CONVERSION FACTORS FROM ENGLISH TO SI UNITS

Length:	1 ft	= 0.3048 m
	1 ft	= 30.48 cm
	1 ft	= 304.8 mm
	1 in.	= 0.0254 m
	1 in.	= 2.54 cm
	1 in.	= 25.4 mm
Area:	1 ft^2	= 929.03×10^{-4} m^2
	1 ft^2	= 929.03 cm^2
	1 ft^2	= 929.03×10^2 mm^2
	1 $in.^2$	= 6.452×10^{-4} m^2
	1 $in.^2$	= 6.452 cm^2
	1 $in.^2$	= 645.16 mm^2
Volume:	1 ft^3	= 28.317×10^{-3} m^3
	1 ft^3	= 28.317×10^3 cm^3
	1 $in.^3$	= 16.387×10^{-6} m^3
	1 $in.^3$	= 16.387 cm^3
Section modulus:	1 $in.^3$	= 0.16387×10^5 mm^3
	1 $in.^3$	= 0.16387×10^{-4} m^3
Hydraulic conductivity:	1 ft/min	= 0.3048 m/min
	1 ft/min	= 30.48 cm/min
	1 ft/min	= 304.8 mm/min
	1 ft/sec	= 0.3048 m/sec
	1 ft/sec	= 304.8 mm/sec
	1 in./min	= 0.0254 m/min
	1 in./sec	= 2.54 cm/sec
	1 in./sec	= 25.4 mm/sec
Coefficient of consolidation:	1 $in.^2$/sec	= 6.452 cm^2/sec
	1 $in.^2$/sec	= 20.346×10^3 m^2/yr
	1 ft^2/sec	= 929.03 cm^2/sec
Force:	1 lb	= 4.448 N
	1 lb	= 4.448×10^{-3} kN
	1 lb	= 0.4536 kgf
	1 kip	= 4.448 kN
	1 U.S. ton	= 8.896 kN
	1 lb	= 0.4536×10^{-3} metric ton
	1 lb/ft	= 14.593 N/m
Stress:	1 lb/ft^2	= 47.88 N/m^2
	1 lb/ft^2	= 0.04788 kN/m^2
	1 U.S. ton/ft^2	= 95.76 kN/m^2
	1 kip/ft^2	= 47.88 kN/m^2
	1 lb/$in.^2$	= 6.895 kN/m^2
Unit weight:	1 lb/ft^3	= 0.1572 kN/m^3
	1 lb/$in.^3$	= 271.43 kN/m^3
Moment:	1 lb-ft	= 1.3558 N · m
	1 lb-in.	= 0.11298 N · m
Energy:	1 ft-lb	= 1.3558 J
Moment of inertia:	1 $in.^4$	= 0.4162×10^6 mm^4
	1 $in.^4$	= 0.4162×10^{-6} m^4

Principles of Geotechnical Engineering

Ninth Edition, SI

BRAJA M. DAS, Dean Emeritus
California State University, Sacramento

KHALED SOBHAN, Professor
Florida Atlantic University

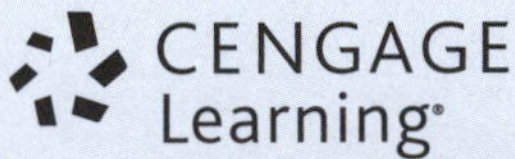

Australia • Brazil • Mexico • Singapore • United Kingdom • United States

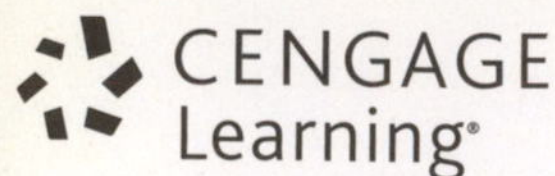

Principles of Geotechnical Engineering, Ninth Edition, SI
Authors: Braja M. Das and Khaled Sobhan

Product Director, Global Engineering: Timothy L. Anderson

Senior Content Developer: Mona Zeftel

Associate Media Content Developer: Ashley Kaupert

Product Assistant: Alexander Sham

Marketing Manager: Kristin Stine

Director, Higher Education Production: Sharon L. Smith

Content Project Manager: Jana Lewis

Production Service: RPK Editorial Services, Inc.

Copyeditor: Lori Martinsek

Proofreader: Harlan James

Indexer: Braja M. Das

Compositor: MPS Limited

Senior Art Director: Michelle Kunkler

Cover and Internal Designer: Harasymczuk Design

Cover Image: Felipe Gabaldon, Getty Images

Intellectual Property

Analyst: Christine Myaskovsky

Project Manager: Sarah Shainwald

Text and Image Permissions Researcher: Kristiina Paul

Manufacturing Planner: Doug Wilke

Library of Congress Control Number: 2016956180

ISBN: 978-1-305-97095-3

Cengage Learning
20 Channel Center Street
Boston, MA 02210
USA

Cengage Learning is a leading provider of customized learning solutions with employees residing in nearly 40 different countries and sales in more than 125 countries around the world. Find your local representative at **www.cengage.com**.

Cengage Learning products are represented in Canada by Nelson Education Ltd.

To learn more about Cengage Learning Solutions, visit **www.cengage.com/engineering**.

Purchase any of our products at your local college store or at our preferred online store **www.cengagebrain.com**.

Printed in the United States of America
Print Number: 01 Print Year: 2016

To Elizabeth Madison, Armaan, and Shaiza

PREFACE TO THE SI EDITION

This edition of *Principles of Geotechnical Engineering* has been adapted to incorporate the International System of Units (*Le Système International d'Unités* or SI) throughout the book.

Le Système International d'Unités

The United States Customary System (USCS) of units uses FPS (foot–pound–second) units (also called English or Imperial units). SI units are primarily the units of the MKS (meter–kilogram–second) system. However, CGS (centimeter–gram–second) units are often accepted as SI units, especially in textbooks.

Using SI Units in this Book

In this book, we have used both MKS and CGS units. USCS (U.S. Customary Units) or FPS (foot-pound-second) units used in the US Edition of the book have been converted to SI units throughout the text and problems. However, in case of data sourced from handbooks, government standards, and product manuals, it is not only extremely difficult to convert all values to SI, it also encroaches upon the intellectual property of the source. Some data in figures, tables, and references, therefore, remains in FPS units. For readers unfamiliar with the relationship between the USCS and the SI systems, a conversion table has been provided inside the front cover.

To solve problems that require the use of sourced data, the sourced values can be converted from FPS units to SI units just before they are to be used in a calculation. To obtain standardized quantities and manufacturers' data in SI units, readers may contact the appropriate government agencies or authorities in their regions.

Instructor Resources

The Instructors' Solution Manual in SI units is available through your Sales Representative or online through the book website at http://login.cengage.com. A digital version of the ISM, Lecture Note PowerPoint slides for the SI text, as well as other resources are available for instructors registering on the book website.

Feedback from users of this SI Edition will be greatly appreciated and will help us improve subsequent editions.

Cengage Learning

PREFACE

Principles of Geotechnical Engineering is intended for use as a text for the introductory course in geotechnical engineering taken by practically all civil engineering students, as well as for use as a reference book for practicing engineers. The book has been revised in 1990, 1994, 1998, 2002, 2006, and 2010. The eighth edition was published in 2014 with coauthor, Khaled Sobhan of Florida Atlantic University. As in the previous editions of the book, this new edition offers a valuable overview of soil properties and mechanics, together with coverage of field practices and basic engineering procedures. It is not the intent of this book to conform to any design codes. The authors appreciate the overwhelming adoptions of this text in various classrooms and are gratified that it has become the market-leading textbook for the course.

New to the Ninth Edition

- This edition includes many new example problems as well as revisions to existing problems. This book now offers more than 185 example problems to ensure understanding. The authors have also added to and updated the book's end-of-chapter problems throughout.
- In Chapter 1 on "Geotechnical Engineering: A Historical Perspective," the list of ISSMGE (International Society for Soil Mechanics and Geotechnical Engineering) technical committees (as of 2013) has been updated. A list of some important geotechnical engineering journals now in publication has been added.
- Chapter 2 on "Origin of Soil and Grain Size" has a more detailed discussion on U.S. sieve sizes. British and Australian standard sieve sizes have also been added.
- Chapter 3 on "Weight-Volume Relationships" now offers an expanded discussion on angularity and the maximum and minimum void ratios of granular soils.
- Students now learn more about the fall cone test used to determine the liquid limit in Chapter 4, which covers "Plasticity and Structure of Soil."
- In Chapter 6 on "Soil Compaction," a newly-developed empirical correlation for maximum dry density and optimum moisture content has been added.
- In Chapter 7 on "Permeability," sections on permeability tests in auger holes, hydraulic conductivity of compacted clay soils, and moisture content-unit weight criteria for clay liner construction have been added.
- Pavlovsky's solution for seepage through an earth dam has been added to Chapter 8 on "Seepage."
- Chapter 10 on "Stresses in a Soil Mass," has new sections on:
    - Vertical stress caused by a horizontal strip load,
    - Westergaard's solution for vertical stress due to a point load, and
    - Stress distribution for Westergaard material.

- An improved relationship for elastic settlement estimation has been incorporated into Chapter 11 on "Compressibility of Soil." This chapter also has a new section on construction time correction (for ramp loading) of consolidation settlement.
- Chapter 12 on "Shear Strength of Soil" now includes some recently-published correlations between drained angle of friction and plasticity index of clayey soil. Additional content has been included on the relationship between undrained shear strength of remolded clay with liquidity index.
- The generalized case for Rankine active and passive pressure (granular backfill) now appears in Chapter 13 on "Lateral Earth Pressure: At-Rest, Rankine, and Coulomb" (Section 13.10). Additional tables for active earth pressure coefficient based on Mononobe-Okabe's equation have been added.
- In Chapter 14 on "Lateral Earth Pressure: Curved Failure Surface," the passive earth pressure coefficient obtained based on the solution by the lower bound theorem of plasticity and the solution by method of characteristics have been summarized. Also, the section on passive force walls with earthquake forces (Section 14.7) has been expanded.
- In Chapter 15 on "Slope Stability," the parameters required for location of the critical failure circle based on Spencer's analysis have been added.
- Chapter 16 on "Soil Bearing Capacity for Shallow Foundations," includes a new section on continuous foundations under eccentrically-inclined load.
- Chapter 18 is a new chapter titled "An Introduction to Geosynthetics," which examines current developments and challenges within this robust and rapidly expanding area of civil engineering.

In the preparation of an engineering text of this type, it is tempting to include many recent developments relating to the behavior of natural soil deposits found in various parts of the world that are available in journals and conference proceedings with the hope that they will prove to be useful to the students in their future practice. However, based on many years of teaching, the authors feel that clarity in explaining the fundamentals of soil mechanics is more important in a first course in this area than filling the book with too many details and alternatives. Many of the fine points can be left to an advanced course in geotechnical engineering. This approach is most likely to nurture students' interest and appreciation in the geotechnical engineering profession at large.

Trusted Features

Principles of Geotechnical Engineering offers more worked-out problems and figures than any other similar text. Unique in the market, these features offer students ample practice and examples, keeping their learning application-oriented, and helping them prepare for work as practicing civil engineers.

In addition to traditional end-of-chapter exercises, this text provides challenging **critical thinking problems**. These problems encourage deeper analyses and drive students to extend their understanding of the subjects covered within each chapter.

A generous **16-page color insert** features distinctive photographs of rocks and rock-forming minerals. These images capture the unique coloring that help geotechnical engineers distinguish one mineral from another.

Each chapter begins with an introduction and concludes with a summary to help students identify what is most important in each chapter. These features clearly preview and reinforce content to guide students and assist them in retaining key concepts.

A complete, comprehensive discussion addresses the weathering of rocks. Students learn about both weathering and the formation of sedimentary and metamorphic rocks in this thorough presentation.

A detailed explanation focuses on the variation of the maximum and minimum void ratios of granular soils. Students examine variations due to grain size, shape, and non-plastic fine contents.

Resource Materials

A detailed Instructor's Solutions Manual containing solutions to all end-of-chapter problems and Lecture Note PowerPoint Slides are available via a secure, password-protected Instructor Resource Center at http://sso.cengage.com.

Principles of Geotechnical Engineering is also available through **MindTap**, Cengage Learning's digital course platform. See the following section on pages xi and xii for more details about this exciting new addition to the book.

Acknowledgments

- We are deeply grateful to Janice Das for her assistance in completing the revision. She has been the driving force behind this textbook since the preparation of the first edition.
- Thanks to Professor Jiliang Li of Purdue University North Central for providing several important review comments on the eighth edition.

The authors would like to thank all of the reviewers and instructors who have provided feedback over the years. In addition we wish to acknowledge and thank our Global Engineering team at Cengage Learning for their dedication to this new book: Timothy Anderson, Product Director; Mona Zeftel, Senior Content Developer; Jana Lewis, Content Project Manager; Kristin Stine, Marketing Manager; Elizabeth Brown and Brittany Burden, Learning Solutions Specialists; Ashley Kaupert, Associate Media Content Developer; Teresa Versaggi and Alexander Sham, Product Assistants; and Rose P. Kernan of RPK Editorial Services. They have skillfully guided every aspect of this text's development and production to successful completion.

Braja M. Das
Khaled Sobhan

MINDTAP ONLINE COURSE

Principles of Geotechnical Engineering is also available through **MindTap**, Cengage Learning's digital course platform. The carefully-crafted pedagogy and exercises in this market-leading textbook are made even more effective by an interactive, customizable eBook, automatically graded assessments, and a full suite of study tools.

As an instructor using MindTap, you have at your fingertips the full text and a unique set of tools, all in an interface designed to save you time. MindTap makes it easy for instructors to build and customize their course, so you can focus on the most relevant material while also lowering costs for your students. Stay connected and informed through real-time student tracking that provides the opportunity to adjust your course as needed based on analytics of interactivity and performance. **Algorithmically generated problem sets** allow your students maximum practice while you can be assured that each student is being tested by unique problems. Videos of real world situations, geotechnical instruments, and soil and rock materials provide students with knowledge of future field experiences.

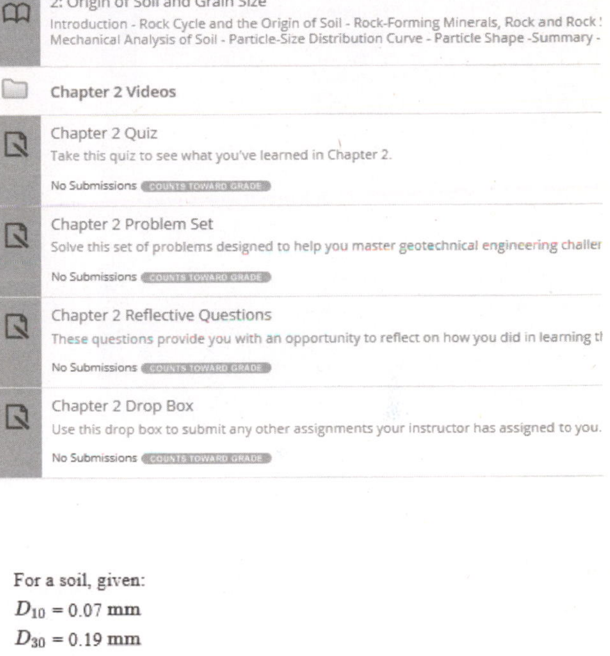

How does MindTap benefit instructors?

- You can build and personalize your course by integrating your own content into the **MindTap Reader** (like lecture notes or problem sets to download) or pull from sources such as RSS feeds, YouTube videos, websites, and more. Control what content students see with a built-in learning path that can be customized to your syllabus.

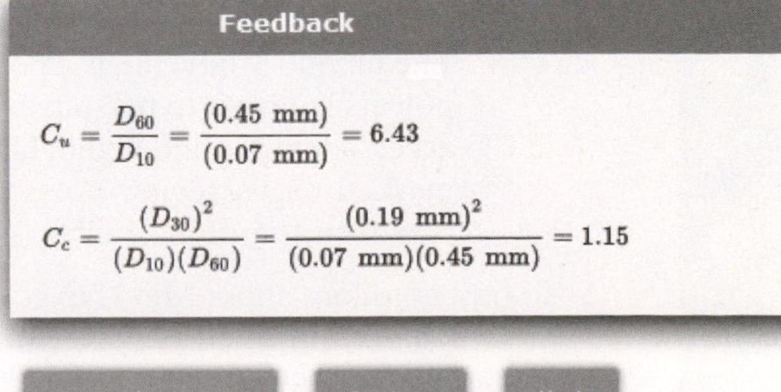

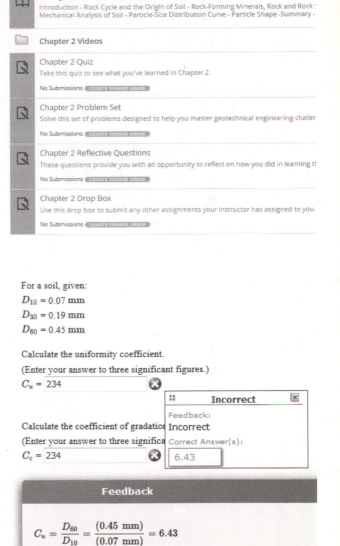

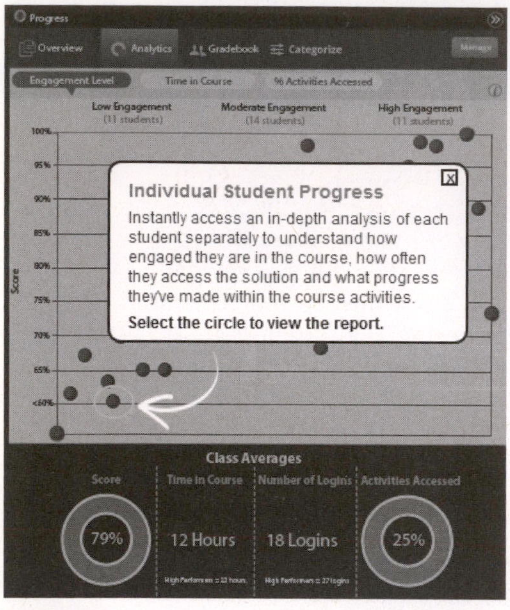

- MindTap saves you time by providing you and your students with **automatically graded assignments and quizzes**, including **algorithmically generated problem sets**. These problems include immediate, specific feedback, so students know exactly where they need more practice.
- The **Message Center** helps you to quickly and easily contact students directly from MindTap. Messages are communicated directly to each student via the communication medium (email, social media, or even text message) designated by the student.
- **StudyHub** is a valuable studying tool that allows you to deliver important information and empowers your students to personalize their experience. Instructors can choose to annotate the text with **notes** and **highlights**, share content from the MindTap Reader, and create **flashcards** to help their students focus and succeed.
- The **Progress App** lets you know exactly how your students are doing (and where they might be struggling) with live analytics. You can see overall class engagement and drill down into individual student performance, enabling you to adjust your course to maximize student success.

How does MindTap benefit your students?

- The **MindTap Reader** adds the abilities to have the content read aloud, to print from the reader, and to take notes and highlights while also capturing them within the linked **StudyHub App**.
- The **MindTap Mobile App** keeps students connected with alerts and notifications while also providing them with on-the-go study tools like Flashcards and quizzing, helping them manage their time efficiently.
- **Flashcards** are pre-populated to provide a jump start on studying, and students and instructors can also create customized cards as they move through the course.
- The **Progress App** allows students to monitor their individual grades, as well as their level compared to the class average. This not only helps them stay on track in the course but also motivates them to do more, and ultimately to do better.
- The unique **StudyHub** is a powerful single-destination studying tool that empowers students to personalize their experience. They can quickly and easily access all notes and highlights marked in the MindTap Reader, locate bookmarked pages, review notes and Flashcards shared by their instructor, and create custom study guides.

To find out more about MindTap go to: www.cengage.com/mindtap.

For more information about MindTap for Engineering, or to schedule a demonstration, please call (800) 354-9706 or email higheredcs@cengage.com. For those instructors outside the United States, please visit http://www.cengage.com/contact/ to locate your regional office.

ABOUT THE AUTHORS

Professor Braja Das is Dean Emeritus of the College of Engineering and Computer Science at California State University, Sacramento. He received his M.S. in Civil Engineering from the University of Iowa and his Ph.D. in the area of Geotechnical Engineering from the University of Wisconsin. He is the author of several geotechnical engineering texts and reference books and has authored more than 300 technical papers in the area of geotechnical engineering. His primary areas of research include shallow foundations, earth anchors, and geosynthetics. He is a Fellow and Life Member of the American Society of Civil Engineers, Life Member of the American Society for Engineering Education, and an Emeritus Member of the Stabilization of Geometrical Materials and Recycled Materials Committee of the Transportation Research Board of the National Research Council (Washington, D.C.). He has previously served as a member of the editorial board of the *Journal of Geotechnical Engineering* of ASCE, a member of the *Lowland Technology International* journal (Japan), associate editor of the *International Journal of Offshore and Polar Engineering* (ISOPE), and co-editor of the *Journal of Geotechnical and Geological Engineering* (Springer, The Netherlands). Presently he is the editor-in-chief of the *International Journal of Geotechnical Engineering* (Taylor and Francis, U.K.). Dr. Das has received numerous awards for teaching excellence, including the AMOCO Foundation Award, AT&T Award for Teaching Excellence from the American Society for Engineering Education, the Ralph Teetor Award from the Society of Automotive Engineers, and the Distinguished Achievement Award for Teaching Excellence from the University of Texas at El Paso.

Dr. Khaled Sobhan is a Professor of Civil, Environmental and Geomatics Engineering at Florida Atlantic University. He received his M.S. degree from The Johns Hopkins University, and his Ph.D. degree from Northwestern University, both in the area of Geotechnical Engineering. His primary research areas include ground improvement, geotechnology of soft soils, experimental soil mechanics, and geotechnical aspects of pavement engineering. He served as the Chair of the Chemical and Mechanical Stabilization committee (AFS90) of the Transportation Research Board (2005–2011), and co-authored the TRB Circular titled *Evaluation of Chemical Stabilizers: State-of-the-Practice Report* (E-C086). He is currently serving as an Associate Editor of ASCE's *Journal of Materials in Civil Engineering*, and in the editorial board of the *ASTM Geotechnical Testing Journal, Geotechnical and Geological Engineering* (Springer, The Netherlands), and the *International Journal of Geotechnical Engineering*. He is a recipient of the distinguished *Award for Excellence and Innovation in Undergraduate Teaching* (2006), and the *Excellence in Graduate Mentoring Award* (2009) from Florida Atlantic University. He has authored/co-authored over 100 technical articles and reports in the area of geotechnical engineering.

CONTENTS

14 Lateral Earth Pressure: Curved Failure Surface 606

15 Slope Stability 638

CHAPTER 1

Geotechnical Engineering—A Historical Perspective

1.1 Introduction

For engineering purposes, *soil* is defined as the uncemented aggregate of mineral grains and decayed organic matter (solid particles) with liquid and gas in the empty spaces between the solid particles. Soil is used as a construction material in various civil engineering projects, and it supports structural foundations. Thus, civil engineers must study the properties of soil, such as its origin, grain-size distribution, ability to drain water, compressibility, strength, and its ability to support structures and resist deformation. *Soil mechanics* is the branch of science that deals with the study of the physical properties of soil and the behavior of soil masses subjected to various types of forces. *Soils engineering* is the application of the principles of soil mechanics to practical problems. *Geotechnical engineering* is the subdiscipline of civil engineering that involves natural materials found close to the surface of the earth. It includes the application of the principles of soil mechanics and rock mechanics to the design of foundations, retaining structures, and earth structures.

1.2 Geotechnical Engineering Prior to the 18th Century

The record of a person's first use of soil as a construction material is lost in antiquity. In true engineering terms, the understanding of geotechnical engineering as it is known today began early in the 18th century (Skempton, 1985). For years, the art of geotechnical engineering was based on only past experiences through a succession

Table 1.1 Major Pyramids in Egypt

Pyramid/Pharaoh	Location	Reign of Pharaoh
Djoser	Saqqara	2630–2612 B.C.
Sneferu	Dashur (North)	2612–2589 B.C.
Sneferu	Dashur (South)	2612–2589 B.C.
Sneferu	Meidum	2612–2589 B.C.
Khufu	Giza	2589–2566 B.C.
Djedefre	Abu Rawash	2566–2558 B.C.
Khafre	Giza	2558–2532 B.C.
Menkaure	Giza	2532–2504 B.C.

of experimentation without any real scientific character. Based on those experimentations, many structures were built—some of which have crumbled, while others are still standing.

Recorded history tells us that ancient civilizations flourished along the banks of rivers, such as the Nile (Egypt), the Tigris and Euphrates (Mesopotamia), the Huang Ho (Yellow River, China), and the Indus (India). Dykes dating back to about 2000 B.C. were built in the basin of the Indus to protect the town of Mohenjo Dara (in what became Pakistan after 1947). During the Chan dynasty in China (1120 B.C. to 249 B.C.) many dykes were built for irrigation purposes. There is no evidence that measures were taken to stabilize the foundations or check erosion caused by floods (Kerisel, 1985). Ancient Greek civilization used isolated pad footings and strip-and-raft foundations for building structures. Beginning around 2700 B.C., several pyramids were built in Egypt, most of which were built as tombs for the country's Pharaohs and their consorts during the Old and Middle Kingdom periods. Table 1.1 lists some of the major pyramids identified through the Pharaoh who ordered it built. As of 2008, a total of 138 pyramids have been discovered in Egypt. Figure 1.1

Figure 1.1 A view of the pyramids at Giza. (*Courtesy of Janice Das, Henderson, Nevada*)

shows a view of the three pyramids at Giza. The construction of the pyramids posed formidable challenges regarding foundations, stability of slopes, and construction of underground chambers. With the arrival of Buddhism in China during the Eastern Han dynasty in 68 A.D., thousands of pagodas were built. Many of these structures were constructed on silt and soft clay layers. In some cases the foundation pressure exceeded the load-bearing capacity of the soil and thereby caused extensive structural damage.

One of the most famous examples of problems related to soil-bearing capacity in the construction of structures prior to the 18th century is the Leaning Tower of Pisa in Italy (See Figure 1.2). Construction of the tower began in 1173 A.D. when the Republic of Pisa was flourishing and continued in various stages for over 200 years. The structure weighs about 15,700 metric tons and is supported by a circular base having a diameter of 20 m. The tower has tilted in the past to the east, north, west, and, finally, to the south. Recent investigations showed that a weak clay layer existed at a depth of about 11 m below the ground surface compression of which caused the tower to tilt. It became more than 5 m out of plumb with the 54 m height

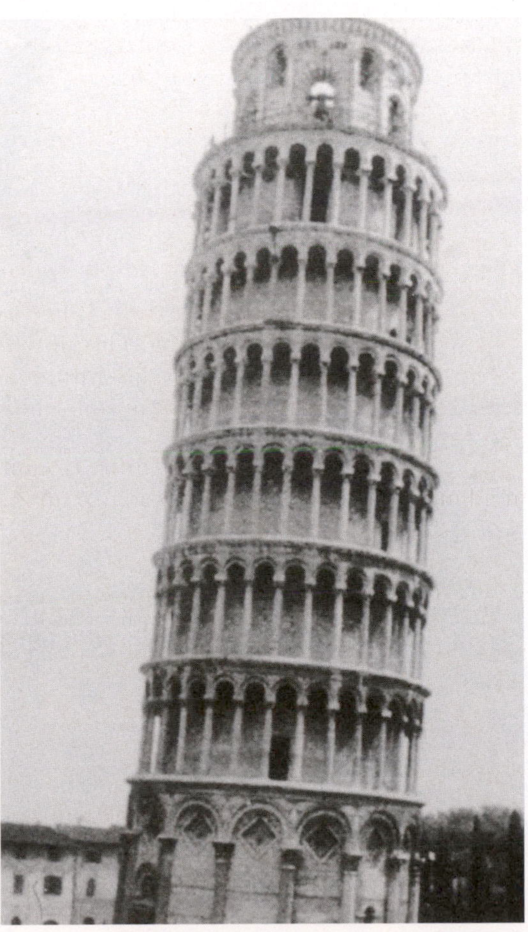

Figure 1.2 Leaning Tower of Pisa, Italy (*Courtesy of Braja M. Das, Henderson, Nevada*)

(about a 5.5 degree tilt). The tower was closed in 1990 because it was feared that it would either fall over or collapse. It recently has been stabilized by excavating soil from under the north side of the tower. About 70 metric tons of earth were removed in 41 separate extractions that spanned the width of the tower. As the ground gradually settled to fill the resulting space, the tilt of the tower eased. The tower now leans 5 degrees. The half-degree change is not noticeable, but it makes the structure considerably more stable. Figure 1.3 is an example of a similar problem. The towers shown in Figure 1.3 are located in Bologna, Italy, and they were built in the 12th century. The tower on the left is usually referred to as the *Garisenda Tower*. It is 48 m in height and weighs about 4210 metric tons. It has tilted about 4 degrees. The tower on the right is the Asinelli Tower, which is 97 m high and weighs 7300 metric tons. It has tilted about 1.3 degrees.

After encountering several foundation-related problems during construction over centuries past, engineers and scientists began to address the properties and

Figure 1.3 Tilting of Garisenda Tower (left) and Asinelli Tower (right) in Bologna, Italy (*Courtesy of Braja M. Das, Henderson, Nevada*)

behaviors of soils in a more methodical manner starting in the early part of the 18th century. Based on the emphasis and the nature of study in the area of geotechnical engineering, the time span extending from 1700 to 1927 can be divided into four major periods (Skempton, 1985):

1. Preclassical (1700 to 1776 A.D.)
2. Classical soil mechanics—Phase I (1776 to 1856 A.D.)
3. Classical soil mechanics—Phase II (1856 to 1910 A.D.)
4. Modern soil mechanics (1910 to 1927 A.D.)

Brief descriptions of some significant developments during each of these four periods are presented below.

1.3 Preclassical Period of Soil Mechanics (1700–1776)

This period concentrated on studies relating to natural slope and unit weights of various types of soils, as well as the semiempirical earth pressure theories. In 1717, a French royal engineer, Henri Gautier (1660–1737), studied the natural slopes of soils when tipped in a heap for formulating the design procedures of retaining walls. The *natural slope* is what we now refer to as the *angle of repose*. According to this study, the natural slope of *clean dry sand* and *ordinary earth* were 31° and 45°, respectively. Also, the unit weight of clean dry sand and ordinary earth were recommended to be 18.1 kN/m^3 and 13.4 kN/m^3, respectively. No test results on clay were reported. In 1729, Bernard Forest de Belidor (1671–1761) published a textbook for military and civil engineers in France. In the book, he proposed a theory for lateral earth pressure on retaining walls that was a follow-up to Gautier's (1717) original study. He also specified a soil classification system in the manner shown in the following table.

Classification	Unit weight kN/m^3
Rock	—
Firm or hard sand, compressible sand	16.7 to 18.4
Ordinary earth (as found in dry locations)	13.4
Soft earth (primarily silt)	16.0
Clay	18.9
Peat	—

The first laboratory model test results on a 76-mm-high retaining wall built with sand backfill were reported in 1746 by a French engineer, Francois Gadroy (1705–1759), who observed the existence of slip planes in the soil at failure. Gadroy's study was later summarized by J. J. Mayniel in 1808. Another notable contribution during this period is that by the French engineer Jean Rodolphe Perronet (1708–1794), who studied slope stability around 1769 and distinguished between intact ground and fills.

1.4 Classical Soil Mechanics—Phase I (1776–1856)

During this period, most of the developments in the area of geotechnical engineering came from engineers and scientists in France. In the preclassical period, practically all theoretical considerations used in calculating lateral earth pressure on retaining walls were based on an arbitrarily based failure surface in soil. In his famous paper presented in 1776, French scientist Charles Augustin Coulomb (1736–1806) used the principles of calculus for maxima and minima to determine the true position of the sliding surface in soil behind a retaining wall. In this analysis, Coulomb used the laws of friction and cohesion for solid bodies. In 1790, the distinguished French civil engineer, Gaspard Clair Marie Riche de Prony (1755–1839) included Coulomb's theory in his leading textbook, *Nouvelle Architecture Hydraulique* (Vol. 1). In 1820, special cases of Coulomb's work were studied by French engineer Jacques Frederic Francais (1775–1833) and by French applied mechanics professor Claude Louis Marie Henri Navier (1785–1836). These special cases related to inclined backfills and backfills supporting surcharge. In 1840, Jean Victor Poncelet (1788–1867), an army engineer and professor of mechanics, extended Coulomb's theory by providing a graphical method for determining the magnitude of lateral earth pressure on vertical and inclined retaining walls with arbitrarily broken polygonal ground surfaces. Poncelet was also the first to use the symbol ϕ for soil friction angle. He also provided the first ultimate bearing-capacity theory for shallow foundations. In 1846 Alexandre Collin (1808–1890), an engineer, provided the details for deep slips in clay slopes, cutting, and embankments. Collin theorized that in all cases the failure takes place when the mobilized cohesion exceeds the existing cohesion of the soil. He also observed that the actual failure surfaces could be approximated as arcs of cycloids.

The end of Phase I of the classical soil mechanics period is generally marked by the year (1857) of the first publication by William John Macquorn Rankine (1820–1872), a professor of civil engineering at the University of Glasgow. This study provided a notable theory on earth pressure and equilibrium of earth masses. Rankine's theory is a simplification of Coulomb's theory.

1.5 Classical Soil Mechanics—Phase II (1856–1910)

Several experimental results from laboratory tests on sand appeared in the literature in this phase. One of the earliest and most important publications is one by French engineer Henri Philibert Gaspard Darcy (1803–1858). In 1856, he published a study on the permeability of sand filters. Based on those tests, Darcy defined the term *coefficient of permeability* (or hydraulic conductivity) of soil, a very useful parameter in geotechnical engineering to this day.

Sir George Howard Darwin (1845–1912), a professor of astronomy, conducted laboratory tests to determine the overturning moment on a hinged wall retaining sand in loose and dense states of compaction. Another noteworthy contribution, which was published in 1885 by Joseph Valentin Boussinesq (1842–1929), was the development of the theory of stress distribution under load bearing areas in a homogeneous, semiinfinite, elastic, and isotropic medium. In 1887, Osborne Reynolds

(1842–1912) demonstrated the phenomenon of dilatancy in sand. Other notable studies during this period are those by John Clibborn (1847–1938) and John Stuart Beresford (1845–1925) relating to the flow of water through sand bed and uplift pressure. Clibborn's study was published in the *Treatise on Civil Engineering, Vol. 2: Irrigation Work in India,* Roorkee, 1901 and also in *Technical Paper No. 97*, Government of India, 1902. Beresford's 1898 study on uplift pressure on the Narora Weir on the Ganges River has been documented in *Technical Paper No. 97*, Government of India, 1902.

1.6 Modern Soil Mechanics (1910–1927)

In this period, results of research conducted on clays were published in which the fundamental properties and parameters of clay were established. The most notable publications are described next.

Around 1908, Albert Mauritz Atterberg (1846–1916), a Swedish chemist and soil scientist, defined *clay-size fractions* as the percentage by weight of particles smaller than 2 microns in size. He realized the important role of clay particles in a soil and the plasticity thereof. In 1911, he explained the consistency of cohesive soils by defining liquid, plastic, and shrinkage limits. He also defined the plasticity index as the difference between liquid limit and plastic limit (see Atterberg, 1911).

In October 1909, the 17-m high earth dam at Charmes, France, failed. It was built between 1902 and 1906. A French engineer, Jean Fontard (1884–1962), carried out investigations to determine the cause of failure. In that context, he conducted undrained double-shear tests on clay specimens (0.77 m^2 in area and 200 mm thick) under constant vertical stress to determine their shear strength parameters (see Frontard, 1914). The times for failure of these specimens were between 10 to 20 minutes.

Arthur Langley Bell (1874–1956), a civil engineer from England, worked on the design and construction of the outer seawall at Rosyth Dockyard. Based on his work, he developed relationships for lateral pressure and resistance in clay as well as bearing capacity of shallow foundations in clay (see Bell, 1915). He also used shear-box tests to measure the undrained shear strength of undisturbed clay specimens.

Wolmar Fellenius (1876–1957), an engineer from Sweden, developed the stability analysis of undrained saturated clay slopes (that is, $\phi = 0$ condition) with the assumption that the critical surface of sliding is the arc of a circle. These were elaborated upon in his papers published in 1918 and 1926. The paper published in 1926 gave correct numerical solutions for the *stability numbers* of circular slip surfaces passing through the toe of the slope.

Karl Terzaghi (1883–1963) of Austria (Figure 1.4) developed the theory of consolidation for clays as we know today. The theory was developed when Terzaghi was teaching at the American Robert College in Istanbul, Turkey. His study spanned a five-year period from 1919 to 1924. Five different clay soils were used. The liquid limit of those soils ranged between 36 and 67, and the plasticity index was in the range of 18 to 38. The consolidation theory was published in Terzaghi's celebrated book *Erdbaumechanik* in 1925.

Figure 1.4 Karl Terzaghi (1883–1963) (*SSPL via Getty Images*)

1.7 Geotechnical Engineering after 1927

The publication of *Erdbaumechanik auf Bodenphysikalisher Grundlage* by Karl Terzaghi in 1925 gave birth to a new era in the development of soil mechanics. Karl Terzaghi is known as the father of modern soil mechanics, and rightfully so. Terzaghi was born on October 2, 1883 in Prague, which was then the capital of the Austrian province of Bohemia. In 1904 he graduated from the Technische Hochschule in Graz, Austria, with an undergraduate degree in mechanical engineering. After graduation he served one year in the Austrian army. Following his army service, Terzaghi studied one more year, concentrating on geological subjects. In January 1912, he received the degree of Doctor of Technical Sciences from his alma mater in Graz. In 1916, he accepted a teaching position at the Imperial School of Engineers in Istanbul. After the end of World War I, he accepted a lectureship at the American Robert College in Istanbul (1918–1925). There he began his research work on the behavior of soil and settlement of clay and on the failure due to piping in sand under dams. The publication *Erdbaumechanik* is primarily the result of this research.

In 1925, Terzaghi accepted a visiting lectureship at Massachusetts Institute of Technology, where he worked until 1929. During that time, he became recognized as the leader of the new branch of civil engineering called soil mechanics. In October 1929, he returned to Europe to accept a professorship at the Technical University of Vienna, which soon became the nucleus for civil engineers interested in soil mechanics. In 1939, he returned to the United States to become a professor at Harvard University.

The first conference of the International Society of Soil Mechanics and Foundation Engineering (ISSMFE) was held at Harvard University in 1936 with Karl Terzaghi presiding. The conference was possible due to the conviction and efforts of Professor Arthur Casagrande of Harvard University. About 200 individuals representing 21 countries attended this conference. It was through the inspiration and guidance of Terzaghi over the preceding quarter-century that papers were brought to that conference covering a wide range of topics, such as

- Effective stress
- Shear strength
- Testing with Dutch cone penetrometer
- Consolidation
- Centrifuge testing
- Elastic theory and stress distribution
- Preloading for settlement control
- Swelling clays
- Frost action
- Earthquake and soil liquefaction
- Machine vibration
- Arching theory of earth pressure

For the next quarter-century, Terzaghi was the guiding spirit in the development of soil mechanics and geotechnical engineering throughout the world. To that effect, in 1985, Ralph Peck wrote that "few people during Terzaghi's lifetime would have disagreed that he was not only the guiding spirit in soil mechanics, but that he was the clearing house for research and application throughout the world. Within the next few years he would be engaged on projects on every continent save Australia and Antarctica." Peck continued with, "Hence, even today, one can hardly improve on his contemporary assessments of the state of soil mechanics as expressed in his summary papers and presidential addresses." In 1939, Terzaghi delivered the 45th James Forrest Lecture at the Institution of Civil Engineers, London. His lecture was entitled "Soil Mechanics—A New Chapter in Engineering Science." In it, he proclaimed that most of the foundation failures that occurred were no longer "acts of God."

Following are some highlights in the development of soil mechanics and geotechnical engineering that evolved after the first conference of the ISSMFE in 1936:

- Publication of the book *Theoretical Soil Mechanics* by Karl Terzaghi in 1943 (Wiley, New York)
- Publication of the book *Soil Mechanics in Engineering Practice* by Karl Terzaghi and Ralph Peck in 1948 (Wiley, New York)

- Publication of the book *Fundamentals of Soil Mechanics* by Donald W. Taylor in 1948 (Wiley, New York)
- Start of the publication of *Geotechnique,* the international journal of soil mechanics in 1948 in England

After a brief interruption for World War II, the second conference of ISSMFE was held in Rotterdam, The Netherlands, in 1948. There were about 600 participants, and seven volumes of proceedings were published. In this conference, A. W. Skempton presented the landmark paper on $\phi = 0$ concept for clays. Following Rotterdam, ISSMFE conferences have been organized about every four years in different parts of the world. The aftermath of the Rotterdam conference saw the growth of regional conferences on geotechnical engineering, such as

- European Regional Conference on Stability of Earth Slopes, Stockholm (1954)
- First Australia–New Zealand Conference on Shear Characteristics of Soils (1952)
- First Pan American Conference, Mexico City (1960)
- Research conference on Shear Strength of Cohesive Soils, Boulder, Colorado, (1960)

Two other important milestones between 1948 and 1960 are (1) the publication of A. W. Skempton's paper on *A* and *B* pore pressure parameters, which made effective stress calculations more practical for various engineering works, and (2) publication of the book entitled *The Measurement of Soil Properties in the Triaxial Text* by A. W. Bishop and B. J. Henkel (Arnold, London) in 1957.

By the early 1950s, computer-aided finite difference and finite element solutions were applied to various types of geotechnical engineering problems. When the projects become more sophisticated with complex boundary conditions, it is no longer possible to apply closed-form solutions. Numerical modeling, using a finite element (e.g. Abaqus, Plaxis) or finite difference (e.g. Flac) software, is becoming increasingly popular in the profession. The dominance of numerical modeling in geotechnical engineering will continue in the next few decades due to new challenges and advances in the modelling techniques. Since the early days, the profession of geotechnical engineering has come a long way and has matured. It is now an established branch of civil engineering, and thousands of civil engineers declare geotechnical engineering to be their preferred area of speciality.

In 1997, the ISSMFE was changed to ISSMGE (International Society of Soil Mechanics and Geotechnical Engineering) to reflect its true scope. These international conferences have been instrumental for exchange of information regarding new developments and ongoing research activities in geotechnical engineering. Table 1.2 gives the location and year in which each conference of ISSMFE/ISSMGE was held.

In 1960, Bishop, Alpan, Blight, and Donald provided early guidelines and experimental results for the factors controlling the strength of partially saturated cohesive soils. Since that time advances have been made in the study of the behavior of unsaturated soils as related to strength and compressibility and other factors affecting construction of earth-supported and earth-retaining structures.

ISSMGE has several technical committees, and these committees organize or cosponsor several conferences around the world. A list of these technical committees

Table 1.2 Details of ISSMFE (1936–1997) and ISSMGE (1997–present) Conferences

Conference	Location	Year
I	Harvard University, Boston, U.S.A.	1936
II	Rotterdam, the Netherlands	1948
III	Zurich, Switzerland	1953
IV	London, England	1957
V	Paris, France	1961
VI	Montreal, Canada	1965
VII	Mexico City, Mexico	1969
VIII	Moscow, U.S.S.R.	1973
IX	Tokyo, Japan	1977
X	Stockholm, Sweden	1981
XI	San Francisco, U.S.A.	1985
XII	Rio de Janeiro, Brazil	1989
XIII	New Delhi, India	1994
XIV	Hamburg, Germany	1997
XV	Istanbul, Turkey	2001
XVI	Osaka, Japan	2005
XVII	Alexandria, Egypt	2009
XVIII	Paris, France	2013
XIX	Seoul, Korea	2017 (scheduled)

(2010–2013) is given in Table 1.3. ISSMGE also conducts International Seminars (formerly known as Touring Lectures), which have proved to be an important activity; these seminars bring together practitioners, contractors, and academics, both on stage and in the audience, to their own benefit irrespective of the region, size, or wealth of the Member Society, thus fostering a sense of belonging to the ISSMGE.

Soils are heterogeneous materials that can have substantial variability within a few meters. The design parameters for all geotechnical projects have to come from a site investigation exercise that includes field tests, collecting soil samples at various locations and depths, and carrying out laboratory tests on these samples. The laboratory and field tests on soils, as for any other materials, are carried out as per standard methods specified by ASTM International (known as American Society for Testing and Materials before 2001). ASTM standards (http://www.astm.org) cover a wide range of materials in more than 80 volumes. The test methods for soils, rocks, and aggregates are bundled into the two volumes—04.08 and 04.09.

Geotechnical engineering is a relatively young discipline that has witnessed substantial developments in the past few decades, and it is still growing. These new developments and most cutting-edge research findings are published in peer reviewed international journals before they find their way into textbooks. Some of these geotechnical journals are (in alphabetical order):

- Canadian Geotechnical Journal (NRC Research Press in cooperation with the Canadian Geotechnical Society)

Table 1.3 List of ISSMGE Technical Committees (November, 2013)

Category	Technical committee number	Technical committee name
Fundamentals	TC101	Laboratory Stress Strength Testing of Geomaterials
	TC102	Ground Property Characterization from In-Situ Tests
	TC103	Numerical Methods in Geomechanics
	TC104	Physical Modelling in Geotechnics
	TC105	Geo-Mechanics from Micro to Macro
	TC106	Unsaturated Soils
Applications	TC201	Geotechnical Aspects of Dykes and Levees, Shore Protection and Land Reclamation
	TC202	Transportation Geotechnics
	TC203	Earthquake Geotechnical Engineering and Associated Problems
	TC204	Underground Construction in Soft Ground
	TC205	Safety and Surviability in Geotechnical Engineering
	TC206	Interactive Geotechnical Design
	TC207	Soil-Structure Interaction and Retaining Walls
	TC208	Slope Stability in Engineering Practice
	TC209	Offshore Geotechnics
	TC210	Dams and Embankments
	TC211	Ground Improvement
	TC212	Deep Foundations
	TC213	Scour and Erosion
	TC214	Foundation Engineering for Difficult Soft Soil Conditions
	TC215	Environmental Geotechnics
	TC216	Frost Geotechnics
Impact on Society	TC301	Preservation of Historic Sites
	TC302	Forensic Geotechnical Engineering
	TC303	Coastal and River Disaster Mitigation and Rehabilitation
	TC304	Engineering Practice of Risk Assessment and Management
	TC305	Geotechnical Infrastructure for Megacities and New Capitals

- Geotechnical and Geoenvironmental Engineering (American Society of Civil Engineers)
- Geotechnical and Geological Engineering (Springer, Germany)
- Geotechnical Testing Journal (ASTM International, USA)
- Geotechnique (Institute of Civil Engineers, UK)
- International Journal of Geomechanics (American Society of Civil Engineers)

- International Journal of Geotechnical Engineering (Taylor and Francis, UK)
- Soils and Foundations (Elsevier on behalf of the Japanese Geotechnical Society)

For a thorough literature review on a research topic, these journals and the proceedings of international conferences (e.g. ICSMGE, see Table 1.2) would be very valuable. The references cited in each chapter in this book are listed at the end of the chapter.

1.8 End of an Era

In Section 1.7, a brief outline of the contributions made to modern soil mechanics by pioneers such as Karl Terzaghi, Arthur Casagrande, Donald W. Taylor, Alec W. Skempton, and Ralph B. Peck was presented. The last of the early giants of the profession, Ralph B. Peck, passed away on February 18, 2008, at the age of 95.

Professor Ralph B. Peck (Figure 1.5) was born in Winnipeg, Canada to American parents Orwin K. and Ethel H. Peck on June 23, 1912. He received B.S. and Ph.D. degrees in 1934 and 1937, respectively, from Rensselaer Polytechnic Institute, Troy, New York. During the period from 1938 to 1939, he took courses from Arthur Casagrande at Harvard University in a new subject called "soil mechanics." From

Figure 1.5 Ralph B. Peck *(Photo courtesy of Ralph B. Peck)*

1939 to 1943, Dr. Peck worked as an assistant to Karl Terzaghi, the "father" of modern soil mechanics, on the Chicago Subway Project. In 1943, he joined the University of Illinois at Champaign–Urbana and was a professor of foundation engineering from 1948 until he retired in 1974. After retirement, he was active in consulting, which included major geotechnical projects in 44 states in the United States and 28 other countries on five continents. Some examples of his major consulting projects include

- Rapid transit systems in Chicago, San Francisco, and Washington, D.C.
- Alaskan pipeline system
- James Bay Project in Quebec, Canada
- Heathrow Express Rail Project (U.K.)
- Dead Sea dikes

His last project was the Rion-Antirion Bridge in Greece. On March 13, 2008, *The Times* of the United Kingdom wrote, "Ralph B. Peck was an American civil engineer who invented a controversial construction technique that would be used on some of the modern engineering wonders of the world, including the Channel Tunnel. Known as 'the godfather of soil mechanics,' he was directly responsible for a succession of celebrated tunneling and earth dam projects that pushed the boundaries of what was believed to be possible."

Dr. Peck authored more than 250 highly distinguished technical publications. He was the president of the ISSMGE from 1969 to 1973. In 1974, he received the National Medal of Science from President Gerald R. Ford. Professor Peck was a teacher, mentor, friend, and counselor to generations of geotechnical engineers in every country in the world. The 16th ISSMGE Conference in Osaka, Japan (2005) was the last major conference of its type that he would attend.

This is truly the end of an era.

References

Atterberg, A. M. (1911). "Über die physikalische Bodenuntersuchung, und über die Plastizität de Tone," International Mitteilungen für Bodenkunde, *Verlag für Fachliteratur.* G.m.b.H. Berlin, Vol. 1, 10–43.

Belidor, B. F. (1729). *La Science des Ingenieurs dans la Conduite des Travaux de Fortification et D'Architecture Civil,* Jombert, Paris.

Bell, A. L. (1915). "The Lateral Pressure and Resistance of Clay, and Supporting Power of Clay Foundations," *Min. Proceeding of Institute of Civil Engineers,* Vol. 199, 233–272.

Bishop, A. W., Alpan, I., Blight, G. E., and Donald, I. B. (1960). "Factors Controlling the Strength of Partially Saturated Cohesive Soils." *Proceedings.* Research Conference on Shear Strength of Cohesive Soils, ASCE, 502–532.

Bishop, A. W. and Henkel, B. J. (1957). *The Measurement of Soil Properties in the Triaxial Test,* Arnold, London.

Boussinesq, J. V. (1885). *Application des Potentiels â L'Etude de L'Équilibre et du Mouvement des Solides Élastiques,* Gauthier-Villars, Paris.

Collin, A. (1846). *Recherches Expérimentales sur les Glissements Spontanés des Terrains Argileux Accompagnées de Considérations sur Quelques Principes de la Mécanique Terrestre*, Carilian-Goeury, Paris.

Coulomb, C. A. (1776). "Essai sur une Application des Règles de Maximis et Minimis à Quelques Problèmes de Statique Relatifs à L'Architecture," *Mèmoires de la Mathèmatique et de Phisique,* présentés à l'Académie Royale des Sciences, par divers savans, et lûs dans sés Assemblées, De L'Imprimerie Royale, Paris, Vol. 7, Annee 1793, 343–382.

Darcy, H. P. G. (1856). *Les Fontaines Publiques de la Ville de Dijon,* Dalmont, Paris.

Darwin, G. H. (1883). "On the Horizontal Thrust of a Mass of Sand," *Proceedings,* Institute of Civil Engineers, London, Vol. 71, 350–378.

Fellenius, W. (1918). "Kaj-och Jordrasen I Göteborg," *Teknisk Tidskrift.* Vol. 48, 17–19.

Francais, J. F. (1820). "Recherches sur la Poussée de Terres sur la Forme et Dimensions des Revêtments et sur la Talus D'Excavation," *Mémorial de L'Officier du Génie,* Paris, Vol. IV, 157–206.

Frontard, J. (1914). "Notice sur L'Accident de la Digue de Charmes," *Anns. Ponts et Chaussées 9th Ser.,* Vol. 23, 173–292.

Gadroy, F. (1746). *Mémoire sur la Poussée des Terres,* summarized by Mayniel, 1808.

Gautier, H. (1717). *Dissertation sur L'Epaisseur des Culées des Ponts . . . sur L'Effort et al Pesanteur des Arches . . . et sur les Profiles de Maconnerie qui Doivent Supporter des Chaussées, des Terrasses, et des Remparts.* Cailleau, Paris.

Kerisel, J. (1985). "The History of Geotechnical Engineering up until 1700," *Proceedings,* XI International Conference on Soil Mechanics and Foundation Engineering, San Francisco, Golden Jubilee Volume, A. A. Balkema, 3–93.

Mayniel, J. J. (1808). *Traité Experimentale, Analytique et Pratique de la Poussé des Terres.* Colas, Paris.

Navier, C. L. M. (1839). *Leçons sur L'Application de la Mécanique à L'Establissement des Constructions et des Machines,* 2nd ed., Paris.

Peck, R. B. (1985). "The Last Sixty Years," *Proceedings,* XI International Conference on Soil Mechanics and Foundation Engineering, San Francisco, Golden Jubilee Volume, A. A. Balkema, 123–133.

Poncelet, J. V. (1840). *Mémoire sur la Stabilité des Revêtments et de seurs Fondations,* Bachelier, Paris.

Prony, G. C. M. L. R. (1790), *Nouvelle Architecture Hydraulique, contenant l' art d'élever l'eau au moyen de différentes machines, de construire dans ce fluide, de le diriger, et généralement de l'appliquer, de diverses manières, aux besoins de la sociétè,* FirminDidot, Paris.

Rankine, W. J. M. (1857). "On the Stability of Loose Earth," *Philosophical Transactions,* Royal Society, Vol. 147, London.

Reynolds, O. (1887). "Experiments Showing Dilatency, a Property of Granular Material Possibly Connected to Gravitation," *Proceedings,* Royal Society, London, Vol. 11, 354–363.

Skempton, A. W. (1948). "The $\phi = 0$ Analysis of Stability and Its Theoretical Basis," *Proceedings,* II International Conference on Soil Mechanics and Foundation Engineering, Rotterdam, Vol. 1, 72–78.

Skempton, A. W. (1954). "The Pore Pressure Coefficients A and B," *Geotechnique,* Vol. 4, 143–147.

Skempton, A. W. (1985). "A History of Soil Properties, 1717–1927," *Proceedings,* XI International Conference on Soil Mechanics and Foundation Engineering, San Francisco, Golden Jubilee Volume, A. A. Balkema, 95–121.

Taylor, D. W. (1948). *Fundamentals of Soil Mechanics,* John Wiley, New York.

Terzaghi, K. (1925). *Erdbaumechanik auf Bodenphysikalisher Grundlage,* Deuticke, Vienna.

Terzaghi, K. (1939). "Soil Mechanics—A New Chapter in Engineering Science," *Institute of Civil Engineers Journal,* London, Vol. 12, No. 7, 106–142.

Terzaghi, K. (1943). *Theoretical Soil Mechanics,* John Wiley, New York.

Terzaghi, K., and Peck, R. B. (1948). *Soil Mechanics in Engineering Practice,* John Wiley, New York.

CHAPTER 2

Origin of Soil and Grain Size

2.1 Introduction

In general, soils are formed by weathering of rocks. The physical properties of soil are dictated primarily by the minerals that constitute the soil particles and, hence, the rock from which it is derived. In this chapter we will discuss the following:

- The formation of various types of rocks, the origins of which are the solidification of molten magma in the mantle of the earth
- Formation of soil by mechanical and chemical weathering of rock
- Determination of the distribution of particle sizes in a given soil mass
- Composition of the clay minerals. The clay minerals provide the plastic properties of a soil mass
- The shape of various particles in a soil mass

2.2 Rock Cycle and the Origin of Soil

The mineral grains that form the solid phase of a soil aggregate are the product of rock weathering. The size of the individual grains varies over a wide range. Many of the physical properties of soil are dictated by the size, shape, and chemical composition of the grains. To better understand these factors, one must be familiar with the basic types of rock that form the earth's crust, the rock-forming minerals, and the weathering process.

On the basis of their mode of origin, rocks can be divided into three basic types: *igneous, sedimentary*, and *metamorphic*. Figure 2.1 shows a diagram of the formation cycle of different types of rock and the processes associated with them. This is called the *rock cycle*. Brief discussions of each element of the rock cycle follow.

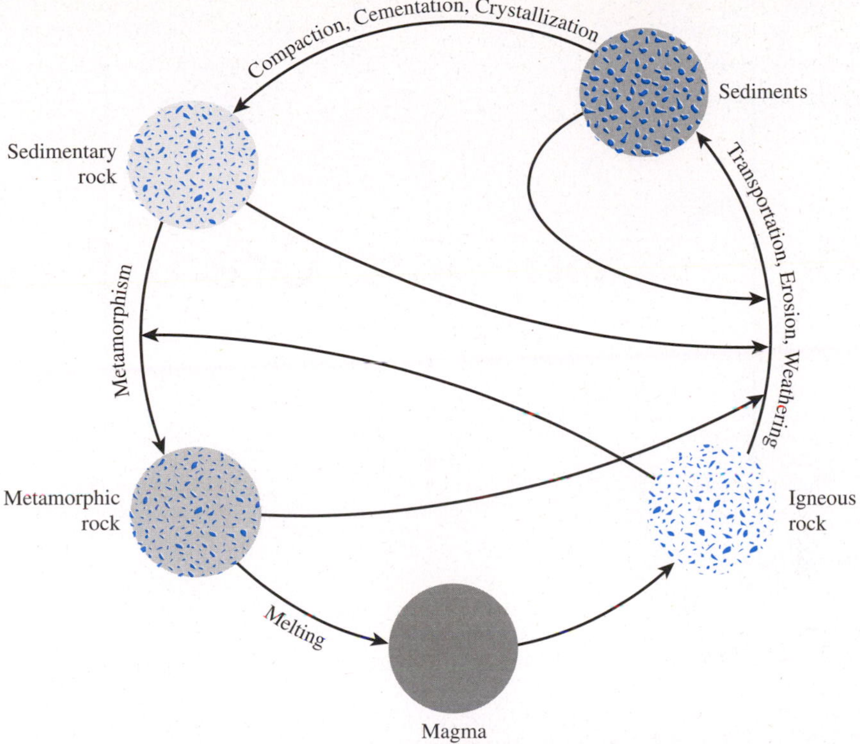

Figure 2.1 Rock cycle

Igneous rock

Igneous rocks are formed by the solidification of molten *magma* ejected from deep within the earth's mantle. After ejection by either *fissure eruption* or *volcanic eruption*, some of the molten magma cools on the surface of the earth. Sometimes magma ceases its mobility below the earth's surface and cools to form intrusive igneous rocks that are called *plutons*. Intrusive rocks formed in the past may be exposed at the surface as a result of the continuous process of erosion of the materials that once covered them.

The types of igneous rock formed by the cooling of magma depend on factors such as the composition of the magma and the rate of cooling associated with it. After conducting several laboratory tests, Bowen (1922) was able to explain the relation of the rate of magma cooling to the formation of different types of rock. This explanation—known as *Bowen's reaction principle*—describes the sequence by which new minerals are formed as magma cools. The mineral crystals grow larger and some of them settle. The crystals that remain suspended in the liquid react with the remaining melt to form a new mineral at a lower temperature. This process continues until the entire body of melt is solidified. Bowen classified these reactions into two groups: (1) *discontinuous ferromagnesian reaction series*, in which

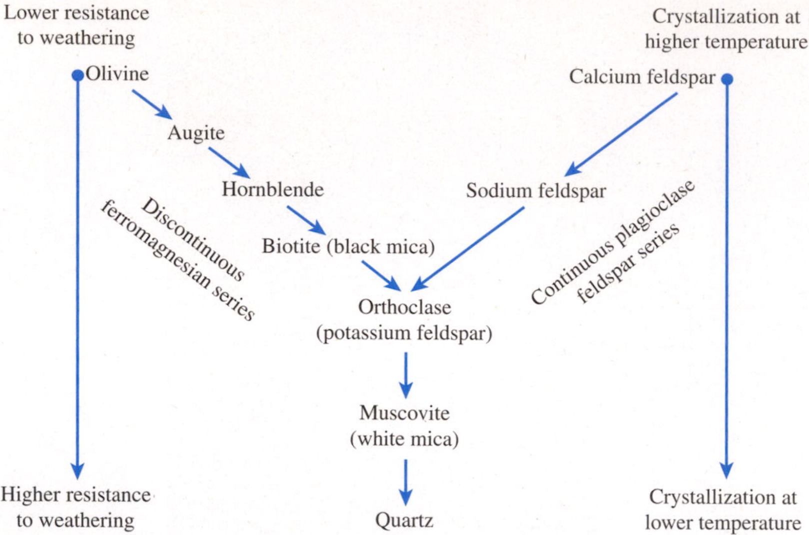

Figure 2.2 Bowen's reaction series

the minerals formed are different in their chemical composition and crystalline structure, and (2) *continuous plagioclase feldspar reaction series*, in which the minerals formed have different chemical compositions with similar crystalline structures. Figure 2.2 shows Bowen's reaction series. The chemical compositions of the minerals are given in Table 2.1. Figure 2.3 is a scanning electron micrograph of a fractured surface of quartz showing glass-like fractures with no discrete planar cleavage. Figure 2.4 is a scanning electron micrograph that shows basal cleavage of individual mica grains.

Thus, depending on the proportions of minerals available, different types of igneous rock are formed. Granite, gabbro, and basalt are some of the common types of igneous rock generally encountered in the field. Table 2.2 shows the general composition of some igneous rocks.

Table 2.1 Composition of Minerals Shown in Bowen's Reaction Series

Mineral		Composition
Olivine		$(Mg, Fe)_2SiO_4$
Augite		$Ca, Na(Mg, Fe, Al)(Al, Si_2O_6)$
Hornblende		Complex ferromagnesian silicate of Ca, Na, Mg, Ti, and Al
Biotite (black mica)		$K(Mg, Fe)_3AlSi_3O_{10}(OH)_2$
Plagioclase	calcium feldspar	$Ca(Al_2Si_2O_8)$
	sodium feldspar	$Na(AlSi_3O_8)$
Orthoclase (potassium feldspar)		$K(AlSi_3O_8)$
Muscovite (white mica)		$KAl_3Si_3O_{10}(OH)_2$
Quartz		SiO_2

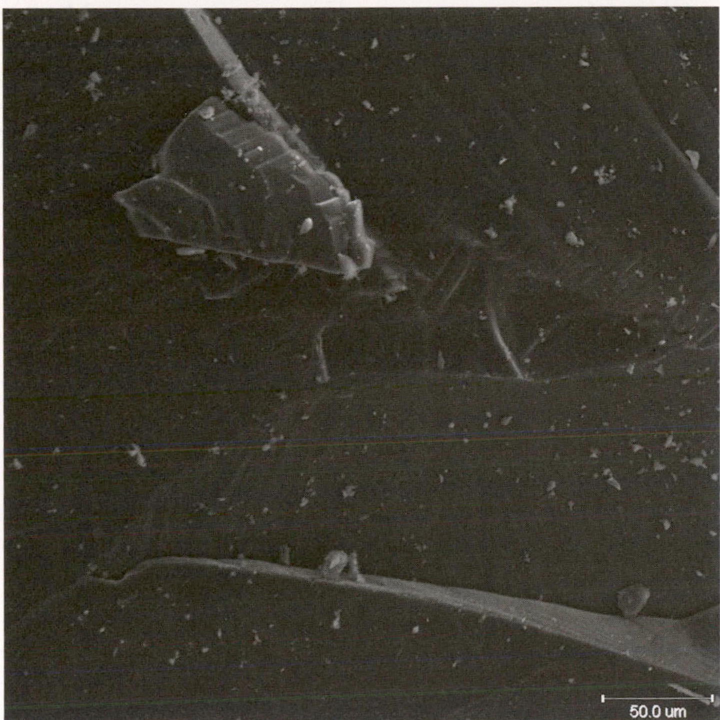

Figure 2.3 Scanning electron micrograph of fractured surface of quartz showing glass-like fractures with no discrete planar surface (*Courtesy of David J. White, Iowa State University, Ames, Iowa*)

Figure 2.4 Scanning electron micrograph showing basal cleavage of individual mica grains (*Courtesy of David J. White, Iowa State University, Ames, Iowa*)

Table 2.2 Composition of Some Igneous Rocks

Name of rock	Mode of occurrence	Texture	Abundant minerals	Less abundant minerals
Granite	Intrusive	Coarse	Quartz, sodium feldspar, potassium feldspar	Biotite, muscovite, hornblende
Rhyolite	Extrusive	Fine		
Gabbro	Intrusive	Coarse	Plagioclase, pyroxines, olivine	Hornblende, biotite, magnetite
Basalt	Extrusive	Fine		
Diorite	Intrusive	Coarse	Plagioclase, hornblende	Biotite, pyroxenes (quartz usually absent)
Andesite	Extrusive	Fine		
Syenite	Intrusive	Coarse	Potassium feldspar	Sodium feldspar, biotite, hornblende
Trachyte	Extrusive	Fine		
Peridotite	Intrusive	Coarse	Olivine, pyroxenes	Oxides of iron

In Table 2.2, the modes of occurrence of various rocks are classified as *intrusive* or *extrusive*. The intrusive rocks are those formed by the cooling of lava beneath the surface. Since the cooling process is very slow, intrusive rocks have very large crystals (coarse grained) and can be seen by the naked eye. When the lava cools on the surface (extrusive rocks), the process is fast. Grains are fine; thus they are difficult to identify by the naked eye.

Weathering

Weathering is the process of breaking down rocks by *mechanical* and *chemical processes* into smaller pieces. Mechanical weathering may be caused by the expansion and contraction of rocks from the continuous gain and loss of heat, which results in ultimate disintegration. Frequently, water seeps into the pores and existing cracks in rocks. As the temperature drops, the water freezes and expands. The pressure exerted by ice because of volume expansion is strong enough to break down even large rocks. Other physical agents that help disintegrate rocks are glacier ice, wind, the running water of streams and rivers, and ocean waves. It is important to realize that, in mechanical weathering, large rocks are broken down into smaller pieces without any change in the chemical composition. Figure 2.5 shows several examples of mechanical erosion due to ocean waves and wind at Yehliu in Taiwan. This area is located at a long and narrow sea cape at the northwest side of Keelung, about 15 kilometers between the north coast of Chin Shan and Wanli. Figure 2.6 shows another example of mechanical weathering in the Precambrian granite outcrop in the Elephant Rocks State Park in southeast Missouri. The freezing and thawing action of water on the surface fractures the rock and creates large cracks and a drainage pattern in the rock (Figure 2.6a). Over a period of time, unweathered rock is transformed into large boulders (Figure 2.6b). Figure 2.7 shows another photograph of *in situ* weathering of granite.

In chemical weathering, the original rock minerals are transformed into new minerals by chemical reaction. Water and carbon dioxide from the atmosphere form carbonic acid, which reacts with the existing rock minerals to form new minerals and soluble salts. Soluble salts present in the groundwater and organic acids formed from

(text continues on page 24)

Figure 2.5 Mechanical erosion due to ocean waves and wind at Yehliu, Taiwan *(Courtesy of Braja Das, Henderson, Nevada)*

Figure 2.5 (*Continued*)

(a)

(b)

Figure 2.6
Mechanical weathering of granite: (a) development of large cracks due to freezing and thawing followed by a drainage pattern, (b) transformation of unweathered rock into large boulders (*Courtesy of Janice Das, Henderson, Nevada*)

Figure 2.7 *In situ* mechanical weathering of granite (*Courtesy of Richard L. Handy, Iowa State University, Ames, Iowa*)

decayed organic matter also cause chemical weathering. An example of the chemical weathering of orthoclase to form clay minerals, silica, and soluble potassium carbonate follows:

$$H_2O + CO_2 \rightarrow \underset{\text{Carbonic acid}}{H_2CO_3} \rightarrow H^+ + (HCO_3)^-$$

$$\underset{\text{Orthoclase}}{2K(AlSi_3O_8)} + 2H^+ + H_2O \rightarrow 2K^+ + \underset{\text{Silica}}{4SiO_2} + \underset{\substack{\text{Kaolinite} \\ \text{(Clay mineral)}}}{Al_2Si_2O_5(OH)_4}$$

Most of the potassium ions released are carried away in solution as potassium carbonate is taken up by plants.

The chemical weathering of plagioclase feldspars is similar to that of orthoclase in that it produces clay minerals, silica, and different soluble salts. Ferromagnesian minerals also form the decomposition products of clay minerals, silica, and soluble salts. Additionally, the iron and magnesium in ferromagnesian minerals result in other products such as hematite and limonite. Quartz is highly resistant to weathering and only slightly soluble in water. Figure 2.2 shows the susceptibility of rock-forming minerals to weathering. The minerals formed at higher temperatures in Bowen's reaction series are less resistant to weathering than those formed at lower temperatures.

The weathering process is not limited to igneous rocks. As shown in the rock cycle (Figure 2.1), sedimentary and metamorphic rocks also weather in a similar manner.

Thus, from the preceding brief discussion, we can see how the weathering process changes solid rock masses into smaller fragments of various sizes that can range from large boulders to very small clay particles. Uncemented aggregates of these small grains in various proportions form different types of soil. The clay minerals, which are a product of chemical weathering of feldspars, ferromagnesians, and micas, give the plastic property to soils. There are three important clay minerals: (1) *kaolinite*, (2) *illite*, and (3) *montmorillonite*. (We discuss these clay minerals later in this chapter.)

Transportation of weathering products

The products of weathering may stay in the same place or may be moved to other places by ice, water, wind, and gravity.

The soils formed by the weathered products at their place of origin are called *residual soils*. An important characteristic of residual soil is the gradation of particle size. Fine-grained soil is found at the surface, and the grain size increases with depth. At greater depths, angular rock fragments may also be found.

The transported soils may be classified into several groups, depending on their mode of transportation and deposition:

1. *Glacial soils*—formed by transportation and deposition of glaciers
2. *Alluvial soils*—transported by running water and deposited along streams
3. *Lacustrine soils*—formed by deposition in quiet lakes
4. *Marine soils*—formed by deposition in the seas
5. *Aeolian soils*—transported and deposited by wind
6. *Colluvial soils*—formed by movement of soil from its original place by gravity, such as during landslides

Sedimentary rock

The deposits of gravel, sand, silt, and clay formed by weathering may become compacted by overburden pressure and cemented by agents like iron oxide, calcite, dolomite, and quartz. Cementing agents are generally carried in solution by groundwater. They fill the spaces between particles and form sedimentary rock. Rocks formed in this way are called *detrital sedimentary* rocks.

All detrital rocks have a *clastic* texture. The following are some examples of detrital rocks with clastic texture.

Particle size	Sedimentary rock
Granular or larger (grain size 2 mm–4 mm or larger)	Conglomerate
Sand	Sandstone
Silt and clay	Mudstone and shale

In the case of conglomerates, if the particles are more angular, the rock is called *breccia*. In sandstone, the particle sizes may vary between $\frac{1}{16}$ mm and 2 mm. When the grains in sandstone are practically all quartz, the rock is referred to as *orthoquartzite*. In mudstone and shale, the size of the particles are generally less than $\frac{1}{16}$ mm. Mudstone has a blocky aspect; whereas, in the case of shale, the rock is split into platy slabs.

Sedimentary rock also can be formed by chemical processes. Rocks of this type are classified as *chemical sedimentary rock*. These rocks can have *clastic* or *nonclastic texture*. The following are some examples of chemical sedimentary rock.

Composition	Rock
Calcite ($CaCO_3$)	Limestone
Halite (NaCl)	Rock salt
Dolomite [$CaMg(CO_3)$]	Dolomite
Gypsum ($CaSO_4 \cdot 2H_2O$)	Gypsum

Limestone is formed mostly of calcium carbonate deposited either by organisms or by an inorganic process. Most limestones have a clastic texture; however, nonclastic textures also are found commonly. Figure 2.8 shows the scanning electron micrograph of a fractured surface of limestone. Individual grains of calcite show rhombohedral cleavage. Chalk is a sedimentary rock made in part from biochemically derived calcite, which are skeletal fragments of microscopic plants and animals. Dolomite is formed either by chemical deposition of mixed carbonates or by the

Figure 2.8 Scanning electron micrograph of the fractured surface of limestone (*Courtesy of David J. White, Iowa State University, Ames, Iowa*)

reaction of magnesium in water with limestone. Gypsum and anhydrite result from the precipitation of soluble $CaSO_4$ due to evaporation of ocean water. They belong to a class of rocks generally referred to as *evaporites*. Rock salt (NaCl) is another example of an evaporite that originates from the salt deposits of seawater.

Sedimentary rock may undergo weathering to form sediments or may be subjected to the process of *metamorphism* to become metamorphic rock.

Metamorphic rock

Metamorphism is the process of changing the composition and texture of rocks (without melting) by heat and pressure. During metamorphism, new minerals are formed, and mineral grains are sheared to give a foliated texture to metamorphic rock. Gneiss is a metamorphic rock derived from high-grade regional metamorphism of igneous rocks, such as granite, gabbro, and diorite. Low-grade metamorphism of shales and mudstones results in slate. The clay minerals in the shale become chlorite and mica by heat; hence, slate is composed primarily of mica flakes and chlorite. Phyllite is a metamorphic rock, which is derived from slate with further metamorphism being subjected to heat greater than 250 to 300°C. Schist is a type of metamorphic rock derived from several igneous, sedimentary, and low-grade metamorphic rocks with a well-foliated texture and visible flakes of platy and micaceous minerals. Metamorphic rock generally contains large quantities of quartz and feldspar as well.

Marble is formed from calcite and dolomite by recrystallization. The mineral grains in marble are larger than those present in the original rock. Green marbles are colored by hornblends, serpentine, or talc. Black marbles contain bituminous material, and brown marbles contain iron oxide and limonite. Quartzite is a metamorphic rock formed from quartz-rich sandstones. Silica enters into the void spaces between the quartz and sand grains and acts as a cementing agent. Quartzite is one of the hardest rocks. Under extreme heat and pressure, metamorphic rocks may melt to form magma, and the cycle is repeated.

2.3 Rock-Forming Minerals, Rock and Rock Structures

In the preceding section we were introduced to the process of the formation of igneous rocks from rock-forming minerals, weathering and formation of sedimentary rocks, and metamorphism and formation of metamorphic rocks. Color insert CI.1 shows some common rock-forming minerals, such as quartz, orthoclase, plagioclase, muscovite, biotite, andradite, garnet, calcite, dolomite, and chlorite. Some common types of rocks that geotechnical engineers may encounter in the field, such as granite, basalt, rhyolite, sandstone, limestone, conglomerate, marble, slate, and schist, are shown in the color insert CI.2. Color insert CI.2j shows an example of *folded schist* from the James Cook University Rock Garden on its campus in Townsville, Australia. Shear stresses and metamorphism involving high temperature and pressure caused the layers to buckle and fold. Color insert CI.3 shows some structures constructed on rock. Figures CI.1 through CI.3 are given after 40.

There are large structures built several centuries ago around the world with, or in/on rock, that are still intact and undergoing partial weathering. The Parthenon (Figure CI.3a), built on the Acropolis in Athens, Greece, in the second half of the 5th century B.C., is made of marble and built on a limestone hill underlain by phyllite, a fine-grained metamorphic rock containing large quantities of mica and resembling slate or schist.

Figure CI.3b shows the Corinth Canal in Greece. The Corinth Canal crosses the Isthmus of Corinth, a narrow strip of land that connects Peloponnesus to the mainland of Greece, thus linking the Saronic Gulf in the Aegean Sea (eastern part of Greece) with the Gulf of Corinth (a deep inlet of the Ionian Sea in western Greece). The canal was completed in 1893. The canal consists of a single channel 8 m deep excavated at sea level (thus requiring no locks) measuring 6346 m long and is 24.6 m wide at the top and 21.3 m wide at the bottom. The canal slopes have an inclination of 3V:1H to 5V:1H. The central part of the canal, where the excavated slopes are highest, consists of Plio-Pleistocene marls with thin interlayers of marly sands and marly limestone. The marls in the upper part of the slopes are whitish yellow to light brown, while those in the middle and lower parts are yellow gray to bluish gray.

2.4 Soil-Particle Size

As discussed in the preceding section, the sizes of particles that make up soil vary over a wide range. Soils generally are called *gravel, sand, silt,* or *clay,* depending on the predominant size of particles within the soil. To describe soils by their particle size, several organizations have developed particle-size classifications. Table 2.3 shows the particle-size classifications developed by the Massachusetts Institute of Technology, the U.S. Department of Agriculture, the American Association of State Highway and Transportation Officials, and the U.S. Army Corps of Engineers and U.S. Bureau of Reclamation. In this table, the MIT system is presented for illustration purposes only. This system is important in the history of the development of the size limits of particles present in soils; however, the Unified Soil Classification System is now almost universally accepted and has been adopted by the American Society for Testing and Materials (ASTM).

Table 2.3 Particle-Size Classifications

	Grain size (mm)			
Name of organization	**Gravel**	**Sand**	**Silt**	**Clay**
Massachusetts Institute of Technology (MIT)	>2	2 to 0.06	0.06 to 0.002	<0.002
U.S. Department of Agriculture (USDA)	>2	2 to 0.05	0.05 to 0.002	<0.002
American Association of State Highway and Transportation Officials (AASHTO)	76.2 to 2	2 to 0.075	0.075 to 0.002	<0.002
Unified Soil Classification System (U.S. Army Corps of Engineers, U.S. Bureau of Reclamation, and American Society for Testing and Materials)	76.2 to 4.75	4.75 to 0.075	Fines (i.e., silts and clays) <0.075	

Note: Sieve openings of 4.75 mm are found on a U.S. No. 4 sieve; 2-mm openings on a U.S. No. 10 sieve; 0.075-mm openings on a U.S. No. 200 sieve. See Table 2.5.

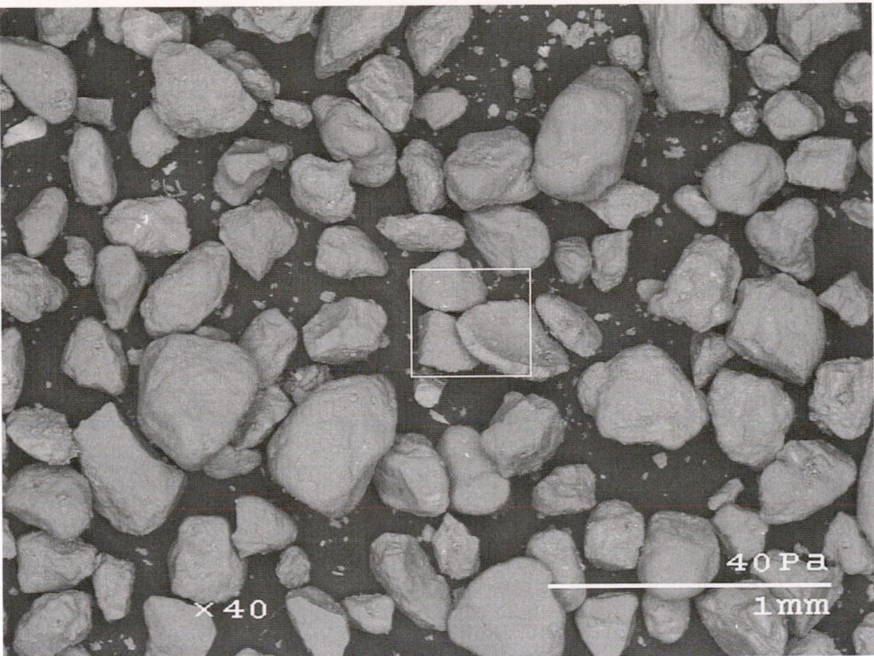

Figure 2.9 Scanning electron micrograph of some sand grains *(Courtesy of David J. White, Iowa State University, Ames, Iowa)*

Gravels are pieces of rocks with occasional particles of quartz, feldspar, and other minerals. *Sand* particles are made of mostly quartz and feldspar. Other mineral grains also may be present at times. Figure 2.9 shows the scanning electron micrograph of some sand grains. Note that the larger grains show rounding that can occur as a result of wear during intermittent transportation by wind and/or water. Figure 2.10 is a higher magnification of the grains highlighted in Figure 2.9, and it reveals a few small clay particles adhering to larger sand grains. *Silts* are the microscopic soil fractions that consist of very fine quartz grains and some flake-shaped particles that are fragments of micaceous minerals. *Clays* are mostly flake-shaped microscopic and submicroscopic particles of mica, clay minerals, and other minerals.

As shown in Table 2.3, clays generally are defined as particles smaller than 0.002 mm. However, in some cases, particles between 0.002 and 0.005 mm in size also are referred to as clay. Particles classified as clay on the basis of their size may not necessarily contain clay minerals. Clays have been defined as those particles "which develop plasticity when mixed with a limited amount of water" (Grim, 1953). (Plasticity is the putty-like property of clays that contain a certain amount of water.) Nonclay soils can contain particles of quartz, feldspar, or mica that are small enough to be within the clay classification. Hence, it is appropriate for soil particles smaller than 2 microns (2 μm), or 5 microns (5 μm) as defined under different systems, to be called clay-sized particles rather than clay. Clay particles are mostly in the colloidal size range (<1 μm), and 2 μm appears to be the upper limit.

Figure 2.10 Higher magnification of the sand grains highlighted in Figure 2.9 *(Courtesy of David J. White, Iowa State University, Ames, Iowa)*

2.5 Clay Minerals

Clay minerals are complex aluminum silicates composed of two basic units: (1) *silica tetrahedron* and (2) *alumina octahedron*. Each tetrahedron unit consists of four oxygen atoms surrounding a silicon atom (Figure 2.11a). The combination of tetrahedral silica units gives a *silica sheet* (Figure 2.11b). Three oxygen atoms at the base of each tetrahedron are shared by neighboring tetrahedra. The octahedral units consist of six hydroxyls surrounding an aluminum atom (Figure 2.11c), and the combination of the octahedral aluminum hydroxyl units gives an *octahedral sheet*. (This also is called a *gibbsite sheet*—Figure 2.11d.) Sometimes magnesium replaces the aluminum atoms in the octahedral units; in this case, the octahedral sheet is called a *brucite sheet*.

In a silica sheet, each silicon atom with a positive charge of four is linked to four oxygen atoms with a total negative charge of eight. But each oxygen atom at the base of the tetrahedron is linked to two silicon atoms. This means that the top oxygen atom of each tetrahedral unit has a negative charge of one to be counterbalanced. When the silica sheet is stacked over the octahedral sheet, as shown in Figure 2.11e, these oxygen atoms replace the hydroxyls to balance their charges.

Of the three important clay minerals, *kaolinite* consists of repeating layers of elemental silica-gibbsite sheets in a 1:1 lattice, as shown in Figures 2.12 and 2.13a. Each layer is about 7.2 Å thick. The layers are held together by hydrogen bonding.

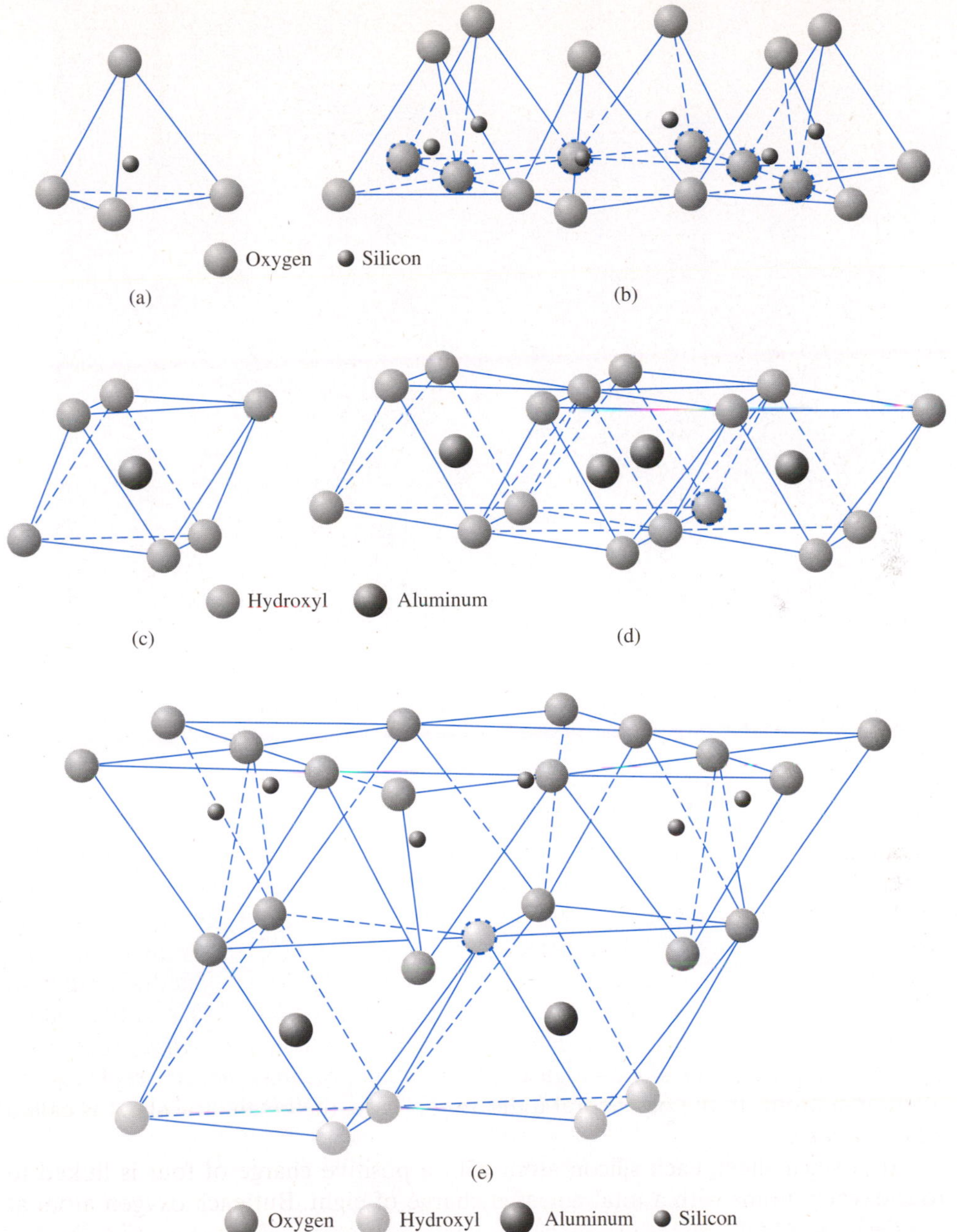

Figure 2.11 (a) Silica tetrahedron; (b) silica sheet; (c) alumina octahedron; (d) octahedral (gibbsite) sheet; (e) elemental silica-gibbsite sheet (After Grim, 1959) (*From Grim, "Physico-Chemical Properties of Soils: Clay Minerals,"* Journal of the Soil Mechanics and Foundations Division, *ASCE, Vol. 85, No. SM2, 1959, pp. 1–17. With permission from ASCE.*)

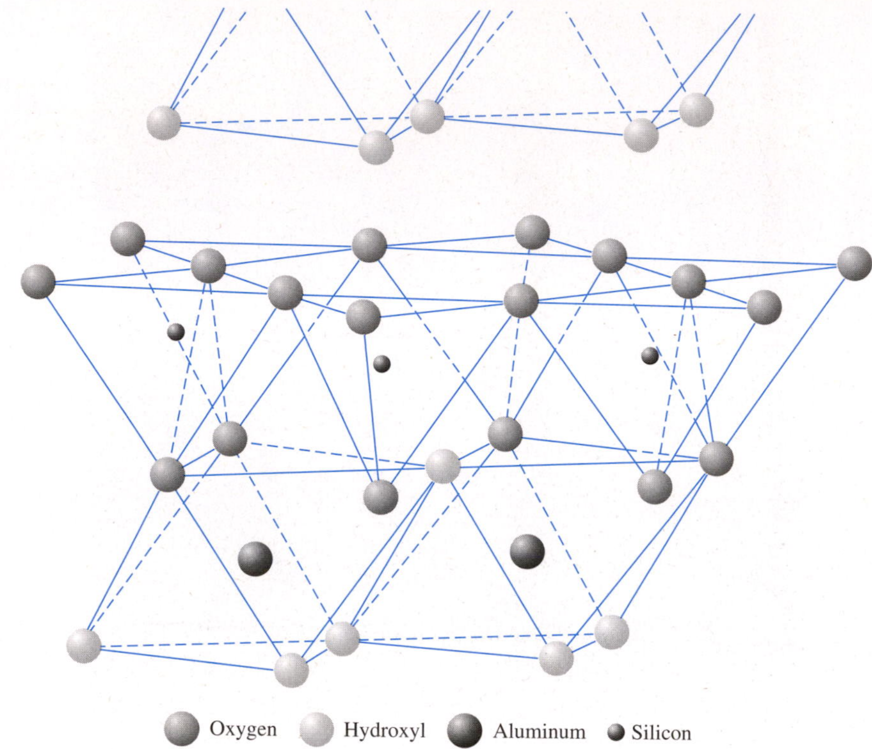

Figure 2.12 Atomic structure of kaolinite (*After Grim, 1959. With permission from ASCE.*)

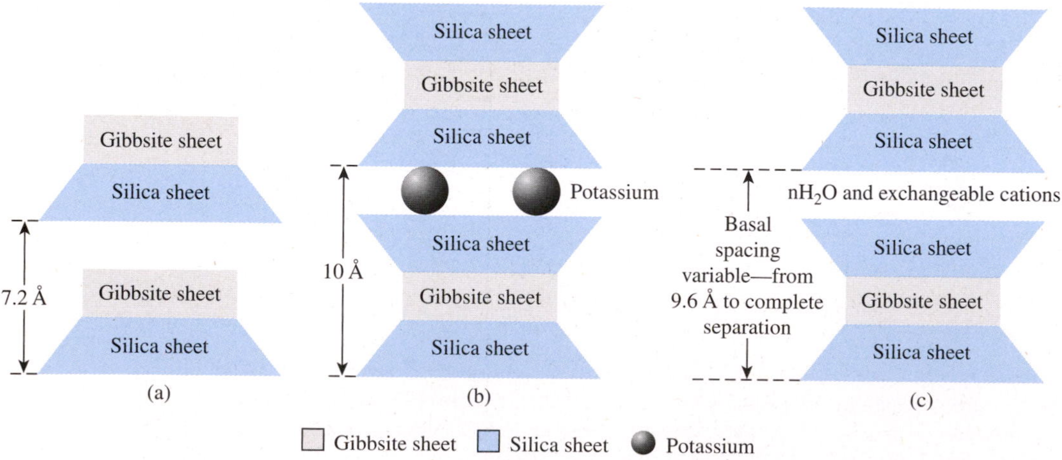

Figure 2.13 Diagram of the structures of (a) kaolinite; (b) illite; (c) montmorillonite (*Note*: 1 Å = 10^{-10} m)

Figure 2.14 Scanning electron micrograph of a kaolinite specimen (*Courtesy of David J. White, Iowa State University, Ames, Iowa*)

Kaolinite occurs as platelets, each with a lateral dimension of 1000 to 20,000 Å and a thickness of 100 to 1000 Å. The surface area of the kaolinite particles per unit mass is about 15 m^2/g. The surface area per unit mass is defined as *specific surface*. Figure 2.14 shows a scanning electron micrograph of a kaolinite specimen.

Illite consists of a gibbsite sheet bonded to two silica sheets—one at the top and another at the bottom (Figures 2.15 and 2.13b). It is sometimes called *clay mica*. The illite layers are bonded by potassium ions. The negative charge to balance the potassium ions comes from the substitution of aluminum for some silicon in the tetrahedral sheets. Substitution of one element for another with no change in the crystalline form is known as *isomorphous substitution*. Illite particles generally have lateral dimensions ranging from 1000 to 5000 Å and thicknesses from 50 to 500 Å. The specific surface of the particles is about 80 m^2/g.

Montmorillonite has a structure similar to that of illite—that is, one gibbsite sheet sandwiched between two silica sheets. (See Figures 2.16 and 2.13c.) In montmorillonite there is isomorphous substitution of magnesium and iron for aluminum in the octahedral sheets. Potassium ions are not present as in illite, and a large amount of water is attracted into the space between the layers. Particles of montmorillonite have lateral dimensions of 1000 to 5000 Å and thicknesses of 10 to 50 Å. The specific surface is about 800 m^2/g. Figure 2.17 is a scanning electron micrograph showing the fabric of montmorillonite.

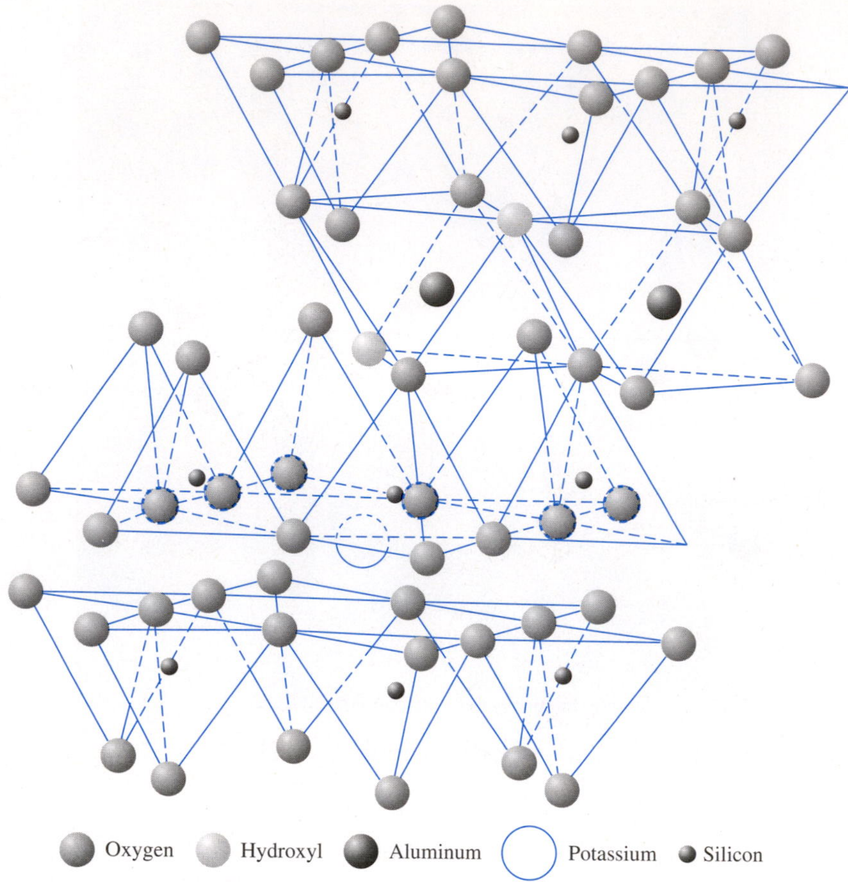

Figure 2.15 Atomic structure of illite *(After Grim, 1959. With permission from ASCE.)*

Besides kaolinite, illite, and montmorillonite, other common clay minerals generally found are chlorite, halloysite, vermiculite, and attapulgite.

The clay particles carry a net negative charge on their surfaces. This is the result both of isomorphous substitution and of a break in continuity of the structure at its edges. Larger negative charges are derived from larger specific surfaces. Some positively charged sites also occur at the edges of the particles. A list of the reciprocal of the average surface densities of the negative charges on the surfaces of some clay minerals follows (Yong and Warkentin, 1966):

Clay mineral	Reciprocal of average surface density of charge (Å^2/electronic charge)
Kaolinite	25
Clay mica and chlorite	50
Montmorillonite	100
Vermiculite	75

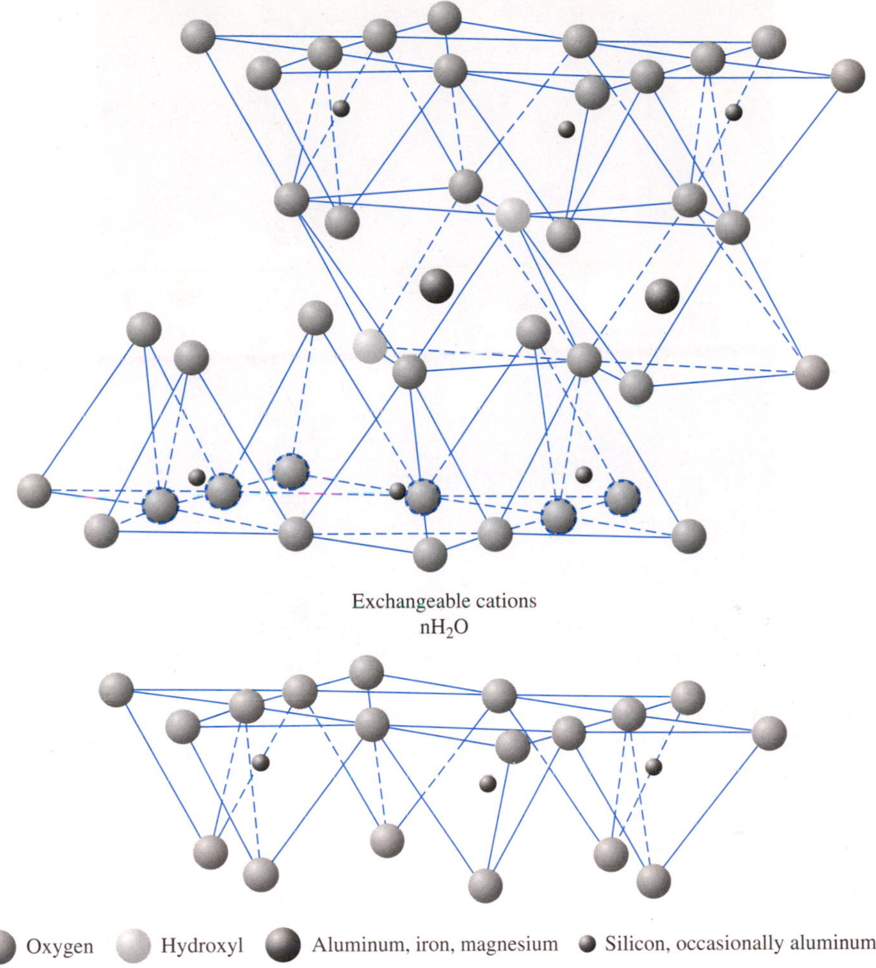

Figure 2.16 Atomic structure of montmorillonite *(After Grim, 1959. With permission from ASCE.)*

In dry clay, the negative charge is balanced by exchangeable cations like Ca^{2+}, Mg^{2+}, Na^+, and K^+ surrounding the particles being held by electrostatic attraction. When water is added to clay, these cations and a few anions float around the clay particles. This configuration is referred to as a *diffuse double layer* (Figure 2.18a). The cation concentration decreases with the distance from the surface of the particle (Figure 2.18b).

Water molecules are dipolar. Hydrogen atoms are not axisymmetric around an oxygen atom; instead, they occur at a bonded angle of 105° (Figure 2.19). As a result, a water molecule has a positive charge at one side and a negative charge at the other side. It is known as a *dipole*.

Dipolar water is attracted both by the negatively charged surface of the clay particles and by the cations in the double layer. The cations, in turn, are attracted to

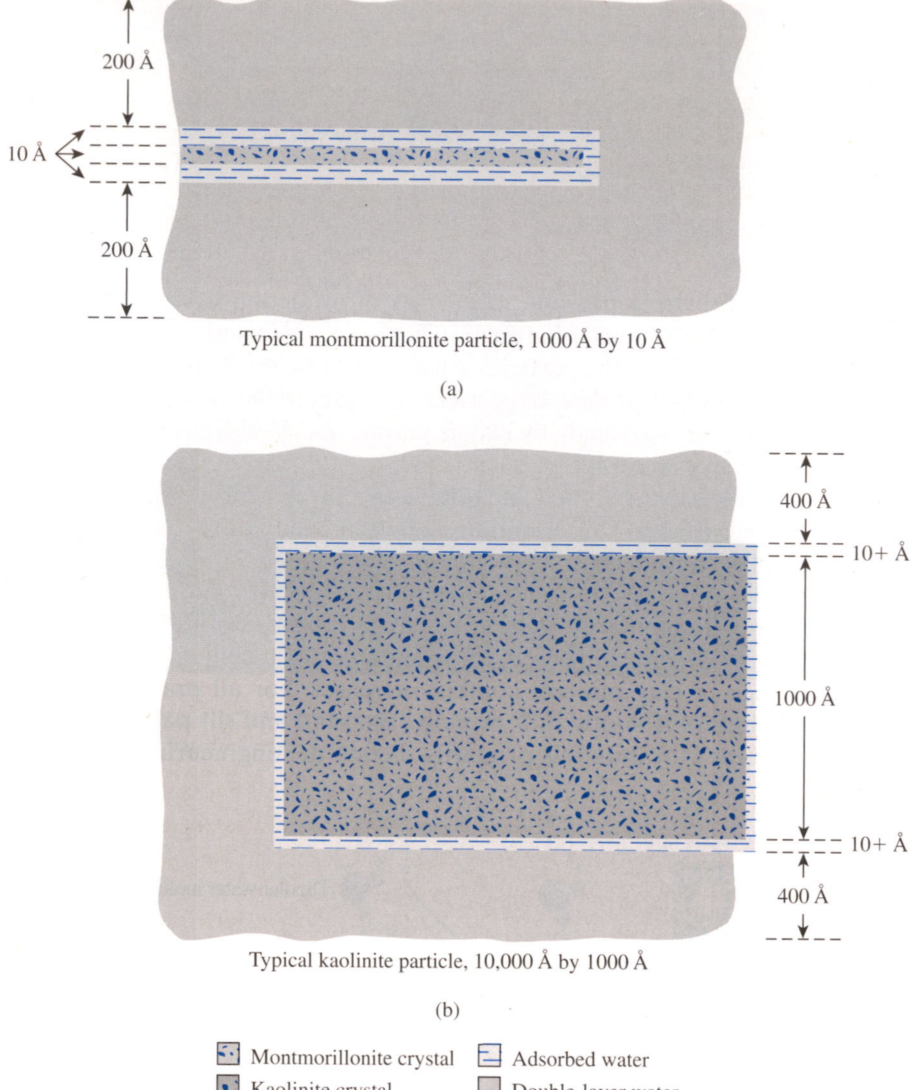

Figure 2.21 Clay water (*Redrawn after Lambe, 1958. With permission from ASCE.*)

2.6 Specific Gravity (G_s)

Specific gravity is defined as the ratio of the unit weight of a given material to the unit weight of water. The specific gravity of soil solids is often needed for various calculations in soil mechanics. It can be determined accurately in the laboratory. Table 2.4 shows the specific gravity of some common minerals found in soils. Most of the values fall within a range of 2.6 to 2.9. The specific gravity of solids of light-colored sand,

Table 2.4 Specific Gravity of Common Minerals

Mineral	Specific gravity, G_s
Quartz	2.65
Kaolinite	2.6
Illite	2.8
Montmorillonite	2.65–2.80
Halloysite	2.0–2.55
Potassium feldspar	2.57
Sodium and calcium feldspar	2.62–2.76
Chlorite	2.6–2.9
Biotite	2.8–3.2
Muscovite	2.76–3.1
Hornblende	3.0–3.47
Limonite	3.6–4.0
Olivine	3.27–3.7

which is mostly made of quartz, may be estimated to be about 2.65; for clayey and silty soils, it may vary from 2.6 to 2.9.

2.7 Mechanical Analysis of Soil

Mechanical analysis is the determination of the size range of particles present in a soil, expressed as a percentage of the total dry weight. Two methods generally are used to find the particle-size distribution of soil: (1) *sieve analysis*—for particle sizes larger than 0.075 mm in diameter, and (2) *hydrometer analysis*—for particle sizes smaller than 0.075 mm in diameter. The basic principles of sieve analysis and hydrometer analysis are described briefly in the following two sections.

Sieve analysis

Sieve analysis consists of shaking the soil sample through a set of sieves that have progressively smaller openings. The current size designation for U. S. sieves uses 100 mm to 6.3 mm, and they are listed in Table 2.5.

Table 2.5 U.S. Sieves by Size Designation

100.0 mm	25.0 mm
75.0 mm	19.0 mm
63.0 mm	16.0 mm
50.0 mm	12.5 mm
45.0 mm	9.5 mm
37.5 mm	8.0 mm
31.5 mm	6.3 mm

Table 2.6 U.S. Sieve Sizes with Number Designation

Sieve no.	Opening (mm)	Sieve no.	Opening (mm)
4	4.75	45	0.355
5	4.00	50	0.300
6	3.35	60	0.250
7	2.80	70	0.212
8	2.36	80	0.180
10	2.00	100	0.150
12	1.70	120	0.125
14	1.40	140	0.106
16	1.18	170	0.090
18	1.00	200	0.075
20	0.85	230	0.063
25	0.71	270	0.053
30	0.60	325	0.045
35	0.500	400	0.038
40	0.425		

After the 6.3-mm size designation, a number designation is used, i.e., No. 4 to No. 400. These are shown in Table 2.6.

The opening for the *i*th sieve given in Table 2.6 can be approximately given as

$$\text{Opening for the } i\text{th sieve} = \frac{\text{Opening for the } (i-1)\text{th sieve}}{(2)^{0.25}} \tag{2.1}$$

For example,

$$\text{The opening for the No. 5 sieve} = \frac{\text{Opening for the No. 4 sieve}}{(2)^{0.25}}$$

$$= \frac{4.75 \text{ mm}}{1.1892} = 3.994 \text{ mm} \approx 4.00 \text{ mm}$$

Similarly,

$$\text{The opening for the No. 50 sieve} = \frac{\text{Opening for the No. 45 sieve}}{(2)^{0.25}}$$

$$= \frac{0.335 \text{ mm}}{1.1892} = 0.2985 \text{ mm} \approx 0.300 \text{ mm}$$

Several other countries have their own sieve sizes that are commonly referred to by their aperture sizes. As an example, the British and Australian standard sieve sizes that have size designation are given in Tables 2.7 and 2.8, respectively.

(a)

(b)

Figure CI.1 Some typical rock-forming minerals: (a) quartz; (b) orthoclase.
(*Courtesy of S. K. Shukla, Edith Cowan University, Perth, Australia*)

(c)

Figure CI.1 Some typical rock-forming minerals: (c) plagioclase. (*Courtesy of S. K. Shukla, Edith Cowan University, Perth, Australia*)

(d)

Figure CI.1 Some typical rock-forming minerals: (d) muscovite. *(Courtesy of S. K. Shukla, Edith Cowan University, Perth, Australia)*

(e)

Figure CI.1 Some typical rock-forming minerals: (e) biotite. (*Courtesy of S. K. Shukla, Edith Cowan University, Perth, Australia*)

(f)

Figure CI.1 Some typical rock-forming minerals: (f) andradite garnet. (*Courtesy of S. K. Shukla, Edith Cowan University, Perth, Australia*)

(g)

Figure CI.1 Some typical rock-forming minerals: (g) calcite. (*Courtesy of S. K. Shukla, Edith Cowan University, Perth, Australia*)

(h)

Figure CI.1 Some typical rock-forming minerals: (h) dolomite. (*Courtesy of S. K. Shukla, Edith Cowan University, Perth, Australia*)

(i)

Figure CI.1 Some typical rock-forming minerals: (i) chlorite. (*Courtesy of S. K. Shukla, Edith Cowan University, Perth, Australia*)

(a)

Figure CI.2 Some typical rocks: (a) granite. (*Figure (a) courtesy of S. K. Shukla, Edith Cowan University, Perth, Australia*)

(b)

Figure CI.3 (b) Corinth Canal in Greece. (*Courtesy of N. Sivakugan, James Cook University, Townsville, Australia*)

Table 2.7 British Standard Sieves

75 mm	3.35 mm
63 mm	2 mm
50 mm	1.18 mm
37.5 mm	0.600 mm
28 mm	0.425 mm
20 mm	0.300 mm
14 mm	0.212 mm
10 mm	0.15 mm
6.3 mm	0.063 mm
5.0 mm	

Table 2.8 Australian Standard Sieves

75.0 mm	2.36 mm
63.0 mm	2 mm
37.5 mm	1.18 mm
26.5 mm	0.600 mm
19.0 mm	0.425 mm
13.2 mm	0.300 mm
9.50 mm	0.212 mm
6.70 mm	0.15 mm
4.75 mm	0.063 mm

In the U.S., for sandy and fine-grained soils, generally sieve Nos. 4, 10, 20, 30, 40, 60, 140, and 200 are used.

The sieves used for soil analysis are generally 203 mm in diameter. To conduct a sieve analysis, one must first oven-dry the soil and then break all lumps into small particles. The soil then is shaken through a stack of sieves with openings of decreasing size from top to bottom (a pan is placed below the stack). Figure 2.22 shows a set of sieves in a shaker used for conducting the test in the laboratory. The smallest-sized sieve that should be used for this type of test is the U.S. No. 200 sieve. After the soil is shaken, the mass of soil retained on each sieve is determined. When cohesive soils are analyzed, breaking the lumps into individual particles may be difficult. In this case, the soil may be mixed with water to make a slurry and then washed through the sieves. Portions retained on each sieve are collected separately and oven-dried before the mass retained on each sieve is measured.

1. Determine the mass of soil retained on each sieve (i.e., $M_1, M_2, \cdots M_n$) and in the pan (i.e., M_p).
2. Determine the total mass of the soil: $M_1 + M_2 + \cdots + M_i + \cdots + M_n + M_p = \Sigma M$.
3. Determine the cumulative mass of soil retained above each sieve. For the ith sieve, it is $M_1 + M_2 + \cdots + M_i$.

Figure 2.22 A set of sieves for a test in the laboratory (*Courtesy of Braja M. Das, Henderson, Nevada*)

4. The mass of soil passing the ith sieve is $\Sigma M - (M_1 + M_2 + \cdots + M_i)$.
5. The percent of soil passing the ith sieve (or *percent finer*) is

$$F = \frac{\Sigma M - (M_1 + M_2 + \cdots + M_i)}{\Sigma M} \times 100$$

Once the percent finer for each sieve is calculated (step 5), the calculations are plotted on semilogarithmic graph paper (Figure 2.23) with percent finer as the ordinate (arithmetic scale) and sieve opening size as the abscissa (logarithmic scale). This plot is referred to as the *particle-size distribution curve*.

Hydrometer analysis

Hydrometer analysis is based on the principle of sedimentation of soil grains in water. When a soil specimen is dispersed in water, the particles settle at different

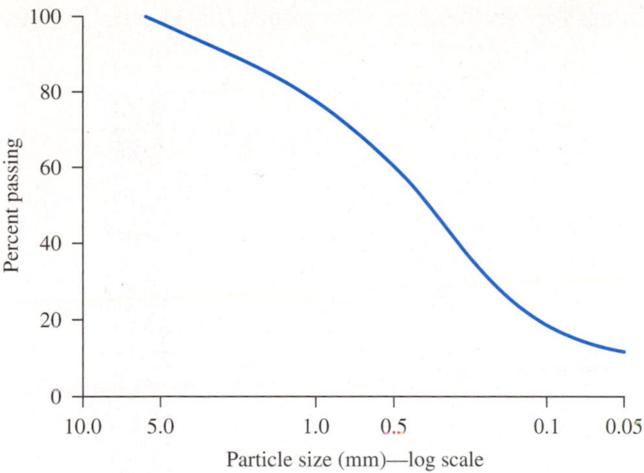

Figure 2.23 Particle-size distribution curve

velocities, depending on their shape, size, weight, and the viscosity of the water. For simplicity, it is assumed that all the soil particles are spheres and that the velocity of soil particles can be expressed by *Stokes' law*, according to which

$$v = \frac{\rho_s - \rho_w}{18\eta} D^2 \tag{2.2}$$

where v = velocity
ρ_s = density of soil particles
ρ_w = density of water
η = viscosity of water
D = diameter of soil particles

Thus, from Eq. (2.2),

$$D = \sqrt{\frac{18\eta v}{\rho_s - \rho_w}} = \sqrt{\frac{18\eta}{\rho_s - \rho_w}}\sqrt{\frac{L}{t}} \tag{2.3}$$

where $v = \dfrac{\text{Distance}}{\text{Time}} = \dfrac{L}{t}$.

Note that

$$\rho_s = G_s\rho_w \tag{2.4}$$

Thus, combining Eqs. (2.3) and (2.4) gives

$$D = \sqrt{\frac{18\eta}{(G_s - 1)\rho_w}}\sqrt{\frac{L}{t}} \tag{2.5}$$

If the units of η are (g · sec)/cm^2, ρ_w is in g/cm^3, L is in cm, t is in min, and D is in mm, then

$$\frac{D(\text{mm})}{10} = \sqrt{\frac{18\eta\,[(\text{g}\cdot\text{sec})/\text{cm}^2]}{(G_s - 1)\rho_w(\text{g/cm}^3)}}\sqrt{\frac{L\text{ (cm)}}{t\text{ (min)} \times 60}}$$

or

$$D = \sqrt{\frac{30\eta}{(G_s - 1)\rho_w}}\sqrt{\frac{L}{t}}$$

Assume ρ_w to be approximately equal to 1 g/cm^3, so that

$$D\text{ (mm)} = K\sqrt{\frac{L\text{ (cm)}}{t\text{ (min)}}} \tag{2.6}$$

where

$$K = \sqrt{\frac{30\eta}{(G_s - 1)}} \tag{2.7}$$

Note that the value of K is a function of G_s and η, which are dependent on the temperature of the test. Table 2.9 gives the variation of K with the test temperature and the specific gravity of soil solids.

In the laboratory, the hydrometer test is conducted in a sedimentation cylinder usually with 50 g of oven-dried sample. Sometimes 100-g samples also can be used. The sedimentation cylinder is 457 mm high and 63.5 mm in diameter. It is marked for a volume of 1000 ml. Sodium hexametaphosphate generally is used as the *dispersing agent*. The volume of the dispersed soil suspension is increased to 1000 ml by adding distilled water. An ASTM 152H type hydrometer (Figure 2.24) is then placed in the sedimentation cylinder (Figure 2.25).

When a hydrometer is placed in the soil suspension at a time t, measured from the start of sedimentation it measures the specific gravity in the vicinity of its bulb at a depth L (Figure 2.26). The specific gravity is a function of the amount of soil particles present per unit volume of suspension at that depth. Also, at a time t, the soil particles in suspension at a depth L will have a diameter smaller than D as calculated in Eq. (2.6). The larger particles would have settled beyond the zone of measurement. Hydrometers are designed to give the amount of soil, in grams, that is still in suspension. They are calibrated for soils that have a specific gravity, G_s, of 2.65; for soils of other specific gravity, a correction must be made.

By knowing the amount of soil in suspension, L, and t, we can calculate the percentage of soil by weight finer than a given diameter. Note that L is the depth measured from the surface of the water to the center of gravity of the hydrometer bulb at which the density of the suspension is measured. The value of L will change with time t. Hydrometer analysis is effective for separating soil fractions down to a

Table 2.9 Values of K from Eq. (2.7)[a]

Temperature (°C)	G_s							
	2.45	2.50	2.55	2.60	2.65	2.70	2.75	2.80
16	0.01510	0.01505	0.01481	0.01457	0.01435	0.01414	0.01394	0.01374
17	0.01511	0.01486	0.01462	0.01439	0.01417	0.01396	0.01376	0.01356
18	0.01492	0.01467	0.01443	0.01421	0.01399	0.01378	0.01359	0.01339
19	0.01474	0.01449	0.01425	0.01403	0.01382	0.01361	0.01342	0.01323
20	0.01456	0.01431	0.01408	0.01386	0.01365	0.01344	0.01325	0.01307
21	0.01438	0.01414	0.01391	0.01369	0.01348	0.01328	0.01309	0.01291
22	0.01421	0.01397	0.01374	0.01353	0.01332	0.01312	0.01294	0.01276
23	0.01404	0.01381	0.01358	0.01337	0.01317	0.01297	0.01279	0.01261
24	0.01388	0.01365	0.01342	0.01321	0.01301	0.01282	0.01264	0.01246
25	0.01372	0.01349	0.01327	0.01306	0.01286	0.01267	0.01249	0.01232
26	0.01357	0.01334	0.01312	0.01291	0.01272	0.01253	0.01235	0.01218
27	0.01342	0.01319	0.01297	0.01277	0.01258	0.01239	0.01221	0.01204
28	0.01327	0.01304	0.01283	0.01264	0.01244	0.01225	0.01208	0.01191
29	0.01312	0.01290	0.01269	0.01249	0.01230	0.01212	0.01195	0.01178
30	0.01298	0.01276	0.01256	0.01236	0.01217	0.01199	0.01182	0.01169

[a] *After ASTM (2014). Copyright ASTM INTERNATIONAL. Reprinted with permission.*

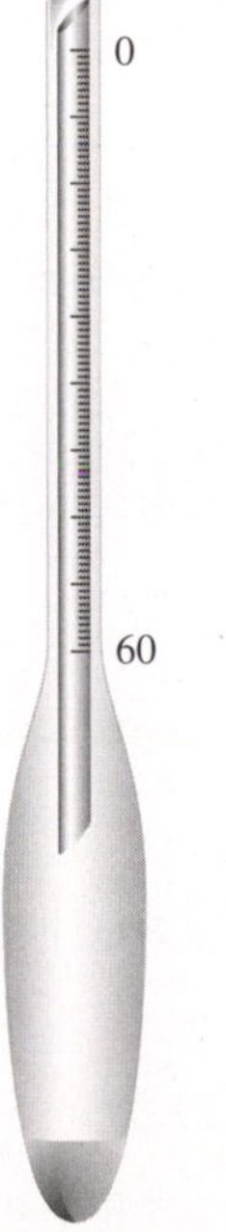

Figure 2.24 ASTM 152H hydrometer (*Courtesy of ELE International*)

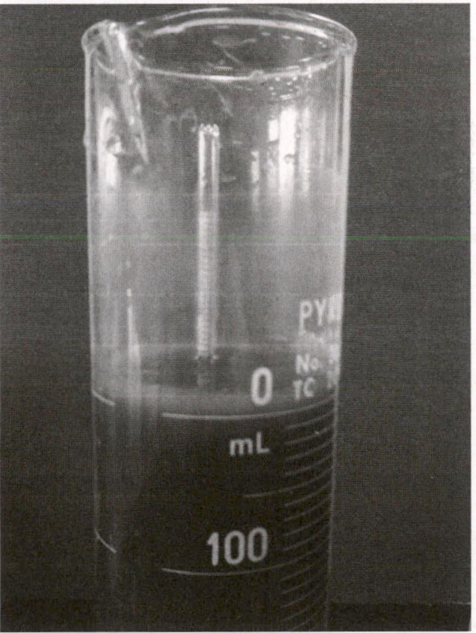

Figure 2.25 ASTM 152H type of hydrometer placed inside the sedimentation cylinder (*Courtesy of Khaled Sobhan, Florida Atlantic University, Boca Raton, Florida*)

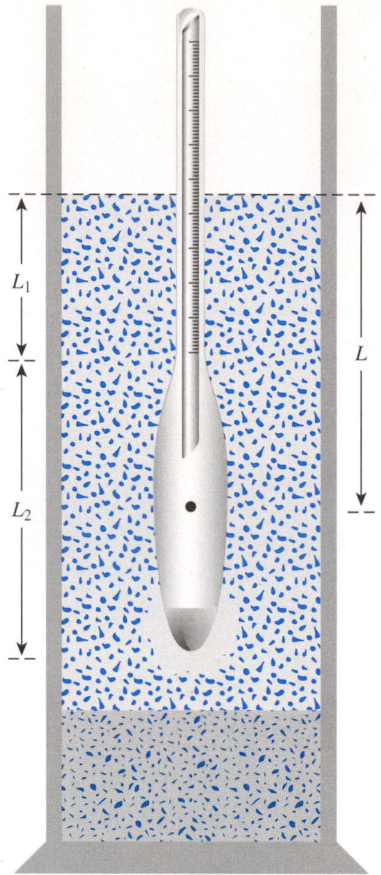

Figure 2.26 Definition of L in hydrometer test

size of about 0.5 μm. The value of L (cm) for the ASTM 152H hydrometer can be given by the expression (see Figure 2.26)

$$L = L_1 + \frac{1}{2}\left(L_2 - \frac{V_B}{A}\right) \tag{2.8}$$

where L_1 = distance along the stem of the hydrometer from the top of the bulb to the mark for a hydrometer reading (cm)
L_2 = length of the hydrometer bulb = 14 cm
V_B = volume of the hydrometer bulb = 67 cm^3
A = cross-sectional area of the sedimentation cylinder = 27.8 cm^2

The value of L_1 is 10.5 cm for a reading of $R = 0$ and 2.3 cm for a reading of $R = 50$. Hence, for any reading R,

$$L_1 = 10.5 - \frac{(10.5 - 2.3)}{50}R = 10.5 - 0.164R \text{ (cm)}$$

Thus, from Eq. (2.8),

$$L = 10.5 - 0.164R + \frac{1}{2}\left(14 - \frac{67}{27.8}\right) = 16.29 - 0.164R \tag{2.9}$$

where R = hydrometer reading corrected for the meniscus.

On the basis of Eq. (2.9), the variations of L with the hydrometer readings R are given in Table 2.10.

Table 2.10 Variation of L with Hydrometer Reading—ASTM 152H Hydrometer

Hydrometer reading, R	L (cm)	Hydrometer reading, R	L (cm)
0	16.3	31	11.2
1	16.1	32	11.1
2	16.0	33	10.9
3	15.8	34	10.7
4	15.6	35	10.6
5	15.5	36	10.4
6	15.3	37	10.2
7	15.2	38	10.1
8	15.0	39	9.9
9	14.8	40	9.7
10	14.7	41	9.6
11	14.5	42	9.4
12	14.3	43	9.2
13	14.2	44	9.1
14	14.0	45	8.9
15	13.8	46	8.8
16	13.7	47	8.6
17	13.5	48	8.4
18	13.3	49	8.3
19	13.2	50	8.1
20	13.0	51	7.9
21	12.9	52	7.8
22	12.7	53	7.6
23	12.5	54	7.4
24	12.4	55	7.3
25	12.2	56	7.1
26	12.0	57	7.0
27	11.9	58	6.8
28	11.7	59	6.6
29	11.5	60	6.5
30	11.4		

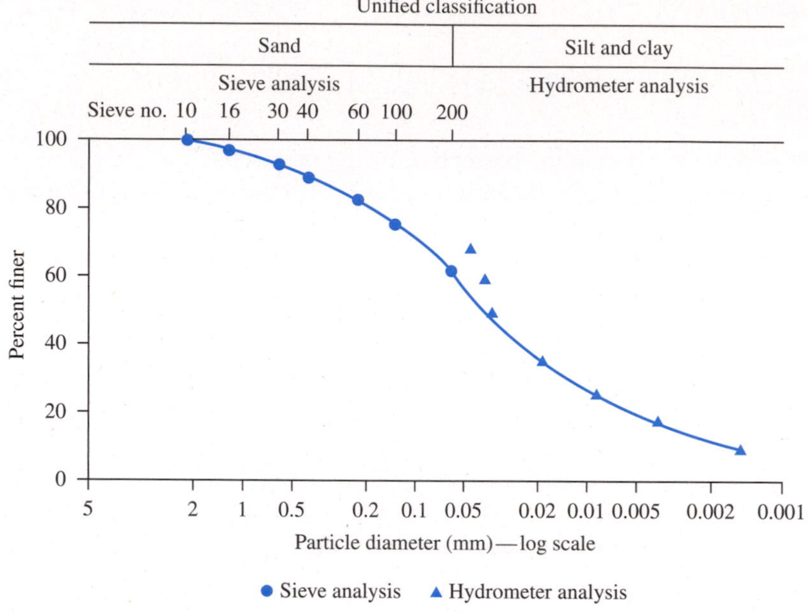

Figure 2.27 Particle-size distribution curve—sieve analysis and hydrometer analysis

In many instances, the results of sieve analysis and hydrometer analysis for finer fractions for a given soil are combined on one graph, such as the one shown in Figure 2.27. When these results are combined, a discontinuity generally occurs in the range where they overlap. This discontinuity occurs because soil particles are generally irregular in shape. Sieve analysis gives the intermediate dimensions of a particle; hydrometer analysis gives the diameter of an equivalent sphere that would settle at the same rate as the soil particle.

2.8 Particle-Size Distribution Curve

A particle-size distribution curve can be used to determine the following four parameters for a given soil (Figure 2.28):

1. *Effective size* (D_{10}): This parameter is the diameter in the particle-size distribution curve corresponding to 10% finer. The effective size of a granular soil is a good measure to estimate the hydraulic conductivity and drainage through soil.
2. *Uniformity coefficient* (C_u): This parameter is defined as

$$C_u = \frac{D_{60}}{D_{10}} \tag{2.10}$$

where D_{60} = diameter corresponding to 60% finer.

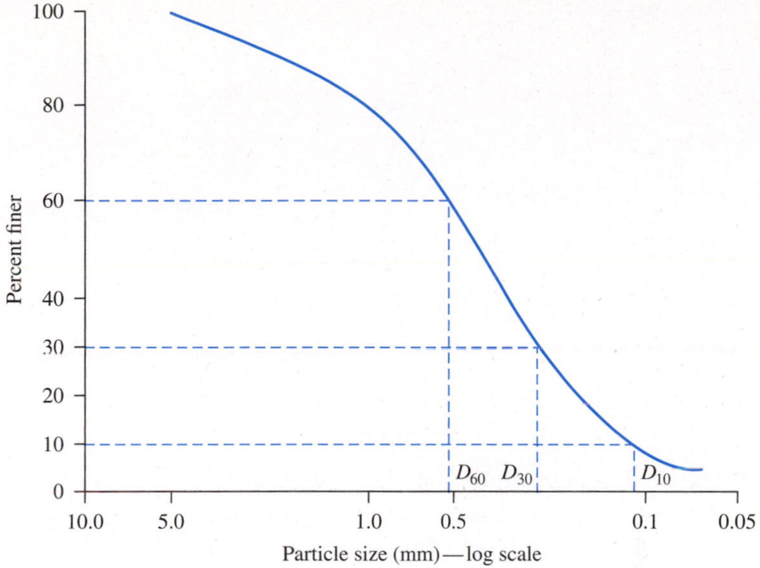

Figure 2.28 Definition of D_{60}, D_{30}, and D_{10}

3. *Coefficient of gradation* (C_c): This parameter is defined as

$$C_c = \frac{D_{30}^2}{D_{60} \times D_{10}} \tag{2.11}$$

The percentages of gravel, sand, silt, and clay-size particles present in a soil can be obtained from the particle-size distribution curve. As an example, we will use the particle-size distribution curve shown in Figure 2.27 to determine the gravel, sand, silt, and clay size particles as follows (according to the Unified Soil Classification System—see Table 2.3):

Size (mm)	Percent finer	Soil type (%)
76.2	100	
		100 − 100 = 0% gravel
4.75	100	
		100 − 62 = 38% sand
0.075	62	
		62 − 0 = 62% silt and clay
—	0	

The particle-size distribution curve shows not only the range of particle sizes present in a soil, but also the type of distribution of various-size particles. Such types of distributions are demonstrated in Figure 2.29. Curve I represents a type of soil in which most of the soil grains are the same size. This is called *poorly graded* soil. Curve II represents a soil in which the particle sizes are distributed over a wide range, termed *well graded*. A well-graded soil has a uniformity coefficient greater

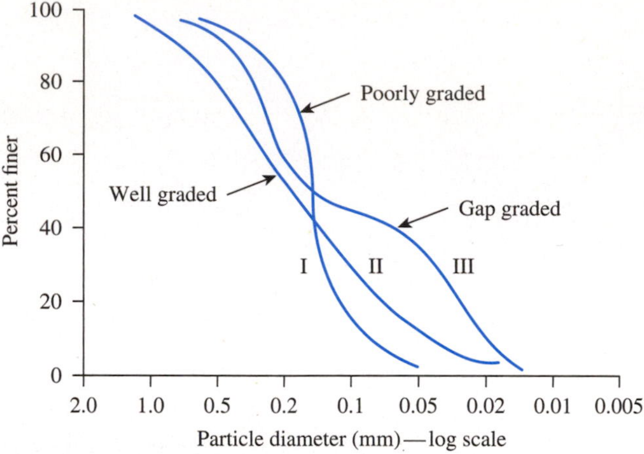

Figure 2.29 Different types of particle-size distribution curves

than about 4 for gravels and 6 for sands, and a coefficient of gradation between 1 and 3 (for gravels and sands). A soil might have a combination of two or more uniformly graded fractions. Curve III represents such a soil. This type of soil is termed *gap graded*.

Example 2.1

The following are the results of a sieve analysis:

U.S. sieve no.	Mass of soil retained on each sieve (g)
4	0
10	21.6
20	49.5
40	102.6
60	89.1
100	95.6
200	60.4
Pan	31.2

a. Perform the necessary calculations and plot a grain-size distribution curve.
b. Determine D_{10}, D_{30}, and D_{60} from the grain-size distribution curve.
c. Calculate the uniformity coefficient, C_u.
d. Calculate the coefficient of gradation, C_c.

Solution

Part a

The following table can now be prepared for obtaining the percent finer.

U.S. sieve (1)	Opening (mm) (2)	Mass retained on each sieve (g) (3)	Cumulative mass retained above each sieve (g) (4)	Percent finer[a] (5)
4	4.75	0	0	100
10	2.00	21.6	21.6	95.2
20	0.850	49.5	71.1	84.2
40	0.425	102.6	173.7	61.4
60	0.250	89.1	262.8	41.6
100	0.150	95.6	358.4	20.4
200	0.075	60.4	418.8	6.9
Pan	—	31.2	$450 = \Sigma M$	

[a] $\frac{\Sigma M - \text{col.4}}{\Sigma M} \times 100 = \frac{450 - \text{col.4}}{450} \times 100$

The particle-size distribution curve is shown in Figure 2.30.

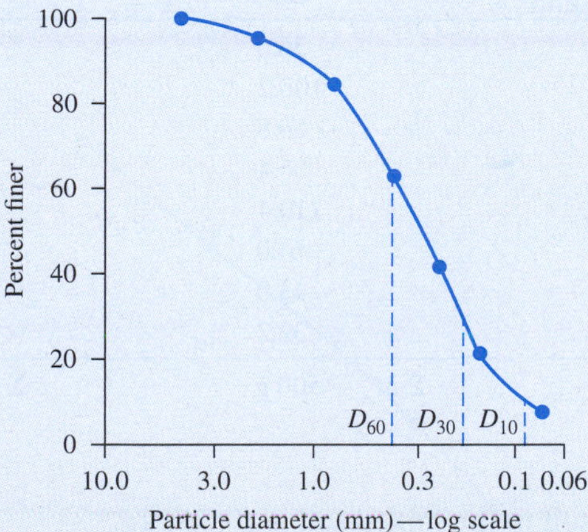

Figure 2.30

Part b

From Figure 2.30,

$$D_{60} = \mathbf{0.41\ mm}$$

$$D_{30} = \mathbf{0.185\ mm}$$

$$D_{10} = \mathbf{0.09\ mm}$$

Example 2.3

Following are the results of a sieve analysis and a hydrometer analysis on a given soil. Plot a combined grain-size distribution curve. From the plot, determine the percent of gravel, sand, silt, and clay based on the ASSHTO Classification System (Table 2.3).

Sieve Analysis

U.S. sieve no.	Sieve opening (mm)	Percent passing
4	4.75	100
10	2.0	92
20	0.850	80
30	0.600	75
40	0.425	68
60	0.250	62
100	0.106	43
200	0.075	31

Hydrometer Analysis

Grain diameter (mm)	Percent finer
0.08	38
0.05	31
0.025	21
0.013	16
0.004	11
0.0017	9

Solution

Figure 2.32 shows the plot of percent passing versus the particle size: passing 2 mm = 92%; passing 0.075 mm = 31%; passing 0.002 mm = 10%. So,

Gravel: 100 − 92 = **8**%

Sand: 92 − 31 = **61**%

Silt: 31 − 10 = **21**%

Clay: **10**%

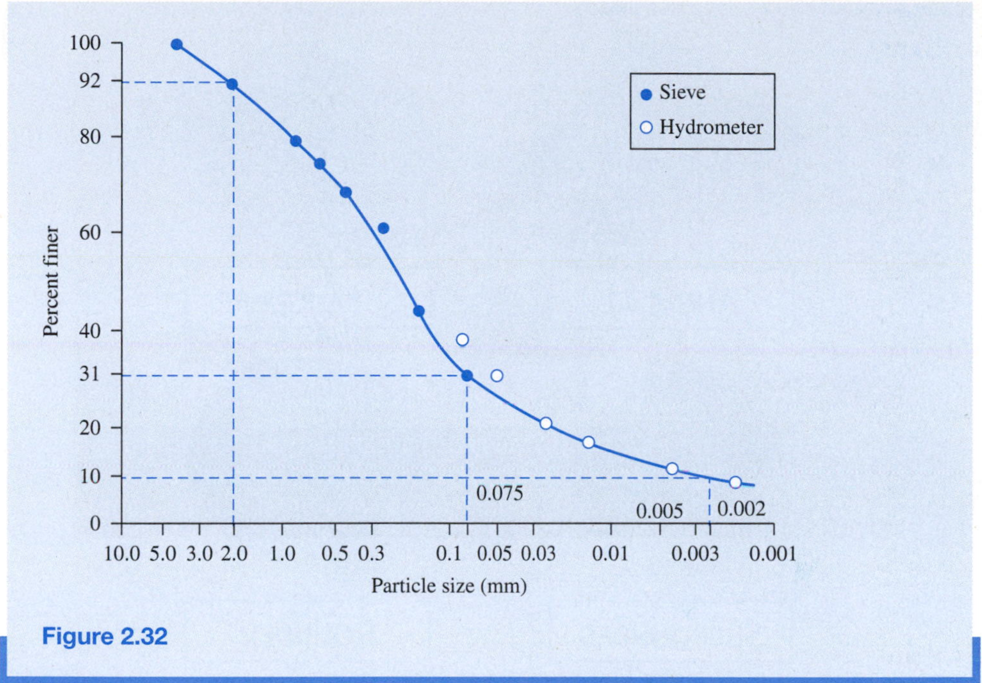

Figure 2.32

2.9 Particle Shape

The shape of particles present in a soil mass is equally as important as the particle-size distribution because it has significant influence on the physical properties of a given soil. However, not much attention is paid to particle shape because it is more difficult to measure. The particle shape generally can be divided into three major categories:

1. Bulky
2. Flaky
3. Needle shaped

Bulky particles are formed mostly by mechanical weathering of rock and minerals. Geologists use such terms as *angular, subangular, subrounded*, and *rounded* to describe the shapes of bulky particles. These shapes are shown qualitatively in Figure 2.33. Small sand particles located close to their origin are generally very angular. Sand particles carried by wind and water for a long distance can be subangular to rounded in shape. The shape of granular particles in a soil mass has a great influence on the physical properties of the soil, such as maximum and minimum void ratios, shear strength parameters, compressibility, etc.

The *angularity*, A, is defined as

$$A = \frac{\text{Average radius of corners and edges}}{\text{Radius of the maximum inscribed sphere}} \tag{2.12}$$

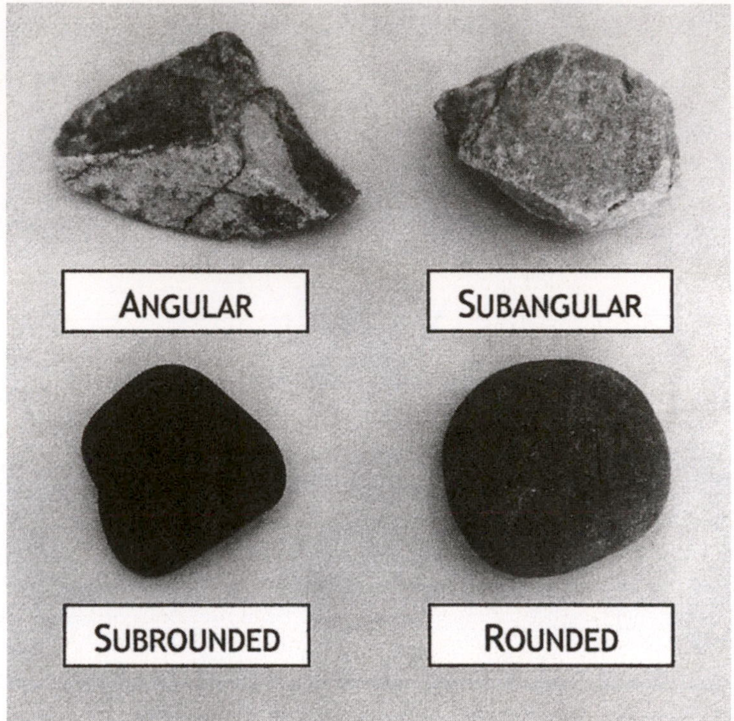

Figure 2.33 Shape of bulky particles (*Courtesy of Janice Das, Henderson, Nevada*)

The *sphericity* of bulky particles is defined as

$$S = \frac{D_e}{L_p} \tag{2.13}$$

where D_e = equivalent diameter of the partilce = $\sqrt[3]{\frac{6V}{\pi}}$

V = volume of particle

L_p = length of particle

Flaky particles have very low sphericity—usually 0.01 or less. These particles are predominantly clay minerals.

Needle-shaped particles are much less common than the other two particle types. Examples of soils containing needle-shaped particles are some coral deposits and attapulgite clays.

2.10 Summary

In this chapter, we discussed the rock cycle, the origin of soil by weathering, the particle-size distribution in a soil mass, the shape of particles, and clay minerals. Some important points include the following:

1. Rocks can be classified into three basic categories: (a) igneous, (b) sedimentary, and (c) metamorphic.
2. Soils are formed by chemical and mechanical weathering of rocks.
3. Based on the size of the particles, soil can be classified as gravel, sand, silt, and clay. According to the Unified Soil Classification System, which is now universally accepted, the grain-size limits of gravel, sand, and fines (silt and clay) are as follows:

Gravel:	76.2 mm–4.75 mm
Sand:	4.75 mm–0.075 mm
Fines (silt and clay):	<0.075 mm

4. Clays are flake-shaped microscopic and submicroscopic particles of mica, clay minerals, and other minerals.
5. Clay minerals are complex aluminum silicates.
6. Clay particles carry a net negative charge on their surfaces. When water is added, a diffuse double layer of water is developed around the clay particles that is responsible for providing plasticity to clay soils.
7. Mechanical analysis is a process of determining the size range of particles present in a soil mass. It consists of two parts—sieve analysis (for particles >0.075 mm) and hydrometer analysis (for particles <0.075 mm)
8. In a sieve analysis,

$$\begin{matrix}\text{Percent finer than} \\ \text{a given sieve size}\end{matrix} = 100 - \left(\frac{\text{Mass of soil passing a given sieve}}{\text{Total mass of soil}}\right)(100)$$

9. In hydrometer analysis, the percent finer than a given particle size (D) can be determined using the hydrometer reading (L) and Eq. (2.6) at a given time.

Problems

2.1 For a gravel with $D_{60} = 0.48$ mm, $D_{30} = 0.25$ mm, and $D_{10} = 0.11$ mm, calculate the uniformity coefficient and the coefficient of gradation. Is it a well-graded or a poorly graded soil?

2.2 The following values for a sand are given: $D_{10} = 0.3$ mm, $D_{30} = 0.41$ mm, and $D_{60} = 0.77$ mm. Determine C_u and C_c, and state if it is a well-graded or a poorly-graded soil.

2.3 The grain-size distribution curves for three different sands (A, B, and C) are shown in Figure 2.34.

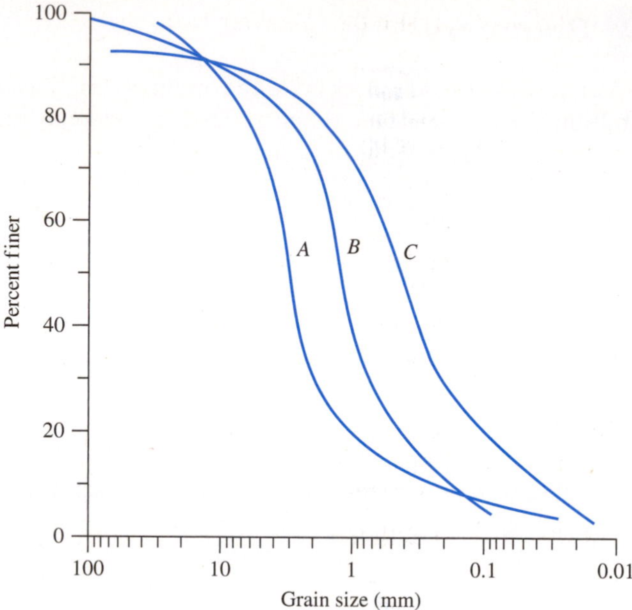

Figure 2.34

a. Determine the uniformity coefficient and the coefficient of gradation for each soil.

b. Identify if the soils are well graded or poorly graded based on Part a.

2.4 The following are the results of a sieve analysis.

U.S. sieve no.	Mass of soil retained (g)
4	0
10	18.5
20	53.2
40	90.5
60	81.8
100	92.2
200	58.5
Pan	26.5

a. Determine the percent finer than each sieve and plot a grain-size distribution curve.

b. Determine D_{10}, D_{30}, and D_{60} for each soil.

c. Calculate the uniformity coefficient C_u.

d. Calculate the coefficient of gradation C_c.

2.5 Repeat Problem 2.4 with the following data.

U.S. sieve no.	Mass of soil retained on each sieve (g)
4	0
6	30.0
10	48.7
20	127.3
40	96.8
60	76.6
100	55.2
200	43.4
Pan	22.0

2.6 Repeat Problem 2.4 with the following data.

U.S. sieve no.	Mass of soil retained on each sieve (g)
4	0
10	44
20	56
40	82
60	51
80	106
100	92
200	85
Pan	35

2.7 Repeat Problem 2.4 with the following data.

U.S. sieve no.	Mass of soil retained on each sieve (g)
4	0
6	0
10	0
20	9.1
40	249.4
60	179.8
100	22.7
200	15.5
Pan	23.5

2.8 The following are the results of a sieve and hydrometer analysis.

Analysis	Sieve number/grain size	Percent finer
Sieve	40	100
	80	97
	170	92
	200	90
Hydrometer	0.04 mm	74
	0.015 mm	42
	0.008 mm	27
	0.004 mm	17
	0.002 mm	11

a. Draw the grain-size distribution curve.
b. Determine the percentages of gravel, sand, silt and clay according to the MIT system.
c. Repeat Part b according to the USDA system.
d. Repeat Part b according to the AASHTO system.

2.9 Repeat Problem 2.8 using the following data.

Analysis	Sieve number/grain size	Percent finer
Sieve	40	100
	80	96
	170	85
	200	80
Hydrometer	0.04 mm	59
	0.02 mm	39
	0.01 mm	26
	0.005 mm	15
	0.0015 mm	8

2.10 Repeat Problem 2.8 with the following data.

Analysis	Sieve number/grain size	Percent finer
Sieve	20	100
	30	96
	40	90
	60	76
	80	65
	200	34
Hydrometer	0.05 mm	27
	0.03 mm	19
	0.015 mm	11
	0.006 mm	7
	0.004 mm	6
	0.0015 mm	5

2.11 The grain-size characteristics of a soil are given in the following table.

Size (mm)	Percent finer
0.425	100
0.1	79
0.04	57
0.02	48
0.01	40
0.002	35
0.001	33

a. Draw the grain-size distribution curve.
b. Determine the percentages of gravel, sand, silt, and clay according to the MIT system.
c. Repeat Part b using the USDA system.
d. Repeat Part b using the AASHTO system.

2.12 Repeat Problem 2.11 with the following data.

Size (mm)	Percent finer
0.425	100.0
0.033	92.1
0.018	81.3
0.01	68.9
0.0062	60.8
0.0035	49.5
0.0018	41.2
0.0005	32.6

2.13 Repeat Problem 2.11 with the following data.

Size (mm)	Percent finer
0.425	100
0.1	92
0.052	84
0.02	62
0.01	46
0.004	32
0.001	22

2.14 A hydrometer test has the following result: $G_s = 2.65$, temperature of water = 26° C, and $L = 10.4$ cm at 45 minutes after the start of sedimentation (see Figure 2.25). What is the diameter D of the smallest-size particles that have settled beyond the zone of measurement at that time (that is, $t = 45$ min)?

2.15 Repeat Problem 2.14 with the following values: $G_s = 2.75$, temperature of water = 21°C, $t = 88$ min, and $L = 11.7$ cm.

Critical Thinking Problems

2.C.1 Three groups of students from the Geotechnical Engineering class collected soil-aggregate samples for laboratory testing from a recycled aggregate processing plant in Palm Beach County, Florida. Three samples, denoted by Soil *A*, Soil *B*, and Soil *C*, were collected from three locations of the aggregate stockpile, and sieve analyses were conducted (see Figure 2.35).

Figure 2.35 (a) Soil-aggregate stockpile; (b) sieve analysis (*Courtesy of Khaled Sobhan, Florida Atlantic University, Boca Raton, Florida*)

a. Determine the coefficient of uniformity and the coefficient of gradation for Soils *A*, *B*, and *C*.

b. Which one is coarser: Soil *A* or Soil *C*? Justify your answer.

c. Although the soils are obtained from the same stockpile, why are the curves so different? (*Hint*: Comment on particle segregation and representative field sampling.)

d. Determine the percentages of gravel, sand and fines according to Unified Soil Classification System.

2.C.2 Refer to Problem 2.C.1. Results of the sieve analysis for Soils *A*, *B*, and *C* are given below. To obtain a more representative sample for further geotechnical testing, a ternary blend is created by uniformly mixing 8000 kg of each soil. Answer the following questions.

Sieve size (mm)	Mass retained Soil *A*, m_A(g)	Mass retained Soil *B*, m_B(g)	Mass retained Soil *C*, m_A(g)
25.0	0.0	0	0
19.0	60	10	30
12.7	130	75	75
9.5	65	80	45
4.75	100	165	90
2.36	50	25	65
0.6	40	60	75
0.075	50	70	105
Pan	5	15	15

a. If a sieve analysis is conducted on the mixture using the same set of sieves as shown above, compute the mass retained (as a percentage) and cumulative percent passing in each sieve.

b. What would be the uniformity coefficient (C_u) and the coefficient of gradation (C_c) of the mixture?

References

American Society for Testing and Materials (2014). *ASTM Book of Standards*, Sec. 4, Vol. 04.08, West Conshohocken, Pa.

Bowen, N. L. (1922). "The Reaction Principles in Petrogenesis," *Journal of Geology*, Vol. 30, 177–198.

Grim, R. E. (1953). *Clay Mineralogy*, McGraw-Hill, New York.

Grim, R. E. (1959). "Physico-Chemical Properties of Soils: Clay Minerals," *Journal of the Soil Mechanics and Foundations Division*, ASCE, Vol. 85, No. SM2, 1–17.

Lambe, T. W. (1958). "The Structure of Compacted Clay," *Journal of the Soil Mechanics and Foundations Division*, ASCE, Vol. 84, No. SM2, 1655–1 to 1655–35.

Yong, R. N., and Warkentin, B. P. (1966). *Introduction of Soil Behavior,* Macmillan, New York.

CHAPTER 3

Weight–Volume Relationships

3.1 Introduction

Chapter 2 presented the geologic processes by which soil is formed, the description of the limits on the sizes of soil particles, and the mechanical analysis of soils. A given volume of soil in natural occurrence consists of solid particles and the void spaces between the particles. The void space may be filled with air and/or water; hence, soil is a three-phase system. If there is no water in the void space, it is a dry soil. If the entire void space is filled with water, it is referred to as a saturated soil. However, if the void is partially filled with water, it is a moist soil. Hence it is important in all geotechnical engineering works to establish relationships between weight and volume in a given soil mass. In this chapter we will discuss the following:

- Define and develop nondimensional volume relationships such as void ratio, porosity, and degree of saturation.
- Define and develop weight relationships such as moisture content and unit weight (dry, saturated, and moist) in combination with the volume relationships.

3.2 Weight–Volume Relationships

Figure 3.1a shows an element of soil of volume V and weight W as it would exist in a natural state. To develop the weight–volume relationships, we must separate the three phases (that is, solid, water, and air) as shown in Figure 3.1b. Thus, the total

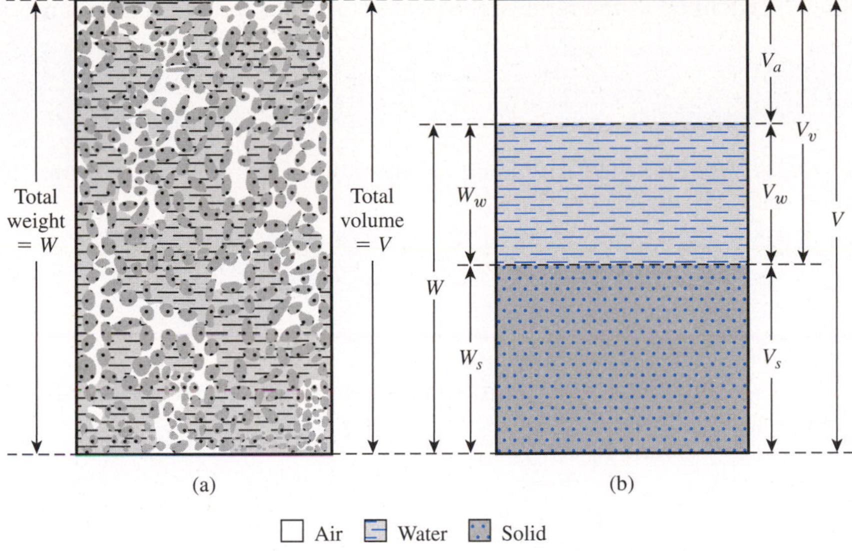

Figure 3.1 (a) Soil element in natural state; (b) three phases of the soil element

volume of a given soil sample can be expressed as

$$V = V_s + V_v = V_s + V_w + V_a \tag{3.1}$$

where V_s = volume of soil solids
V_v = volume of voids
V_w = volume of water in the voids
V_a = volume of air in the voids

Assuming that the weight of the air is negligible, we can express the total weight of the sample as

$$W = W_s + W_w \tag{3.2}$$

where W_s = weight of soil solids
W_w = weight of water

The *volume relationships* commonly used for the three phases in a soil element are *void ratio, porosity*, and *degree of saturation*. *Void ratio* (e) is defined as the ratio of the volume of voids to the volume of solids. Thus,

$$e = \frac{V_v}{V_s} \tag{3.3}$$

Porosity (n) is defined as the ratio of the volume of voids to the total volume, or

$$n = \frac{V_v}{V} \tag{3.4}$$

The *degree of saturation* (S) is defined as the ratio of the volume of water to the volume of voids, or

$$S = \frac{V_w}{V_v} \tag{3.5}$$

It is commonly expressed as a percentage.

The relationship between void ratio and porosity can be derived from Eqs. (3.1), (3.3), and (3.4) as follows:

$$e = \frac{V_v}{V_s} = \frac{V_v}{V - V_v} = \frac{\left(\frac{V_v}{V}\right)}{1 - \left(\frac{V_v}{V}\right)} = \frac{n}{1 - n} \tag{3.6}$$

Also, from Eq. (3.6),

$$n = \frac{e}{1 + e} \tag{3.7}$$

The common terms used for *weight relationships* are *moisture content* and *unit weight. Moisture content* (w) is also referred to as *water content* and is defined as the ratio of the weight of water to the weight of solids in a given volume of soil:

$$w = \frac{W_w}{W_s} \tag{3.8}$$

Unit weight (γ) is the weight of soil per unit volume. Thus,

$$\gamma = \frac{W}{V} \tag{3.9}$$

The unit weight can also be expressed in terms of the weight of soil solids, the moisture content, and the total volume. From Eqs. (3.2), (3.8), and (3.9),

$$\gamma = \frac{W}{V} = \frac{W_s + W_w}{V} = \frac{W_s\left[1 + \left(\frac{W_w}{W_s}\right)\right]}{V} = \frac{W_s(1 + w)}{V} \tag{3.10}$$

Soils engineers sometimes refer to the unit weight defined by Eq. (3.9) as the *moist unit weight.*

Often, to solve earthwork problems, one must know the weight per unit volume of soil, excluding water. This weight is referred to as the *dry unit weight*, γ_d. Thus,

$$\gamma_d = \frac{W_s}{V} \tag{3.11}$$

From Eqs. (3.10) and (3.11), the relationship of unit weight, dry unit weight, and moisture content can be given as

$$\gamma_d = \frac{\gamma}{1 + w} \tag{3.12}$$

Unit weight is expressed in English units (a gravitational system of measurement) as pounds per cubic foot (lb/ft^3). In SI (Système International), the unit used is kiloNewton per cubic meter (kN/m^3). Because the Newton is a derived unit, working with mass densities (ρ) of soil may sometimes be convenient. The SI unit of mass density is kilograms per cubic meter (kg/m^3). We can write the density equations [similar to Eqs. (3.9) and (3.11)] as

$$\rho = \frac{M}{V} \tag{3.13}$$

and

$$\rho_d = \frac{M_s}{V} \tag{3.14}$$

where ρ = density of soil (kg/m^3)
ρ_d = dry density of soil (kg/m^3)
M = total mass of the soil sample (kg)
M_s = mass of soil solids in the sample (kg)

The unit of total volume, V, is m^3.
The unit weight in kN/m^3 can be obtained from densities in kg/m^3 as

$$\gamma\,(\text{kN/m}^3) = \frac{g\rho\,(\text{kg/m}^3)}{1000}$$

and

$$\gamma_d\,(\text{kN/m}^3) = \frac{g\rho_d(\text{kg/m}^3)}{1000}$$

where g = acceleration due to gravity = 9.81 m/sec^2.

Note that unit weight of water (γ_w) is equal to 9.81 kN/m^3 or 1000 kgf/m^3.

Some typical values of void ratio, moisture content, and dry unit weight in a natural state are given in Table 3.1.

Table 3.1 Void Ratio, Moisture Content, and Dry Unit Weight for Some Typical Soils in a Natural State

Type of soil	Void ratio, e	Natural moisture content in a saturated state (%)	Dry unit weight, γ_d kN/m^3
Loose uniform sand	0.8	30	14.5
Dense uniform sand	0.45	16	18
Loose angular-grained silty sand	0.65	25	16
Dense angular-grained silty sand	0.4	15	19
Stiff clay	0.6	21	17
Soft clay	0.9–1.4	30–50	11.5–14.5
Loess	0.9	25	13.5
Soft organic clay	2.5–3.2	90–120	6–8
Glacial till	0.3	10	21

3.3 Relationships among Unit Weight, Void Ratio, Moisture Content, and Specific Gravity

To obtain a relationship among unit weight (or density), void ratio, and moisture content, let us consider a volume of soil in which the volume of the soil solids is one, as shown in Figure 3.2. If the volume of the soil solids is 1, then the volume of voids

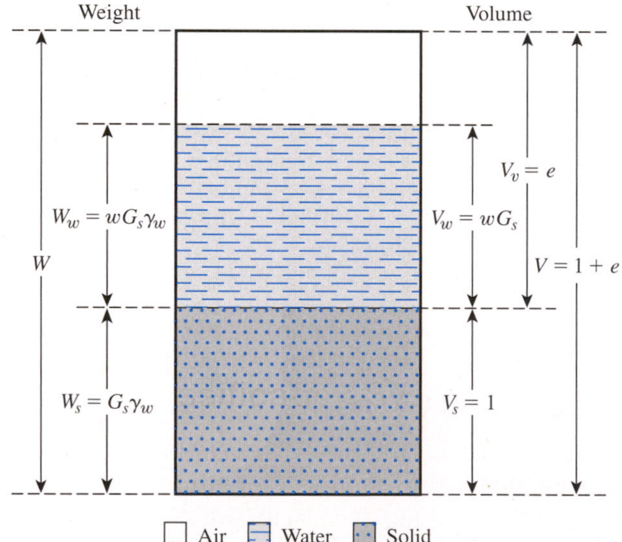

Figure 3.2 Three separate phases of a soil element with volume of soil solids equal to 1

is numerically equal to the void ratio, *e* [from Eq. (3.3)]. The weights of soil solids and water can be given as

$$W_s = G_s\gamma_w$$

$$W_w = wW_s = wG_s\gamma_w$$

where G_s = specific gravity of soil solids
w = moisture content
γ_w = unit weight of water

Specific gravity of soil solids (G_s) was defined in Section 2.6 of Chapter 2. It can be expressed as

$$G_s = \frac{W_s}{V_s\gamma_w} \tag{3.15}$$

Now, using the definitions of unit weight and dry unit weight [Eqs. (3.9) and (3.11)], we can write

$$\gamma = \frac{W}{V} = \frac{W_s + W_w}{V} = \frac{G_s\gamma_w + wG_s\gamma_w}{1+e} = \frac{(1+w)\,G_s\gamma_w}{1+e} \tag{3.16}$$

and

$$\gamma_d = \frac{W_s}{V} = \frac{G_s\gamma_w}{1+e} \tag{3.17}$$

or

$$e = \frac{G_s\gamma_w}{\gamma_d} - 1 \tag{3.18}$$

Because the weight of water for the soil element under consideration is $wG_s\gamma_w$, the volume occupied by water is

$$V_w = \frac{W_w}{\gamma_w} = \frac{wG_s\gamma_w}{\gamma_w} = wG_s$$

Hence, from the definition of degree of saturation [Eq. (3.5)],

$$S = \frac{V_w}{V_v} = \frac{wG_s}{e}$$

or

$$Se = wG_s \tag{3.19}$$

This equation is useful for solving problems involving three-phase relationships.

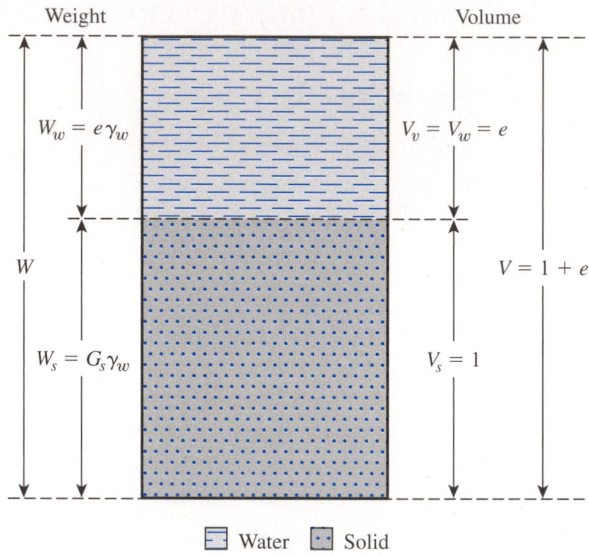

Figure 3.3 Saturated soil element with volume of soil solids equal to one

If the soil sample is *saturated*—that is, the void spaces are completely filled with water (Figure 3.3)—the relationship for saturated unit weight (γ_{sat}) can be derived in a similar manner:

$$\gamma_{sat} = \frac{W}{V} = \frac{W_s + W_w}{V} = \frac{G_s\gamma_w + e\gamma_w}{1 + e} = \frac{(G_s + e)\gamma_w}{1 + e} \tag{3.20}$$

Also, from Eq. (3.19) with $S = 1$,

$$e = wG_s \tag{3.21}$$

As mentioned before, due to the convenience of working with densities in the SI system, the following equations, similar to unit–weight relationships given in Eqs. (3.16), (3.17), and (3.20), will be useful:

$$\text{Density} = \rho = \frac{(1 + w)G_s\rho_w}{1 + e} \tag{3.22}$$

$$\text{Dry density} = \rho_d = \frac{G_s\rho_w}{1 + e} \tag{3.23}$$

$$\text{Saturated density} = \rho_{sat} = \frac{(G_s + e)\,\rho_w}{1 + e} \tag{3.24}$$

where ρ_w = density of water = 1000 kg/m^3.

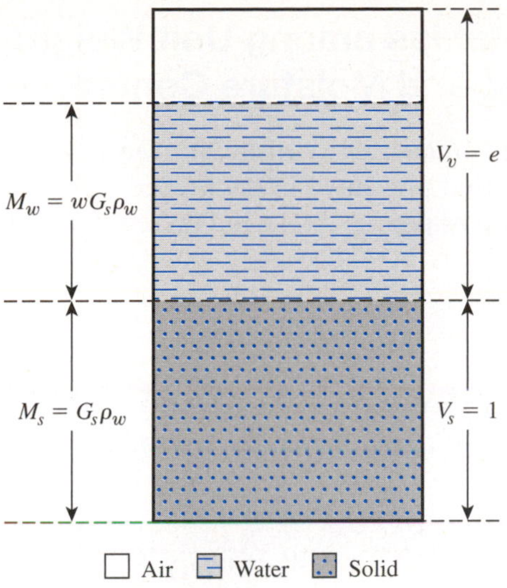

Figure 3.4 Three separate phases of a soil element showing mass–volume relationship

Equation (3.22) may be derived by referring to the soil element shown in Figure 3.4, in which the volume of soil solids is equal to 1 and the volume of voids is equal to e. Hence, the mass of soil solids, M_s, is equal to $G_s\rho_w$. The moisture content has been defined in Eq. (3.8) as

$$w = \frac{W_w}{W_s} = \frac{(\text{mass of water}) \cdot g}{(\text{mass of solid}) \cdot g}$$

$$= \frac{M_w}{M_s}$$

where M_w = mass of water.

Since the mass of soil in the element is equal to $G_s\rho_w$, the mass of water

$$M_w = wM_s = wG_s\rho_w$$

From Eq. (3.13), density

$$\rho = \frac{M}{V} = \frac{M_s + M_w}{V_s + V_v} = \frac{G_s\rho_w + wG_s\rho_w}{1 + e}$$

$$= \frac{(1 + w)G_s\rho_w}{1 + e}$$

Equations (3.23) and (3.24) can be derived similarly.

3.4 Relationships among Unit Weight, Porosity, and Moisture Content

The relationship among *unit weight, porosity*, and *moisture content* can be developed in a manner similar to that presented in the preceding section. Consider a soil that has a total volume equal to one, as shown in Figure 3.5. From Eq. (3.4),

$$n = \frac{V_v}{V}$$

If V is equal to 1, then V_v is equal to n, so $V_s = 1 - n$. The weight of soil solids (W_s) and the weight of water (W_w) can then be expressed as follows:

$$W_s = G_s\gamma_w(1 - n) \tag{3.25}$$

$$W_w = wW_s = wG_s\gamma_w(1 - n) \tag{3.26}$$

So, the dry unit weight equals

$$\gamma_d = \frac{W_s}{V} = \frac{G_s\gamma_w(1 - n)}{1} = G_s\gamma_w(1 - n) \tag{3.27}$$

The moist unit weight equals

$$\gamma = \frac{W_s + W_w}{V} = G_s\gamma_w(1 - n)(1 + w) \tag{3.28}$$

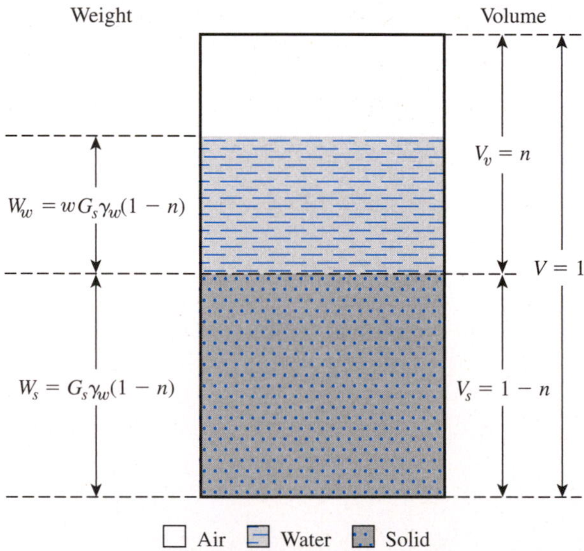

Figure 3.5 Soil element with total volume equal to one

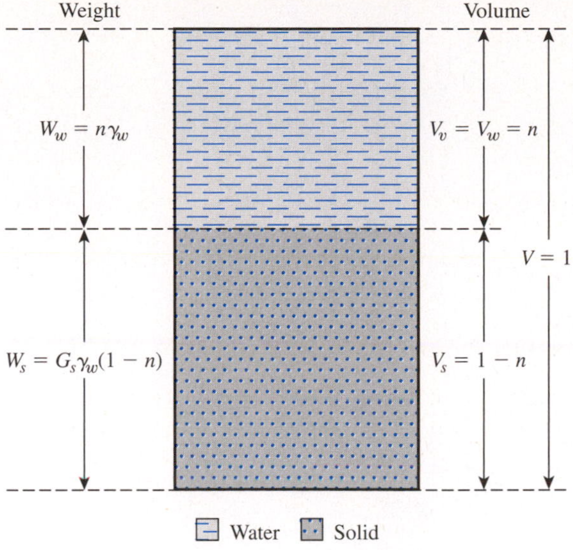

Figure 3.6 Saturated soil element with total volume equal to 1

Figure 3.6 shows a soil sample that is saturated and has $V = 1$. According to this figure,

$$\gamma_{sat} = \frac{W_s + W_w}{V} = \frac{(1 - n)G_s\gamma_w + n\gamma_w}{1} = [(1 - n)G_s + n]\gamma_w \tag{3.29}$$

The moisture content of a saturated soil sample can be expressed as

$$w_{sat} = \frac{W_w}{W_s} = \frac{n\gamma_w}{(1 - n)\gamma_w G_s} = \frac{n}{(1 - n)G_s} \tag{3.30}$$

Example 3.1

For a saturated soil, show that

$$\gamma_{sat} = \left(\frac{1 + w_{sat}}{1 + w_{sat}G_s}\right)G_s\gamma_w$$

Solution

$$\gamma_{sat} = \frac{W}{V} = \frac{W_w + W_s}{V} = \frac{w_{sat}W_s + W_s}{V} = (1 + w_{sat})\frac{W_s}{V} \tag{a}$$

From Eq. (3.15),

$$W_s = G_s V_s \gamma_w \tag{b}$$

Also, from Eq. (3.3),

$$e + 1 = \frac{V_v + V_s}{V_s} = \frac{V}{V_s} \tag{c}$$

Substituting Eqs. (b) and (c) into Eq. (a),

$$\gamma_{\text{sat}} = (1 + w_{\text{sat}}) \frac{G_s \gamma_w}{1 + e} \tag{d}$$

From Eq. (3.21),

$$e = w_{\text{sat}} G_s \tag{e}$$

Substituting (e) into (d) gives

$$\boldsymbol{\gamma_{\text{sat}} = \left(\frac{1 + w_{\text{sat}}}{1 + w_{\text{sat}} G_s}\right) G_s \gamma_w}$$

Example 3.2

For a moist soil sample, the following are given.

Total volume: $V = 1.2\ \text{m}^3$
Total mass: $M = 2350$ kg
Moisture content: $w = 8.6\%$
Specific gravity of soil solids: $G_s = 2.71$

Determine the following.

a. Moist density
b. Dry density
c. Void ratio
d. Porosity
e. Degree of saturation
f. Volume of water in the soil sample

Solution

Part a
From Eq. (3.13),

$$\rho = \frac{M}{V} = \frac{2350}{1.2} = \mathbf{1958.3\ kg/m^3}$$

Part b

From Eq. (3.14),

$$\rho_d = \frac{M_s}{V} = \frac{M}{(1+w)V} = \frac{2350}{\left(1+\frac{8.6}{100}\right)(1.2)} = \mathbf{1803.3\ kg/m^3}$$

Part c

From Eq. (3.23),

$$\rho_d = \frac{G_s\rho_w}{1+e}$$

$$e = \frac{G_s\rho_w}{\rho_d} - 1 = \frac{(2.71)(1000)}{1803.3} - 1 = \mathbf{0.503}$$

Part d

From Eq. (3.7),

$$n = \frac{e}{1+e} = \frac{0.503}{1+0.503} = \mathbf{0.335}$$

Part e

From Eq. (3.19),

$$S = \frac{wG_s}{e} = \frac{\left(\frac{8.6}{100}\right)(2.71)}{0.503} = 0.463 = \mathbf{46.3\%}$$

Part f

The volume of water is

$$\frac{M_w}{\rho_w} = \frac{M - M_s}{\rho_w} = \frac{M - \frac{M}{1+w}}{\rho_w} = \frac{2350 - \left(\frac{2350}{1+\frac{8.6}{100}}\right)}{1000} = \mathbf{0.186\ m^3}$$

Alternate Solution

Refer to Figure 3.7.

Part a

$$\rho = \frac{M}{V} = \frac{2350}{1.2} = \mathbf{1958.3\ kg/m^3}$$

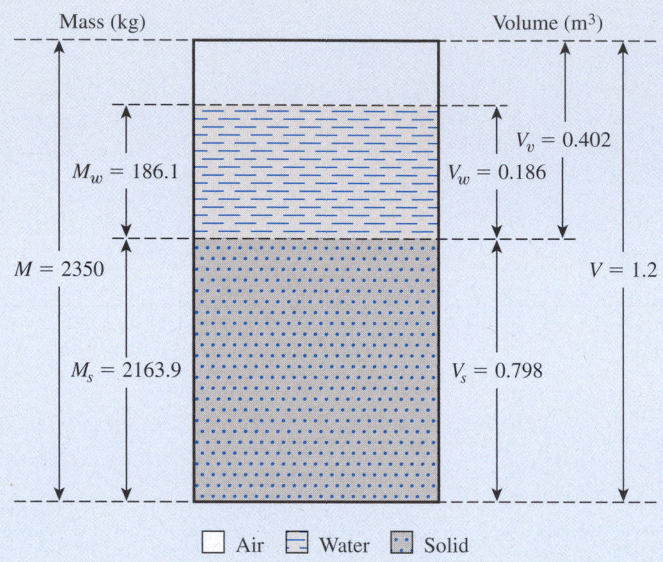

Figure 3.7

Part b

$$M_s = \frac{M}{1 + w} = \frac{2350}{1 + \frac{8.6}{100}} = 2163.9 \text{ kg}$$

$$\rho_d = \frac{M_s}{V} = \frac{M}{(1 + w)V} = \frac{2350}{\left(1 + \frac{8.6}{100}\right)(1.2)} = \mathbf{1803.3\ kg/m^3}$$

Part c

The volume of solids: $\dfrac{M_s}{G_s\rho_w} = \dfrac{2163.9}{(2.71)(1000)} = 0.798 \text{ m}^3$

The volume of voids: $V_v = V - V_s = 1.2 - 0.798 = 0.402 \text{ m}^3$

Void ratio: $e = \dfrac{V_v}{V_s} = \dfrac{0.402}{0.798} = \mathbf{0.503}$

Part d

$$\text{Porosity: } n = \frac{V_v}{V} = \frac{0.402}{1.2} = \mathbf{0.335}$$

Part e

$$S = \frac{V_w}{V_v}$$

$$\text{Volume of water: } V_w = \frac{M_w}{\rho_w} = \frac{186.1}{1000} = 0.186 \text{ m}^3$$

Hence,

$$S = \frac{0.186}{0.402} = 0.463 = \mathbf{46.3\%}$$

Part f
From Part e,

$$V_w = \mathbf{0.186 \ m^3}$$

Example 3.3

The following data are given for a soil:

Porosity: $n = 0.4$
Specific gravity of the soil solids: $G_s = 2.68$
Moisture content: $w = 12\%$

Determine the mass of water to be added to 10 m³ of soil for full saturation.

Solution

Equation (3.28) can be rewritten in terms of density as

$$\rho = G_s \rho_w (1 - n)(1 + w)$$

Similarly, from Eq. (3.29)

$$\rho_{\text{sat}} = [(1 - n)G_s + n]\rho_w$$

Thus,

$$\rho = (2.68)(1000)(1 - 0.4)(1 + 0.12) = 1800.96 \text{ kg/m}^3$$

$$\rho_{\text{sat}} = [(1 - 0.4)(2.68) + 0.4](1000) = 2008 \text{ kg/m}^3$$

Mass of water needed per cubic meter equals

$$\rho_{\text{sat}} - \rho = 2008 - 1800.96 = 207.04 \text{ kg}$$

So, total mass of water to be added equals

$$207.04 \times 10 = \mathbf{2070.4 \ kg}$$

Example 3.4

A saturated soil has a dry unit weight of 16.19 kN/m^3. Its moisture content is 23%.

Determine:

a. Saturated unit weight, γ_{sat}
b. Specific gravity, G_s
c. Void ratio, e

Solution

Part a: Saturated Unit Weight
From Eq. (3.12),

$$\gamma_{sat} = \gamma_d(1 + w) = (16.19)\left(1 + \frac{23}{100}\right) = \mathbf{19.91\ kN/m^3}$$

Part b: Specific Gravity, G_s
From Eq. (3.17),

$$\gamma_d = \frac{G_s\gamma_w}{1 + e}$$

Also from Eq. (3.21) for saturated soils, $e = wG_s$. Thus,

$$\gamma_d = \frac{G_s\gamma_w}{1 + wG_s}$$

So,

$$16.19 = \frac{G_s(9.81)}{1 + (0.23)(G_s)}$$

or

$$16.19 + 3.72G_s = 9.81G_s$$

$$G_s = \mathbf{2.66}$$

Part c: Void Ratio, e
For saturated soils,

$$e = wG_s = (0.23)(2.66) = \mathbf{0.61}$$

Example 3.5

The dry density of a sand with a porosity of 0.387 is 1600 kg/m^3. Determine the void ratio of the soil and the specific gravity of soil solids.

Solution

From Eq. (3.6),

$$\text{Void ratio, } e = \frac{n}{1-n} = \frac{0.387}{1-0.387} = \mathbf{0.63}$$

From Eq. (3.23),

$$\text{Dry density: } \rho_d = \frac{G_s \rho_w}{1+e}$$

$$1600 = \frac{(G_s)(1000)}{1+0.63}; \; G_s = \mathbf{2.61}$$

Example 3.6

Figure 3.8 shows the cross section of an embankment to be constructed. For the embankment, $\gamma = 17.29$ kN/m^3. The soil for the embankment has to be brought from a borrow pit. The soil at the borrow pit has the following: $e = 0.68$, $G_s = 2.68$, and $w = 10\%$. Determine the volume of soil from the borrow pit that will be required to construct the embankment 305 m long.

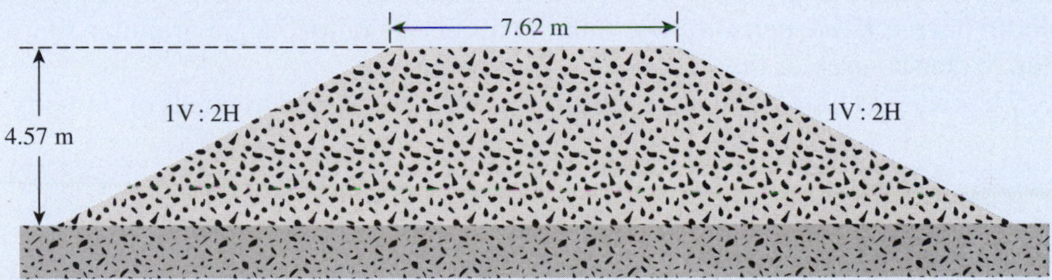

Figure 3.8

Solution

At the borrow pit,

$$\text{Dry unit weight, } \gamma_d = \frac{G_s \gamma_w}{1+e} = \frac{(2.68)(9.81)}{1+0.68} = 15.65 \text{ kN/m}^3$$

$$\text{Total volume of embankment} = \left[(7.62)(4.57) + (2)\left(\frac{1}{2} \times 4.57 \times 9.14\right)\right](305)$$

$$= 23{,}360.8 \text{ m}^3$$

$$\text{Volume of soil from borrow pit} = (23{,}360.8)\left(\frac{\gamma_{d\text{-embankment}}}{\gamma_{d\text{-borrow pit}}}\right)$$

$$= (23{,}360.8)\left(\frac{17.29}{15.65}\right) \sim \mathbf{25{,}809\ m^3}$$

3.5 Relative Density

The term *relative density* is commonly used to indicate the *in situ* denseness or looseness of granular soil. It is defined as

$$D_r = \frac{e_{max} - e}{e_{max} - e_{min}} \tag{3.31}$$

where D_r = relative density, usually given as a percentage
e = *in situ* void ratio of the soil
e_{max} = void ratio of the soil in the loosest state
e_{min} = void ratio of the soil in the densest state

The values of D_r may vary from a minimum of 0% for very loose soil to a maximum of 100% for very dense soils. Soils engineers qualitatively describe the granular soil deposits according to their relative densities, as shown in Table 3.2. In-place soils seldom have relative densities less than 20 to 30%. Compacting a granular soil to a relative density greater than about 85% is difficult.

The relationships for relative density can also be defined in terms of porosity, or

$$e_{max} = \frac{n_{max}}{1 - n_{max}} \tag{3.32}$$

$$e_{min} = \frac{n_{min}}{1 - n_{min}} \tag{3.33}$$

Table 3.2 Qualitative Description of Granular Soil Deposits

Relative density (%)	Description of soil deposit
0–15	Very loose
15–50	Loose
50–70	Medium
70–85	Dense
85–100	Very dense

$$e = \frac{n}{1 - n} \tag{3.34}$$

where n_{max} and n_{min} = porosity of the soil in the loosest and densest conditions, respectively. Substituting Eqs. (3.32), (3.33), and (3.34) into Eq. (3.31), we obtain

$$D_r = \frac{(1 - n_{min})(n_{max} - n)}{(n_{max} - n_{min})(1 - n)} \tag{3.35}$$

By using the definition of dry unit weight given in Eq. (3.17), we can express relative density in terms of maximum and minimum possible dry unit weights. Thus,

$$D_r = \frac{\left[\frac{1}{\gamma_{d(min)}}\right] - \left[\frac{1}{\gamma_d}\right]}{\left[\frac{1}{\gamma_{d(min)}}\right] - \left[\frac{1}{\gamma_{d(max)}}\right]} = \left[\frac{\gamma_d - \gamma_{d(min)}}{\gamma_{d(max)} - \gamma_{d(min)}}\right]\left[\frac{\gamma_{d(max)}}{\gamma_d}\right] \tag{3.36}$$

where $\gamma_{d(min)}$ = dry unit weight in the loosest condition (at a void ratio of e_{max})
γ_d = *in situ* dry unit weight (at a void ratio of e)
$\gamma_{d(max)}$ = dry unit weight in the densest condition (at a void ratio of e_{min})

In terms of density, Eq. (3.36) can be expressed as

$$D_r = \left[\frac{\rho_d - \rho_{d(min)}}{\rho_{d(max)} - \rho_{d(min)}}\right]\frac{\rho_{d(max)}}{\rho_d} \tag{3.37}$$

ASTM Test Designations D-4253 and D-4254 (2014) provide a procedure for determining the maximum and minimum dry unit weights of granular soils so that they can be used in Eq. (3.36) to measure the relative density of compaction in the field. For sands, this procedure involves using a mold with a volume of 2830 cm^3. For a determination of the *minimum dry unit weight*, sand is poured loosely into the mold from a funnel with a 12.7 mm diameter spout. The average height of the fall of sand into the mold is maintained at about 25.4 mm. The value of $\gamma_{d(min)}$ then can be calculated by using the following equation:

$$\gamma_{d(min)} = \frac{W_{s(mold)}}{V_m} \tag{3.38}$$

where $W_{s(mold)}$ = weight of sand required to fill the mold
V_m = volume of the mold

The *maximum dry unit weight* is determined by vibrating sand in the mold for 8 min. A surcharge of 14 kN/m^2 is added to the top of the sand in the mold. The mold is placed on a table that vibrates at a frequency of 3600 cycles/min and that has an amplitude of vibration of 0.635 mm. The value of $\gamma_{d(max)}$ can be determined at the end of the vibrating period with knowledge of the weight and volume of the sand. Figure 3.9 shows the equipment needed to conduct the test for determination of e_{min}.

Figure 3.9 Laboratory equipment for determination of minimum and maximum dry densities of granular soil *(Courtesy of K. Reddy, University of Illinois, Chicago)*

Several factors control the magnitude of $\gamma_{d(\max)}$: the magnitude of acceleration, the surcharge load, and the geometry of acceleration. Hence, one can obtain a larger-value $\gamma_{d(\max)}$ than that obtained by using the ASTM standard method described earlier.

Example 3.7

For a given sandy soil, $e_{\max} = 0.75$ and $e_{\min} = 0.4$. Let $G_s = 2.68$. In the field, the soil is compacted to a moist unit weight of 1794.9 kg/m³ at a moisture content of 12%. Determine the relative density of compaction.

Solution

From Eq. (3.16),

$$\rho = \frac{(1 + w)G_s\gamma_w}{1 + e}$$

or

$$e = \frac{G_s\rho_w(1 + w)}{\rho} - 1 = \frac{(2.68)(1000)(1 + 0.12)}{1794.9} - 1 = 0.67$$

From Eq. (3.31),

$$D_r = \frac{e_{max} - e}{e_{max} - e_{min}} = \frac{0.75 - 0.67}{0.75 - 0.4} = \mathbf{0.229 = 22.9\%}$$

3.6 Comments on e_{max} and e_{min}

The maximum and minimum void ratios for granular soils described in Section 3.5 depend on several factors, such as

- Grain size
- Grain shape
- Nature of the grain-size distribution curve
- Fine contents, F_c (that is, fraction smaller than 0.075 mm)

Youd (1973) analyzed the variation of e_{max} and e_{min} of several sand samples and provided relationships between angularity A (see Section 2.9) of sand particles and the uniformity coefficient ($C_u = D_{60}/D_{10}$; see Section 2.8). The qualitative descriptions of sand particles with the range of angularity as provide by Youd (1973) are given below.

- **Very angular**—the particles that have unworn fractured surfaces and multiple sharp corners and edges. The value of A varies within a range of 0.12–0.17 with a mean value of 0.14.
- **Angular**—the particles with sharp corners having approximately prismoidal or tetrahedral shapes with A = 0.17–0.25 with a mean value of 0.21.
- **Subangular**—The particles have blunted or slightly rounded corners and edges with A = 0.25–0.35 with a mean value of about 0.30.
- **Subrounded**—The particles have well rounded edges and corners. The magnitude of A varies in the range of 0.35–0.49 with a mean value of 0.41.
- **Rounded**—The particles are irregularly shaped and rounded with no distinct corners or edges for which A = 0.49–0.79 with a mean value of 0.59.
- **Well-rounded**—The particles are spherical or ellipsoidal shape with A = 0.7–1.0 with a mean value of about 0.84.

The variations of e_{max} and e_{min} with criteria described above are given in Figure 3.10. Note that, for a given value of C_u, the maximum and minimum void ratios increase with the decrease in angularity. Also, for a given value of A, the magnitudes of e_{max} and e_{min} decrease with an increase in C_u. The amount of *nonplastic fines* present in a given granular soil has a great influence on e_{max} and e_{min}.

Lade et al. (1998) conducted several tests by mixing sand with nonplastic fines (passing 0.075 mm–U.S. No. 200 sieve) at different proportions by volume to determine e_{max} and e_{min} in two types of sand (Nevada 50/80 and Nevada 80/200) along with one type of nonplastic fine. The median grain size of the sand samples ($D_{50\text{-sand}}$) and the fines ($D_{50\text{-fine}}$) are given in Table 3.3.

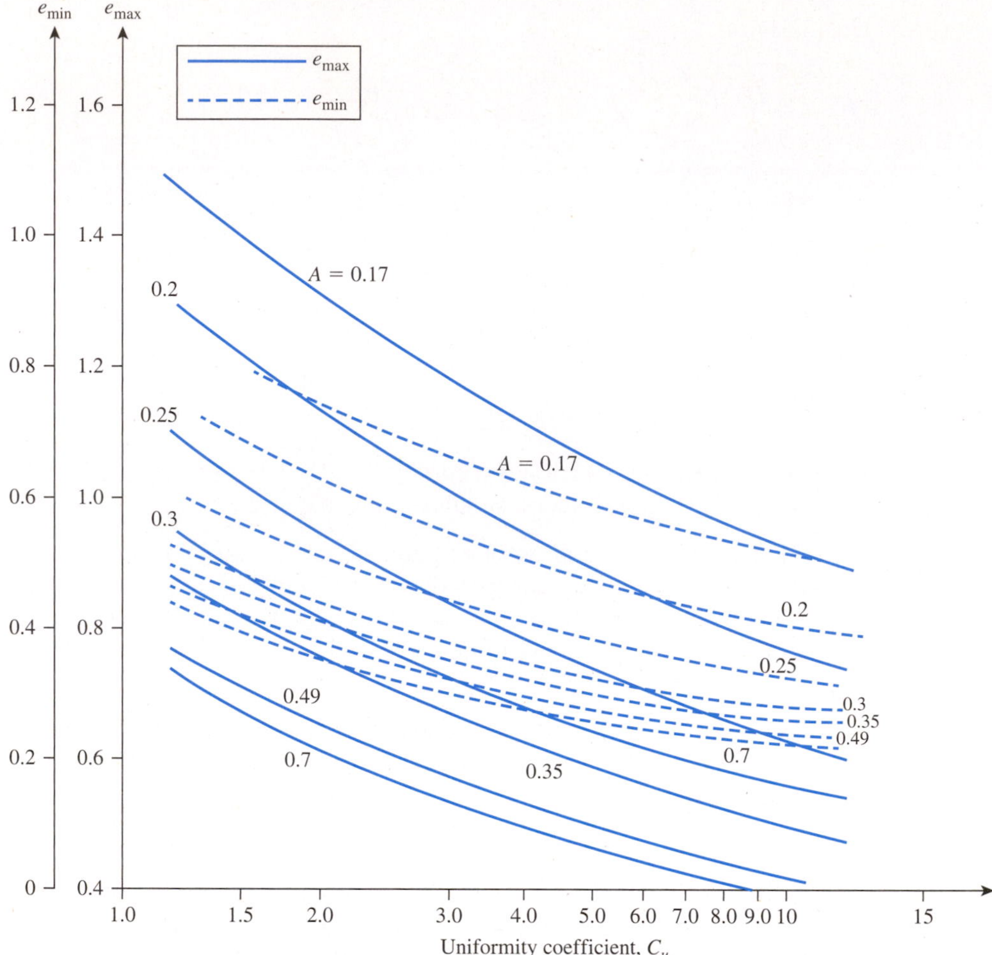

Figure 3.10 Variation of e_{max} and e_{min} with A and C_u *(Adapted after Youd, 1973)*

Figure 3.11 shows the variation of e_{max} and e_{min} with percent of fine by volume for (a) Nevada 50/80 sand and fines and (b) Nevada 80/200 sand and fines. From this figure, it can be seen that

- For a given sand and fine mixture, the e_{max} and e_{min} decrease with the increase in the volume of fines from zero to about 30%. This is the *filling-of-the-void phase*, where fines tend to fill the void spaces between the larger sand particles.

Table 3.3 $D_{50\text{-sand}}$ and $D_{50\text{-fine}}$ of the soils used by Lade et al. (1998)

Sand description	$D_{50\text{-sand}}$ (mm)	$D_{50\text{-fine}}$ (mm)	$\dfrac{D_{50\text{-sand}}}{D_{50\text{-fine}}}$
Nevada 50/80	0.211	0.050	4.22
Nevada 80/200	0.120	0.050	2.4

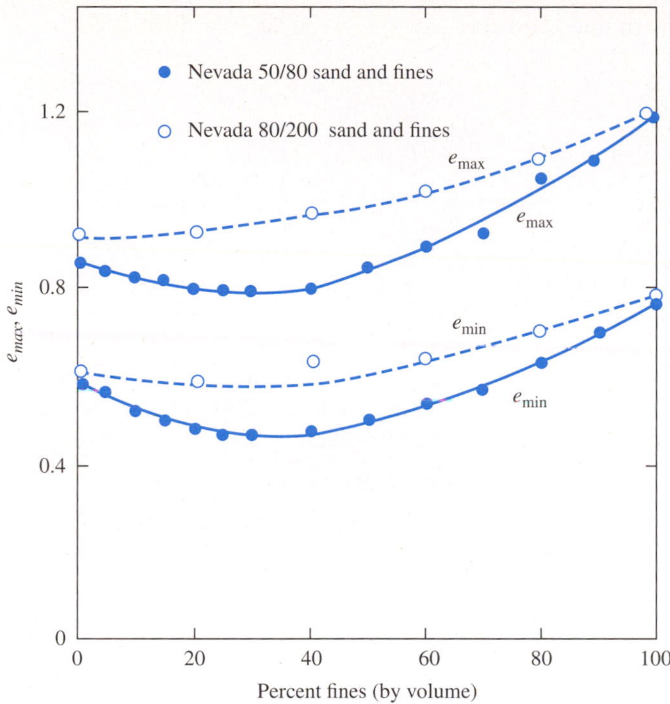

Figure 3.11 Variation of e_{max} and e_{min} with percent of nonplastic fines (Based on the test results of Lade et al., 1998). *Note*: For 50/80 sand and fines, $D_{50\text{-sand}}/D_{50\text{-fine}} = 4.22$ and for 80/200 sand and fines, $D_{50\text{-sand}}/D_{50\text{-fine}} = 2.4$

- There is a *transition zone*, where the percentage of fines is between 30% to 40%.
- For percentage of fines greater than about 40%, the magnitudes of e_{max} and e_{min} start increasing. This is the *replacement-of-solids phase*, where larger-sized solid particles are pushed out and gradually replaced by fines.

3.7 Correlations between e_{max}, e_{min}, $e_{max} - e_{min}$, and Median Grain Size (D_{50})

Cubrinovski and Ishihara (2002) studied the variation of e_{max} and e_{min} for a much larger number of soils. Based on the best-fit linear-regression lines, they provided the following relationships.

- Clean sand ($F_c = 0$ to 5%)

$$e_{max} = 0.072 + 1.53\, e_{min} \tag{3.39}$$

- Sand with fines ($5 < F_c \leq 15\%$)

$$e_{max} = 0.25 + 1.37\, e_{min} \tag{3.40}$$

- Sand with fines and clay ($15 < F_c \le 30\%$; $P_c = 5$ to 20%)

$$e_{max} = 0.44 + 1.21\, e_{min} \tag{3.41}$$

- Silty soils ($30 < F_c \le 70\%$; $P_c = 5$ to 20%)

$$e_{max} = 0.44 + 1.32\, e_{min} \tag{3.42}$$

where F_c = fine fraction for which grain size is smaller than 0.075 mm
P_c = clay-size fraction (<0.005 mm)

Figure 3.12 shows a plot of $e_{max} - e_{min}$ versus the mean grain size (D_{50}) for a number of soils (Cubrinovski and Ishihara, 1999 and 2002). From this figure, the average plot for sandy and gravelly soils can be given by the relationship

$$e_{max} - e_{min} = 0.23 + \frac{0.06}{D_{50}\,(\text{mm})} \tag{3.43}$$

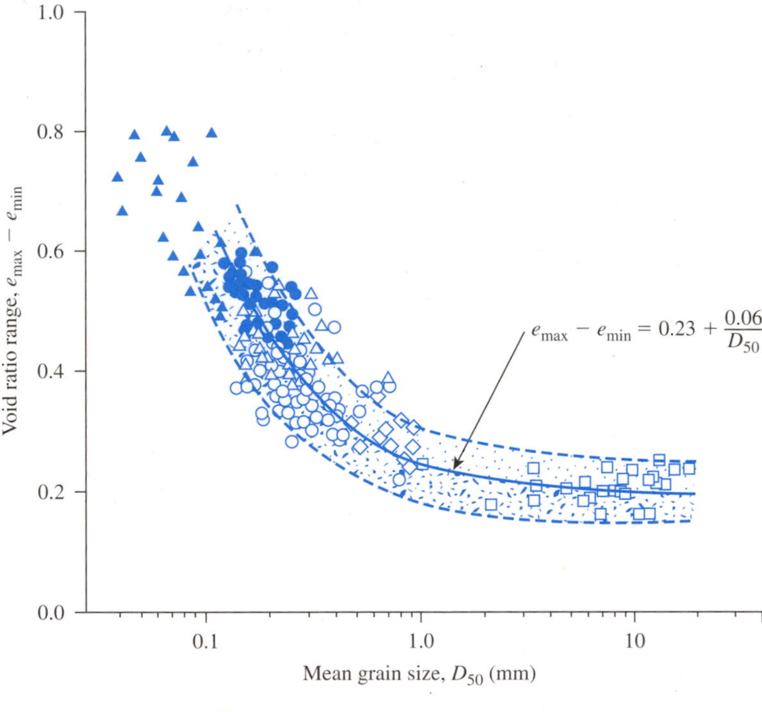

Figure 3.12 Plot of $e_{max} - e_{min}$ versus the mean grain size *(Cubrinovski and Ishihara, 2002)*

Example 3.8

The median grain size (D_{50}) of a clean sand is 0.5 mm. The sand is compacted in the field to a dry unit weight of 15.72 kN/m^3. Estimate the relative density of compaction. Given: G_s for the sand is 2.66.

Solution

We will use the correlations provided by Cubrinovski and Ishihara. From Eq. (3.39)

$$e_{max} = 0.072 + 1.53e_{min}$$

or

$$e_{min} = \frac{e_{max} - 0.072}{1.53} \tag{a}$$

From Eq. (3.43),

$$e_{max} - e_{min} = 0.23 + \frac{0.06}{D_{50}} \tag{b}$$

Combining Eqs. (a) and (b),

$$e_{max} - \left(\frac{e_{max} - 0.072}{1.53}\right) = 0.23 + \frac{0.06}{0.5}$$

$$e_{max} - 0.6536e_{max} + 0.04706 = 0.35$$

$$e_{max} = \frac{0.35 - 0.04706}{1 - 0.6536} = 0.875$$

From Eq. (a),

$$e_{min} = \frac{0.875 - 0.072}{1.53} = 0.525$$

From Eq. (3.18),

$$e_{field} = \frac{G_s\gamma_w}{\gamma_d} - 1 = \frac{(2.66)(9.81)}{15.72} - 1 = 0.66$$

Hence, the relative density of compaction in the field is

$$D_r = \frac{e_{max} - e_{field}}{e_{max} - e_{min}} = \frac{0.875 - 0.66}{0.875 - 0.525} = 0.614 = \mathbf{61.4\%}$$

3.8 Summary

In this chapter, we discussed weight–volume relationships of soils. Following is a summary of the subjects covered:

- Volume relationships consist of void ratio (e), porosity (n), and degree of saturation (S), or

$$\text{Void ratio, } e = \frac{\text{Volume of void}}{\text{Volume of solid}}$$

$$\text{Porosity, } n = \frac{\text{Volume of void}}{\text{Total volume}}$$

$$\text{Degree of saturation, } S = \frac{\text{Volume of water in void}}{\text{Total volume of void}}$$

- Weight relationships consist of moisture content (w) and unit weight ($\gamma_d, \gamma, \gamma_{sat}$).

$$\text{Moisture content, } w = \frac{\text{Weight of water in void}}{\text{Weight of solid}}$$

The relationships of dry, moist, and saturated unit weights are given, respectively, by Eqs. (3.17), (3.16), and (3.20).
- Relative density (D_r) is a measure of denseness of granular soil in the field and is defined by Eqs. (3.31) and (3.36).
- Approximate empirical relationships between maximum void ratio (e_{max}) and minimum void ratio (e_{min}) for granular soils with varying fine contents and clay-size fraction are given in Eqs. (3.39)–(3.42).
- The magnitude of $e_{max} - e_{min}$ for sandy and gravelly soils can be correlated to the median grain size (D_{50}) via Eq. (3.43).

Problems

3.1 For a given soil, show that,

a. $\gamma_{sat} = \gamma_d + \left(\frac{e}{1+e}\right)\gamma_w$

b. $\gamma_d = \frac{eS\gamma_w}{(1+e)w}$

c. $e = \frac{\gamma_{sat} - \gamma_d}{\gamma_d - \gamma_{sat} + \gamma_w}$

d. $w_{sat} = \frac{n\gamma_w}{\gamma_{sat} - n\gamma_w}$

3.2 The moist unit weight of a soil is 17.8 kN/m^3 and the moisture content is 14%. If the specific gravity of the soil solids is 2.69, calculate the following:

a. Dry unit weight
b. Void ratio
c. Degree of saturation

3.3 Refer to Problem 3.2. For a unit volume of the soil, determine the various quantities of the phase diagram shown in Figure 3.13.

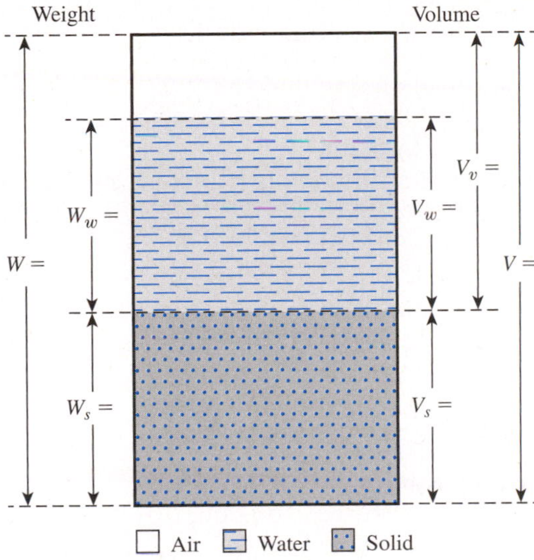

Figure 3.13

3.4 During a compaction test in the geotechnical laboratory, the students compacted a clayey soil into a cylindrical mold 101.6 mm in diameter and 116.4 mm in height. The compacted soil in the mold had a mass of 1814 grams, and it had a moisture content of 12%. If G_s = 2.72, determine the following:

a. Dry density
b. Void ratio
c. Degree of saturation
d. Additional water (in kg) needed to achieve 100% saturation in the soil sample

3.5 Two undisturbed soil samples, each having a volume of 2832 cm^3, are collected from different depths of the same soil layer. For sample *A*, located above the groundwater table, mass = 4990 g and w = 9%. Sample *B* is located below the groundwater table. If G_s = 2.68, determine

a. Void ratio of *A*
b. Degree of saturation of *A*
c. Water content of *B*
d. Total weight of *B*

3.6 A saturated clay soil has a moisture content of 40%. Given that $G_s = 2.73$, determine the following:

a. Porosity
b. Dry unit weight
c. Saturated unit weight

3.7 The moist mass of 2832 cm^3 of soil is 5670 g. If the moisture content is 14% and the specific gravity of soil solids is 2.71, determine the following:

a. Moist unit weight
b. Dry unit weight
c. Void ratio
d. Porosity
e. Degree of saturation
f. Volume occupied by water

3.8 The dry unit weight of a soil sample is 14.8 kN/m^3. Given that $G_s = 2.72$ and $w = 17\%$, determine:

a. Void ratio
b. Moist unit weight
c. Degree of saturation
d. Unit weight when the sample is fully saturated

3.9 Refer to Problem 3.8. Determine the mass of water (in kg) to be added per cubic meter (m^3) of soil for

a. 90% degree of saturation
b. 100% degree of saturation

3.10 The void ratio of an undisturbed soil sample is 0.55 and the moisture content is 11%. If $G_s = 2.68$, determine:

a. Moist unit weight
b. Dry unit weight
c. Degree of saturation
d. Moisture content when the sample is fully saturated

3.11 During a subsurface exploration, an undisturbed soil sample was collected from the field using a split-spoon sampler for laboratory evaluation (see Figure 17.6 in Chapter 17). The sample has a diameter of 34.9 mm, length of 457.2 mm, and a moist mass of 840 g. If the oven-dried mass was 680 g and $G_s = 2.74$, calculate the following:

a. Moist unit weight
b. Moisture content
c. Dry unit weight
d. Void ratio
e. Degree of saturation

3.12 Refer to Problem 3.11. A 76.2 mm long specimen was cut from the split-spoon sampler for performing a shear strength test. If the specimen is required to be 100% saturated for the test, determine:

a. Saturated unit weight
b. Moisture content at 100% saturation
c. Amount of water (in kg) needed to achieve full saturation

3.13 When the moisture content of a soil is 26%, the degree of saturation is 72% and the moist unit weight is 16.97 kN/m^3. Determine:

a. Specific gravity of soil solids
b. Void ratio
c. Saturated unit weight

3.14 For a given soil, the following are known: G_s = 2.74, moist unit weight, γ = 20.6 kN/m^3, and moisture content, w = 16.6%. Determine:

a. Dry unit weight
b. Void ratio
c. Porosity
d. Degree of saturation

3.15 Refer to Problem 3.14. Determine the mass of water, in kg, to be added per cubic meter (m^3) of soil for

a. 90% degree of saturation
b. 100% degree of saturation

3.16 The moist density of a soil is 1935 kg/m^3. Given w = 18% and G_s = 2.7, determine:

a. Dry density
b. Porosity
c. Degree of saturation
d. Mass of water, in kg/m^3, to be added to reach full saturation

3.17 For a moist soil, given the following: V = 7080 cm^3; M = 13.95 kg; w = 9.8%; and G_s = 2.66. Determine:

a. Dry unit weight
b. Void ratio
c. Volume occupied by water

3.18 For a given soil, ρ_d = 1750 kg/m^3 and n = 0.36. Determine:

a. Void ratio
b. Specific gravity of soil solids

3.19 The moisture content of a soil sample is 22% and the dry unit weight is 15.65 kN/m^3. If G_s = 2.67, what is the degree of saturation?

3.20 For a given soil, w = 14.8%, G_s = 2.71, and S = 72%. Determine:

a. Moist unit weight in kN/m^3
b. Volume occupied by water

3.21 The degree of saturation of a soil is 55% and the moist unit weight is 16.63 kN/m^3. When the moist unit weight increased to 17.88 kN/m^3, the degree of saturation increased to 82.2%. Determine:

a. G_s
b. Void ratio

3.22 Refer to Figure 3.14. After the construction of a concrete retaining wall, backfill material from a nearby borrow pit was brought into the excavation behind the wall and compacted to a final void ratio of 0.8. Given that the soil in the borrow pit has void ratio of 1.1, determine the volume of borrow material needed to construct 1 m^3 of compacted backfill.

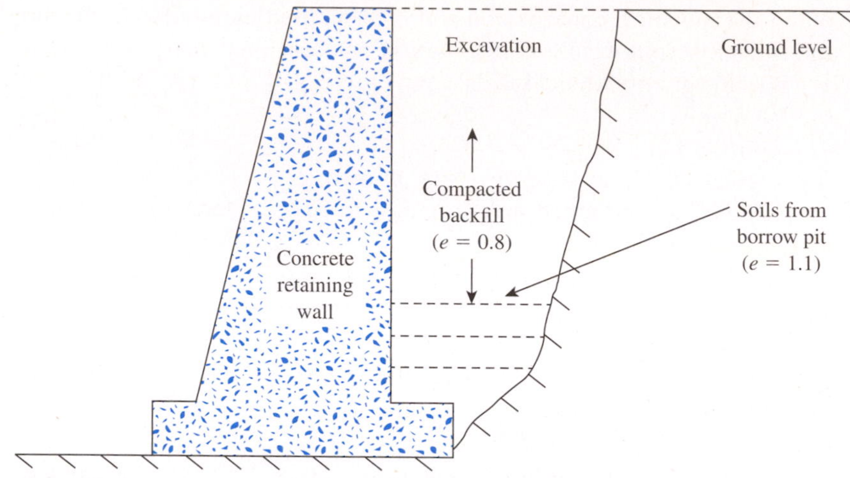

Figure 3.14

3.23 Refer to Problem 3.22. Given that the borrow pit soil has a moisture content of 11% and $G_s = 2.7$, determine

- **a.** Moist unit weight of the borrow soil
- **b.** Degree of saturation of the borrow soil
- **c.** Moist unit weight of the compacted backfill

3.24 Refer to the 4.57 m high embankment shown in Figure 3.8. Embankments are generally constructed in several lifts or layers that are compacted according to geotechnical specifications. Each lift thickness is 0.91 m and must have a dry unit weight of 18.51 kN/m^3. It is known that the soil at the borrow pit has a moist unit weight of 17.41 kN/m^3, moisture content of 23%, and $G_s = 2.67$. Perform the following tasks.

- **a.** Determine the moist weight of borrow soil needed to construct the first lift (bottom layer) per m of the embankment.
- **b.** On the day of the construction, there was a heavy rain that caused the borrow pit to reach a near saturated condition. Recalculate the moist weight of the borrow soil needed to construct the first lift.

3.25 For a given sandy soil, $e_{max} = 0.75$ and $e_{min} = 0.52$. If $G_s = 2.67$ and $D_r = 65\%$, determine the void ratio and the dry unit weight.

3.26 For a given sandy soil, the maximum and minimum void ratios are 0.77 and 0.41, respectively. If $G_s = 2.66$ and $w = 9\%$, what is the moist unit weight of compaction (kN/m^3) in the field if $D_r = 90\%$?

3.27 In a construction project, the field moist unit weight was 17.5 kN/m^3 and the moisture content was 11%. If maximum and minimum dry unit weights determined in the laboratory were 19.2 kN/m^3 and 14.1 kN/m^3, respectively, what was the field relative density?

3.28 In a highway project, the granular sub-base layer is compacted to a moist unit weight of 19.14 kN/m^3 at a moisture content of 16%. What is the relative density of the compacted sub-base? Given: $e_{max} = 0.85$, $e_{min} = 0.42$, and $G_s = 2.68$.

3.29 Refer to Problem 3.28. To improve the bearing capacity of the same sub-base, the field engineers decided to increase the relative density to 88% by additional compaction. What would be the final dry unit weight of the compacted sub-base?

Critical Thinking Problems

3.C.1 It is known that the natural soil at a construction site has a void ratio of 0.92. At the end of compaction, the in-place void ratio was found to be 0.65. If the moisture content remains unchanged, determine the following:

a. Percent decrease in the total volume of the soil due to compaction
b. Percent increase in the field dry unit weight
c. Percent change in the degree of saturation

3.C.2 A 3-m high sandy fill material was placed loosely at a relative density of 55%. Laboratory studies indicated that the maximum and minimum void ratios of the fill material are 0.94 and 0.66, respectively. Construction specifications required that the fill be compacted to a relative density of 85%. If $G_s = 2.65$, determine:

a. Dry unit weight of the fill before and after compaction
b. Final height of the fill after compaction

3.C.3 In a certain beach restoration project involving mixing and compaction of various sandy soils, the engineers studied the role of median grain size, D_{50}, on compacted density. Binary granular mixes of coarse and fine materials were synthetically prepared by mixing different volume percentages of finer soils with coarser soils at three different median grain size ratios; $D_{50\text{-coarse}}/D_{50\text{-fine}} = 1.67, 3$, and 6.

The table below shows all mixes used in this study. For each binary mix, the maximum dry unit weight was determined by compacting the mix in the Proctor mold using the same compactive energy. Perform the following tasks.

a. On the same graph, plot the variation of dry unit weight with the volume percent of finer soil for each median grain size ratio.
b. What can you conclude about the role of $D_{50\text{-coarse}}/D_{50\text{-fine}}$ ratios on compacted density of granular binary mixes?

Fine soil by volume (%)	Dry unit weight, γ_d(kN/m^3)		
	$\frac{D_{50\text{-coarse}}}{D_{50\text{-fine}}} = 1.67$	$\frac{D_{50\text{-coarse}}}{D_{50\text{-fine}}} = 3.0$	$\frac{D_{50\text{-coarse}}}{D_{50\text{-fine}}} = 6.0$
10	16.61	16.3	16.78
20	16.1	16.17	17.10
30	16.42	16.52	17.37
40	16.72	16.78	17.59
50	16.6	16.68	17.2
80	16.0	16.4	16.64

References

American Society for Testing and Materials (2014). *Annual Book of ASTM Standards*, Sec. 4, Vol. 04.08. West Conshohocken, Pa.

Cubrinovski, M., and Ishihara, K. (1999). "Empirical Correlation Between SPT N-Value and Relative Density for Sandy Soils," *Soils and Foundations*. Vol. 39, No. 5, 61–71.

Cubrinovski, M., and Ishihara, K. (2002). "Maximum and Minimum Void Ratio Characteristics of Sands," *Soils and Foundations*. Vol. 42, No. 6, 65–78.

Lade, P. V., Liggio, C. D., and Yamamuro, J. A. (1998). "Effects of Non-Plastic Fines on Minimum and Maximum Void Ratios of Sand," *Geotechnical Testing Journal*, ASTM. Vol. 21, No. 4, 336–347.

Youd, T. L. (1973). "Factors controlling Maximum and Minimum Densities of Sand," *Evaluation of Relative Density and Its Role in Geotechnical Projects Involving Cohesionless Soils*, STP 523, ASTM, 98–122.

CHAPTER 4

Plasticity and Structure of Soil

4.1 Introduction

When clay minerals are present in fine-grained soil, the soil can be remolded in the presence of some moisture without crumbling. This cohesive nature is caused by the adsorbed water surrounding the clay particles. In the early 1900s, a Swedish scientist named Atterberg developed a method to describe the consistency of fine-grained soils with varying moisture contents. At a very low moisture content, soil behaves more like a solid. When the moisture content is very high, the soil and water may flow like a liquid. Hence, on an arbitrary basis, depending on the moisture content, the behavior of soil can be divided into four basic states—*solid, semisolid, plastic,* and *liquid*—as shown in Figure 4.1.

The moisture content, in percent, at which the transition from solid to semisolid state takes place is defined as the *shrinkage limit*. The moisture content at the point of transition from semisolid to plastic state is the *plastic limit*, and from plastic to liquid state is the *liquid limit*. These parameters are also known as *Atterberg limits*. This chapter describes the procedures to determine the Atterberg limits. Also discussed in this chapter are soil structure and geotechnical parameters, such as activity and liquidity index, which are related to Atterberg limits.

4.2 Liquid Limit (*LL*)

Percussion cup method

The percussion method was developed by Casagrande (1932) and used throughout the world. This is the only method adopted by ASTM (Test Designation D-4318) to determine the liquid limit of cohesive soils. A schematic diagram (side view) of

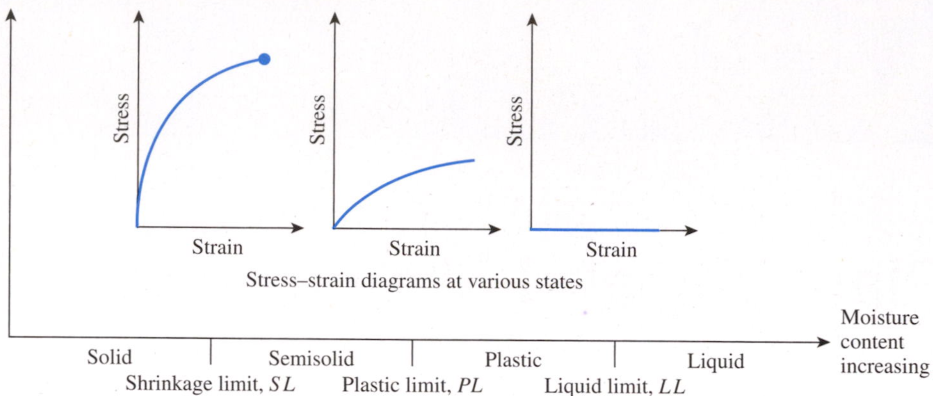

Figure 4.1 Atterberg limits

a liquid limit device is shown in Figure 4.2a. This device consists of a brass cup and a hard rubber base. The brass cup can be dropped onto the base by a cam operated by a crank. To perform the liquid limit test, one must place a soil paste in the cup. A groove is then cut at the center of the soil pat with the standard grooving tool (Figures 4.2b and 4.2c). Note that there are two types of grooving tools in use. They are flat grooving tools (Figure 4.2b) and wedge grooving tools (Figure 4.2c). By the use of the crank-operated cam, the cup is lifted and dropped from a height of 10 mm. The moisture content, in percent, required to close a distance of 12.5 mm along the bottom of the groove (see Figures 4.2d and 4.2e) after 25 blows is defined as the *liquid limit.*

It is difficult to adjust the moisture content in the soil to meet the required 12.5 mm closure of the groove in the soil pat at 25 blows. Hence, at least three tests for the same soil are conducted at varying moisture contents, with the number of blows, N, required to achieve closure varying between 15 and 35. Figure 4.3 shows a photograph of a liquid limit test device and grooving tools. Figure 4.4 shows photographs of the soil pat in the liquid limit device before and after the test. The moisture content of the soil, in percent, and the corresponding number of blows are plotted on semilogarithmic graph paper (Figure 4.5). The relationship between moisture content and log N is approximated as a straight line. This line is referred to as the *flow curve*. The moisture content corresponding to $N = 25$, determined from the flow curve, gives the liquid limit of the soil. The slope of the flow line is defined as the *flow index* and may be written as

$$I_F = \frac{w_1 - w_2}{\log\left(\dfrac{N_2}{N_1}\right)} \tag{4.1}$$

where I_F = flow index
w_1 = moisture content of soil, in percent, corresponding to N_1 blows
w_2 = moisture content corresponding to N_2 blows

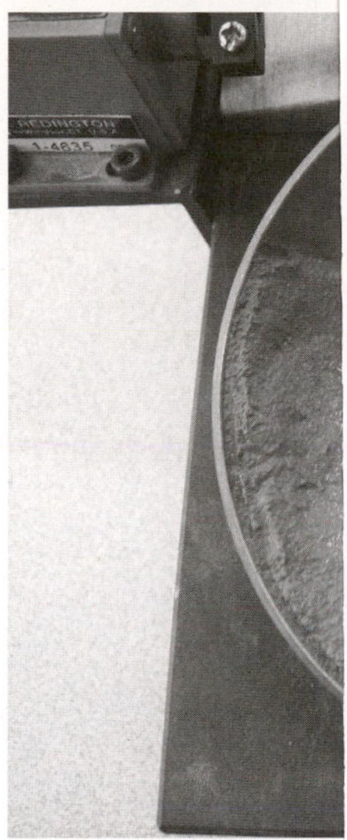

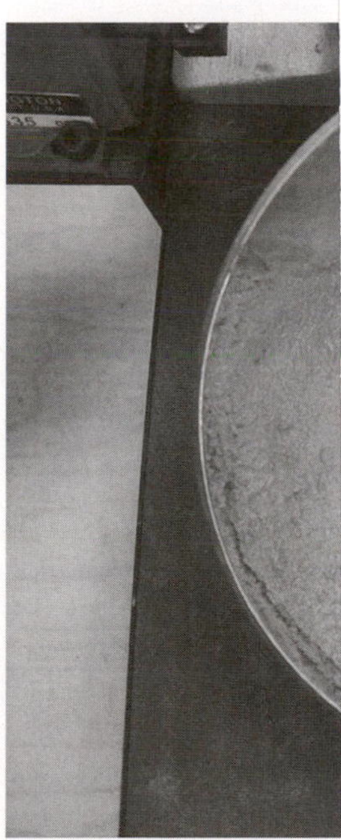

Figure 4.4 Photographs showi
(b) after test [*Note:* The 12.5 m
(*Courtesy of Khaled Sobhan, Florida*

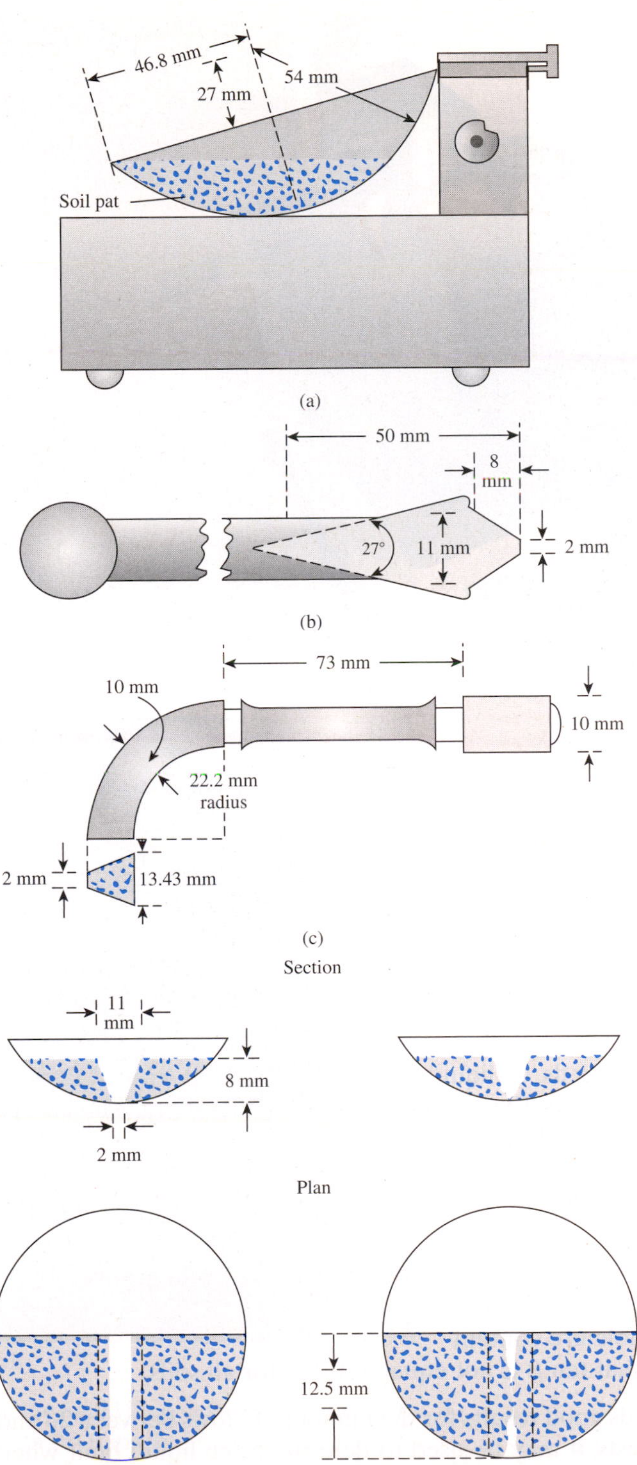

Figure 4.2 Liquid limit test: (a) liquid limit device; (b) flat grooving tool; (c) wedge grooving tool; (d) soil pat before test; (e) soil pat after test

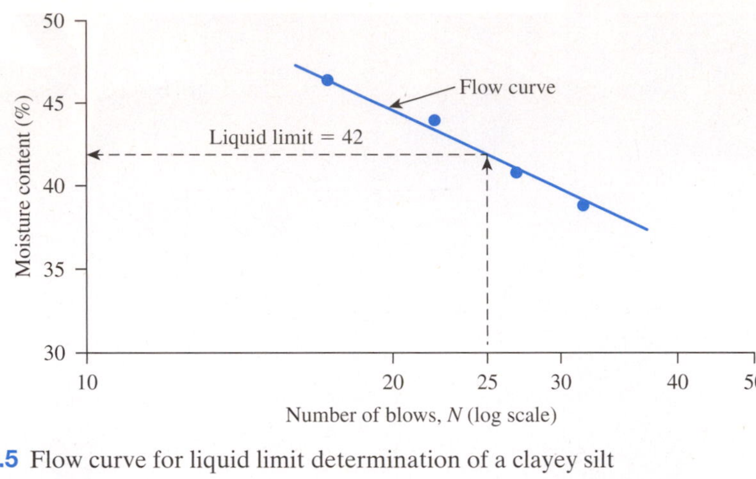

Figure 4.5 Flow curve for liquid limit determination of a clayey silt

Table 4.1 Values of $\left(\frac{N}{25}\right)^{0.121}$

N	$\left(\frac{N}{25}\right)^{0.121}$	N	$\left(\frac{N}{25}\right)^{0.121}$
20	0.973	26	1.005
21	0.979	27	1.009
22	0.985	28	1.014
23	0.990	29	1.018
24	0.995	30	1.022
25	1.000		

method and was also adopted by ASTM under designation D-4318. The reason that the one-point method yields fairly good results is that a small range of moisture content is involved when $N = 20$ to $N = 30$. Table 4.1 shows the values of the term $(\frac{N}{25})^{0.121}$ given in Eq. (4.3) for $N = 20$ to $N = 30$.

Example 4.1

Following are the results of a test conducted in the laboratory. Determine the liquid limit (LL) and the flow index (I_F).

Number of blows, N	Moisture content (%)
15	42.0
20	40.8
28	39.3

Solution

The plot of w against N (log scale) is shown in Figure 4.6. For $N = 25$, $w = \mathbf{39.5\%} = \boldsymbol{LL}$.

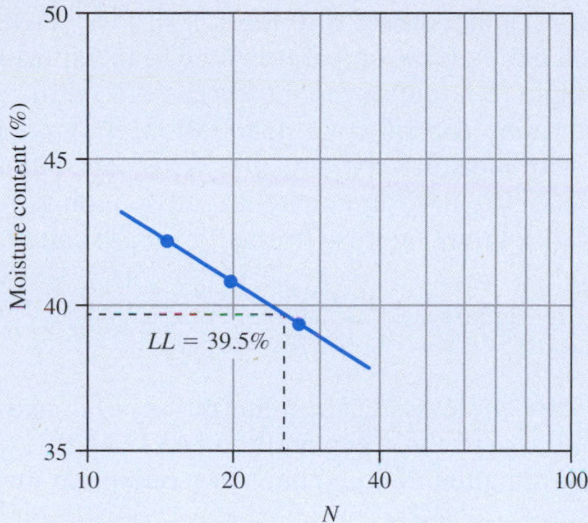

Figure 4.6

From Eq. (4.1),

$$I_F = \frac{w_1 - w_2}{\log\left(\frac{N_2}{N_1}\right)} = \frac{42 - 39.3}{\log\left(\frac{28}{15}\right)} = \mathbf{9.96}$$

Example 4.2

For the soil discussed in Example 4.1, assume that only one liquid limit was conducted, i.e., $N = 20$ and the moisture content = 40.8%. Estimate the liquid limit of the soil by the one-point method.

Solution

From Eq. (4.3),

$$LL = w_N\left(\frac{N}{25}\right)^{0.121} = (40.8)\left(\frac{20}{25}\right)^{0.121} = \mathbf{39.7}$$

Fall cone method

Another method of determining liquid limit that is popular in Europe and Asia is the *fall cone method* (British Standard—BS1377). In this test the liquid limit is defined as the moisture content at which a standard cone of apex angle 30° and weight of 0.78 N (80 gf) will penetrate a distance $d = 20$ mm in 5 seconds when allowed to drop from a position of point contact with the soil surface (Figure 4.7a). Figure 4.8 shows the photograph of a fall cone apparatus. Due to the difficulty in achieving the liquid limit from a single test, four or more tests can be conducted at various moisture contents to determine the fall cone penetration, d. A semilogarithmic graph can then be plotted with moisture content (w) versus cone penetration d. The plot results in a straight line. The moisture content corresponding to $d = 20$ mm is the liquid limit (Figure 4.7b). From Figure 4.7b, the *flow index* can be defined as

$$I_{FC} = \frac{w_2(\%) - w_1(\%)}{\log d_2 - \log d_1} \tag{4.4}$$

where w_1, w_2 = moisture contents at cone penetrations of d_1 and d_2, respectively.

As in the case of the percussion cup method (ASTM 4318), attempts have been made to develop the estimation of liquid limit by a one-point method. They are

- Nagaraj and Jayadeva (1981)

$$LL = \frac{w}{0.77 \log d} \tag{4.5}$$

$$LL = \frac{w}{0.65 + 0.0175d} \tag{4.6}$$

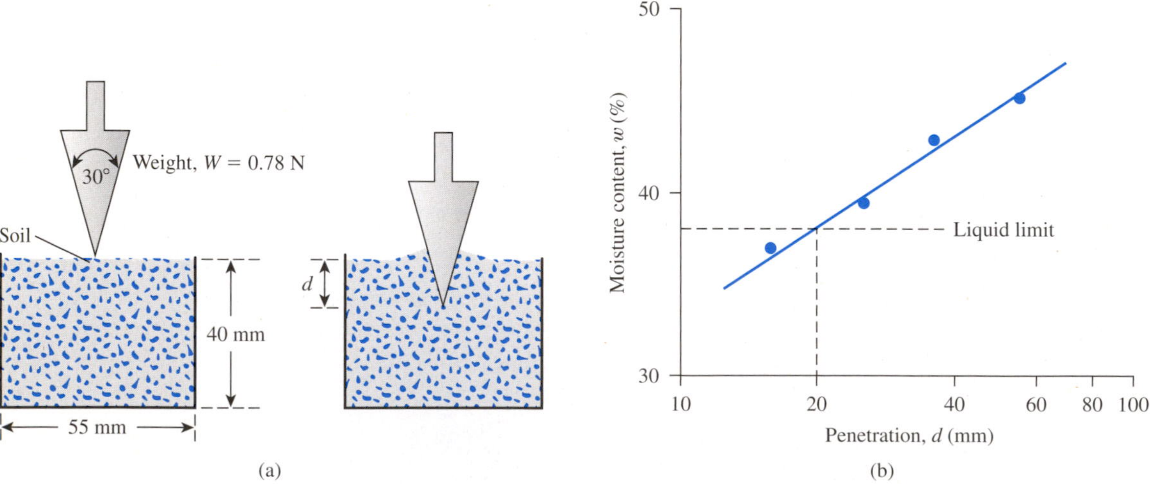

Figure 4.7 (a) Fall cone test (b) plot of moisture content vs. cone penetration for determination of liquid limit

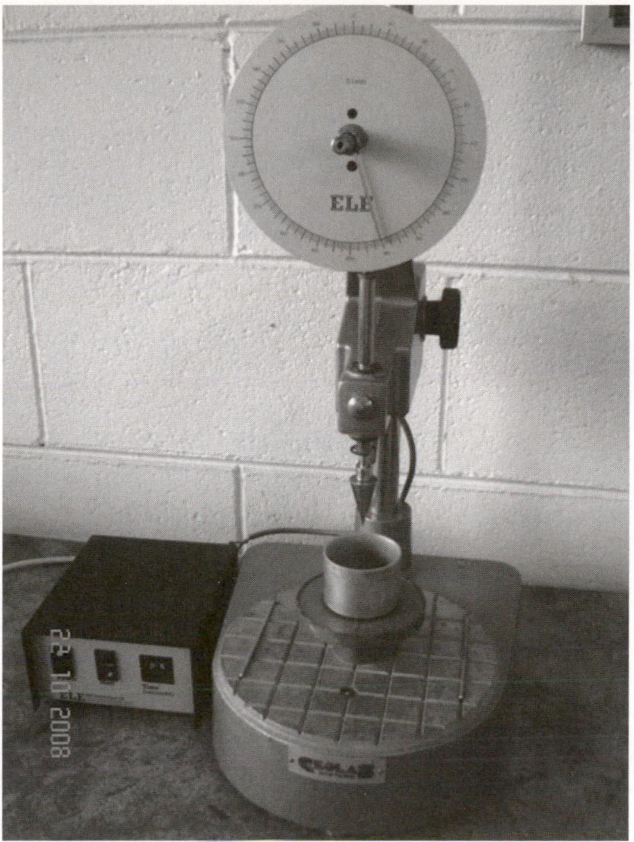

Figure 4.8 Fall cone apparatus (*Courtesy of N. Sivakugan, James Cook University, Australia*)

- Feng (2001)

$$LL = w\left(\frac{20}{d}\right)^{0.33} \tag{4.7}$$

where w (%) is the moisture content for a cone penetration d (mm) falling between 15 mm to 25 mm.

Example 4.3

Following are the results of a liquid limit test using a fall cone. Estimate the liquid limit.

Cone penetration, d (mm)	Moisture content (%)
15	29.5
26	35.5
34	38.5
43	41.5

Solution

Figure 4.9 shows the moisture content versus d (mm). From this plot, the moisture content can be determined to be **32.5**.

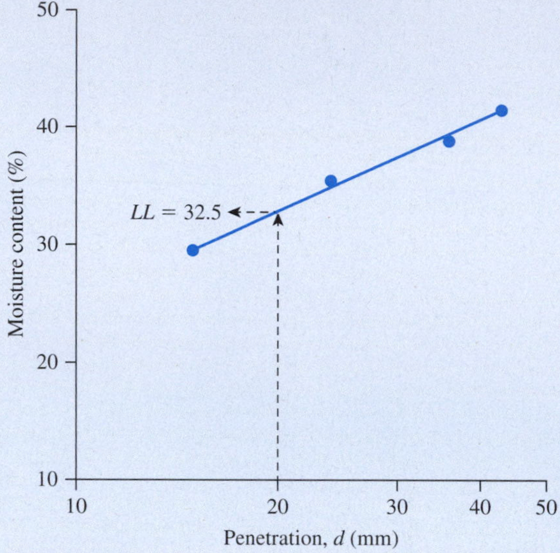

Figure 4.9

Example 4.4

Let us assume that only one liquid limit test is conducted using the fall cone for the soil reported in Example 4.3; i.e., $w = 29.5\%$ at $d = 15$ mm. Estimate the liquid limit using Eqs. (4.5), (4.6) and (4.7).

Solution

From Eq. (4.5),

$$LL = \frac{w}{0.77 \log d} = \frac{29.5}{(0.77)(\log 15)} = \mathbf{32.58}$$

From Eq. (4.6),

$$LL = \frac{w}{0.65 + 0.0175d} = \frac{29.5}{0.65 + (0.0175)(15)} = \mathbf{32.33}$$

From Eq. (4.7),

$$LL = w\left(\frac{20}{d}\right)^{0.33} = (29.5)\left(\frac{20}{15}\right)^{0.33} = \mathbf{32.43}$$

Table 4.2 Summary of Main Differences among Fall Cones (Summarized from Budhu, 1985)

Country	Cone details	Penetration for liquid limit (mm)
Russia	Cone angle = 30° Cone mass = 76 g	10
Britian, France	Cone angle = 30° Cone mass = 80 g	20
India	Cone angle = 31° Cone mass = 148 g	20.4
Sweden, Canada (Quebec)	Cone angle = 60° Cone mass = 60 g	10

Note: Duration of penetration is 5 s in all cases.

General comments

The dimensions of the cone tip angle, cone weight, and the penetration (mm) at which the liquid limit is determined varies from country to country. Table 4.2 gives a summary of different fall cones used in various countries.

A number of major studies have shown that the undrained shear strength of the soil at liquid limit varies between 1.7 to 2.3 kN/m^2. Based on tests conducted on a large number of soil samples, Feng (2001) has given the following correlation between the liquid limits determined according to ASTM D4318 and British Standard BS1377.

$$LL_{(BS)} = 2.6 + 0.94[LL_{(ASTM)}] \tag{4.8}$$

4.3 Plastic Limit (*PL*)

The *plastic limit* is defined as the moisture content in percent, at which the soil crumbles, when rolled into threads of 3.2 mm in diameter. The plastic limit is the lower limit of the plastic stage of soil. The plastic limit test is simple and is performed by repeated rollings of an ellipsoidal-sized soil mass by hand on a ground glass plate (Figure 4.10). The procedure for the plastic limit test is given by ASTM in Test Designation D-4318.

As in the case of liquid limit determination, the fall cone method can be used to obtain the plastic limit. This can be achieved by using a cone of similar geometry but with a mass of 2.35 N (240 gf). Three to four tests at varying moisture contents of soil are conducted, and the corresponding cone penetrations (d) are determined. The moisture content corresponding to a cone penetration of $d = 20$ mm is the plastic limit. Figure 4.11 shows the liquid and plastic limit determination of Cambridge Gault clay reported by Wroth and Wood (1978).

Table 4.3 gives the ranges of liquid limit, plastic limit, and activity (Section 4.7) of some clay minerals (Mitchell, 1976; Skempton, 1953).

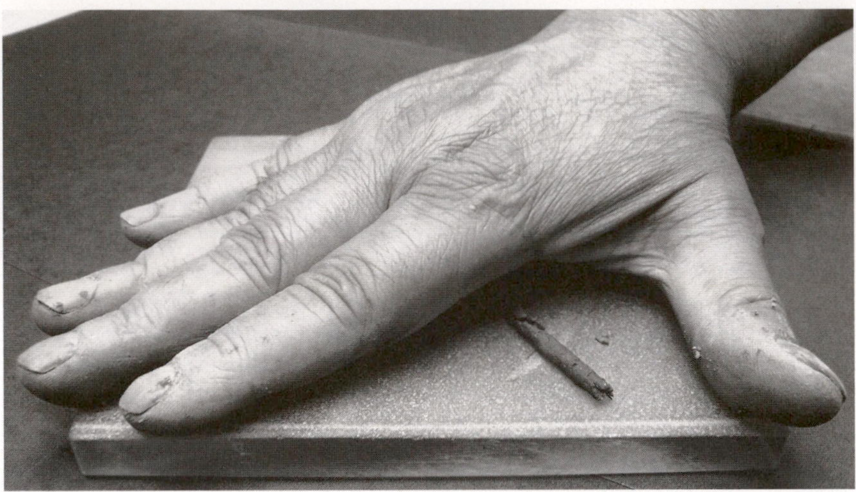

Figure 4.10 Rolling of soil mass on ground glass plate to determine plastic limit *(Courtesy of Braja M. Das, Henderson, Nevada)*

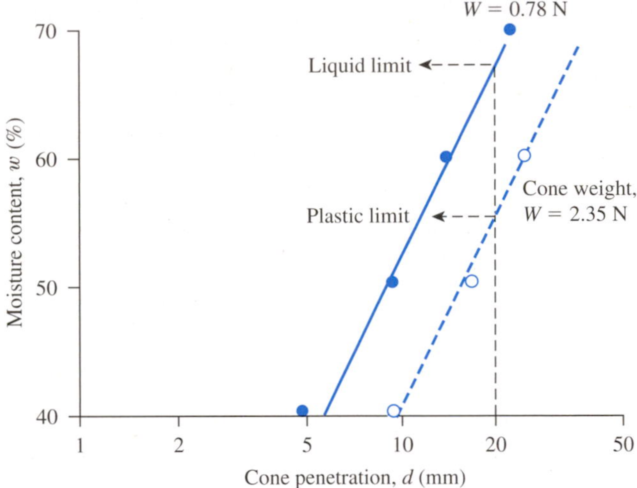

Figure 4.11 Liquid and plastic limits for Cambridge Gault clay determined by fall cone test

Table 4.3 Typical Values of Liquid Limit, Plastic Limit, and Activity of Some Clay Minerals

Mineral	Liquid limit, *LL*	Plastic limit, *PL*	Activity, *A*
Kaolinite	35–100	20–40	0.3–0.5
Illite	60–120	35–60	0.5–1.2
Montmorillonite	100–900	50–100	1.5–7.0
Halloysite (hydrated)	50–70	40–60	0.1–0.2
Halloysite (dehydrated)	40–55	30–45	0.4–0.6
Attapulgite	150–250	100–125	0.4–1.3
Allophane	200–250	120–150	0.4–1.3

4.4 Plasticity Index

The *plasticity index* (*PI*) is the difference between the liquid limit and the plastic limit of a soil, or

$$PI = LL - PL \tag{4.9}$$

Burmister (1949) classified the plasticity index in a qualitative manner as follows:

PI	Description
0	Nonplastic
1–5	Slightly plastic
5–10	Low plasticity
10–20	Medium plasticity
20–40	High plasticity
>40	Very high plasticity

The plasticity index is important in classifying fine-grained soils. It is fundamental to the Casagrande plasticity chart (presented in Section 4.8), which is currently the basis for the Unified Soil Classification System. (See Chapter 5.)

Sridharan et al. (1999) showed that the plasticity index can be correlated to the flow index as obtained from the liquid limit tests (Section 4.2). According to their study (Figure 4.12a),

$$PI\,(\%) = 4.12 I_F\,(\%) \tag{4.10}$$

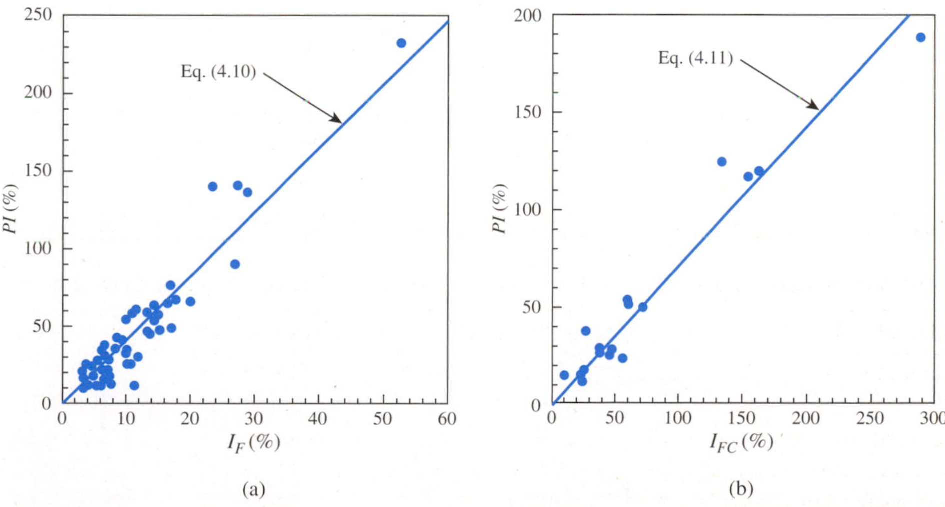

Figure 4.12 Variation of *PI* with (a) I_F; and (b) I_{FC} [*Adapted after Sridharan et al. (1999). With Permission from ASTM*]

and (Figure 4.12 b)

$$PI\ (\%) = 0.74 I_{FC}\ (\%) \quad (4.11)$$

4.5 Shrinkage Limit (*SL*)

Soil shrinks as moisture is gradually lost from it. With continuing loss of moisture, a stage of equilibrium is reached at which more loss of moisture will result in no further volume change (Figure 4.13). The moisture content, in percent, at which the volume of the soil mass ceases to change is defined as the *shrinkage limit*.

Shrinkage limit tests are performed in the laboratory with a porcelain dish about 44 mm in diameter and about 12.7 mm high. The inside of the dish is coated with petroleum jelly and is then filled completely with wet soil. Excess soil standing above the edge of the dish is struck off with a straightedge. The mass of the wet soil inside the dish is recorded. The soil pat in the dish is then oven-dried. The volume of the oven-dried soil pat is then determined.

By reference to Figure 4.13, the shrinkage limit can be determined as

$$SL = w_i\ (\%) - \Delta w\ (\%) \quad (4.12)$$

where w_i = initial moisture content when the soil is placed in the shrinkage limit dish
Δw = change in moisture content (that is, between the initial moisture content and the moisture content at the shrinkage limit)

However,

$$w_i\ (\%) = \frac{M_1 - M_2}{M_2} \times 100 \quad (4.13)$$

where M_1 = mass of the wet soil pat in the dish at the beginning of the test (g)
M_2 = mass of the dry soil pat (g) (see Figure 4.14)

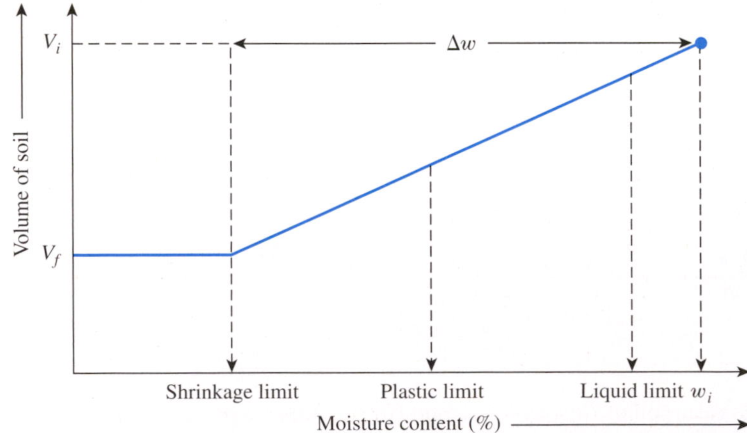

Figure 4.13 Definition of shrinkage limit

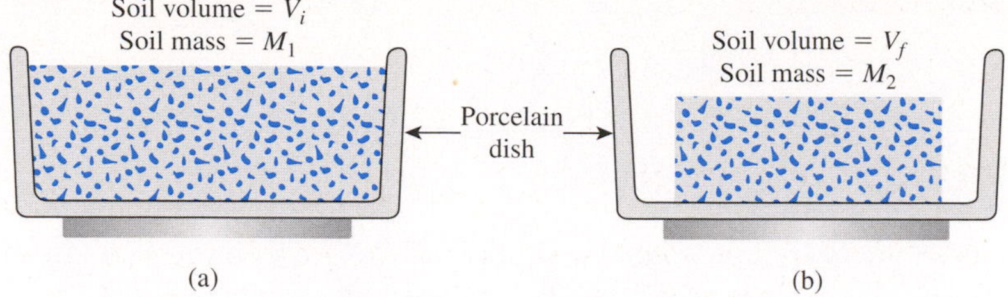

Figure 4.14 Shrinkage limit test: (a) soil pat before drying; (b) soil pat after drying

Figure 4.15 shows photographs of the soil pat in the shrinkage limit dish before and after drying.
Also,

$$\Delta w\ (\%) = \frac{(V_i - V_f)\rho_w}{M_2} \times 100 \tag{4.14}$$

where V_i = initial volume of the wet soil pat (that is, inside volume of the dish, cm^3)
V_f = volume of the oven-dried soil pat (cm^3)
ρ_w = density of water (g/cm^3)

Finally, combining Eqs. (4.12), (4.13), and (4.14) gives

$$SL = \left(\frac{M_1 - M_2}{M_2}\right)(100) - \left(\frac{V_i - V_f}{M_2}\right)(\rho_w)\ (100) \tag{4.15}$$

ASTM (2014) Test Designation D-4943 describes a method where volume V_i is determined by filling the shrinkage limit dish with water, or

$$V_i = \frac{\text{Mass of water to fill the dish (g)}}{\rho_w\ (\text{g/cm}^3)} \tag{4.16}$$

In order to determine V_f, the dry soil pat is dipped in a molten pot of wax and cooled. The mass of the dry soil and wax is determined in air and in submerged water. Thus

$$M_5 = M_3 - M_4 \tag{4.17}$$

where M_3 = mass of dry soil pat and wax in air (g)
M_4 = mass of dry soil pat and wax in water (g)
M_5 = mass of water displaced by dry soil pat and wax (g)

The volume of the dry soil pat and wax can be calculated as

$$V_{fwx}\ (\text{cm}^3) = \frac{M_5\ (\text{g})}{\rho_w\ (\text{g/cm}^3)} \tag{4.18}$$

The mass of wax (M_6) coating the dry soil pat is then obtained as

$$M_6\ (\text{g}) = M_3\ (\text{g}) - M_2\ (\text{g}) \tag{4.19}$$

(a)

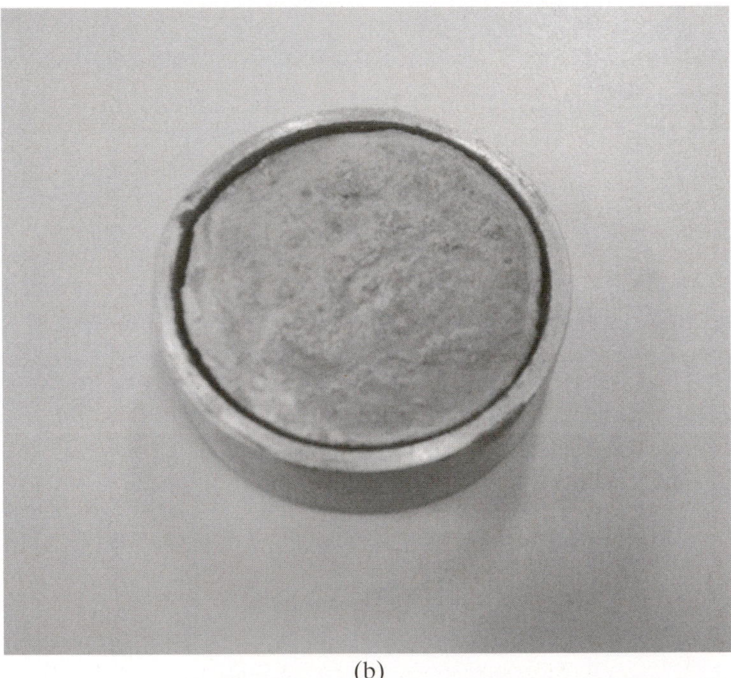

(b)

Figure 4.15 Photograph of soil pat in the shrinkage limit dish: (a) before drying; (b) after drying *(Courtesy of Braja Das, Henderson, Nevada)*

Thus the volume of wax coating (V_{wx}) is

$$V_{wx}\,(\text{cm}^3) = \frac{M_6\,(\text{g})}{G_{wx}\,\rho_w\,(\text{g/cm}^3)} \tag{4.20}$$

where G_{wx} = specific gravity of wax

Finally, the volume of the dry soil pat (V_f) can be obtained as

$$V_f\,(\text{cm}^3) = V_{fwx} - V_{wx} \tag{4.21}$$

Equations (4.16) and (4.21) can be substituted into Eq. (4.15) to obtain the shrinkage limit.

Another parameter that can be determined from a shrinkage limit test is the *shrinkage ratio*, which is the ratio of the volume change of soil as a percentage of the dry volume to the corresponding change in moisture content, or

$$SR = \frac{\left(\dfrac{\Delta V}{V_f}\right)}{\left(\dfrac{\Delta M}{M_2}\right)} = \frac{\left(\dfrac{\Delta V}{V_f}\right)}{\left(\dfrac{\Delta V \rho_w}{M_2}\right)} = \frac{M_2}{V_f \rho_w} \tag{4.22}$$

where ΔV = change in volume

ΔM = corresponding change in the mass of moisture

It can also be shown that

$$G_s = \frac{1}{\dfrac{1}{SR} - \left(\dfrac{SL}{100}\right)} \tag{4.23}$$

where G_s = specific gravity of soil solids.

If desired, the maximum expected volumetric shrinkage and linear shrinkage at given moisture contents (w) can be calculated as

$$VS\,(\%) = SR[w(\%) - SL] \tag{4.24}$$

where VS = volumetric shrinkage, and

$$LS\,(\%) = 100\left[1 - \left(\frac{100}{VS(\%) + 100}\right)^{\frac{1}{3}}\right] \tag{4.25}$$

where LS = linear shrinkage

Typical values of shrinkage limit for some clay minerals are as follows (Mitchell, 1976).

Mineral	Shrinkage limit
Montmorillonite	8.5–15
Illite	15–17
Kaolinite	25–29

Example 4.5

Following are the results of a shrinkage limit test:

- Initial volume of soil in a saturated state = 24.6 cm^3
- Final volume of soil in a dry state = 15.9 cm^3
- Initial mass in a saturated state = 44.0 g
- Final mass in a dry state = 30.1 g

Determine the shrinkage limit of the soil.

Solution

From Eq. (4.15),

$$SL = \left(\frac{M_1 - M_2}{M_2}\right)(100) - \left(\frac{V_i - V_f}{M_2}\right)(\rho_w)(100)$$

$$M_1 = 44.0 \text{ g} \qquad V_i = 24.6 \text{ cm}^3 \qquad \rho_w = 1 \text{ g/cm}^3$$

$$M_2 = 30.1 \text{ g} \qquad V_f = 15.9 \text{ cm}^3$$

$$SL = \left(\frac{44.0 - 30.1}{30.1}\right)(100) - \left(\frac{24.6 - 15.9}{30.1}\right)(1)(100)$$

$$= 46.18 - 28.9 = \mathbf{17.28\%}$$

Example 4.6

Refer to Example 4.5. Determine the shrinkage ratio of the soil. Also estimate the specific gravity of the soil solids.

Solution

From Eq. (4.22),

$$SR = \frac{M_2}{V_f \rho_w} = \frac{30.1 \text{ g}}{(15.9 \text{ cm}^3)(1 \text{ g/cm}^3)} = \mathbf{1.89}$$

Also, from Eq. (4.23),

$$G_s = \frac{1}{\dfrac{1}{SR} - \left(\dfrac{SL}{100}\right)} = \frac{1}{\left(\dfrac{1}{1.89}\right) - \left(\dfrac{17.28}{100}\right)} \approx \mathbf{2.81}$$

Example 4.7

Refer to Example 4.5. If the soil is at a moisture content of 28%, estimate the maximum volumetric shrinkage (VS) and the linear shrinkage (LS).

Solution

From Eq. (4.24),

$$VS\ (\%) = SR[w(\%) - SL]$$

From Example 4.6, $SR = 1.89$. So

$$VS = (1.89)(28 - 17.28) = \mathbf{20.26\%}$$

Again, from Eq. (4.25),

$$LS\ (\%) = 100\left[1 - \left(\frac{100}{VS(\%) + 100}\right)^{\frac{1}{3}}\right] = 100\left[1 - \left(\frac{100}{20.26 + 100}\right)^{\frac{1}{3}}\right] \approx \mathbf{5.96\%}$$

4.6 Liquidity Index and Consistency Index

The relative consistency of a cohesive soil in the natural state can be defined by a ratio called the *liquidity index*, which is given by

$$LI = \frac{w - PL}{LL - PL} \tag{4.26}$$

where w = *in situ* moisture content of soil.

The *in situ* moisture content for a sensitive clay may be greater than the liquid limit. In this case (Figure 4.16),

$$LI > 1$$

These soils, when remolded, can be transformed into a viscous form to flow like a liquid.

Soil deposits that are heavily overconsolidated may have a natural moisture content less than the plastic limit. In this case (Figure 4.16),

$$LI < 0$$

Another index that is commonly used for engineering purposes is the *consistency index* (CI), which may be defined as

$$CI = \frac{LL - w}{LL - PL} \tag{4.27}$$

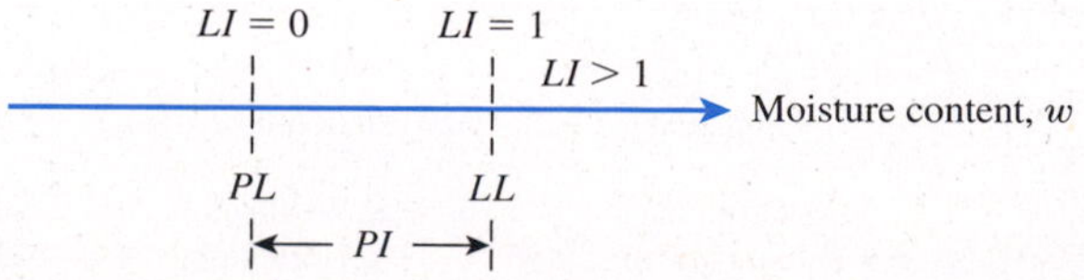

Figure 4.16 Liquidity index

Table 4.4 Approximate Correlation between *CI* and Unconfined Compression Strength of Clay

CI	Unconfined compression strength kN/m^2
<0.5	<25
0.5–0.75	25–80
0.75–1.0	80–150
1.0–1.5	150–400
>1.5	>400

where w = *in situ* moisture content. If w is equal to the liquid limit, the consistency index is zero. Again, if $w = PL$, then $CI = 1$. Table 4.4 gives an approximate correlation between *CI* and the unconfined compression strength of clay (see Chapter 12).

4.7 Activity

Because the plasticity of soil is caused by the adsorbed water that surrounds the clay particles, we can expect that the type of clay minerals and their proportional amounts in a soil will affect the liquid and plastic limits. Skempton (1953) observed that the plasticity index of a soil increases linearly with the percentage of clay-size fraction (% finer than 2 μm by weight) present (Figure 4.17). The correlations of *PI* with the clay-size fractions for different clays plot separate lines. This difference is due to the diverse plasticity characteristics of the various types of clay minerals. On the basis of

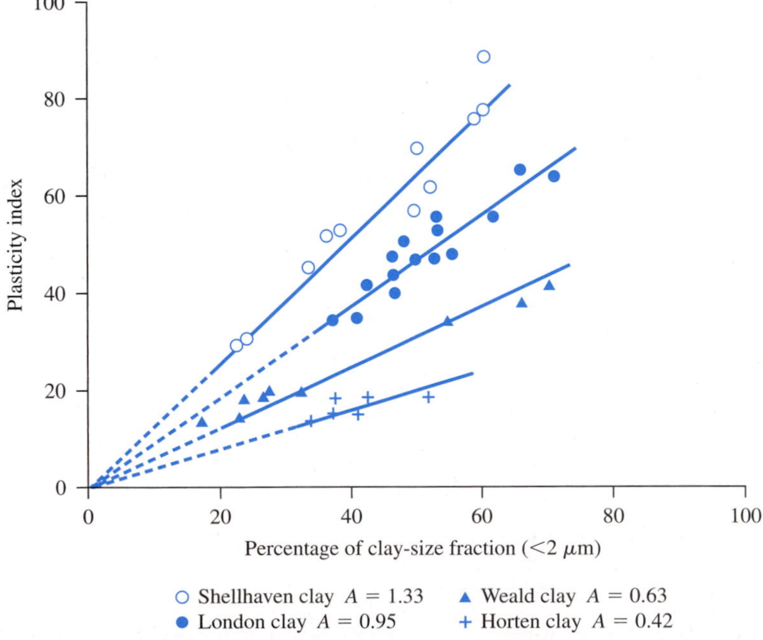

Figure 4.17 Activity (*Based on Skempton, 1953*)

these results, Skempton defined a quantity called *activity*, which is the slope of the line correlating PI and % finer than 2 μm. This activity may be expressed as

$$A = \frac{PI}{(\%\text{of clay-size fraction, by weight})} \tag{4.28}$$

where A = activity. Activity is used as an index for identifying the swelling potential of clay soils. Typical values of activities for various clay minerals are given in Table 4.3.

Seed, Woodward, and Lundgren (1964a) studied the plastic property of several artificially prepared mixtures of sand and clay. They concluded that, although the relationship of the plasticity index to the percentage of clay-size fraction is linear (as observed by Skempton), it may not always pass through the origin. This is shown in Figures 4.18 and 4.19. Thus, the activity can be redefined as

$$A = \frac{PI}{\%\text{of clay-size fraction} - C'} \tag{4.29}$$

where C' is a constant for a given soil.

For the experimental results shown in Figures 4.18 and 4.19, $C' = 9$.

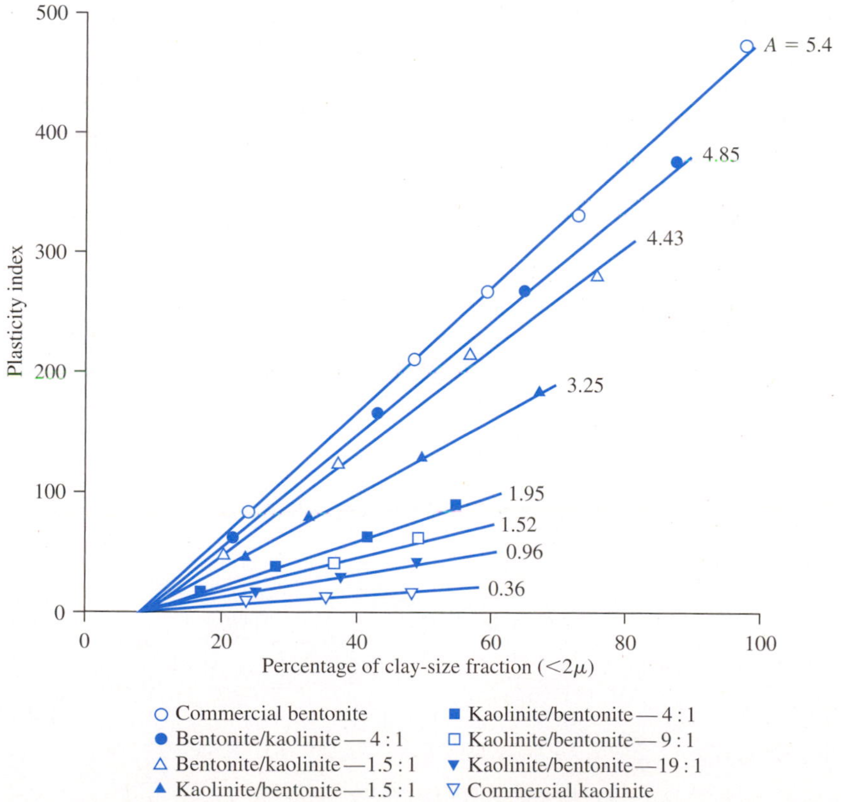

Figure 4.18 Relationship between plasticity index and clay-size fraction by weight for kaolinite/bentonite clay mixtures (*After Seed, Woodward, and Lundgren, 1964a. With permission from ASCE.*)

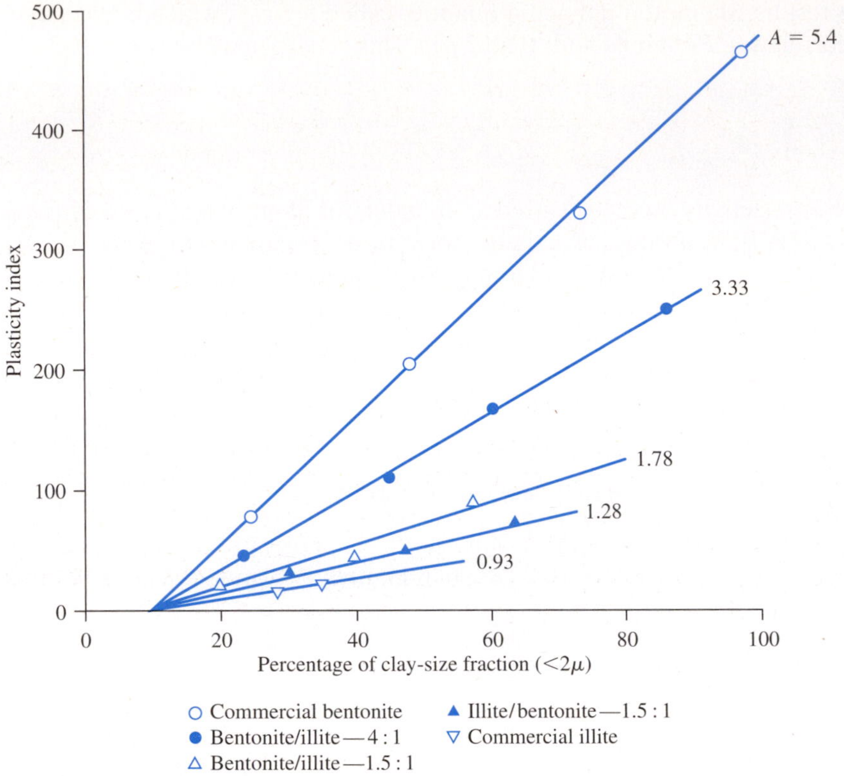

Figure 4.19 Relationship between plasticity index and clay-size fraction by weight for illite/bentonite clay mixtures *(After Seed, Woodward, and Lundgren, 1964a. With permission from ASCE.)*

Further works of Seed, Woodward, and Lundgren (1964b) have shown that the relationship of the plasticity index to the percentage of clay-size fractions present in a soil can be represented by two straight lines. This is shown qualitatively in Figure 4.20. For clay-size fractions greater than 40%, the straight line passes through the origin when it is projected back.

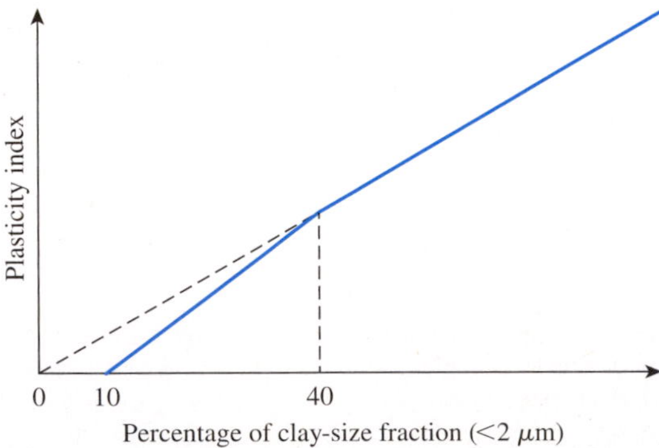

Figure 4.20 Simplified relationship between plasticity index and percentage of clay-size fraction by weight *(After Seed, Woodward, and Lundgren, 1964b. With permission from ASCE.)*

4.8 Plasticity Chart

Liquid and plastic limits are determined by relatively simple laboratory tests that provide information about the nature of cohesive soils. Engineers have used the tests extensively for the correlation of several physical soil parameters as well as for soil identification. Casagrande (1932) studied the relationship of the plasticity index to the liquid limit of a wide variety of natural soils. On the basis of the test results, he proposed a plasticity chart as shown in Figure 4.21. The important feature of this chart is the empirical A-line that is given by the equation $PI = 0.73(LL - 20)$. An A-line separates the inorganic clays from the inorganic silts. Inorganic clay values lie above the A-line, and values for inorganic silts lie below the A-line. Organic silts plot in the same region (below the A-line and with LL ranging from 30 to 50) as the inorganic silts of medium compressibility. Organic clays plot in the same region as inorganic silts of high compressibility (below the A-line and LL greater than 50). The information provided in the plasticity chart is of great value and is the basis for the classification of fine-grained soils in the Unified Soil Classification System. (See Chapter 5.)

Note that a line called the U-line lies above the A-line. The U-line is approximately the upper limit of the relationship of the plasticity index to the liquid limit for any currently known soil. The equation for the U-line can be given as

$$PI = 0.9(LL - 8) \tag{4.30}$$

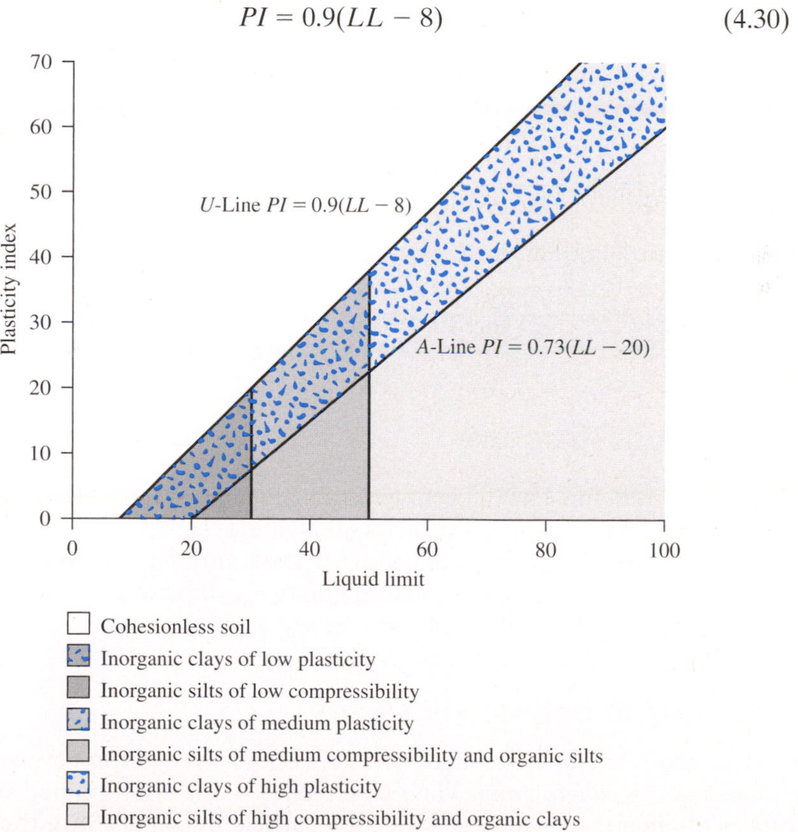

Figure 4.21 Plasticity chart

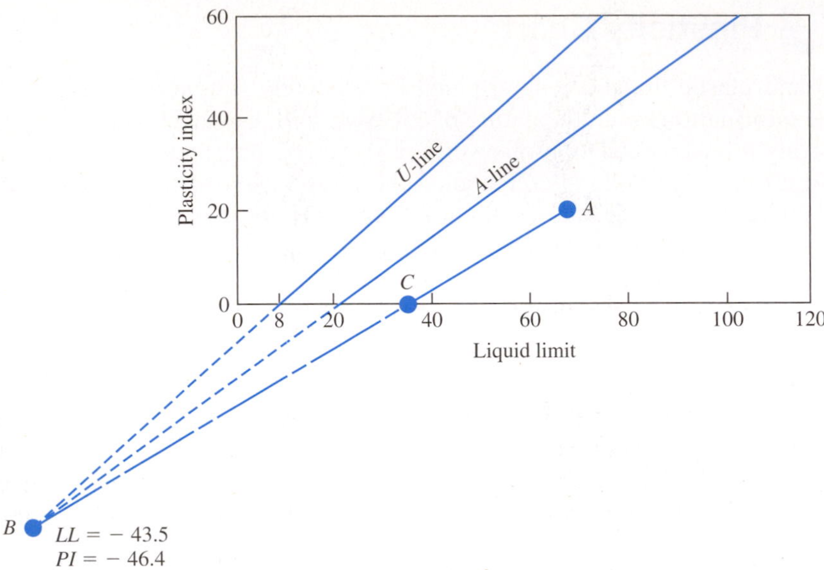

Figure 4.22 Estimation of shrinkage from plasticity chart (*Adapted from Holtz and Kovacs, 1981*)

There is another use for the *A*-line and the *U*-line. Casagrande has suggested that the shrinkage limit of a soil can be approximately determined if its plasticity index and liquid limit are known (see Holtz and Kovacs, 1981). This can be done in the following manner with reference to Figure 4.22.

a. Plot the plasticity index against the liquid limit of a given soil such as point *A* in Figure 4.22.

b. Project the *A*-line and the *U*-line downward to meet at point *B*. Point *B* will have the coordinates of $LL = -43.5$ and $PI = -46.4$.

c. Join points *B* and *A* with a straight line. This will intersect the liquid limit axis at point *C*. The abscissa of point *C* is the estimated shrinkage limit.

4.9 Soil Structure

Soil structure is defined as the geometric arrangement of soil particles with respect to one another. Among the many factors that affect the structure are the shape, size, and mineralogical composition of soil particles, and the nature and composition of soil water. In general, soils can be placed into two groups: cohesionless and cohesive. The structures found in soils in each group are described next.

Structures in cohesionless soil

The structures generally encountered in cohesionless soils can be divided into two major categories: *single-grained* and *honeycombed*. In single-grained structures, soil particles are in stable positions, with each particle in contact with the surrounding

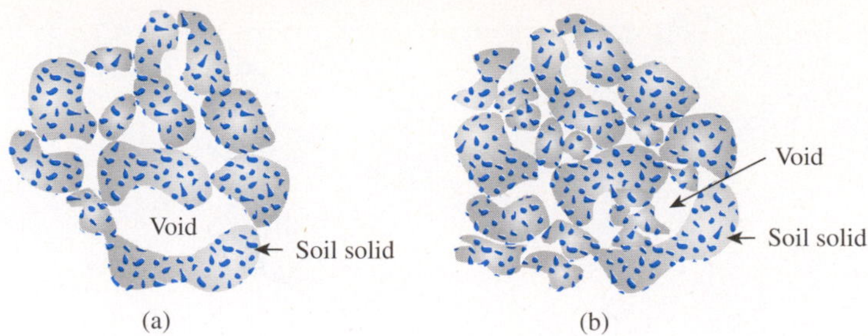

Figure 4.23 Single-grained structure: (a) loose; (b) dense

ones. The shape and size distribution of the soil particles and their relative positions influence the denseness of packing (Figure 4.23); thus, a wide range of void ratios is possible. To get an idea of the variation of void ratios caused by the relative positions of the particles, let us consider the mode of packing of equal spheres shown in Figures 4.24 and 4.25.

Figure 4.24a shows the case of a very loose state of packing. If we isolate a cube with each side measuring d, which is equal to the diameter of each sphere as shown in the figure, the void ratio can be calculated as

$$e = \frac{V_v}{V_s} = \frac{V - V_s}{V_s}$$

where V = volume of the cube = d^3

V_s = volume of sphere (i.e., solid) inside the cube

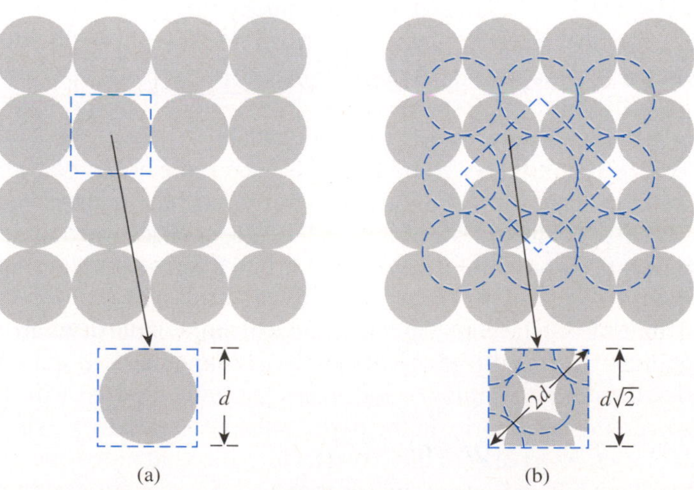

Figure 4.24 Mode of packing of equal spheres (plan views): (a) very loose packing ($e = 0.91$); (b) very dense packing ($e = 0.35$)

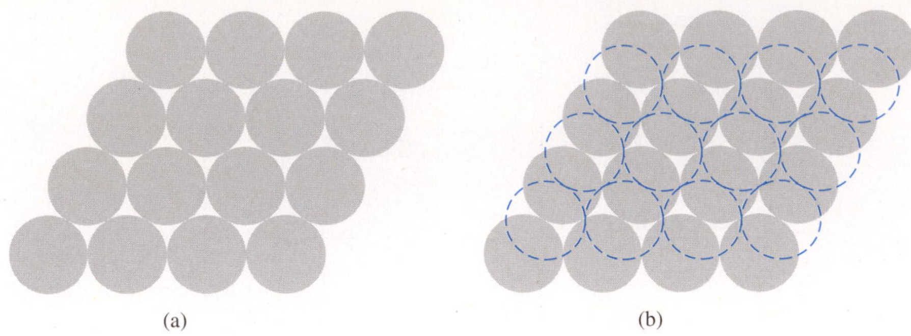

Figure 4.25 Packing of equal spheres: (a) simple stagger; (b) double stagger

Noting that $V = d^3$ and $V_s = \pi d^3/6$ yields

$$e = \frac{d^3 - \left(\frac{\pi d^3}{6}\right)}{\left(\frac{\pi d^3}{6}\right)} = 0.91$$

The type of packing shown in Figure 4.24a is called *cubical* or *simple cubical* packing.

Similarly, Figure 4.24b shows the case of a very dense state of packing. Figure 4.24b also shows an isolated cube, for which each side measures $d\sqrt{2}$. It can be shown that, for this case, $e = 0.35$. This is referred to as *pyramidal packing*.

There can be other types of packing of equal spheres between the loosest and densest states, and these are shown in Figure 4.25. Figure 4.25a shows a *simple stagger packing*. In this pattern, each sphere touches six neighboring spheres in its own layer, and the spheres in different layers are stacked directly on top of each other. The void ratio for the single stagger pattern is 0.65. Figure 4.25b shows a *double stagger packing*. This is similar to the single stagger pattern, except that each sphere in one layer has slid over and down to contact two spheres in the second layer. The void ratio for the double stagger arrangement is 0.43.

McGeary (1961) conducted some tests (also see Lade et al., 1998) by depositing equal-sized steel spheres into a container to determine the average minimum void ratio, which was 0.6. In those tests about 20% of the spheres were in double stagger arrangement ($e = 0.43$) and about 80% of the spheres were in single stagger arrangement ($e = 0.65$).

Real soil differs from the equal-spheres model in that soil particles are neither equal in size nor spherical. The smaller-size particles may occupy the void spaces between the larger particles, thus the void ratio of soils is decreased compared with that for equal spheres. However, the irregularity in the particle shapes generally yields an increase in the void ratio of soils. As a result of these two factors, the void ratios encountered in real soils have approximately the same range as those obtained in equal spheres.

In the honeycombed structure (Figure 4.26), relatively fine sand and silt form small arches with chains of particles. Soils that exhibit a honeycombed structure

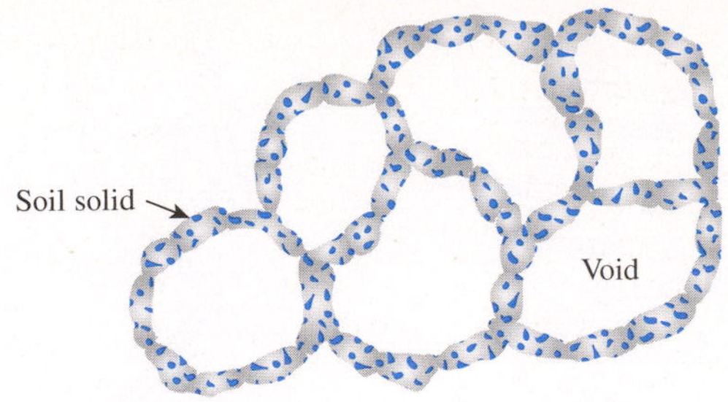

Figure 4.26 Honeycombed structure

have large void ratios, and they can carry an ordinary static load. However, under a heavy load or when subjected to shock loading, the structure breaks down, which results in a large amount of settlement.

Structures in cohesive soils

To understand the basic structures in cohesive soils, we need to know the types of forces that act between clay particles suspended in water. In Chapter 2, we discussed the negative charge on the surface of the clay particles and the diffuse double layer surrounding each particle. When two clay particles in suspension come close to each other, the tendency for interpenetration of the diffuse double layers results in repulsion between the particles. At the same time, an attractive force exists between the clay particles that is caused by van der Waals forces and is independent of the characteristics of water. Both repulsive and attractive forces increase with decreasing distance between the particles, but at different rates. When the spacing between the particles is very small, the force of attraction is greater than the force of repulsion. These are the forces treated by colloidal theories.

The fact that local concentrations of positive charges occur at the edges of clay particles was discussed in Chapter 2. If the clay particles are very close to each other, the positively charged edges can be attracted to the negatively charged faces of the particles.

Let us consider the behavior of clay in the form of a dilute suspension. When the clay is initially dispersed in water, the particles repel one another. This repulsion occurs because with larger interparticle spacing, the forces of repulsion between the particles are greater than the forces of attraction (van der Waals forces). The force of gravity on each particle is negligible. Thus, the individual particles may settle very slowly or remain in suspension, undergoing *Brownian motion* (a random zigzag motion of colloidal particles in suspension). The sediment formed by the settling of the individual particles has a dispersed structure, and all particles are oriented more or less parallel to one another (Figure 4.27a).

If the clay particles initially dispersed in water come close to one another during random motion in suspension, they might aggregate into visible flocs with edge-to-face contact. In this instance, the particles are held together by electrostatic attraction of positively charged edges to negatively charged faces. This aggregation is known as *flocculation*. When the flocs become large, they settle under the force of gravity. The sediment formed in this manner has a flocculent structure (Figure 4.27b).

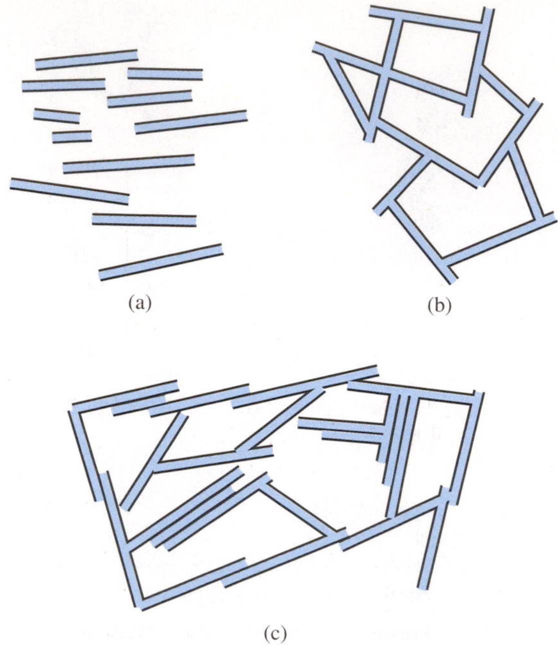

Figure 4.27 Sediment structures: (a) dispersion; (b) nonsalt flocculation; (c) salt flocculation (*Adapted from Lambe, 1958*)

When salt is added to a clay–water suspension that has been initially dispersed, the ions tend to depress the double layer around the particles. This depression reduces the interparticle repulsion. The clay particles are attracted to one another to form flocs and settle. The flocculent structure of the sediments formed is shown in Figure 4.27c. In flocculent sediment structures of the salt type, the particle orientation approaches a large degree of parallelism, which is due to van der Waals forces.

Clays that have flocculent structures are lightweight and possess high void ratios. Clay deposits formed in the sea are highly flocculent. Most of the sediment deposits formed from freshwater possess an intermediate structure between dispersed and flocculent.

A deposit of pure clay minerals is rare in nature. When a soil has 50% or more particles with sizes of 0.002 mm or less, it is generally termed *clay*. Studies with scanning electron microscopes (Collins and McGown, 1974; Pusch, 1978; Yong and Sheeran, 1973) have shown that individual clay particles tend to be aggregated or flocculated in submicroscopic units. These units are referred to as *domains*. The domains then group together, and these groups are called *clusters*. Clusters can be seen under a light microscope. This grouping to form clusters is caused primarily by interparticle forces. The clusters, in turn, group to form *peds*. Peds can be seen without a microscope. Groups of peds are macrostructural features along with joints and fissures. Figure 4.28a shows the arrangement of the peds and macropore spaces. The arrangement of domains and clusters with silt-size particles is shown in Figure 4.28b.

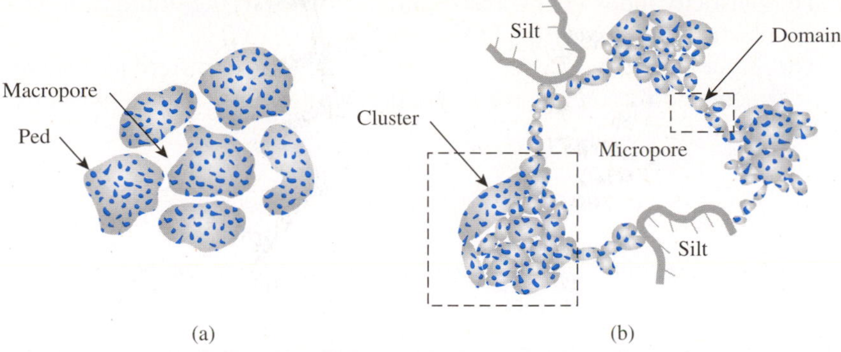

Figure 4.28 Soil structure: (a) arrangement of peds and macropore spaces; (b) arrangement of domains and clusters with silt-sized particles

Table 4.5 Structure of Clay Soils

Item	Remarks
Dispersed structures	Formed by settlement of individual clay particles; more or less parallel orientation (see Figure 4.27a)
Flocculent structures	Formed by settlement of flocs of clay particles (see Figures 4.27b and 4.27c)
Domains	Aggregated or flocculated submicroscopic units of clay particles
Clusters	Domains group to form clusters; can be seen under light microscope
Peds	Clusters group to form peds; can be seen without microscope

From the preceding discussion, we can see that the structure of cohesive soils is highly complex. Macrostructures have an important influence on the behavior of soils from an engineering viewpoint. The microstructure is more important from a fundamental viewpoint. Table 4.5 summarizes the macrostructures of clay soils.

4.10 Summary

Following is a summary of the materials presented in this chapter.

- The consistency of fine-grained soils can be described by three parameters: the liquid limit, plastic limit, and shrinkage limit. These are referred to as Atterberg limits.
- The liquid (LL), plastic (PL), and shrinkage (SL) limits are, respectively, the moisture contents (%) at which the consistency of soil changes from liquid to plastic stage, plastic to semisolid stage, and semisolid to solid stage.

- The plasticity index (PI) is the difference between the liquid limit (LL) and the plastic limit (PL) [Eq. (4.9)].
- The liquidity index of soil (LI) is the ratio of the difference between the *in situ* moisture content (%) and the plastic limit to the plasticity index [Eq. (4.26)], or

$$LI = \frac{w - PL}{LL - PL}$$

- Activity, A, is defined as the ratio of plasticity index to the percent of clay-size fraction by weight in a soil [Eq. (4.28)].
- The structure of cohesionless soils can be single-grained or honeycombed. Soils with honeycombed structure have large void ratios that may break down under heavy load and dynamic loading.
- Dispersion, nonsalt flocculation, and salt flocculation of clay soils were discussed in Section 4.9. Also discussed in this section is the structure of fine-grained soil as it relates to the arrangement of peds and micropore spaces and the arrangement of domains and clusters with silt-size particles.

Problems

4.1 During Atterberg limit tests in the soil mechanics laboratory, the students obtained the following results from a clayey soil.
Liquid limit tests:

Number of blows, N	Moisture content (%)
14	38.4
16	36.5
20	33.1
28	27.0

Plastic limit tests: Students conducted two trials and found that PL = 17.2% for the first trial and PL = 17.8% for the second trial.

a. Draw the flow curve and obtain the liquid limit.

b. What is the plasticity index of the soil? Use an average value of PL from the two plastic limit trails.

4.2 Refer to the soil in Problem 4.1. A second group of students conducted only one test and found that the groove on the soil sample closed 12.5 mm when $N = 21$ and $w = 30.4\%$. Estimate the liquid limit by the one-point method.

4.3 Refer to the soil in Problem 4.1.

a. Determine the flow index.

b. Determine the liquidity index of the soil if the *in situ* moisture content is 21%.

4.4 Results from a liquid limit test conducted on a soil are given below.

Number of blows, N	Moisture content (%)
12	35.2
19	29.5
27	25.4
37	21

a. Determine the liquid limit of the soil.
b. If it is known that the $PI = 6.5$, what would be the plastic limit of the soil?
c. Determine the liquidity index of the soil if $w_{\text{in situ}} = 23.8\%$

4.5 The following data were obtained by conducting liquid limit and plastic limit tests on a soil collected from the site.
Liquid limit tests:

Number of blows, N	Moisture content (%)
15	39.5
21	37.9
29	36.4
38	35.1

Plastic limit test: $PL = 19.3\%$
a. Draw the flow curve and determine the liquid limit.
b. Using the Casagrande plasticity chart (Figure 4.21), determine the soil type.

4.6 Refer to the soil in Problem 4.5. Using the Casagrande plasticity chart, graphically estimate the shrinkage limit of the soil as shown in Figure 4.22.

4.7 Following results are obtained for a liquid limit test using a fall cone device. Estimate the liquid limit of the soil and the flow index.

Cone penetration, d (mm)	Moisture content (%)
13	26.3
19	31.9
26	39.3
31	42.6

4.8 Refer to the same soil in Problem 4.7. A single test was conducted with the fall cone device and the following results were obtained: $d = 17$ mm and $w = 28.5\%$. Using Eqs. (4.5), (4.6), and (4.7), estimate the liquid limit by the one-point method.

4.9 Refer to the liquid limit determined in Problem 4.5 using the percussion cup method (ASTM 4318). Estimate the liquid limit for the same soil if the fall cone method (BS 1377) were used. Use Eq. (4.8).

4.10 During a shrinkage limit test, a 19.3 cm³ saturated clay sample with a mass of 37 g was placed in a porcelain dish and dried in the oven. The oven-dried sample had a mass of 28 g with a final volume of 16 cm³. Determine the shrinkage limit and the shrinkage ratio.

4.11 The following data were recorded during a shrinkage limit test on a clay soil pat: $V_i = 20.6\ \text{cm}^3$, $V_f = 13.8\ \text{cm}^3$, $M_1 = 47.5$ g, and mass of dry soil, $M_2 = 34.6$ g. Determine the shrinkage limit and the shrinkage ratio.

4.12 In a shrinkage limit test, a sample of saturated clay was dried in the oven. The dry mass of the soil was 22.5 g. As shown in Figure 4.13, when the moisture content is at the shrinkage limit, the soil reaches a constant total volume, V_f. If $V_f = 10.3\ \text{cm}^3$, calculate the shrinkage limit of the soil. Given: $G_s = 2.72$

Critical Thinking Problems

4.C.1 The properties of seven different clayey soils are shown below (Skempton and Northey, 1952). Investigate the relationship between the strength and plasticity characteristics by performing the following tasks:

a. Estimate the plasticity index for each soil using Skempton's definition of activity [Eq. (4.28)].

b. Estimate the probable mineral composition of the clay soils based on *PI* and *A* (use Table 4.3)

c. Sensitivity (S_t) refers to the loss of strength when the soil is remolded or disturbed. It is defined as the ratio of the undisturbed strength ($\tau_{f\text{-undisturbed}}$) to the remolded strength ($\tau_{f\text{-remolded}}$) at the same moisture content [Eq. (12.49)]. From the given data, estimate $\tau_{f\text{-remolded}}$ for the clay soils.

d. Plot the variations of undisturbed and remolded shear strengths with the activity, *A*, and explain the observed behavior.

Soil	% Clay fraction (< 2μm)	Activity, A	Undisturbed shear strength (kN/m²)	Sensitivity, S_t
Beauharnois	79	0.52	18	14
Detroit I	36	0.36	17	2.5
Horten	40	0.42	41	17
Gosport	55	0.89	29	2.2
Mexico City	90	4.5	46	5.3
Shellhaven	41	1.33	36	7.6
St. Thuribe	36	0.33	38	150

4.C.2 Liquidity index, *LI,* defined by Eq. (4.26), can indicate probable engineering behavior depending on the natural or current state of moisture content. For example, the material behavior can vary from a brittle solid ($LI < 1$) to viscous fluid ($LI > 1$), with an intermediate plastic state ($0 < LI < 1$). From the plasticity characteristics and ranges of moisture contents listed in the following table,

a. Determine the range of liquidity index for each soil over the range of moisture content.

b. Comment on the probable engineering behavior of each soil as the moisture content changes (refer to Figure 4.1).

Soil	% Clay fraction ($< 2\mu m$)	Natural moisture content, w_n (%)	Liquid limit, LL (%)	Plastic limit, PL (%)
1	34	59–67	49	26
2	44	29–36	37	21
3	54	51–56	61	26
4	81	61–70	58	24
5	28	441–600	511	192
6	67	98–111	132	49
7	72	51–65	89	31

References

American Society for Testing and Materials (2014). *Annual Book of ASTM Standards*, Sec. 4, Vol. 04.08, West Conshohocken, Pa.

BS:1377 (1990). *British Standard Methods of Tests for Soil for Engineering Purposes*, Part 2, BSI, London.

Budhu, M. (1985). "The Effect of Clay Content on Liquid Limit from a Fall Cone and the British Cup Device," *Geotechnical Testing Journal*, ASTM, Vol. 8, No. 2, 91–95.

Burmister, D. M. (1949). "Principles and Techniques of Soil Identification," *Proceedings*, Annual Highway Research Board Meeting, National Research Council, Washington, D.C., Vol. 29, 402–434.

Casagrande, A. (1932). "Research of Atterberg Limits of Soils," *Public Roads*, Vol. 13, No. 8, 121–136.

Collins, K., and McGown, A. (1974). "The Form and Function of Microfabric Features in a Variety of Natural Soils," *Geotechnique*, Vol. 24, No. 2, 223–254.

Feng, T. W. (2001). "A Linear log d − log w Model for the Determination of Consistency Limits of Soils," *Canadian Geotechnical Journal*, Vol. 38, No. 6, 1335–1342.

Holtz, R. D., and Kovacs, W. D. (1981). *An Introduction to Geotechnical Engineering*, Prentice-Hall, Englewood Cliffs, NJ.

Lade, P. V., Liggio, C. D., and Yamamuro, J. A. (1998). "Effects of Non-Plastic Fines on Minimum and Maximum Void Ratios of Sand," *Geotechnical Testing Journal*, ASTM. Vol. 21, No. 4, 336–347.

Lambe, T. W. (1958). "The Structure of Compacted Clay," *Journal of the Soil Mechanics and Foundations Division*, ASCE, Vol. 85, No. SM2, 1654-1 to 1654-35.

Mitchell, J. K. (1976). *Fundamentals of Soil Behavior*, Wiley, New York.

McGeary, R. K. (1961). "Mechanical Packing of Spherical Particles," *Journal of the American Ceramic Society*, Vol. 44, No. 11, 513–522.

Nagaraj, T. S., and Jayadeva, M. S. (1981). "Re-examination of One-Point Methods of Liquid Limit Determination," *Geotechnique*, Vol. 31, No. 3, 413–425.

Pusch, R. (1978). "General Report on Physico-Chemical Processes Which Affect Soil Structure and Vice Versa," *Proceedings*, International Symposium on Soil Structure, Gothenburg, Sweden, Appendix, 33.

Seed, H. B., Woodward, R. J., and Lundgren, R. (1964a). "Clay Mineralogical Aspects of Atterberg Limits," *Journal of the Soil Mechanics and Foundations Division*, ASCE, Vol. 90, No. SM4, 107–131.

Seed, H. B., Woodward, R. J., and Lundgren, R. (1964b). "Fundamental Aspects of the Atterberg Limits," *Journal of the Soil Mechanics and Foundations Division*, ASCE, Vol. 90, No. SM6, 75–105.

Skempton, A. W. (1953). "The Colloidal Activity of Clays," *Proceedings*, 3rd International Conference on Soil Mechanics and Foundation Engineering, London, Vol. 1, 57–61.

Skempton, A. W., and Northey, R. D. (1952). "The Sensitivity of Clays," *Geotechnique*, Vol. 3, No. 1, 30–53.

Sridharan, A., Nagaraj, H. B., and Prakash, K. (1999). "Determination of the Plasticity Index from Flow Index," *Geotechnical Testing Journal*, ASTM, Vol. 22, No. 2, 175–181.

U.S. Army Corps of Engineers (1949). *Technical Memo 3-286,* U.S. Waterways Experiment Station, Vicksburg, Miss.

Wroth, C. P., and Wood, D. M. (1978). "The Correlation of Index Properties with Some Basic Engineering Properties of Soils," *Canadian Geotechnical Journal,* Vol. 15, No. 2, 137–145.

Yong, R. N., and Sheeran, D. E. (1973). "Fabric Unit Interaction and Soil Behavior," *Proceedings*, International Symposium on Soil Structure, Gothenburg, Sweden, 176–184.

CHAPTER 5

Classification of Soil

5.1 Introduction

Different soils with similar properties may be classified into groups and subgroups according to their engineering behavior. Classification systems provide a common language to concisely express the general characteristics of soils, which are infinitely varied, without detailed descriptions. Most of the soil classification systems that have been developed for engineering purposes are based on simple index properties such as particle-size distribution and plasticity. Although several classification systems are now in use, none is totally definitive of any soil for all possible applications because of the wide diversity of soil properties.

In general, there are two major categories into which the classification systems developed in the past can be grouped.

1. The textural classification is based on the particle-size distribution of the percent of sand, silt, and clay-size fractions present in a given soil. In this chapter, we will discuss the textural classification system developed by the U.S. Department of Agriculture.
2. The other major category is based on the engineering behavior of soil and takes into consideration the particle-size distribution and the plasticity (i.e., liquid limit and plasticity index). Under this category, there are two major classification systems in extensive use now:
    a. The AASHTO classification system, and
    b. The Unified classification system.

The guidelines for classifying soil according to both of the aforementioned systems will be discussed in detail in the chapter.

5.2 Textural Classification

In a general sense, *texture* of soil refers to its surface appearance. Soil texture is influenced by the size of the individual particles present in it. Table 2.3 divided soils into gravel, sand, silt, and clay categories on the basis of particle size. In most cases, natural soils are mixtures of particles from several size groups. In the textural classification system, the soils are named after their principal components, such as sandy clay, silty clay, and so forth.

A number of textural classification systems were developed in the past by different organizations to serve their needs, and several of those are in use today. Figure 5.1 shows the textural classification systems developed by the U.S. Department of Agriculture (USDA). This classification method is based on the particle-size limits as described under the USDA system in Table 2.3; that is

- *Sand size:* 2.0 to 0.05 mm in diameter
- *Silt size:* 0.05 to 0.002 mm in diameter
- *Clay size:* smaller than 0.002 mm in diameter

The use of this chart can best be demonstrated by an example. If the particle-size distribution of soil *A* shows 30% sand, 40% silt, and 30% clay-size particles, its textural classification can be determined by proceeding in the manner indicated by the arrows in Figure 5.1. This soil falls into the zone of *clay loam*. Note that this chart

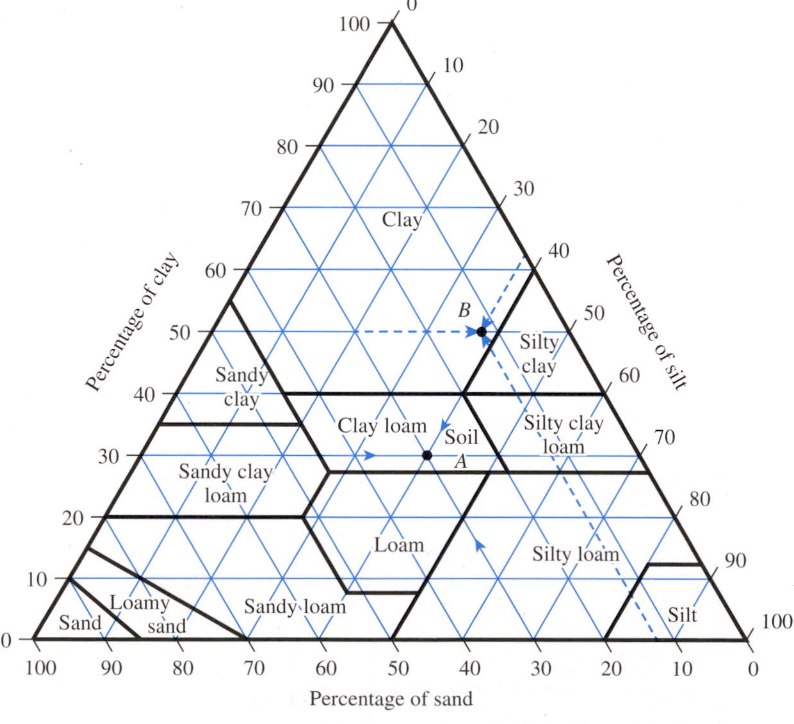

Figure 5.1 U.S. Department of Agriculture textural classification (USDA)

is based on only the fraction of soil that passes through the No. 10 sieve. Hence, if the particle-size distribution of a soil is such that a certain percentage of the soil particles is larger than 2 mm in diameter, a correction will be necessary. For example, if soil *B* has a particle-size distribution of 20% gravel, 10% sand, 30% silt, and 40% clay, the modified textural compositions are

$$\textit{Sand size: } \frac{10 \times 100}{100 - 20} = 12.5\%$$

$$\textit{Silt size: } \frac{30 \times 100}{100 - 20} = 37.5\%$$

$$\textit{Clay size: } \frac{40 \times 100}{100 - 20} = 50.0\%$$

On the basis of the preceding modified percentages, the USDA textural classification is *clay* (see Figure 5.1). However, because of the large percentage of gravel, it may be called *gravelly clay*.

Several other textural classification systems are also used, but they are no longer useful for civil engineering purposes.

Example 5.1

Classify the following soils according to the USDA textural classification system.

Particle-size distribution (%)	Soil			
	A	*B*	*C*	*D*
Gravel	12	18	0	12
Sand	25	31	15	22
Silt	32	30	30	26
Clay	31	21	55	40

Solution

Step 1. Calculate the modified percentages of sand, gravel, and silt as follows:

$$\text{Modified \% sand} = \frac{\%\text{sand}}{100 - \%\text{gravel}} \times 100$$

$$\text{Modified \% silt} = \frac{\%\text{silt}}{100 - \%\text{gravel}} \times 100$$

$$\text{Modified \% clay} = \frac{\%\text{clay}}{100 - \%\text{gravel}} \times 100$$

Thus, the following table results:

Particle-size distribution (%)	Soil			
	A	*B*	*C*	*D*
Sand	28.4	37.8	15	25
Silt	36.4	36.6	30	29.5
Clay	35.2	25.6	55	45.5

Step 2. With the modified composition calculated, refer to Figure 5.1 to determine the zone into which each soil falls. The results are as follows:

Classification of soil			
A	*B*	*C*	*D*
Gravelly clay loam	Gravelly loam	Clay	Gravelly clay

Note: The word *gravelly* was added to the classification of soils *A*, *B*, and *D* because of the large percentage of gravel present in each.

5.3 Classification by Engineering Behavior

Although the textural classification of soil is relatively simple, it is based entirely on the particle-size distribution. The amount and type of clay minerals present in fine-grained soils dictate to a great extent their physical properties. Hence, the soils engineer must consider *plasticity*, which results from the presence of clay minerals, to interpret soil characteristics properly. Because textural classification systems do not take plasticity into account and are not totally indicative of many important soil properties, they are inadequate for most engineering purposes. Currently, two more elaborate classification systems are commonly used by soils engineers. Both systems take into consideration the particle-size distribution and Atterberg limits. They are the American Association of State Highway and Transportation Officials (AASHTO) classification system and the Unified Soil Classification System. The AASHTO classification system is used mostly by state and county highway departments. Geotechnical engineers generally prefer the Unified system.

5.4 AASHTO Classification System

The AASHTO system of soil classification was developed in 1929 as the Public Road Administration classification system. It has undergone several revisions, with the present version proposed by the Committee on Classification of Materials for Subgrades and Granular Type Roads of the Highway Research Board in 1945 (ASTM designation D-3282; AASHTO method M145).

Table 5.1 Classification of Highway Subgrade Materials

General classification	Granular materials (35% or less of total sample passing No. 200)						
	A-1			A-2			
Group classification	**A-1-a**	**A-1-b**	**A-3**	**A-2-4**	**A-2-5**	**A-2-6**	**A-2-7**
Sieve analysis (percentage passing)							
No. 10	50 max.						
No. 40	30 max.	50 max.	51 min.				
No. 200	15 max.	25 max.	10 max.	35 max.	35 max.	35 max.	35 max.
Characteristics of fraction passing No. 40							
Liquid limit				40 max.	41 min.	40 max.	41 min.
Plasticity index	6 max.		NP	10 max.	10 max.	11 min.	11 min.
Usual types of significant constituent materials	Stone, fragments, gravel and sand		Fine sand	Silty or clayey gravel, and sand			
General subgrade rating	Excellent to good						

General classification	Silt-clay materials (more than 35% of total sample passing No. 200)			
Group classification	**A-4**	**A-5**	**A-6**	**A-7** **A-7-5**[a] **A-7-6**[b]
Sieve analysis (percentage passing)				
No. 10				
No. 40				
No. 200	36 min.	36 min.	36 min.	36 min.
Characteristics of fraction passing No. 40				
Liquid limit	40 max.	41 min.	40 max.	41 min.
Plasticity index	10 max.	10 max.	11 min.	11 min.
Usual types of significant constituent materials	Silty soils		Clayey soils	
General subgrade rating	Fair to poor			

[a]For A-7-5, $PI \leq LL - 30$

[b]For A-7-6, $PI > LL - 30$

The AASHTO (See AASHTO, 1982) classification in present use is given in Table 5.1. According to this system, soil is classified into seven major groups: A-1 through A-7. Soils classified under groups A-1, A-2, and A-3 are granular materials of which 35% or less of the particles pass through the No. 200 sieve. Soils of which more than 35% pass through the No. 200 sieve are classified under groups A-4, A-5, A-6, and A-7. These soils are mostly silt and clay-type materials. This classification system is based on the following criteria:

1. *Grain size*
   a. *Gravel:* fraction passing the 75-mm sieve and retained on the No. 10 (2-mm) U.S. sieve

   b. *Sand:* fraction passing the No. 10 (2-mm) U.S. sieve and retained on the No. 200 (0.075-mm) U.S. sieve
   c. *Silt and clay:* fraction passing the No. 200 U.S. sieve
2. *Plasticity:* The term *silty* is applied when the fine fractions of the soil have a plasticity index of 10 or less. The term *clayey* is applied when the fine fractions have a plasticity index of 11 or more.
3. If cobbles and *boulders* (size larger than 75 mm) are encountered, they are excluded from the portion of the soil sample from which classification is made. However, the percentage of such material is recorded.

To classify a soil according to Table 5.1, one must apply the test data from left to right. By process of elimination, the first group from the left into which the test data fit is the correct classification. Figure 5.2 shows a plot of the range of the liquid limit and the plasticity index for soils that fall into groups A-2, A-4, A-5, A-6, and A-7.

To evaluate the quality of a soil as a highway subgrade material, one must also incorporate a number called the *group index (GI)* with the groups and subgroups of the soil. This index is written in parentheses after the group or subgroup designation. The group index is given by the equation

$$GI = (F_{200} - 35)[0.2 + 0.005(LL - 40)] + 0.01(F_{200} - 15)(PI - 10) \quad (5.1)$$

where F_{200} = percentage passing through the No. 200 sieve
LL = liquid limit
PI = plasticity index

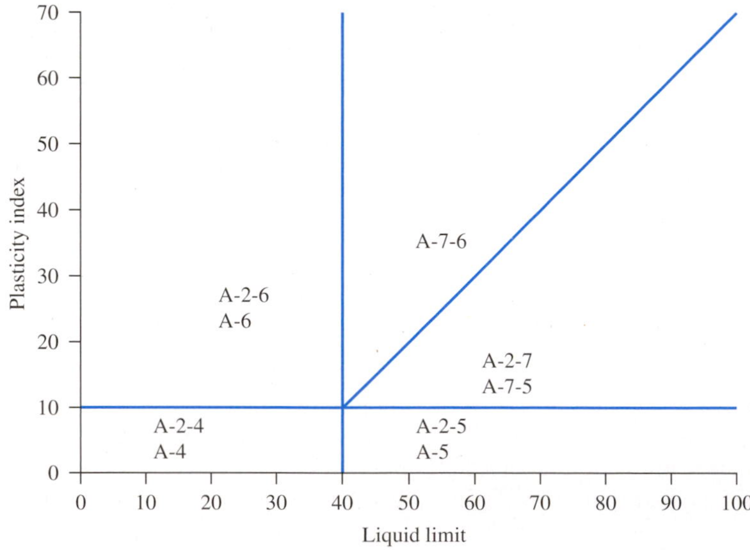

Figure 5.2 Range of liquid limit and plasticity index for soils in groups A-2, A-4, A-5, A-6, and A-7

The first term of Eq. (5.1)—that is, $(F_{200} - 35)[0.2 + 0.005(LL - 40)]$—is the partial group index determined from the liquid limit. The second term—that is, $0.01(F_{200} - 15)(PI - 10)$—is the partial group index determined from the plasticity index. Following are some rules for determining the group index:

1. If Eq. (5.1) yields a negative value for GI, it is taken as 0.
2. The group index calculated from Eq. (5.1) is rounded off to the nearest whole number (for example, $GI = 3.4$ is rounded off to 3; $GI = 3.5$ is rounded off to 4).
3. There is no upper limit for the group index.
4. The group index of soils belonging to groups A-1-a, A-1-b, A-2-4, A-2-5, and A-3 is always 0.
5. When calculating the group index for soils that belong to groups A-2-6 and A-2-7, use the partial group index for PI, or

$$GI = 0.01(F_{200} - 15)(PI - 10) \tag{5.2}$$

In general, the quality of performance of a soil as a subgrade material is inversely proportional to the group index.

Example 5.2

The results of the particle-size analysis of a soil are as follows:

- Percent passing the No. 10 sieve = 42
- Percent passing the No. 40 sieve = 35
- Percent passing the No. 200 sieve = 20

The liquid limit and plasticity index of the minus No. 40 fraction of the soil are 25 and 20, respectively. Classify the soil by the AASHTO system.

Solution

Since 20% (i.e., less than 35%) of soil is passing No. 200 sieve, it is a granular soil. Hence it can be A-1, A-2, or A-3. Refer to Table 5.1. Starting from the left of the table, the soil falls under A-1-b (see the table below).

Parameter	Specifications in Table 5.1	Parameters of the given soil
Percent passing sieve		
No. 10	—	
No. 40	50 max	35
No. 200	25 max	20
Plasticity index (PI)	6 max	$PI = LL - PL = 25 - 20 = 5$

The group index of the soil is 0. So, the soil is **A-1-b(0)**.

Example 5.3

Ninety-five percent of a soil passes through the No. 200 sieve and has a liquid limit of 60 and plasticity index of 40. Classify the soil by the AASHTO system.

Solution

Ninety-five percent of the soil (which is $\geq 36\%$) is passing through No. 200 sieve. So it is a silty-clay material. Now refer to Table 5.1. Starting from the left of the table, it falls under A-7-6 (see the table below).

Parameter	Specifications in Table 5.1	Parameters of the given soil
Percent passing No. 200 sieve	36 min.	95
Liquid limit (LL)	41 min.	60
Plasticity index (PI)	11 min.	40
PI	$> LL - 30$	$PI = 40 > LL - 30 = 60 - 30 = 30$

$$GI = (F_{200} - 35)[0.2 + 0.005(LL - 40)] + 0.01(F_{200} - 15)(PI - 10)$$

$$= (95 - 35)[0.2 + 0.005(60 - 40)] + (0.01)(95 - 15)(40 - 10)$$

$$= 42$$

So, the classification is **A-7-6(42)**.

5.5 Unified Soil Classification System

The original form of this system was proposed by Casagrande in 1942 for use in the airfield construction works undertaken by the Army Corps of Engineers during World War II. In cooperation with the U.S. Bureau of Reclamation, this system was revised in 1952. At present, it is used widely by engineers (ASTM Test Designation D-2487). The Unified classification system is presented in Table 5.2.

This system classifies soils into two broad categories:

1. Coarse-grained soils that are gravelly and sandy in nature with less than 50% passing through the No. 200 sieve. The group symbols start with a prefix of G or S. G stands for gravel or gravelly soil, and S for sand or sandy soil.
2. Fine-grained soils are with 50% or more passing through the No. 200 sieve. The group symbols start with prefixes of M, which stands for inorganic silt, C for inorganic clay, or O for organic silts and clays. The symbol Pt is used for peat, muck, and other highly organic soils.

Table 5.2 Unified Soil Classification System (Based on Material Passing 76.2-mm Sieve)

Criteria for assigning group symbols				Group symbol
Coarse-grained soils More than 50% retained on No. 200 sieve	**Gravels** More than 50% of coarse fraction retained on No. 4 sieve	Clean Gravels Less than 5% fines[a]	$C_u \geq 4$ and $1 \leq C_c \leq 3$[c]	GW
			$C_u < 4$ and/or $C_c < 1$ or $C_c > 3$[c]	GP
		Gravels with Fines More than 12% fines[a,d]	$PI < 4$ or plots below "A" line (Figure 5.3)	GM
			$PI > 7$ and plots on or above "A" line (Figure 5.3)	GC
	Sands 50% or more of coarse fraction passes No. 4 sieve	Clean Sands Less than 5% fines[b]	$C_u \geq 6$ and $1 \leq C_c \leq 3$[c]	SW
			$C_u < 6$ and/or $C_c < 1$ or $C_c > 3$[c]	SP
		Sands with Fines More than 12% fines[b,d]	$PI < 4$ or plots below "A" line (Figure 5.3)	SM
			$PI > 7$ and plots on or above "A" line (Figure 5.3)	SC
Fine-grained soils 50% or more passes No. 200 sieve	**Silts and clays** Liquid limit less than 50	Inorganic	$PI > 7$ and plots on or above "A" line (Figure 5.3)[e]	CL
			$PI < 4$ or plots below "A" line (Figure 5.3)[e]	ML
		Organic	$\frac{\text{Liquid limit—oven dried}}{\text{Liquid limit—not dried}} < 0.75$; see Figure 5.3; OL zone	OL
	Silts and clays Liquid limit 50 or more	Inorganic	PI plots on or above "A" line (Figure 5.3)	CH
			PI plots below "A" line (Figure 5.3)	MH
		Organic	$\frac{\text{Liquid limit—oven dried}}{\text{Liquid limit—not dried}} < 0.75$; see Figure 5.3; OH zone	OH
Highly organic soils	Primarily organic matter, dark in color, and organic odor			Pt

[a]Gravels with 5 to 12% fine require dual symbols: GW-GM, GW-GC, GP-GM, GP-GC.

[b]Sands with 5 to 12% fines require dual symbols: SW-SM, SW-SC, SP-SM, SP-SC.

[c]$C_u = \frac{D_{60}}{D_{10}}$; $C_c = \frac{(D_{30})^2}{D_{60} \times D_{10}}$

[d]If $4 \leq PI \leq 7$ and plots in the hatched area in Figure 5.3, use dual symbol GC-GM or SC-SM.

[e]If $4 \leq PI \leq 7$ and plots in the hatched area in Figure 5.3, use dual symbol CL-ML.

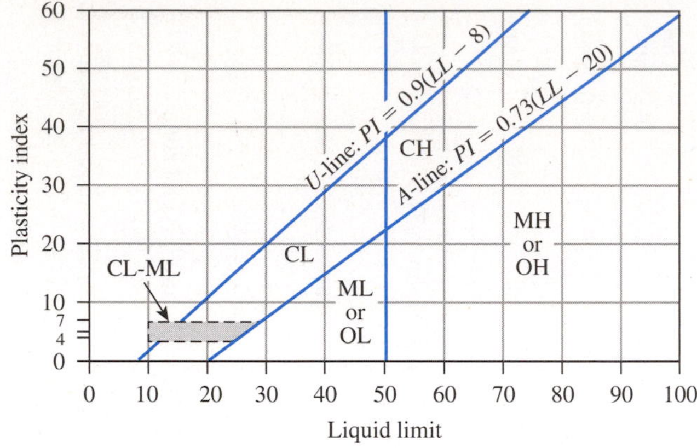

Figure 5.3 Plasticity chart

Other symbols used for the classification are

- W—well graded
- P—poorly graded
- L—low plasticity (liquid limit less than 50)
- H—high plasticity (liquid limit more than 50)

For proper classification according to this system, some or all of the following information must be known:

1. Percent of gravel—that is, the fraction passing the 76.2-mm sieve and retained on the No. 4 sieve (4.75-mm opening)
2. Percent of sand—that is, the fraction passing the No. 4 sieve (4.75-mm opening) and retained on the No. 200 sieve (0.075-mm opening)
3. Percent of silt and clay—that is, the fraction finer than the No. 200 sieve (0.075-mm opening)
4. Uniformity coefficient (C_u) and the coefficient of gradation (C_c)
5. Liquid limit and plasticity index of the portion of soil passing the No. 40 sieve

The group symbols for coarse-grained gravelly soils are GW, GP, GM, GC, GC-GM, GW-GM, GW-GC, GP-GM, and GP-GC. Similarly, the group symbols for fine-grained soils are CL, ML, OL, CH, MH, OH, CL-ML, and Pt.

More recently, ASTM designation D-2487 created an elaborate system to assign *group names* to soils. These names are summarized in Figures 5.4, 5.5, and 5.6. In using these figures, one needs to remember that, in a given soil,

- Fine fraction = percent passing No. 200 sieve
- Coarse fraction = percent retained on No. 200 sieve
- Gravel fraction = percent retained on No. 4 sieve
- Sand fraction = (percent retained on No. 200 sieve) − (percent retained on No. 4 sieve)

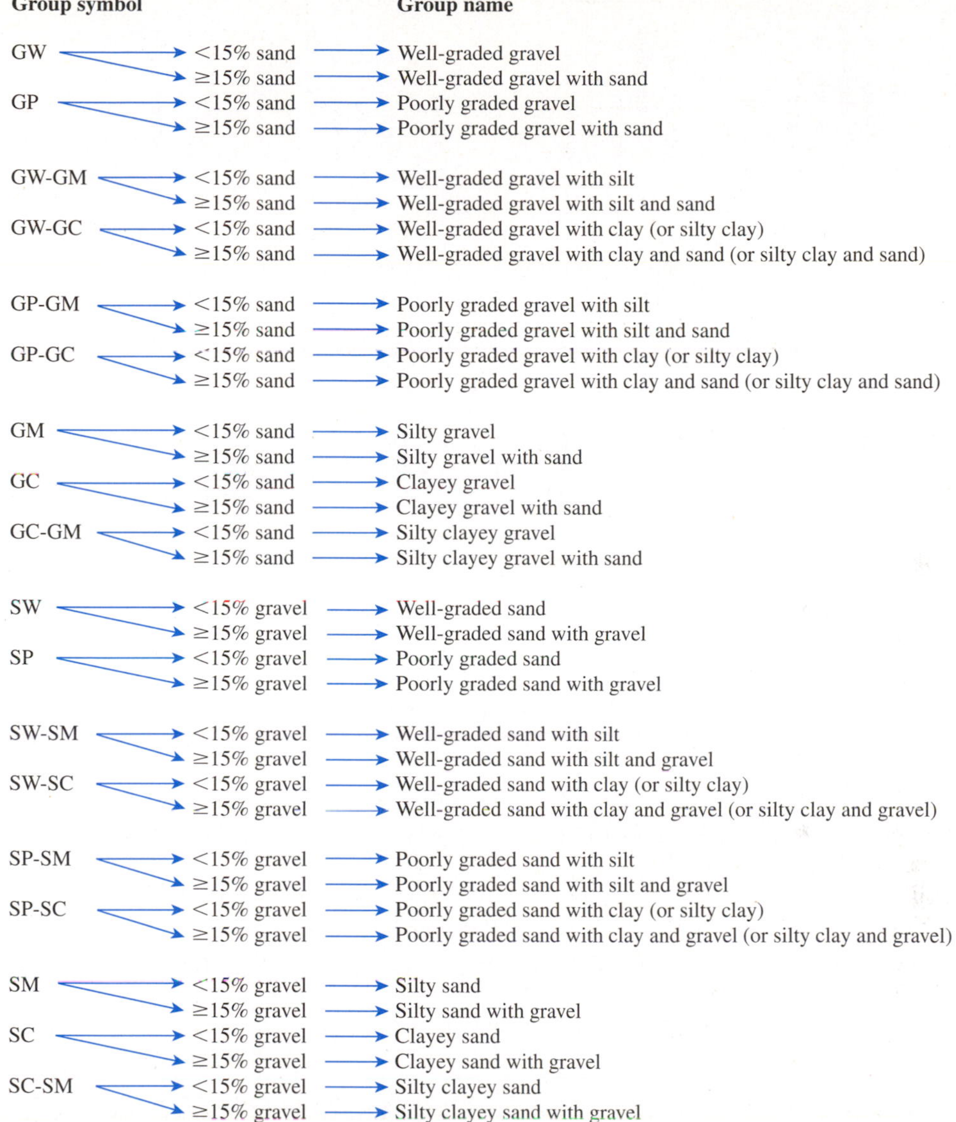

Figure 5.4 Flowchart group names for gravelly and sandy soil (*Source:* From "Annual Book of ASTM Standards, 04.08, 2014." Copyright ASTM INTERNATIONAL. Reprinted with permission.)

5.6 Comparison between the AASHTO and Unified Systems

Both soil classification systems, AASHTO and Unified, are based on the texture and plasticity of soil. Also, both systems divide the soils into two major categories, coarse grained and fine grained, as separated by the No. 200 sieve. According to

(text continues on page 142)

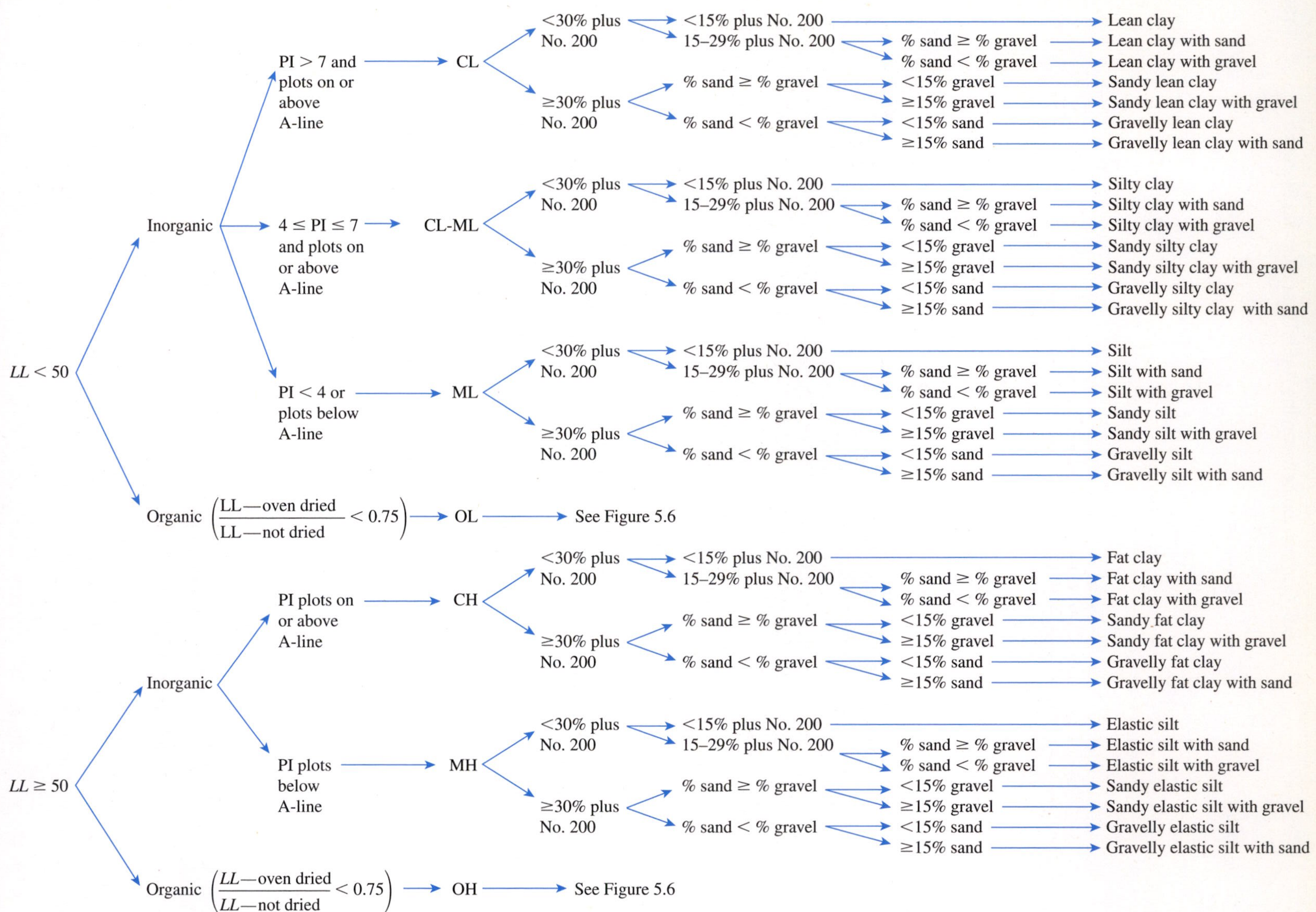

Figure 5.5 Flowchart group names for inorganic silty and clayey soils (*Source:* From "Annual Book of ASTM Standards, 04.08." Copyright ASTM INTERNATIONAL, 2014. Reprinted with permission.)

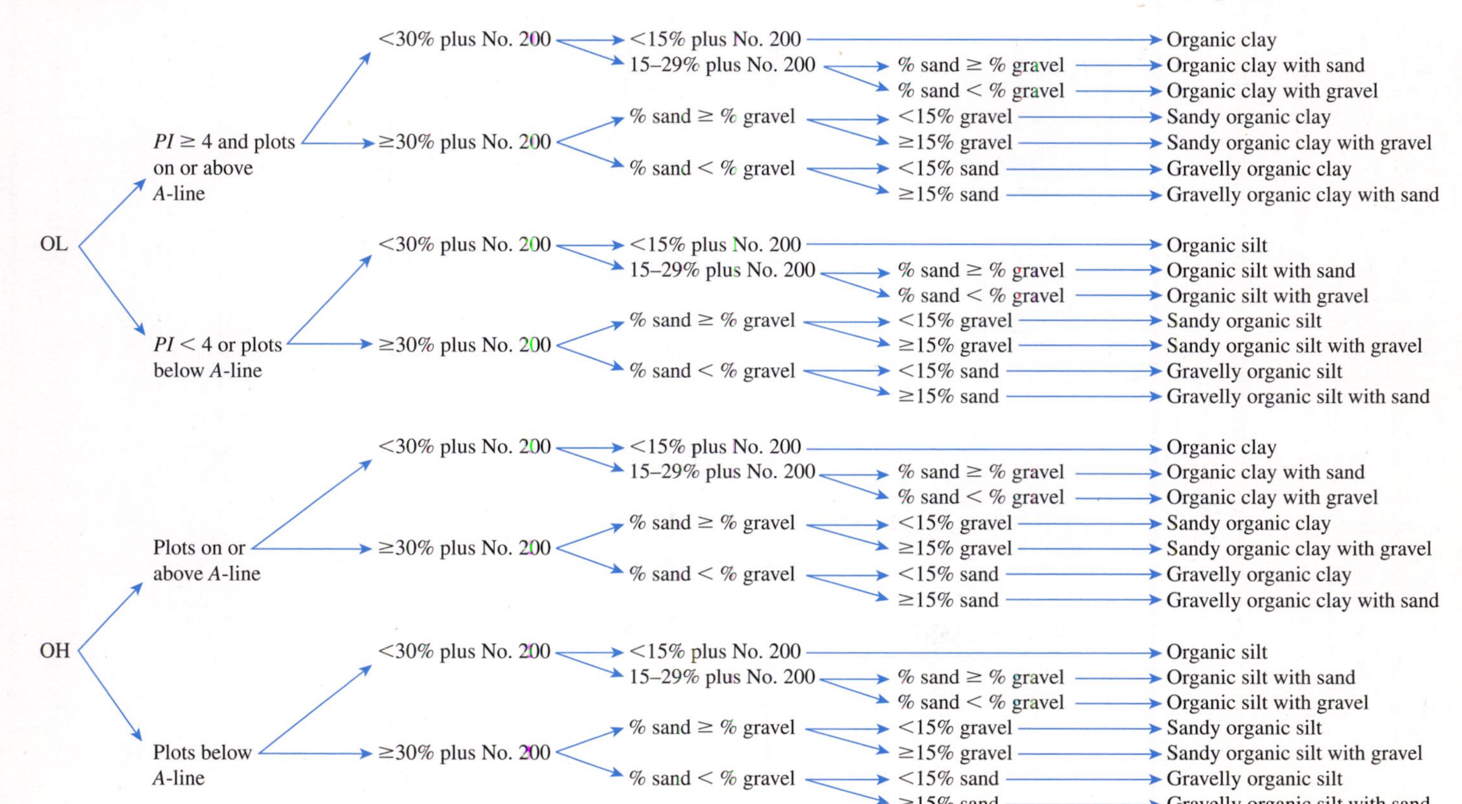

Figure 5.6 Flowchart group names for organic silty and clayey soils (*Source:* From "Annual Book of ASTM Standards, 04.08." Copyright ASTM INTERNATIONAL, 2014. Reprinted with permission.)

the AASHTO system, a soil is considered fine grained when more than 35% passes through the No. 200 sieve. According to the Unified system, a soil is considered fine grained when more than 50% passes through the No. 200 sieve. A coarse-grained soil that has about 35% fine grains will behave like a fine-grained material. This is because enough fine grains exist to fill the voids between the coarse grains and hold them apart. In this respect, the AASHTO system appears to be more appropriate. In the AASHTO system, the No. 10 sieve is used to separate gravel from sand; in the Unified system, the No. 4 sieve is used. From the point of view of soil-separate size limits, the No. 10 sieve is the more accepted upper limit for sand. This limit is used in concrete and highway base-course technology.

Example 5.4

The results of the particle-size analysis of a soil are as follows:

Percent passing through the No. 10 sieve = 100
Percent passing through the No. 40 sieve = 80
Percent passing through the No. 200 sieve = 58

The liquid limit and plasticity index of the minus No. 40 fraction of the soil are 30 and 10, respectively. Classify the soil by the Unified classification system.

Solution

Refer to Table 5.2. Since 58% of the soil passes through the No. 200 sieve, it is a fine-grained soil. Referring to the plasticity chart in Figure 5.3, for $LL = 30$ and $PI = 10$, it can be classified (group symbol) as **CL**.

In order to determine the group name, we refer to Figure 5.5 and Figure 5.7, which is taken from Figure 5.5. The percent retained on No. 200 sieve is more than 30%. Percent of gravel = 0; percent of sand = (100 − 58) − (0) = 42. Hence, percent sand > percent gravel. Also, percent gravel is less than 15%. Hence the group name is **sandy lean clay**.

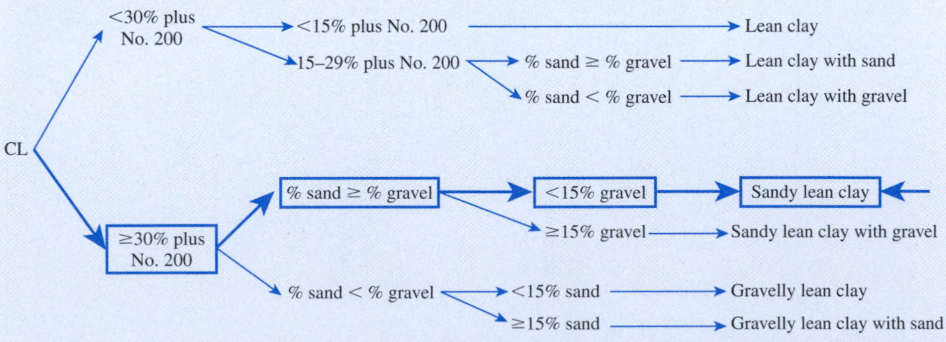

Figure 5.7 Determination of group name for the soil in Example 5.4

Example 5.5

Classify the soil given in Example 5.4 according to the AASHTO Classification System.

Solution

Refer to Table 5.1. Since 58% is passing No. 200 sieve, it is a silt-clay material.

Given:

$$LL = 30$$
$$PI = 10$$

Referring to Table 5.1, the soil is **A-4**.

From Eq. (5.1),

$$GI = (F_{200} - 35)[0.2 + 0.005(LL - 40)] + 0.01(F_{200} - 15)(PI - 10)$$

or

$$GI = (58 - 35)[0.2 + 0.005(30 - 40)] + 0.01(58 - 15)(10 - 10)$$
$$= 3.45 \approx 4$$

So, the soil is **A-4(4)**.

Note: Compare this with Table 5.4, according to which a CL soil may be A-4.

Example 5.6

For a given soil, the following are known:

Percent passing through No. 4 sieve = 70
Percent passing through No. 200 sieve = 30
Liquid limit = 33
Plastic limit = 12

Classify the soil using the Unified Soil Classification System. Give the group symbol and the group name.

Solution

Refer to Table 5.2. The percentage passing No. 200 sieve is 30%, which is less than 50%. So it is a coarse-grained soil. Thus

$$\text{Coarse fraction} = 100 - 30 = 70\%$$

$$\text{Gravel fraction} = \text{percent retained on No. 4 sieve} = 100 - 70 = 30\%$$

Hence, more than 50% of the coarse fraction is passing No. 4 sieve. Thus, it is a sandy soil. Since more than 12% is passing No. 200 sieve, it is SM or SC. For this soil, $PI = 33 - 12 = 21$ (which is greater than 7). With $LL = 33$ and $PI = 21$, it plots above the A-line in Figure 5.3. Thus the group symbol is **SC**.

For the group name, refer to Figure 5.4 and Figure 5.8 (which is taken from Figure 5.4). Since the percentage of gravel is more than 15%, it is **clayey sand with gravel**.

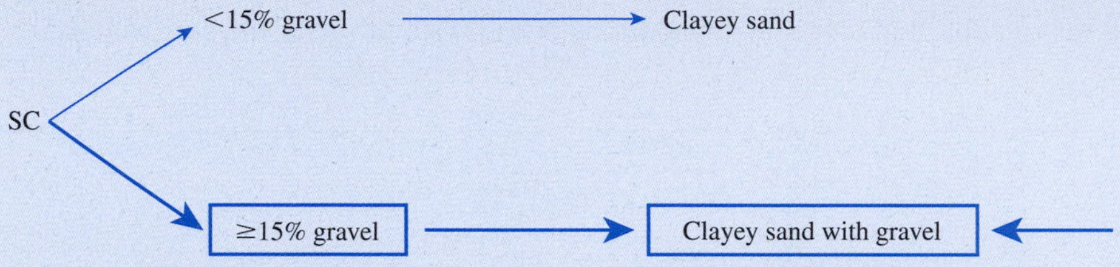

Figure 5.8 Determination of group name for the soil in Example 5.6

Example 5.7

Figure 5.9 gives the grain-size distribution of two soils. The liquid and plastic limits of minus No. 40 sieve fraction of the soil are as follows:

	Soil A	Soil B
Liquid limit	30	26
Plastic limit	22	20

Determine the group symbols and group names according to the Unified Soil Classification System.

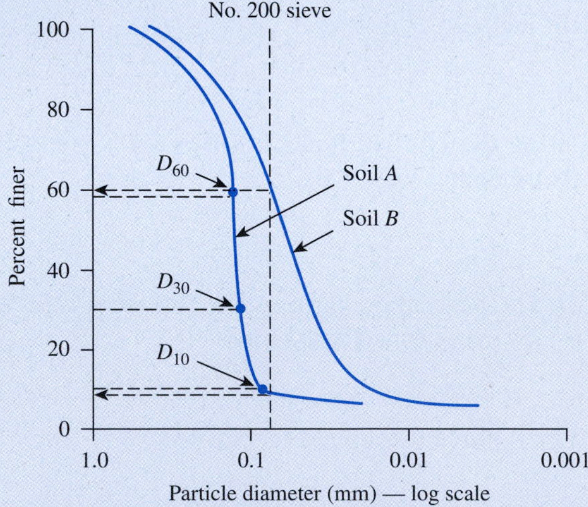

Figure 5.9 Particle-size distribution of two soils

Solution

Soil *A*

The grain-size distribution curve (Figure 5.9) indicates that percent passing No. 200 sieve is 8. According to Table 5.2, it is a coarse-grained soil. Also, from Figure 5.9, the percent retained on No. 4 sieve is zero. Hence, it is a sandy soil.

From Figure 5.9, $D_{10} = 0.085$ mm, $D_{30} = 0.12$ m, and $D_{60} = 0.135$ mm. Thus,

$$C_u = \frac{D_{60}}{D_{10}} = \frac{0.135}{0.085} = 1.59 < 6$$

$$C_c = \frac{D_{30}^2}{D_{60} \times D_{10}} = \frac{(0.12)^2}{(0.135)(0.085)} = 1.25 > 1$$

With $LL = 30$ and $PI = 30 - 22 = 8$ (which is greater than 7), it plots above the A-line in Figure 5.3. Hence, the group symbol is **SP-SC**.

In order to determine the group name, we refer to Figure 5.4 and Figure 5.10.

$$\text{Percent of gravel} = 0 \text{ (which is } < 15\%)$$

Figure 5.10 Determination of group name for soil A in Example 5.7

So, the group name is **poorly graded sand with clay**.

Soil *B*

The grain-size distribution curve in Figure 5.9 shows that percent passing No. 200 sieve is 61 (>50%); hence, it is a fine-grained soil. Given: $LL = 26$ and $PI = 26 - 20 = 6$. In Figure 5.3, the PI plots in the hatched area. So, from Table 5.2, the group symbol is **CL-ML**.

For group name (assuming that the soil is inorganic), we go to Figure 5.5 and obtain Plus No. 200 sieve = 100 − 61 = 39 (which is greater than 30).

$$\text{Percent of gravel} = 0; \text{ percent of sand} = 100 - 61 = 39$$

Thus, because the percent of sand is greater than the percent of gravel, the soil is **sandy silty clay** as shown in Figure 5.11.

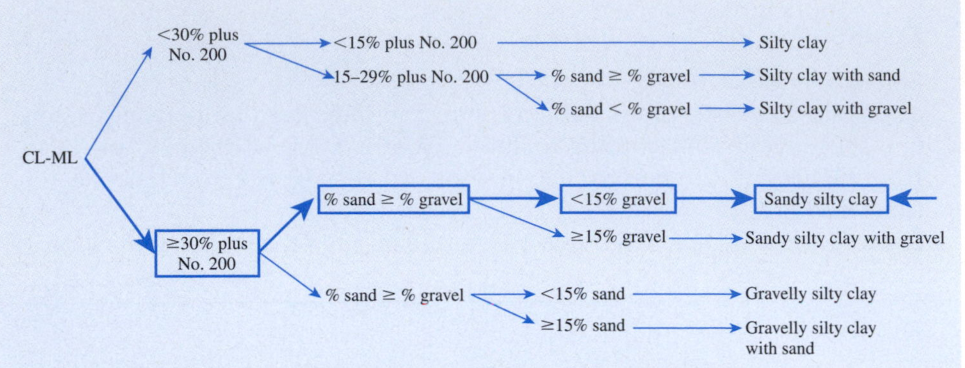

Figure 5.11 Determination of group name for soil *B* in Example 5.7

Example 5.8

For a given inorganic soil, the following are known:

Percent passing No. 4 sieve	100
Percent passing No. 200 sieve	77
Liquid limit	63
Plasticity index	25

Classify the soil using the Unified Soil Classification System. Give the group symbol and the group name.

Solution

Refer to Table 5.2. For the soil, 77% is passing No. 200 sieve. So, it is a fine-grained soil (i.e., CL, ML, MH or CH). Given:

$$LL = 63$$

$$PI = 25$$

Referring to Figure 5.3, it is **MH**.
Referring to Figure 5.5,

- Soil has less than 30% (100 − 77 = 23%) plus No. 200
- % sand (100 − 77 = 23%) > % gravel (0%)

So the group name is **elastic silt with sand**.

Example 5.9

The grain-size analysis for a soil is given next:

Sieve no.	% passing
4	94
10	63
20	21
40	10
60	7
100	5
200	3

Given that the soil is nonplastic, classifiy the soil using the Unified Soil Classification System.

Solution

Refer to Table 5.2. The soil has 3% passing No. 200 sieve (i.e., less than 5% fines) and 94% passing No. 4 sieve. This is a nonplastic soil; therefore, it is a sandy soil (i.e., SW or SP).

The grain-size distribution is shown in Figure 5.12. From this figure, we obtain

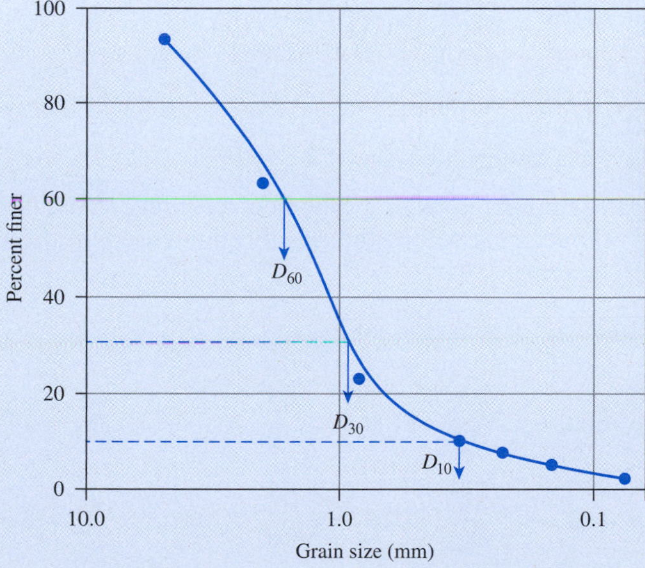

Figure 5.12

$$D_{60} = 1.41 \text{ mm} \qquad D_{30} = 0.96 \text{ mm} \qquad D_{10} = 0.41 \text{ mm}$$

Thus,

$$C_u = \frac{D_{60}}{D_{10}} = \frac{1.41}{0.41} = 3.44$$

$$C_c = \frac{(D_{30})^2}{D_{60} \times D_{10}} = \frac{0.96^2}{1.41 \times 0.41} = 1.59$$

From Table 5.2, we see that the group symbol is **SP**.

Now refer to Figure 5.4. Since the gravel portion is 100 − 94 = 6% (i.e., less than 15%), the group name is **poorly graded sand**.

Example 5.10

Fuller and Thompson (1907) developed the following relationship for proportioning aggregate for maximum density:

$$p = \sqrt{\frac{D}{D_{max}}} \times 100$$

where p = percent passing
D = grain size
D_{max} = maximum grain size of the soil

If an aggregate is prepared by proportioning according to the above relation with D_{max} = 40 mm, what will be the classification based on the Unified Soil Classification System?

Solution

$$\text{For } p = 60 = \sqrt{\frac{D_{60}}{D_{max}}} \times 100 = \sqrt{\frac{D_{60}}{40}} \times 100, \text{ or } \quad D_{60} = 14.4 \text{ mm}$$

$$\text{For } p = 30 = \sqrt{\frac{D_{30}}{D_{max}}} \times 100 = \sqrt{\frac{D_{30}}{40}} \times 100, \text{ or } \quad D_{30} = 3.6 \text{ mm}$$

$$\text{For } p = 10 = \sqrt{\frac{D_{10}}{D_{max}}} \times 100 = \sqrt{\frac{D_{10}}{40}} \times 100, \text{ or } \quad D_{10} = 0.4 \text{ mm}$$

So

$$C_u = \frac{D_{60}}{D_{10}} = \frac{14.4}{0.4} = 36;$$

$$C_c = \frac{D_{30}^2}{D_{60} \times D_{10}} = \frac{3.6^2}{14.4 \times 0.4} = 2.25$$

Now, referring to Table 5.2, the material is **GW**.

In the Unified system, the gravelly and sandy soils clearly are separated; in the AASHTO system, they are not. The A-2 group, in particular, contains a large variety of soils. Symbols like GW, SM, CH, and others that are used in the Unified system are more descriptive of the soil properties than the A symbols used in the AASHTO system.

The classification of organic soils, such as OL, OH, and Pt, is provided in the Unified system. Under the AASHTO system, there is no place for organic soils. Peats usually have a high moisture content, low specific gravity of soil solids, and low unit weight.

Liu (1967) compared the AASHTO and Unified systems. The results of his study are presented in Tables 5.3 and 5.4.

Table 5.3 Comparison of the AASHTO System with the Unified System*

Soil group in AASHTO system	Comparable soil groups in Unified system		
	Most probable	**Possible**	**Possible but improbable**
A-1-a	GW, GP	SW, SP	GM, SM
A-1-b	SW, SP, GM, SM	GP	—
A-3	SP	—	SW, GP
A-2-4	GM, SM	GC, SC	GW, GP, SW, SP
A-2-5	GM, SM	—	GW, GP, SW, SP
A-2-6	GC, SC	GM, SM	GW, GP, SW, SP
A-2-7	GM, GC, SM, SC	—	GW, GP, SW, SP
A-4	ML, OL	CL, SM, SC	GM, GC
A-5	OH, MH, ML, OL	—	SM, GM
A-6	CL	ML, OL, SC	GC, GM, SM
A-7-5	OH, MH	ML, OL, CH	GM, SM, GC, SC
A-7-6	CH, CL	ML, OL, SC	OH, MH, GC, GM, SM

*After Liu (1967)

Source: From A Review of Engineering Soil Classification Systems. In Highway Research Record 156, Highway Research Board, National Research Council, Washington, D.C., 1967, Table 5, p. 16. Reproduced with permission of the Transportation Research Board.

Table 5.4 Comparison of the Unified System with the AASHTO System*

Soil group in Unified system	Comparable soil groups in AASHTO system		
	Most probable	**Possible**	**Possible but improbable**
GW	A-1-a	—	A-2-4, A-2-5, A-2-6, A-2-7
GP	A-1-a	A-1-b	A-3, A-2-4, A-2-5, A-2-6, A-2-7
GM	A-1-b, A-2-4, A-2-5, A-2-7	A-2-6	A-4, A-5, A-6, A-7-5, A-7-6, A-1-a
GC	A-2-6, A-2-7	A-2-4	A-4, A-6, A-7-6, A-7-5
SW	A-1-b	A-1-a	A-3, A-2-4, A-2-5, A-2-6, A-2-7
SP	A-3, A-1-b	A-1-a	A-2-4, A-2-5, A-2-6, A-2-7
SM	A-1-b, A-2-4, A-2-5, A-2-7	A-2-6, A-4	A-5, A-6, A-7-5, A-7-6, A-1-a
SC	A-2-6, A-2-7	A-2-4, A-6, A-4, A-7-6	A-7-5
ML	A-4, A-5	A-6, A-7-5, A-7-6	—
CL	A-6, A-7-6	A-4	—
OL	A-4, A-5	A-6, A-7-5, A-7-6	—
MH	A-7-5, A-5	—	A-7-6
CH	A-7-6	A-7-5	—
OH	A-7-5, A-5	—	A-7-6
Pt	—	—	—

*After Liu (1967)

Source: From A Review of Engineering Soil Classification Systems. In Highway Research Record 156, Highway Research Board, National Research Council, Washington, D.C., 1967, Table 6, p. 17. Reproduced with permission of the Transportation Research Board.

5.7 Summary

In this chapter we have discussed the following:

1. Textural classification is based on naming soils based on their principal components such as sand, silt, and clay-size fractions determined from particle-size distribution. The USDA textural classification system is described in detail in Section 5.2.
2. The AASHTO soil classification system is based on sieve analysis (i.e., percent finer than No. 10, 40, and 200 sieves), liquid limit, and plasticity index (Table 5.1). Soils can be classified under categories
   - A-1, A-2, and A-3 (granular soils)
   - A-4, A-5, A-6, and A-7 (silty and clayey soils)

   Group index [Eqs. (5.1) and (5.2)] is added to the soil classification which evaluates the quality of soil as a subgrade material.
3. Unified soil classification is based on sieve analysis (i.e., percent finer than No. 4 and No. 200 sieves), liquid limit, and plasticity index (Table 5.2 and Figure 5.3). It uses classification symbols such as
   - GW, GP, GM, GC, GW-GM, GW-GC, GP-GM, GP-GC, GC-GM, SW, SP, SM, SC, SW-SM, SW-SC, SP-SM, SP-SC, and SC-SM (for coarse-grained soils)
   - CL, ML, CL-ML, OL, CH, MH, and OH (for fine-grained soils)

4. In addition to group symbols, the group names under the Unified classification system can be determined using Figures 5.4, 5.5, and 5.6. The group name is primarily based on percent retained on No. 200 sieve, percent of gravel (i.e., percent retained on No. 4 sieve), and percent of sand (i.e., percent passing No. 4 sieve but retained on No. 200 sieve).

Problems

5.1 Classify the following soils using the U.S. Department of Agriculture textural classification chart.

	Particle-size distribution (%)				Particle-size distribution (%)		
Soil	**Sand**	**Silt**	**Clay**	**Soil**	**Sand**	**Silt**	**Clay**
A	20	20	60	F	30	58	12
B	55	5	40	G	40	25	35
C	45	35	20	H	30	25	45
D	50	15	35	I	5	45	50
E	70	15	15	J	45	45	10

5.2 The gravel, sand, silt and clay contents of five different soils are given below. Classify the soils using the U.S. Department of Agriculture textural classification chart.

	Particle-size distribution (%)			
Soil	**Gravel**	**Sand**	**Silt**	**Clay**
A	18	51	22	9
B	10	20	41	29
C	21	12	35	32
D	0	18	24	58
E	12	22	26	40

5.3 Classify the following soils by the AASHTO classification system. Give the group index for each soil.

	Sieve analysis—Percent finer				
Soil	**No. 10**	**No. 40**	**No. 200**	**Liquid limit**	**Plasticity index**
A	90	74	32	28	9
B	86	56	8	NP	
C	42	28	12	18	13
D	92	68	30	42	18
E	90	48	22	31	5

5.4 The sieve analysis of ten soils and the liquid and plastic limits of the fraction passing through the No. 40 sieve are given below. Classify the soils by the AASHTO classification system and give the group index for each soil.

	Sieve analysis—Percent finer				
Soil	**No. 10**	**No. 40**	**No. 200**	**Liquid limit**	**Plasticity index**
1	98	80	50	38	29
2	100	92	80	56	23
3	100	88	65	37	22
4	85	55	45	28	20
5	92	75	62	43	28
6	97	60	30	25	16
7	100	55	8	—	NP
8	94	80	63	40	21
9	83	48	20	20	15
10	100	92	86	70	38

5.5 Determine the group symbols for the fine-grained soils given in Problem 5.4 by the Unified soil classification system.

5.6 For an inorganic soil, the following grain-size analysis is given.

U.S. sieve no.	Percent passing
4	100
10	90
20	64
40	38
80	18
200	13

For this soil, $LL = 23$ and $PL = 19$. Classify the soil by using

a. AASHTO soil classification system. Give the group index.

b. Unified soil classification system. Give group symbol and group name.

5.7 Classify the following soils using the Unified soil classification system. Give the group symbols and the group names.

	Percent passing				
Sieve size	**A**	**B**	**C**	**D**	**E**
No. 4	94	98	100	100	100
No. 10	63	86	100	100	100
No. 20	21	50	98	100	100
No. 40	10	28	93	99	94
No. 60	7	18	88	95	82
No. 100	5	14	83	90	66

Sieve size	Percent passing A	B	C	D	E
No. 200	3	10	77	86	45
0.01 mm	—	—	65	42	26
0.002 mm	—	—	60	17	21
Liquid limit	—	—	63	55	36
Plasticity index	NP	NP	25	28	22

5.8 Classify the following soils using the Unified soil classification system. Give the group symbols and the group names.

Soil	Sieve analysis—Percent finer No. 4	No. 200	Liquid limit	Plasticity index
A	92	48	30	8
B	60	40	26	4
C	99	76	60	32
D	90	60	41	12
E	80	35	24	2

5.9 Classify the following soils using the Unified soil classification system. Give the group symbols and the group names.

Soil	Sieve analysis—Percent finer No. 4	No. 200	Liquid limit	Plasticity index	C_u	C_c
1	70	30	33	21		
2	48	20	41	22		
3	95	70	52	28		
4	100	82	30	19		
5	100	74	35	21		
6	87	26	38	18		
7	88	78	69	38		
8	99	57	54	26		
9	71	11	32	16	4.8	2.9
10	100	2		NP	7.2	2.2
11	89	65	44	21		
12	90	8	39	31	3.9	2.1

5.10 9% of a soil is retained on No. 4 sieve, and 11% passes the No. 200 sieve. It is also known that 10%, 30%, and 60% of the soil is smaller than 0.1 mm, 0.8 mm, and 1.9 mm, respectively. When Atterberg limit tests are conducted, it is found that the liquid limit is 32% and the plastic limit is 27%. Classify this soil according to the Unified soil classification system and give group symbol and group name.

Critical Thinking Problem

5.C.1 The subsurface characteristics for a highway pavement rehabilitation project in the southeastern United States are shown in a "boring log" in Figure 5.13. The highway structure consists of the asphalt pavement underlain by four different soil strata up to a depth of 6 m, after which the boring was terminated. Some data on the grain size and plasticity characteristics are also provided for each stratum. Perform the following tasks:

1. Determine the AASHTO soil classification and the *group index* (*GI*) for each layer.
2. Determine the "most probable" group symbols and group names for the various layers according to the Unified soil classification system. Use Table 5.3 and the soil characteristics given in the boring log.

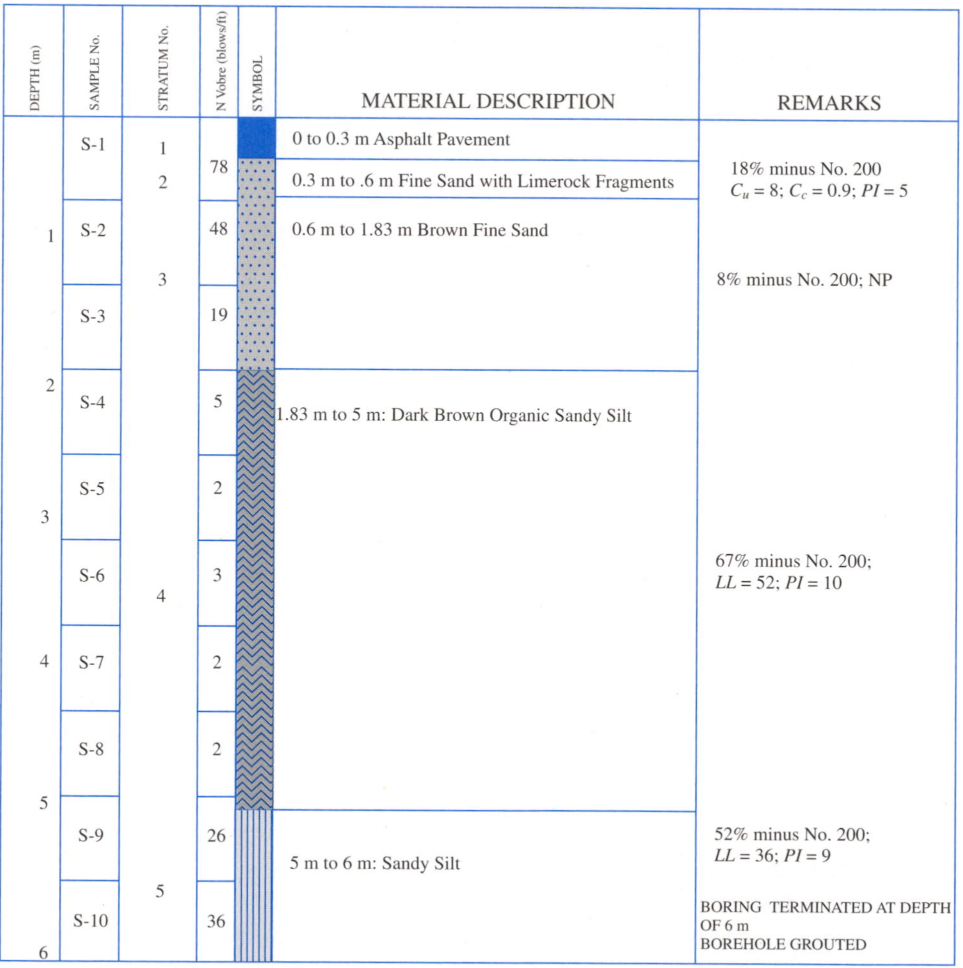

Figure 5.13 Soil boring log for a highway rehabilitation project

References

American Association of State Highway and Transportation Officials (1982). *AASHTO Materials, Part I, Specifications,* Washington, D.C.

American Society for Testing and Materials (2010). *Annual Book of ASTM Standards,* Sec. 4, Vol. 04.08, West Conshohoken, Pa.

Casagrande, A. (1948). "Classification and Identification of Soils," *Transactions*, ASCE, Vol. 113, 901–930.

Fuller, W. B. and Thompson, S.E. (1907). "The Law of Proportioning Concrete," *Transactions*, ASCE, Vol. 59, 67–118.

Liu, T. K. (1967). "A Review of Engineering Soil Classification Systems," *Highway Research Record No. 156,* National Academy of Sciences, Washington, D.C., 1–22.

CHAPTER 6

Soil Compaction

6.1 Introduction

In the construction of highway embankments, earth dams, and many other engineering structures, loose soils must be compacted to increase their unit weights. Compaction increases the strength characteristics of soils, which increase the bearing capacity of foundations constructed over them. Compaction also decreases the amount of undesirable settlement of structures and increases the stability of slopes of embankments. Smooth-wheel rollers, sheepsfoot rollers, rubber-tired rollers, and vibratory rollers are generally used in the field for soil compaction. Vibratory rollers are used mostly for the densification of granular soils. Vibroflot devices are also used for compacting granular soil deposits to a considerable depth. Compaction of soil in this manner is known as *vibroflotation*. This chapter discusses in some detail the principles of soil compaction in the laboratory and in the field.

This chapter includes elaboration of the following:

- Laboratory compaction test methods
- Factors affecting compaction in general
- Empirical relationships related to compaction
- Structure and properties of compacted cohesive soils
- Field compaction
- Tests for quality control of field compaction
- Special compaction techniques in the field

6.2 Compaction—General Principles

Compaction, in general, is the densification of soil by removal of air, which requires mechanical energy. The degree of compaction of a soil is measured in terms of its dry unit weight. When water is added to the soil during compaction, it acts as a softening agent on the soil particles. The soil particles slip over each other and move into a densely packed position. The dry unit weight after compaction first increases as the moisture content increases. (See Figure 6.1.) Note that at a moisture content $w = 0$, the moist unit weight (γ) is equal to the dry unit weight (γ_d), or

$$\gamma = \gamma_{d(w=0)} = \gamma_1$$

When the moisture content is gradually increased and the same compactive effort is used for compaction, the weight of the soil solids in a unit volume gradually increases. For example, at $w = w_1$,

$$\gamma = \gamma_2$$

However, the dry unit weight at this moisture content is given by

$$\gamma_{d(w=w_1)} = \gamma_{d(w=0)} + \Delta\gamma_d$$

Beyond a certain moisture content $w = w_2$ (Figure 6.1), any increase in the moisture content tends to reduce the dry unit weight. This phenomenon occurs because the water takes up the spaces that would have been occupied by the solid particles. The moisture content at which the maximum dry unit weight is attained is generally referred to as the *optimum moisture content*.

The laboratory test generally used to obtain the maximum dry unit weight of compaction and the optimum moisture content is called the *Proctor compaction test*

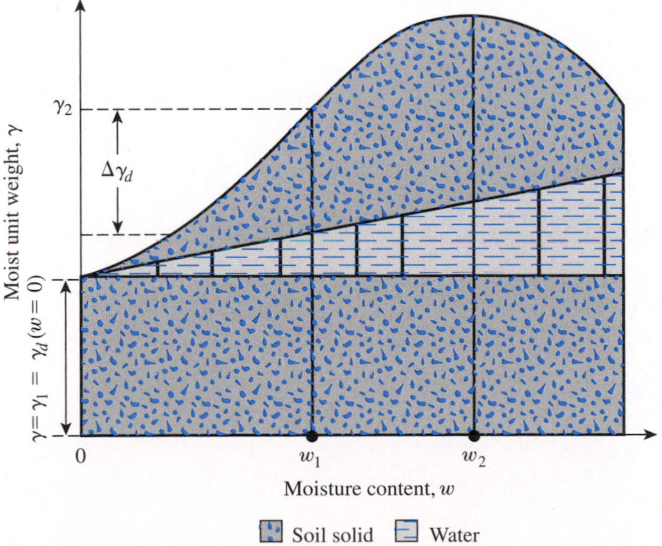

Figure 6.1 Principles of compaction

(Proctor, 1933). The procedure for conducting this type of test is described in the following section.

6.3 Standard Proctor Test

In the Proctor test, the soil is compacted in a mold that has a volume of 944 cm^3. The diameter of the mold is 101.6 mm. During the laboratory test, the mold is attached to a baseplate at the bottom and to an extension at the top (Figure 6.2a). The soil is mixed with varying amounts of water and then compacted in three equal layers by a hammer (Figure 6.2b) that delivers 25 blows to each layer. The hammer has a mass of 2.5 kg and has a drop of 305 mm. Figure 6.2c is a photograph of the laboratory equipment required for conducting a standard Proctor test.

For each test, the moist unit weight of compaction, γ, can be calculated as

$$\gamma = \frac{W}{V_m} \tag{6.1}$$

where W = weight of the compacted soil in the mold
V_m = volume of the mold [944 cm^3]

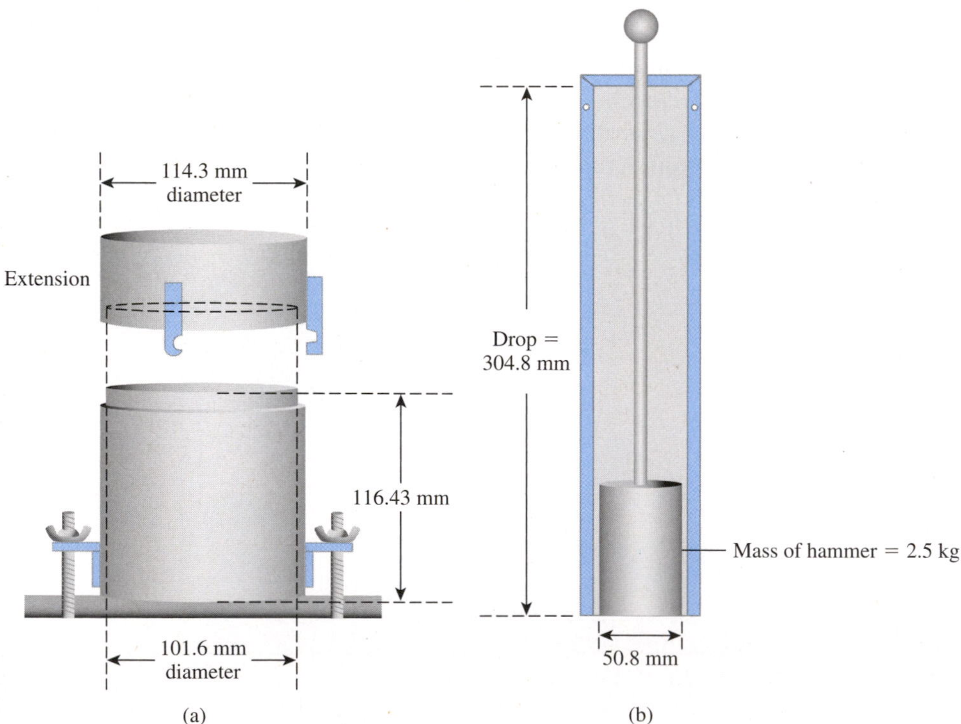

Figure 6.2 Standard Proctor test equipment: (a) mold; (b) hammer; (c) photograph of laboratory equipment used for test *(Courtesy of Braja M. Das, Henderson, Nevada)*

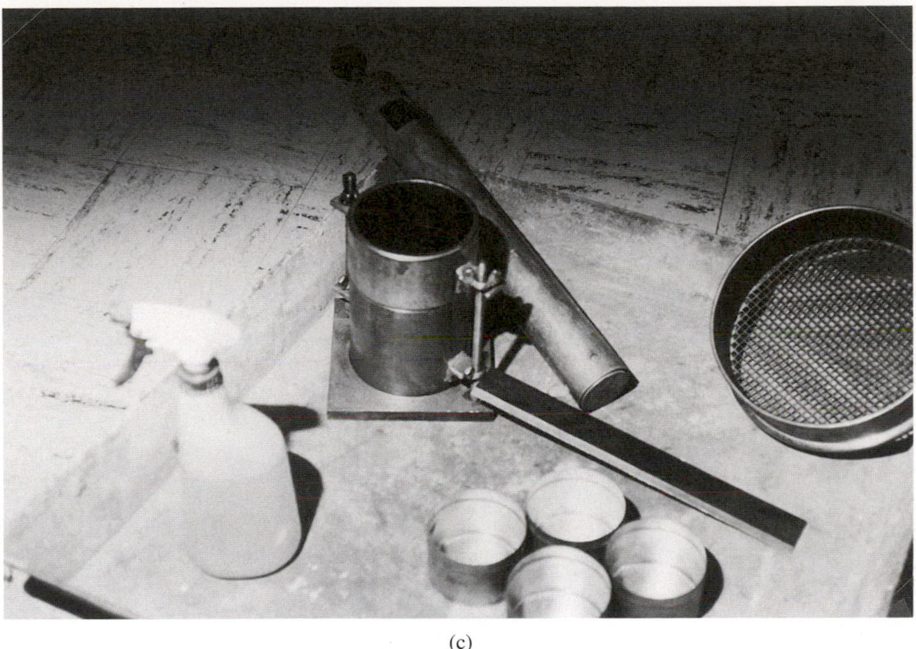

(c)

Figure 6.2 (*Continued*)

For each test, the moisture content of the compacted soil is determined in the laboratory. With the known moisture content, the dry unit weight can be calculated as

$$\gamma_d = \frac{\gamma}{1 + \dfrac{w\,(\%)}{100}} \tag{6.2}$$

where $w\,(\%)$ = percentage of moisture content.

The values of γ_d determined from Eq. (6.2) can be plotted against the corresponding moisture contents to obtain the maximum dry unit weight and the optimum moisture content for the soil. Figure 6.3 shows such a plot for a silty-clay soil.

The procedure for the standard Proctor test is elaborated in ASTM Test Designation D-698 (ASTM, 2014) and AASHTO Test Designation T-99 (AASHTO, 1982).

In order to avoid a large number of compaction tests, it is desirable to begin the first test at a moisture content that is about 4 to 5% below the approximate optimum moisture content. Figure 6.4 may be used to estimate the approximate optimum moisture content (Johnson and Sallberg, 1962) if the liquid and plastic limits of the soil are known. As an example, for a given soil (if the liquid limit is 50 and plastic limit is 20), the approximate average optimum moisture content will be 19.

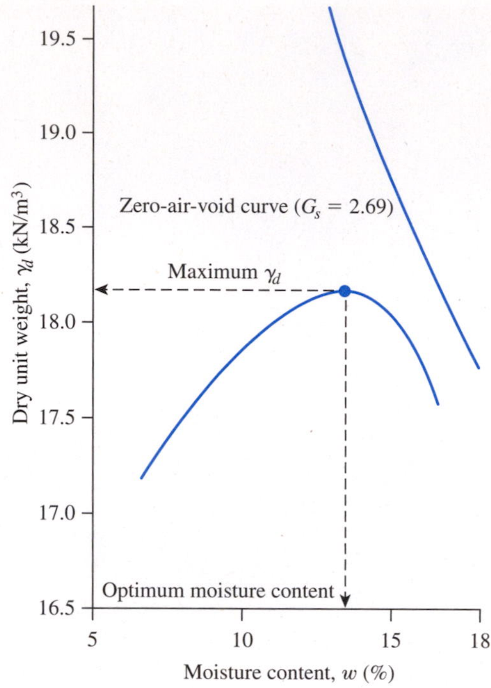

Figure 6.3 Standard Proctor compaction test results for a silty clay

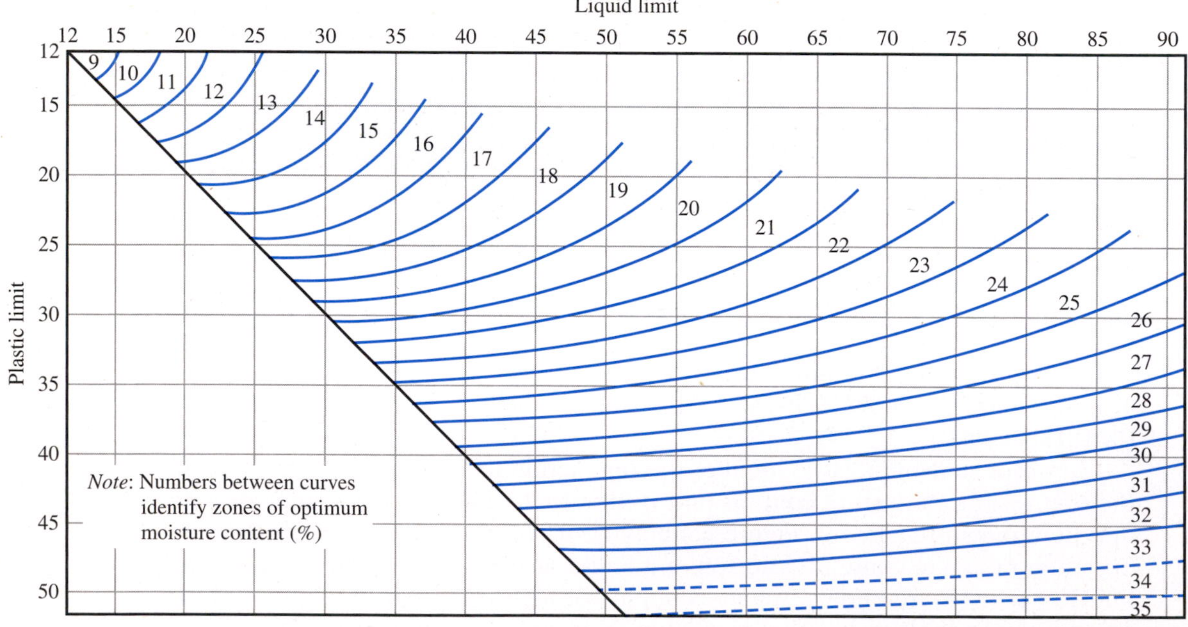

Figure 6.4 Approximate optimum moisture content for a soil using the standard Proctor compaction test (*After Johnson and Sallberg, 1962*)

For a given *moisture content* w and *degree of saturation* S, the dry unit weight of compaction can be calculated as follows. From Chapter 3 [Eq. (3.17)], for any soil,

$$\gamma_d = \frac{G_s \gamma_w}{1 + e}$$

where G_s = specific gravity of soil solids
γ_w = unit weight of water
e = void ratio
and, from Eq. (3.19),

$$Se = G_s w$$

or

$$e = \frac{G_s w}{S}$$

Thus,

$$\gamma_d = \frac{G_s \gamma_w}{1 + \frac{G_s w}{S}} \tag{6.3}$$

For a given moisture content, the theoretical maximum dry unit weight is obtained when no air is in the void spaces—that is, when the degree of saturation equals 100%. Hence, the maximum dry unit weight at a given moisture content with zero air voids can be obtained by substituting $S = 1$ into Eq. (6.3), or

$$\gamma_{zav} = \frac{G_s \gamma_w}{1 + wG_s} = \frac{\gamma_w}{w + \frac{1}{G_s}} \tag{6.4}$$

where γ_{zav} = zero-air-void unit weight.

To obtain the variation of γ_{zav} with moisture content, use the following procedure:

1. Determine the specific gravity of soil solids.
2. Know the unit weight of water (γ_w).
3. Assume several values of w, such as 5%, 10%, 15%, and so on.
4. Use Eq. (6.4) to calculate γ_{zav} for various values of w.

Figure 6.3 also shows the variation of γ_{zav} with moisture content and its relative location with respect to the compaction curve. Under no circumstances should any part of the compaction curve lie to the right of the zero-air-void curve.

6.4 Factors Affecting Compaction

The preceding section showed that moisture content has a strong influence on the degree of compaction achieved by a given soil. Besides moisture content, other important factors that affect compaction are soil type and compaction effort (energy per unit volume). The importance of each of these two factors is described in more detail in the following two sections.

Effect of soil type

The soil type—that is, grain-size distribution, shape of the soil grains, specific gravity of soil solids, and amount and type of clay minerals present—has a great influence on the maximum dry unit weight and optimum moisture content. Figure 6.5 shows typical compaction curves obtained from four soils. The laboratory tests were conducted in accordance with ASTM Test Designation D-698.

Note also that the bell-shaped compaction curve shown in Figure 6.3 is typical of most clayey soils. Figure 6.5 shows that for sands, the dry unit weight has

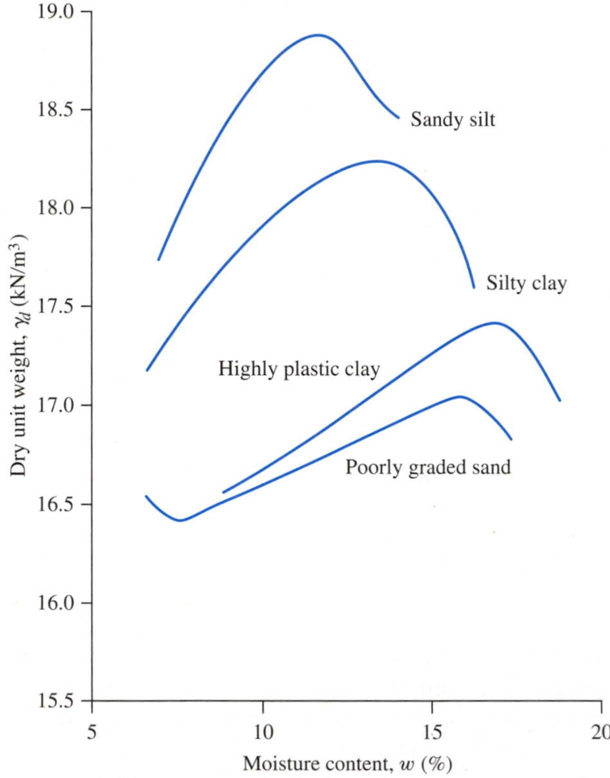

Figure 6.5 Typical compaction curves for four soils (ASTM D-698)

a general tendency first to decrease as moisture content increases and then to increase to a maximum value with further increase of moisture. The initial decrease of dry unit weight with increase of moisture content can be attributed to the capillary tension effect. At lower moisture contents, the capillary tension in the pore water inhibits the tendency of the soil particles to move around and be compacted densely.

Lee and Suedkamp (1972) studied compaction curves for 35 soil samples. They observed that four types of compaction curves can be found. These curves are shown in Figure 6.6. The following table is a summary of the type of compaction curves encountered in various soils with reference to Figure 6.6.

Type of compaction curve (Figure 6.6)	Description of curve	Liquid limit
A	Bell shaped	Between 30 to 70
B	1-1/2 peak	Less than 30
C	Double peak	Less than 30 and those greater than 70
D	Odd shaped	Greater than 70

Effect of compaction effort

The compaction energy per unit volume used for the standard Proctor test described in Section 6.3 can be given as

$$E = \frac{\begin{pmatrix}\text{Number}\\ \text{of blows}\\ \text{per layer}\end{pmatrix} \times \begin{pmatrix}\text{Number}\\ \text{of}\\ \text{layers}\end{pmatrix} \times \begin{pmatrix}\text{Weight}\\ \text{of}\\ \text{hammer}\end{pmatrix} \times \begin{pmatrix}\text{Height of}\\ \text{drop of}\\ \text{hammer}\end{pmatrix}}{\text{Volume of mold}} \tag{6.5}$$

or, in SI units,

$$E = \frac{(25)(3)\left(\frac{2.5 \times 9.81}{1000}\text{ kN}\right)(0.305\text{ m})}{944 \times 10^{-6}\text{ m}^3} = 594\text{ kN-m/m}^3 \approx 600\text{ kN-m/m}^3$$

If the compaction effort per unit volume of soil is changed, the moisture–unit weight curve also changes. This fact can be demonstrated with the aid of Figure 6.7, which shows four compaction curves for a sandy clay. The standard Proctor mold and hammer were used to obtain these compaction curves. The number of layers of soil used for compaction was three for all cases. However, the number of hammer blows per each layer varied from 20 to 50, which varied the energy per unit volume.

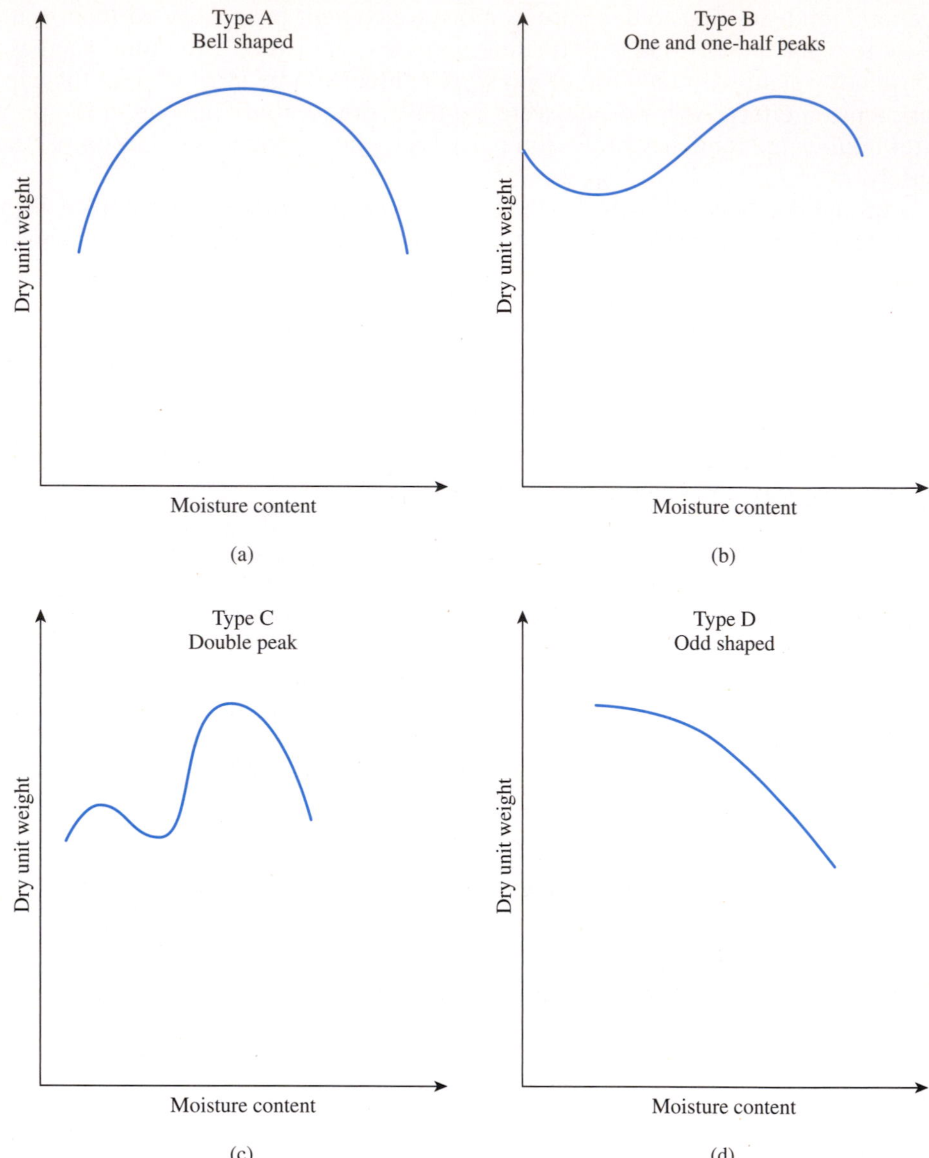

Figure 6.6 Various types of compaction curves encountered in soils

From the preceding observation and Figure 6.7, we can see that

1. As the compaction effort is increased, the maximum dry unit weight of compaction is also increased.
2. As the compaction effort is increased, the optimum moisture content is decreased to some extent.

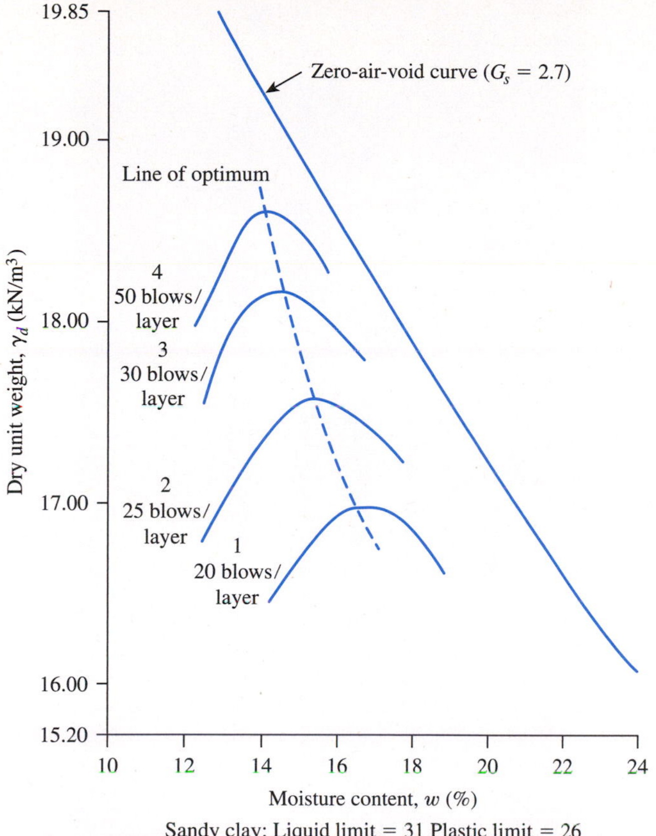

Figure 6.7 Effect of compaction energy on the compaction of a sandy clay

The preceding statements are true for all soils. Note, however, that the degree of compaction is not directly proportional to the compaction effort.

6.5 Modified Proctor Test

With the development of heavy rollers and their use in field compaction, the standard Proctor test was modified to better represent field conditions. This revised version sometimes is referred to as the *modified Proctor test* (ASTM Test Designation D-1557 and AASHTO Test Designation T-180). For conducting the modified Proctor test, the same mold is used with a volume of 944 cm^3, as in the case of the standard Proctor test. However, the soil is compacted in five layers by a hammer that has a mass of 4.54 kg. The drop of the hammer is 457 mm. The number of hammer blows for each layer is kept at 25 as in the case of the standard Proctor test. Figure 6.8 shows a comparison between the hammers used in standard and modified Proctor tests.

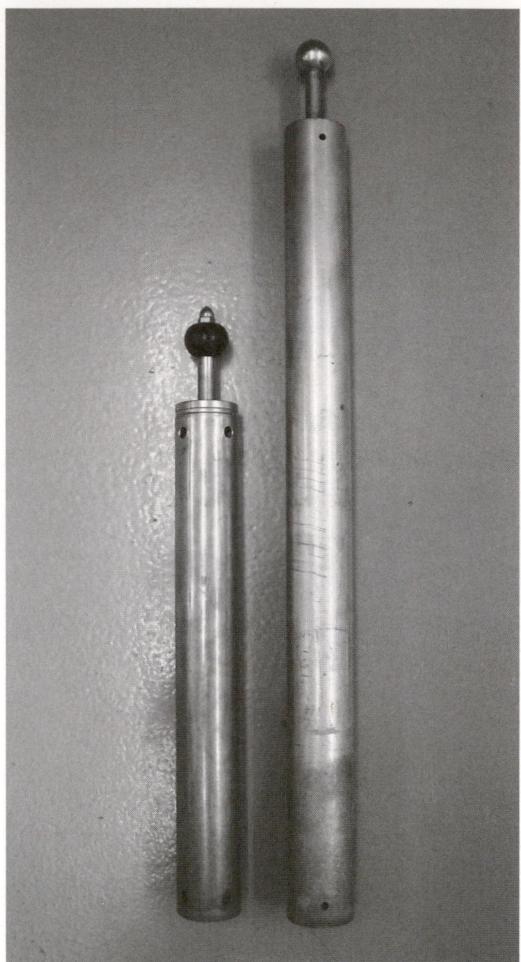

Figure 6.8 Comparison between standard Proctor hammer (left) and modified Proctor hammer (right) *(Courtesy of Braja M. Das, Henderson, Nevada)*

The compaction energy for this type of compaction test can be calculated as 2700 kN-m/m^3.

Because it increases the compactive effort, the modified Proctor test results in an increase in the maximum dry unit weight of the soil. The increase in the maximum dry unit weight is accompanied by a decrease in the optimum moisture content.

In the preceding discussions, the specifications given for Proctor tests adopted by ASTM and AASHTO regarding the volume of the mold and the number of blows are generally those adopted for fine-grained soils that pass through the U.S. No. 4 sieve. However, under each test designation, there are three suggested methods that reflect the mold size, the number of blows per layer, and the maximum particle size in a soil aggregate used for testing. A summary of the test methods is given in Table 6.1.

Table 6.1 Summary of Standard and Modified Proctor Compaction Test Specifications (ASTM D-698 and D-1557)

	Description	Method A	Method B	Method C
Physical data for the tests	Material	Passing No. 4 sieve	Passing 9.5 mm sieve	Passing 19 mm sieve
	Use	Used if 20% or less by weight of material is retained on No. 4 (4.75 mm) sieve	Used if more than 20% by weight of material is retained on No. 4 (4.75 mm) sieve and 20% or less by weight of material is retained on 9.5 mm sieve	Used if more than 20% by weight of material is retained on 9.5 mm sieve and less than 30% by weight of material is retained on 19 mm sieve
	Mold volume	944 cm^3	944 cm^3	2124 cm^3
	Mold diameter	101.6 mm	101.6 mm	152.4 mm
	Mold height	116.4 mm	116.4 mm	116.4 mm
Standard Proctor test	Weight of hammer	24.4 N	24.4 N	24.4 N
	Height of drop	305 mm	305 mm	305 mm
	Number of soil layers	3	3	3
	Number of blows/layer	25	25	56
Modified Proctor test	Weight of hammer	44.5 N	44.5 N	44.5 N
	Height of drop	457 mm	457 mm	457 mm
	Number of soil layers	5	5	5
	Number of blows/layer	25	25	56

6.6 Empirical Relationships

Omar et al. (2003) presented the results of modified Proctor compaction tests on 311 soil samples. Of these samples, 45 were gravelly soil (GP, GP-GM, GW, GW-GM, and GM), 264 were sandy soil (SP, SP-SM, SW-SM, SW, SC-SM, SC, and SM), and two were clay with low plasticity (CL). All compaction tests were conducted using ASTM D-1557 method C to avoid over-size correction. Based on the tests, the following correlations were developed.

$$\rho_{d(\max)}\,(\text{kg/m}^3) = [4{,}804{,}574G_s - 195.55(LL)^2 + 156{,}971\,(\text{R\#4})^{0.5} - 9{,}527{,}830]^{0.5} \tag{6.6}$$

$$\ln(w_{\text{opt}}) = 1.195 \times 10^{-4}\,(LL)^2 - 1.964G_s - 6.617 \times 10^{-5}\,(\text{R\#4}) + 7.651 \tag{6.7}$$

where $\rho_{d(\max)}$ = maximum dry density (kg/m^3)
w_{opt} = optimum moisture content(%)

G_s = specific gravity of soil solids
LL = liquid limit, in percent
R#4 = percent retained on No. 4 sieve

Mujtaba et al. (2013) conducted laboratory compaction tests on 110 sandy soil samples (SM, SP-SM, SP, SW-SM, and SW). Based on the test results, the following correlations were provided for $\gamma_{d\,(\max)}$ and w_{opt} (optimum moisture content):

$$\gamma_{d\,(\max)}(\text{kN/m}^3) = 4.49 \log(C_u) + 1.51 \log(E) + 10.2 \tag{6.8}$$

$$\log w_{opt}(\%) = 1.67 - 0.193 \log(C_u) - 0.153 \log(E) \tag{6.9}$$

where C_u = uniformity coefficient
E = compaction energy (kN-m/m^3)

For granular soils with less than 12% fines (i.e., finer than No. 200 sieve), relative density may be a better indicator for end product compaction specification in the field. Based on laboratory compaction tests on 55 clean sands (less than 5% finer than No. 200 sieve), Patra et al. (2010) provided the following relationships:

$$D_r = AD_{50}^{-B} \tag{6.10}$$

$$A = 0.216 \ln E - 0.850 \tag{6.11}$$

$$B = -0.03 \ln E + 0.306 \tag{6.12}$$

where D_r = maximum relative density of compaction achieved with compaction energy E (kN-m/m^3)
D_{50} = median grain size (mm)

Gurtug and Sridharan (2004) proposed correlations for optimum moisture content and maximum dry unit weight with the plastic limit (PL) of cohesive soils. These correlations can be expressed as

$$w_{opt}(\%) = [1.95 - 0.38(\log\ E)](PL) \tag{6.13}$$

$$\gamma_{d(\max)}\ (\text{kN/m}^3) = 22.68e^{-0.0183 w_{opt}(\%)} \tag{6.14}$$

where PL = plastic limit (%)
E = compaction energy (kN-m/m^3)

For a modified Proctor test, $E = 2700$ kN-m/m^3. Hence,

$$w_{opt}(\%) \approx 0.65\ (PL)$$

and

$$\gamma_{d(\max)}\ (\text{kN/m}^3) = 22.68e^{-0.012(PL)}$$

Osman et al. (2008) analyzed a number of laboratory compaction test results on fine-grained (cohesive) soil, including those provided by Gurtug and Sridharan (2004). Based on this study, the following correlations were developed:

$$w_{opt}(\%) = (1.99 - 0.165 \ln E)(PI) \tag{6.15}$$

and

$$\gamma_{d(\max)}\ (\text{kN/m}^3) = L - Mw_{\text{opt}} \tag{6.16}$$

where

$$L = 14.34 + 1.195 \ln E \tag{6.17}$$

$$M = -0.19 + 0.073 \ln E \tag{6.18}$$

w_{opt} = optimum water content (%)
PI = plasticity index (%)
$\gamma_{d(\max)}$ = maximum dry unit weight (kN/m³)
E = compaction energy (kN-m/m³)

Matteo et al. (2009) analyzed the results of 71 fine-grained soils and provided the following correlations for optimum water content (w_{opt}) and maximum dry unit weight [$\gamma_{d(\max)}$] for modified Proctor tests (E = 2700 kN-m/m³):

$$w_{\text{opt}}(\%) = -0.86(LL) + 3.04\left(\frac{LL}{G_s}\right) + 2.2 \tag{6.19}$$

and

$$\gamma_{d(\max)}(\text{kN/m}^3) = 40.316(w_{\text{opt}}^{-0.295})(PI^{0.032}) - 2.4 \tag{6.20}$$

where LL = liquid limit (%)
PI = plasticity index (%)
G_s = specific gravity of soil solids

Example 6.1

The laboratory test results of a standard Proctor test are given in the following table.

Volume of mold (cm³)	Weight of moist soil in mold (N)	Moisture content, w (%)
944	16.81	10
944	17.84	12
944	18.41	14
944	18.33	16
944	17.84	18
944	17.35	20

a. Determine the maximum dry unit weight of compaction and the optimum moisture content.
b. Calculate and plot γ_d versus the moisture content for degree of saturation, S = 80, 90, and 100% (i.e., γ_{zav}). Given: G_s = 2.7.

Solution

Part a

The following table can be prepared.

Volume of mold V_m (cm^3)	Weight of soil, W (N)	Moist unit weight, γ (kN/m^3)[a]	Moisture content, w (%)	Dry unit weight, γ_d (kN/m^3)[b]
944	16.81	17.81	10	16.19
944	17.84	18.90	12	16.87
944	18.41	19.50	14	17.11
944	18.33	19.42	16	16.74
944	17.84	18.90	18	16.02
944	17.35	18.38	20	15.32

[a] $\gamma = \dfrac{W}{V_m}$

[b] $\gamma_d = \dfrac{\gamma}{1 + \dfrac{w\%}{100}}$

The plot of γ_d versus w is shown at the bottom of Figure 6.9. From the plot, we see that the maximum dry unit weight $\gamma_{d(\max)}$ = **17.15 kN/m³** and the optimum moisture content is **14.4%**.

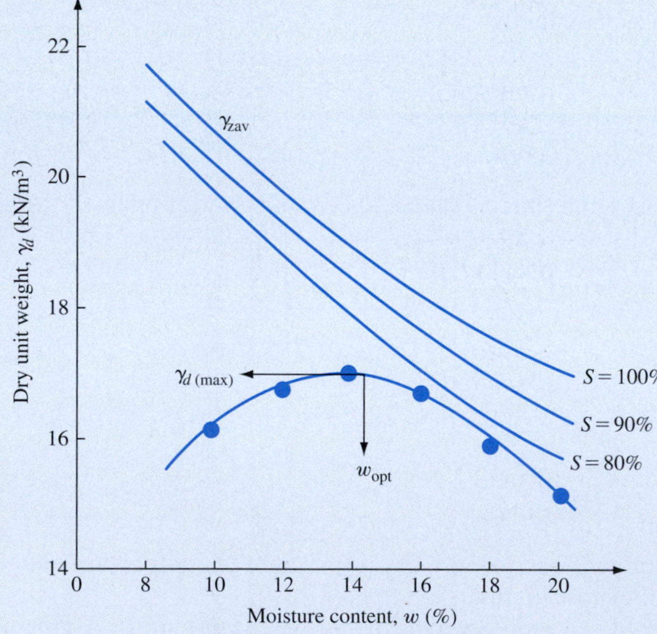

Figure 6.9

Part b

From Eq. (6.3),

$$\gamma_d = \frac{G_s \gamma_w}{1 + \dfrac{G_s w}{S}}$$

The following table can be prepared.

		γ_d (kN/m³)		
G_s	w (%)	$S = 80\%$	$S = 90\%$	$S = 100\%$
2.7	8	20.84	21.37	21.79
2.7	10	19.81	20.37	20.86
2.7	12	18.85	19.48	20.01
2.7	14	17.99	18.65	19.23
2.7	16	17.20	17.89	18.50
2.7	18	16.48	17.20	17.83
2.7	20	15.82	16.55	17.20

The plot of γ_d versus w for the various degrees of saturation is also shown in Figure 6.9.

Example 6.2

A modified Proctor compaction test was carried out on a clayey sand in a cylindrical mold that has a volume of 944 cm³. The specific gravity of the soil grains is 2.68. The moisture content and the mass of the six compacted specimens are given below.

Moisture content (%)	5.0	7.0	9.5	11.8	14.1	17.0
Mass (moist) of specimen in the mold (g)	1776	1890	2006	2024	2005	1977

a. Using the compaction test data determine the optimum moisture content and the maximum dry unit weight.
b. Plot the zero air void curve and check whether it intersects the compaction curve.
c. Plot the void ratio and the degree of saturation against the moisture content.
d. What are the void ratio and degree of saturation at the optimum moisture content?

Solution

For $w = 5.0\%$, mass of moist specimen = 1776 g; volume is 944 cm³.

$$\text{Moist unit weight, } \gamma = \frac{1776}{944} \times 9.81 = 18.46 \text{ kN/m}^3$$

Part d

From Figure 6.11, at an optimum moisture content of 10%,

Void ratio = **0.38**

$$\text{Degree of saturation, } \frac{wG_s}{e} = \frac{(0.1)(2.68)}{0.38} = 0.7 = \mathbf{70\%}$$

Example 6.3

For a granular soil, the following are given:

- $G_s = 2.6$
- Liquid limit on the fraction passing No. 40 sieve = 20
- Percent retained on No. 4 sieve = 20

Using Eqs. (6.6) and (6.7), estimate the maximum dry density of compaction and the optimum moisture content based on the modified Proctor test.

Solution

From Eq. (6.6),

$$\begin{aligned}\rho_{d(\max)}\ (\text{kg/m}^3) &= [4{,}804{,}574G_s - 195.55(LL)^2 + 156{,}971(R\#4)^{0.5} - 9{,}527{,}830]^{0.5}\\ &= [4{,}804{,}574(2.6) - 195.55(20)^2 + 156{,}971(20)^{0.5} - 9{,}527{,}830]^{0.5}\\ &= \mathbf{1894\ kg/m^3}\end{aligned}$$

From Eq. (6.7),

$$\begin{aligned}\ln(w_{\text{opt}}) &= 1.195 \times 10^{-4}(LL)^2 - 1.964G_s - 6.617 \times 10^{-5}(R\#4) + 7{,}651\\ &= 1.195 \times 10^{-4}(20)^2 - 1.964(2.6) - 6.617 \times 10^{-5}(20) + 7{,}651\\ &= 2.591\\ w_{\text{opt}} &= \mathbf{13.35\%}\end{aligned}$$

Example 6.4

For a sand with 4% finer than No. 200 sieve, estimate the maximum relative density of compaction that may be obtained from a modified Proctor test. Given $D_{50} = 1.4$ mm. Use Eq. (6.10).

Solution

For modified Proctor test, $E = 2696$ kN-m/m^3.

From Eq. (6.11),

$$A = 0.216 \ln E - 0.850 = (0.216)(\ln 2696) - 0.850 = 0.856$$

From Eq. (6.12),

$$B = -0.03 \ln E + 0.306 = -(0.03)(\ln 2696) + 0.306 = 0.069$$

From Eq. (6.10),

$$D_r = AD_{50}^{-B} = (0.856)(1.4)^{-0.069} = 0.836 = \mathbf{83.6\%}$$

Example 6.5

For a silty clay soil given $LL = 43$ and $PL = 18$. Estimate the maximum dry unit weight of compaction that can be achieved by conducting a modified Proctor test. Use Eq. (6.16).

Solution

For modified Proctor test, $E = 2696$ kN-m/m^3.

From Eqs. (6.17) and (6.18),

$$L = 14.34 + 1.195 \ln E = 14.34 + 1.195 \ln (2696) = 23.78$$

$$M = -0.19 + 0.073 \ln E = -0.19 + 0.073 \ln (2696) = 0.387$$

From Eq. (6.15),

$$\begin{aligned} w_{\text{opt}}(\%) &= (1.99 - 0.165 \ln E)(PI) \\ &= [1.99 - 0.165 \ln (2696)](43 - 18) \\ &= 17.16\% \end{aligned}$$

From Eq. (6.16),

$$\gamma_{d(\max)} = L - Mw_{\text{opt}} = 23.78 - (0.387)(17.16) = \mathbf{17.14\ kN/m^3}$$

Example 6.6

Refer to the silty clay soil given in Example 6.5. Using Eqs. (6.13) and (6.14), estimate w_{opt} and $\gamma_{d(max)}$ that can be obtained from a modified compaction test.

Solution

Given $PL = 18$. For a modified compaction test, $E = 2700$ kN-m/m^3. So, from Eqs. (6.13) and (6.14),

$$w_{opt}(\%) = 0.65(PL)$$

and

$$\gamma_{d(max)}(\text{kN/m}^3) = 22.68e^{-0.012(PL)}$$

Hence,

$$w_{opt}(\%) = (0.65)(18) = \mathbf{11.7\%}$$

$$\gamma_{d(max)} = 22.68e^{-0.012(18)} = \mathbf{18.27\ kN/m^3}$$

Example 6.7

Refer to Example 6.5. Estimate w_{opt} and $\gamma_{d(max)}$ using Eqs. (6.19) and (6.20). Use $G_s = 2.66$.

Solution

From Eq. (6.19),

$$w_{opt}(\%) = -0.86(LL) + 3.04\left(\frac{LL}{G_s}\right) + 2.2$$

Given $LL = 43$ and $Gs = 2.66$,

$$w_{opt}(\%) = -0.86(43) + 3.04\left(\frac{43}{2.66}\right) + 2.2 = \mathbf{14.36\%}$$

From Eq. (6.20),

$$\begin{aligned}\gamma_{d(max)} &= 40.316(w_{opt}^{-0.295})(PI^{0.032}) - 2.4 \\ &= (40.316)(14.36)^{-0.295}(43 - 18)^{0.032} - 2.4 = \mathbf{17.97 kN/m^3}\end{aligned}$$

6.7 Structure of Compacted Clay Soil

Lambe (1958a) studied the effect of compaction on the structure of clay soils, and the results of his study are illustrated in Figure 6.12. If clay is compacted with a moisture content on the dry side of the optimum, as represented by point A, it will possess a flocculent structure. This type of structure results because, at low moisture content, the diffuse double layers of ions surrounding the clay particles cannot be fully developed; hence, the interparticle repulsion is reduced. This reduced repulsion results in a more random particle orientation and a lower dry unit weight. When the moisture content of compaction is increased, as shown by point B, the diffuse double layers around the particles expand, which increases the repulsion between the clay particles and gives a lower degree of flocculation and a higher dry unit weight. A continued increase in moisture content from B to C expands the double layers more. This expansion results in a continued increase of repulsion between the particles and thus a still greater degree of particle orientation and a more or less dispersed structure. However, the dry unit weight decreases because the added water dilutes the concentration of soil solids per unit volume.

At a given moisture content, higher compactive effort yields a more parallel orientation to the clay particles, which gives a more dispersed structure. The particles are closer and the soil has a higher unit weight of compaction. This phenomenon can be seen by comparing point A with point E in Figure 6.12.

Figure 6.13 shows the variation in the degree of particle orientation with molding moisture content for compacted Boston blue clay. Works of Seed and Chan (1959) have shown similar results for compacted kaolin clay.

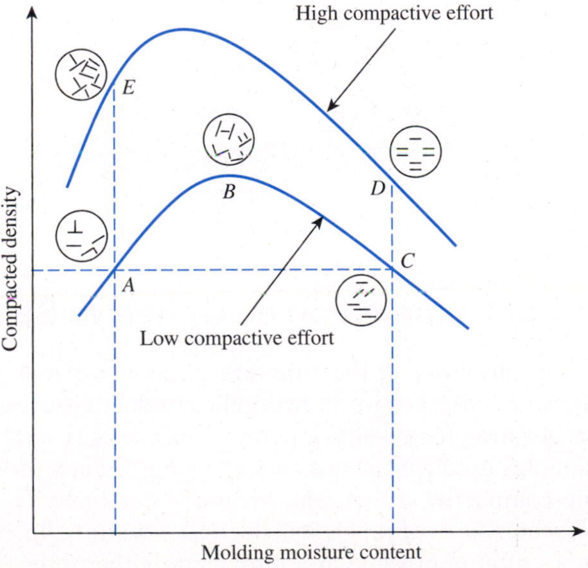

Figure 6.12 Effect of compaction on structure of clay soils (*Redrawn after Lambe, 1958a. With permission from ASCE.*)

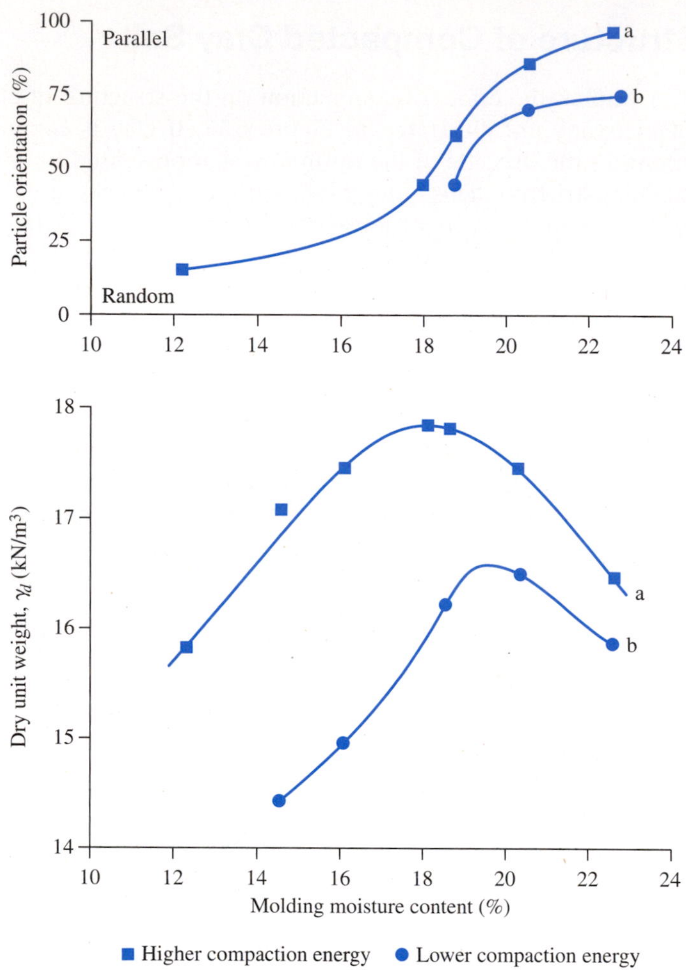

Figure 6.13 Orientation against moisture content for Boston blue clay (*After Lambe, 1958a. With permission from ASCE.*)

6.8 Effect of Compaction on Cohesive Soil Properties

Compaction induces variations in the structure of cohesive soils. Results of these structural variations include changes in hydraulic conductivity, compressibility, and strength. Figure 6.14 shows the results of permeability tests (Chapter 7) on Jamaica sandy clay. The samples used for the tests were compacted at various moisture contents by the same compactive effort. The hydraulic conductivity, which is a measure of how easily water flows through soil, decreases with the increase of moisture content. It reaches a minimum value at approximately the optimum moisture content. Beyond the optimum moisture content, the hydraulic conductivity increases slightly. The high value of the hydraulic conductivity on the dry side of the optimum

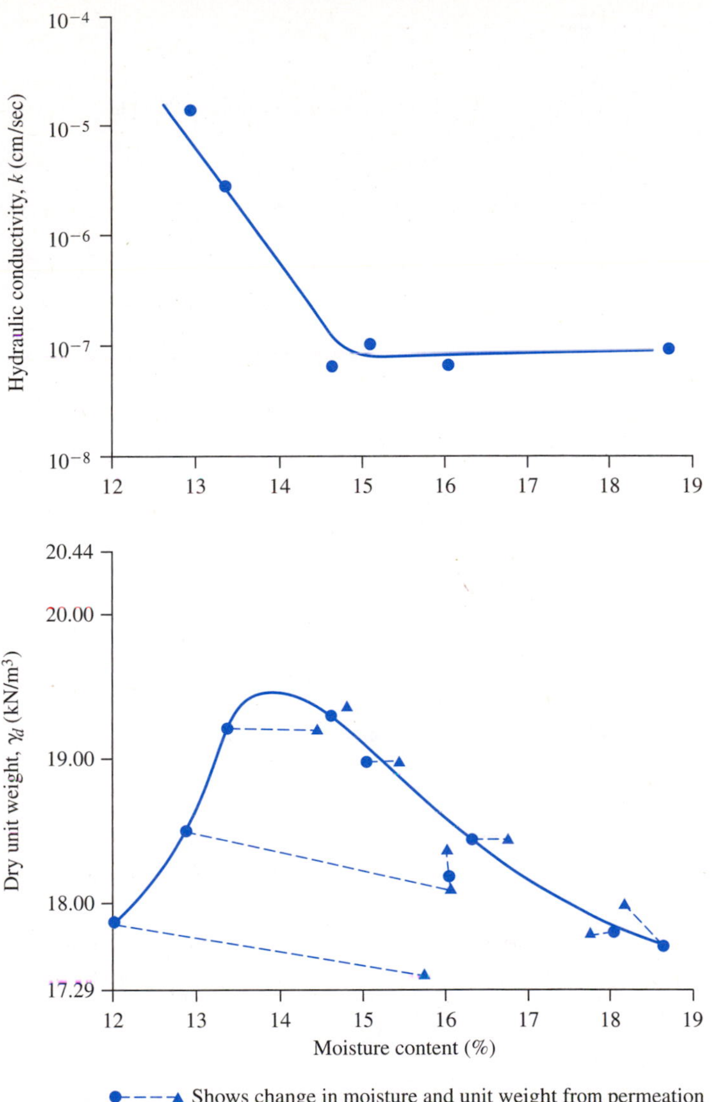

Figure 6.14 Effect of compaction on hydraulic conductivity of clayey soil *(Redrawn after Lambe, 1958b. With permission from ASCE.)*

moisture content is due to the random orientation of clay particles that results in larger pore spaces.

One-dimensional compressibility characteristics (Chapter 11) of clay soils compacted on the dry side of the optimum and compacted on the wet side of the optimum are shown in Figure 6.15. Under lower pressure, a soil that is compacted on the wet side of the optimum is more compressible than a soil that is compacted on the dry side of the optimum. This is shown in Figure 6.15a. Under high pressure, the

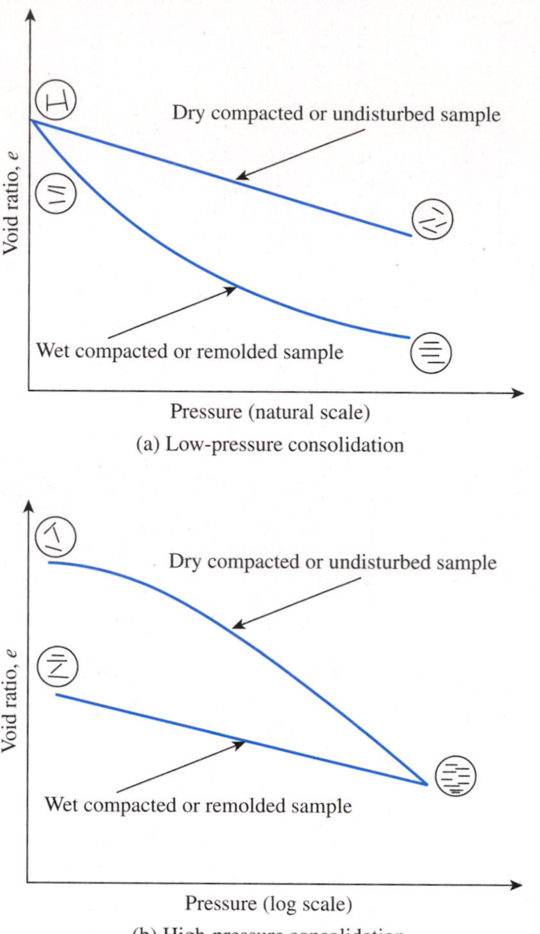

Figure 6.15 Effect of compaction on one-dimensional compressibility of clayey soil (*Redrawn after Lambe, 1958b. With permission from ASCE.*)

trend is exactly the opposite, and this is shown in Figure 6.15b. For samples compacted on the dry side of the optimum, the pressure tends to orient the particles normal to its direction of application. The space between the clay particles is also reduced at the same time. However, for samples compacted on the wet side of the optimum, pressure merely reduces the space between the clay particles. At very high pressure, it is possible to have identical structures for samples compacted on the dry and wet sides of optimum.

The strength of compacted clayey soils (Chapter 12) generally decreases with the molding moisture content. This is shown in Figure 6.16, which is the result of several unconfined compression-strength tests on compacted specimens of a silty clay soil. The test specimens were prepared by kneading compaction. The insert in Figure 6.16 shows the relationship between dry unit weight and moisture content for the soil. Note that specimens A, B, and C have been compacted, respectively,

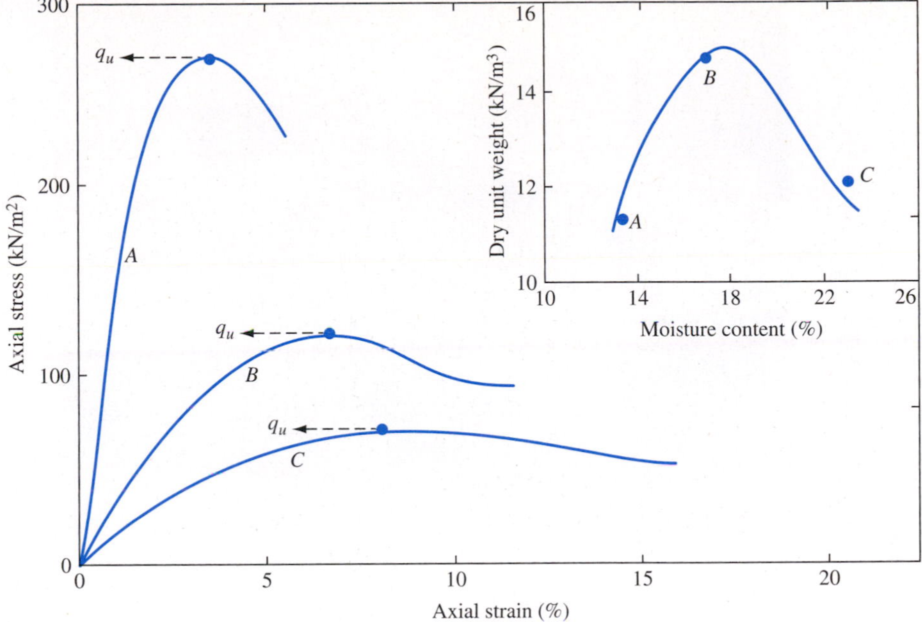

Figure 6.16 Unconfined compression test on compacted specimens of a silty clay

on the dry side of the optimum moisture content, near optimum moisture content, and on the wet side of the optimum moisture content. The unconfined compression strength, q_u, is greatly reduced for the specimen compacted on the wet side of the optimum moisture content.

6.9 Field Compaction

Compaction equipment

Most of the compaction in the field is done with rollers. The four most common types of rollers are

1. Smooth-wheel rollers (or smooth-drum rollers)
2. Pneumatic rubber-tired rollers
3. Sheepsfoot rollers
4. Vibratory rollers

Smooth-wheel rollers (Figure 6.17) are suitable for proof rolling subgrades and for finishing operation of fills with sandy and clayey soils. These rollers provide 100% coverage under the wheels, with ground contact pressures as high as 310 to 380 kN/m^2. They are not suitable for producing high unit weights of compaction when used on thicker layers.

Pneumatic rubber-tired rollers (Figure 6.18) are better in many respects than the smooth-wheel rollers. The former are heavily loaded with several rows of tires.

Figure 6.17 Smooth-wheel roller (*Ingram Compaction LLC*)

Figure 6.18 Pneumatic rubber-tired roller (*Ingram Compaction LLC*)

Figure 6.19 Sheepsfoot roller (*SuperStock/Alamy*)

These tires are closely spaced—four to six in a row. The contact pressure under the tires can range from 600 to 700 kN/m^2, and they produce about 70 to 80% coverage. Pneumatic rollers can be used for sandy and clayey soil compaction. Compaction is achieved by a combination of pressure and kneading action.

Sheepsfoot rollers (Figure 6.19) are drums with a large number of projections. The area of each projection may range from 25 to 85 cm^2. These rollers are most effective in compacting clayey soils. The contact pressure under the projections can range from 1400 to 7000 kN/m^2. During compaction in the field, the initial passes compact the lower portion of a lift. Compaction at the top and middle of a lift is done at a later stage.

Vibratory rollers are extremely efficient in compacting granular soils. Vibrators can be attached to smooth-wheel, pneumatic rubber-tired, or sheepsfoot rollers to provide vibratory effects to the soil. Figure 6.20 demonstrates the principles of vibratory rollers. The vibration is produced by rotating off-center weights.

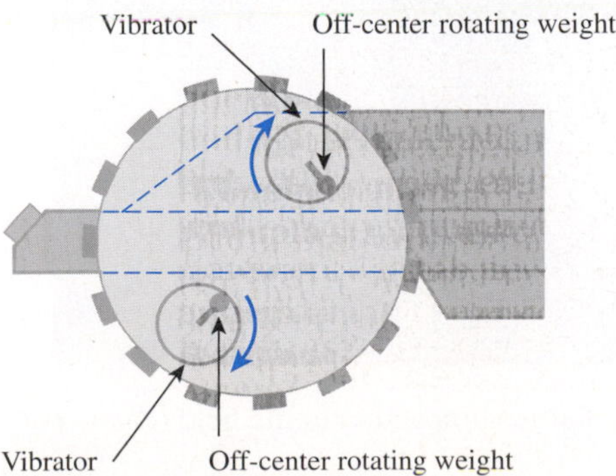

Figure 6.20 Principles of vibratory rollers

Handheld vibrating plates can be used for effective compaction of granular soils over a limited area. Vibrating plates are also gang-mounted on machines. These plates can be used in less restricted areas.

Field compaction and factors affecting field compaction

For field compaction, soil is spread in layers and a predetermined amount of water is sprayed (Figure 6.21) on each layer (lift) of soil, after which compaction is initiated by a desired roller.

In addition to soil type and moisture content, other factors must be considered to achieve the desired unit weight of compaction in the field. These factors include the thickness of lift, the intensity of pressure applied by the compacting equipment, and the area over which the pressure is applied. These factors are important because the pressure applied at the surface decreases with depth, which results in a decrease in the degree of soil compaction. During compaction, the dry unit weight of soil also is affected by the number of roller passes. Figure 6.22 shows the growth curves for a silty clay soil. The dry unit weight of a soil at a given moisture content increases to a certain point with the number of roller passes. Beyond this point, it remains approximately constant. In most cases, about 10 to 15 roller passes yield the maximum dry unit weight economically attainable.

Figure 6.23a shows the variation in the unit weight of compaction with depth for a poorly graded dune sand for which compaction was achieved by a vibratory drum roller. Vibration was produced by mounting an eccentric weight on a single rotating

Figure 6.21 Spraying of water on each lift of soil before compaction in the field *(Courtesy of N. Sivakugan, James Cook University, Australia)*

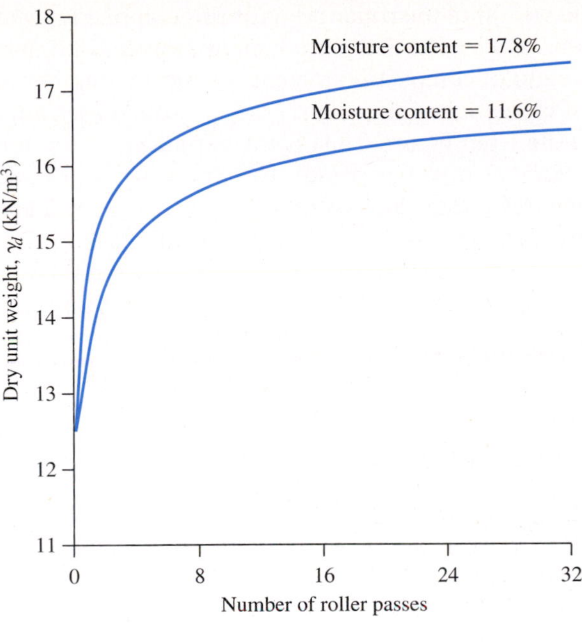

Figure 6.22 Growth curves for a silty clay—relationship between dry unit weight and number of passes of 84.5 kN three-wheel roller when the soil is compacted in 229 mm loose layers at different moisture contents (*From Full-Scale Field Tests on 3-Wheel Power Rollers. In Highway Research Bulletin 272, Highway Research Board, National Research Council, Washington, D.C., 1960, Figure 15, p. 23. Reproduced with permission of the Transportation Research Board.*)

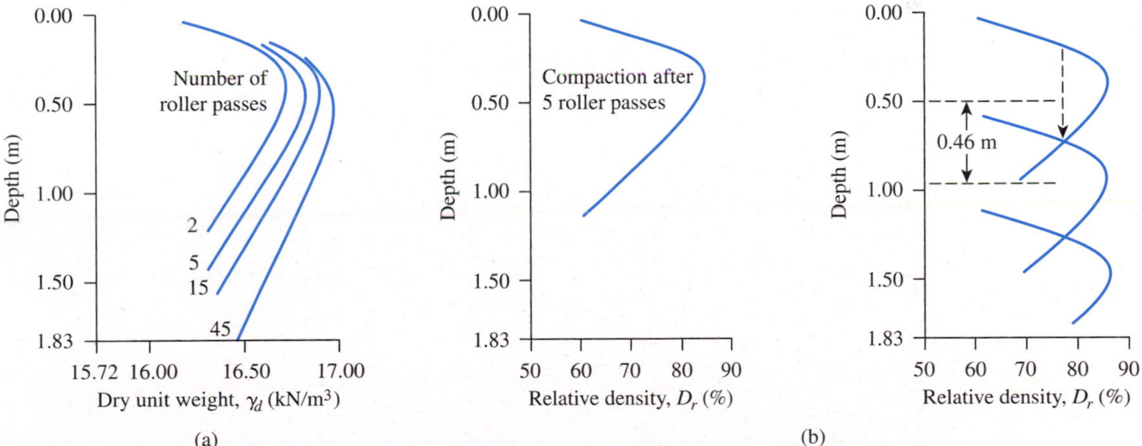

Figure 6.23 (a) Vibratory compaction of a sand—variation of dry unit weight with number of roller passes; thickness of lift = 2.44 m; (b) estimation of compaction lift thickness for minimum required relative density of 75% with five roller passes (*After D' Appolonia, Whitman, and D' Appolonia, 1969. With permission from ASCE.*)

shaft within the drum cylinder. The weight of the roller used for this compaction was 55.6 kN, and the drum diameter was 1.19 m. The lifts were kept at 2.44 m. Note that, at any given depth, the dry unit weight of compaction increases with the number of roller passes. However, the rate of increase in unit weight gradually decreases after about 15 passes. Another fact to note from Figure 6.23a is the variation of dry unit weight with depth for any given number of roller passes. The dry unit weight and hence the relative density, D_r, reach maximum values at a depth of about 0.5 m and gradually decrease at lesser depths. This decrease occurs because of the lack of confining pressure toward the surface. Once the relationship between depth and relative density (or dry unit weight) for a given soil with a given number of roller passes is determined, estimating the approximate thickness of each lift is easy. This procedure is shown in Figure 6.23b (D' Appolonia et al., 1969).

6.10 Specifications for Field Compaction

In most specifications for earthwork, the contractor is instructed to achieve a compacted field dry unit weight of 90 to 95% of the maximum dry unit weight determined in the laboratory by either the standard or modified Proctor test. This is a specification for relative compaction, which can be expressed as

$$R(\%) = \frac{\gamma_{d(\text{field})}}{\gamma_{d(\text{max-lab})}} \times 100 \tag{6.21}$$

where R = relative compaction.

For the compaction of granular soils, specifications sometimes are written in terms of the required relative density D_r or the required relative compaction. Relative density should not be confused with relative compaction. From Chapter 3, we can write

$$D_r = \left[\frac{\gamma_{d(\text{field})} - \gamma_{d(\text{min})}}{\gamma_{d(\text{max})} - \gamma_{d(\text{min})}}\right]\left[\frac{\gamma_{d(\text{max})}}{\gamma_{d(\text{field})}}\right] \tag{6.22}$$

Comparing Eqs. (6.21) and (6.22), we see that

$$R = \frac{R_0}{1 - D_r(1 - R_0)} \tag{6.23}$$

where

$$R_0 = \frac{\gamma_{d(\text{min})}}{\gamma_{d(\text{max})}} \tag{6.24}$$

The specification for field compaction based on relative compaction or on relative density is an end product specification. The contractor is expected to achieve a minimum dry unit weight regardless of the field procedure adopted. The most

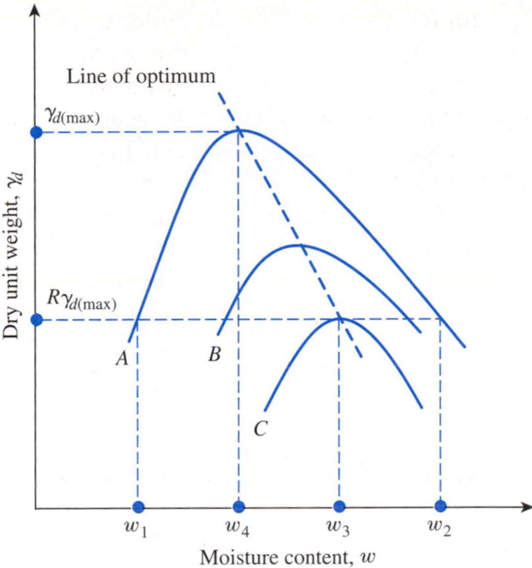

Figure 6.24 Most economical compaction condition

economical compaction condition can be explained with the aid of Figure 6.24. The compaction curves A, B, and C are for the same soil with varying compactive effort. Let curve A represent the conditions of maximum compactive effort that can be obtained from the existing equipment. Let the contractor be required to achieve a minimum dry unit weight of $\gamma_{d(\text{field})} = R\gamma_{d(\max)}$. To achieve this, the contractor must ensure that the moisture content w falls between w_1 and w_2. As can be seen from compaction curve C, the required $\gamma_{d(\text{field})}$ can be achieved with a lower compactive effort at a moisture content $w = w_3$. However, for most practical conditions, a compacted field unit weight of $\gamma_{d(\text{field})} = R\gamma_{d(\max)}$ cannot be achieved by the minimum compactive effort. Hence, equipment with slightly more than the minimum compactive effort should be used. The compaction curve B represents this condition. Now we can see from Figure 6.24 that the most economical moisture content is between w_3 and w_4. Note that $w = w_4$ is the optimum moisture content for curve A, which is for the maximum compactive effort.

The concept described in the preceding paragraph, along with Figure 6.24, is attributed historically to Seed (1964) and is elaborated on in more detail in Holtz and Kovacs (1981).

Example 6.8

The maximum and minimum unit weights of a sand collected from the field were determined in the laboratory as 18.38 kN/m^3 and 15.99 kN/m^3, respectively. It is required that the sand in the field be compacted to a relative density of 85%. Determine what would be the relative compaction in the field.

Solution

From Eq. (6.24),

$$R_o = \frac{\gamma_{d(\min)}}{\gamma_{d(\max)}} = \frac{15.99}{18.38} = 0.87$$

From Eq. (6.23),

$$R = \frac{R_o}{1 - D_r(1 - R_o)} = \frac{0.87}{1 - (0.85)(1 - 0.87)} = 0.978 = \mathbf{97.8\%}$$

6.11 Determination of Field Unit Weight of Compaction

When the compaction work is progressing in the field, knowing whether the specified unit weight has been achieved is useful. The standard procedures for determining the field unit weight of compaction include

1. Sand cone method
2. Rubber balloon method
3. Nuclear method

Following is a brief description of each of these methods.

Sand cone method (ASTM Designation D-1556)

The sand cone device consists of a glass or plastic jar with a metal cone attached at its top (Figure 6.25). The jar is filled with uniform dry Ottawa sand. The combined weight of the jar, the cone, and the sand filling the jar is determined (W_1). In the field, a small hole is excavated in the area where the soil has been compacted. If the weight of the moist soil excavated from the hole (W_2) is determined and the moisture content of the excavated soil is known, the dry weight of the soil can be obtained as

$$W_3 = \frac{W_2}{1 + \frac{w\,(\%)}{100}} \tag{6.25}$$

where w = moisture content.

After excavation of the hole, the cone with the sand-filled jar attached to it is inverted and placed over the hole (Figure 6.26). Sand is allowed to flow out of the jar to fill the hole and the cone. After that, the combined weight of the jar, the cone, and the remaining sand in the jar is determined (W_4), so

$$W_5 = W_1 - W_4 \tag{6.26}$$

where W_5 = weight of sand to fill the hole and cone.

Figure 6.25 Glass jar filled with Ottawa sand with sand cone attached *(Courtesy of Braja M. Das, Henderson, Nevada)*

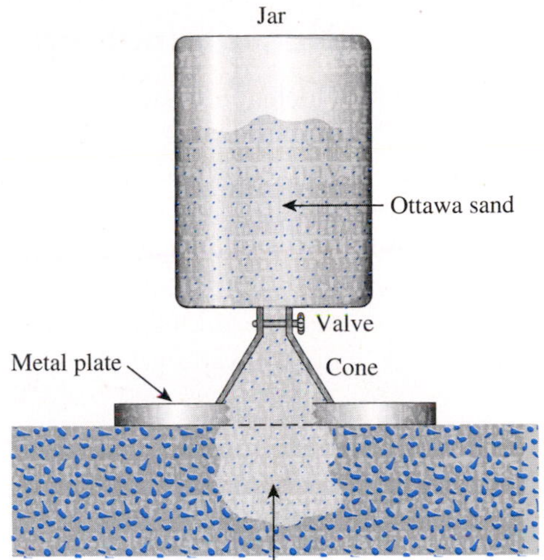

Figure 6.26 Field unit weight determined by sand cone method

The volume of the excavated hole can then be determined as

$$V = \frac{W_5 - W_c}{\gamma_{d(\text{sand})}} \tag{6.27}$$

where W_c = weight of sand to fill the cone only

$\gamma_{d(\text{sand})}$ = dry unit weight of Ottawa sand used

The values of W_c and $\gamma_{d(\text{sand})}$ are determined from the calibration done in the laboratory. The dry unit weight of compaction made in the field then can be determined as follows:

$$\gamma_d = \frac{\text{Dry weight of the soil excavated from the hole}}{\text{Volume of the hole}} = \frac{W_3}{V} \tag{6.28}$$

Rubber balloon method (ASTM Designation D-2167)

The procedure for the rubber balloon method is similar to that for the sand cone method; a test hole is made and the moist weight of soil removed from the hole and its moisture content are determined. However, the volume of the hole is determined by introducing into it a rubber balloon filled with water from a calibrated vessel, from which the volume can be read directly. The dry unit weight of the compacted soil can be determined by using Eq. (6.28). Figure 6.27 shows a calibrated vessel that would be used with a rubber balloon.

Nuclear method

Nuclear density meters are often used for determining the compacted dry unit weight of soil. The density meters operate either in drilled holes or from the ground surface. It uses a radioactive isotope source. The isotope gives off Gamma rays that radiate back to the meter's detector. Dense soil absorbs more radiation than loose soil. The instrument measures the weight of wet soil per unit volume and the weight of water present in a unit volume of soil. The dry unit weight of compacted soil can be determined by subtracting the weight of water from the moist unit weight of soil. Figure 6.28 shows a photograph of a nuclear density meter.

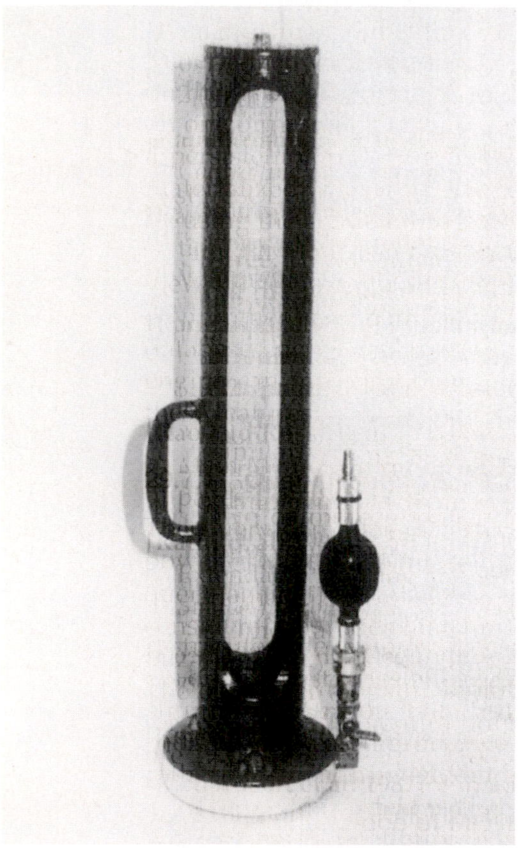

Figure 6.27 Calibrated vessel used with rubber balloon (not shown) (*Courtesy of John Hester, Carterville, Illinois*)

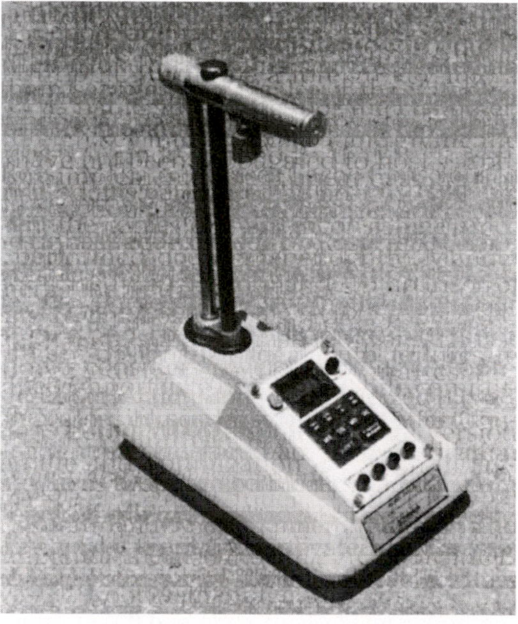

Figure 6.28 Nuclear density meter (*Courtesy of Braja M. Das, Henderson, Nevada*)

Example 6.9

Laboratory compaction test results for a clayey silt are given in the following table.

Moisture content (%)	Dry unit weight (kN/m^3)
6	14.80
8	17.45
9	18.52
11	18.9
12	18.5
14	16.9

Following are the results of a field unit-weight determination test performed on the same soil by means of the sand cone method:

- Calibrated dry density of Ottawa sand = 1570 kg/m^3
- Calibrated mass of Ottawa sand to fill the cone = 0.545 kg
- Mass of jar + cone + sand (before use) = 7.59 kg
- Mass of jar + cone + sand (after use) = 4.78 kg
- Mass of moist soil from hole = 3.007 kg
- Moisture content of moist soil = 10.2%

Determine:

a. Dry unit weight of compaction in the field
b. Relative compaction in the field

Solution

Part a

In the field,

$$\text{Mass of sand used to fill the hole and cone} = 7.59\text{ kg} - 4.78\text{ kg} = 2.81\text{ kg}$$

$$\text{Mass of sand used to fill the hole} = 2.81\text{ kg} - 0.545\text{ kg} = 2.265\text{ kg}$$

$$\text{Volume of the hole } (V) = \frac{2.265\text{ kg}}{\text{Dry density of Ottawa sand}}$$

$$= \frac{2.265\text{ kg}}{1570\text{ kg/m}^3} = 0.0014426\text{ m}^3$$

$$\text{Moist density of compacted soil} = \frac{\text{Mass of moist soil}}{\text{Volume of hole}}$$

$$= \frac{3.007}{0.0014426} = 2.084.4\text{ kg/m}^3$$

$$\text{Moist unit weight of compacted soil} = \frac{(2084.4)(9.81)}{1000} = 20.45\text{ kN/m}^3$$

Hence,

$$\gamma_d = \frac{\gamma}{1 + \frac{w(\%)}{100}} = \frac{20.45}{1 + \frac{10.2}{100}} = \mathbf{18.56\ kN/m^3}$$

Part b

The results of the laboratory compaction test are plotted in Figure 6.29. From the plot, we see that $\gamma_{d(\max)} = 19\ \text{kN/m}^3$. Thus, from Eq. (6.21),

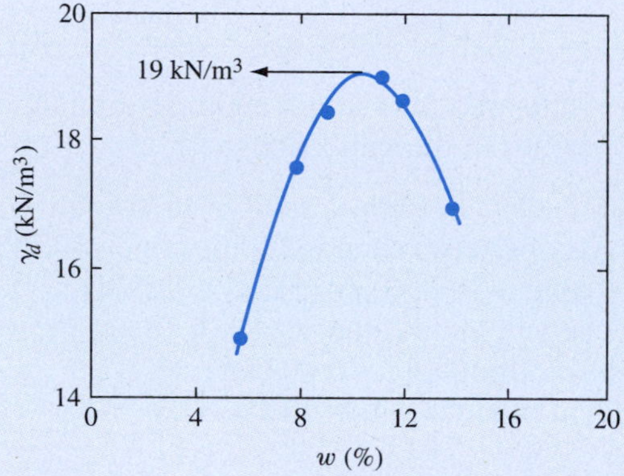

Figure 6.29 Plot of laboratory-compaction test results

$$R = \frac{\gamma_{d(\text{field})}}{\gamma_{d(\max)}} = \frac{18.56}{19.0} = \mathbf{97.7\%}$$

Example 6.10

For a given soil ($G_s = 2.72$), following are the results of compaction tests conducted in the laboratory.

Moisture content (%)	Dry unit weight γ_d (kN/m³)
12	16.34
14	16.93
16	17.24
18	17.20
20	16.75
22	16.23

After compaction of the soil in the field, sand cone tests (control tests) were conducted at five separate locations. Following are the results:

Location	Moisture content (%)	Moist density, ρ (kg/m^3)
1	15.2	2055
2	16.4	2060
3	17.2	1971
4	18.8	1980
5	21.1	2104

The specifications require that

a. γ_d must be at least 0.95 $\gamma_{d(\max)}$.
b. Moisture content w should be within $\pm 2\%$ of w_{opt}.

Make necessary calculations to see if the control tests meet the specifications.

Solution

From Eq. (6.4),

$$\gamma_{\text{zav}} = \frac{\gamma_w}{w + \dfrac{1}{G_s}}$$

Given $G_s = 2.72$. Now the following table can be prepared.

w (%)	γ_{zav} (kN/m^3)
12	20.12
14	19.33
16	18.59
18	17.91
20	17.28
22	16.70

Figure 6.30 shows the plot of γ_d and γ_{zav}. From the plot, it can be seen that

$$\gamma_{d(\max)} = 17.4 \text{ kN/m}^3$$

$$w_{\text{opt}} = 16.8\%$$

Based on the specifications, γ_d must be at least $0.95\gamma_{d(\max)} = (0.95)(17.4) = 16.54 \text{ kN/m}^3$ with a moisture content of $16.8\% \pm 2\% = 14.8\%$ to 18.8%. This zone is shown in Figure 6.30.

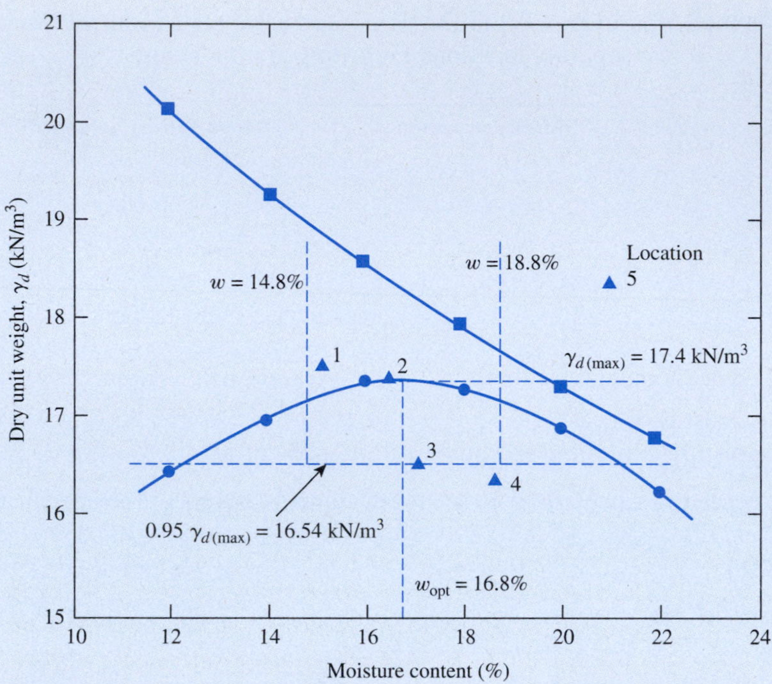

Figure 6.30

For the control tests, the following table can be prepared.

Location	w (%)	ρ (kg/m³)	γ_d* (kN/m³)
1	15.2	2055	17.5
2	16.4	2060	17.36
3	17.2	1971	16.51
4	18.8	1980	16.35
5	21.1	2104	18.41

$$^*\gamma_d(\text{kN/m}^3) = \left[\frac{\rho(\text{kg/m}^3)}{1 + \dfrac{w\,(\%)}{100}}\right]\left(\frac{9.81}{1000}\right)$$

The results of the control tests are also plotted in Figure 6.30. From the plot, it appears that the **tests at locations 1 and 2 meet the specifications.** The test at location 3 is a borderline case. Also note that there is some error for the test in location 5, since it falls above the zero-air-void line.

6.12 Evaluation of Soils as Compaction Material

Table 6.2 provides a general summary of the evaluation of various types of soils as fill material as they relate to roller type, maximum dry unit weight of compaction based on standard Proctor tests, and compaction characteristics. The compressibility and expansion characteristics on compacted soils are as follow (Sowers, 1979):

GW, GP, SW, SP	Practically none
GM, GC, SM, SC	Slight
ML	Slight to medium
MH	High
CL	Medium
CH	Very high

6.13 Special Compaction Techniques

Several special types of compaction techniques have been developed for deep compaction of in-place soils, and these techniques are used in the field for large-scale compaction works. Among these, the popular methods are vibroflotation, dynamic compaction, and blasting. Details of these methods are provided in the following sections.

Table 6.2 Summary of Evaluation of Fill Materials for Compaction Based on Sowers (1979) and Highway Research Board (1962)

			Maximum dry unit weight—standard Proctor compaction	
Soil type	**Unified classification**	**Roller(s) for best results**	**kN/m³**	**Compaction characteristics**
Gravelly	GW	Rubber-tired, steel wheel, vibratory	18.9–20.4	Good
	GP	Rubber-tired, steel wheel, vibratory	18.1–18.9	Good
	GM	Rubber-tired, sheepsfoot	18.9–20.4	Good to fair
	GC	Rubber-tired, sheepsfoot	18.1–19.7	Good to fair
Sandy	SW	Rubber-tired, vibratory	18.1–19.7	Good
	SP	Rubber-tired, vibratory	16.5–18.1	Good
	SM	Rubber-tired, sheepsfoot	17.3–18.9	Good to fair
	SC	Rubber-tired, sheepsfoot	16.5–18.9	Good to fair
Silty	ML	Rubber-tired, sheepsfoot	15.7–17.3	Good to poor
	MH	Rubber-tired, sheepsfoot	13.4–15.7	Fair to poor
Clayey	CL	Rubber-tired, sheepsfoot	14.1–18.1	Fair to poor
	CH	Sheepsfoot	13.4–16.5	Fair to poor

Vibroflotation

Vibroflotation is a technique for *in situ* densification of thick layers of loose granular soil deposits. It was developed in Germany in the 1930s. The first vibroflotation device was used in the United States about 10 years later. The process involves the use of a *Vibroflot* unit (also called the *vibrating unit*), which is about 2.1 m long. (As shown in Figure 6.31.) This vibrating unit has an eccentric weight inside it and can

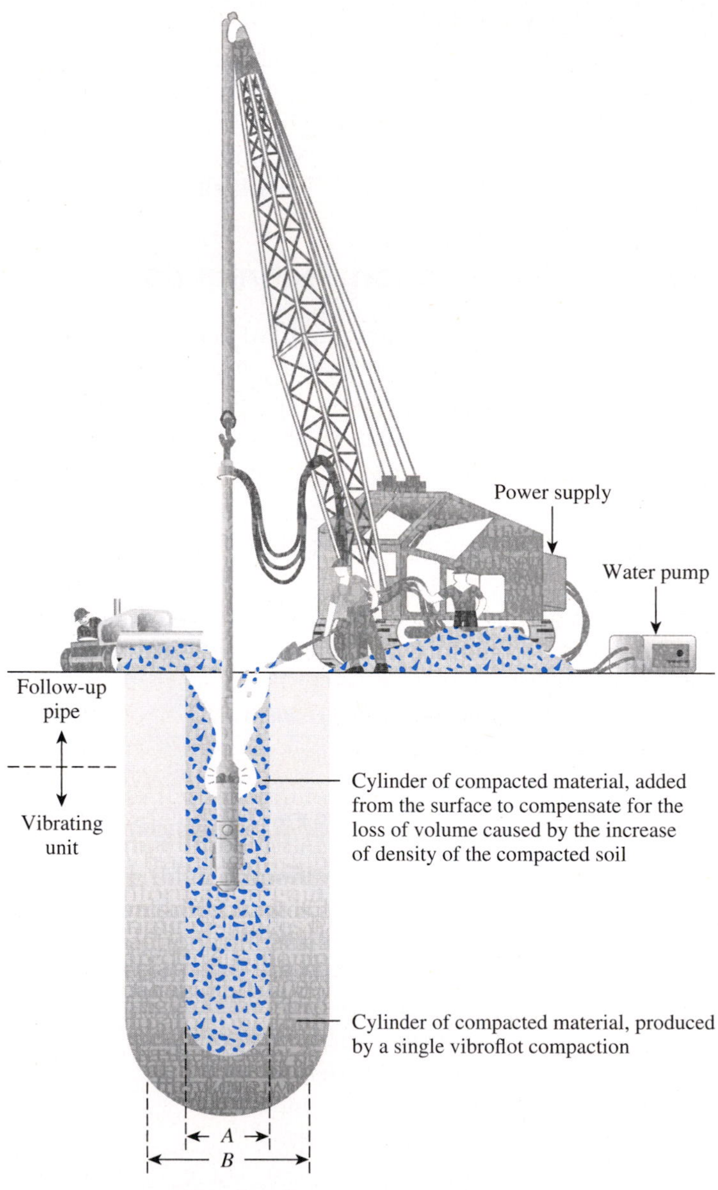

Figure 6.31 Vibroflotation unit (*After Brown, 1977. With permission from ASCE.*)

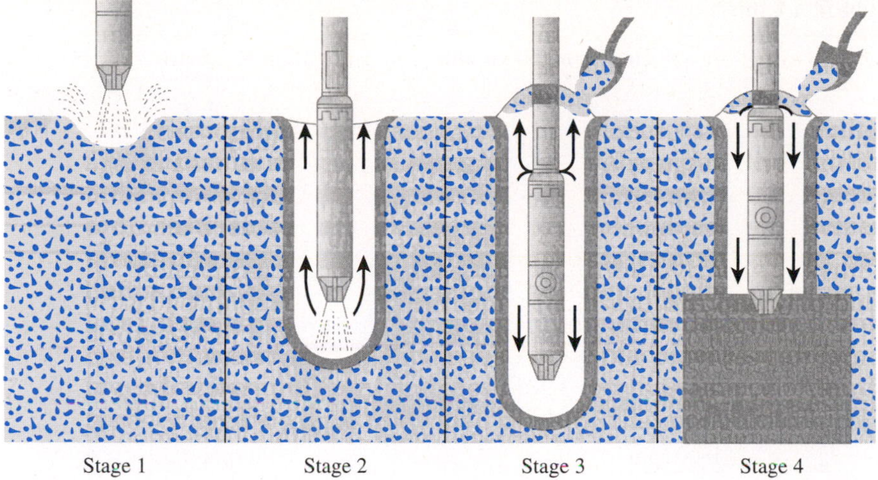

Figure 6.32 Compaction by vibroflotation process (*After Brown, 1977. With permission from ASCE.*)

develop a centrifugal force, which enables the vibrating unit to vibrate horizontally. There are openings at the bottom and top of the vibrating unit for water jets. The vibrating unit is attached to a follow-up pipe. Figure 6.31 shows the entire assembly of equipment necessary for conducting the field compaction.

The entire vibroflotation compaction process in the field can be divided into four stages (Figure 6.32):

Stage 1. The jet at the bottom of the Vibroflot is turned on and lowered into the ground.

Stage 2. The water jet creates a quick condition in the soil and it allows the vibrating unit to sink into the ground.

Stage 3. Granular material is poured from the top of the hole. The water from the lower jet is transferred to the jet at the top of the vibrating unit. This water carries the granular material down the hole.

Stage 4. The vibrating unit is gradually raised in about 0.3 m lifts and held vibrating for about 30 seconds at each lift. This process compacts the soil to the desired unit weight.

The details of various types of Vibroflot units used in the United States are given in Table 6.3. Note that 23 kW electric units have been used since the latter part of the 1940s. The 75 kW units were introduced in the early 1970s.

The zone of compaction around a single probe varies with the type of Vibroflot used. The cylindrical zone of compaction has a radius of about 2 m for a 23 kW unit. This radius can extend to about 3 m for a 75 kW unit.

Typical patterns of Vibroflot probe spacings are shown in Figure 6.33. Square and rectangular patterns generally are used to compact soil for isolated, shallow foundations. Equilateral triangular patterns generally are used to compact large areas. The capacity for successful densification of *in situ* soil depends on several factors, the most important of which is the grain-size distribution of the soil and the type

Table 6.3 Types of Vibroflot Units*

Motor type	75 kW electric and hydraulic	23 kW electric
a. Vibrating tip		
Length	2.1 m	1.86 m
Diameter	406 mm	381 mm
Weight	17.8 kN	17.8 kN
Maximum movement when full Centrifugal force	12.5 mm 160 kN	7.6 mm 89 kN
b. Eccentric		
Weight	1.2 kN	0.76 kN
Offset	38 mm	32 mm
Length	610 mm	390 mm
Speed	1800 rpm	1800 rpm
c. Pump		
Operating flow rate	0–1.6 m^3/min	0–0.6 m^3/min
Pressure	700–1050 kN/m^2	700–1050 kN/m^2
d. Lower follow-up pipe and extensions		
Diameter	305 mm	305 mm
Weight	3.65 kN/m	3.65 kN/m

**After Brown 1977. With permission from ASCE.*

of backfill used to fill the holes during the withdrawal period of the Vibroflot. The range of the grain-size distribution of *in situ* soil marked Zone 1 in Figure 6.34 is most suitable for compaction by vibroflotation. Soils that contain excessive amounts of fine sand and silt-size particles are difficult to compact, and considerable effort is needed to reach the proper relative density of compaction. Zone 2 in Figure 6.34 is the approximate lower limit of grain-size distribution for which compaction by vibroflotation is effective. Soil deposits whose grain-size distributions fall in Zone 3 contain appreciable amounts of gravel. For these soils, the rate of probe penetration may be slow and may prove uneconomical in the long run.

The grain-size distribution of the backfill material is an important factor that controls the rate of densification. Brown (1977) has defined a quantity called the *suitability number* for rating backfill as

$$S_N = 1.7\sqrt{\frac{3}{(D_{50})^2} + \frac{1}{(D_{20})^2} + \frac{1}{(D_{10})^2}} \tag{6.29}$$

where D_{50}, D_{20}, and D_{10} are the diameters (in mm) through which, respectively, 50, 20, and 10% of the material passes.

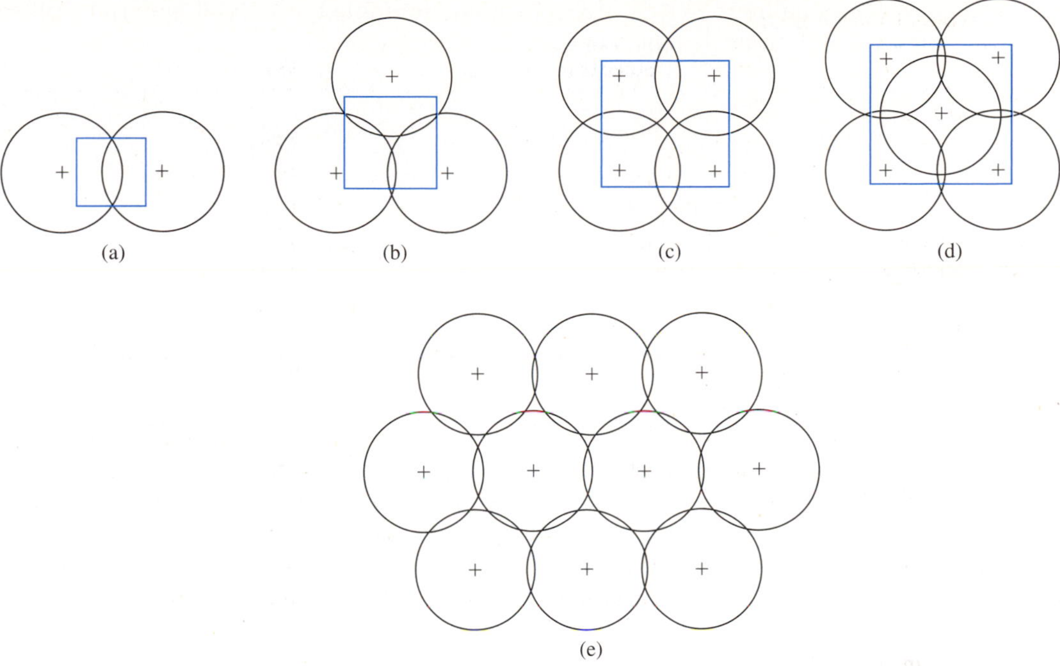

Figure 6.33 Typical patterns of Vibroflot probe spacings for a column foundation (a, b, c, and d) and for compaction over a large area (e)

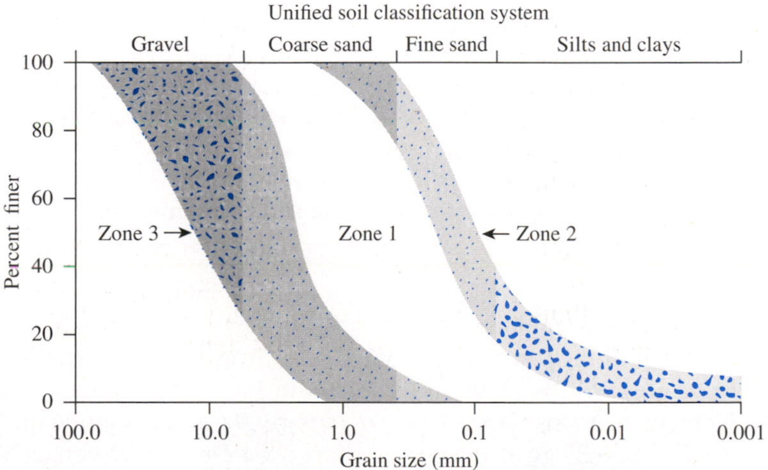

Figure 6.34 Effective range of grain-size distribution of soil for vibroflotation

The smaller the value of S_N, the more desirable the backfill material. Following is a backfill rating system proposed by Brown:

Range of S_N	Rating as backfill
0–10	Excellent
10–20	Good
20–30	Fair
30–50	Poor
>50	Unsuitable

Dynamic compaction

Dynamic compaction is a technique that has gained popularity in the United States for the densification of granular soil deposits. This process consists primarily of dropping a heavy weight repeatedly on the ground at regular intervals. The weight of the hammer used varies over a range of 80 to 360 kN, and the height of the hammer drop varies between 7.5 and 30.5 m. The stress waves generated by the hammer drops aid in the densification. The degree of compaction achieved at a given site depends on the following three factors:

1. Weight of hammer
2. Height of hammer drop
3. Spacing of locations at which the hammer is dropped

Figure 6.35 shows a dynamic compaction in progress. Leonards, Cutter, and Holtz (1980) suggested that the significant depth of influence for compaction can be approximated by using the equation

$$DI \simeq \left(\tfrac{1}{2}\right)\sqrt{W_H h} \tag{6.30}$$

where DI = significant depth of densification (m)
W_H = dropping weight (metric ton) (*Note*: 1 metric ton = 1000 kgf = 9.81 kN)
h = height of drop (m)

In 1992, Poran and Rodriguez suggested a rational method for conducting dynamic compaction for granular soils in the field. According to their method, for a hammer of width D having a weight W_H and a drop h, the approximate shape of the densified area will be of the type shown in Figure 6.36 (i.e., a semiprolate spheroid). Note that in this figure $b = DI$ (where DI is the significant depth of densification). Figure 6.37 gives the design chart for a/D and b/D versus $NW_H h/Ab$ (D = width of the hammer if not circular in cross section; A = area of cross section of the hammer; and N = number of required hammer drops). This method uses the following steps.

Step 1. Determine the required significant depth of densification, DI (= b).

Figure 6.35 Dynamic compaction in progress (*Courtesy of Khaled Sobhan, Florida Atlantic University, Boca Raton, Florida*)

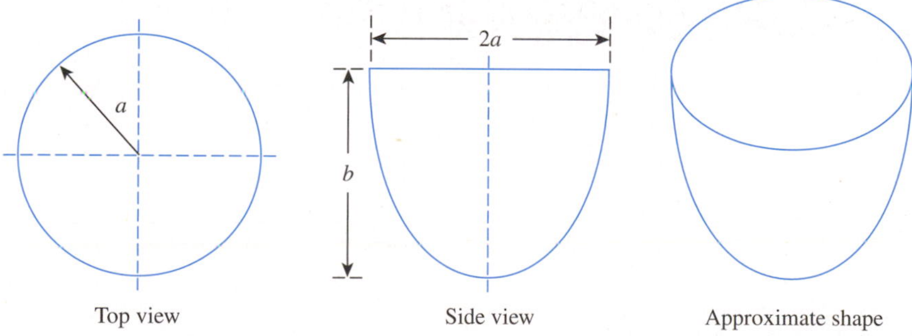

Figure 6.36 Approximate shape of the densified area due to dynamic compaction

Step 2. Determine the hammer weight (W_H), height of drop (h), dimensions of the cross section, and thus, the area A and the width D.

Step 3. Determine $DI/D = b/D$.

Step 4. Use Figure 6.37 and determine the magnitude of NW_Hh/Ab for the value of b/D obtained in step 3.

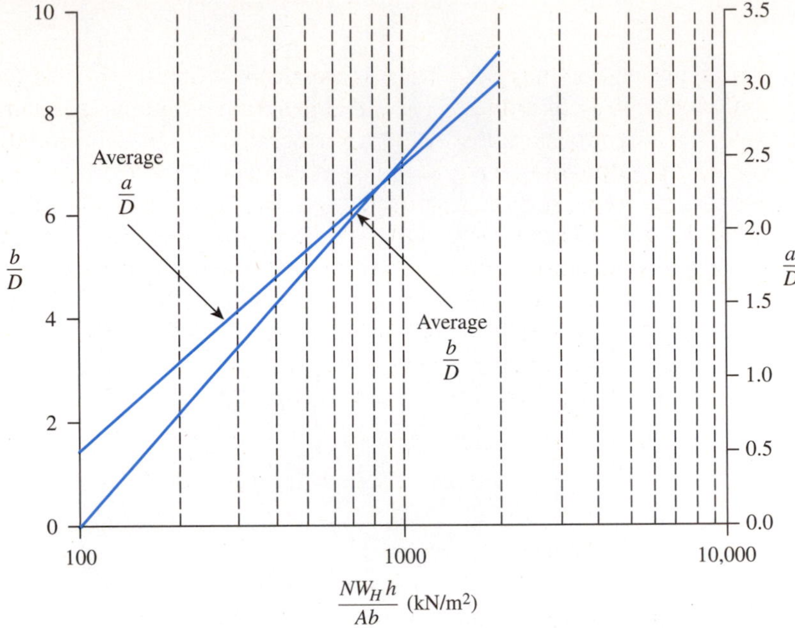

Figure 6.37 Poran and Rodriguez chart for a/D, b/D versus $NW_H h/Ab$

Step 5. Since the magnitudes of W_H, h, A, and b are known (or assumed) from step 2, the number of hammer drops can be estimated from the value of $NW_H h/Ab$ obtained from step 4.

Step 6. With known values of $NW_H h/Ab$, determine a/D and thus a from Figure 6.37.

Step 7. The grid spacing, S_g, for dynamic compaction may now be assumed to be equal to or somewhat less than a. (See Figure 6.38.)

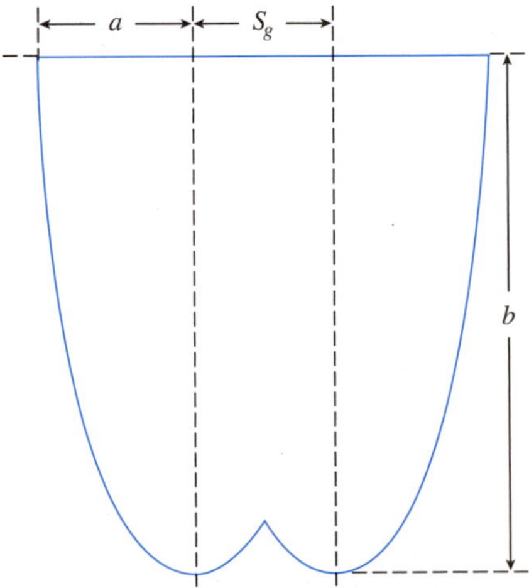

Figure 6.38 Approximate grid spacing for dynamic compaction

Blasting

Blasting is a technique that has been used successfully in many projects (Mitchell, 1970) for the densification of granular soils. The general soil grain sizes suitable for compaction by blasting are the same as those for compaction by vibroflotation. The process involves the detonation of explosive charges, such as 60% dynamite at a certain depth below the ground surface in saturated soil. The lateral spacing of the charges varies from about 3 to 9 m. Three to five successful detonations are usually necessary to achieve the desired compaction. Compaction (up to a relative density of about 80%) up to a depth of about 18 m over a large area can easily be achieved by using this process. Usually, the explosive charges are placed at a depth of about two-thirds of the thickness of the soil layer desired to be compacted. The sphere of influence of compaction by a 60% dynamite charge can be given as follows (Mitchell, 1970):

$$r = \sqrt{\frac{W_{\text{EX}}}{C}} \tag{6.31}$$

where r = sphere of influence
W_{EX} = weight of explosive—60% dynamite
C = 0.0122 when W_{EX} is in kg and r is in m

Figure 6.39 shows the test results of soil densification by blasting in an area measuring 15 m by 9 m (Mitchell, 1970). For these tests, twenty 2.09 kg (4.6 lb) charges of Gelamite No. 1 (Hercules Powder Company, Wilmington, Delaware) were used.

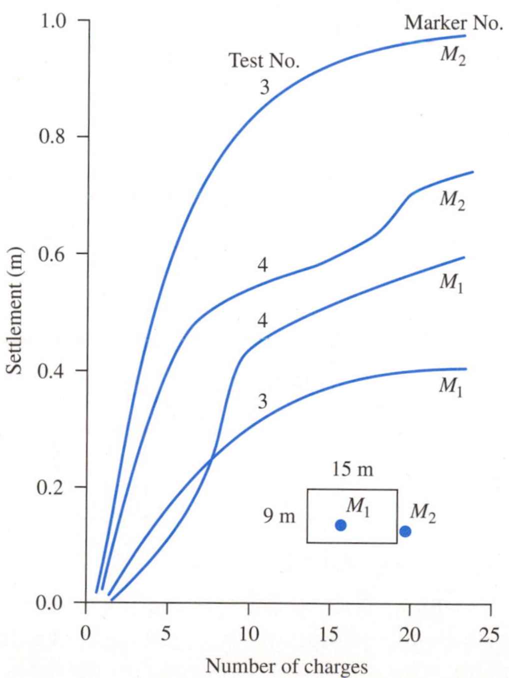

Figure 6.39 Ground settlement as a function of number of explosive charges

Example 6.11

Following are the details for the backfill material used in a vibroflotation project:

- $D_{10} = 0.36$ mm
- $D_{20} = 0.52$ mm
- $D_{50} = 1.42$ mm

Determine the suitability number S_N. What would be its rating as a backfill material?

Solution

From Eq. (6.29),

$$S_N = 1.7\sqrt{\frac{3}{(D_{50})^2} + \frac{1}{(D_{20})^2} + \frac{1}{(D_{10})^2}}$$

$$= 1.7\sqrt{\frac{3}{(1.42)^2} + \frac{1}{(0.52)^2} + \frac{1}{(0.36)^2}}$$

$$= \mathbf{6.1}$$

Rating: **Excellent**

Example 6.12

Consider the case of a dynamic compaction in the field. Given:

- Weight of hammer = 11.34 metric ton
- Height of drop = 10.67 m

Determine the significant depth of densification, DI.

Solution

From Eq. (6.31),

$$DI = 0.5\sqrt{W_H h} = (0.5)(11.34 \times 10.67)^{0.5} \simeq \mathbf{5.5\ m}$$

6.14 Summary

In this chapter, we have discussed the following:

- Standard and modified Proctor compaction tests are conducted in the laboratory to determine the maximum dry unit weight of compaction $[\gamma_{d(\max)}]$ and optimum moisture content (w_{opt}) (Sections 6.3 and 6.5).

- $\gamma_{d(max)}$ and w_{opt} are functions of the energy of compaction E.
- Several empirical relations have been presented to estimate $\gamma_{d(max)}$ and w_{opt} for cohesionless and cohesive soils (Section 6.6). Also included in this section is an empirical relationship to estimate the relative density of compaction (D_r) with known median grain size (D_{50}) and energy of compaction (E).
- For a given energy of compaction (E) in a cohesive soil, the hydraulic conductivity and unconfined compression strength are functions of molding moisture content.
- Field compaction is generally carried out by rollers such as smooth-wheel, rubber-tired, sheepsfoot, and vibratory (Section 6.9).
- Control tests to determine the quality of field compaction can be done by using the sand cone method, rubber balloon method, and nuclear method.
- Vibroflotation, dynamic compaction, and blasting are special techniques used for large-scale compaction in the field (Section 6.13).

Laboratory standard and modified Proctor compaction tests described in this chapter are essentially for *impact* or *dynamic* compaction of soil; however, in the laboratory, *static compaction* and *kneading compaction* also can be used. It is important to realize that the compaction of clayey soils achieved by rollers in the field is essentially the kneading type. The relationships of dry unit weight (γ_d) and moisture content (w) obtained by dynamic and kneading compaction are not the same. Proctor compaction test results obtained in the laboratory are used primarily to determine whether the roller compaction in the field is sufficient. The structures of compacted cohesive soil at a similar dry unit weight obtained by dynamic and kneading compaction may be different. This difference, in turn, affects physical properties, such as hydraulic conductivity, compressibility, and strength.

For most fill operations, the final selection of the borrow site depends on such factors as the soil type and the cost of excavation and hauling.

Problems

6.1 The maximum dry unit weight and the optimum moisture content of a soil are 16.8 kN/m^3 and 17%, respectively. If G_s is 2.73, what is the degree of saturation at optimum moisture content?

6.2 For a soil with G_s = 2.7, calculate and plot the variation of dry density (in kg/m^3) at w = 8, 12, 16, and 20% and for the degree of saturation at S = 55, 70, 85, and 100%, respectively.

6.3 Calculate the zero-air-void unit weights (kN/m^3) for a soil with G_s = 2.66 at moisture contents of 7, 11, 15, 19, and 23%.

6.4 The results of a standard Proctor test are given in the following table.

a. Determine the maximum dry unit weight of compaction and the optimum moisture content. Given: Mold volume = 943.3 cm^3.

b. Determine the void ratio and the degree of saturation at the optimum moisture content. Given: $G_s = 2.69$.

Trial no.	Mass of moist soil in the mold (kg)	Moisture content (%)
1	1.68	8.6
2	1.88	10.6
3	2.13	12.5
4	2.10	14.9
5	1.83	16.7
6	1.65	18.3

6.5 The laboratory test in Problem 6.4 is used to develop field compaction specification for a highway project. A field unit weight determination during the construction revealed that the *in situ* moist unit weight is 19.46 kN/m³ and the moisture content is 13.7%. Determine the relative compaction in the field.

6.6 Repeat Problem 6.4 with the following data (use $G_s = 2.73$):

Trial no.	Mass of moist soil in the mold (g)	Moisture content (%)
1	1668	7
2	1723	8.9
3	1800	12.3
4	1850	14.8
5	1873	17.3
6	1868	18.5

6.7 Results of a standard Proctor compaction test on a silty sand are shown in Figure 6.40.

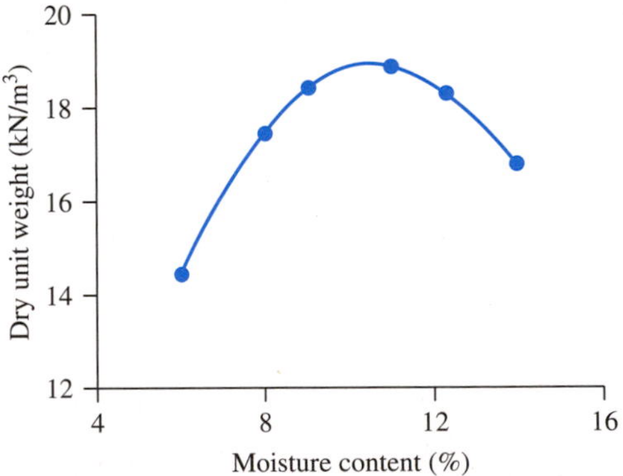

Figure 6.40

a. Find the maximum dry unit weight and optimum moisture content.
b. What is the moist unit weight at optimum moisture content?

c. What is the degree of saturation at optimum moisture content? Given: $G_s = 2.69$.

d. If the required field dry unit weight is 18.5 kN/m^3, what is the relative compaction?

e. What should be the range of compaction moisture contents in the field to achieve the above relative compaction?

f. If the minimum and maximum void ratios are 0.31 and 0.82, respectively, what is the relative density of compaction in the field?

6.8 A standard Proctor test was conducted on a silty clay soil collected from a proposed construction site. The results are shown in the following table.

Trial no.	Mass of moist soil in the mold (g)	Moisture content (%)
1	1689	12.7
2	1752	15.0
3	1800	17.8
4	1845	20.6
5	1844	23.8

a. Determine the maximum dry density (kg/m^3) of compaction and the optimum moisture content. Given: mold volume = 943.3 cm^3.

b. If specification calls for 99% relative compaction in the field, what would be the field dry density and the range of acceptable moisture content?

6.9 Refer to the silty clay soil at the construction site in Problem 6.8. As part of a quality control program, the field inspection engineer conducted a sand cone test to determine the field density. The following data were recorded using the sand cone method.

- Calibrated dry density of Ottawa sand = 1667 kg/m^3
- Mass of Ottawa sand to fill the cone = 0.117 kg
- Mass of jar + cone + sand (before use) = 6.1 kg
- Mass of jar + cone + sand (after use) = 2.83 kg
- Mass of moist soil from hole = 3.35 kg
- Moisture content of moist soil = 16.1%

a. Determine the dry unit weight of compaction in the field.

b. What is the relative compaction in the field?

c. Was the compaction specification stated in Problem 6.8 met?

6.10 The *in situ* moist unit weight of a soil is 16.6 kN/m^3, and the moisture content is 19%. The specific gravity of soil solids is 2.69. This soil is to be excavated and transported to a construction site for use in a compacted fill. If the specification calls for the soil to be compacted to a minimum dry unit weight of 19.5 kN/m^3 at the same moisture content of 19%, how many cubic meters of soil from the excavation site are needed to produce 2500 m^3 of compacted fill? How many 20-ton ($\approx$ 18,144 kgf) truckloads are needed to transport the excavated soil?

6.11 A proposed embankment fill requires 7500 m^3 of compacted soil. The void ratio of the compacted fill is specified as 0.7. Soil can be transported from one of the four borrow pits, as described in the following table. The void ratio, specific gravity of soil solids, and the cost per cubic meter for moving the soil to the proposed construction site are provided in the table.

a. Determine the volume of each borrow pit soil required to meet the specification of the embankment site.

b. Make the necessary calculations to select the borrow pit which would be most cost-effective.

Borrow pit	Void ratio	G_s	Cost (\$/m^3)
I	0.85	2.66	11
II	0.92	2.69	8
III	1.21	2.71	9
IV	0.89	2.73	10

6.12 The maximum and minimum dry unit weights of a sand were determined in the laboratory to be 16.9 kN/m^3 and 14.2 kN/m^3, respectively. What is the relative compaction in the field if the relative density is 82%?

6.13 The relative compaction of a silty sand in the field is 94%. Given that $\gamma_{d(\max)}$ = 17 kN/m^3 and $\gamma_{d(\min)}$ = 13.8 kN/m^3, determine the dry unit weight in the field and the relative density of compaction.

6.14 The relative compaction of a clayey sand in the field is 90%. The maximum and minimum dry unit weights of the sand are 18 kN/m^3 and 14.6 kN/m^3, respectively. Determine:

a. Dry unit weight in the field

b. Relative density of compaction

c. Moist unit weight at a moisture content of 18%

6.15 Refer to the field compaction of the clayey sand in Problem 6.14. If the soil layer before compaction had a void ratio of 0.97 and a thickness of 1.75 m, what would be the final thickness after compaction? Assume G_s = 2.67

6.16 For a dynamic compaction test, the weight of the hammer was 16 metric ton and the height of the hammer drop was 11 m. Estimate the significant depth of densification.

6.17 Vibroflotation is being considered for *in situ* densification of a thick deposit of granular soils at a particular site. The results of the sieve analysis of the proposed backfill material is shown in Figure 6.41.

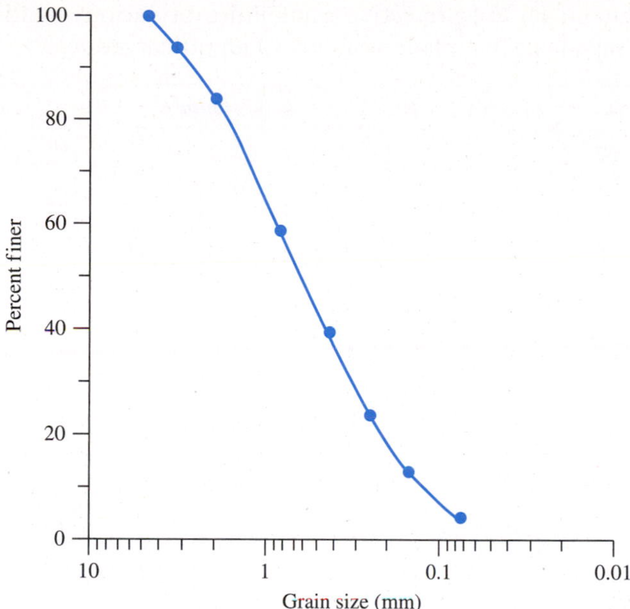

Figure 6.41

Determine the suitability number, S_N, and rate it as a backfill material.

Critical Thinking Problem

6.C.1 Since laboratory or field experiments are generally expensive and time consuming, geotechnical engineers often have to rely on empirical relationships to predict design parameters. Section 6.6 presents such relationships for predicting optimum moisture content and maximum dry unit weight. Let us use some of these equations and compare our results with known experimental data. The following table presents the results from laboratory compaction tests conducted on a wide range of fine-grained soils using various compactive efforts (E). Based on the soil data given in the table, determine the optimum moisture content and maximum dry unit weight using the empirical relationships presented in Section 6.6.

a. Use the Osman et al. (2008) method [Eqs. (6.15) through (6.18)].

b. Use the Gurtug and Sridharan (2004) method [Eqs. (6.13) and (6.14)].

c. Use the Matteo et al. (2009) method [Eqs. (6.19) and (6.20)].

d. Plot the calculated w_{opt} against the experimental w_{opt}, and the calculated $\gamma_{d(max)}$ with the experimental $\gamma_{d(max)}$. Draw a 45° *line of equality* on each plot.

e. Comment on the predictive capabilities of various methods. What can you say about the inherent nature of empirical models?

Soil	G_s	LL (%)	PL (%)	E (kN-m/m³)	w_{opt} (%)	$\gamma_{d(max)}$ (kN/m³)
1[a]	2.67	17	16	2700[b]	8	20.72
				600[c]	10	19.62
				354[d]	10	19.29
2[a]	2.73	68	21	2700	20	16.00
				600	28	13.80
				354	31	13.02
3	2.68	56	14	2700	15	18.25
				1300[e]	16	17.5
				600	17	16.5
				275[f]	19	15.75
4	2.68	66	27	600	21	15.89
5	2.67	25	21	600	18	16.18
6	2.71	35	22	600	17	16.87
7	2.69	23	18	600	12	18.63
8	2.72	29	19	600	15	17.65

Note:

[a] Tschebotarioff (1951)
[b] Modified Proctor test
[c] Standard Proctor test
[d] Standard Proctor mold and hammer; drop: 305 mm; layers: 3; blows/layer: 15
[e] Modified Proctor mold and hammer; drop: 457 mm; layers: 5; blows/layer: 26
[f] Modified Proctor mold; standard Proctor hammer; drop: 305 mm; layers: 3; blows/layer: 25

References

American Association of State Highway and Transportation Officials (1982). *AASHTO Materials, Part II*, Washington, D.C.

American Society for Testing and Materials (2014). *Annual Book of ASTM Standards*, Vol 04.08, West Conshohocken, Pa.

Brown, E. (1977). "Vibroflotation Compaction of Cohesionless Soils," *Journal of the Geo technical Engineering Division*, ASCE, Vol. 103, No. GT12, 1437–1451.

D' Appolonia, D. J., Whitman, R. V., and D' Appolonia, E. D. (1969). "Sand Compaction with Vibratory Rollers," *Journal of the Soil Mechanics and Foundations Division*, ASCE, Vol. 95, No. SM1, 263–284.

Gurtug, Y., and Sridharan, A. (2004). "Compaction Behaviour and Prediction of Its Characteristics of Fine Grained Soils with Particular Reference to Compaction Energy," *Soils and Foundations*, Vol. 44, No. 5, 27–36.

Highway Research Board (1962). *Factors influencing Compaction Test Results*, Bulletin 319, Washington, D.C.

Holtz, R. D., and Kovacs, W. D. (1981). *An Introduction to Geotechnical Engineering*, Prentice-Hall, Englewood Cliffs, N.J.

Johnson, A. W., and Sallberg, J. R. (1960). "Factors That Influence Field Compaction of Soil," Highway Research Board, *Bulletin No. 272*.

Johnson, A. W,. and Sallberg, J.R. (1962). "Factors Influencing Compaction Test Results," *Bulletin 319*, Highway Research Board, National Research Council, 146pp.

Lambe, T. W. (1958a). "The Structure of Compacted Clay," *Journal of the Soil Mechanics and Foundations Division*, ASCE, Vol. 84, No. SM2, 1654–1 to 1654–35.

Lambe, T. W. (1958b). "The Engineering Behavior of Compacted Clay," *Journal of the Soil Mechanics and Foundations Division*, ASCE, Vol. 84, No. SM2, 1655–1 to 1655–35.

Lee, P. Y., and Suedkamp, R. J. (1972). "Characteristics of Irregularly Shaped Compaction Curves of Soils," *Highway Research Record No. 381*, National Academy of Sciences, Washington, D.C., 1–9.

Leonards, G. A., Cutter, W. A., and Holtz, R. D. (1980). "Dynamic Compaction of Granular Soils," *Journal of the Geotechnical Engineering Division*, ASCE, Vol. 106, No. GT1, 35–44.

Matteo, L. D., Bigotti, F., and Ricco, R. (2009). "Best-Fit Model to Estimate Proctor Properties of Compacted Soil," *Journal of Geotechnical and Geoenvironmental Engineering*, ASCE, Vol. 135, No. 7, 992–996.

Mitchell, J. K. (1970). "In-Place Treatment of Foundation Soils," *Journal of the Soil Mechanics and Foundations Division*, ASCE, Vol. 96, No. SM1, 73–110.

Mujtaba, H., Farooq, K., Sivakugan, N., and Das, B. M. (2013). "Correlation between Gradation Parameters and Compaction Characteristics of Sandy Soil," *International Journal of Geotechnical Engineering*, Maney Publishing, U.K,. Vol. 7, No. 4, 395–401.

Omar, M., Abdallah, S., Basma, A., and Barakat, S. (2003). "Compaction Characteristics of Granular Soils in United Arab Emirates," *Geotechnical and Geological Engineering*, Vol. 21, No. 3, 283–295.

Osman, S., Togrol, E., and Kayadelen, C. (2008). "Estimating Compaction Behavior of Fine-Grained Soils Based on Compaction Energy," *Canadian Geotechnical Journal*, Vol. 45, No. 6, 877–887.

Patra, C. R., Sivakugan, N., Das, B. M., and Rout, S. K. (2010). "Correlation of Relative Density of Clean Sand with Median Grain Size and Compaction Energy," *International Journal of Geotechnical Engineering*, Vol. 4, No. 2, 196–203.

Poran, C. J., and Rodriguez, J. A. (1992). "Design of Dynamic Compaction," *Canadian Geotechnical Journal*, Vol. 2, No. 5, 796–802.

Proctor, R. R. (1933). "Design and Construction of Rolled Earth Dams," *Engineering News Record*, Vol. 3, 245–248, 286–289, 348–351, 372–376.

Seed, H. B. (1964). Lecture Notes, CE 271, Seepage and Earth Dam Design, University of California, Berkeley.

Seed, H. B., and Chan, C. K. (1959). "Structure and Strength Characteristics of Compacted Clays," *Journal of the Soil Mechanics and Foundations Division*, ASCE, Vol. 85, No. SM5, 87–128.

Sowers, G. F. (1979). *Introductory Soil Mechanics and Foundations: Geotechnical Engineering*, Macmillan, New York.

Tschebotarioff, G. P. (1951). *Soil Mechanics, Foundations, and Earth Structures*, McGraw-Hill, New York.

CHAPTER 7

Permeability

7.1 Introduction

Soils are permeable due to the existence of interconnected voids through which water can flow from points of high energy to points of low energy. The study of the flow of water through permeable soil media is important in soil mechanics. It is necessary for estimating the quantity of underground seepage under various hydraulic conditions, for investigating problems involving the pumping of water for underground construction, and for making stability analyses of earth dams and earth-retaining structures that are subject to seepage forces.

One of the major physical parameters of a soil that controls the rate of seepage through it is *hydraulic conductivity*, otherwise known as the *coefficient of permeability*. In this chapter, we will study the following:

- Definition of hydraulic conductivity and its magnitude in various soils
- Laboratory determination of hydraulic conductivity
- Empirical relationship to estimate hydraulic conductivity
- Equivalent hydraulic conductivity in stratified soil based on the direction of the flow of water
- Hydraulic conductivity determination from field tests

7.2 Bernoulli's Equation

From fluid mechanics, we know that, according to Bernoulli's equation, the total head at a point in water under motion can be given by the sum of the pressure, velocity, and elevation heads, or

$$h = \frac{u}{\gamma_w} + \frac{v^2}{2g} + Z \qquad (7.1)$$

(Pressure head: $\frac{u}{\gamma_w}$; Velocity head: $\frac{v^2}{2g}$; Elevation head: Z)

where h = total head
u = pressure
v = velocity
g = acceleration due to gravity
γ_w = unit weight of water

Note that the elevation head, Z, is the vertical distance of a given point above or below a datum plane. The pressure head is the water pressure, u, at that point divided by the unit weight of water, γ_w.

If Bernoulli's equation is applied to the flow of water through a porous soil medium, the term containing the velocity head can be neglected because the seepage velocity is small, and the total head at any point can be adequately represented by

$$h = \frac{u}{\gamma_w} + Z \qquad (7.2)$$

Figure 7.1 shows the relationship among pressure, elevation, and total heads for the flow of water through soil. Open standpipes called *piezometers* are installed at points A and B. The levels to which water rises in the piezometer tubes situated at points A and B are known as the *piezometric levels* of points A and B, respectively. The pressure head at a point is the height of the vertical column of water in the piezometer installed at that point.

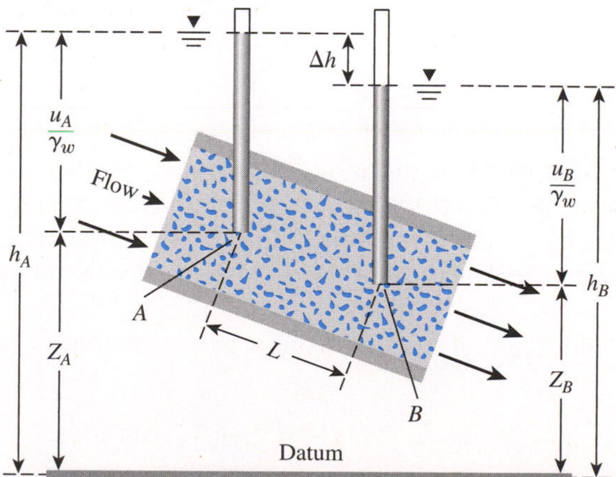

Figure 7.1 Pressure, elevation, and total heads for flow of water through soil

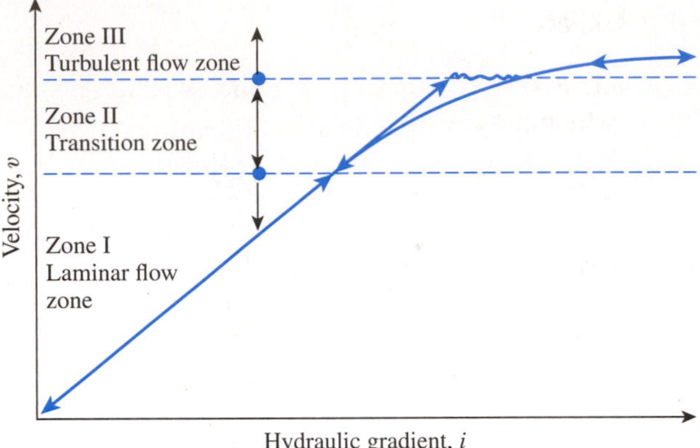

Figure 7.2 Nature of variation of v with hydraulic gradient, i

The loss of head between two points, A and B, can be given by

$$\Delta h = h_A - h_B = \left(\frac{u_A}{\gamma_w} + Z_A\right) - \left(\frac{u_B}{\gamma_w} + Z_B\right) \tag{7.3}$$

The head loss, Δh, can be expressed in a nondimensional form as

$$i = \frac{\Delta h}{L} \tag{7.4}$$

where i = hydraulic gradient
L = distance between points A and B—that is, the length of flow over which the loss of head occurred

In general, the variation of the velocity v with the hydraulic gradient i is as shown in Figure 7.2. This figure is divided into three zones:

1. Laminar flow zone (Zone I)
2. Transition zone (Zone II)
3. Turbulent flow zone (Zone III)

When the hydraulic gradient is increased gradually, the flow remains laminar in Zones I and II, and the velocity, v, bears a linear relationship to the hydraulic gradient. At a higher hydraulic gradient, the flow becomes turbulent (Zone III). When the hydraulic gradient is decreased, laminar flow conditions exist only in Zone I.

In most soils, the flow of water through the void spaces can be considered laminar; thus,

$$v \propto i \tag{7.5}$$

In fractured rock, stones, gravels, and very coarse sands, turbulent flow conditions may exist, and Eq. (7.5) may not be valid.

7.3 Darcy's Law

In 1856, Darcy published a simple equation for the discharge velocity of water through saturated soils, which may be expressed as

$$v = ki \tag{7.6}$$

where v = *discharge velocity*, which is the quanity of water flowing in unit time through a unit gross cross-sectional area of soil at right angles to the direction of flow

k = hydraulic conductivity (otherwise known as the coefficient of permeability)

This equation was based primarily on Darcy's observations about the flow of water through clean sands. Note that Eq. (7.6) is similar to Eq. (7.5); both are valid for laminar flow conditions and applicable for a wide range of soils.

In Eq. (7.6), v is the discharge velocity of water based on the gross cross-sectional area of the soil. However, the actual velocity of water (that is, the seepage velocity) through the void spaces is greater than v. A relationship between the discharge velocity and the seepage velocity can be derived by referring to Figure 7.3, which shows a soil of length L with a gross cross-sectional area A. If the quantity of water flowing through the soil in unit time is q, then

$$q = vA = A_v v_s \tag{7.7}$$

where v_s = seepage velocity

A_v = area of void in the cross section of the specimen

However,

$$A = A_v + A_s \tag{7.8}$$

where A_s = area of soil solids in the cross section of the specimen.

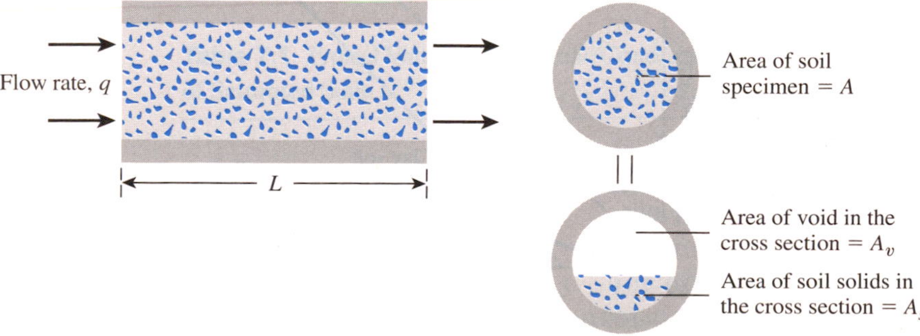

Figure 7.3 Derivation of Eq. (7.10)

Solution

From Eq. (7.31),

$$\frac{k_1}{k_2} = \frac{\frac{e_1^3}{1+e_1}}{\frac{e_2^3}{1+e_2}}$$

$$\frac{0.047}{k_2} = \frac{\frac{(0.8)^3}{1+0.8}}{\frac{(0.5)^3}{1+0.5}}$$

$$k_2 = \mathbf{0.014\ cm/sec}$$

Example 7.8

The grain-size distribution curve for a sand is shown in Figure 7.11. Estimate the hydraulic conductivity using Eq. (7.30). Given: The void ratio of the sand is 0.6. Use $SF = 7$.

Solution

From Figure 7.11, the following table can be prepared.

Sieve no.	Sieve opening (cm)	Percent passing	Fraction of particles between two consecutive sieves (%)
30	0.06	100	
			4
40	0.0425	96	
			12
60	0.02	84	
			34
100	0.015	50	
			50
200	0.0075	0	

For fraction between Nos. 30 and 40 sieves,

$$\frac{f_i}{D_{li}^{0.404} \times D_{si}^{0.595}} = \frac{4}{(0.06)^{0.404} \times (0.0425)^{0.595}} = 81.62$$

For fraction between Nos. 40 and 60 sieves,

$$\frac{f_i}{D_{li}^{0.404} \times D_{si}^{0.595}} = \frac{12}{(0.0425)^{0.404} \times (0.02)^{0.595}} = 440.76$$

Similarly, for fraction between Nos. 60 and 100 sieves,

$$\frac{f_i}{D_{li}^{0.404} \times D_{si}^{0.595}} = \frac{34}{(0.02)^{0.404} \times (0.015)^{0.595}} = 2009.5$$

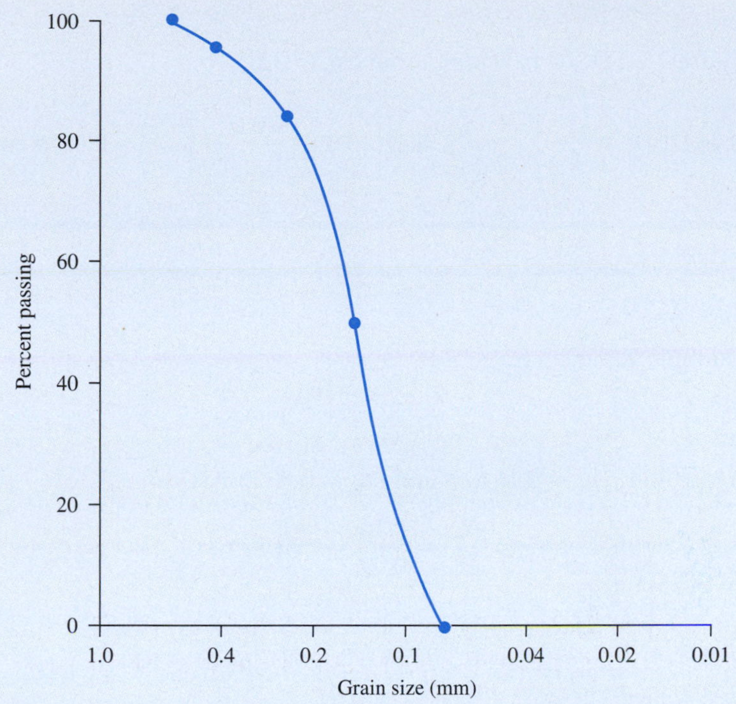

Figure 7.11

And, for between Nos. 100 and 200 sieves,

$$\frac{f_i}{D_{li}^{0.404} \times D_{si}^{0.595}} = \frac{50}{(0.015)^{0.404} \times (0.0075)^{0.595}} = 5013.8$$

$$\frac{100\%}{\sum \frac{f_i}{D_{li}^{0.404} \times D_{si}^{0.595}}} = \frac{100}{81.62 + 440.76 + 2009.5 + 5013.8} \approx 0.0133$$

From Eq. (7.30),

$$k = (1.99 \times 10^4)(0.0133)^2 \left(\frac{1}{7}\right)^2 \left(\frac{0.6^3}{1 + 0.6}\right) = \mathbf{0.0097\ cm/s}$$

Example 7.9

Solve Example 7.8 using Eq. (7.32).

Solution

From Figure 7.11, $D_{10} = 0.09$ mm. From Eq. (7.32),

$$k = 2.4622\left[D_{10}^2 \frac{e^3}{1+e}\right]^{0.7825} = 2.4622\left[(0.09)^2 \frac{0.6^3}{1+0.6}\right]^{0.7825} = \mathbf{0.0119\ cm/sec}$$

Example 7.10

Solve Example 7.8 using Eq. (7.34).

Solution

From Figure 7.11, $D_{60} = 0.16$ mm and $D_{10} = 0.09$ mm. Thus,

$$C_u = \frac{D_{60}}{D_{10}} = \frac{0.16}{0.09} = 1.78$$

From Eq. (7.34),

$$k = 35\left(\frac{e^3}{1+e}\right)C_u^{0.6}(D_{10})^{2.32} = 35\left(\frac{0.6^3}{1+0.6}\right)(1.78)^{0.6}(0.09)^{2.32} = \mathbf{0.025\ cm/sec}$$

7.7 Relationships for Hydraulic Conductivity—Cohesive Soils

The Kozeny–Carman equation [Eq. (7.24)] has been used in the past to see if it will hold good for cohesive soil. Olsen (1961) conducted hydraulic conductivity tests on sodium illite and compared the results with Eq. (7.24). This comparison is shown in Figure 7.12. The marked degrees of variation between the theoretical and experimental values arise from several factors, including deviations from Darcy's law, high viscosity of the pore water, and unequal pore sizes.

Taylor (1948) proposed a linear relationship between the logarithm of k and the void ratio as

$$\log k = \log k_o - \frac{e_o - e}{C_k} \tag{7.36}$$

where k_o = in situ hydraulic conductivity at a void ratio e_o
k = hydraulic conductivity at a void ratio e
C_k = hydraulic conductivity change index

The preceding equation is a good correlation for e_o less than about 2.5. In this equation, the value of C_k may be taken to be about $0.5e_o$ (see Figure 7.13).

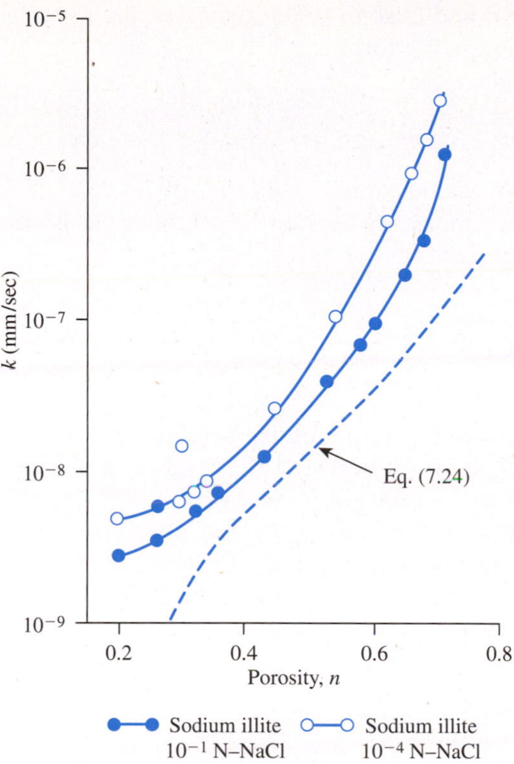

Figure 7.12 Coefficient of permeability for sodium illite (*Based on Olsen, 1961*)

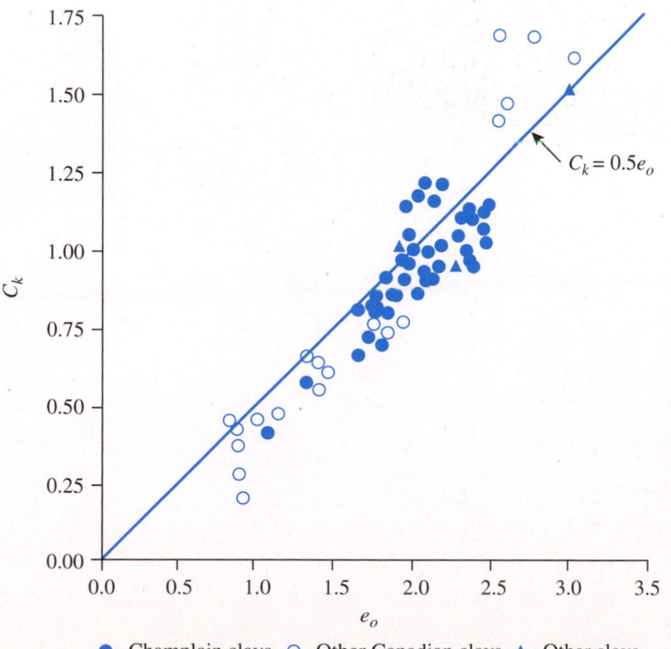

Figure 7.13 Basis of the relationship as given in Eq. (7.36) (*Tavenas, F., Jean, P., Leblond, P., and Leroueil, S. (1983). "The Permeabilty of Natural Soft Clays. Part II: Permeability Characteristics," Canadian Geotechnical Journal, Vol. 20, No. 4, pp. 645–660. Figure 17, p. 658. © 2008 Canadian Science Publishing or its licensors. Reproduced with permission.*)

For a wide range of void ratio, Mesri and Olson (1971) suggested the use of a linear relationship between log k and log e in the form

$$\log k = A' \log e + B' \tag{7.37}$$

where A' and B' are experimentally derived constants.

Samarasinghe et al. (1982) conducted laboratory tests on New Liskeard clay and proposed that, for normally consolidated clays,

$$k = C\left(\frac{e^n}{1 + e}\right) \tag{7.38}$$

where C and n are constants to be determined experimentally (see Figure 7.14).

Tavenas et al. (1983) also gave a correlation between the void ratio and the hydraulic conductivity of clayey soil. This correlation is shown in Figure 7.15. An important point to note, however, is that in Figure 7.15, *PI*, the plasticity index, and *CF*, the clay-size fraction in the soil, are in *fraction* (decimal) form. One should keep in mind, however, that any empirical relationship of this type is for estimation only, because the magnitude of k is a highly variable parameter and depends on several factors.

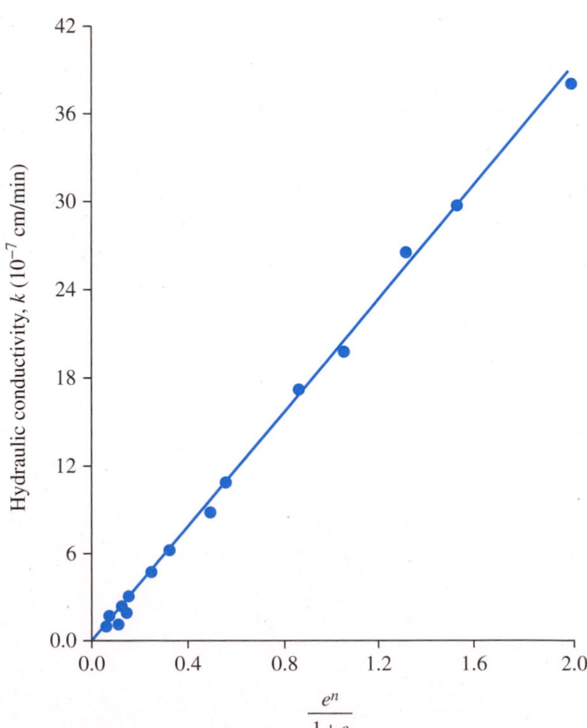

Figure 7.14 Variation of k with $e^n/(1 + e)$ for normally consolidated New Liskeard clay *(After Samarasinghe, Huang, and Drnevich, 1982. With permission from ASCE)*

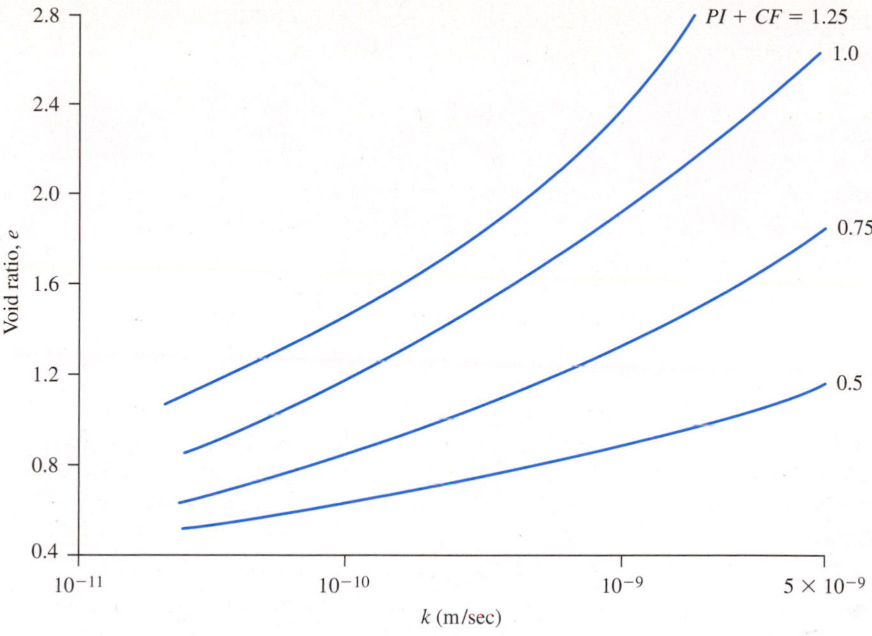

Figure 7.15 Variation of void ratio with hydraulic conductivity of clayey soils (*Based on Tavenas et al, 1983*)

Example 7.11

For a normally consolidated clay soil, the following values are given:

Void ratio	k (cm/sec)
1.1	0.302×10^{-7}
0.9	0.12×10^{-7}

Estimate the hydraulic conductivity of the clay at a void ratio of 0.75. Use Eq. (7.38).

Solution

From Eq. (7.38),

$$k = C\left(\frac{e^n}{1 + e}\right)$$

$$\frac{k_1}{k_2} = \frac{\left(\dfrac{e_1^n}{1 + e_1}\right)}{\left(\dfrac{e_2^n}{1 + e_2}\right)}$$

$$\frac{0.302 \times 10^{-7}}{0.12 \times 10^{-7}} = \frac{\dfrac{(1.1)^n}{1 + 1.1}}{\dfrac{(0.9)^n}{1 + 0.9}}$$

$$2.517 = \left(\frac{1.9}{2.1}\right)\left(\frac{1.1}{0.9}\right)^n$$

$$2.782 = (1.222)^n$$

$$n = \frac{\log (2.782)}{\log (1.222)} = \frac{0.444}{0.087} = 5.1$$

so

$$k = C\left(\frac{e^{5.1}}{1 + e}\right)$$

To find C,

$$0.302 \times 10^{-7} = C\left[\frac{(1.1)^{5.1}}{1 + 1.1}\right] = \left(\frac{1.626}{2.1}\right)C$$

$$C = \frac{(0.302 \times 10^{-7})(2.1)}{1.626} = 0.39 \times 10^{-7}$$

Hence,

$$k = (0.39 \times 10^{-7} \text{ cm/sec})\left(\frac{e^n}{1 + e}\right)$$

At a void ratio of 0.75,

$$k = (0.39 \times 10^{-7})\left(\frac{0.75^{5.1}}{1 + 0.75}\right) = \mathbf{0.514 \times 10^{-8}\ cm/sec}$$

Example 7.12

A soft saturated clay has the following:

Percent less than 0.002 mm = 32%

Plasticity index = 21

Saturated unit weight, $\gamma_{sat} = 19.4 \text{ kN/m}^3$

Specific gravity of soil solids = 2.76

Estimate the hydraulic conductivity of the clay. Use Figure 7.15.

Solution

Given:

PI (in fraction) = 0.21

Clay-size fraction, $CF = 0.32$

$$CF + PI = 0.32 + 0.21 = 0.53$$

$$\gamma_{sat} = \frac{(G_s + e)\gamma_w}{1 + e} = \frac{(2.76 + e)(9.81)}{1 + e}; e = 0.8$$

Now, from Figure 7.15, for $e = 0.8$ and $CF + PI = 0.53$, the value of

$$k \approx 3.59 \times 10^{-10} \text{ m/sec} = \mathbf{3.59 \times 10^{-8} \text{ cm/sec}}$$

EXAMPLE 7.13

The void ratio and hydraulic conductivity relation for a normally consolidated clay are given here.

Void ratio	k (cm/sec)
1.2	0.6×10^{-7}
1.52	1.519×10^{-7}

Estimate the value of k for the same clay with a void ratio of 1.4. Use Eq. (7.37).

Solution

From Eq. (7.37),

$$\log k = A' \log e + B'$$

So,

$$\log (0.6 \times 10^{-7}) = A' \log (1.2) + B' \tag{a}$$

$$\log (1.519 \times 10^{-7}) = A' \log (1.52) + B' \tag{b}$$

From Eqs. (a) and (b),

$$\log \left(\frac{0.6 \times 10^{-7}}{1.519 \times 10^{-7}}\right) = A' \log \left(\frac{1.2}{1.52}\right)$$

$$A' = \frac{-0.4034}{-0.1027} = 3.928 \tag{c}$$

From Eqs. (a) and (c),

$$B' = \log (0.6 \times 10^{-7}) - (3.928)(\log 1.2) = -7.531$$

Thus,

$$\log k = 3.928 \log e - 7.531$$

With $e = 1.4$,

$$\log k = 3.928 \log (1.4) - 7.531 = -6.957$$

Hence,

$$k = \mathbf{1.1 \times 10^{-7}\ cm/sec}$$

7.8 Directional Variation of Permeability

Most soils are not isotropic with respect to permeability. In a given soil deposit, the magnitude of k changes with respect to the direction of flow. Figure 7.16 shows a soil layer through which water flows in a direction inclined at an angle α with the vertical. Let the hydraulic conductivity in the vertical ($\alpha = 0$) and horizontal ($\alpha = 90°$) directions be k_V and k_H, respectively. The magnitudes of k_V and k_H in a given soil depend on several factors, including the method of deposition in the field.

There are several published results for fine-grained soils that show that the ratio of k_H/k_V varies over a wide range. Table 7.3 provides a summary of some of those studies.

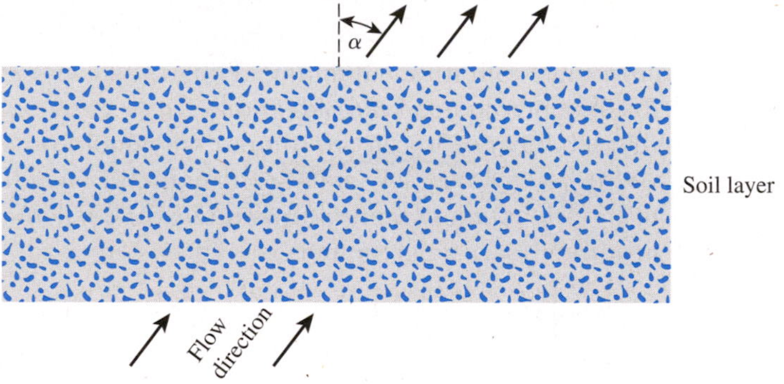

Figure 7.16 Directional variation of permeability

Table 7.3 k_H/k_V for Fine-Grained Soils—Summary of Several Studies

Soil type	k_H/k_V	Reference
Organic silt with peat	1.2 to 1.7	Tsien (1955)
Plastic marine clay	1.2	Lumb and Holt (1968)
Soft clay	1.5	Basett and Brodie (1961)
Varved clay	1.5 to 1.7	Chan and Kenney (1973)
Varved clay	1.5	Kenney and Chan (1973)
Varved clay	3 to 15	Wu et al. (1978)
Varved clay	4 to 40	Casagrande and Poulos (1969)

7.9 Equivalent Hydraulic Conductivity in Stratified Soil

In a stratified soil deposit where the hydraulic conductivity for flow in a given direction changes from layer to layer, an equivalent hydraulic conductivity can be computed to simplify calculations. The following derivations relate to the equivalent hydraulic conductivities for flow in vertical and horizontal directions through multilayered soils with horizontal stratification.

Figure 7.17 shows n layers of soil with flow in the *horizontal direction*. Let us consider a cross section of unit length passing through the n layer and perpendicular to the direction of flow. The total flow through the cross section in unit time can be written as

$$
\begin{aligned}
q &= v \cdot 1 \cdot H \\
&= v_1 \cdot 1 \cdot H_1 + v_2 \cdot 1 \cdot H_2 + v_3 \cdot 1 \cdot H_3 + \cdots + v_n \cdot 1 \cdot H_n
\end{aligned} \tag{7.39}
$$

where v = average discharge velocity

$v_1, v_2, v_3, \ldots, v_n$ = discharge velocities of flow in layers denoted by the subscripts

If $k_{H_1}, k_{H_2}, k_{H_3}, \ldots, k_{H_n}$ are the hydraulic conductivities of the individual layers in the horizontal direction and $k_{H(\text{eq})}$ is the equivalent hydraulic conductivity in the horizontal direction, then, from Darcy's law,

$$v = k_{H(\text{eq})} i_{\text{eq}}; \quad v_1 = k_{H_1} i_1; \quad v_2 = k_{H_2} i_2; \quad v_3 = k_{H_3} i_3; \quad \ldots \; v_n = k_{H_n} i,$$

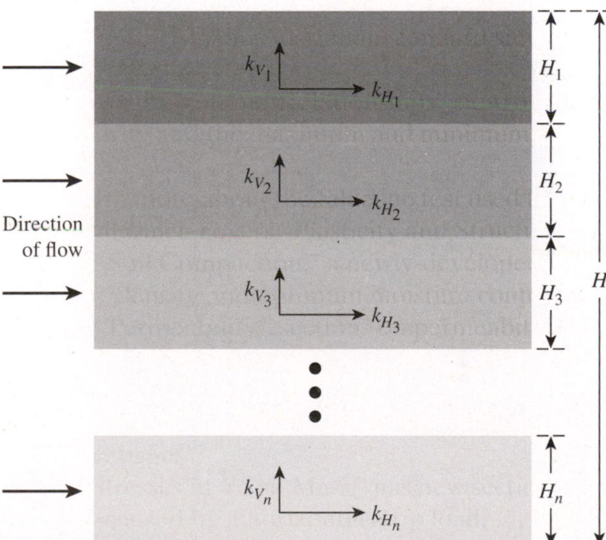

Figure 7.17 Equivalent hydraulic conductivity determination—horizontal flow in stratified soil

Substituting the preceding relations for velocities into Eq. (7.39) and noting that $i_{eq} = i_1 = i_2 = i_3 = \cdots = i_n$ results in

$$k_{H(eq)} = \frac{1}{H}(k_{H_1}H_1 + k_{H_2}H_2 + k_{H_3}H_3 + \cdots + k_{H_n}H_n) \tag{7.40}$$

Figure 7.18 shows n layers of soil with flow in the vertical direction. In this case, the velocity of flow through all the layers is the same. However, the total head loss, h, is equal to the sum of the head losses in all layers. Thus,

$$v = v_1 = v_2 = v_3 = \cdots = v_n \tag{7.41}$$

and

$$h = h_1 + h_2 + h_3 + \cdots + h_n \tag{7.42}$$

Using Darcy's law, we can rewrite Eq. (7.41) as

$$k_{V(eq)}\left(\frac{h}{H}\right) = k_{V_1}i_1 = k_{V_2}i_2 = k_{V_3}i_3 = \cdots = k_{V_n}i_n \tag{7.43}$$

where $k_{V_1}, k_{V_2}, k_{V_3}, \ldots, k_{V_n}$ are the hydraulic conductivities of the individual layers in the vertical direction and $k_{V(eq)}$ is the equivalent hydraulic conductivity.

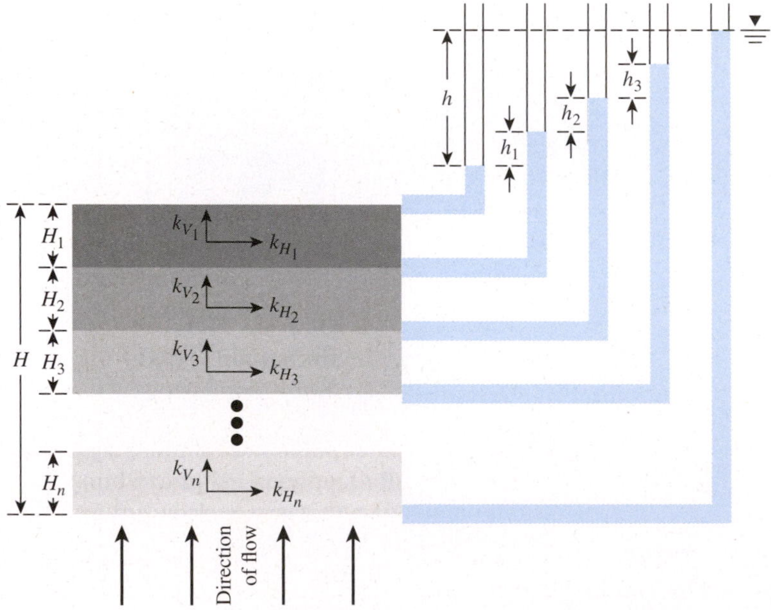

Figure 7.18 Equivalent hydraulic conductivity determination—vertical flow in stratified soil

Again, from Eq. (7.42),

$$h = H_1 i_1 + H_2 i_2 + H_3 i_3 + \cdots + H_n i_n \tag{7.44}$$

Solving Eqs. (7.43) and (7.44) gives

$$k_{v(\text{eq})} = \frac{H}{\left(\frac{H_1}{k_{V_1}}\right) + \left(\frac{H_2}{k_{V_2}}\right) + \left(\frac{H_3}{k_{V_3}}\right) + \cdots + \left(\frac{H_n}{k_{V_n}}\right)} \tag{7.45}$$

An excellent example of naturally deposited layered soil is *varved soil*, which is a rhythmically layered sediment of coarse and fine minerals. Varved soils result from annual seasonal fluctuation of sediment conditions in glacial lakes. Figure 7.19 shows the variation of moisture content and grain-size distribution in New Liskeard, Canada, varved soil. Each varve is about 41 to 51 mm (1.6 to 2.0 in.) thick and consists of two homogeneous layers of soil—one coarse and one fine—with a transition layer between.

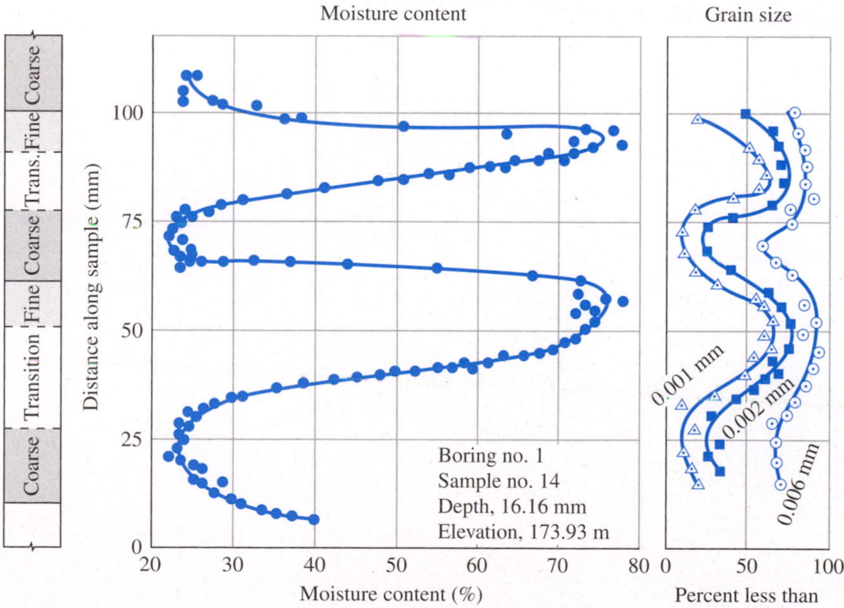

Figure 7.19 Variation of moisture content and grain-size distribution in New Liskeard varved soil. (*Source:* After "Laboratory Investigation of Permeability Ratio of New Liskeard Varved Clay," by H. T. Chan and T. C. Kenney, 1973, *Canadian Geotechnical Journal*, 10(3), p. 453–472. © 2008 NRC Canada or its licensors. Reproduced with permission.)

Example 7.14

A layered soil is shown in Figure 7.20. Given:

- $H_1 = 1$ m $\quad k_1 = 10^{-4}$ cm/sec
- $H_2 = 1.5$ m $\quad k_2 = 3.2 \times 10^{-2}$ cm/sec
- $H_3 = 2$ m $\quad k_3 = 4.1 \times 10^{-5}$ cm/sec

Estimate the ratio of equivalent hydraulic conductivity,

$$\frac{k_{H(\mathrm{eq})}}{k_{V(\mathrm{eq})}}$$

Solution

From Eq. (7.40),

$$k_{H(\mathrm{eq})} = \frac{1}{H}(k_{H_1}H_1 + k_{H_2}H_2 + k_{H_3}H_3)$$

$$= \frac{1}{(1 + 1.5 + 2)}[(10^{-4})(1) + (3.2 \times 10^{-2})(1.5) + (4.1 \times 10^{-5})(2)]$$

$$= 107.07 \times 10^{-4} \text{ cm/sec}$$

Again, from Eq. (7.45),

$$k_{v(\mathrm{eq})} = \frac{H}{\left(\dfrac{H_1}{k_{V_1}}\right) + \left(\dfrac{H_2}{k_{V_2}}\right) + \left(\dfrac{H_3}{k_{V_3}}\right)}$$

$$= \frac{1 + 1.5 + 2}{\left(\dfrac{1}{10^{-4}}\right) + \left(\dfrac{1.5}{3.2 \times 10^{-2}}\right) + \left(\dfrac{2}{4.1 \times 10^{-5}}\right)}$$

$$= 0.765 \times 10^{-4} \text{ cm/sec}$$

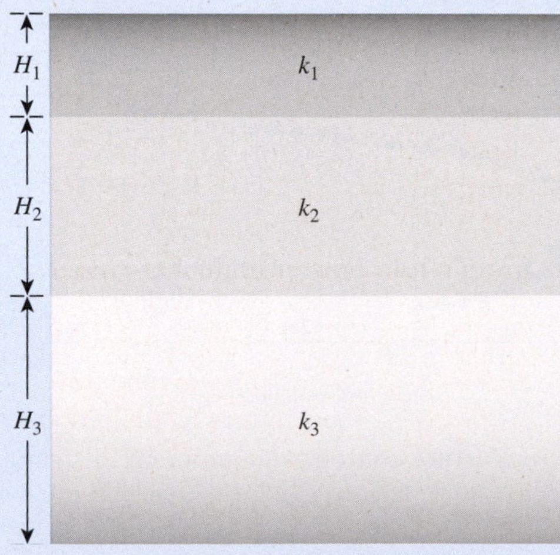

Figure 7.20 A layered soil profile

Hence,

$$\frac{k_{H(\text{eq})}}{k_{V(\text{eq})}} = \frac{107.07 \times 10^{-4}}{0.765 \times 10^{-4}} \approx \mathbf{140}$$

Example 7.15

Figure 7.21 shows three layers of soil in a tube that is 100 mm × 100 mm in cross section. Water is supplied to maintain a constant-head difference of 300 mm across the sample. The hydraulic conductivities of the soils in the direction of flow through them are as follows:

Soil	*k* (cm/sec)
A	10^{-2}
B	3×10^{-3}
C	4.9×10^{-4}

Find the rate of water supply in cm³/hr.

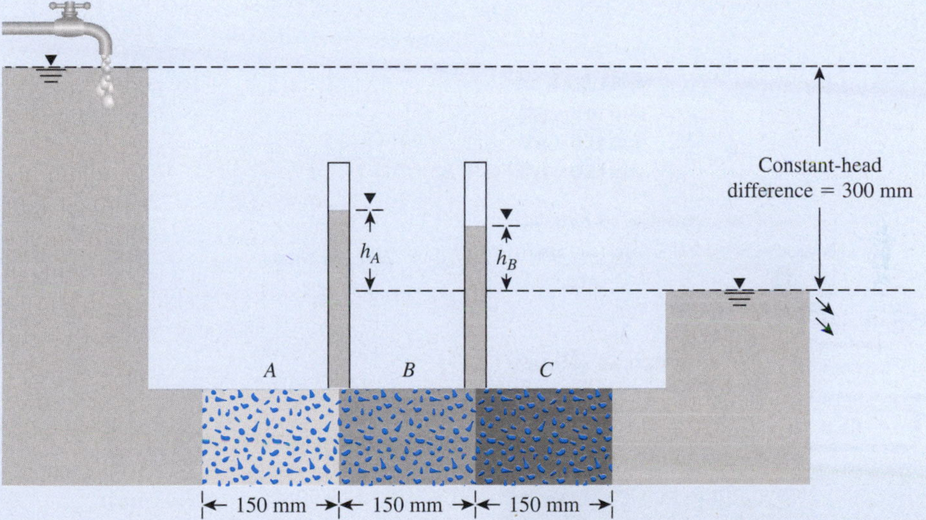

Figure 7.21 Three layers of soil in a tube 100 mm × 100 mm in cross section

Solution

From Eq. (7.45),

$$k_{V(\text{eq})} = \frac{H}{\left(\frac{H_1}{k_1}\right) + \left(\frac{H_2}{k_2}\right) + \left(\frac{H_3}{k_3}\right)} = \frac{450}{\left(\frac{150}{10^{-2}}\right) + \left(\frac{150}{3 \times 10^{-3}}\right) + \left(\frac{150}{4.9 \times 10^{-4}}\right)}$$

$$= 0.001213 \text{ cm/sec}$$

$$q = k_{V(\text{eq})} i A = (0.001213)\left(\frac{300}{450}\right)\left(\frac{100}{10} \times \frac{100}{10}\right)$$

$$= 0.0809 \text{ cm}^3/\text{sec} = \mathbf{291.24\ cm^3/hr}$$

EXAMPLE 7.16

Refer to Example 7.15 and Figure 7.21. Determine the magnitudes of h_A and h_B.

Solution

The loss of head during flow through Soil A can be calculated as

$$q = k_A i_A A = k_A \frac{\Delta h_A A}{L_A}$$

where Δh_A and L_A are, respectively, the head loss in Soil A and the length of Soil A. Hence,

$$\Delta h_A = \frac{q L_A}{k_A A}$$

From Example 7.15, $q = 0.0809$ cm³/sec, $L_A = 15$ cm, and $k_A = 10^{-2}$ cm/sec. Thus,

$$\Delta h_A = \frac{(0.0809)(15)}{(0.01)(10 \times 10 \text{ cm}^2)} = 1.2135 \text{ cm} \approx 12.14 \text{ mm}$$

Hence,

$$h_A = 300 - 12.14 = \mathbf{287.86\ mm}$$

Similarly, for Soil B,

$$\Delta h_B = \frac{q L_B}{k_B A} = \frac{(0.0809)(15)}{(0.003)(10 \times 10)} = 4.045 \text{ cm} = 40.45 \text{ mm}$$

Hence,

$$h_B = 300 - \Delta h_A - \Delta h_B = 300 - 12.14 - 40.45 = \mathbf{247.41\ mm}$$

7.10 Permeability Test in the Field by Pumping from Wells

In the field, the average hydraulic conductivity of a soil deposit in the direction of flow can be determined by performing pumping tests from wells. Figure 7.22 shows a case where the top permeable layer, whose hydraulic conductivity has

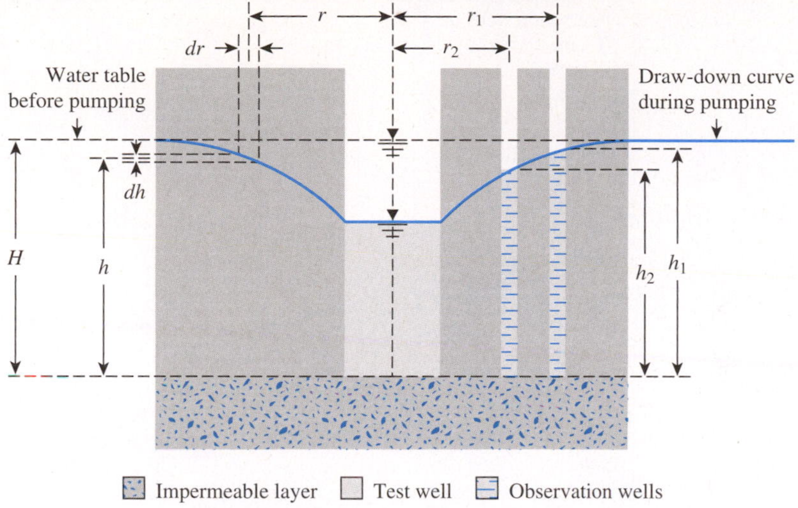

Figure 7.22 Pumping test from a well in an unconfined permeable layer underlain by an impermeable stratum.

to be determined, is unconfined and underlain by an impermeable layer. During the test, water is pumped out at a constant rate from a test well that has a perforated casing. Several observation wells at various radial distances are made around the test well. Continuous observations of the water level in the test well and in the observation wells are made after the start of pumping, until a steady state is reached. The steady state is established when the water level in the test and observation wells becomes constant. The expression for the rate of flow of groundwater into the well, which is equal to the rate of discharge from pumping, can be written as

$$q = k\left(\frac{dh}{dr}\right)2\pi rh \tag{7.46}$$

or

$$\int_{r_2}^{r_1} \frac{dr}{r} = \left(\frac{2\pi k}{q}\right)\int_{h_2}^{h_1} h\,dh$$

Thus,

$$k = \frac{2.303q\ \log_{10}\left(\frac{r_1}{r_2}\right)}{\pi(h_1^2 - h_2^2)} \tag{7.47}$$

From field measurements, if q, r_1, r_2, h_1, and h_2 are known, the hydraulic conductivity can be calculated from the simple relationship presented in Eq. (7.47).

Ahmad et al. (1975) have reported the results of a field pumping test in southwestern India. For this case, $H = 30.49$ m (see Figure 7.22 for definition of H). Several observation wells were located along three radial lines from the test

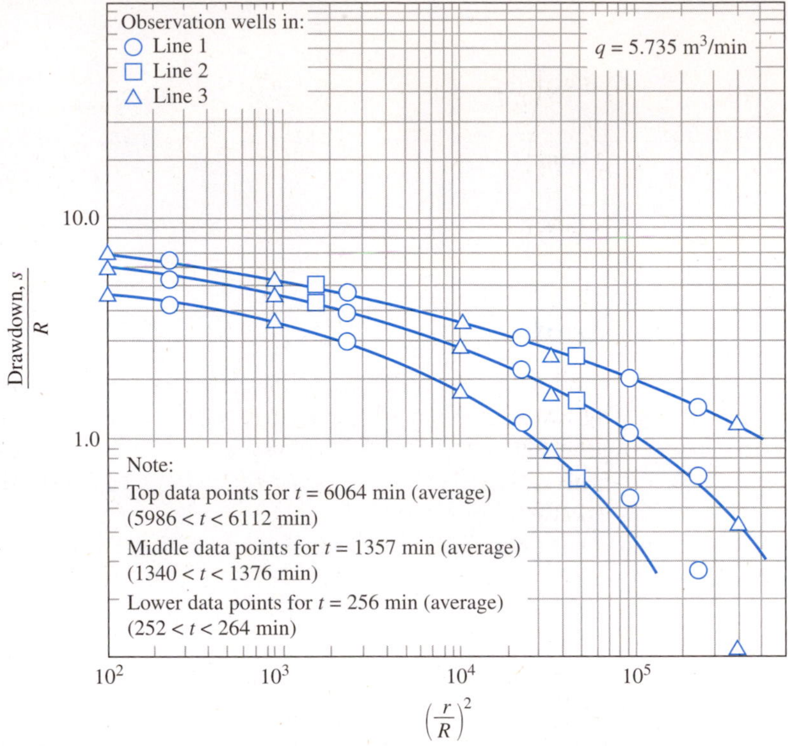

Figure 7.23 Plot of drawdown versus r^2 in a field pumping test. *Note*: R = reference distance. $R = 0.305$ m when r and s are m. (*Based on Ahmad, Lacroix, and Steinback, 1975*)

well. During pumping, the drawdown, s, at each observation well was measured. The results of the observed drawdown, s, versus radial distance, r, for the observations wells are shown in a nondimensional form in Figure 7.23. From the plots, it appears that the steady state was reached at time $t \geqslant 6064$ min. With this, the hydraulic conductivity of the permeable layer can be calculated as follows:

$\left(\frac{r}{R^*}\right)^2$	r (m)	$\frac{s}{R^*}$	s (m)	h (m)
1000	9.65	5	1.525	30.49 – 1.525 = 28.965
10,000	30.5	3.5	1.068	30.49 – 1.068 = 29.422

Note: $R = 0.305$ m

From Eq. (7.47),

$$k = \frac{2.303q \log_{10}\left(\frac{r_1}{r_2}\right)}{\pi\left(h_1^2 - h_2^2\right)} = \frac{(2.303)\left(\frac{5.735}{60}\ \text{m}^3/\text{sec}\right)\log_{10}\left(\frac{30.5}{9.65}\right)}{\pi[(29.422)^2 - (28.965)^2]}$$

$$= 0.00131\ \text{m/sec} = \mathbf{0.131\ cm/sec}$$

Pumping from a confined aquifer

The average hydraulic conductivity for a confined aquifer can also be determined by conducting a pumping test from a well with a perforated casing that penetrates the full depth of the aquifer and by observing the piezometric level in a number of observation wells at various radial distances (Figure 7.24). Pumping is continued at a uniform rate q until a steady state is reached.

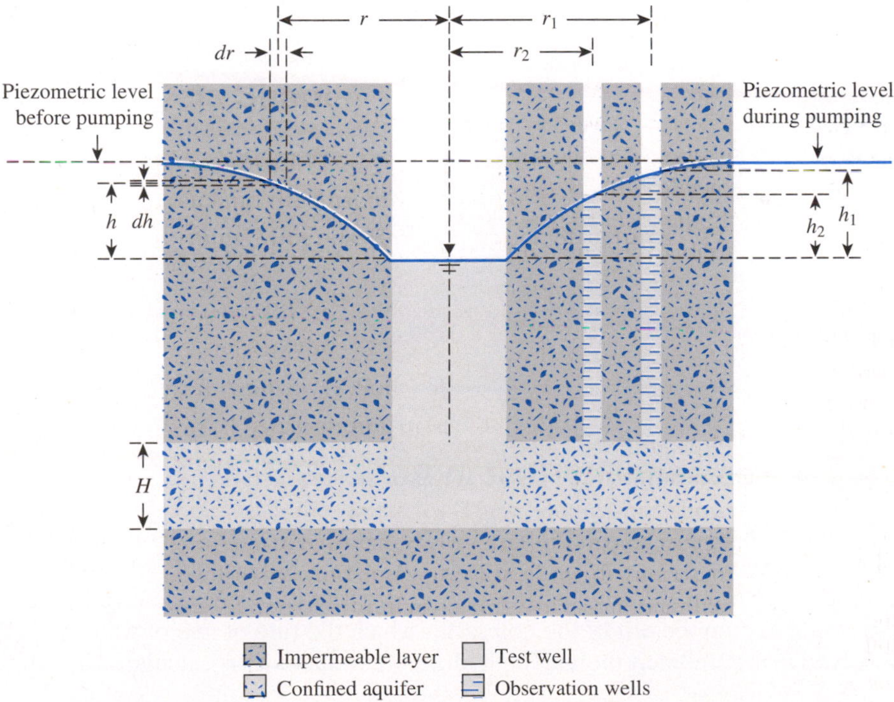

Figure 7.24 Pumping test from a well penetrating the full depth in a confined aquifer

Because water can enter the test well only from the aquifer of thickness H, the steady state of discharge is

$$q = k\left(\frac{dh}{dr}\right)2\pi rH \tag{7.48}$$

or

$$\int_{r_2}^{r_1}\frac{dr}{r} = \int_{h_2}^{h_1}\frac{2\pi kH}{q}\,dh$$

This gives the hydraulic conductivity in the direction of flow as

$$k = \frac{q\,\log_{10}\left(\dfrac{r_1}{r_2}\right)}{2.727H(h_1 - h_2)} \tag{7.49}$$

EXAMPLE 7.17

A pumping test from a confined aquifer yielded the following results: $q = 0.303\ \text{m}^3/\text{min}$, $h_1 = 2.44$ m, $h_2 = 1.52$ m, $r_1 = 18.3$ m, $r_2 = 9.15$ m, and $H = 3.05$ m. Refer to Figure 7.24 and determine the magnitude of k of the permeable layer.

Solution

From Eq. (7.49),

$$k = \frac{q\log_{10}\left(\dfrac{r_1}{r_2}\right)}{2.727H(h_1 - h_2)} = \frac{(0.303)\log_{10}\left(\dfrac{18.3}{9.15}\right)}{(2.727)(3.05)(2.44 - 1.52)}$$

$$= 0.01192\ \text{m/min} \approx \mathbf{0.0199\ cm/sec}$$

7.11 Permeability Test in Auger Holes

Van Bavel and Kirkham (1948) suggested a method to determine k from an auger hole (Figure 7.25). In this method, an auger hole is made in the ground that should extend to a depth of 10 times the diameter of the hole or to an impermeable layer, whichever is less. Water is pumped out of the hole, after which the rate of rise of water with time is observed in several increments. The hydraulic conductivity is calculated as

$$k = 0.617\frac{r_w}{Sd}\frac{dh}{dt} \tag{7.50}$$

where r_w = the radius of the auger hole
d = the depth of the hole below the water table

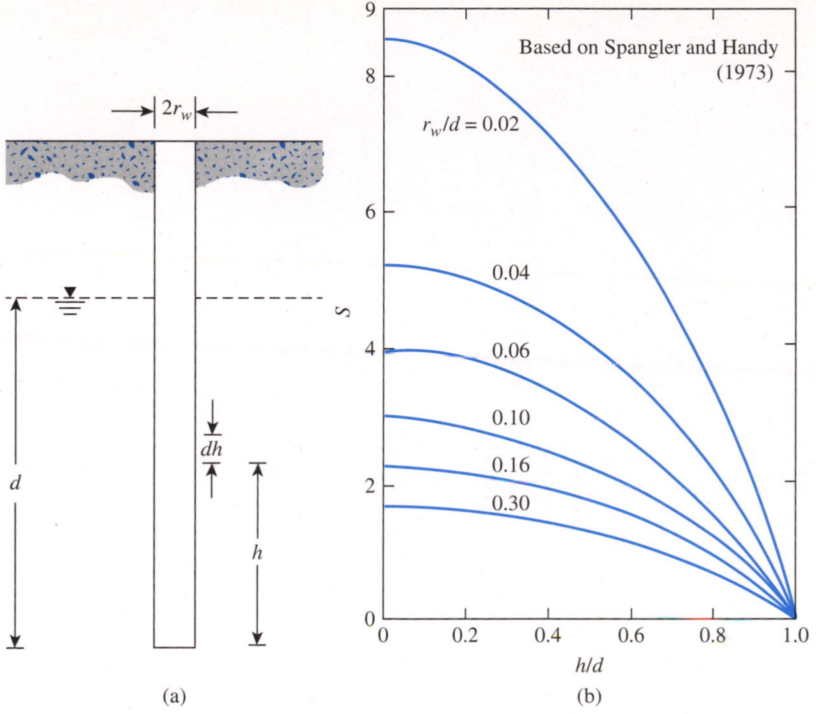

Figure 7.25 Auger hole test: (a) auger hole; (b) plot of S with h/d and r_w/d

S = the shape factor for auger hole

dh/dt = the rate of increase of water table at a depth h measured from the bottom of the hole

The variation of S with r_w/d and h/d is given in Figure 7.25b (Spangler and Handy, 1973).

EXAMPLE 7.18

A 100-mm diameter auger hole was made to a depth of 3 m. The ground water level is located at a depth of 1.2 m below the ground surface. Water was bailed out several times from the auger hole. Referring to Figure 7.25a, when h was equal to 1.5 m, the water table in the auger hole rose 3 cm in a time period of 10 min. Estimate k.

Solution

Referring to Figure 7.25a,

$$d = 3 - 1.2 = 1.8 \text{ m}$$

$$h = 1.5 \text{ m}$$

$$dh = 3 \text{ cm}$$
$$dt = 10 \text{ min}$$
$$r_w = \frac{100}{2} = 50 \text{ mm}$$
$$\frac{h}{d} = \frac{1.5}{1.8} = 0.833$$
$$\frac{r_w}{d} = \frac{\left(\frac{50}{1000}\text{ m}\right)}{1.8 \text{ m}} = 0.0278$$

From Figure 7.25, $S \approx 2$.

From Eq. (7.50),

$$k = 0.617\frac{r_w}{Sd}\frac{dh}{dt} = (0.617)\left(\frac{0.05 \text{ m}}{2 \times 1.8 \text{ m}}\right)\left(\frac{0.03 \text{ m}}{10 \times 60 \text{ sec}}\right)$$
$$= 4.28 \times 10^{-7} \text{ m/sec} = \mathbf{4.28 \times 10^{-5} \text{ cm/sec}}$$

7.12 Hydraulic Conductivity of Compacted Clayey Soils

It was shown in Chapter 6 (Section 6.7) that when a clay is compacted at a lower moisture content it possesses a flocculent structure. Approximately at optimum moisture content of compaction, the clay particles have a lower degree of flocculation. A further increases in the moisture content at compaction provides a greater degree of particle orientation; however, the dry unit weight decreases because the added water dilutes the concentration of soil solids per unit volume.

Figure 7.26 shows the results of laboratory compaction tests on a clay soil as well as the variation of hydraulic conductivity on the compacted clay specimens. The compaction tests and thus the specimens for hydraulic conductivity tests were prepared from clay clods that were 19 mm and 4.8 mm. From the laboratory test results shown, the following observations can be made.

1. For similar compaction effort and molding moisture content, the magnitude of k decreases with the decrease in clod size.
2. For a given compaction effort, the hydraulic conductivity decreases with the increase in molding moisture content, reaching a minimum value at about the optimum moisture content (that is, approximately where the soil has a higher unit weight with the clay particles having a lower degree of flocculation). Beyond the optimum moisture content, the hydraulic conductivity increases slightly.
3. For similar compaction effort and dry unit weight, a soil will have a lower hydraulic conductivity when it is compacted on the wet side of the optimum moisture content. This fact is further illustrated in Figure 7.27, which shows a summary of hydraulic conductivity test results on a silty clay (Mitchell, Hopper, and Campanella, 1965).

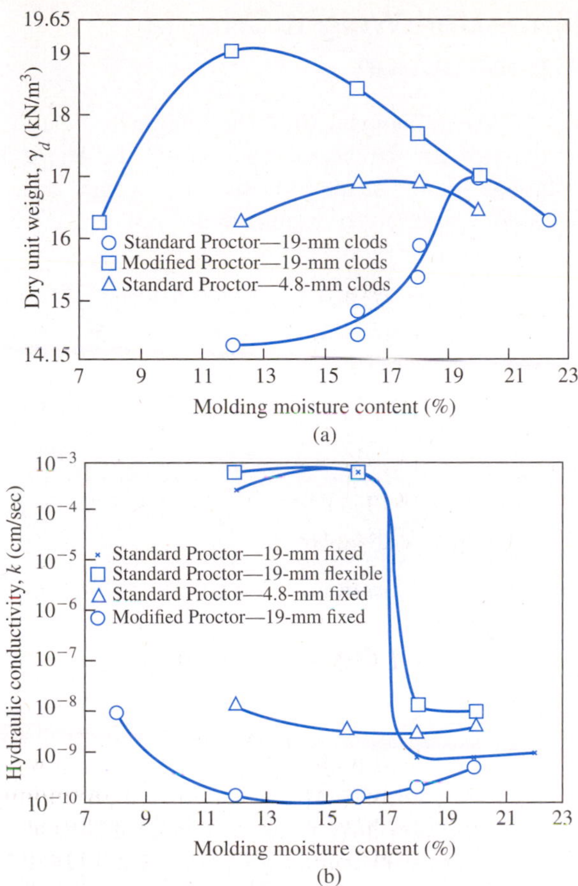

Figure 7.26 Tests on a clay soil: (a) Standard and modified Proctor compaction curves; (b) variation of k with molding moisture content (*Source*: After "Influence of Clods on Hydraulic Conductivity of Compacted Clay," by C. H. Benson and D. E. Daniel, 1990, *Journal of Geotechnical Engineering, 116*(8), p. 1231–1248. Copyright © 1990 American Society of Civil Engineers. Used by permission.)

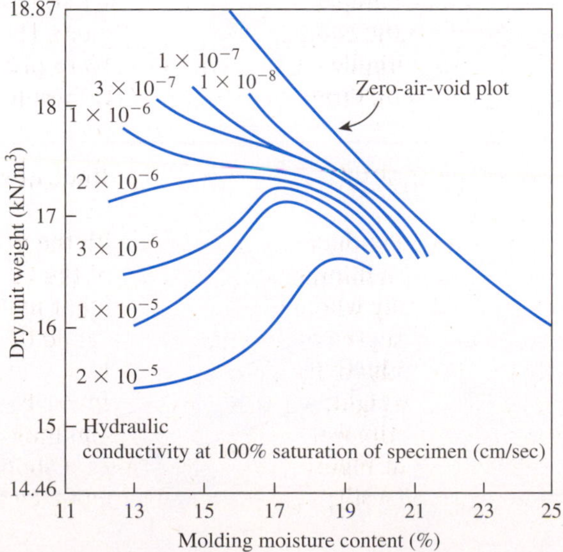

Figure 7.27 Contours of hydraulic conductivity for a silty clay (*Source*: After "Permeability of Compacted Clay," by J. K. Mitchell, D. R. Hooper, and R. B. Campenella, 1965, *Journal of the Soil Mechanics and Foundations Divisions, 91* (SM4), p. 41–65. Copyright © 1965 American Society of Civil Engineers. Used by permission.)

7.13 Moisture Content—Unit Weight Criteria for Clay Liner Construction

For construction of clay liners for solid-waste disposal sites, the compacted clay is required to have a hydraulic conductivity of 10^{-7} cm/sec or less. Daniel and Benson (1990) developed a procedure to establish the moisture content—unit weight criteria for clayey soils to meet the hydraulic conductivity requirement. Following is a step-by-step procedure to develop the criteria.

1. Conduct *modified, standard,* and *reduced* Proctor tests to establish the dry unit weight versus molding moisture content relationships (Figure 7.28a). Modified

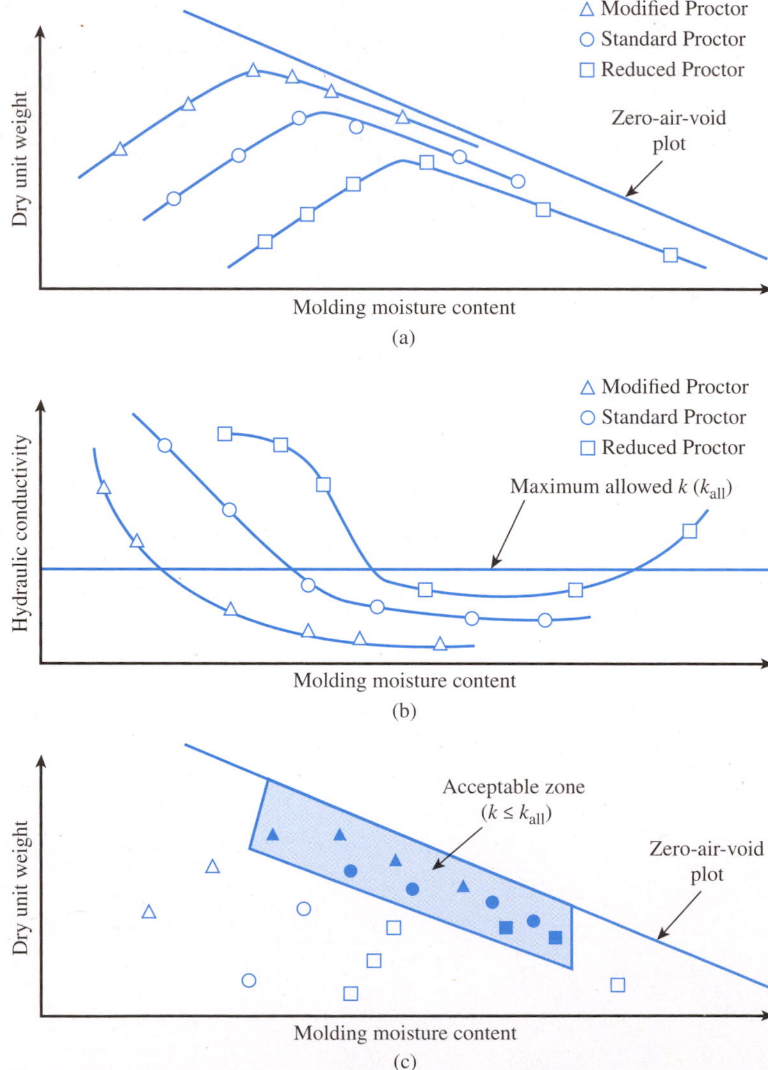

Figure 7.28 (a) Proctor curves; (b) variation of hydraulic conductivity of compacted specimens; (c) determination of acceptable zone (*Source*: After "Water Content-Density Criteria for Compacted Soil Linkers," by D. E. Daniel and C. H. Benson, 1990, *Journal of Geo-technical Engineering, 116*(12), pp. 1811–1830. Copyright © 1990 American Society of Civil Engineers. Used by permission.)

and standard Proctor tests were discussed in Chapter 6. The *reduced* Proctor test is similar to the standard Proctor test, except the hammer is dropped only 15 times per lift instead of the usual 25 times. Modified, standard, and reduced Proctor efforts represent, respectively, the upper, medium, and minimum levels of compaction energy for a typical clayey soil liner.

2. Conduct permeability tests on the compacted soil specimens (from step 1), and plot the results, as shown in Figure 7.28b. In this figure, also plot the maximum allowable value of k (that is, k_{all}).
3. Replot the dry unit weight–moisture content points (Figure 7.28c) with different symbols to represent the compacted specimens with $k > k_{all}$ and $k \leq k_{all}$.
4. Plot the acceptable zone for which k is less than or equal to k_{all} (Figure 7.28c).

7.14 Summary

Following is a summary of the important subjects covered in this chapter.

- Darcy's law can be expressed as

$$\underset{\substack{\text{discharge}\\ \text{velocity}}}{\underset{\uparrow}{v}} = \underset{\substack{\text{hydraulic}\\ \text{conductivity}}}{\underset{\uparrow}{k}} \quad \underset{\substack{\text{hydraulic}\\ \text{gradient}}}{\underset{\uparrow}{i}}$$

- Seepage velocity (v_s) of water through the void spaces can be given as

$$v_s = \frac{\text{discharge velocity}}{\text{porosity of soil}}$$

- Hydraulic conductivity is a function of viscosity (and hence temperature) of water.
- Constant-head and falling-head types of tests are conducted to determine the hydraulic conductivity of soils in the laboratory (Section 7.5).
- There are several empirical correlations for hydraulic conductivity in granular and cohesive soil. Some of those are given in Sections 7.6 and 7.7. It is important, however, to realize that these are only approximations, since hydraulic conductivity is a highly variable quantity.
- For layered soil, depending on the direction of flow, an equivalent hydraulic conductivity relation can be developed to estimate the quantity of flow [Eqs. (7.40) and (7.45)].
- Hydraulic conductivity in the field can be determined by pumping from wells (Section 7.10).

The hydraulic conductivity of saturated cohesive soils also can be determined by laboratory consolidation tests. The actual value of the hydraulic conductivity in the field also may be somewhat different than that obtained in the laboratory because of the non-homogeneity of the soil. Hence, proper care should be taken in assessing the order of the magnitude of k for all design considerations.

Problems

7.1 A permeable soil layer is underlain by an impervious layer as shown in Figure 7.29. Knowing that $k = 6 \times 10^{-3}$ cm/sec for the permeable layer, calculate the rate of seepage through this layer in m³/hr/m width. Given: $H = 5.4$ m and $\alpha = 7°$.

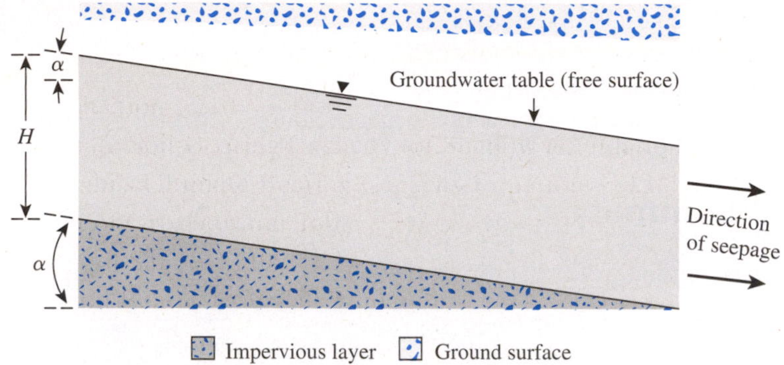

Figure 7.29

7.2 Redo Problem 7.1 for $\alpha = 5°$ and $\alpha = 9°$. All other site conditions remaining the same, what impact does the slope angle have on the rate of seepage?

7.3 Seepage is occurring through the sandy layer underneath the concrete dam as shown in Figure 7.30.

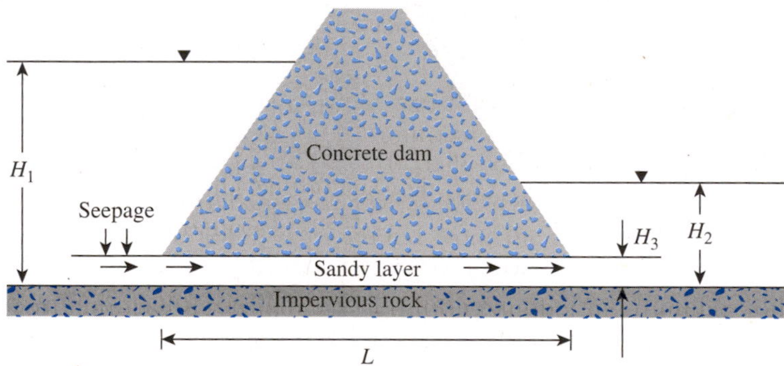

Figure 7.30

Given: upstream water level, $H_1 = 16$ m; downstream water level, $H_2 = 2.3$ m; thickness of the sandy layer, $H_3 = 0.75$ m; hydraulic conductivity of the sandy layer, $k = 0.009$ cm/sec; void ratio of sand, $e = 0.8$; and $L = 45$ m. Determine:

a. Rate of seepage per unit length of the dam (in m³/hr/m)
b. Seepage velocity
c. Quantity of seepage per day if the dam is 350 m long

7.4 Redo Problem 7.3 with the following information: $H_1 = 12$ m; $H_2 = 2$ m; $H_3 = 0.5$ m; $k = 6.3 \times 10^{-3}$ cm/sec; $e = 1.22$; $L = 52$ m; and dam length = 410 m.

7.5 A pervious soil layer is sandwiched between two impervious layers as shown in Figure 7.31. Find the rate of flow in m³/sec/m (at right angles to the cross section) through the pervious soil layer. Given: $H = 3.5$ m, $H_1 = 1.75$ m, $h = 2.5$ m, $S = 28$ m, $\alpha = 12°$, and $k = 0.055$ cm/sec.

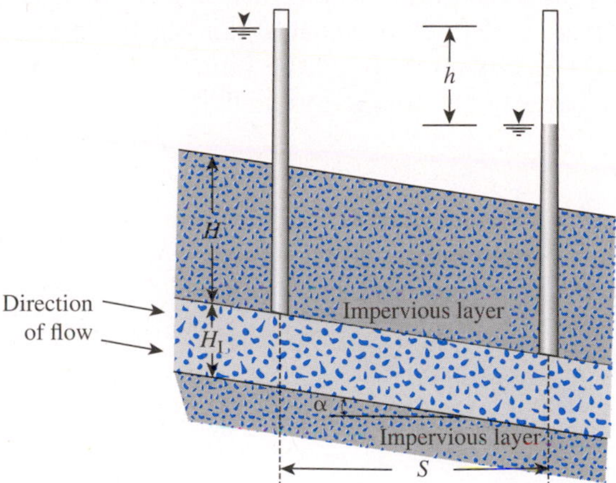

Figure 7.31

7.6 The results of a constant-head permeability test for a fine sand sample having a diameter of 70 mm and a length of 140 mm are as follows (refer to Figure 7.5):

- Constant-head difference = 550 mm
- Water collected in 7 min = 450 cm³
- Void ratio of sand = 0.8

Determine:

a. Hydraulic conductivity, k (cm/sec)
b. Seepage velocity

7.7 In a constant-head permeability test, the length of the specimen is 200 mm and the cross sectional area is 78.5 cm². If $k = 2.1 \times 10^{-2}$ cm/sec, and a rate of flow of 130 cm³/min has to be maintained during the test, what should be the head difference across the specimen?

7.8 The following data are for a falling-head permeability test:

- Length of the soil sample = 140 mm
- Diameter of soil sample = 70 mm
- Area of the standpipe = 19.6 mm²
- At time $t = 0$, head difference = 500 mm
- At time $t = 7$ min, head difference = 350 mm

a. Determine the hydraulic conductivity of the soil (cm/sec)
b. What was the head difference at $t = 5$ min?

7.9 The following data are for a falling-head permeability test:

- Length of the soil sample = 400 mm
- Area of the soil sample = 7854 mm²
- Diameter of the standpipe = 11 mm

- At time $t = 0$, head difference = 450 mm
- At time $t = 8$ min, head difference = 200 mm

If the test was conducted at 20°C at which $\gamma_w = 9.789$ kN/m^3 and $\eta = 1.005 \times 10^{-3}$ N · s/m^2,

a. Determine the absolute permeability of the soil (cm/sec).

b. What was the head difference at $t = 4$ min?

7.10 Figure 7.32 shows the cross section of a levee which is 650 m long and is underlain by a 2.5-m-thick permeable sand layer. It was observed that the quantity of water flowing through the sand layer into the collection ditch is 13.5 m^3/hr. What is the hydraulic conductivity of the sand layer?

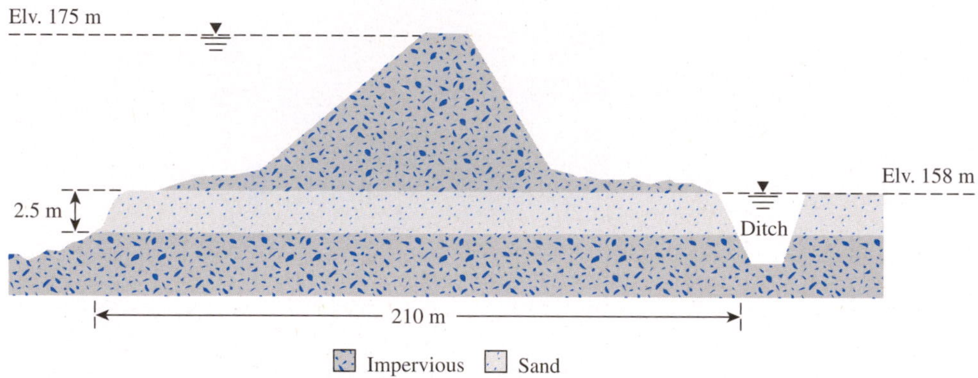

Figure 7.32

7.11 The hydraulic conductivity of a sandy soil is 0.011 cm/sec at a room temperature of 24°C. What would be the hydraulic conductivity at 20°C? Use Eq. (7.15).

7.12 The hydraulic conductivity of a sand at a void ratio of 0.85 is 0.08 cm/sec. Estimate its hydraulic conductivity at a void ratio of 0.68. Use Eq. (7.31).

7.13 For a sandy soil, the following are given:

- Maximum void ratio = 0.75
- Minimum void ratio = 0.39
- Effective size, $D_{10} = 0.32$ mm

Determine the hydraulic conductivity of the sand at a relative density of 80%. Use Eq. (7.32).

7.14 For a sandy soil, the following are given:

- Maximum void ratio = 0.86
- Minimum void ratio = 0.4
- Hydraulic conductivity at a relative density of 70% = 0.003 cm/sec

Determine the hydraulic conductivity of the sand at a relative density of 50%. Use Eq. (7.31).

7.15 For a sand, the porosity $n = 0.28$ and $k = 0.058$ cm/sec. Determine k when $n = 0.45$. Use Eq. (7.31).

7.16 The maximum dry unit weight of a quartz sand determined in the laboratory is 18.5 kN/m^3. If the relative compaction in the field is 88%, determine the hydraulic conductivity of the sand in the field compaction condition. Given: $G_s = 2.66$, $D_{10} = 0.28$ mm, and $C_u = 4.2$. Use Eq. (7.34).

7.17 The grain-size analysis data for a sand is given in the following table. Estimate the hydraulic conductivity of the sand at a void ratio of 0.77. Use Eq. (7.30) and $SF = 6.5$.

U.S. sieve no.	Percent passing
30	100
40	85
60	64
100	32
200	3

7.18 For a normally consolidated clay, the following values are given.

e	k (cm/sec)
0.78	0.45×10^{-6}
1.1	0.88×10^{-6}

Estimate k at a void ratio of 0.97. Use Eq. (7.38).

7.19 Redo Problem 7.18 using Mesri and Olson's (1971) procedure given by Eq. (7.37).

7.20 Estimate the hydraulic conductivity of a saturated clay having a clay-size fraction, $CF = 42\%$, and plasticity index, $PI = 27\%$. Given: $\gamma_{sat} = 18.8$ kN/m^3 and $G_s = 2.73$. Use Tavenas et al.'s (1983) method illustrated in Figure 7.15.

7.21 A layered soil is shown in Figure 7.33. Given:

- $H_1 = 1.5$ m $k_1 = 9 \times 10^{-4}$ cm/sec
- $H_2 = 2.5$ m $k_2 = 7.8 \times 10^{-3}$ cm/sec
- $H_3 = 3.5$ m $k_3 = 4.5 \times 10^{-5}$ cm/sec

Estimate the ratio of equivalent permeability, $k_{H(eq)}/k_{V(eq)}$.

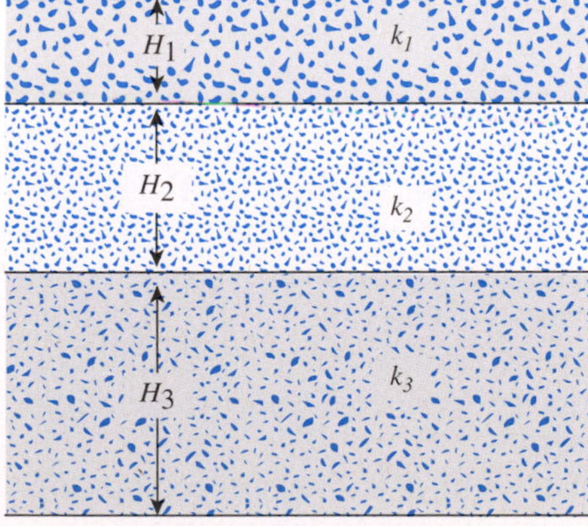

Figure 7.33

7.22 Refer to Figure 7.24. The following data were collected during the field permeability measurement of a confined aquifer using a pumping test. Determine the hydraulic conductivity of the permeable layer. Use Eq. (7.49).

Thickness of the aquifer, $H = 4.5$ m

Piezometric level and radial distance of the first observation well:

$h_1 = 2.9$ m; $r_1 = 17.8$ m

Piezometric level and radial distance of the second observation well:

$h_2 = 1.8$ m; $r_2 = 8.1$ m

Rate of discharge from pumping, $q = 0.5$ m^3/min

7.23 Refer to Figure 7.25. During an auger hole test to determine field permeability, it was observed that the water table inside the hole rose by 5 cm in 8 min, when $h = 2$ m. Given: diameter of the auger hole = 150 mm, length of auger hole = 4 m, and depth of the ground water table from the surface = 1 m. Estimate k. Use Eq. (7.50).

Critical Thinking Problems

7.C.1 Section 7.2 described the importance of total head and hydraulic gradient on the seepage of water through permeable soil media. In this problem, we will study the variations of head along the axis of a soil specimen through which seepage is occurring. Consider the setup shown in Figure 7.34 (similar to Example 7.15) in which three different soil layers, each 200 mm in length, are located inside a cylindrical tube of diameter 150 mm. A constant-head difference of 470 mm is maintained across the soil sample. The porosities and hydraulic conductivities of the three soils in the direction of the flow are given here.

Soil	n	k (cm/sec)
I	0.5	5×10^{-3}
II	0.6	4.2×10^{-2}
III	0.33	3.9×10^{-4}

Perform the following tasks.

a. Determine the quantity of water flowing through the sample per hour.

b. Denoting the downstream water level (Y–Y) to be the datum, determine the elevation head (Z), pressure head (u/γ_w) and the total head (h) at the entrance and exit of each soil layer.

c. Plot the variation of the elevation head, pressure head and the total head with the horizontal distance along the sample axis (X–X).

d. Plot the variations of discharge velocity and the seepage velocity along the sample axis.

e. What will be the height of the vertical columns of water inside piezometers A and B installed on the sample axis?

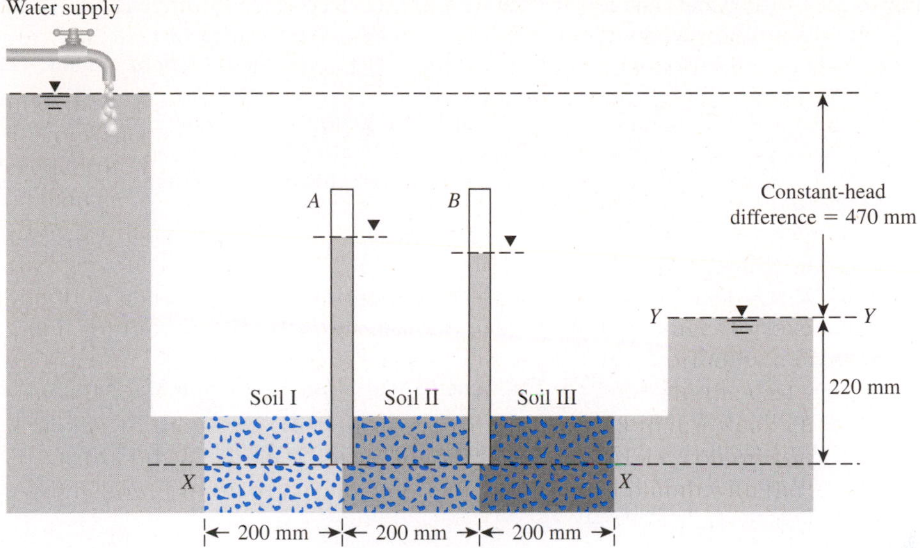

Figure 7.34

References

Ahmad, S., Lacroix, Y., and Steinback, J. (1975). "Pumping Tests in an Unconfined Aquifer," *Proceedings*, Conference on *in situ* Measurement of Soil Properties, ASCE, Vol. 1, 1–21.

Amer, A. M., and Awad, A. A. (1974). "Permeability of Cohesionless Soils," *Journal of the Geotechnical Engineering Division*, ASCE, Vol. 100, No. GT12, 1309–1316.

Basett, D. J., and Brodie, A. F. (1961). "A Study of Matabitchuan Varved Clay," *Ontario Hydroelectric Research News*, Vol. 13, No. 4, 1–6.

Benson, C. H., and Daniel, D. E. (1990). "Influence of Clods on Hydraulic Conductivity of Compacted Clay," *Journal of Geotechnical Engineering,* ASCE, Vol. 116, No. 8, 1231–1248.

Carman, P. C. (1938). "The Determination of the Specific Surface of Powders." *J. Soc. Chem. Ind. Trans.*, Vol. 57, 225.

Carman, P. C. (1956). *Flow of Gases through Porous Media*, Butterworths Scientific Publications, London.

Carrier III, W. D. (2003). "Goodbye, Hazen; Hello, Kozeny-Carman," *Journal of Geotechnical and Geoenvironmental Engineering,* ASCE, Vol. 129, No. 11, 1054–1056.

Casagrande, L., and Poulos, S. J. (1969). "On the Effectiveness of Sand Drains," *Canadian Geotechnical Journal,* Vol. 6, No. 3, 287–326.

Chan, H. T., and Kenney, T. C. (1973). "Laboratory Investigation of Permeability Ratio of New Liskeard Varved Soil," *Canadian Geotechnical Journal*, Vol. 10, No. 3, 453–472.

Chapuis, R. P. (2004). "Predicting the Saturated Hydraulic Conductivity of Sand and Gravel Using Effective Diameter and Void Ratio," *Canadian Geotechnical Journal*, Vol. 41, No. 5, 787–795.

Daniel, D. E., and Benson, C. H. (1990). "Water Content-Density Criteria for Compacted Soil Liners," *Journal of Geotechnical Engineering*, ASCE, Vol. 116, No. 12, 1811–1830.

Darcy, H. (1856). *Les Fontaines Publiques de la Ville de Dijon,* Dalmont, Paris.

Hansbo, S. (1960). "Consolidation of Clay with Special Reference to Influence of Vertical Sand Drains," Swedish Geotechnical Institute, *Proc. No.* 18, 41–61.

Hazen, A. (1930). "Water Supply," in *American Civil Engineers Handbook,* Wiley, New York.

Kenney, T. C., and Chan, H. T. (1973). "Field Investigation of Permeability Ratio of New Liskeard Varved Soil," *Canadian Geotechnical Journal*, Vol. 10, No. 3, 473–488.

Kenney, T. C., Lau, D., and Ofoegbu, G. I. (1984). "Permeability of Compacted Granular Materials," *Canadian Geotechnical Journal,* Vol. 21, No. 4, 726–729.

Kozeny, J. (1927). "Ueber kapillare Leitung des Wassers im Boden." *Wien, Akad. Wiss.*, Vol. 136, No. 2a, 271.

Krumbein, W. C., and Monk, G. D. (1943). "Permeability as a Function of the Size Parameters of Unconsolidated Sand," *Transactions*, AIMME (Petroleum Division), Vol. 151, 153–163.

Lumb, P., and Holt, J. K. (1968). "The Undrained Shear Strength of a Soft Marine Clay from Hong Kong," *Geotechnique,* Vol. 18, 25–36.

Mesri, G., and Olson, R. E. (1971). "Mechanism Controlling the Permeability of Clays," *Clay and Clay Minerals,* Vol. 19, 151–158.

Mitchell, J. K. (1976). *Fundamentals of Soil Behavior,* Wiley, New York.

Olsen, H. W. (1961). "Hydraulic Flow Through Saturated Clay," Sc.D. Thesis, Massachusetts Institute of Technology.

Samarasinghe, A. M., Huang, Y. H., and Drnevich, V. P. (1982). "Permeability and Consolidation of Normally Consolidated Soils," *Journal of the Geotechnical Engineering Division*, ASCE, Vol. 108, No. GT6, 835–850.

Spangler. M. G. and Handy, R. L. (1973). *Soil Engineering*, 3rd Ed., Intext Educational, New York.

Tavenas, F., Jean, P., Leblond, F. T. P., and Leroueil, S. (1983). "The Permeability of Natural Soft Clays. Part II: Permeability Characteristics," *Canadian Geotechnical Journal*, Vol. 20, No. 4, 645–660.

Taylor, D. W. (1948). *Fundamentals of Soil Mechanics*, Wiley, New York.

Tsien, S. I. (1955). "Stabilization of Marsh Deposit," *Highway Research Record*, Bulletin 115, 15–43.

U.S. Army Corps of Engineers (1953). "Filter Experiments and Design Criteria," *Technical Memorandum No. 3–360,* U.S. Army Waterways Experiment Station, Vicksburg, Ms.

U.S. Department of Navy (1986). "Soil Mechanics Design Manual 7.01," U.S. Government Printing Office, Washington, D.C.

Van Bavel, C. H. M. and Kirkham, D. (1948). "Field Measurement of Soil Permeability Using Auger Holes," *Proceedings*, Soil Science Society of America, Vol. 13, 90–96.

Wu, T. H., Chang, N. Y., and Ali, E. M. (1978). "Consolidation and Strength Properties of a Clay," *Journal of the Geotechnical Engineering Division,* ASCE, Vol. 104, No. GT7, 899–905.

CHAPTER 8

Seepage

8.1 Introduction

In the preceding chapter, we considered some simple cases for which direct application of Darcy's law was required to calculate the flow of water through soil. In many instances, the flow of water through soil is not in one direction only, nor is it uniform over the entire area perpendicular to the flow. In such cases, the groundwater flow is generally calculated by the use of graphs referred to as *flow nets*. The concept of the flow net is based on *Laplace's equation of continuity*, which governs the steady flow condition for a given point in the soil mass.

In this chapter, we will discuss the following:

- Derivation of Laplace's equation of continuity and some simple applications of the equation
- Procedure to construct flow nets and calculation of seepage in isotropic and anisotropic soils
- Seepage through earth dams

8.2 Laplace's Equation of Continuity

To derive the Laplace differential equation of continuity, let us consider a single row of sheet piles that have been driven into a permeable soil layer, as shown in Figure 8.1a. The row of sheet piles is assumed to be impervious. The steady-state flow of water from the upstream to the downstream side through the permeable layer is a two-dimensional flow. For flow at a point A, we consider an elemental soil block. The block has dimensions dx, dy, and dz (length dy is perpendicular to the plane of the paper); it is shown

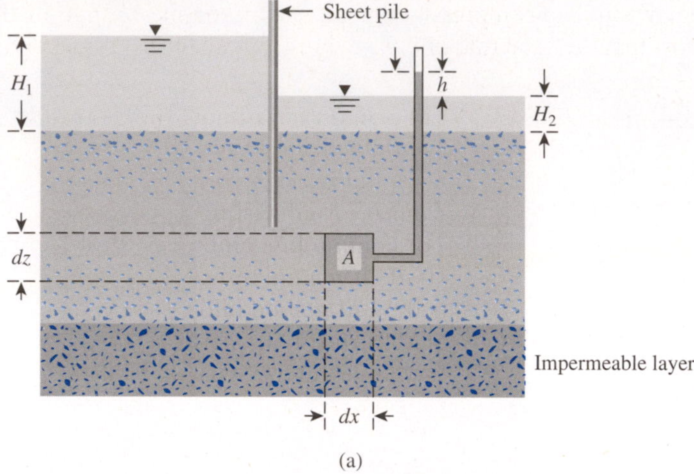

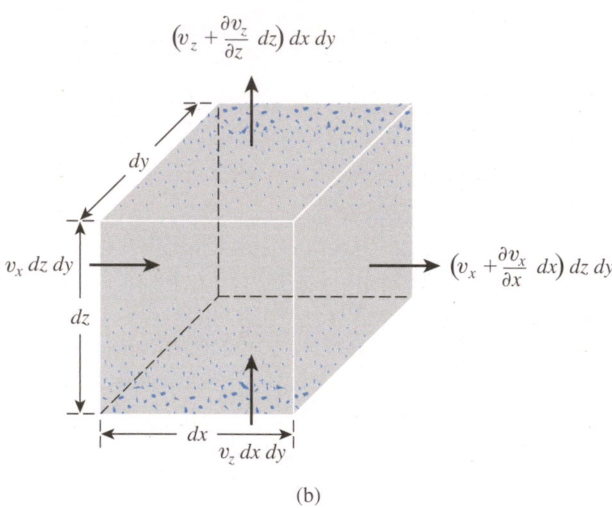

Figure 8.1 (a) Single-row sheet piles driven into permeable layer; (b) flow at A

in an enlarged scale in Figure 8.1b. Let v_x and v_z be the components of the discharge velocity in the horizontal and vertical directions, respectively. The rate of flow of water into the elemental block in the horizontal direction is equal to $v_x\, dz\, dy$, and in the vertical direction it is $v_z\, dx\, dy$. The rates of outflow from the block in the horizontal and vertical directions are, respectively,

$$\left(v_x + \frac{\partial v_x}{\partial x} dx\right) dz\, dy$$

and

$$\left(v_z + \frac{\partial v_z}{\partial z} dz\right) dx\, dy$$

Assuming that water is incompressible and that no volume change in the soil mass occurs, we know that the total rate of inflow should equal the total rate of outflow. Thus,

$$\left[\left(v_x + \frac{\partial v_x}{\partial x}dx\right)dz\,dy + \left(v_z + \frac{\partial v_z}{\partial z}dz\right)dx\,dy\right] - [v_x dz\,dy + v_z\,dx\,dy] = 0$$

or

$$\frac{\partial v_x}{\partial x} + \frac{\partial v_z}{\partial z} = 0 \tag{8.1}$$

With Darcy's law, the discharge velocities can be expressed as

$$v_x = k_x i_x = k_x \frac{\partial h}{\partial x} \tag{8.2}$$

and

$$v_z = k_z i_z = k_z \frac{\partial h}{\partial z} \tag{8.3}$$

where k_x and k_z are the hydraulic conductivities in the horizontal and vertical directions, respectively.

From Eqs. (8.1), (8.2), and (8.3), we can write

$$k_x \frac{\partial^2 h}{\partial x^2} + k_z \frac{\partial^2 h}{\partial z^2} = 0 \tag{8.4}$$

If the soil is isotropic with respect to the hydraulic conductivity—that is, $k_x = k_z$—the preceding continuity equation for two-dimensional flow simplifies to

$$\frac{\partial^2 h}{\partial x^2} + \frac{\partial^2 h}{\partial z^2} = 0 \tag{8.5}$$

8.3 Flow Nets

The continuity equation [Eq. (8.5)] in an isotropic medium represents *two orthogonal families of curves*—that is, the flow lines and the equipotential lines. A *flow line* is a line along which a water particle will travel from upstream to the downstream side in the permeable soil medium. An *equipotential line* is a line along which the potential head at all points is equal. Thus, if piezometers are placed at different points along an equipotential line, the water level will rise to the same elevation in all of them. Figure 8.2a demonstrates the definition of flow and equipotential lines for flow in the permeable soil layer around the row of sheet piles shown in Figure 8.1 (for $k_x = k_z = k$).

A combination of a number of flow lines and equipotential lines is called a *flow net*. As mentioned in the introduction, flow nets are constructed for the calculation

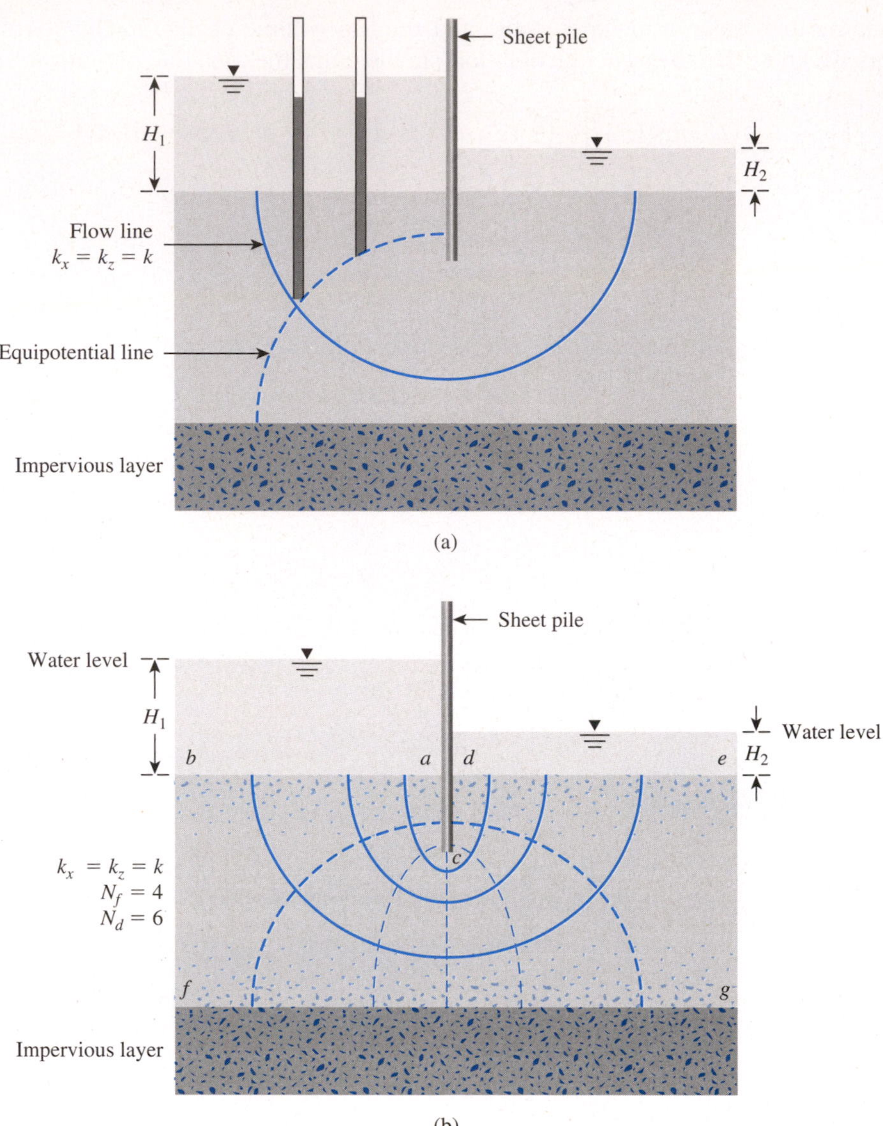

Figure 8.2 (a) Definition of flow lines and equipotential lines; (b) completed flow net

of groundwater flow and the evaluation of heads in the media. To complete the graphic construction of a flow net, one must draw the flow and equipotential lines in such a way that

1. The equipotential lines intersect the flow lines at right angles.
2. The flow elements formed are approximate squares.

Figure 8.2b shows an example of a completed flow net. One more example of flow net in isotropic permeable layer is given in Figure 8.3. In these figures, N_f is the

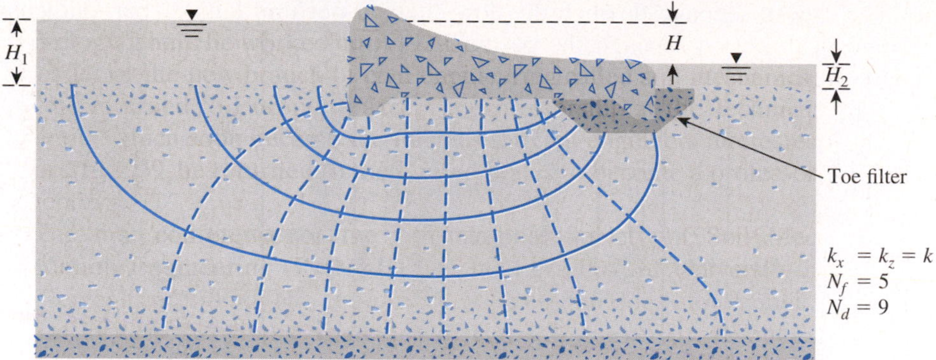

Figure 8.3 Flow net under a dam with toe filter

number of flow channels in the flow net, and N_d is the number of potential drops (defined later in this chapter).

Drawing a flow net takes several trials. While constructing the flow net, keep the boundary conditions in mind. For the flow net shown in Figure 8.2b, the following four boundary conditions apply:

Condition 1. The upstream and downstream surfaces of the permeable layer (lines *ab* and *de*) are equipotential lines.

Condition 2. Because *ab* and *de* are equipotential lines, all the flow lines intersect them at right angles.

Condition 3. The boundary of the impervious layer—that is, line *fg*—is a flow line, and so is the surface of the impervious sheet pile, line *acd*.

Condition 4. The equipotential lines intersect *acd* and *fg* at right angles.

8.4 Seepage Calculation from a Flow Net

In any flow net, the strip between any two adjacent flow lines is called a *flow channel*. Figure 8.4 shows a flow channel with the equipotential lines forming square elements. Let $h_1, h_2, h_3, h_4, \ldots, h_n$ be the piezometric levels corresponding to the equipotential

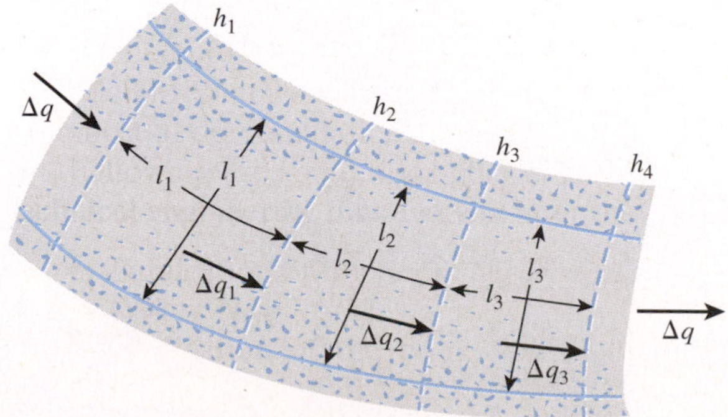

Figure 8.4 Seepage through a flow channel with square elements

lines. The rate of seepage through the flow channel per unit length (perpendicular to the vertical section through the permeable layer) can be calculated as follows. Because there is no flow across the flow lines,

$$\Delta q_1 = \Delta q_2 = \Delta q_3 = \cdots = \Delta q \tag{8.6}$$

From Darcy's law, the flow rate is equal to kiA. Thus, Eq. (8.6) can be written as

$$\Delta q = k\left(\frac{h_1 - h_2}{l_1}\right)l_1 = k\left(\frac{h_2 - h_3}{l_2}\right)l_2 = k\left(\frac{h_3 - h_4}{l_3}\right)l_3 = \cdots \tag{8.7}$$

Equation (8.7) shows that if the flow elements are drawn as approximate squares, the drop in the piezometric level between any two adjacent equipotential lines is the same. This is called the *potential drop*. Thus,

$$h_1 - h_2 = h_2 - h_3 = h_3 - h_4 = \cdots = \frac{H}{N_d} \tag{8.8}$$

and

$$\Delta q = k\frac{H}{N_d} \tag{8.9}$$

where H = head difference between the upstream and downstream sides
N_d = number of potential drops

In Figure 8.2b, for any flow channel, $H = H_1 - H_2$ and $N_d = 6$.

If the number of flow channels in a flow net is equal to N_f, the total rate of flow through all the channels per unit length can be given by

$$q = k\frac{HN_f}{N_d} \tag{8.10}$$

Although drawing square elements for a flow net is convenient, it is not always necessary. Alternatively, one can draw a rectangular mesh for a flow channel, as shown in Figure 8.5, provided that the width-to-length ratios for all the rectangular elements in the flow net are the same. In this case, Eq. (8.7) for rate of flow through the channel can be modified to

$$\Delta q = k\left(\frac{h_1 - h_2}{l_1}\right)b_1 = k\left(\frac{h_2 - h_3}{l_2}\right)b_2 = k\left(\frac{h_3 - h_4}{l_3}\right)b_3 = \cdots \tag{8.11}$$

If $b_1/l_1 = b_2/l_2 = b_3/l_3 = \cdots = n$ (i.e., the elements are not square), Eqs. (8.9) and (8.10) can be modified to

$$\Delta q = kH\left(\frac{n}{N_d}\right) \tag{8.12}$$

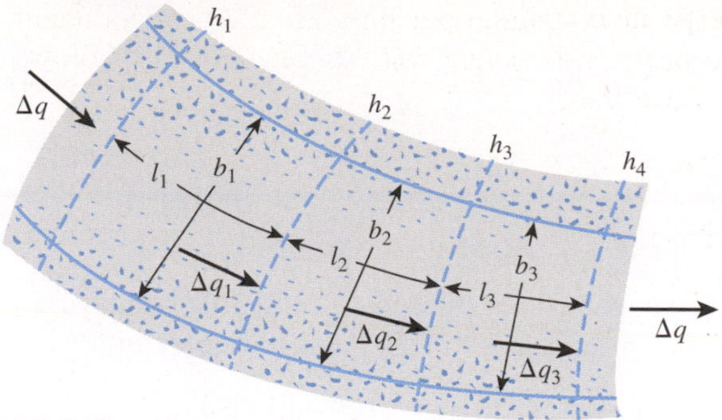

Figure 8.5 Seepage through a flow channel with rectangular elements

and

$$q = kH\left(\frac{N_f}{N_d}\right)n \tag{8.13}$$

Figure 8.6 shows a flow net for seepage around a single row of sheet piles. Note that flow channels 1 and 2 have square elements. Hence, the rate of flow through these two channels can be obtained from Eq. (8.9):

$$\Delta q_1 + \Delta q_2 = \frac{k}{N_d}H + \frac{k}{N_d}H = \frac{2kH}{N_d}$$

However, flow channel 3 has rectangular elements. These elements have a width-to-length ratio of about 0.38; hence, from Eq. (8.12),

$$\Delta q_3 = \frac{k}{N_d}H(0.38)$$

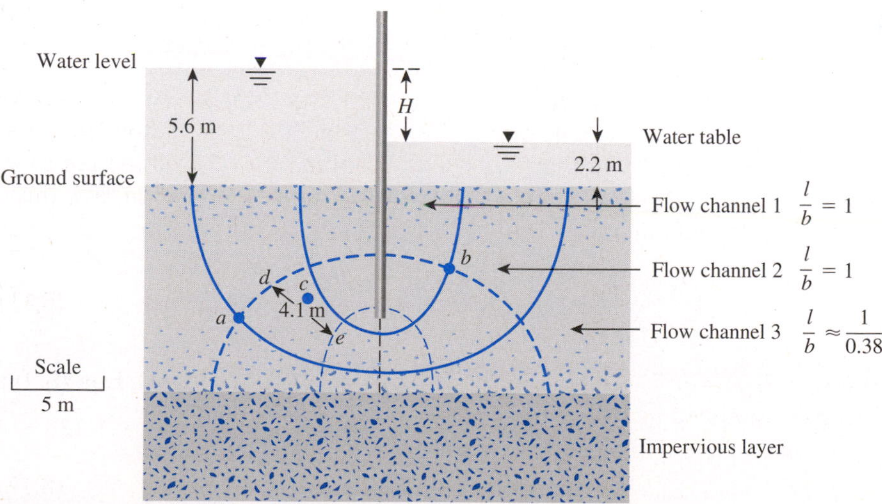

Figure 8.6 Flow net for seepage around a single row of sheet piles

So, the total rate of seepage can be given as

$$q = \Delta q_1 + \Delta q_2 + \Delta q_3 = 2.38\frac{kH}{N_d} \tag{8.14}$$

Example 8.1

A flow net for flow around a single row of sheet piles in a permeable soil layer is shown in Figure 8.6. Given that $k_x = k_z = k = 5 \times 10^{-3}$ cm/sec, determine

a. How high (above the ground surface) the water will rise if piezometers are placed at points *a* and *b*
b. The total rate of seepage through the permeable layer per unit length
c. The approximate average hydraulic gradient at *c*

Solution

Part a

From Figure 8.6, we have $N_d = 6$, $H_1 = 5.6$ m, and $H_2 = 2.2$ m. So the head loss of each potential drop is

$$\Delta H = \frac{H_1 - H_2}{N_d} = \frac{5.6 - 2.2}{6} = 0.567 \text{ m}$$

At point *a*, we have gone through one potential drop. So the water in the piezometer will rise to an elevation of

$$(5.6 - 0.567) = \textbf{5.033 m above the ground surface}$$

At point *b*, we have five potential drops. So the water in the piezometer will rise to an elevation of

$$[5.6 - (5)(0.567)] = \textbf{2.765 m above the ground surface}$$

Part b

From Eq. (8.14),

$$q = 2.38\frac{k(H_1 - H_2)}{N_d} = \frac{(2.38)(5 \times 10^{-5} \text{ m/sec})(5.6 - 2.2)}{6}$$

$$= \mathbf{6.74 \times 10^{-5}\ m^3/sec/m}$$

Part c

The average hydraulic gradient at *c* can be given as

$$i = \frac{\text{head loss}}{\text{average length of flow between } d \text{ and } e} = \frac{\Delta H}{\Delta L} = \frac{0.567 \text{ m}}{4.1 \text{ m}} = \mathbf{0.138}$$

(*Note:* The average length of flow has been scaled.)

Example 8.2

Seepage takes place around a retaining wall shown in Figure 8.7. The hydraulic conductivity of the sand is 1.5×10^{-3} cm/s. The retaining wall is 50 m long. Determine the quantity of seepage across the entire wall per day.

Solution

For the flow net shown in Figure 8.7, $N_f = 3$ and $N_d = 10$. The total head loss from right to left, $H = 5.0$ m. The flow rate is given by [Eq. (8.10)],

$$q = kH\frac{N_f}{N_d} = (1.5 \times 10^{-5} \text{ m/s})(5.0)\left(\frac{3}{10}\right) = 2.25 \times 10^{-5} \text{m}^3\text{/s/m}$$

Seepage across the entire wall,

$$Q = 2.25 \times 10^{-5} \times 50.0 \times 24 \times 3600 \text{ m}^3\text{/day} = \mathbf{97.2 \ m^3/day}$$

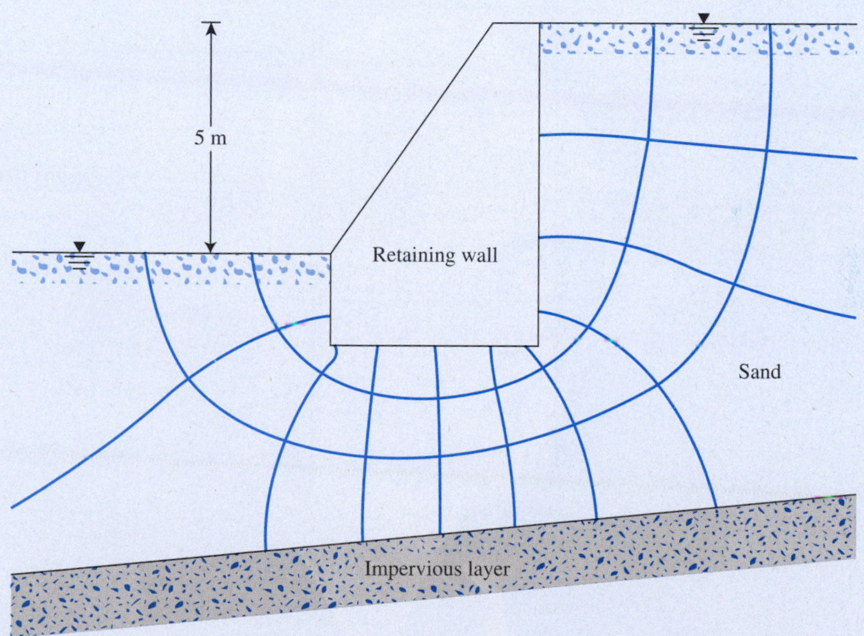

Figure 8.7

Example 8.3

Two sheet piles were driven 4 m apart into clayey sand, as shown in Figure 8.8, and a 2-m depth of soil between the two sheet piles was removed. To facilitate some proposed construction work, the region between the sheet piles is being dewatered where the water level is lowered to the excavation level by pumping out water continuously. Some equipotential lines have been drawn. Complete the flow net.

Assuming the hydraulic conductivity of the clayey sand as 2×10^{-4} cm/s, estimate the quantity of water that has to be pumped out per meter length per day.

Solution

By symmetry, it is possible to analyze only one half of the configuration shown in Figure 8.8. The flow net for the left half is shown in Figure 8.9. Here, $N_f \approx 2.9$ (≈ 3), $N_d = 10$, and $H = 4.5$ m.

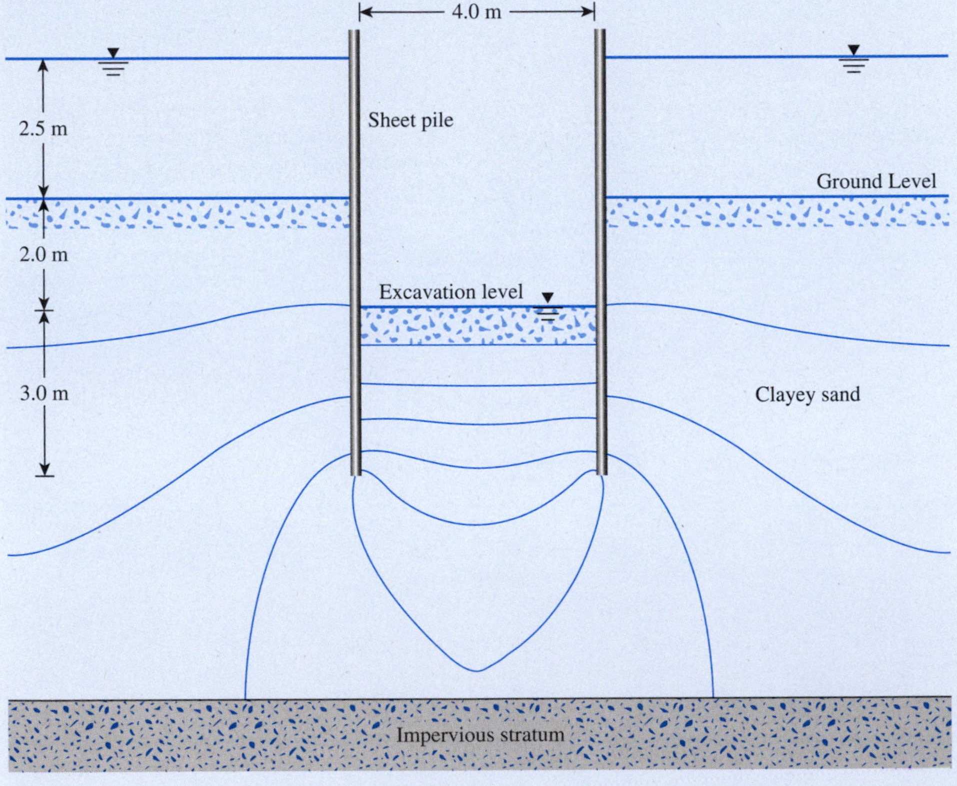

Figure 8.8

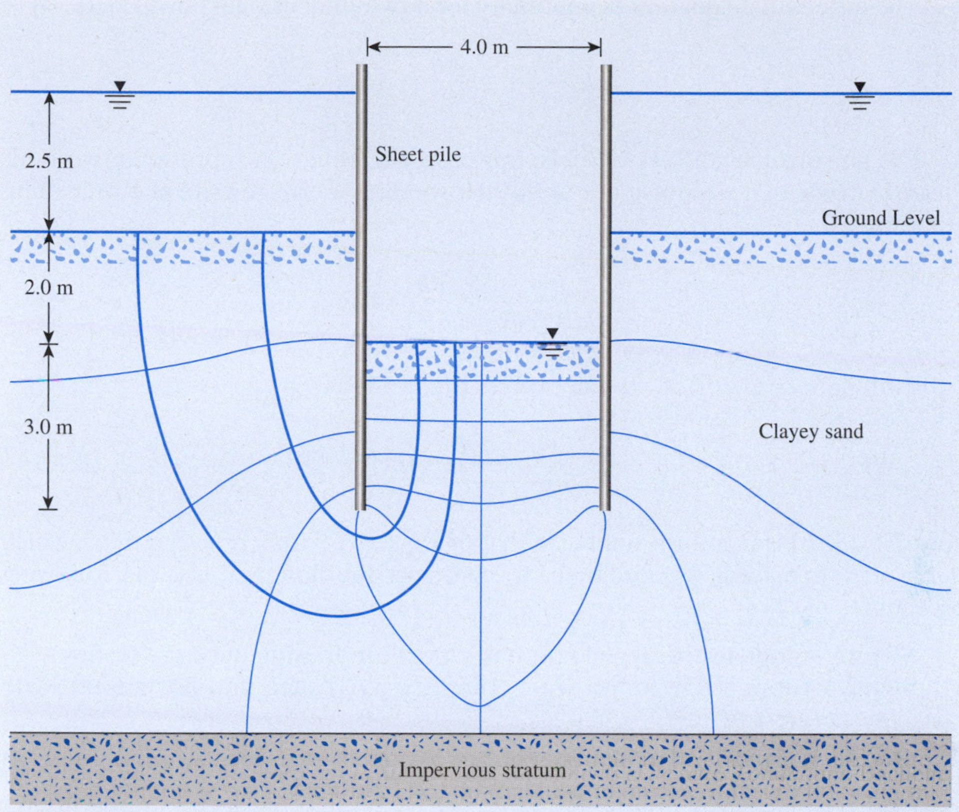

Figure 8.9

The flow rate in the left half can be given by

$$q = kH\frac{N_f}{N_d} = (2 \times 10^{-6})(4.5)\left(\frac{2.9}{10}\right)(24 \times 3600) = 0.226 \text{ m}^3\text{/day/m}$$

Considering the two halves, **the flow rate is 0.452 m³/day/m.**

8.5 Flow Nets in Anisotropic Soil

The flow-net construction described thus far and the derived Eqs. (8.10) and (8.13) for seepage calculation have been based on the assumption that the soil is isotropic. However, in nature, most soils exhibit some degree of anisotropy. To account for soil anisotropy with respect to hydraulic conductivity, we must modify the flow net construction.

The differential equation of continuity for a two-dimensional flow [Eq. (8.4)] is

$$k_x \frac{\partial^2 h}{\partial x^2} + k_z \frac{\partial^2 h}{\partial z^2} = 0$$

For anisotropic soils, $k_x \neq k_z$. In this case, the equation represents two families of curves that do not meet at 90°. However, we can rewrite the preceding equation as

$$\frac{\partial^2 h}{(k_z/k_x)\,\partial x^2} + \frac{\partial^2 h}{\partial z^2} = 0 \tag{8.15}$$

Substituting $x' = \sqrt{k_z/k_x}\,x$, we can express Eq. (8.15) as

$$\frac{\partial^2 h}{\partial x'^2} + \frac{\partial^2 h}{\partial z^2} = 0 \tag{8.16}$$

Now Eq. (8.16) is in a form similar to that of Eq. (8.5), with x replaced by x', which is the new transformed coordinate. To construct the flow net, use the following procedure:

Step 1. Adopt a vertical scale (that is, z axis) for drawing the cross section.
Step 2. Adopt a horizontal scale (that is, x axis) such that horizontal scale $= \sqrt{k_z/k_x} \times$ vertical scale.
Step 3. With scales adopted as in steps 1 and 2, plot the vertical section through the permeable layer parallel to the direction of flow.
Step 4. Draw the flow net for the permeable layer on the section obtained from step 3, with flow lines intersecting equipotential lines at right angles and the elements as approximate squares.

The rate of seepage per unit length can be calculated by modifying Eq. (8.10) to

$$q = \sqrt{k_x k_z}\,\frac{HN_f}{N_d} \tag{8.17}$$

where H = total head loss
N_f and N_d = number of flow channels and potential drops, respectively (from flow net drawn in step 4)

Note that when flow nets are drawn in transformed sections (in anisotropic soils), the flow lines and the equipotential lines are orthogonal. However, when they are redrawn in a true section, these lines are not at right angles to each other. This fact is shown in Figure 8.10. In this figure, it is assumed that $k_x = 6k_z$. Figure 8.10a shows a flow element in a transformed section. The flow element has been redrawn in a true section in Figure 8.10b.

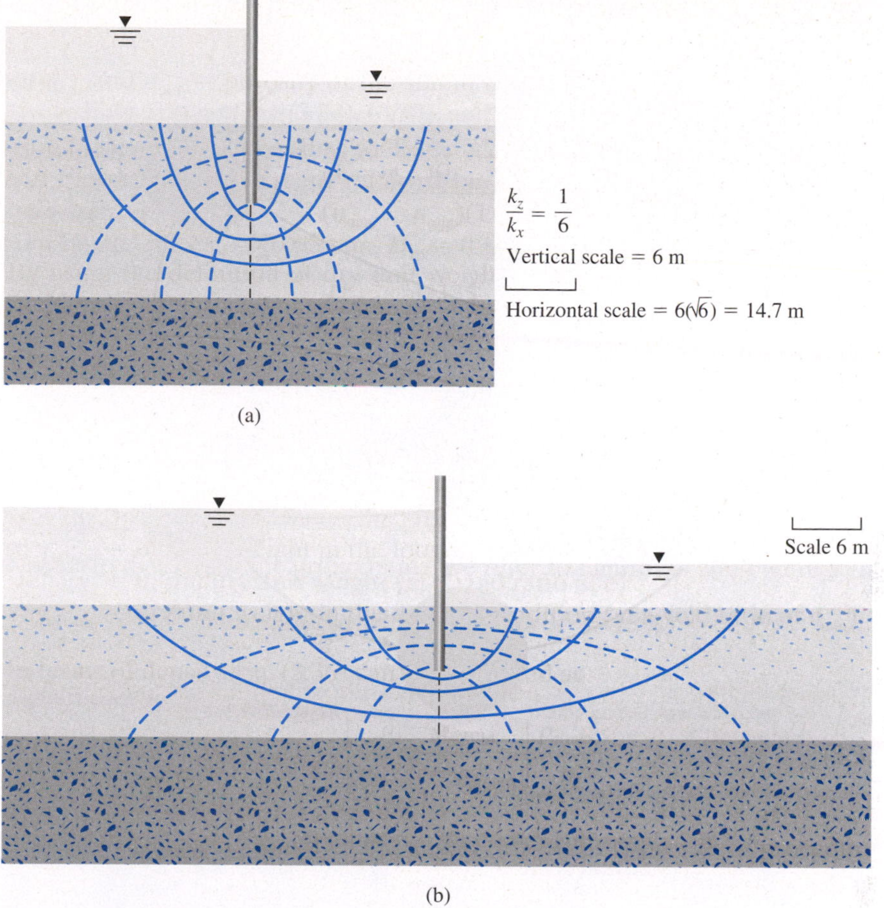

Figure 8.10 A flow element in anisotropic soil: (a) in transformed section; (b) in true section

Example 8.4

A dam section is shown in Figure 8.11a. The hydraulic conductivity of the permeable layer in the vertical and horizontal directions are 2×10^{-2} mm/s and 4×10^{-2} mm/s, respectively. Draw a flow net and calculate the seepage loss of the dam in m^3/day/m

Solution

From the given data,

$$k_z = 2 \times 10^{-2} \text{ mm/s} = 1.728 \text{ m/day}$$

$$k_x = 4 \times 10^{-2} \text{ mm/s} = 3.456 \text{ m/day}$$

and $H = 6.1$ m. For drawing the flow net,

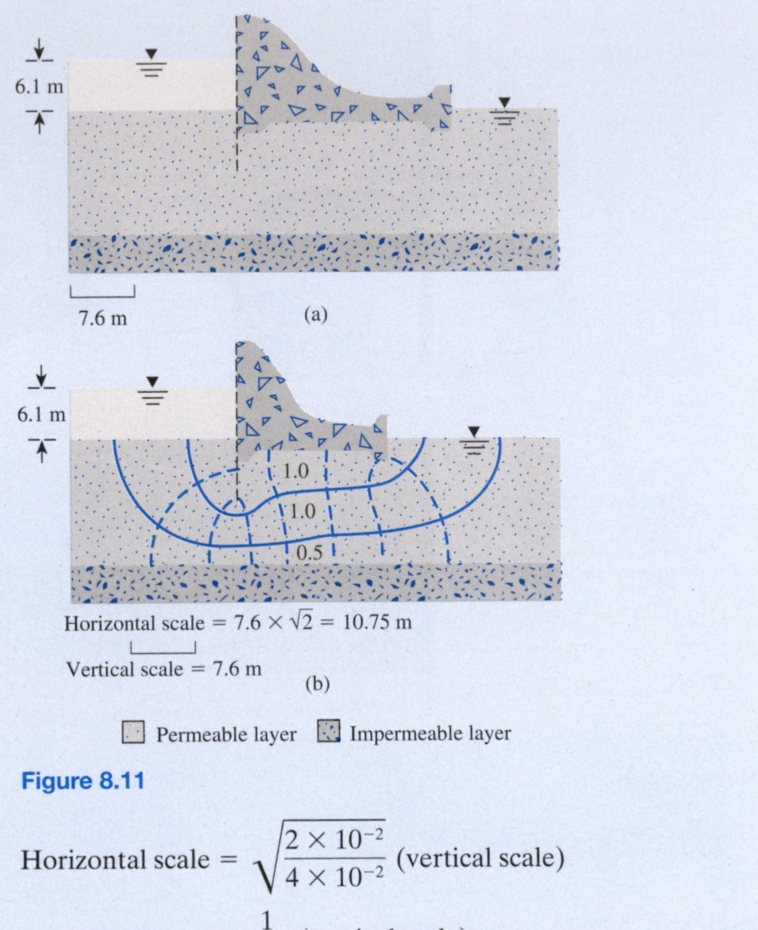

Figure 8.11

$$\text{Horizontal scale} = \sqrt{\frac{2 \times 10^{-2}}{4 \times 10^{-2}}}\ (\text{vertical scale})$$

$$= \frac{1}{\sqrt{2}}\ (\text{vertical scale})$$

On the basis of this, the dam section is replotted, and the flow net drawn as in Figure 8.11b. The rate of seepage is given by $q = \sqrt{k_x k_z}\, H(N_f/N_d)$. From Figure 8.11b, $N_d = 8$ and $N_f = 2.5$ (the lowermost flow channel has a width-to-length ratio of 0.5). So,

$$q = \sqrt{(1.728)(3.456)}(6.1)(2.5/8) = \mathbf{4.66\ m^3/day/m}$$

8.6 Mathematical Solution for Seepage

The seepage under several simple hydraulic structures can be solved mathematically. Harr (1962) has analyzed many such conditions. Figure 8.12 shows a nondimensional plot for the rate of seepage around a single row of sheet piles. In a similar manner, Figure 8.13 is a nondimensional plot for the rate of seepage under a dam. In Figures 8.12 and 8.13, the depth of penetration of the sheet pile is S, and the thickness of the permeable soil layer is T'.

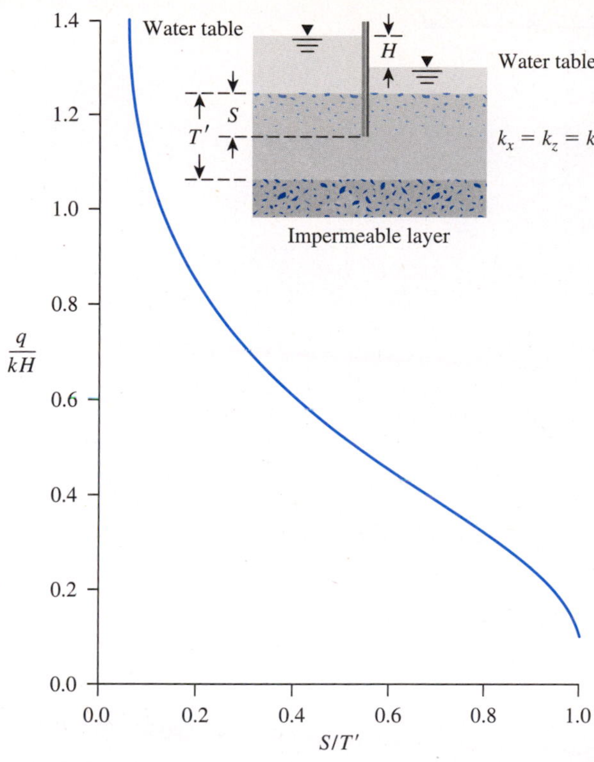

Figure 8.12 Plot of q/kH against S/T' for flow around a single row of sheet piles (*After Harr, 1962. By permission of Dover Publications, Inc.*)

Example 8.5

Refer to Figure 8.13. Given; the width of the dam, $B = 6$ m; length of the dam, $L = 120$ m; $S = 3$ m; $T' = 6$ m; $x = 2.4$ m; and $H_1 - H_2 = 5$ m. If the hydraulic conductivity of the permeable layer is 0.008 cm/sec, estimate the seepage under the dam (Q) in m³/day.

Solution

Given that $B = 6$ m, $T' = 6$ m, and $S = 3$ m, so $b = B/2 = 3$ m.

$$\frac{b}{T'} = \frac{3}{6} = 0.5$$

$$\frac{S}{T'} = \frac{3}{6} = 0.5$$

$$\frac{x}{b} = \frac{2.4}{3} = 0.8$$

From Figure 8.13, for $b/T' = 0.5$, $S/T' = 0.5$, and $x/b = 0.8$, the value of $q/kH \approx 0.378$.

Thus,

$$Q = q\,L = 0.378\,k\,HL = (0.378)(0.008 \times 10^{-2} \times 60 \times 60 \times 24 \text{ m/day})(5)(120)$$

$$= \mathbf{1567.64\ m^3/day}$$

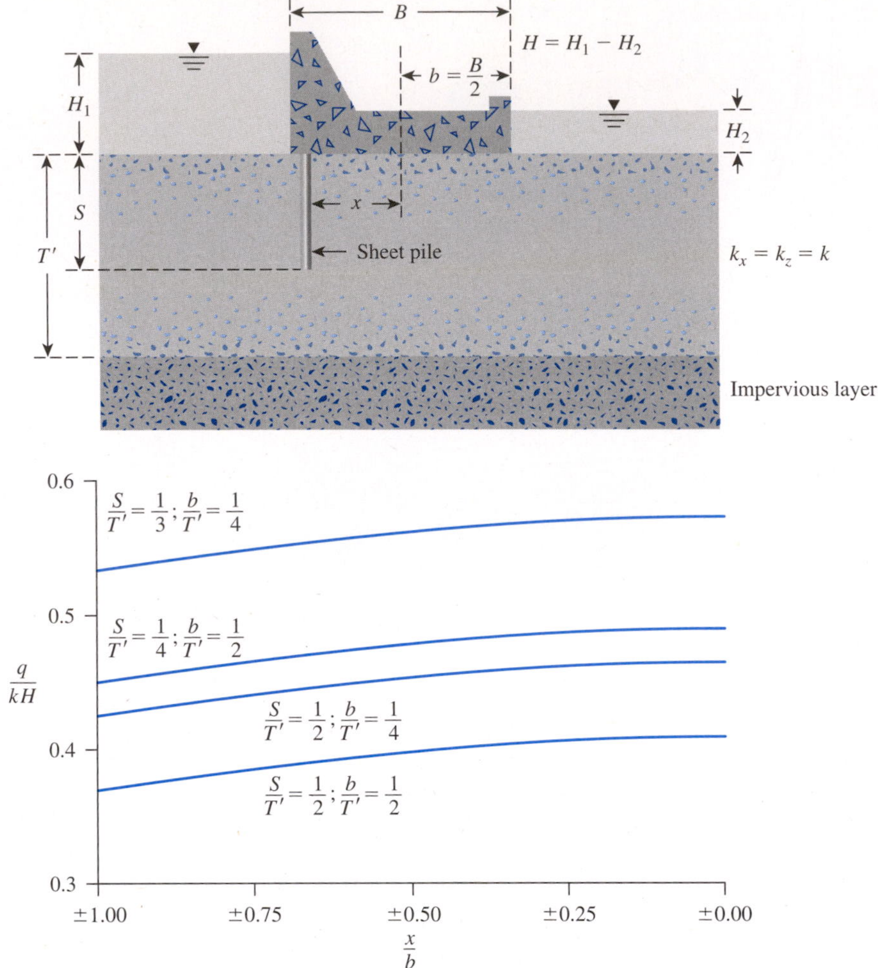

Figure 8.13 Seepage under a dam (*After Harr, 1962. By permission of Dover Publications, Inc.*)

8.7 Uplift Pressure under Hydraulic Structures

Flow nets can be used to determine the uplift pressure at the base of a hydraulic structure. This general concept can be demonstrated by a simple example. Figure 8.14a shows a weir, the base of which is 2 m below the ground surface. The necessary flow net also has been drawn (assuming that $k_x = k_z = k$). The pressure distribution diagram at the base of the weir can be obtained from the equipotential lines as follows.

There are seven equipotential drops (N_d) in the flow net, and the difference in the water levels between the upstream and downstream sides is $H = 7$ m. The head loss for each potential drop is $H/7 = 7/7 = 1$ m. The uplift pressure at

$$a \text{ (left corner of the base)} = (\text{Pressure head at } a) \times (\gamma_w)$$
$$= [(7 + 2) - 1]\gamma_w = 8\gamma_w$$

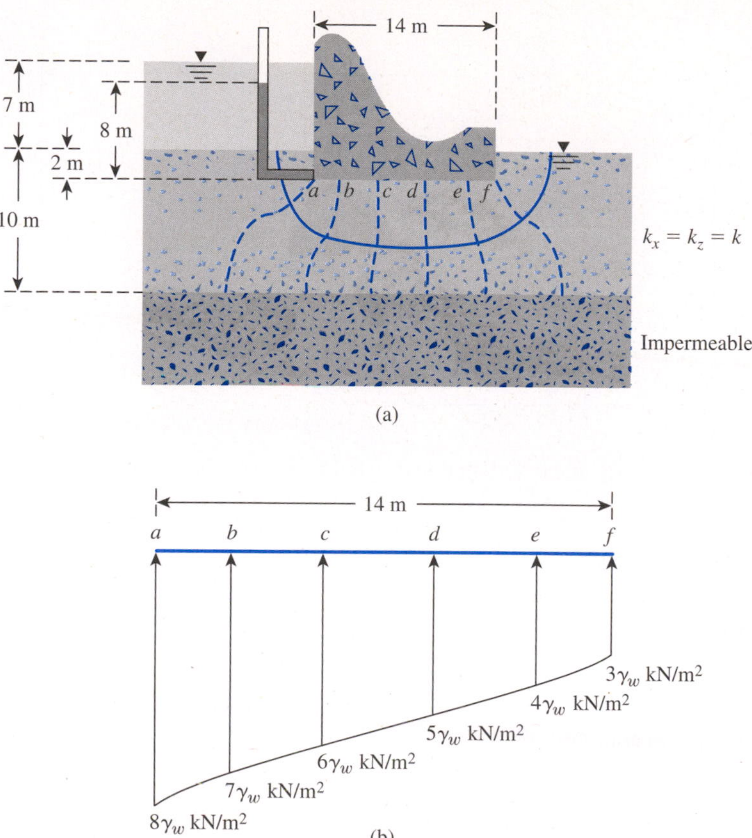

Figure 8.14 (a) A weir; (b) uplift force under a hydraulic structure

Similarly, the uplift pressure at

$$b = [9 - (2)(1)]\gamma_w = 7\gamma_w$$

and at

$$f = [9 - (6)(1)]\gamma_w = 3\gamma_w$$

The uplift pressures have been plotted in Figure 8.14b. The uplift force per unit length measured along the axis of the weir can be calculated by finding the area of the pressure diagram.

8.8 Seepage through an Earth Dam on an Impervious Base

Figure 8.15 shows a homogeneous earth dam resting on an impervious base. Let the hydraulic conductivity of the compacted material of which the earth dam is made be equal to k. The free surface of the water passing through the dam is given by *abcd*. It is assumed that $a'bc$ is parabolic. The slope of the free surface can be assumed to

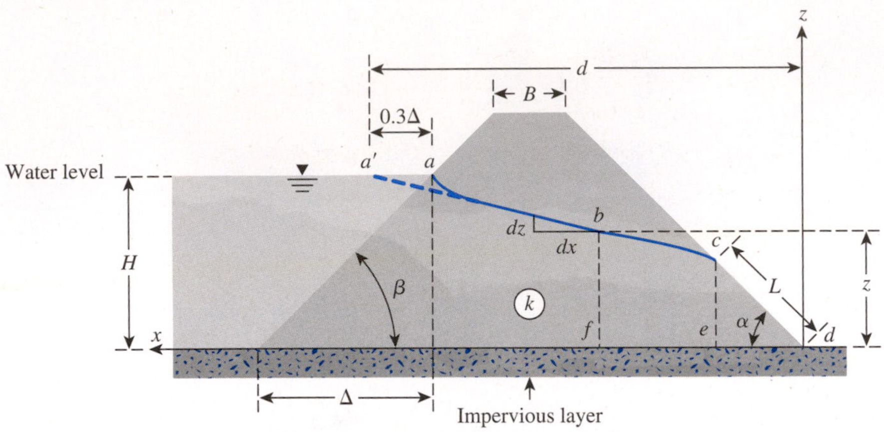

Figure 8.15 Flow through an earth dam constructed over an impervious base

be equal to the hydraulic gradient. It is also is assumed that, because this hydraulic gradient is constant with depth (Dupuit, 1863),

$$i \simeq \frac{dz}{dx} \tag{8.18}$$

Considering the triangle *cde*, we can give the rate of seepage per unit length of the dam (at right angles to the cross section shown in Figure 8.15) as

$$q = kiA$$

$$i = \frac{dz}{dx} = \tan \alpha$$

$$A = (\overline{ce})(1) = L \sin \alpha$$

So

$$q = k(\tan \alpha)(L \sin \alpha) = kL \tan \alpha \sin \alpha \tag{8.19}$$

Again, the rate of seepage (per unit length of the dam) through the section *bf* is

$$q = kiA = k\left(\frac{dz}{dx}\right)(z \times 1) = kz\frac{dz}{dx} \tag{8.20}$$

For continuous flow,

$$q_{\text{Eq. (8.19)}} = q_{\text{Eq. (8.20)}}$$

or

$$kz\frac{dz}{dx} = kL \tan \alpha \sin \alpha$$

or

$$\int_{z=L\sin\alpha}^{z=H} kz\,dz = \int_{x=L\cos\alpha}^{x=d} (kL\tan\alpha\sin\alpha)\,dx$$

$$\tfrac{1}{2}(H^2 - L^2\sin^2\alpha) = L\tan\alpha\sin\alpha(d - L\cos\alpha)$$

$$\frac{H^2}{2} - \frac{L^2\sin^2\alpha}{2} = Ld\left(\frac{\sin^2\alpha}{\cos\alpha}\right) - L^2\sin^2\alpha$$

$$\frac{H^2\cos\alpha}{2\sin^2\alpha} - \frac{L^2\cos\alpha}{2} = Ld - L^2\cos\alpha$$

or

$$L^2\cos\alpha - 2Ld + \frac{H^2\cos\alpha}{\sin^2\alpha} = 0$$

So,

$$L = \frac{d}{\cos\alpha} - \sqrt{\frac{d^2}{\cos^2\alpha} - \frac{H^2}{\sin^2\alpha}} \tag{8.21}$$

Following is a step-by-step procedure to obtain the seepage rate q (per unit length of the dam):

Step 1. Obtain α.
Step 2. Calculate Δ (see Figure 8.15) and then 0.3Δ.
Step 3. Calculate d.
Step 4. With known values of α, H and d, calculate L from Eq. (8.21).
Step 5. With known value of L, calculate q from Eq. (8.19).

The preceding solution generally is referred to as Schaffernak's solution (1917) with Casagrande's (1937) correction, since Casagrande experimentally showed that the parabolic free surface starts from a', not a (Figure 8.15).

Example 8.6

Refer to the earth dam shown in Figure 8.15. Given that $\beta = 45°$, $\alpha = 30°$, $B = 3$ m, $H = 6$ m, height of dam = 7.6 m, and $k = 61 \times 10^{-6}$ m/min, calculate the seepage rate, q, in m³/day/m length.

Solution

We know that $\beta = 45°$ and $\alpha = 30°$. Thus,

$$\Delta = \frac{H}{\tan\beta} = \frac{6}{\tan 45°} = 6 \text{ m} \quad 0.3\Delta = (0.3)(6) = 1.8 \text{ m}$$

$$d = 0.3\Delta + \frac{(7.6 - 6)}{\tan\beta} + B + \frac{7.6}{\tan\alpha}$$

$$= 1.8 + \frac{(7.6 - 6)}{\tan 45°} + 3 + \frac{7.6}{\tan 30} = 23.36 \text{ m}$$

From Eq. (8.21),

$$L = \frac{d}{\cos\alpha} - \sqrt{\frac{d^2}{\cos^2\alpha} - \frac{H^2}{\sin^2\alpha}}$$

$$= \frac{23.36}{\cos 30} - \sqrt{\left(\frac{23.36}{\cos 30}\right)^2 - \left(\frac{6}{\sin 30}\right)^2} = 2.81 \text{ m}$$

From Eq. (8.19)

$$q = kL\ \tan\alpha \sin\alpha = (61 \times 10^{-6})(2.81)(\tan\ 30)(\sin\ 30)$$

$$= 49.48 \times 10^{-6} \text{ m}^3\text{/min/m} = \mathbf{7.13 \times 10^{-2}\ m^3/day/m}$$

8.9 L. Casagrande's Solution for Seepage through an Earth Dam

Equation (8.21) is derived on the basis of Dupuit's assumption (i.e., $i \approx dz/dx$). It was shown by Casagrande (1932) that, when the downstream slope angle α in Figure 8.15 becomes greater than 30°, deviations from Dupuit's assumption become more noticeable. Thus (see Figure 8.15), L. Casagrande (1932) suggested that

$$i = \frac{dz}{ds} = \sin\alpha \tag{8.22}$$

where $ds = \sqrt{dx^2 + dz^2}$.

Now Eq. (8.19) can be modified as

$$q = kiA = k\sin\alpha(L\sin\alpha) = kL\sin^2\alpha \tag{8.23}$$

Again,

$$q = kiA = k\left(\frac{dz}{ds}\right)(1 \times z) \tag{8.24}$$

Combining Eqs. (8.23) and (8.24) yields

$$\int_{L\sin\alpha}^{H} z\,dz = \int_{L}^{S} L\sin^2\alpha\,ds \tag{8.25}$$

where s = length of curve $a'bc$

$$\frac{1}{2}(H^2 - L^2 \sin^2 \alpha) = L \sin^2 \alpha(s - L)$$

or

$$L = s - \sqrt{s^2 - \frac{H^2}{\sin^2 \alpha}} \tag{8.26}$$

With about 4 to 5% error, we can write

$$s = \sqrt{d^2 + H^2} \tag{8.27}$$

Combining Eqs. (8.26) and (8.27) yields

$$L = \sqrt{d^2 + H^2} - \sqrt{d^2 - H^2 \cot^2 \alpha} \tag{8.28}$$

Once the magnitude of L is known, the rate of seepage can be calculated from Eq. (8.23) as

$$q = kL \sin^2 \alpha$$

In order to avoid the approximation introduced in Eqs. (8.27) and (8.28), a solution was provided by Gilboy (1934). This is shown in a graphical form in Figure 8.16. Note, in this graph,

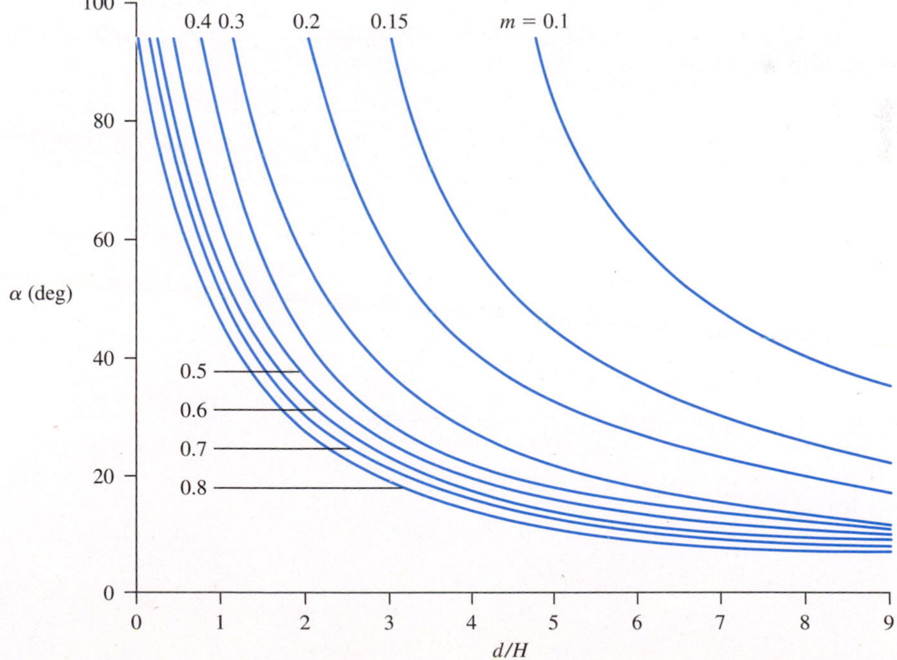

Figure 8.16 Chart for solution by L. Casagrande's method based on Gilboy's solution

$$m = \frac{L \sin \alpha}{H} \tag{8.29}$$

In order to use the graph,

Step 1. Determine d/H.
Step 2. For a given d/H and α, determine m.
Step 3. Calculate $L = \dfrac{mH}{\sin \alpha}$.
Step 4. Calculate $q = kL \sin^2 \alpha$.

8.10 Pavlovsky's Solution for Seepage through an Earth Dam

Pavlovksy (1931; also see Harr, 1962) gave a solution for calculating seepage through an earth dam. This can be explained with reference to Figure 8.17. The dam section can be divided into three zones, and the rate of seepage through each zone can be calculated as follows.

Zone I (Area *agOf*)

In Zone I the seepage lines are actually curved, but Pavlovsky assumed that they can be replaced by horizontal lines. The rate of seepage through an elemental strip of thickness dz then can be given by

$$dq = ki\,(dA)$$

$$dA = (dz)(1) = dz$$

$$i = \frac{\text{Loss of head, } l_1}{\text{Lenght of flow}} = \frac{l_1}{(H_d - z)\cot \beta}$$

So,

$$q = \int dq = \int_0^{h_1} \frac{kl_1}{(H_d - z)\cot \beta} dz = \frac{kl_1}{\cot \beta} \ln \frac{H_d}{H_d - h_1}$$

However, $l_1 = H - h_1$. So,

$$q = \frac{k(H - h_1)}{\cot \beta} \ln \frac{H_d}{H_d - h_1} \tag{8.30}$$

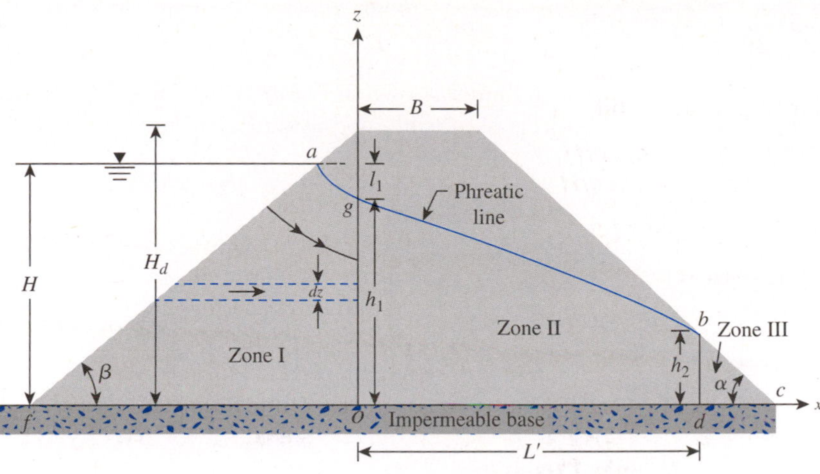

Figure 8.17 Pavlovsky's solution for seepage through an earth dam

Zone II (Area *Ogbd*)

The flow in Zone II can be given by the equation derived by Dupuit [Eq. (8.18)] or

$$q = \frac{k}{2L'}(h_1^2 - h_2^2) \tag{8.31}$$

where

$$L' = B + (H_d - h_2)\cot\alpha \tag{8.32}$$

Zone III (Area *bcd*)

As in Zone I, the stream lines in Zone III are also assumed to be horizontal:

$$q = k\int_0^{h_2}\frac{dz}{\cot\alpha} = \frac{kh_2}{\cot\alpha} \tag{8.33}$$

Combining Eqs. (8.30) through (8.32),

$$h_2 = \frac{B}{\cot\alpha} + H_d - \sqrt{\left(\frac{B}{\cot\alpha} + H_d\right)^2 - h_1^2} \tag{8.34}$$

From Eqs. (8.30) and (8.33),

$$\frac{H - h_1}{\cot\beta}\ln\frac{H_d}{H_d - h_1} = \frac{h_2}{\cot\alpha} \tag{8.35}$$

Equations (8.34) and (8.35) contain two unknowns, h_1 and h_2, which can be solved graphically. Once these are known, the rate of seepage per unit length of the dam can be obtained from any one of the Eqs. (8.30), (8.31), and (8.33).

Example 8.7

The cross section of an earth dam is shown in Figure 8.18. Calculate the rate of seepage through the dam [q is in $m^3/(min \cdot m)$] using Pavlovksy's method.

Solution

From Figure 8.18, $\alpha = \beta = \tan^{-1}(\frac{1}{2}) = 26.57°$; $H_d = 30$ m; $H = 25$ m; $B = 5$ m. From Eqs. (8.34) and (8.35),

$$h_2 = \frac{B}{\cot\alpha} + H_d - \sqrt{\left(\frac{B}{\cot\alpha} + H_d\right)^2 - h_1^2}$$

and

$$\frac{H - h_1}{\cot\beta}\ln\frac{H_d}{H_d - h_1} = \frac{h_2}{\cot\alpha}$$

Hence, from Eq. (8.34),

$$h_2 = \frac{5}{2} + 30 - \sqrt{\left(\frac{5}{2} + 30\right)^2 - h_1^2}$$

or

$$h_2 = 32.5 - \sqrt{1056.25 - h_1^2} \qquad \text{(a)}$$

Similarly, from Eq. (8.35),

$$\frac{25 - h_1}{2}\ln\frac{30}{30 - h_1} = \frac{h_2}{2}$$

or

$$h_2 = (25 - h_1)\ln\frac{30}{30 - h_1} \qquad \text{(b)}$$

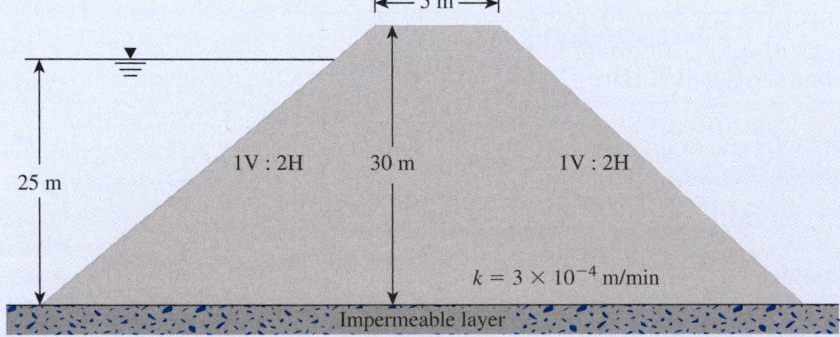

Figure 8.18

Equations (a) and (b) must be solved by trial and error:

h_1 (m)	h_2 from Eq. (a) (m)	h_2 from Eq. (b) (m)
2	0.062	1.587
4	0.247	3.005
6	0.559	4.240
8	1.0	5.273
10	1.577	6.082
12	2.297	6.641
14	3.170	6.915
16	4.211	6.859
18	5.400	6.414
20	6.882	5.493

Using the values of h_1 and h_2 calculated in the preceding table, we can plot the graph as shown in Figure 8.19 and, from that, $h_1 = 18.9$ m and $h_2 = 6.06$ m. From Eq. (8.33),

$$q = \frac{kh_2}{\cot \alpha} = \frac{(3 \times 10^{-4})(6.06)}{2} = \mathbf{9.09 \times 10^{-4}\ m^3/(min \cdot m)}$$

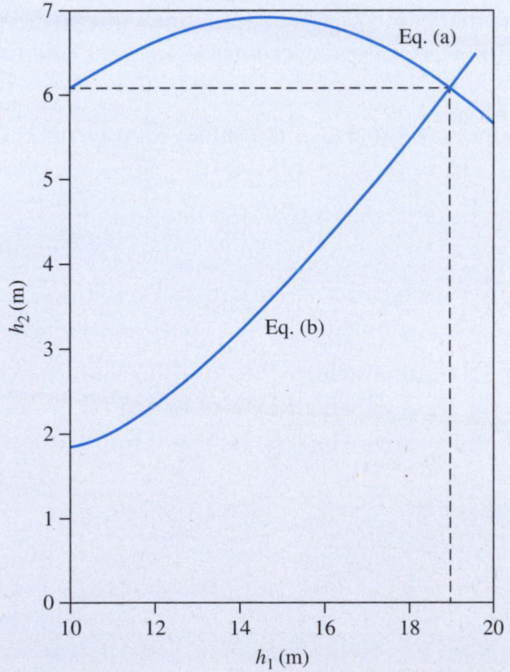

Figure 8.19

8.11 Filter Design

When seepage water flows from a soil with relatively fine grains into a coarser material, there is danger that the fine soil particles may wash away into the coarse material. Over a period of time, this process may clog the void spaces in the coarser material. Hence, the grain-size distribution of the coarse material should be properly manipulated to avoid this situation. A properly designed coarser material is called a *filter*. Figure 8.20 shows the steady-state seepage condition in an earth dam which has a toe filter. For proper selection of the filter material, two conditions should be kept in mind:

Condition 1. The size of the voids in the filter material should be small enough to hold the larger particles of the protected material in place.

Condition 2. The filter material should have a high hydraulic conductivity to prevent buildup of large seepage forces and hydrostatic pressures in the filters.

It can be shown that, if three perfect spheres have diameters greater than 6.5 times the diameter of a smaller sphere, the small sphere can move through the void spaces of the larger ones (Figure 8.21a). Generally speaking, in a given soil, the sizes of the grains vary over a wide range. If the pore spaces in a filter are small enough to hold D_{85} of the soil to be protected, then the finer soil particles also will be protected (Figure 8.21b). This means that the effective diameter of the pore spaces in the filter should be less than D_{85} of the soil to be protected. The effective pore diameter is about $\frac{1}{5}D_{15}$ of the filter. With this in mind and based on the experimental investigation of filters, Terzaghi and Peck (1948) provided the following criteria to satisfy Condition 1:

$$\frac{D_{15(F)}}{D_{85(S)}} \leq 4 \text{ to } 5 \qquad \text{(to satisfy Condition 1)} \tag{8.36}$$

In order to satisfy Condition 2, they suggested that

$$\frac{D_{15(F)}}{D_{15(S)}} \geq 4 \text{ to } 5 \qquad \text{(to satisfy Condition 2)} \tag{8.37}$$

where $D_{15(F)}$ = diameter through which 15% of filter material will pass
$D_{15(S)}$ = diameter through which 15% of soil to be protected will pass
$D_{85(S)}$ = diameter through which 85% of soil to be protected will pass

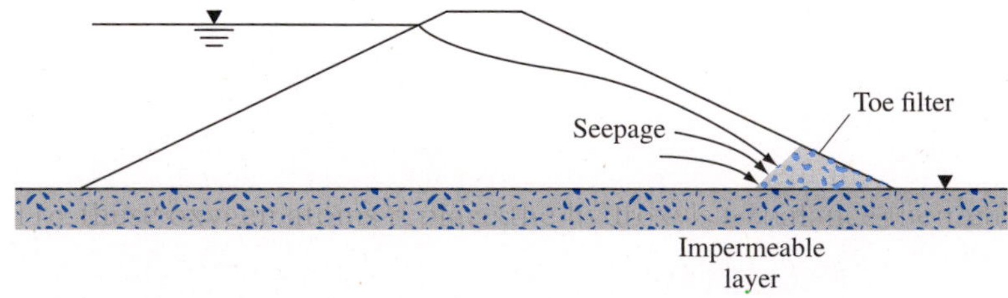

Figure 8.20 Steady-state seepage in an earth dam with a toe filter

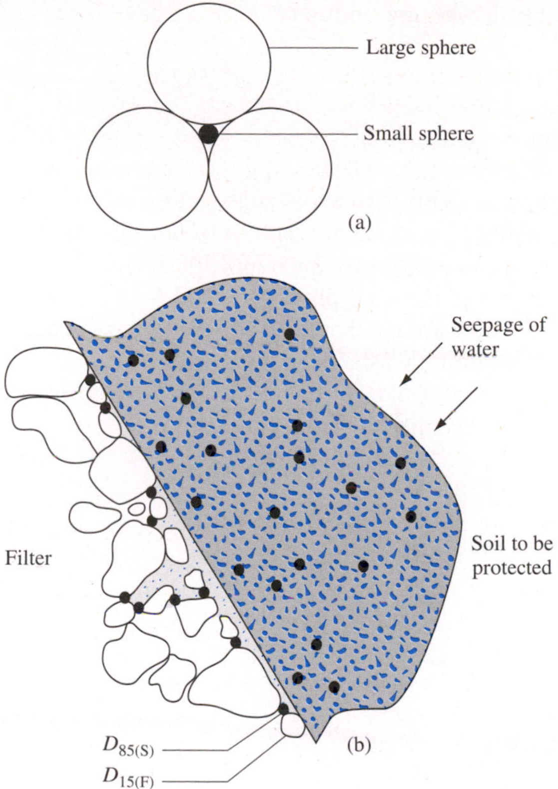

Figure 8.21 (a) Large spheres with diameters of 6.5 times the diameter of the small sphere; (b) boundary between a filter and the soil to be protected

The U.S. Navy (1971) requires the following conditions for the design of filters.

Condition 1. For avoiding the movement of the particles of the protected soil,

$$\frac{D_{15(F)}}{D_{85(S)}} < 5 \tag{8.38}$$

$$\frac{D_{50\,(F)}}{D_{50\,(S)}} < 25 \tag{8.39}$$

$$\frac{D_{15(F)}}{D_{15(S)}} < 20 \tag{8.40}$$

If the uniformity coefficient C_u of the protected soil is less than 1.5, $D_{15(F)}/D_{85(S)}$ may be increased to 6. Also, if C_u of the protected soil is greater than 4, $D_{15(F)}/D_{15(S)}$ may be increased to 40.

Condition 2. For avoiding buildup of large seepage force in the filter,

$$\frac{D_{15(F)}}{D_{15(S)}} > 4 \tag{8.41}$$

Condition 3. The filter material should not have grain sizes greater than 76.2 mm (3 in.). (This is to avoid segregation of particles in the filter.)

Condition 4. To avoid internal movement of fines in the filter, it should have no more than 5% passing a No. 200 sieve.

Condition 5. When perforated pipes are used for collecting seepage water, filters also are used around the pipes to protect the fine-grained soil from being washed into the pipes. To avoid the movement of the filter material into the drain-pipe perforations, the following additional conditions should be met:

$$\frac{D_{85(F)}}{\text{Slot width}} > 1.2 \text{ to } 1.4$$

$$\frac{D_{85(F)}}{\text{Hole diameter}} > 1.0 \text{ to } 1.2$$

Example 8.8

The grain-size distribution of a soil to be protected is shown as curve *a* in Figure 8.22. Given for the soil: $D_{15(S)} = 0.009$ mm, $D_{50(S)} = 0.05$ mm, and $D_{85(S)} = 0.11$ mm. Using Eqs. (8.38) through (8.41), determine the zone of the grain-size distribution of the filter material.

Solution

From Eq. (8.38),

$$\frac{D_{15(F)}}{D_{85(S)}} < 5$$

or

$$D_{15(F)} < 5D_{85(S)} = (5)(0.11) = 0.55 \text{ mm}$$

From Eq. (8.39),

$$\frac{D_{50(F)}}{D_{50(S)}} < 25$$

or

$$D_{50(F)} < 25D_{50(S)} = (25)(0.05) = 1.25 \text{ mm}$$

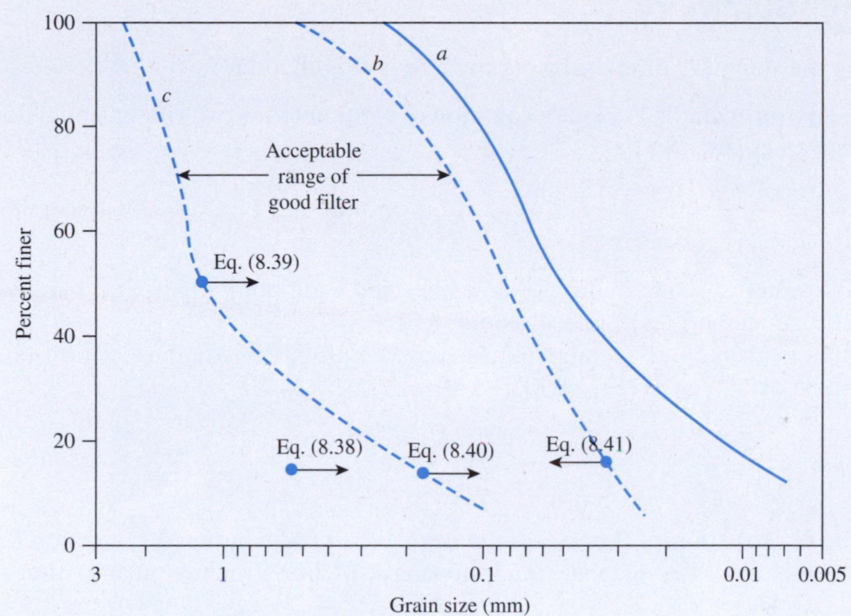

Figure 8.22

From Eq. (8.40),

$$\frac{D_{15(F)}}{D_{15(S)}} < 20$$

or

$$D_{15(F)} < 20D_{15(S)} = (20)(0.009) = 0.18 \text{ mm}$$

From Eq. (8.41),

$$\frac{D_{15(F)}}{D_{85(S)}} > 4$$

or

$$D_{15(F)} > 4D_{15(S)} = (4)(0.009) = 0.036 \text{ mm}$$

The above calculations have been plotted in Figure 8.22. Curves b and c are approximately the same shape as curve a. **The acceptable range of good filter falls between curves *b* and *c*.**

8.12 Summary

Following is a summary of the subjects covered in this chapter.

- In an *isotropic soil*, Laplace's equation of continuity for two-dimensional flow is given as [Eq. (8.5)]:

$$\frac{\partial^2 h}{\partial x^2} + \frac{\partial^2 h}{\partial z^2} = 0$$

- A flow net is a combination of flow lines and equipotential lines that are two orthogonal families of lines (Section 8.3).
- In an isotropic soil, seepage (q) for unit length of the structure in unit time can be expressed as [Eq. (8.13)]

$$q = kH\left(\frac{N_f}{N_d}\right)n$$

- The construction of flow nets in *anisotropic soil* was outlined in Section 8.5. For this case, the seepage for unit length of the structure in unit time is [Eq. (8.17)]

$$q = \sqrt{k_x k_z}\,\frac{HN_f}{N_d}$$

- Seepage through an earth dam on an impervious base was discussed in Section 8.8 (Schaffernak's solution with Casagrande's correction), Section 8.9 (L. Casagrande solution), and Section 8.10 (Pavlovsky's solution).
- The criteria for filter design (Terzaghi and Peck, 1948) are given in Section 8.11 [Eqs. (8.36) and (8.37)], according to which

$$\frac{D_{15(\mathrm{F})}}{D_{85(\mathrm{S})}} \le 4 \text{ to } 5$$

and

$$\frac{D_{15(\mathrm{F})}}{D_{15(\mathrm{S})}} \ge 4 \text{ to } 5$$

Problems

8.1 Refer to Figure 8.23. Given:
- $H_1 = 6$ m
- $H_2 = 1.5$ m
- $D = 3$ m
- $D_1 = 6$ m

Draw a flow net. Calculate the seepage loss per meter length of the sheet pile (at a right angle to the cross section shown).

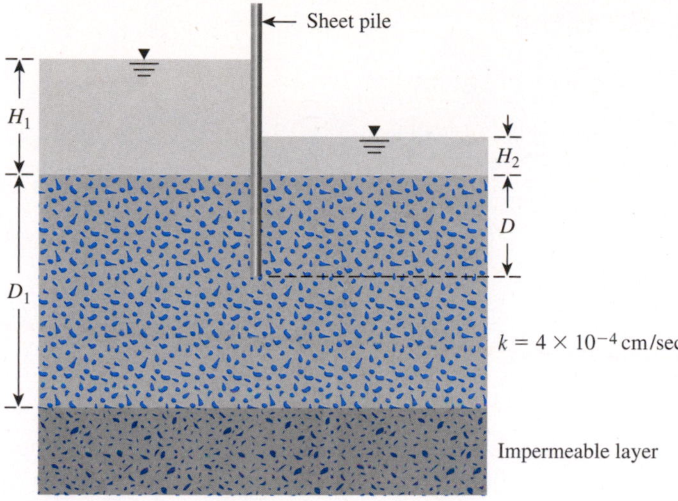

Figure 8.23

8.2 Draw a flow net for the single row of sheet piles driven into a permeable layer as shown in Figure 8.23. Given:

- $H_1 = 3$ m
- $H_2 = 0.5$ m
- $D = 1.5$ m
- $D_1 = 3.75$ m

Calculate the seepage loss per meter length of the sheet pile (at right angles to the cross section shown).

8.3 Refer to Figure 8.23. Given:

- $H_1 = 4$ m
- $H_2 = 1.5$ m
- $D_1 = 6$ m
- $D = 3.6$ m

Calculate the seepage loss in m^3/day per meter length of the sheet pile (at right angles to the cross section shown). Use Figure 8.12.

8.4 For the hydraulic structure shown in Figure 8.24, draw a flow net for flow through the permeable layer and calculate the seepage loss in m^3/day/m.

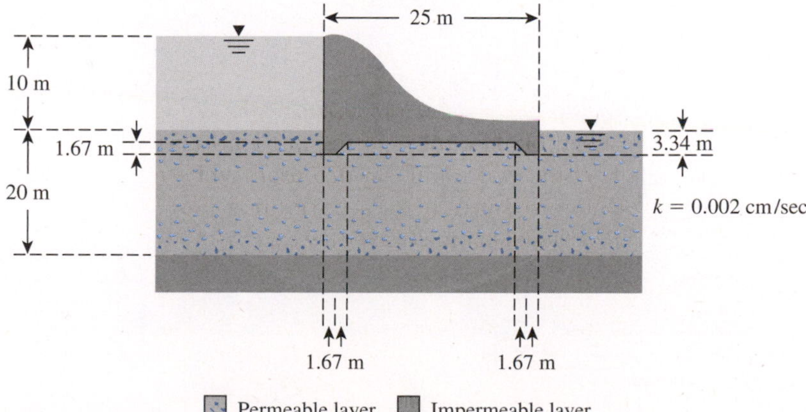

Figure 8.24

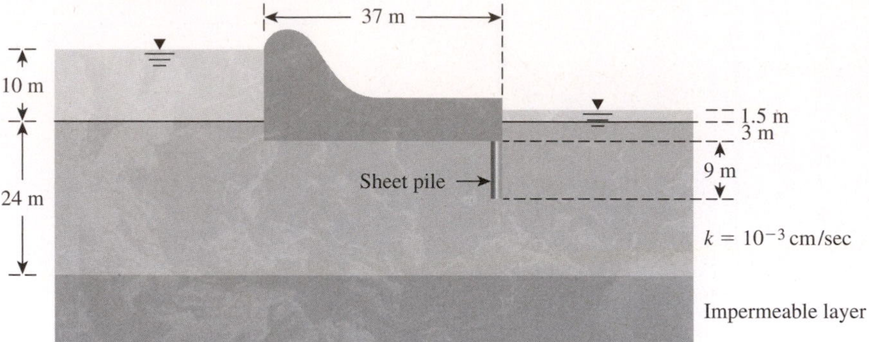

Figure 8.25

8.5 Refer to Problem 8.4. Using the flow net drawn, calculate the hydraulic uplift force at the base of the hydraulic structure per meter length (measured along the axis of the structure).

8.6 Draw a flow net for the weir shown in Figure 8.25. Calculate the rate of seepage under the weir.

8.7 For the weir shown in Figure 8.26, calculate the seepage in the permeable layer in m^3/day/m for (a) $x' = 1$ m and (b) $x' = 2$ m. Use Figure 8.13.

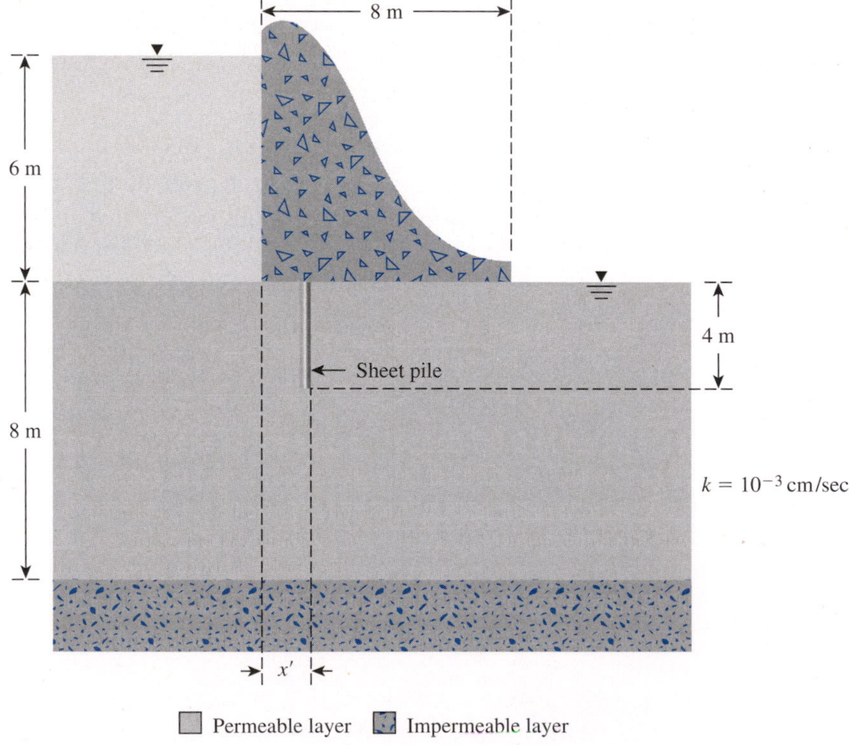

Figure 8.26

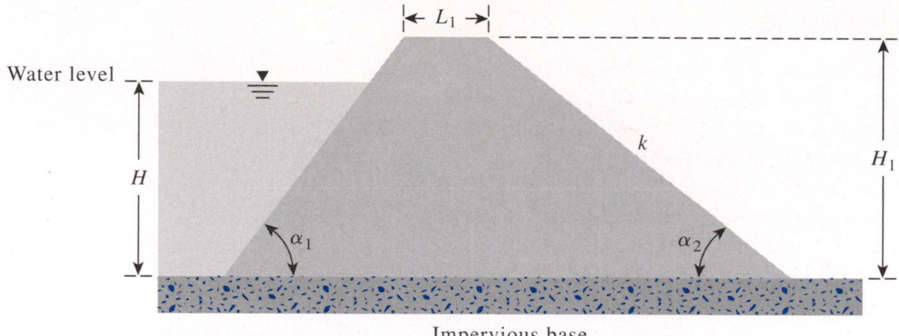

Figure 8.27

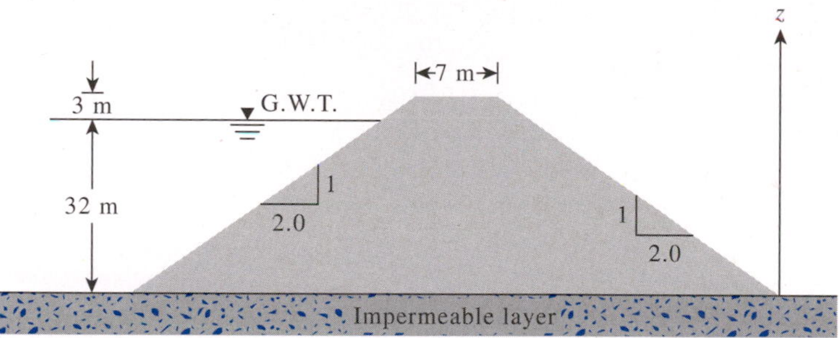

Figure 8.28

8.8 An earth dam is shown in Figure 8.27. Determine the seepage rate, q, in m^3/day/m length. Given: $\alpha_1 = 35°$, $\alpha_2 = 40°$, $L_1 = 5$ m, $H = 7$ m, $H_1 = 10$ m, and $k = 3 \times 10^{-4}$ cm/sec. Use Schaffernak's solution.

8.9 Repeat Problem 8.8 using L. Casagrande's method.

8.10 Refer to the cross section of the earth dam shown in Figure 8.18. Calculate the rate of seepage through the dam (q in m^3/min/m) using Schaffernak's solution.

8.11 Solve Problem 8.10 using L. Casagrande's method.

8.12 An earth dam section is shown in Figure 8.28. Determine the rate of seepage through the earth dam using Pavlovsky's solution. Use $k = 4 \times 10^{-5}$ mm/s.

Critical Thinking Problem

8.C.1 Refer to Problem 8.12. Given $k_x = 4 \times 10^{-5}$ m/min and $k_z = 1 \times 10^{-5}$ m/min, calculate the rate of seepage through the dam (m^3/min/m) using Schaffernak's solution.

References

Casagrande, A. (1937). "Seepage Through Dams," in *Contribution to Soil Mechanics 1925–1940*, Boston Society of Civil Engineers, Boston.

Casagrande, L. (1932). "Naeherungsmethoden zur Bestimmurg von Art und Menge der Sickerung durch geschuettete Daemme," Thesis, Technische Hochschule, Vienna.

Dupuit, J. (1863). *Etudes Theoriques et Practiques sur le Mouvement des Eaux dans les Canaux Decouverts et a Travers les Terrains Permeables*, Dunod, Paris.

Gilboy, G. (1934). "Mechanics of Hydraulic Fill Dams," in *Contributions to Soil Mechanics 1925–1940*, Boston Society of Civil Engineers, Boston.

Harr, M. E. (1962). *Ground Water and Seepage*, McGraw-Hill, New York.

Pavlovksy, N.N. (1931). *Seepage through Earth Dams* (in Russian), Inst. Gidrotekhniki I Melioratsii, Leningrad, Russia.

Schaffernak, F. (1917). "Über die Standicherheit durchlaessiger geschuetteter Dämme," *Allgem. Bauzeitung*.

Terzaghi, K., and Peck, R. B. (1948). *Soil Mechanics in Engineering Practice*, Wiley, New York.

U.S. Department of the Navy, Naval Facilities Engineering Command (1971). "Design Manual—Soil Mechanics, Foundations, and Earth Structures," *NAVFAC DM-7*, Washington, D.C.

CHAPTER 9

In Situ Stresses

9.1 Introduction

As described in Chapter 3, soils are multiphase systems. In a given volume of soil, the solid particles are distributed randomly with void spaces between. The void spaces are continuous and are occupied by water and/or air. To analyze problems (such as compressibility of soils, bearing capacity of foundations, stability of embankments, and lateral pressure on earth-retaining structures), we need to know the nature of the distribution of stress along a given cross section of the soil profile. We can begin the analysis by considering a saturated soil with no seepage.
In this chapter, we will discuss the following:

- Concept of effective stress
- Stresses in saturated soil without seepage, upward seepage, and downward seepage
- Seepage force per unit volume of soil
- Conditions for *heaving* or boiling for seepage under a hydraulic structure
- Use of filter to increase the stability against heaving or boiling
- Effective stress in partially saturated soil

9.2 Stresses in Saturated Soil without Seepage

Figure 9.1a shows a column of saturated soil mass with no seepage of water in any direction. The total stress at the elevation of point A can be obtained from the saturated unit weight of the soil and the unit weight of water above it. Thus,

$$\sigma = H\gamma_w + (H_A - H)\gamma_{sat} \tag{9.1}$$

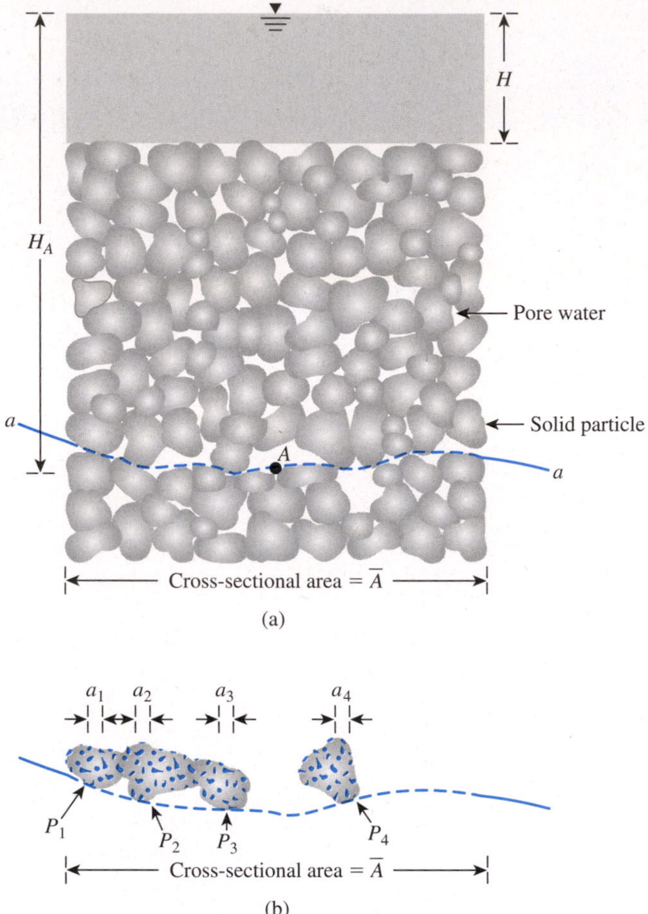

Figure 9.1 (a) Effective stress consideration for a saturated soil column without seepage; (b) forces acting at the points of contact of soil particles at the level of point *A*

where σ = total stress at the elevation of point *A*
γ_w = unit weight of water
γ_{sat} = saturated unit weight of the soil
H = height of water table from the top of the soil column
H_A = distance between point *A* and the water table

The total stress, σ, given by Eq. (9.1) can be divided into two parts:

1. A portion is carried by water in the continuous void spaces. This portion acts with equal intensity in all directions.
2. The rest of the total stress is carried by the soil solids at their points of contact. The sum of the vertical components of the forces developed at the points of contact of the solid particles per unit cross-sectional area of the soil mass is called the *effective stress*.

This can be seen by drawing a wavy line, a–a, through point A that passes only through the points of contact of the solid particles. Let $P_1, P_2, P_3, \ldots, P_n$ be the forces that act at the points of contact of the soil particles (Figure 9.1b). The sum of the vertical components of all such forces over the unit cross-sectional area is equal to the effective stress σ', or

$$\sigma' = \frac{P_{1(v)} + P_{2(v)} + P_{3(v)} + \cdots + P_{n(v)}}{\overline{A}} \tag{9.2}$$

where $P_{1(v)}, P_{2(v)}, P_{3(v)}, \ldots, P_{n(v)}$ are the vertical components of $P_1, P_2, P_3, \ldots, P_n$, respectively, and $\overline{A}$ is the cross-sectional area of the soil mass under consideration.

Again, if a_s is the cross-sectional area occupied by solid-to-solid contacts (that is, $a_s = a_1 + a_2 + a_3 + \cdots + a_n$), then the space occupied by water equals $(\overline{A} - a_s)$. So we can write

$$\sigma = \sigma' + \frac{u(\overline{A} - a_s)}{\overline{A}} = \sigma' + u(1 - a_s') \tag{9.3}$$

where $u = H_A \gamma_w$ = pore water pressure (that is, the hydrostatic pressure at A)
$a_s' = a_s/\overline{A}$ = fraction of unit cross-sectional area of the soil mass occupied by solid-to-solid contacts

The value of a_s' is extremely small and can be neglected for pressure ranges generally encountered in practical problems. Thus, Eq. (9.3) can be approximated by

$$\sigma = \sigma' + u \tag{9.4}$$

where u is also referred to as *neutral stress*. Substitution of Eq. (9.1) for σ in Eq. (9.4) gives

$$\begin{aligned}\sigma' &= [H\gamma_w + (H_A - H)\gamma_{sat}] - H_A\gamma_w \\ &= (H_A - H)(\gamma_{sat} - \gamma_w) \\ &= (\text{Height of the soil column}) \times \gamma' \end{aligned} \tag{9.5}$$

where $\gamma' = \gamma_{sat} - \gamma_w$ equals the submerged unit weight of soil. Thus, we can see that the effective stress at any point A is independent of the depth of water, H, above the submerged soil.

Figure 9.2a shows a layer of submerged soil in a tank where there is no seepage. Figures 9.2b through 9.2d show plots of the variations of the total stress, pore water pressure, and effective stress, respectively, with depth for a submerged layer of soil placed in a tank with no seepage.

The principle of effective stress [Eq. (9.4)] was first developed by Terzaghi (1925, 1936). Skempton (1960) extended the work of Terzaghi and proposed the relationship between total and effective stress in the form of Eq. (9.3).

In summary, effective stress is approximately the force per unit area carried by the soil skeleton. The effective stress in a soil mass controls its volume change

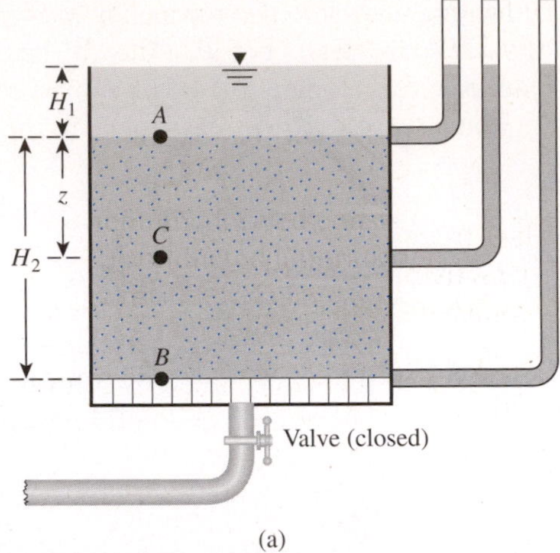

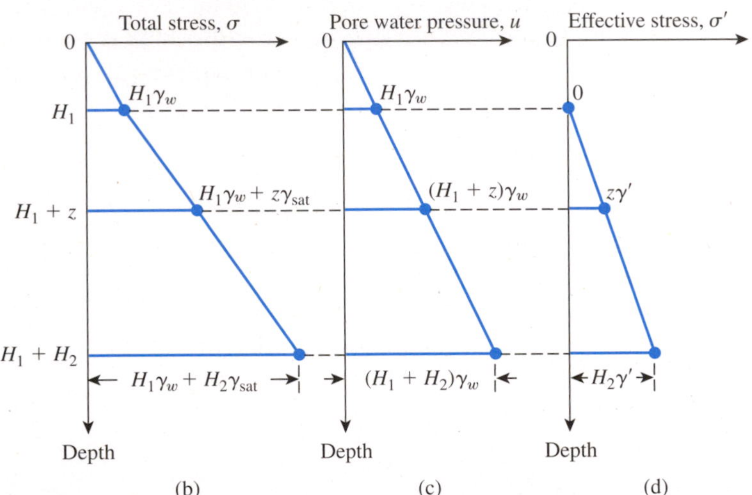

Figure 9.2 (a) Layer of soil in a tank where there is no seepage; variation of (b) total stress, (c) pore water pressure, and (d) effective stress with depth for a submerged soil layer without seepage

and strength. Increasing the effective stress induces soil to move into a denser state of packing.

The effective stress principle is probably the most important concept in geotechnical engineering. The compressibility and shearing resistance of a soil depend to a great extent on the effective stress. Thus, the concept of effective stress is significant in solving geotechnical engineering problems, such as the lateral earth pressure on retaining structures, the load-bearing capacity and settlement of foundations, and the stability of earth slopes.

In Eq. (9.2), the effective stress, σ', is defined as the sum of the vertical components of all intergranular *contact* forces over a unit gross cross-sectional area. This definition is mostly true for granular soils; however, for fine-grained soils, intergranular contact may not physically be there, because the clay particles are surrounded by tightly held water film. In a more general sense, Eq. (9.3) can be rewritten as

$$\sigma = \sigma_{ig} + u(1 - a_s') - A' + R' \tag{9.6}$$

where σ_{ig} = intergranular stress
A' = electrical attractive force per unit cross-sectional area of soil
R' = electrical repulsive force per unit cross-sectional area of soil

For granular soils, silts, and clays of low plasticity, the magnitudes of A' and R' are small. Hence, for all practical purposes,

$$\sigma_{ig} = \sigma' \approx \sigma - u$$

However, if $A' - R'$ is large, then $\sigma_{ig} \neq \sigma'$. Such situations can be encountered in highly plastic, dispersed clay. Many interpretations have been made in the past to distinguish between the intergranular stress and effective stress. In any case, the effective stress principle is an excellent approximation used in solving engineering problems.

Example 9.1

A soil profile is shown in Figure 9.3. Calculate the total stress, pore water pressure, and effective stress at points A, B, and C.

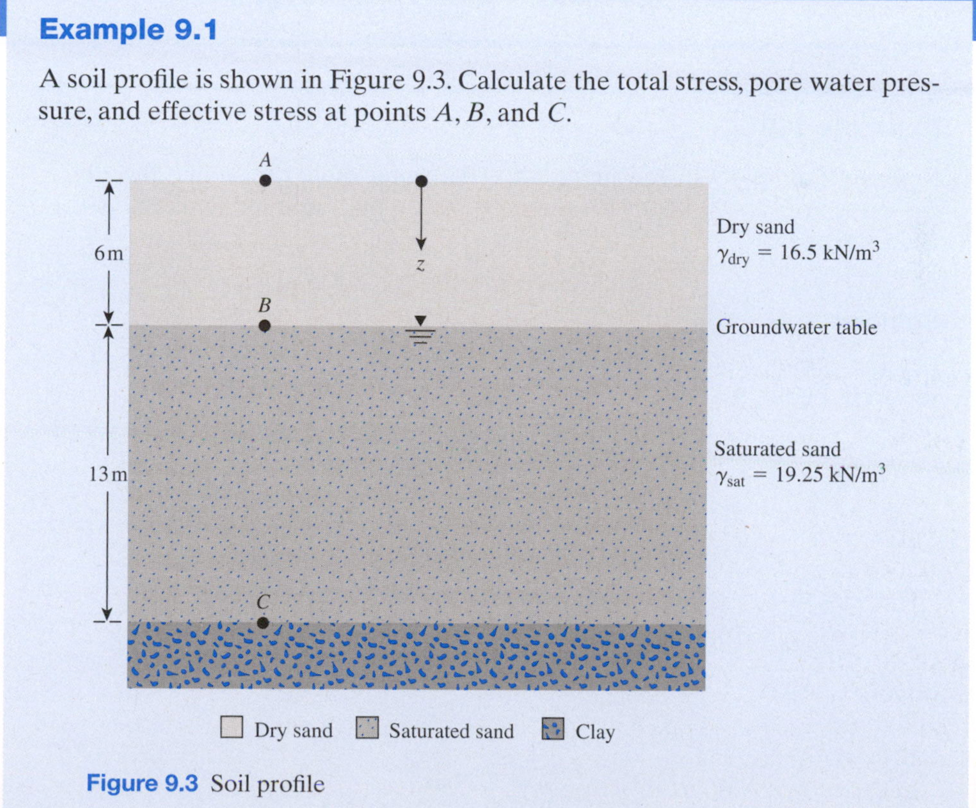

Figure 9.3 Soil profile

Solution

At Point A,

$$\text{Total stress: } \sigma_A = \mathbf{0}$$
$$\text{Pore water pressure: } u_A = \mathbf{0}$$
$$\text{Effective stress: } \sigma'_A = \mathbf{0}$$

At Point B,

$$\sigma_B = 6\gamma_{\text{dry(sand)}} = 6 \times 16.5 = \mathbf{99\ kN/m^2}$$
$$u_B = \mathbf{0\ kN/m^2}$$
$$\sigma'_B = 99 - 0 = \mathbf{99\ kN/m^2}$$

At Point C,

$$\begin{aligned}\sigma_C &= 6\gamma_{\text{dry(sand)}} + 13\gamma_{\text{sat(sand)}}\\ &= 6 \times 16.5 + 13 \times 19.25\\ &= 99 + 250.25 = \mathbf{349.25\ kN/m^2}\end{aligned}$$
$$u_C = 13\gamma_w = 13 \times 9.81 = \mathbf{127.53\ kN/m^2}$$
$$\sigma'_C = 349.25 - 127.53 = \mathbf{221.72\ kN/m^2}$$

Example 9.2

Refer to Example 9.1. How high should the water table rise so that the effective stress at C is 190 kN/m^2? Assume γ_{sat} to be the same for both layers (i.e., 19.25 kN/m^2).

Solution

Let the groundwater table rise be h above the present groundwater table shown in Figure 9.3 with

$$\sigma_C = (6 - h)\gamma_{\text{dry}} + h\gamma_{\text{sat}} + 13\gamma_{\text{sat}}$$
$$u = (h + 13)\gamma_w$$

So

$$\begin{aligned}\sigma'_C = \sigma_c - u &= (6 - h)\gamma_{\text{dry}} + h\gamma_{\text{sat}} + 13\gamma_{\text{sat}} - h\gamma_w - 13\gamma_w\\ &= (6 - h)\gamma_{\text{dry}} + h(\gamma_{\text{sat}} - \gamma_w) + 13(\gamma_{\text{sat}} - \gamma_w)\end{aligned}$$

or

$$190 = (6 - h)16.5 + h(19.25 - 9.81) + 13(19.25 - 9.81)$$
$$h = \mathbf{4.49\ m}$$

9.3 Stresses in Saturated Soil with Upward Seepage

If water is seeping, the effective stress at any point in a soil mass will differ from that in the static case. It will increase or decrease, depending on the direction of seepage.

Figure 9.4a shows a layer of granular soil in a tank where upward seepage is caused by adding water through the valve at the bottom of the tank. The rate of water supply is kept constant. The loss of head caused by upward seepage between the levels of A and B is h. Keeping in mind that the total stress at any point in the soil mass is due solely to the weight of soil and water above it, we find that the effective stress calculations at points A and B are as follows:

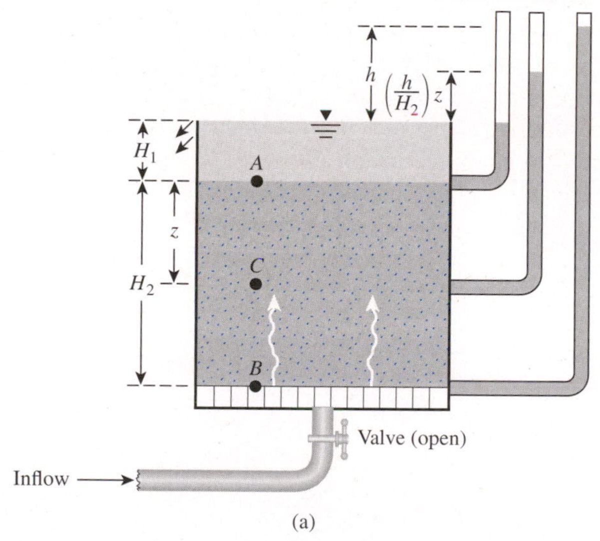

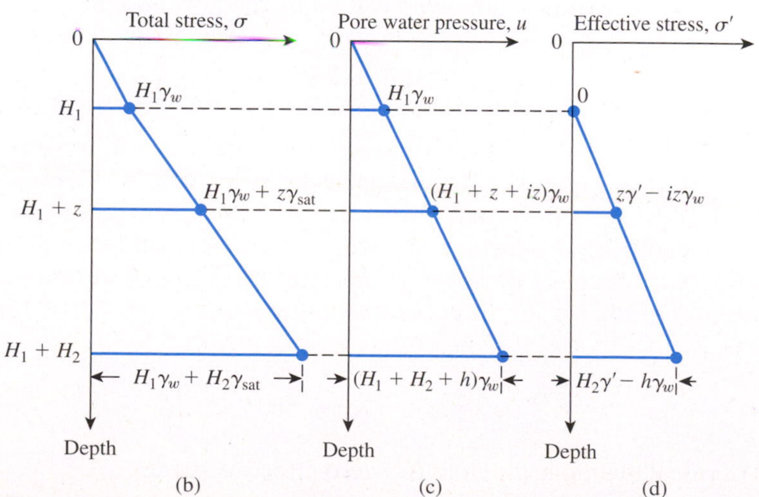

Figure 9.4 (a) Layer of soil in a tank with upward seepage. Variation of (b) total stress; (c) pore water pressure; and (d) effective stress with depth for a soil layer with upward seepage

At A,

- Total stress: $\sigma_A = H_1\gamma_w$
- Pore water pressure: $u_A = H_1\gamma_w$
- Effective stress: $\sigma'_A = \sigma_A - u_A = 0$

At B,

- Total stress: $\sigma_B = H_1\gamma_w + H_2\gamma_{sat}$
- Pore water pressure: $u_B = (H_1 + H_2 + h)\gamma_w$
- Effective stress: $\sigma'_B = \sigma_B - u_B$

$$= H_2(\gamma_{sat} - \gamma_w) - h\gamma_w$$

$$= H_2\gamma' - h\gamma_w$$

Similarly, the effective stress at a point C located at a depth z below the top of the soil surface can be calculated as follows:
At C,

- Total stress: $\sigma_C = H_1\gamma_w + z\gamma_{sat}$
- Pore water pressure: $u_C = \left(H_1 + z + \frac{h}{H_2}z\right)\gamma_w$
- Effective stress: $\sigma'_C = \sigma_C - u_C$

$$= z(\gamma_{sat} - \gamma_w) - \frac{h}{H_2}z\gamma_w$$

$$= z\gamma' - \frac{h}{H_2}z\gamma_w$$

Note that h/H_2 is the hydraulic gradient i caused by the flow, and therefore,

$$\sigma'_C = z\gamma' - iz\gamma_w \tag{9.7}$$

The variations of total stress, pore water pressure, and effective stress with depth are plotted in Figures 9.4b through 9.4d, respectively. A comparison of Figures 9.2d and 9.4d shows that the effective stress at a point located at a depth z measured from the surface of a soil layer is reduced by an amount $iz\gamma_w$ because of upward seepage of water. If the rate of seepage and thereby the hydraulic gradient gradually are increased, a limiting condition will be reached, at which point

$$\sigma'_C = z\gamma' - i_{cr}z\gamma_w = 0 \tag{9.8}$$

where i_{cr} = critical hydraulic gradient (for zero effective stress).

Under such a situation, soil stability is lost. This situation generally is referred to as *boiling,* or a *quick condition.*

From Eq. (9.8),

$$i_{cr} = \frac{\gamma'}{\gamma_w} \tag{9.9}$$

For most soils, the value of i_{cr} varies from 0.9 to 1.1, with an average of 1.

Example 9.3

A 9-m-thick layer of stiff saturated clay is underlain by a layer of sand (Figure 9.5). The sand is under artesian pressure. Calculate the maximum depth of cut H that can be made in the clay.

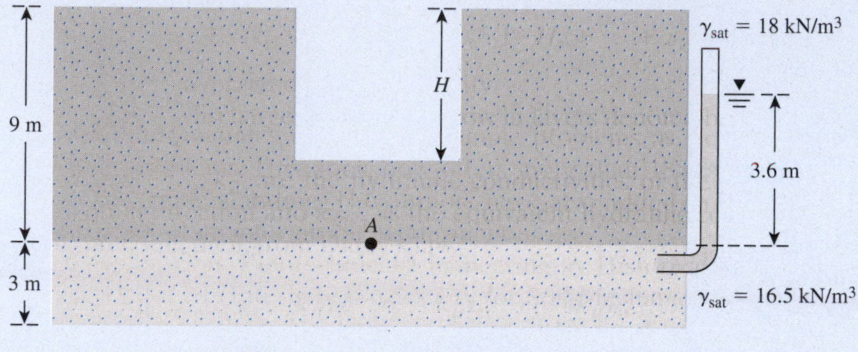

Figure 9.5

Solution

Due to excavation, there will be unloading of the overburden pressure. Let the depth of the cut be H, at which point the bottom will heave. Let us consider the stability of point A at that time:

$$\sigma_A = (9 - H)\gamma_{sat(clay)}$$

$$u_A = 3.6\gamma_w$$

For heave to occur, σ'_A should be 0. So

$$\sigma_A - u_A = (9 - H)\gamma_{sat(clay)} - 3.6\gamma_w$$

or

$$(9 - H)18 - (3.6)9.81 = 0$$

$$H = \frac{(9)18 - (3.6)9.81}{18} = \mathbf{7.04\ m}$$

Example 9.4

A cut is made in a stiff, saturated clay that is underlain by a layer of sand (Figure 9.6). What should be the height of the water, *h*, in the cut so that the stability of the saturated clay is not lost?

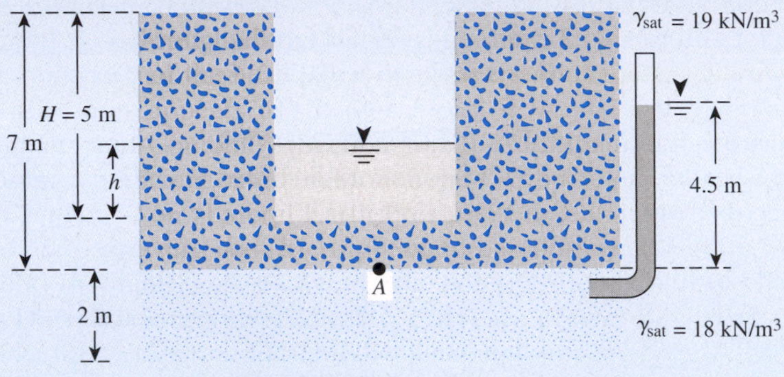

Figure 9.6

Solution

At point *A*,

$$\sigma_A = (7 - 5)\gamma_{sat(clay)} + h\gamma_w = (2)(19) + (h)(9.81) = 38 + 9.81h \text{ (kN/m}^2\text{)}$$

$$u_A = 4.5\gamma_w = (4.5)(9.81) = 44.15 \text{ kN/m}^2$$

For loss of stability, $\sigma_A' = 0$. So,

$$\sigma_A - u_A = 0$$

$$38 + 9.81h - 44.15 = 0$$

$$h = \mathbf{0.63\ m}$$

9.4 Stresses in Saturated Soil with Downward Seepage

The condition of downward seepage is shown in Figure 9.7a on the next page. The water level in the soil tank is held constant by adjusting the supply from the top and the outflow at the bottom.

The hydraulic gradient caused by the downward seepage equals $i = h/H_2$. The total stress, pore water pressure, and effective stress at any point *C* are, respectively,

$$\sigma_C = H_1\gamma_w + z\gamma_{sat}$$

$$u_C = (H_1 + z - iz)\gamma_w$$

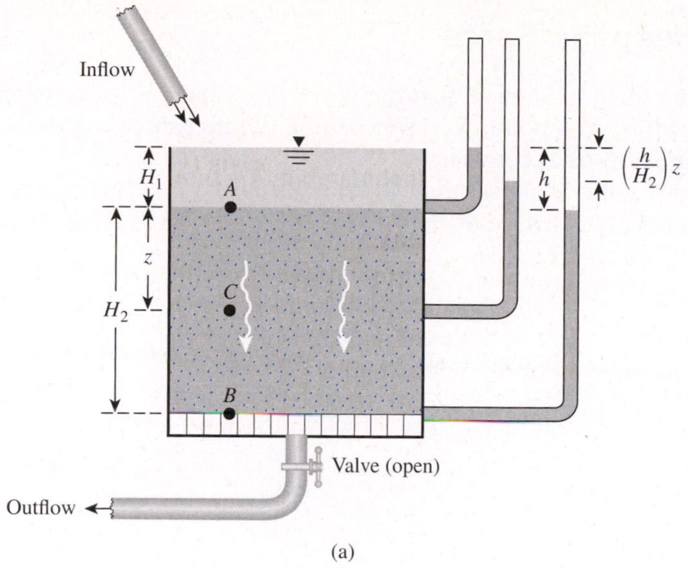

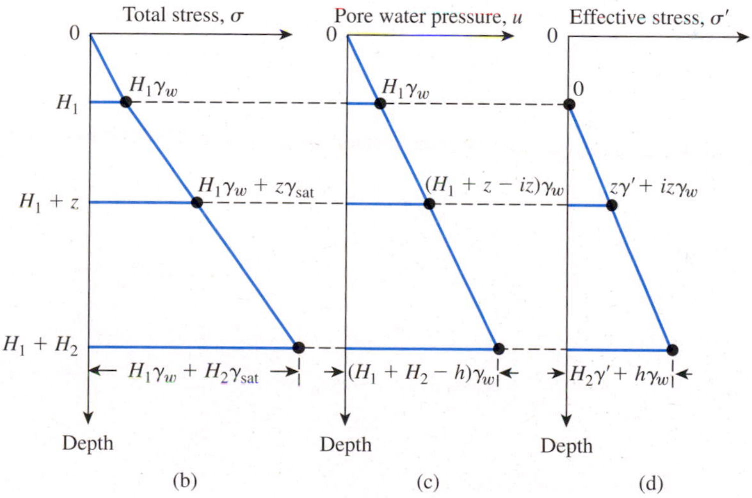

Figure 9.7 (a) Layer of soil in a tank with downward seepage; variation of (b) total stress; (c) pore water pressure; (d) effective stress with depth for a soil layer with downward seepage

$$\sigma'_C = (H_1\gamma_w + z\gamma_{sat}) - (H_1 + z - iz)\gamma_w$$
$$= z\gamma' + iz\gamma_w$$

The variations of total stress, pore water pressure, and effective stress with depth also are shown graphically in Figures 9.7b through 9.7d.

9.5 Seepage Force

The preceding sections showed that the effect of seepage is to increase or decrease the effective stress at a point in a layer of soil. Often, expressing the seepage force per unit volume of soil is convenient.

In Figure 9.2, it was shown that, with no seepage, the effective stress at a depth z measured from the surface of the soil layer in the tank is equal to $z\gamma'$. Thus, the effective force on an area A is

$$P_1' = z\gamma' A \tag{9.10}$$

(The direction of the force P_1' is shown in Figure 9.8a.)

Again, if there is an upward seepage of water in the vertical direction through the same soil layer (Figure 9.4), the effective force on an area A at a depth z can be given by

$$P_2' = (z\gamma' - iz\gamma_w)A \tag{9.11}$$

Hence, the decrease in the total force because of seepage is

$$P_1' - P_2' = iz\gamma_w A \tag{9.12}$$

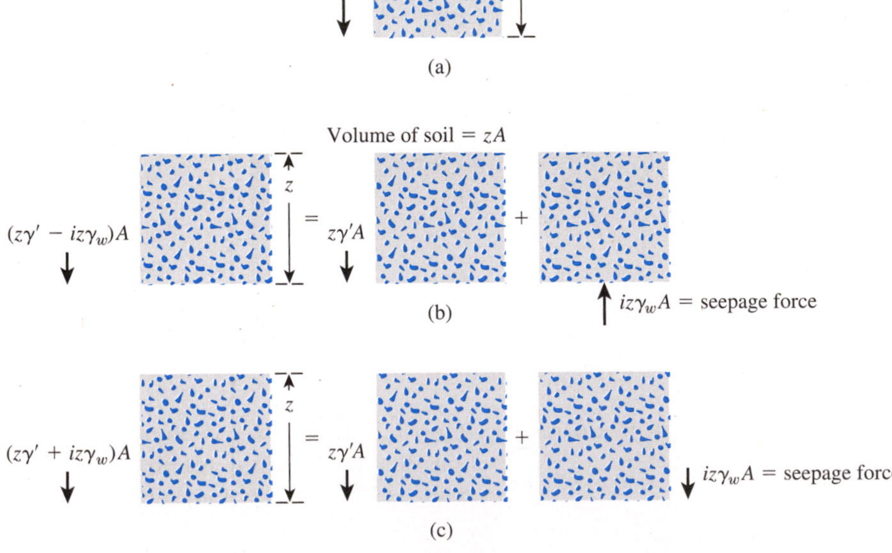

Figure 9.8 Force due to (a) no seepage; (b) upward seepage; (c) downward seepage on a volume of soil

The volume of the soil contributing to the effective force equals zA, so the seepage force per unit volume of soil is

$$\frac{P_1' - P_2'}{(\text{Volume of soil})} = \frac{iz\gamma_w A}{zA} = i\gamma_w \tag{9.13}$$

The force per unit volume, $i\gamma_w$, for this case acts in the upward direction—that is, in the direction of flow. This upward force is demonstrated in Figure 9.8b. Similarly, for downward seepage, it can be shown that the seepage force in the downward direction per unit volume of soil is $i\gamma_w$ (Figure 9.8c).

From the preceding discussions, we can conclude that the seepage force per unit volume of soil is equal to $i\gamma_w$, and in isotropic soils the force acts in the same direction as the direction of flow. This statement is true for flow in any direction. Flow nets can be used to find the hydraulic gradient at any point and, thus, the seepage force per unit volume of soil. The mathematical derivation for a general case is given below.

Figure 9.9 shows a soil mass bounded by two flow lines *ab* and *cd* and two equipotential lines *ef* and *gh*. This is taken from a flow net. The soil mass has a unit thickness at right angles to the section shown. Let h_1 and h_2 be the average piezometric elevations, respectively, along the faces $a'c'$ and $b'd'$ of the flow element. Also let F and $F + \Delta F$ be the forces acting, respectively, on the faces $a'c'$ and $b'd'$. The saturated self-weight of the soil mass $a'c'd'b'$ (of unit thickness) can then be given as

$$W = (l)(l)(1)\gamma_{sat} \tag{9.14}$$

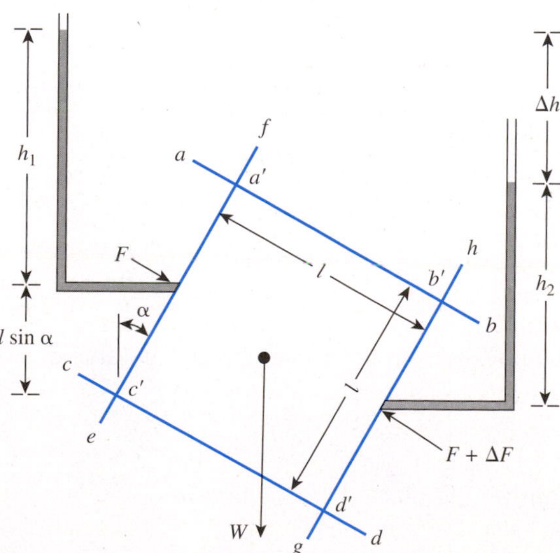

Figure 9.9 Seepage force per unit volume—determination from flow net

The hydrostatic force on the face $a'c'$ is $h_1\gamma_w l$; and, similarly, the hydrostatic force on the face $b'd'$ is $h_2\gamma_w l$. Hence

$$\Delta F = h_1\,\gamma_w l + l^2\,\gamma_{sat}\sin\alpha - h_2\,\gamma_w l \tag{9.15}$$

However,

$$h_2 = h_1 + l\sin\alpha - \Delta h \tag{9.16}$$

Combining Eqs. (9.15) and (9.16),

$$\Delta F = h_1\,\gamma_w l + l^2\,\gamma_{sat}\sin\alpha - (h_1 + l\sin\alpha - \Delta h)\gamma_w l$$

or

$$\begin{aligned}\Delta F &= l^2(\gamma_{sat} - \gamma_w)\sin\alpha + \Delta h\gamma_w l \\ &= \underbrace{l^2\gamma'\sin\alpha}_{\substack{\text{component of}\\\text{the effective}\\\text{weight of soil}\\\text{in direction of}\\\text{flow}}} + \underbrace{\Delta h\gamma_w l}_{\substack{\text{seepage}\\\text{force}}}\end{aligned} \tag{9.17}$$

where $\gamma' = \gamma_{sat} - \gamma_w =$ effective unit weight of soil. Hence

$$\text{Seepage force/unit volume} = \frac{\Delta h\gamma_w l}{l^2} = \gamma_w i \tag{9.18}$$

where $i =$ hydraulic gradient along the direction of flow. Note that Eqs. (9.13) and (9.18) are identical.

Example 9.5

Consider the upward flow of water through a layer of sand in a tank as shown in Figure 9.10. For the sand, the following are given: void ratio (e) = 0.52 and specific gravity of solids = 2.67.

a. Calculate the total stress, pore water pressure, and effective stress at points A and B.
b. What is the upward seepage force per unit volume of soil?

Solution

Part a

The saturated unit weight of sand is calculated as follows:

$$\gamma_{sat} = \frac{(G_s + e)\gamma_w}{1 + e} = \frac{(2.67 + 0.52)9.81}{1 + 0.52} = 20.59\ \text{kN/m}^3$$

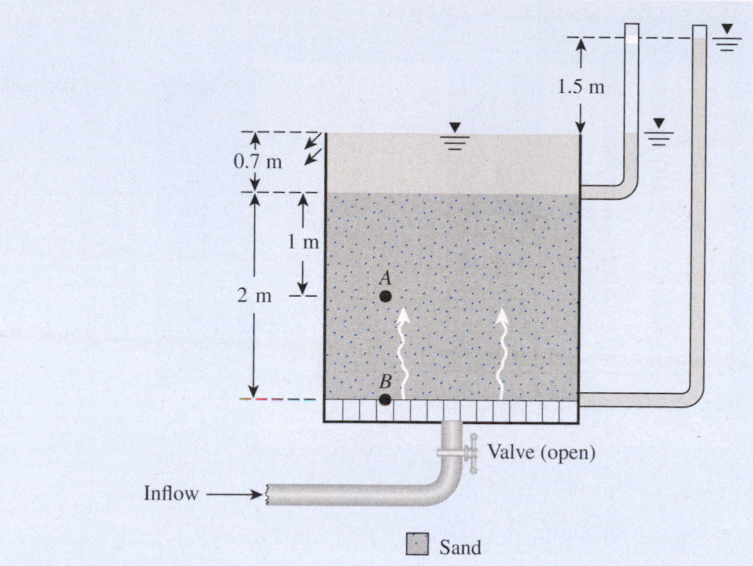

Figure 9.10 Upward flow of water through a layer of sand in a tank

Now, the following table can be prepared:

Point	Total stress, σ (kN/m²)	Pore water pressure, u (kN/m²)	Effective stress, $\sigma' = \sigma - u$ (kN/m²)
A	$0.7\gamma_w + 1\gamma_{sat} = (0.7)(9.81) + (1)(20.59) = \mathbf{27.46}$	$\left[(1 + 0.7) + \left(\frac{1.5}{2}\right)(1)\right]\gamma_w = (2.45)(9.81) = \mathbf{24.03}$	**3.43**
B	$0.7\gamma_w + 2\gamma_{sat} = (0.7)(9.81) + (2)(20.59) = \mathbf{48.05}$	$(2 + 0.7 + 1.5)\gamma_w = (4.2)(9.81) = \mathbf{41.2}$	**6.85**

Part b

Hydraulic gradient (i) = 1.5/2 = 0.75. Thus, the seepage force per unit volume can be calculated as

$$i\gamma_w = (0.75)(9.81) = \mathbf{7.36\ kN/m^3}$$

9.6 Heaving in Soil Due to Flow around Sheet Piles

Seepage force per unit volume of soil can be used for checking possible failure of sheet-pile structures where underground seepage may cause heaving of soil on the downstream side (Figure 9.11a). After conducting several model tests, Terzaghi (1922) concluded that heaving generally occurs within a distance of $D/2$ from the

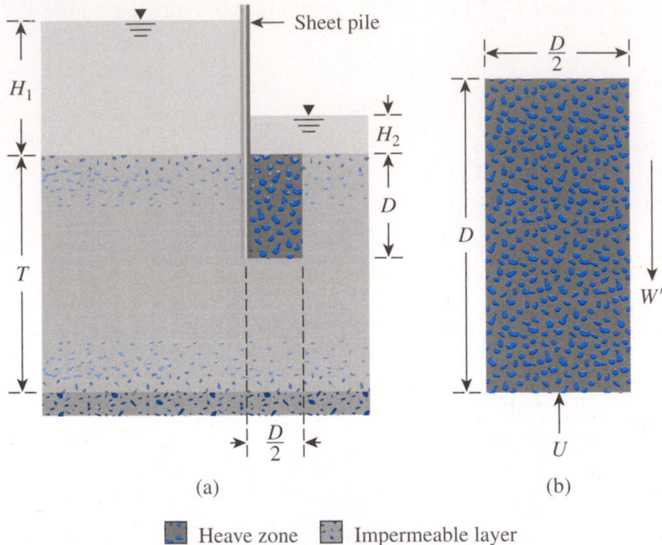

Figure 9.11 (a) Check for heaving on the downstream side for a row of sheet piles driven into a permeable layer; (b) enlargement of heave zone

sheet piles (when D equals depth of embedment of sheet piles into the permeable layer). Therefore, we need to investigate the stability of soil in a zone measuring D by $D/2$ in cross-section as shown in Figure 9.11b.

The factor of safety against heaving can be given by

$$FS = \frac{W'}{U} \tag{9.19}$$

where FS = factor of safety

W' = submerged weight of soil in the heave zone per unit length of sheet pile = $D(D/2)(\gamma_{sat} - \gamma_w) = (\frac{1}{2})D^2\gamma'$

U = uplifting force caused by seepage on the same volume of soil

From Eq. (9.13),

$$U = (\text{Soil volume}) \times (i_{av}\gamma_w) = \tfrac{1}{2}D^2 i_{av}\gamma_w$$

where i_{av} = average hydraulic gradient at the bottom of the block of soil (see Example 9.6).

Substituting the values of W' and U in Eq. (9.19), we can write

$$FS = \frac{\gamma'}{i_{av}\gamma_w} \tag{9.20}$$

For the case of *flow around a sheet pile in a homogeneous soil*, as shown in Figure 9.11, it can be demonstrated that

$$\frac{U}{0.5\gamma_w D(H_1 - H_2)} = C_o \tag{9.21}$$

where C_o is a function of D/T (see Table 9.1). Hence, from Eq. (9.19),

$$FS = \frac{W'}{U} = \frac{0.5D^2\gamma'}{0.5C_o\gamma_w D(H_1 - H_2)} = \frac{D\gamma'}{C_o\gamma_w(H_1 - H_2)} \tag{9.22}$$

Harza (1935) investigated the safety of hydraulic structures against heaving. According to his work, the factor of safety (FS) against heaving (or piping) can be expressed as

$$FS = \frac{i_{cr}}{i_{exit}} \tag{9.23}$$

where i_{cr} = critical hydraulic gradient
i_{exit} = maximum exit gradient

From Eq. (9.9),

$$i_{cr} = \frac{\gamma'}{\gamma_w} = \frac{\left[\dfrac{(G_s - 1)\gamma_w}{1 + e}\right]}{\gamma_w} = \frac{G_s - 1}{1 + e} \tag{9.24}$$

The maximum exit gradient also can be determined from a flow net. Referring to Figure 9.12, the maximum exit gradient is

$$i_{exit} = \frac{\Delta h}{l} = \frac{H}{N_d l} \tag{9.24a}$$

A factor of safety of 3 also is considered adequate for the safe performance of the structure. Harza also presented a chart for i_{exit} for dams constructed over deep homogeneous deposits (Figure 9.13). Using the notations shown in Figure 9.13,

$$i_{exit} = C\frac{H}{B} \tag{9.25}$$

Table 9.1 Variation of C_o with D/T

D/T	C_o
0.1	0.385
0.2	0.365
0.3	0.359
0.4	0.353
0.5	0.347
0.6	0.339
0.7	0.327
0.8	0.309
0.9	0.274

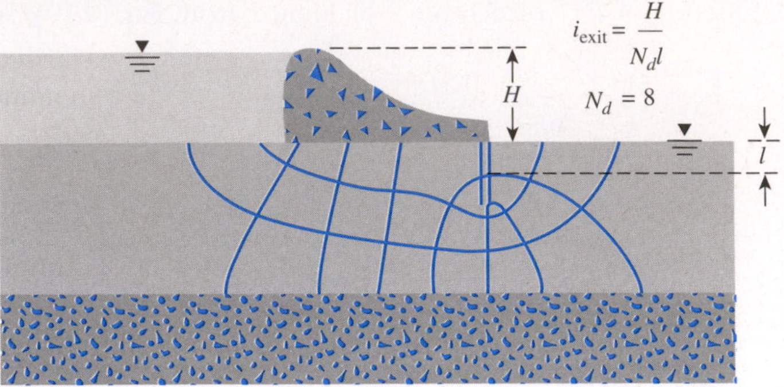

Figure 9.12 Definition of i_{exit} [Eq. (9.24a)]

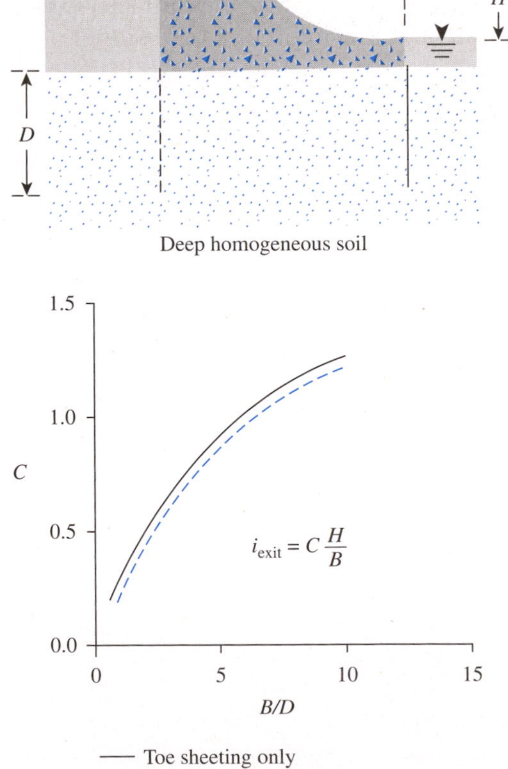

Figure 9.13 Hazra chart for i_{exit} [see Eq. (9.25)] for dams constructed over deep homogeneous deposits

Example 9.6

Figure 9.14 shows the flow net for seepage of water around a single row of sheet piles driven into a permeable layer. Calculate the factor of safety against downstream heave, given that γ_{sat} for the permeable layer = 17.7 kN/m³. (*Note:* Thickness of permeable layer T = 18 m)

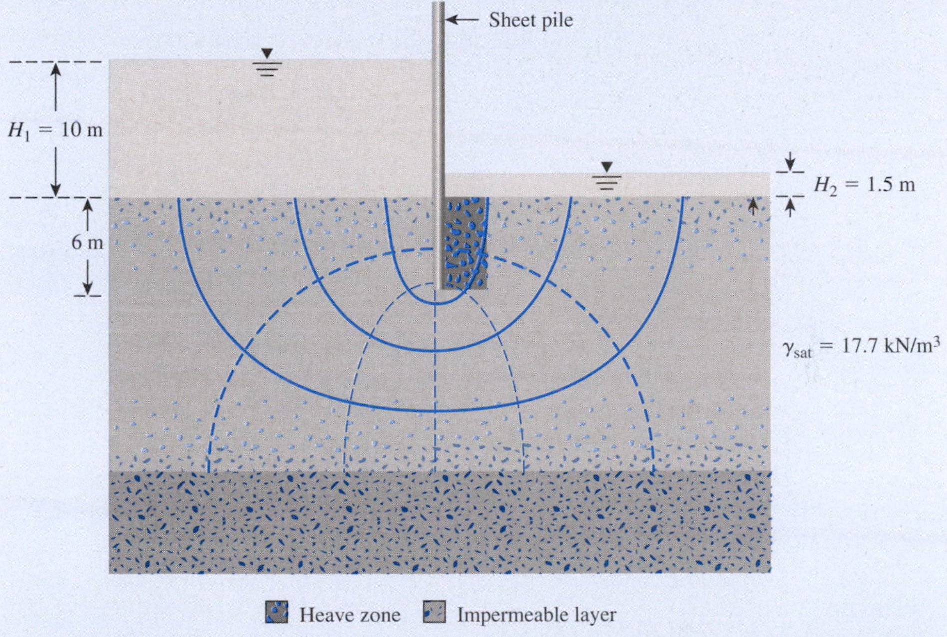

Figure 9.14 Flow net for seepage of water around sheet piles driven into permeable layer

Solution

From the dimensions given in Figure 9.14, the soil prism to be considered is 6 m × 3 m in cross section.

The soil prism is drawn to an enlarged scale in Figure 9.15. By use of the flow net, we can calculate the head loss through the prism.

At b,

$$\text{Driving head} = \frac{3}{6}(H_1 - H_2)$$

At c,

$$\text{Driving head} \approx \frac{1.6}{6}(H_1 - H_2)$$

Similarly, for other intermediate points along bc, the approximate driving heads have been calculated and are shown in Figure 9.15.

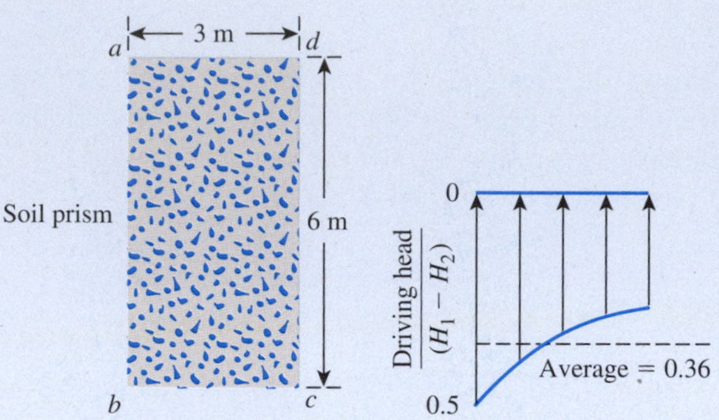

Figure 9.15 Soil prism—enlarged scale

The average value of the head loss in the prism is $0.36(H_1 - H_2)$, and the average hydraulic gradient is

$$i_{\text{av}} = \frac{0.36(H_1 - H_2)}{D}$$

Thus, the factor of safety [Eq. (9.20)] is

$$FS = \frac{\gamma'}{i_{\text{av}}\gamma_w} = \frac{\gamma' D}{0.36(H_1 - H_2)\gamma_w} = \frac{(17.7 - 9.81)6}{0.36(10 - 1.5) \times 9.81} = \mathbf{1.58}$$

Alternate Solution

For this case, $D/T = 1/3$. From Table 9.1, for $D/T = 1/3$, the value of $C_o = 0.357$. Thus, from Eq. (9.22),

$$FS = \frac{D\gamma'}{C_o\gamma_w(H_1 - H_2)} = \frac{(6)(17.7 - 9.81)}{(0.357)(9.81)(10 - 1.5)} = \mathbf{1.59}$$

Example 9.7

Refer to Figure 9.16. For the flow under the weir, estimate the factor of safety against piping.

Solution

We can scale the following:

$$H = 4.2 \text{ m}$$
$$l = 1.65 \text{ m}$$

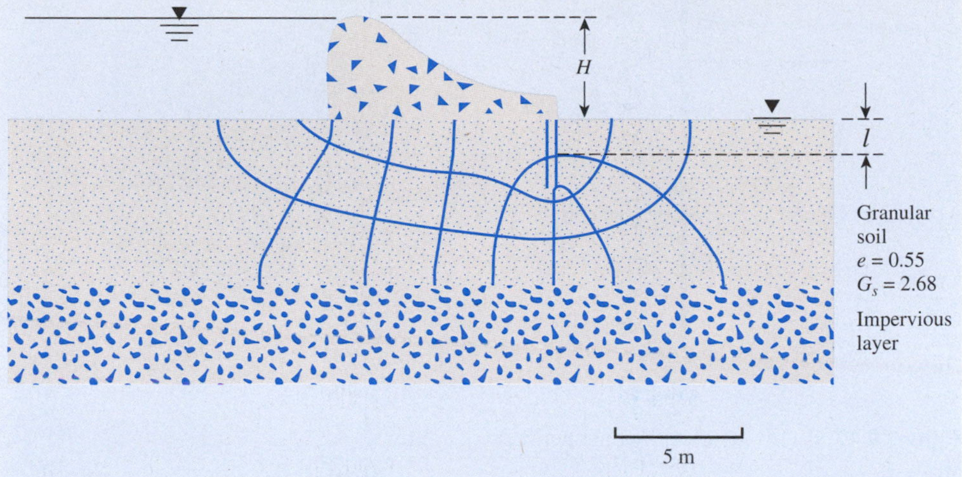

Figure 9.16

From the flow net, note that $N_d = 8$. So

$$\Delta h = \frac{H}{N_d} = \frac{4.2}{8} = 0.525 \text{ m}$$

$$i_{\text{exit}} = \frac{\Delta h}{l} = \frac{0.525}{1.65} = 0.318$$

From Eq. (9.24),

$$i_{\text{cr}} = \frac{G_s - 1}{1 + e} = \frac{2.68 - 1}{1 + 0.55} = 1.08$$

From Eq. (9.23),

$$FS = \frac{i_{\text{cr}}}{i_{\text{exit}}} = \frac{1.08}{0.318} = \mathbf{3.14}$$

9.7 Use of Filters to Increase the Factor of Safety against Heave

The factor of safety against heave as calculated in Example 9.6 is low. In practice, a minimum factor of safety of about 4 to 5 is required for the safety of the structure. Such a high factor of safety is recommended primarily because of the inaccuracies inherent in the analysis. One way to increase the factor of safety against heave is to

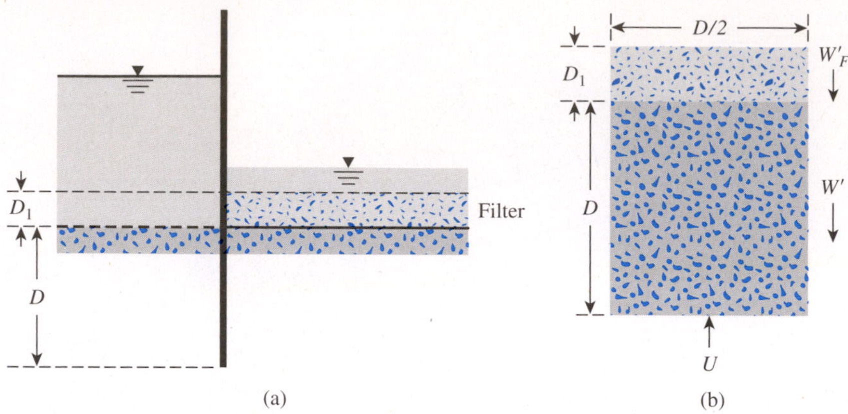

Figure 9.17 Factor of safety against heave, with a filter

use a *filter* in the downstream side of the sheet-pile structure (Figure 9.17a). A filter is a granular material with openings small enough to prevent the movement of the soil particles upon which it is placed and, at the same time, is pervious enough to offer little resistance to seepage through it (see Section 8.11). In Figure 9.17a, the thickness of the filter material is D_1. In this case, the factor of safety against heave can be calculated as follows (Figure 9.17b).

The effective weight of the soil and the filter in the heave zone per unit length of sheet pile $= W' + W'_F$, where

$$W' = (D)\left(\frac{D}{2}\right)(\gamma_{\text{sat}} - \gamma_w) = \frac{1}{2}D^2\gamma'$$

$$W'_F = (D_1)\left(\frac{D}{2}\right)(\gamma'_F) = \frac{1}{2}D_1 D\gamma'_F$$

in which γ'_F = effective unit weight of the filter.

The uplifting force caused by seepage on the same volume of soil is given by

$$U = \frac{1}{2}D^2 i_{\text{av}}\gamma_w$$

The preceding relationship was derived in Section 9.6.

The factor of safety against heave is thus

$$FS = \frac{W' + W'_F}{U} = \frac{\frac{1}{2}D^2\gamma' + \frac{1}{2}D_1 D\gamma'_F}{\frac{1}{2}D^2 i_{\text{av}}\gamma_w} = \frac{\gamma' + \left(\frac{D_1}{D}\right)\gamma'_F}{i_{\text{av}}\gamma_w} \tag{9.26}$$

The principles for selection of filter materials were given in Section 8.11.

If Eq. (9.21) is used,

$$FS = \frac{\frac{1}{2}D^2\gamma' + \frac{1}{2}D_1D\gamma'_F}{0.5C_o\gamma_w D(H_1 - H_2)} = \frac{D\gamma' + D_1\gamma'_F}{C_o\gamma_w(H_1 - H_2)} \tag{9.27}$$

The value of C_o is given in Table 9.1.

Example 9.8

Refer to Example 9.6. If the factor of safety against heaving needs to be increased to 2.5 by laying a filter layer on the downstream side, what should be the thickness of the layer? Given: dry and saturated unit weights of the filter material are 16 kN/m³ and 20 kN/m³, respectively.

Solution

Refer to Figure 9.18. The filter material has a thickness of D_1. The top $(D_1 - 1.5\text{ m})$ of the filter is dry, and the bottom 1.5 m of the filter is submerged. Now, from Eq. (9.27),

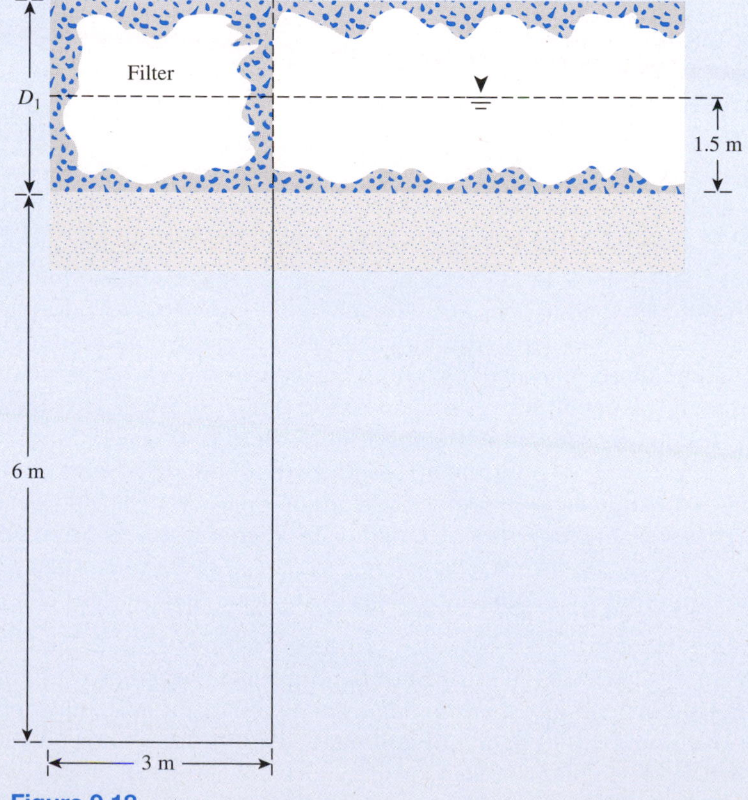

Figure 9.18

$$FS = \frac{D\gamma' + (D_1 - 1.5)\gamma_{d(F)} + 1.5\gamma'_F}{C_o\gamma_w(H_1 - H_2)}$$

or

$$2.5 = \frac{(6)(17.7 - 9.81) + (D_1 - 1.5)(16) + (1.5)(20 - 9.81)}{(0.375)(9.81)(10 - 1.5)}$$

$D_1 \approx$ **2.47 m**

9.8 Effective Stress in Partially Saturated Soil

In partially saturated soil, water in the void spaces is not continuous, and it is a three-phase system—that is, solid, pore water, and pore air (Figure 9.19). Hence, the total stress at any point in a soil profile consists of intergranular, pore air, and pore water pressures. From laboratory test results, Bishop et al. (1960) gave the following equation for effective stress in partially saturated soils:

$$\sigma' = \sigma - u_a + \chi(u_a - u_w) \tag{9.28}$$

where σ' = effective stress
σ = total stress
u_a = pore air pressure
u_w = pore water pressure

In Eq. (9.28), χ represents the fraction of a unit cross-sectional area of the soil occupied by water. For dry soil $\chi = 0$, and for saturated soil $\chi = 1$.

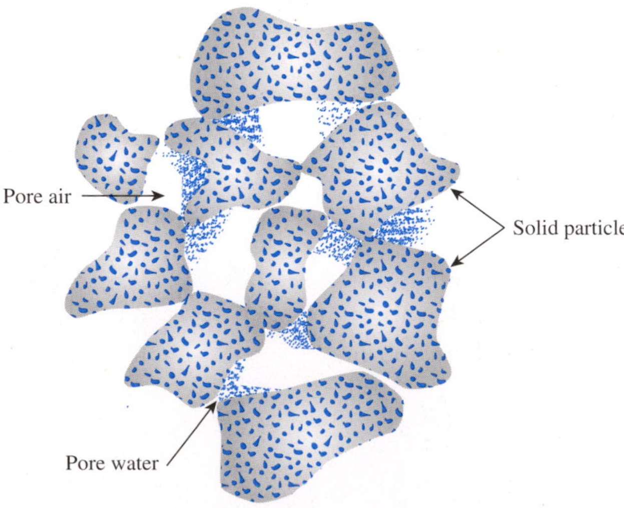

Figure 9.19 Partially saturated soil

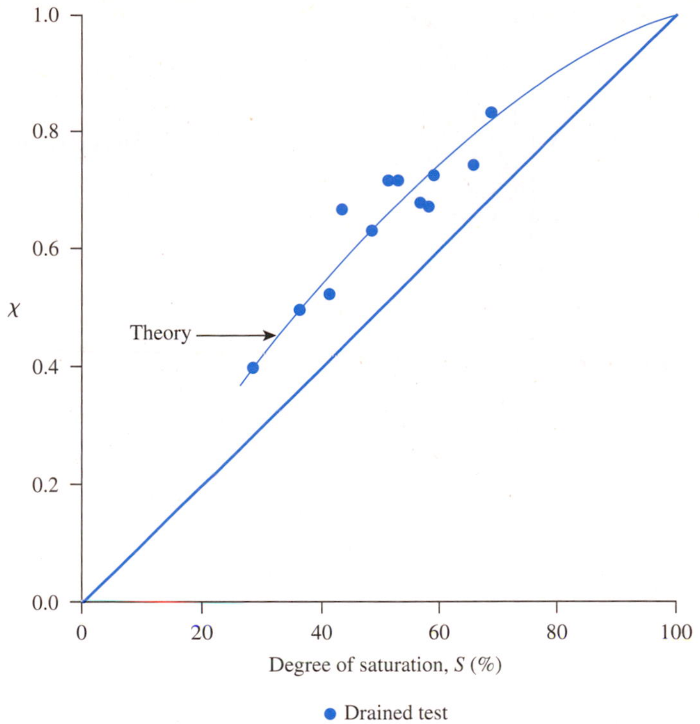

Figure 9.20 Relationship between the parameter χ and the degree of saturation for Bearhead silt (*After Bishop et al., 1960. With permission from ASCE.*)

Bishop et al. (1960) have pointed out that the intermediate values of χ will depend primarily on the degree of saturation S. However, these values also will be influenced by factors such as soil structure. The nature of variation of χ with the degree of saturation for a silt is shown in Figure 9.20.

9.9 Capillary Rise in Soils

The continuous void spaces in soil can behave as bundles of capillary tubes of variable cross section. Because of surface tension force, water may rise above the phreatic surface.

Figure 9.21 shows the fundamental concept of the height of rise in a capillary tube. The height of rise of water in the capillary tube can be given by summing the forces in the vertical direction, or

$$\left(\frac{\pi}{4}d^2\right)h_c\gamma_w = \pi d T \cos\alpha$$

$$h_c = \frac{4T\cos\alpha}{d\gamma_w} \qquad (9.29)$$

$$u = -(S\gamma_w H_2) = -(0.5)(9.81)(0.91) = \mathbf{-4.46\ kN/m^2}\ \text{(immediately below)}$$

$$\sigma' = 31.71 - 0 = \mathbf{31.71\ kN/m^2}\ \text{(immediately above)}$$

$$\sigma' = 31.71 - (-4.46) = \mathbf{36.17\ kN/m^2}\ \text{(immediately below)}$$

At depth $H_1 + H_2$ (i.e., at point C):

$$\sigma = (17.33)(1.83) + (18.97)(0.91) = \mathbf{48.97\ kN/m^2}$$

$$u = \mathbf{0}$$

$$\sigma' = 48.97 - 0 = \mathbf{48.97\ kN/m^2}$$

At depth $H_1 + H_2 + H_3$ (i.e., at point D):

$$\sigma = 48.97 + (17.66)(1.83) = \mathbf{81.29\ kN/m^2}$$

$$u = 1.83\gamma_w = (1.83)(9.81) = \mathbf{17.95\ kN/m^2}$$

$$\sigma' = 81.29 - 17.95 = \mathbf{63.34\ kN/m^2}$$

The plot of the stress variation is shown in Figure 9.24.

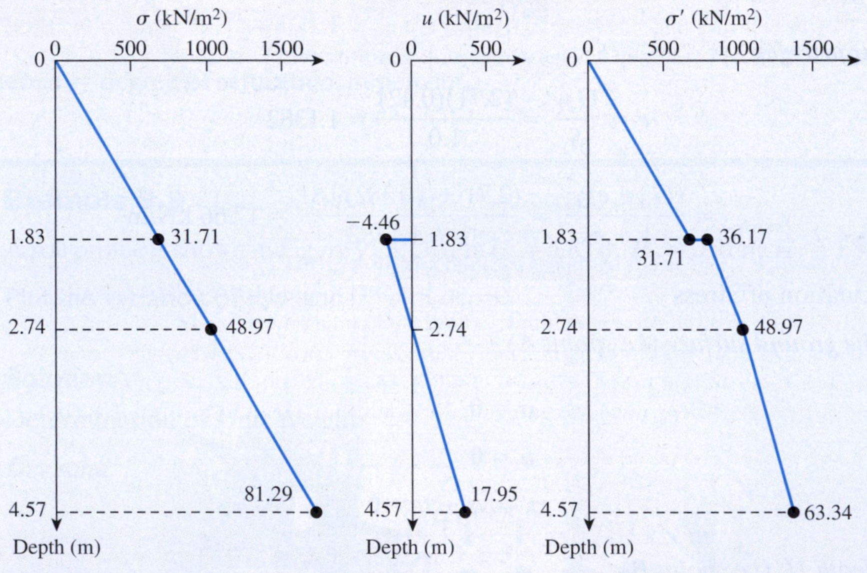

Figure 9.24

9.11 Summary

The effective stress principle is probably the most important concept in geotechnical engineering. The compressibility and shearing resistance of a soil depend to a great extent on the effective stress. Thus, the concept of effective stress is significant in solving geotechnical engineering problems, such as the lateral earth pressure on retaining structures, the load-bearing capacity and settlement of foundations, and the stability of earth slopes.

Following is a summary of the topics discussed in this chapter:

- The total stress (σ) at a point in the soil mass is the sum of effective stress (σ') and pore water pressure (u), or [Eq. (9.4)]

$$\sigma = \sigma' + u$$

- The critical hydraulic gradient (i_{cr}) for *boiling* or *quick condition* is given as

$$i_{cr} = \frac{\gamma'}{\gamma_w} = \frac{\text{effective unit weight of soil}}{\text{unit weight of water}}$$

- Seepage force per unit volume in the direction of flow is equal to $i\gamma_w$ (i = hydraulic gradient in the direction of flow).
- The relationships to check for heaving for flow under a hydraulic structure are discussed in Section 9.6. Also, the possibility of using filters to increase the factor of safety against heaving is discussed in Section 9.7.
- Effective stress at a point in a partially saturated soil can be expressed as [Eq. (9.28)]

$$\sigma' = \sigma - u_a + \chi(u_a - u_w)$$

where σ = total stress
u_a, u_w = pore air and pore water pressure, respectively
χ = a factor which is zero for dry soil and 1 for saturated soil

- Capillary rise in soil has been discussed in Section 9.9. Capillary rise can range from 0.1 m to 0.2 m in coarse sand to 7.5 m to 23 m in clay.

Problems

9.1 Through 9.3 A soil profile consisting of three layers is shown in Figure 9.25. Calculate the values of σ, u, and σ' at points A, B, C, and D for the following cases. In each case, plot the variations of σ, u, and σ' with depth. Characteristics of layers 1, 2, and 3 for each case are given below:

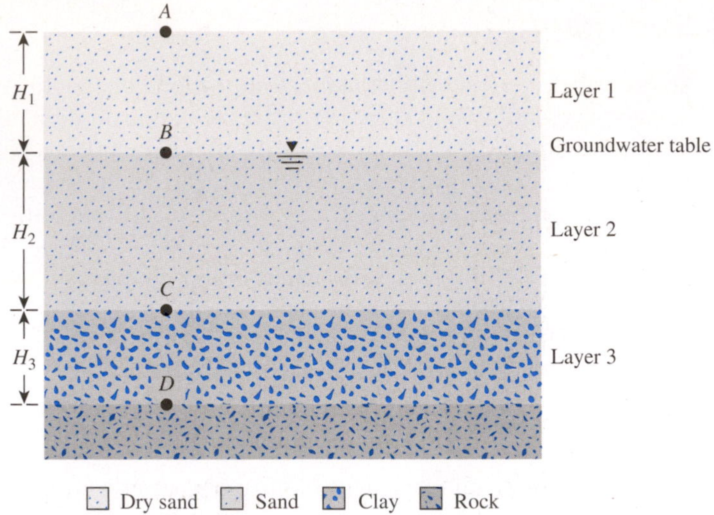

Figure 9.25

Problem	Layer no.	Thickness	Soil parameters
9.1	1	$H_1 = 2.1$ m	$\gamma_d = 17.23$ kN/m^3
	2	$H_2 = 3.66$ m	$\gamma_{sat} = 18.96$ kN/m^3
	3	$H_3 = 1.83$ m	$\gamma_{sat} = 18.5$ kN/m^3
9.2	1	$H_1 = 5$ m	$e = 0.7$; $G_s = 2.69$
	2	$H_2 = 8$ m	$e = 0.55$; $G_s = 2.7$
	3	$H_3 = 3$ m	$w = 38\%$; $e = 1.2$
9.3	1	$H_1 = 3$ m	$\gamma_d = 16$ kN/m^3
	2	$H_2 = 6$ m	$\gamma_{sat} = 18$ kN/m^3
	3	$H_3 = 2.5$ m	$\gamma_{sat} = 17$ kN/m^3

9.4 Consider the soil profile in Problem 9.2. What is the change in effective stress at point C if:

a. the water table drops by 2 m?

b. the water table rises to the surface up to point A?

c. the water level rises 3 m above point A due to flooding?

9.5 Consider the soil profile shown in Figure 9.26:

a. Calculate the variations of σ, u, and σ' at points A, B, and C.

b. How high should the groundwater table rise so that the effective stress at C is 111 kN/m^2?

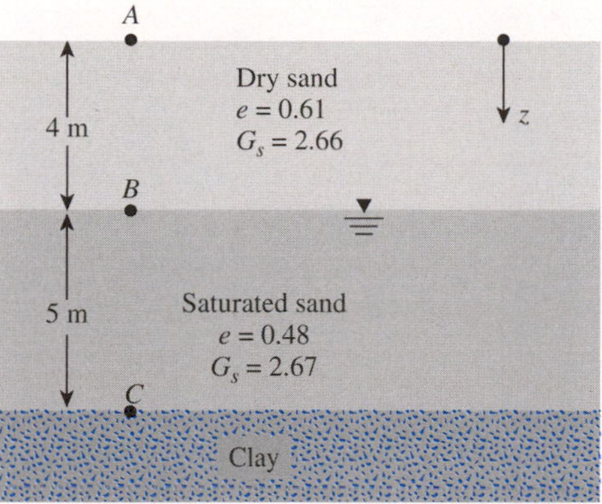

Figure 9.26

9.6 For a sandy soil with $G_s = 2.68$, calculate the critical hydraulic gradient that will cause *boiling* or *quick condition* for $e = 0.38, 0.48, 0.6, 0.7$, and 0.8. Plot the variation of i_{cr} with the void ratio.

9.7 An exploratory drill hole was made in a stiff saturated clay having a moisture content of 29% and $G_s = 2.68$ (Figure 9.27). The sand layer underlying the clay was observed to be under artesian pressure. Water in the drill hole rose to a height of 6 m above the top of the sand layer. If an open excavation is to be made in the clay, determine the safe depth of excavation before the bottom heaves.

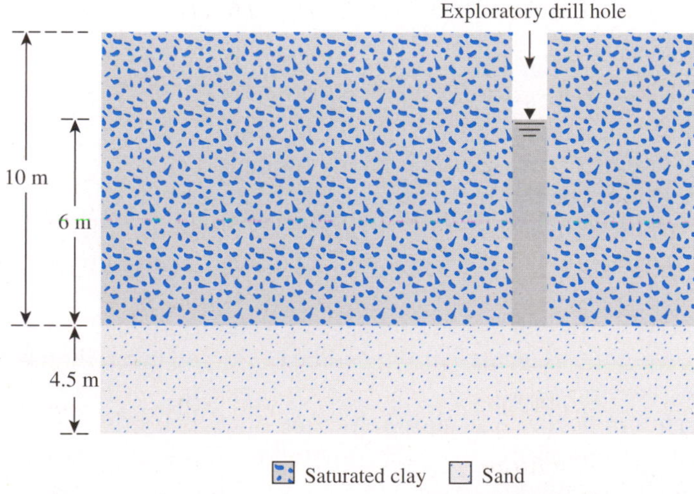

Figure 9.27

9.8 A 10-m-thick layer of stiff saturated clay is underlain by a layer of sand (Figure 9.28). The sand is under artesian pressure. A 5.75-m-deep cut is made in the clay. Determine the factor of safety against heaving at point A.

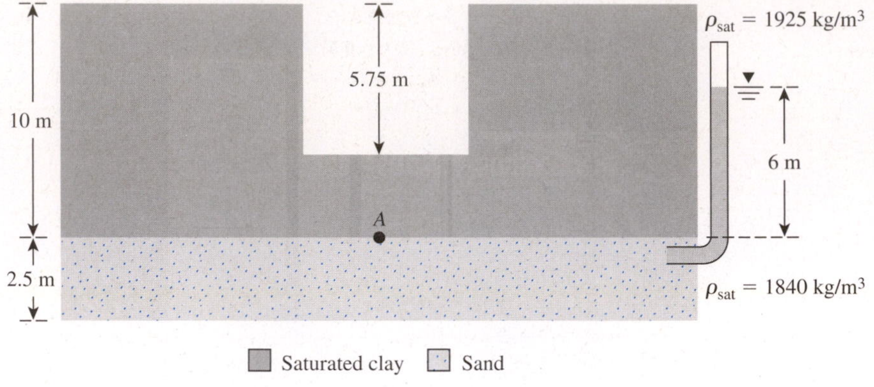

Figure 9.28

9.9 Refer to Figure 9.28. What would be the maximum permissible depth of cut before heaving would occur?

9.10 Refer to Problem 9.9. Water may be introduced into the cut to improve the stability against heaving. Assuming that a cut is made up to the maximum permissible depth calculated in Problem 9.9, what would be the required height of water inside the cut in order to ensure a factor of safety of 1.5?

9.11 Refer to Figure 9.4a in which upward seepage is taking place through a granular soil contained in a tank. Given: $H_1 = 1.5$ m; $H_2 = 2.5$ m; $h = 1.5$ m; area of the tank = 0.62 m^2; void ratio of the soil, $e = 0.49$; $G_s = 2.66$; and hydraulic conductivity of the sand (k) = 0.21 cm/sec.

a. What is the rate of upward seepage?

b. Will boiling occur when $h = 1.5$ m? Explain.

c. What would be the critical value of h to cause boiling?

9.12 Refer to Figure 9.4a. If $H_1 = 0.91$ m, $H_2 = 1.37$ m, $h = 0.46$ m, $\gamma_{sat} = 18.67$ kN/m^3, area of the tank = 0.58 m^2, and hydraulic conductivity of the sand (k) = 0.16 cm/sec,

a. What is the rate of upward seepage of water (m^3/min)?

b. If the point C is located at the middle of the soil layer, then what is the effective stress at C?

9.13 Through 9.14 Figure 9.29 shows the zone of capillary rise within a clay layer above the groundwater table. For the following variables, calculate and plot σ, u, and σ' with depth.

Problem	H_1	H_2	H_3	Degree of saturation in capillary rise zone, S (%)
9.13	3.05 m	2.43 m	4.88 m	40
9.14	4 m	2.5 m	4.5 m	60

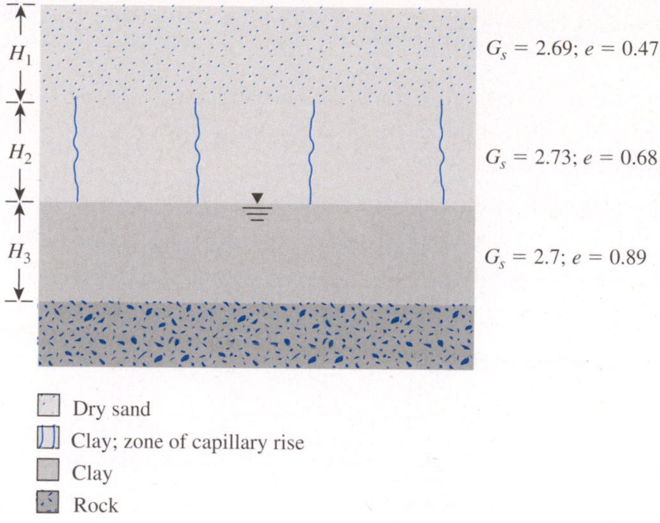

Figure 9.29

9.15 Determine the factor of safety against heave on the downstream side of the single-row sheet pile structure shown in Figure 9.30. Use the following soil and design parameters: $H_1 = 7$ m, $H_2 = 3$ m, thickness of permeable layer $(T) = 12$ m, design depth of penetration of sheet pile $(D) = 4.5$ m, and $\gamma_{sat} = 17$ kN/m^3.

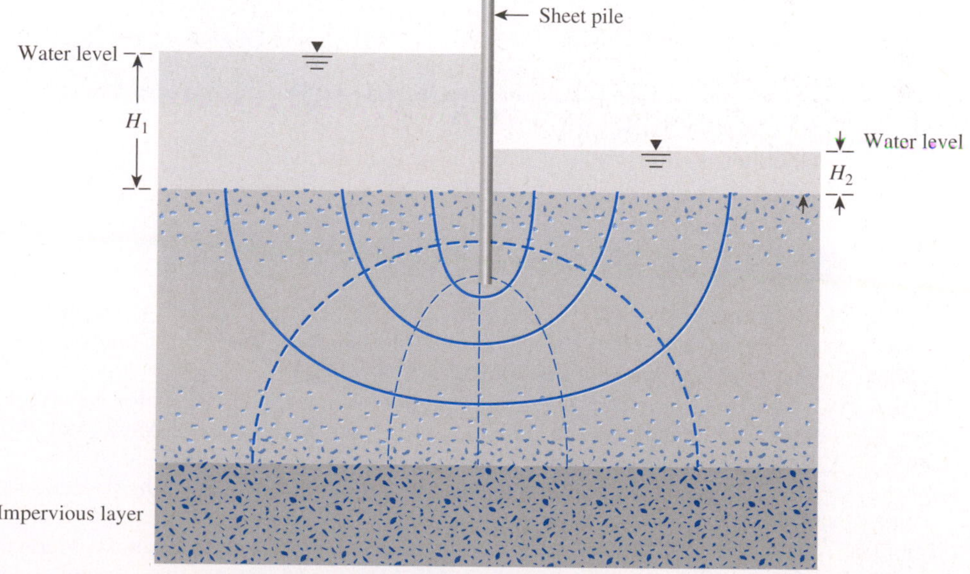

Figure 9.30

Critical Thinking Problem

9.C.1 Figure 9.31 shows a concrete dam. Consider Case 1 without the sheet pile, and Case 2 with the sheet pile along the upstream side.

a. Draw flow nets for both cases.

b. Determine the value of $\frac{q}{k}$ for both cases. (*Note:* q = m³/s/m; k = m/s.)

c. Determine the factor of safety (*FS*) against heaving using Eqs. (9.23), (9.24), and (9.24a), for Cases 1 and 2. Comment on any differences in the magnitude of *FS*.

d. Estimate the seepage force (kN/m³) at point *A* in the direction of seepage for Cases 1 and 2. Comment on any difference in the magnitude of the seepage force.

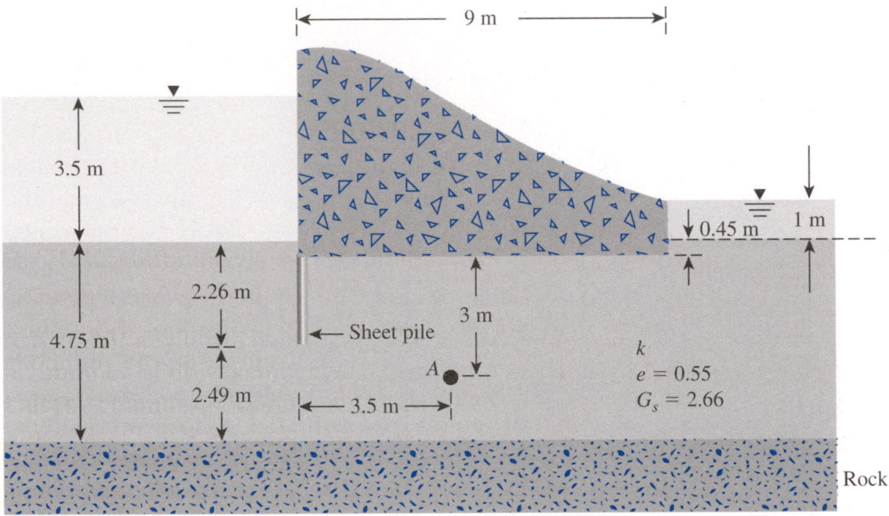

Figure 9.31

References

Bishop, A. W., Alpan, I., Blight, G. E., and Donald, I. B. (1960). "Factors Controlling the Strength of Partially Saturated Cohesive Soils," *Proceedings,* Research Conference on Shear Strength of Cohesive Soils, ASCE, 502–532.

Hazen, A. (1930). "Water Supply," in *American Civil Engineering Handbook*, Wiley, New York.

Harza, L. F. (1935). "Uplift and Seepage Under Dams in Sand." *Transactions*, ASCE, Vol. 100.

Skempton, A. W. (1960). "Correspondence," *Geotechnique*, Vol. 10, No. 4, 186.

Terzaghi, K. (1922). "Der Grundbruch an Stauwerken und seine Verhütung," *Die Wasserkraft*, Vol. 17, 445–449.

Terzaghi, K. (1925). *Erdbaumechanik auf Bodenphysikalischer Grundlage*, Dueticke, Vienna.

Terzaghi, K. (1936). "Relation Between Soil Mechanics and Foundation Engineering: Presidential Address," *Proceedings*, First International Conference on Soil Mechanics and Foundation Engineering, Boston, Vol. 3, 13–18.

CHAPTER 10

Stresses in a Soil Mass

10.1 Introduction

Construction of a foundation causes changes in the stress, usually a net increase. The net stress increase in the soil depends on the load per unit area to which the foundation is subjected, the depth below the foundation at which the stress estimation is desired, and other factors. It is necessary to estimate the net increase of vertical stress in soil that occurs as a result of the construction of a foundation so that settlement can be calculated. The settlement calculation procedure is discussed in more detail in Chapter 11. This chapter discusses the principles of estimation of vertical stress increase in soil caused by various types of loading, based on the theory of elasticity. Topics discussed in this chapter include:

- Determination of normal and shear stresses on an inclined plane with known stresses on a two-dimensional stress element
- Determination of vertical stress increase at a certain depth due to the application of load on the surface. The loading type includes:
  - Point load
  - Line load
  - Uniformly distributed vertical strip load
  - Linearly increasing vertical loading on a strip
  - Embankment type of loading
  - Uniformly loaded circular area
  - Uniformly loaded rectangular area

Although natural soil deposits, in most cases, are not fully elastic, isotropic, or homogeneous materials, calculations for estimating increases in vertical stress yield fairly good results for practical work.

10.2 Normal and Shear Stresses on a Plane

Students in a soil mechanics course are familiar with the fundamental principles of the mechanics of deformable solids. This section is a brief review of the basic concepts of normal and shear stresses on a plane that can be found in any course on the mechanics of materials.

Figure 10.1a shows a two-dimensional soil element that is being subjected to normal and shear stresses ($\sigma_y > \sigma_x$). To determine the normal stress and the shear stress on a plane *EF* that makes an angle θ with the plane *AB*, we need to consider the free body diagram of *EFB* shown in Figure 10.1b. Let σ_n and τ_n be the normal stress and the shear stress, respectively, on the plane *EF*. From geometry, we know that

$$\overline{EB} = \overline{EF}\cos\theta \tag{10.1}$$

and

$$\overline{FB} = \overline{EF}\sin\theta \tag{10.2}$$

Summing the components of forces that act on the element in the direction of *N* and *T*, we have

$$\sigma_n(\overline{EF}) = \sigma_x(\overline{EF})\sin^2\theta + \sigma_y(\overline{EF})\cos^2\theta + 2\tau_{xy}(\overline{EF})\sin\theta\cos\theta$$

or

$$\sigma_n = \sigma_x\sin^2\theta + \sigma_y\cos^2\theta + 2\tau_{xy}\sin\theta\cos\theta$$

or

$$\sigma_n = \frac{\sigma_y + \sigma_x}{2} + \frac{\sigma_y - \sigma_x}{2}\cos 2\theta + \tau_{xy}\sin 2\theta \tag{10.3}$$

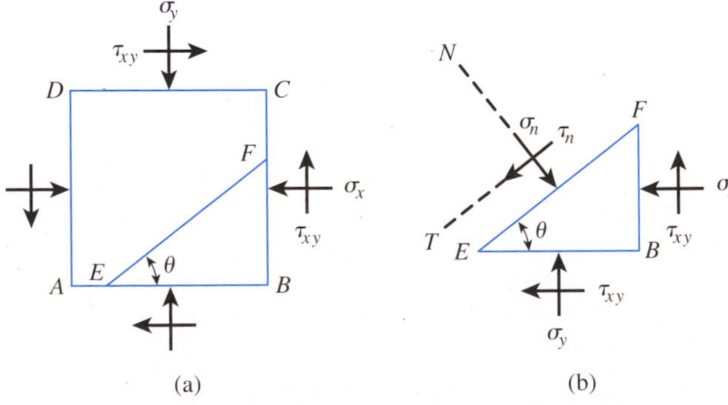

Figure 10.1 (a) A soil element with normal and shear stresses acting on it; (b) free body diagram of *EFB* as shown in (a)

Again,

$$\tau_n(\overline{EF}) = -\sigma_x(\overline{EF}) \sin\theta \cos\theta + \sigma_y(\overline{EF}) \sin\theta \cos\theta$$
$$-\tau_{xy}(\overline{EF}) \cos^2\theta + \tau_{xy}(\overline{EF}) \sin^2\theta$$

or

$$\tau_n = \sigma_y \sin\theta \cos\theta - \sigma_x \sin\theta \cos\theta - \tau_{xy}(\cos^2\theta - \sin^2\theta)$$

or

$$\tau_n = \frac{\sigma_y - \sigma_x}{2} \sin 2\theta - \tau_{xy} \cos 2\theta \tag{10.4}$$

From Eq. (10.4), we can see that we can choose the value of θ in such a way that τ_n will be equal to zero. Substituting $\tau_n = 0$, we get

$$\tan 2\theta = \frac{2\tau_{xy}}{\sigma_y - \sigma_x} \tag{10.5}$$

For given values of τ_{xy}, σ_x, and σ_y, Eq. (10.5) will give two values of θ that are 90° apart. This means that there are two planes that are at right angles to each other on which the shear stress is zero. Such planes are called *principal planes*. The normal stresses that act on the principal planes are referred to as *principal stresses*. The values of principal stresses can be found by substituting Eq. (10.5) into Eq. (10.3), which yields

Major principal stress:

$$\sigma_n = \sigma_1 = \frac{\sigma_y + \sigma_x}{2} + \sqrt{\left[\frac{(\sigma_y - \sigma_x)}{2}\right]^2 + \tau_{xy}^2} \tag{10.6}$$

Minor principal stress:

$$\sigma_n = \sigma_3 = \frac{\sigma_y + \sigma_x}{2} - \sqrt{\left[\frac{(\sigma_y - \sigma_x)}{2}\right]^2 + \tau_{xy}^2} \tag{10.7}$$

The normal stress and shear stress that act on any plane can also be determined by plotting a Mohr's circle, as shown in Figure 10.2. The following sign conventions are used in Mohr's circles: Compressive normal stresses are taken as positive, and shear stresses are considered positive if they act on opposite faces of the element in such a way that they tend to produce a counterclockwise rotation.

For plane AD of the soil element shown in Figure 10.1a, normal stress equals $+\sigma_x$ and shear stress equals $+\tau_{xy}$. For plane AB, normal stress equals $+\sigma_y$ and shear stress equals $-\tau_{xy}$.

The points R and M in Figure 10.2 represent the stress conditions on planes AD and AB, respectively. O is the point of intersection of the normal stress axis with the

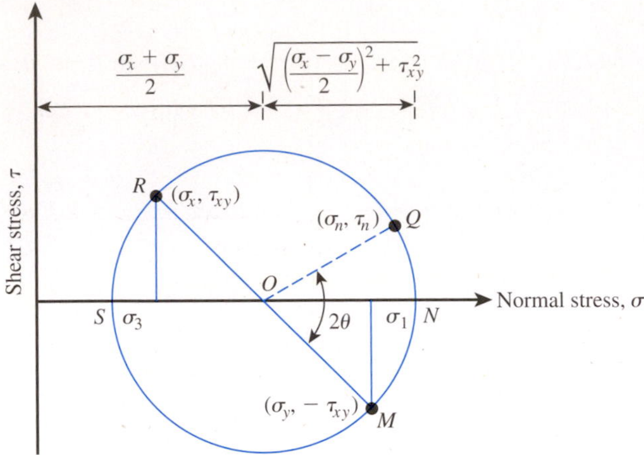

Figure 10.2 Principles of the Mohr's circle

line RM. The circle $MNQRS$ drawn with O as the center and OR as the radius is the Mohr's circle for the stress conditions considered. The radius of the Mohr's circle is equal to

$$\sqrt{\left[\frac{(\sigma_y - \sigma_x)}{2}\right]^2 + \tau_{xy}^2}$$

The stress on plane EF can be determined by moving an angle 2θ (which is twice the angle that the plane EF makes with plane AB in Figure 10.1a) in a counterclockwise direction from point M along the circumference of the Mohr's circle to reach point Q. The abscissa and ordinate of point Q, respectively, give the normal stress σ_n and the shear stress τ_n on plane EF.

Because the ordinates (that is, the shear stresses) of points N and S are zero, they represent the stresses on the principal planes. The abscissa of point N is equal to σ_1 [Eq. (10.6)], and the abscissa for point S is σ_3 [Eq. (10.7)].

As a special case, if the planes AB and AD were major and minor principal planes, the normal stress and the shear stress on plane EF could be found by substituting $\tau_{xy} = 0$. Equations (10.3) and (10.4) show that $\sigma_y = \sigma_1$ and $\sigma_x = \sigma_3$ (Figure 10.3a). Thus,

$$\sigma_n = \frac{\sigma_1 + \sigma_3}{2} + \frac{\sigma_1 - \sigma_3}{2} \cos 2\theta \tag{10.8}$$

$$\tau_n = \frac{\sigma_1 - \sigma_3}{2} \sin 2\theta \tag{10.9}$$

The Mohr's circle for such stress conditions is shown in Figure 10.3b. The abscissa and the ordinate of point Q give the normal stress and the shear stress, respectively, on the plane EF.

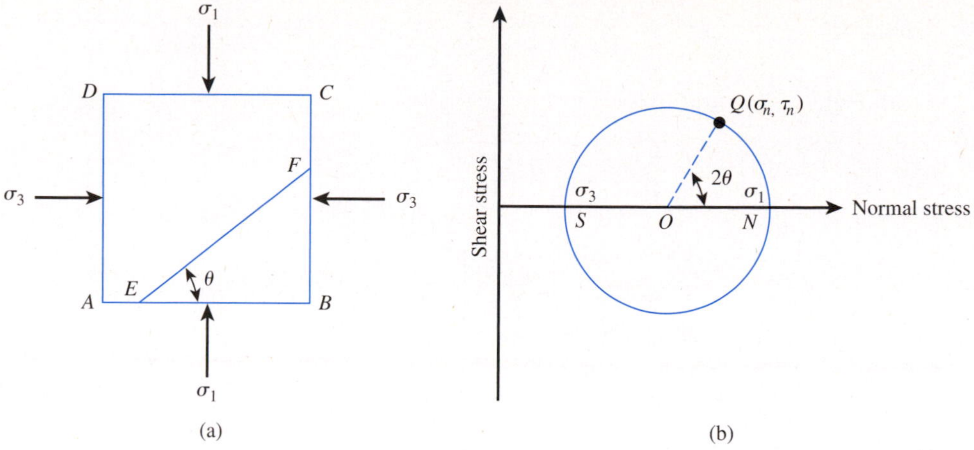

Figure 10.3 (a) Soil element with AB and AD as major and minor principal planes; (b) Mohr's circle for soil element shown in (a)

Example 10.1

A soil element is shown in Figure 10.4. The magnitudes of stresses are $\sigma_x = 120\ \text{kN/m}^2$, $\tau = 40\ \text{kN/m}^2$, $\sigma_y = 300\ \text{kN/m}^2$, and $\theta = 20°$. Determine

a. Magnitudes of the principal stresses.
b. Normal and shear stresses on plane AB. Use Eqs. (10.3), (10.4), (10.6), and (10.7).

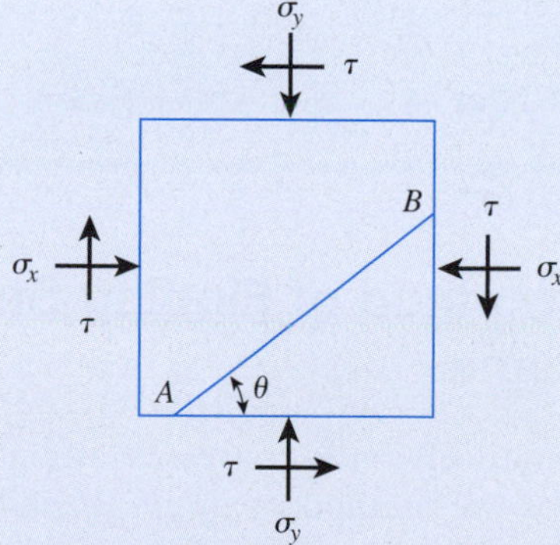

Figure 10.4 Soil element with stresses acting on it

Solution

Part a

From Eqs. (10.6) and (10.7),

$$\left.\begin{matrix}\sigma_3\\ \sigma_1\end{matrix}\right\} = \frac{\sigma_y + \sigma_x}{2} \pm \sqrt{\left[\frac{\sigma_y - \sigma_x}{2}\right]^2 + \tau_{xy}^2}$$

$$= \frac{300 + 120}{2} \pm \sqrt{\left[\frac{300 - 120}{2}\right]^2 + (-40)^2}$$

$$\sigma_1 = \mathbf{308.5\ kN/m^2}$$

$$\sigma_3 = \mathbf{111.5\ kN/m^2}$$

Part b

From Eq. (10.3),

$$\sigma_n = \frac{\sigma_y + \sigma_x}{2} + \frac{\sigma_y - \sigma_x}{2}\cos 2\theta + \tau \sin 2\theta$$

$$= \frac{300 + 120}{2} + \frac{300 - 120}{2}\cos(2 \times 20) + (-40)\sin(2 \times 20)$$

$$= \mathbf{253.23\ kN/m^2}$$

From Eq. (10.4),

$$\tau_n = \frac{\sigma_y - \sigma_x}{2}\sin 2\theta - \tau \cos 2\theta$$

$$= \frac{300 - 120}{2}\sin(2 \times 20) - (-40)\cos(2 \times 20)$$

$$= \mathbf{88.40\ kN/m^2}$$

10.3 The Pole Method of Finding Stresses along a Plane

Another important technique of finding stresses along a plane from a Mohr's circle is the *pole method*, or the method of *origin of planes*. This is demonstrated in Figure 10.5. Figure 10.5a is the same stress element that is shown in Figure 10.1a; Figure 10.5b is the Mohr's circle for the stress conditions indicated. According to the pole method, we draw a line from a known point on the Mohr's circle parallel to the plane on which the state of stress acts. The point of intersection of this line with the Mohr's circle is called the *pole*. This is a unique point for the state of stress under consideration. For example, the point M on the Mohr's circle in Figure 10.5b represents the stresses on the plane AB. The line MP is drawn parallel to AB. So

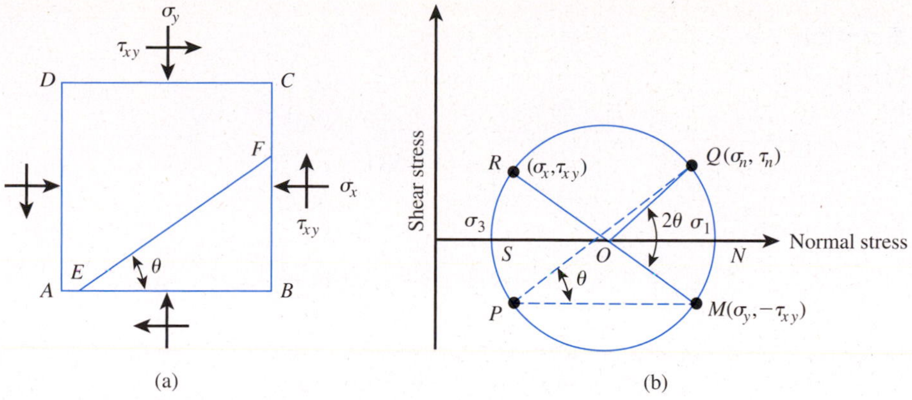

Figure 10.5 (a) Soil element with normal and shear stresses acting on it; (b) use of pole method to find the stresses along a plane

point P is the pole (origin of planes) in this case. If we need to find the stresses on a plane EF, we draw a line from the pole parallel to EF. The point of intersection of this line with the Mohr's circle is Q. The coordinates of Q give the stresses on the plane EF. (*Note:* From geometry, angle QOM is twice the angle QPM.)

Example 10.2

For the stressed soil element shown in Figure 10.6a, determine

a. Major principal stress
b. Minor principal stress
c. Normal and shear stresses on the plane DE

Use the pole method.

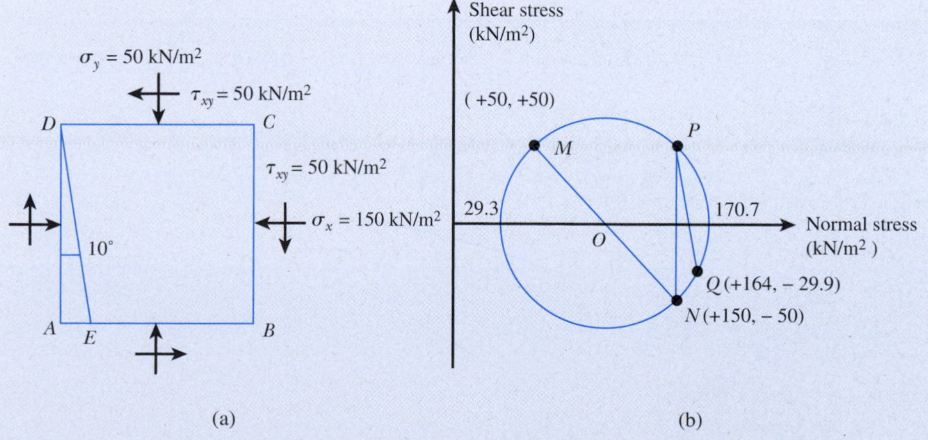

Figure 10.6 (a) Stressed soil element; (b) Mohr's circle for the soil element

Solution

On plane AD:

$$\text{Normal stress} = +150 \text{ kN/m}^2$$
$$\text{Shear stress} = -50 \text{ kN/m}^2$$

On plane AB:

$$\text{Normal stress} = +50 \text{ kN/m}^2$$
$$\text{Shear stress} = +50 \text{ kN/m}^2$$

The Mohr's circle is plotted in Figure 10.6b. From the plot,

a. Major principal stress = **170.7 kN/m²**
b. Minor principal stress = **29.3 kN/m²**
c. NP is the line drawn parallel to the plane CB. P is the pole. PQ is drawn parallel to DE (Figure 10.6a). The coordinates of point Q give the stress on the plane DE. Thus,

$$\text{Normal stress} = \mathbf{164 \text{ kN/m}^2}$$
$$\text{Shear stress} = \mathbf{-29.9 \text{ kN/m}^2}$$

10.4 Stresses Caused by a Point Load

Boussinesq (1883) solved the problem of stresses produced at any point in a homogeneous, elastic, and isotropic medium as the result of a point load applied on the surface of an infinitely large half-space. According to Figure 10.7, Boussinesq's solution for normal stresses at a point caused by the point load P is

$$\Delta\sigma_x = \frac{P}{2\pi}\left\{\frac{3x^2z}{L^5} - (1-2\mu)\left[\frac{x^2-y^2}{Lr^2(L+z)} + \frac{y^2z}{L^3r^2}\right]\right\} \tag{10.10}$$

$$\Delta\sigma_y = \frac{P}{2\pi}\left\{\frac{3y^2z}{L^5} - (1-2\mu)\left[\frac{y^2-x^2}{Lr^2(L+z)} + \frac{x^2z}{L^3r^2}\right]\right\} \tag{10.11}$$

and

$$\Delta\sigma_z = \frac{3P}{2\pi}\frac{z^3}{L^5} = \frac{3P}{2\pi}\frac{z^3}{(r^2+z^2)^{5/2}} \tag{10.12}$$

where $r = \sqrt{x^2+y^2}$
$L = \sqrt{x^2+y^2+z^2} = \sqrt{r^2+z^2}$
μ = Poisson's ratio

Note that Eqs. (10.10) and (10.11), which are the expressions for horizontal normal stresses, depend on the Poisson's ratio of the medium. However, the relationship

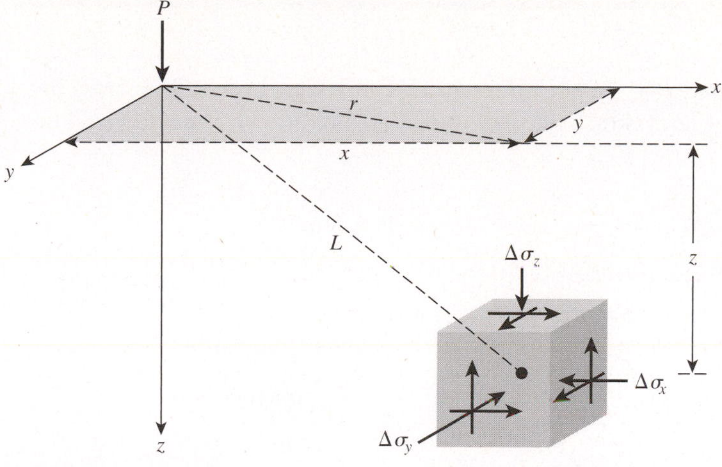

Figure 10.7 Stresses in an elastic medium caused by a point load

for the vertical normal stress, $\Delta\sigma_z$, as given by Eq. (10.12), is independent of Poisson's ratio. The relationship for $\Delta\sigma_z$ can be rewritten as

$$\Delta\sigma_z = \frac{P}{z^2}\left\{\frac{3}{2\pi}\frac{1}{[(r/z)^2+1]^{5/2}}\right\} = \frac{P}{z^2}I_1 \tag{10.13}$$

where

$$I_1 = \frac{3}{2\pi}\frac{1}{[(r/z)^2+1]^{5/2}} \tag{10.14}$$

The variation of I_1 for various values of r/z is given in Table 10.1.

Table 10.1 Variation of I_1 for Various Values of r/z [Eq. (10.14)]

r/z	I_1	r/z	I_1	r/z	I_1
0	0.4775	0.36	0.3521	1.80	0.0129
0.02	0.4770	0.38	0.3408	2.00	0.0085
0.04	0.4765	0.40	0.3294	2.20	0.0058
0.06	0.4723	0.45	0.3011	2.40	0.0040
0.08	0.4699	0.50	0.2733	2.60	0.0029
0.10	0.4657	0.55	0.2466	2.80	0.0021
0.12	0.4607	0.60	0.2214	3.00	0.0015
0.14	0.4548	0.65	0.1978	3.20	0.0011
0.16	0.4482	0.70	0.1762	3.40	0.00085
0.18	0.4409	0.75	0.1565	3.60	0.00066
0.20	0.4329	0.80	0.1386	3.80	0.00051
0.22	0.4242	0.85	0.1226	4.00	0.00040
0.24	0.4151	0.90	0.1083	4.20	0.00032
0.26	0.4050	0.95	0.0956	4.40	0.00026
0.28	0.3954	1.00	0.0844	4.60	0.00021
0.30	0.3849	1.20	0.0513	4.80	0.00017
0.32	0.3742	1.40	0.0317	5.00	0.00014
0.34	0.3632	1.60	0.0200		

Example 10.3

Consider a point load $P = 5$ kN (Figure 10.7). Calculate the vertical stress increase ($\Delta\sigma_z$) at $z = 0, 2$ m, 4 m, 6 m, 10 m, and 20 m. Given $x = 3$ m and $y = 4$ m.

Solution

$$r = \sqrt{x^2 + y^2} = \sqrt{3^2 + 4^2} = 5 \text{ m}$$

The following table can now be prepared.

r (m)	z (m)	$\frac{r}{z}$	I_1	$\Delta\sigma_z = \left(\frac{P}{z^2}\right)I_1$ (kN/m²)
5	0	∞	0	**0**
	2	2.5	0.0034	**0.0043**
	4	1.25	0.0424	**0.0133**
	6	0.83	0.1295	**0.0180**
	10	0.5	0.2733	**0.0137**
	20	0.25	0.4103	**0.0051**

Example 10.4

Refer to Example 10.3. Calculate the vertical stress increase ($\Delta\sigma_z$) at $z = 2$ m; $y = 3$ m; and $x = 0, 1, 2, 3$, and 4 m.

Solution

The following table can now be prepared. *Note*: $r = \sqrt{x^2 + y^2}$; $P = 5$ kN

x (m)	y (m)	r (m)	z (m)	$\frac{r}{z}$	I_1	$\Delta\sigma_z = \left(\frac{P}{z^2}\right)I_1$ (kN/m²)
0	3	3	2	1.5	0.025	**0.031**
1	3	3.16	2	1.58	0.0208	**0.026**
2	3	3.61	2	1.81	0.0126	**0.0158**
3	3	4.24	2	2.1	0.007	**0.009**
4	3	5	2	2.5	0.0034	**0.004**

10.5 Vertical Stress Caused by a Vertical Line Load

Figure 10.8 shows a vertical flexible line load of infinite length that has an intensity q/unit length on the surface of a semi-infinite soil mass. The vertical stress increase, $\Delta\sigma_z$, inside the soil mass can be determined by using the principles of the theory of elasticity, or

$$\Delta\sigma_z = \frac{2qz^3}{\pi(x^2 + z^2)^2} \tag{10.15}$$

This equation can be rewritten as

$$\Delta\sigma_z = \frac{2q}{\pi z[(x/z)^2 + 1]^2}$$

or

$$\frac{\Delta\sigma_z}{(q/z)} = \frac{2}{\pi[(x/z)^2 + 1]^2} \tag{10.16}$$

Note that Eq. (10.16) is in a nondimensional form. Using this equation, we can calculate the variation of $\Delta\sigma_z/(q/z)$ with x/z. This is given in Table 10.2. The value of $\Delta\sigma_z$ calculated by using Eq. (10.16) is the additional stress on soil caused by the line load. The value of $\Delta\sigma_z$ does not include the overburden pressure of the soil above point A.

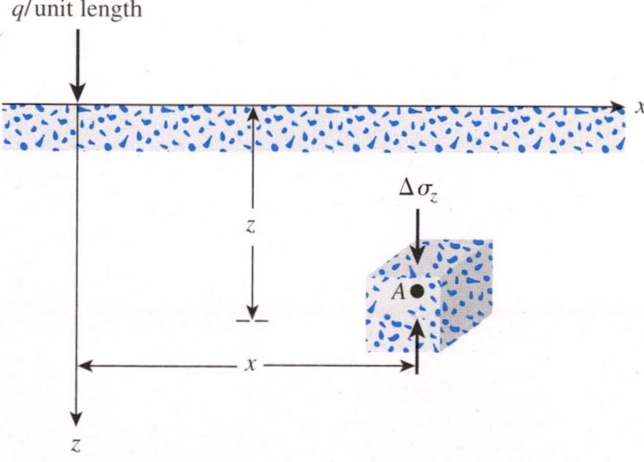

Figure 10.8 Line load over the surface of a semi-infinite soil mass

Table 10.2 Variation of $\Delta\sigma_z/(q/z)$ with x/z [Eq. (10.16)]

x/z	$\Delta\sigma_z/(q/z)$	x/z	$\Delta\sigma_z/(q/z)$
0	0.637	1.3	0.088
0.1	0.624	1.4	0.073
0.2	0.589	1.5	0.060
0.3	0.536	1.6	0.050
0.4	0.473	1.7	0.042
0.5	0.407	1.8	0.035
0.6	0.344	1.9	0.030
0.7	0.287	2.0	0.025
0.8	0.237	2.2	0.019
0.9	0.194	2.4	0.014
1.0	0.159	2.6	0.011
1.1	0.130	2.8	0.008
1.2	0.107	3.0	0.006

Example 10.5

Figure 10.9a shows two line loads on the ground surface. Determine the increase of stress at point A.

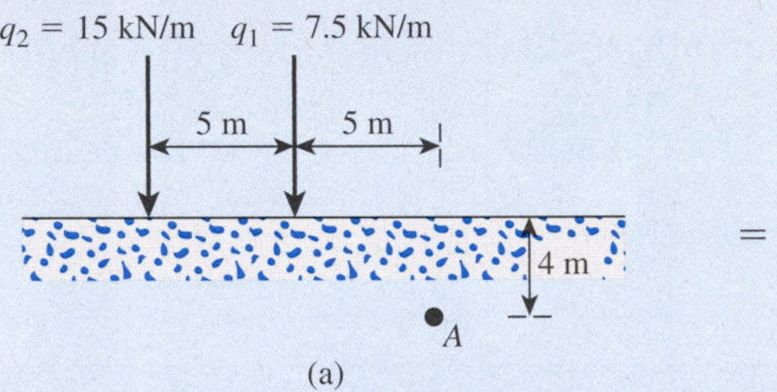

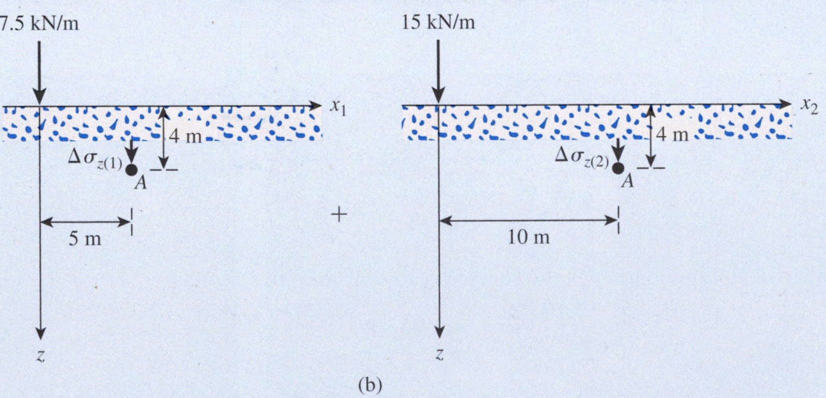

Figure 10.9 (a) Two line loads on the ground surface; (b) use of superposition principle to obtain stress at point A

Solution

Refer to Figure 10.9b. The total stress at A is

$$\Delta\sigma_z = \Delta\sigma_{z(1)} + \Delta\sigma_{z(2)}$$

$$\Delta\sigma_{z(1)} = \frac{2q_1z^3}{\pi(x_1^2 + z^2)^2} = \frac{(2)(7.5)(4)^3}{\pi(5^2 + 4^2)^2} = 0.182 \text{ kN/m}^2$$

$$\Delta\sigma_{z(2)} = \frac{2q_2z^3}{\pi(x_2^2 + z^2)^2} = \frac{(2)(15)(4)^3}{\pi(10^2 + 4^2)^2} = 0.045 \text{ kN/m}^2$$

$$\Delta\sigma_z = 0.182 + 0.045 = \mathbf{0.227\ kN/m^2}$$

10.6 Vertical Stress Caused by a Horizontal Line Load

Figure 10.10 shows a horizontal flexible line load on the surface of a semi-infinite soil mass. The vertical stress increase at point A in the soil mass can be given as

$$\Delta\sigma_z = \frac{2q}{\pi}\frac{xz^2}{(x^2 + z^2)^2} \quad (10.17)$$

Table 10.3 gives the variation of $\Delta\sigma_z/(q/z)$ with x/z.

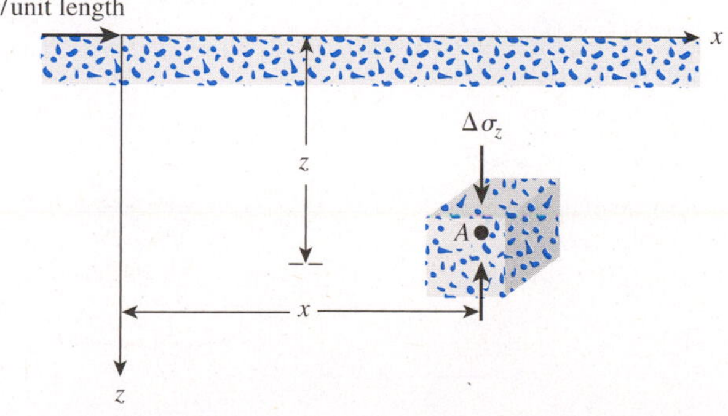

Figure 10.10 Horizontal line load over the surface of a semi-infinite soil mass

Table 10.3 Variation of $\Delta\sigma_z/(q/z)$ with x/z

x/z	$\Delta\sigma_z/(q/z)$	x/z	$\Delta\sigma_z/(q/z)$
0	0	0.7	0.201
0.1	0.062	0.8	0.189
0.2	0.118	0.9	0.175
0.3	0.161	1.0	0.159
0.4	0.189	1.5	0.090
0.5	0.204	2.0	0.051
0.6	0.207	3.0	0.019

Example 10.6

An inclined line load with a magnitude of 10 kN/m is shown in Figure 10.11. Determine the increase of vertical stress $\Delta\sigma_z$ at point A due to the line load.

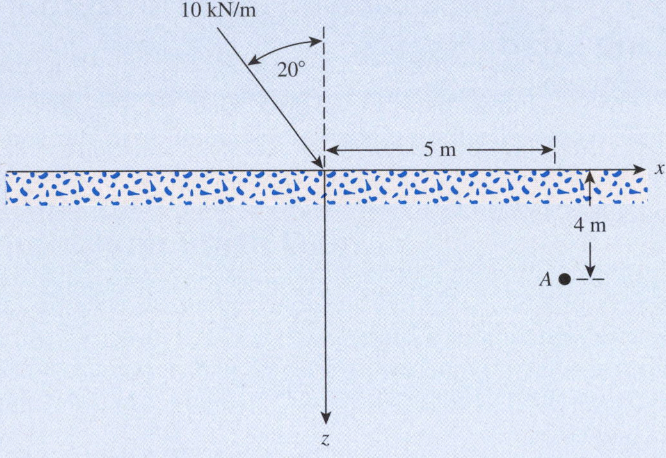

Figure 10.11

Solution

The vertical component of the inclined load $q_V = 10 \cos 20 = 9.4$ kN/m, and the horizontal component $q_H = 10 \sin 20 = 3.42$ kN/m. For point A, $x/z = 5/4 = 1.25$. Using Table 10.2, the vertical stress increase at point A due to q_V is

$$\frac{\Delta\sigma_{z(V)}}{\left(\dfrac{q_V}{z}\right)} = 0.098$$

$$\Delta\sigma_{z(V)} = (0.098)\left(\frac{q_V}{z}\right) = (0.098)\left(\frac{9.4}{4}\right) = 0.23 \text{ kN/m}^2$$

Similarly, using Table 10.3, the vertical stress increase at point A due to q_H is

$$\frac{\Delta\sigma_{z(H)}}{\left(\dfrac{q_H}{z}\right)} = 0.125$$

$$\Delta\sigma_{z(V)} = (0.125)\left(\frac{3.42}{4}\right) = 0.107\ \text{kN/m}^2$$

Thus, the total is

$$\Delta\sigma_z = \Delta\sigma_{z(V)} + \Delta\sigma_{z(H)} = 0.23 + 0.107 = \mathbf{0.337\ kN/m^2}$$

10.7 Vertical Stress Caused by a Vertical Strip Load (Finite Width and Infinite Length)

The fundamental equation for the vertical stress increase at a point in a soil mass as the result of a line load (Section 10.5) can be used to determine the vertical stress at a point caused by a flexible strip load of width B. (See Figure 10.12.) Let the load per unit area of the strip shown in Figure 10.12 be equal to q. If we consider an elemental strip of width dr, the load per unit length of this strip is equal to $q\ dr$.

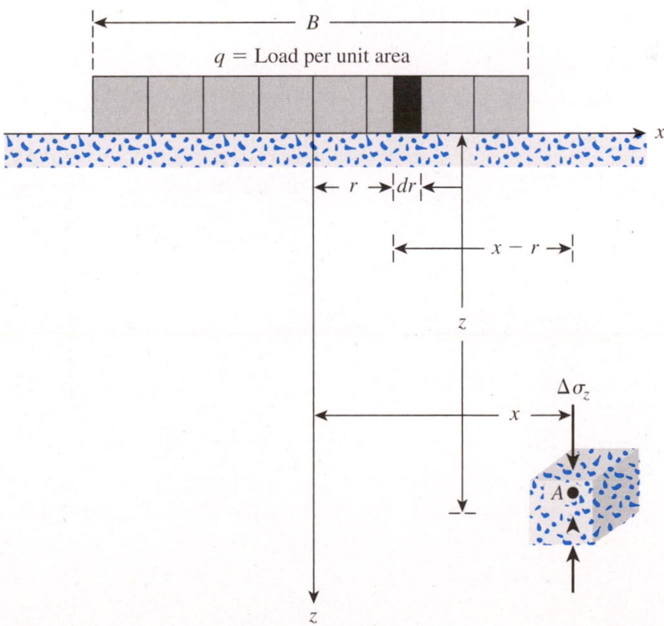

Figure 10.12 Vertical stress caused by a flexible strip load

This elemental strip can be treated as a line load. Equation (10.15) can be used to give the vertical stress increase $d\sigma_z$ at point A inside the soil mass caused by this elemental strip load. To calculate the vertical stress increase, we need to substitute $q\,dr$ for q and $(x - r)$ for x in Eq. (10.15). So,

$$d\sigma_z = \frac{2(q\,dr)z^3}{\pi[(x - r)^2 + z^2]^2} \tag{10.18}$$

The total increase in the vertical stress ($\Delta\sigma_z$) at point A caused by the entire strip load of width B can be determined by integration of Eq. (10.18) with limits of r from $-B/2$ to $+B/2$, or

$$\begin{aligned}\Delta\sigma_z = \int d\sigma_z &= \int_{-B/2}^{+B/2} \left(\frac{2q}{\pi}\right)\left\{\frac{z^3}{[(x - r)^2 + z^2]^2}\right\} dr \\ &= \frac{q}{\pi}\left\{\tan^{-1}\left[\frac{z}{x - (B/2)}\right] - \tan^{-1}\left[\frac{z}{x + (B/2)}\right] \right. \\ &\quad \left. - \frac{Bz[x^2 - z^2 - (B^2/4)]}{[x^2 + z^2 - (B^2/4)]^2 + B^2z^2}\right\}\end{aligned} \tag{10.19}$$

With respect to Eq. (10.19), the following should be kept in mind:

1. $\tan^{-1}\left[\dfrac{z}{x - \left(\dfrac{B}{2}\right)}\right]$ and $\tan^{-1}\left[\dfrac{z}{x + \left(\dfrac{B}{2}\right)}\right]$ are in radians.
2. The magnitude of $\Delta\sigma_z$ is the same value of x/z ($\pm$).
3. Equation (10.19) is valid as shown in Figure 10.12; that is, for point A, $x \geq B/2$.

   However, for $x = 0$ to $x < B/2$, the magnitude of $\tan^{-1}\left[\dfrac{z}{x - \left(\dfrac{B}{2}\right)}\right]$ becomes negative. For this case, that should be replaced by $\pi + \tan^{-1}\left[\dfrac{z}{x - \left(\dfrac{B}{2}\right)}\right]$.

Table 10.4 shows the variation of $\Delta\sigma_z/q$ with $2z/B$ and $2x/B$. This table can be used conveniently for the calculation of vertical stress at a point caused by a flexible strip load. Contours of $\Delta\sigma_z/q$ varying from 0.05 to 0.9 are shown in Figure 10.13.

Table 10.4 Variation of $\Delta\sigma_z/q$ with $2z/B$ and $2x/B$ [Eq. (10.19)]

	$2x/B$										
$2z/B$	0.0	0.1	0.2	0.3	0.4	0.5	0.6	0.7	0.8	0.9	1.0
0.00	1.000	1.000	1.000	1.000	1.000	1.000	1.000	1.000	1.000	1.000	0.000
0.10	1.000	1.000	0.999	0.999	0.999	0.998	0.997	0.993	0.980	0.909	0.500
0.20	0.997	0.997	0.996	0.995	0.992	0.988	0.979	0.959	0.909	0.775	0.500
0.30	0.990	0.989	0.987	0.984	0.978	0.967	0.947	0.908	0.833	0.697	0.499
0.40	0.977	0.976	0.973	0.966	0.955	0.937	0.906	0.855	0.773	0.651	0.498
0.50	0.959	0.958	0.953	0.943	0.927	0.902	0.864	0.808	0.727	0.620	0.497
0.60	0.937	0.935	0.928	0.915	0.896	0.866	0.825	0.767	0.691	0.598	0.495
0.70	0.910	0.908	0.899	0.885	0.863	0.831	0.788	0.732	0.662	0.581	0.492
0.80	0.881	0.878	0.869	0.853	0.829	0.797	0.755	0.701	0.638	0.566	0.489
0.90	0.850	0.847	0.837	0.821	0.797	0.765	0.724	0.675	0.617	0.552	0.485
1.00	0.818	0.815	0.805	0.789	0.766	0.735	0.696	0.650	0.598	0.540	0.480
1.10	0.787	0.783	0.774	0.758	0.735	0.706	0.670	0.628	0.580	0.529	0.474
1.20	0.755	0.752	0.743	0.728	0.707	0.679	0.646	0.607	0.564	0.517	0.468
1.30	0.725	0.722	0.714	0.699	0.679	0.654	0.623	0.588	0.548	0.506	0.462
1.40	0.696	0.693	0.685	0.672	0.653	0.630	0.602	0.569	0.534	0.495	0.455
1.50	0.668	0.666	0.658	0.646	0.629	0.607	0.581	0.552	0.519	0.484	0.448
1.60	0.642	0.639	0.633	0.621	0.605	0.586	0.562	0.535	0.506	0.474	0.440
1.70	0.617	0.615	0.608	0.598	0.583	0.565	0.544	0.519	0.492	0.463	0.433
1.80	0.593	0.591	0.585	0.576	0.563	0.546	0.526	0.504	0.479	0.453	0.425
1.90	0.571	0.569	0.564	0.555	0.543	0.528	0.510	0.489	0.467	0.443	0.417
2.00	0.550	0.548	0.543	0.535	0.524	0.510	0.494	0.475	0.455	0.433	0.409
2.10	0.530	0.529	0.524	0.517	0.507	0.494	0.479	0.462	0.443	0.423	0.401
2.20	0.511	0.510	0.506	0.499	0.490	0.479	0.465	0.449	0.432	0.413	0.393
2.30	0.494	0.493	0.489	0.483	0.474	0.464	0.451	0.437	0.421	0.404	0.385
2.40	0.477	0.476	0.473	0.467	0.460	0.450	0.438	0.425	0.410	0.395	0.378
2.50	0.462	0.461	0.458	0.452	0.445	0.436	0.426	0.414	0.400	0.386	0.370
2.60	0.447	0.446	0.443	0.439	0.432	0.424	0.414	0.403	0.390	0.377	0.363
2.70	0.433	0.432	0.430	0.425	0.419	0.412	0.403	0.393	0.381	0.369	0.355
2.80	0.420	0.419	0.417	0.413	0.407	0.400	0.392	0.383	0.372	0.360	0.348
2.90	0.408	0.407	0.405	0.401	0.396	0.389	0.382	0.373	0.363	0.352	0.341
3.00	0.396	0.395	0.393	0.390	0.385	0.379	0.372	0.364	0.355	0.345	0.334
3.10	0.385	0.384	0.382	0.379	0.375	0.369	0.363	0.355	0.347	0.337	0.327
3.20	0.374	0.373	0.372	0.369	0.365	0.360	0.354	0.347	0.339	0.330	0.321
3.30	0.364	0.363	0.362	0.359	0.355	0.351	0.345	0.339	0.331	0.323	0.315
3.40	0.354	0.354	0.352	0.350	0.346	0.342	0.337	0.331	0.324	0.316	0.308
3.50	0.345	0.345	0.343	0.341	0.338	0.334	0.329	0.323	0.317	0.310	0.302
3.60	0.337	0.336	0.335	0.333	0.330	0.326	0.321	0.316	0.310	0.304	0.297
3.70	0.328	0.328	0.327	0.325	0.322	0.318	0.314	0.309	0.304	0.298	0.291
3.80	0.320	0.320	0.319	0.317	0.315	0.311	0.307	0.303	0.297	0.292	0.285
3.90	0.313	0.313	0.312	0.310	0.307	0.304	0.301	0.296	0.291	0.286	0.280
4.00	0.306	0.305	0.304	0.303	0.301	0.298	0.294	0.290	0.285	0.280	0.275
4.10	0.299	0.299	0.298	0.296	0.294	0.291	0.288	0.284	0.280	0.275	0.270
4.20	0.292	0.292	0.291	0.290	0.288	0.285	0.282	0.278	0.274	0.270	0.265
4.30	0.286	0.286	0.285	0.283	0.282	0.279	0.276	0.273	0.269	0.265	0.260
4.40	0.280	0.280	0.279	0.278	0.276	0.274	0.271	0.268	0.264	0.260	0.256
4.50	0.274	0.274	0.273	0.272	0.270	0.268	0.266	0.263	0.259	0.255	0.251
4.60	0.268	0.268	0.268	0.266	0.265	0.263	0.260	0.258	0.254	0.251	0.247
4.70	0.263	0.263	0.262	0.261	0.260	0.258	0.255	0.253	0.250	0.246	0.243
4.80	0.258	0.258	0.257	0.256	0.255	0.253	0.251	0.248	0.245	0.242	0.239
4.90	0.253	0.253	0.252	0.251	0.250	0.248	0.246	0.244	0.241	0.238	0.235
5.00	0.248	0.248	0.247	0.246	0.245	0.244	0.242	0.239	0.237	0.234	0.231

(continued)

Table 10.4 (*continued*)

	2x/B									
2z/B	**1.1**	**1.2**	**1.3**	**1.4**	**1.5**	**1.6**	**1.7**	**1.8**	**1.9**	**2.0**
0.00	0.000	0.000	0.000	0.000	0.000	0.000	0.000	0.000	0.000	0.000
0.10	0.091	0.020	0.007	0.003	0.002	0.001	0.001	0.000	0.000	0.000
0.20	0.225	0.091	0.040	0.020	0.011	0.007	0.004	0.003	0.002	0.002
0.30	0.301	0.165	0.090	0.052	0.031	0.020	0.013	0.009	0.007	0.005
0.40	0.346	0.224	0.141	0.090	0.059	0.040	0.027	0.020	0.014	0.011
0.50	0.373	0.267	0.185	0.128	0.089	0.063	0.046	0.034	0.025	0.019
0.60	0.391	0.298	0.222	0.163	0.120	0.088	0.066	0.050	0.038	0.030
0.70	0.403	0.321	0.250	0.193	0.148	0.113	0.087	0.068	0.053	0.042
0.80	0.411	0.338	0.273	0.218	0.173	0.137	0.108	0.086	0.069	0.056
0.90	0.416	0.351	0.291	0.239	0.195	0.158	0.128	0.104	0.085	0.070
1.00	0.419	0.360	0.305	0.256	0.214	0.177	0.147	0.122	0.101	0.084
1.10	0.420	0.366	0.316	0.271	0.230	0.194	0.164	0.138	0.116	0.098
1.20	0.419	0.371	0.325	0.282	0.243	0.209	0.178	0.152	0.130	0.111
1.30	0.417	0.373	0.331	0.291	0.254	0.221	0.191	0.166	0.143	0.123
1.40	0.414	0.374	0.335	0.298	0.263	0.232	0.203	0.177	0.155	0.135
1.50	0.411	0.374	0.338	0.303	0.271	0.240	0.213	0.188	0.165	0.146
1.60	0.407	0.373	0.339	0.307	0.276	0.248	0.221	0.197	0.175	0.155
1.70	0.402	0.370	0.339	0.309	0.281	0.254	0.228	0.205	0.183	0.164
1.80	0.396	0.368	0.339	0.311	0.284	0.258	0.234	0.212	0.191	0.172
1.90	0.391	0.364	0.338	0.312	0.286	0.262	0.239	0.217	0.197	0.179
2.00	0.385	0.360	0.336	0.311	0.288	0.265	0.243	0.222	0.203	0.185
2.10	0.379	0.356	0.333	0.311	0.288	0.267	0.246	0.226	0.208	0.190
2.20	0.373	0.352	0.330	0.309	0.288	0.268	0.248	0.229	0.212	0.195
2.30	0.366	0.347	0.327	0.307	0.288	0.268	0.250	0.232	0.215	0.199
2.40	0.360	0.342	0.323	0.305	0.287	0.268	0.251	0.234	0.217	0.202
2.50	0.354	0.337	0.320	0.302	0.285	0.268	0.251	0.235	0.220	0.205
2.60	0.347	0.332	0.316	0.299	0.283	0.267	0.251	0.236	0.221	0.207
2.70	0.341	0.327	0.312	0.296	0.281	0.266	0.251	0.236	0.222	0.208
2.80	0.335	0.321	0.307	0.293	0.279	0.265	0.250	0.236	0.223	0.210
2.90	0.329	0.316	0.303	0.290	0.276	0.263	0.249	0.236	0.223	0.211
3.00	0.323	0.311	0.299	0.286	0.274	0.261	0.248	0.236	0.223	0.211
3.10	0.317	0.306	0.294	0.283	0.271	0.259	0.247	0.235	0.223	0.212
3.20	0.311	0.301	0.290	0.279	0.268	0.256	0.245	0.234	0.223	0.212
3.30	0.305	0.296	0.286	0.275	0.265	0.254	0.243	0.232	0.222	0.211
3.40	0.300	0.291	0.281	0.271	0.261	0.251	0.241	0.231	0.221	0.211
3.50	0.294	0.286	0.277	0.268	0.258	0.249	0.239	0.229	0.220	0.210
3.60	0.289	0.281	0.273	0.264	0.255	0.246	0.237	0.228	0.218	0.209
3.70	0.284	0.276	0.268	0.260	0.252	0.243	0.235	0.226	0.217	0.208
3.80	0.279	0.272	0.264	0.256	0.249	0.240	0.232	0.224	0.216	0.207
3.90	0.274	0.267	0.260	0.253	0.245	0.238	0.230	0.222	0.214	0.206
4.00	0.269	0.263	0.256	0.249	0.242	0.235	0.227	0.220	0.212	0.205
4.10	0.264	0.258	0.252	0.246	0.239	0.232	0.225	0.218	0.211	0.203
4.20	0.260	0.254	0.248	0.242	0.236	0.229	0.222	0.216	0.209	0.202
4.30	0.255	0.250	0.244	0.239	0.233	0.226	0.220	0.213	0.207	0.200
4.40	0.251	0.246	0.241	0.235	0.229	0.224	0.217	0.211	0.205	0.199
4.50	0.247	0.242	0.237	0.232	0.226	0.221	0.215	0.209	0.203	0.197
4.60	0.243	0.238	0.234	0.229	0.223	0.218	0.212	0.207	0.201	0.195
4.70	0.239	0.235	0.230	0.225	0.220	0.215	0.210	0.205	0.199	0.194
4.80	0.235	0.231	0.227	0.222	0.217	0.213	0.208	0.202	0.197	0.192
4.90	0.231	0.227	0.223	0.219	0.215	0.210	0.205	0.200	0.195	0.190
5.00	0.227	0.224	0.220	0.216	0.212	0.207	0.203	0.198	0.193	0.188

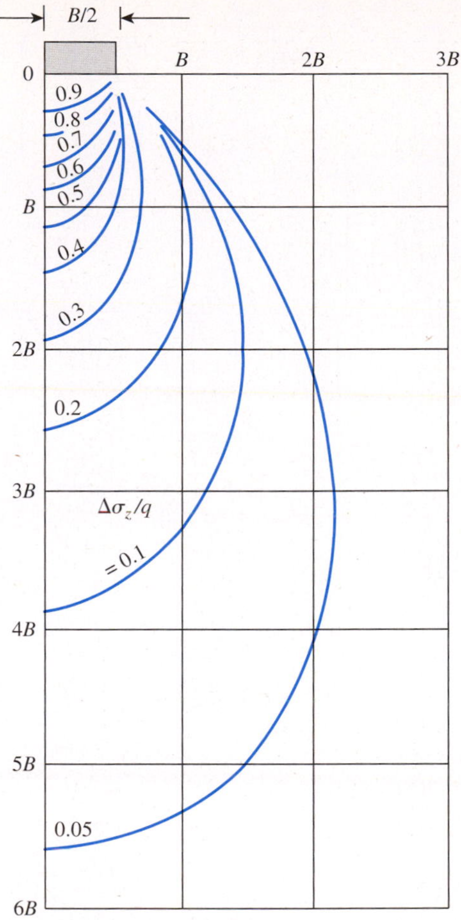

Figure 10.13 Contours of $\Delta\sigma_z/q$ below a strip load

Example 10.7

Refer to Figure 10.12. Given: $B = 4$ m and $q = 100$ kN/m^2. For point A, $z = 1$ m and $x = 1$ m. Determine the vertical stress $\Delta\sigma_z$ at A. Use Eq. (10.19).

Solution

Since $x = 1 \text{ m} < B/2 = 2$ m,

$$\Delta\sigma_z = \frac{q}{\pi}\left\{\tan^{-1}\left[\frac{z}{x - \left(\frac{B}{2}\right)}\right] + \pi - \tan^{-1}\left[\frac{z}{x + \left(\frac{B}{2}\right)}\right]\right.$$

$$- \frac{Bz\left[x^2 - z^2 - \left(\frac{B^2}{4}\right)\right]}{\left[x^2 + z^2 - \left(\frac{B^2}{4}\right)\right]^2 + B^2z^2} \right\}$$

$$\tan^{-1}\left[\frac{z}{x - \left(\frac{B}{2}\right)}\right] = \tan^{-1}\left(\frac{1}{1-2}\right) = -45° = -0.785 \text{ rad}$$

$$\tan^{-1}\left[\frac{z}{x + \left(\frac{B}{2}\right)}\right] = \tan^{-1}\left(\frac{1}{1+2}\right) = 18.43° = 0.322 \text{ rad}$$

$$\frac{Bz\left[x^2 - z^2 - \left(\frac{B^2}{4}\right)\right]}{\left[x^2 + z^2 - \left(\frac{B^2}{4}\right)\right]^2 + B^2z^2} = \frac{(4)(1)\left[(1)^2 - (1)^2 - \left(\frac{16}{4}\right)\right]}{\left[(1)^2 + (1)^2 - \left(\frac{16}{4}\right)\right]^2 + (16)(1)} = -0.8$$

Hence,

$$\frac{\Delta\sigma_z}{q} = \frac{1}{\pi}[-0.785 + \pi - 0.322 - (-0.8)] = 0.902$$

Now, compare with Table 10.4. For this case, $\frac{2x}{B} = \frac{(2)(1)}{4} = 0.5$ and $\frac{2z}{B} = \frac{(2)(1)}{4} = 0.5$.

So, $\frac{\Delta\sigma_z}{q} = 0.902$ (Check)

$$\Delta\sigma_z = 0.902q = (0.902)(100) = \mathbf{90.2\ kN/m^2}$$

10.8 Vertical Stress Caused by a Horizontal Strip Load

Figure 10.14 shows a horizontal, flexible strip load with a width B on a semi-infinite soil mass. The load per unit area is equal to q. The vertical stress $\Delta\sigma_z$ at a point $A(x,z)$ can be given as

$$\Delta\sigma_z = \frac{4bqxz^2}{\pi[(x^2 + z^2 - b^2)^2 + 4b^2z^2]} \tag{10.20}$$

where $b = B/2$.

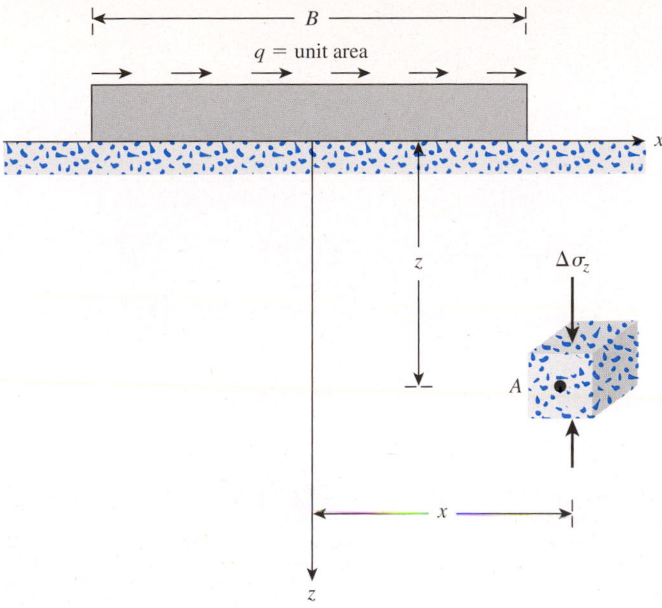

Figure 10.14 Horizontal strip load on a semi-infinite soil mass

Table 10.5 Variation of $\Delta\sigma_z/q$ with z/b and x/b [Eq. (10.20)]

	x/*b*					
z*/*b	**0**	**0.5**	**1.0**	**1.5**	**2.0**	**2.5**
0	—	—	—	—	—	—
0.25	—	0.052	0.313	0.061	0.616	
0.5	—	0.127	0.300	0.147	0.055	0.025
1.0	—	0.159	0.255	0.210	0.131	0.074
1.5	—	0.128	0.204	0.202	0.157	0.110
2.0	—	0.096	0.159	0.175	0.157	0.126
2.5	—	0.072	0.124	0.147	0.144	0.127

Table 10.5 gives the variation of $\Delta\sigma_z/q$ with z/b and x/b.

Example 10.8

Refer to Figure 10.14. Given: $B = 4$ m, $z = 1$ m, and $q = 100$ kN/m^2. Determine $\Delta\sigma_z$ at points ± 1 m.

Solution

From Eq. (10.20),

$$\frac{\Delta\sigma_z}{q} = \frac{4bxz^2}{\pi[(x^2 + z^2 - b^2)^2 + 4b^2z^2]}$$

$$= \frac{(4)\left(\frac{4}{2}\right)(\pm 1)(1)^2}{\pi\{[(\pm 1)^2 + (1)^2 - (2)^2]^2 + (4)(2)^2(1)^2\}}$$

$$= \frac{\pm 8}{\pi[(4) + (16)]} = \pm 0.127$$

Note: Compare this value of $\Delta\sigma_z/q = 0.127$ for $z/b = 1/2 = 0.5$ and $x/b = 1/2 = 0.5$ in Table 10.5. So,

$$\Delta\sigma_z = (0.127)(100) = \mathbf{12.7\ kN/m^2\ at\ } \boldsymbol{x} \mathbf{\ = +1\ m}$$

and

$$\Delta\sigma_z = (-0.127)(100) = \mathbf{-12.7\ kN/m^2\ at\ } \boldsymbol{x} \mathbf{\ = -1\ m}$$

Example 10.9

Consider the inclined strip load shown in Figure 10.15. Determine the vertical stress $\Delta\sigma_z$ at A (x = 2.25 m, z = 3 m) and B (x = –2.25 m, z = 3 m). Given: width of the strip = 3 m.

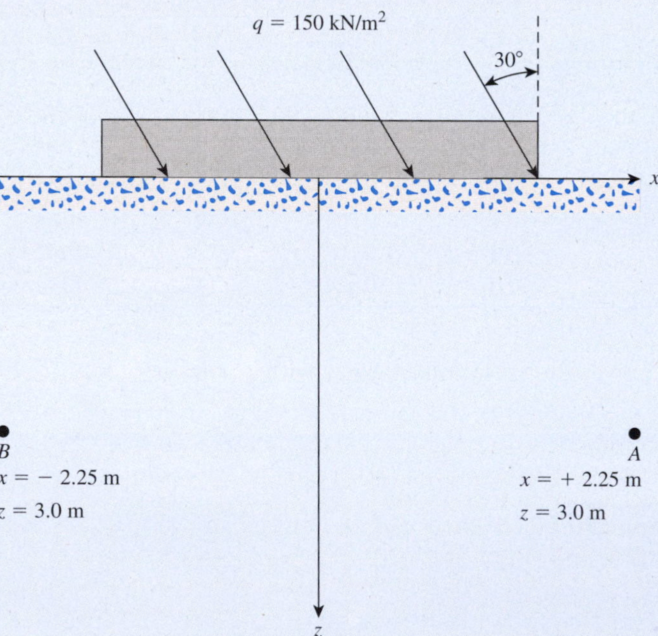

Figure 10.15

Solution

Vertical component of $q = q_v = q \cos 30 = 150 \cos 30 = 129.9 \text{ kN/m}^2$
Horizontal component of $q = q_h = q \sin 30 = 150 \sin 30 = 75 \text{ kN/m}^2$

$\Delta\sigma_z$ due to q_v:

$$\frac{2z}{B} = \frac{(2)(3)}{3} = 2$$

$$\frac{2x}{B} = \frac{(2)(\pm 2.25)}{3} = \pm 1.5$$

From Table 10.4, $\Delta\sigma_z/q_v = 0.288$.

$$\Delta\sigma_{z(v)} = (0.288)(129.9) = 37.4 \text{ kN/m}^2 \text{ (at } A \text{ and at } B\text{)}$$

$\Delta\sigma_z$ due to q_h:

$$b = \frac{B}{2} = \frac{3}{2} = 1.5$$

$$\frac{z}{b} = \frac{3}{1.5} = 2$$

$$\frac{x}{b} = \frac{\pm 2.25}{1.5} = \pm 1.5$$

From Table 10.5, $\Delta\sigma_z/q_h = \pm 0.175$. So at A,

$$\Delta\sigma_z = (+0.175)q_h = (0.175)(75) = 13.13 \text{ kN/m}^2$$

and at B,

$$\Delta\sigma_z = (-0.175)q_h = (-0.175)(75) = -13.13 \text{ kN/m}^2$$

Hence, at A,

$$\Delta\sigma_z = \Delta\sigma_{z(v)} + \Delta\sigma_{z(h)} = 37.4 + 13.13 = \mathbf{50.53 \text{ kN/m}^2}$$

At B,

$$\Delta\sigma_z = \Delta\sigma_{z(v)} + \Delta\sigma_{z(h)} = 37.4 + (-13.13) = \mathbf{24.27 \text{ kN/m}^2}$$

10.9 Linearly Increasing Vertical Loading on an Infinite Strip

Figure 10.16 shows a vertical loading on an infinity strip of width B. The intensity of load increases from zero at $x = 0$ to q/unit area at $x = B$. For the elemental strip of width dr, the load per unit length can be given as $\left(\frac{q}{B}\right)x \cdot dr$. Approximating this as a line load, we can substitute $\left(\frac{q}{B}\right)x \cdot dr$ for q and $(x - r)$ for x in Eq. (10.15) to determine the vertical stress at A (x, z), or

$$\Delta\sigma_z = \int d\sigma_z = \int_0^B \frac{2\left(\frac{q}{B}\right)r\,drz^3}{\pi[(x-r)^2 + z^2]^2} = \left(\frac{1}{B}\right)\left(\frac{2q}{\pi}\right)\int_0^B \frac{z^3 r\,dr}{[(x-r)^2 + z^2]^2}$$

or

$$\Delta\sigma_z = \frac{q}{2\pi}\left(\frac{2x}{B}\alpha - \sin 2\delta\right) \qquad (10.21)$$

In Eq. (10.21), α is in radians. *Also. note the sign for the angle* δ. Table 10.6 shows the variation of $\Delta\sigma_z$ with $2x/B$ and $2z/B$.

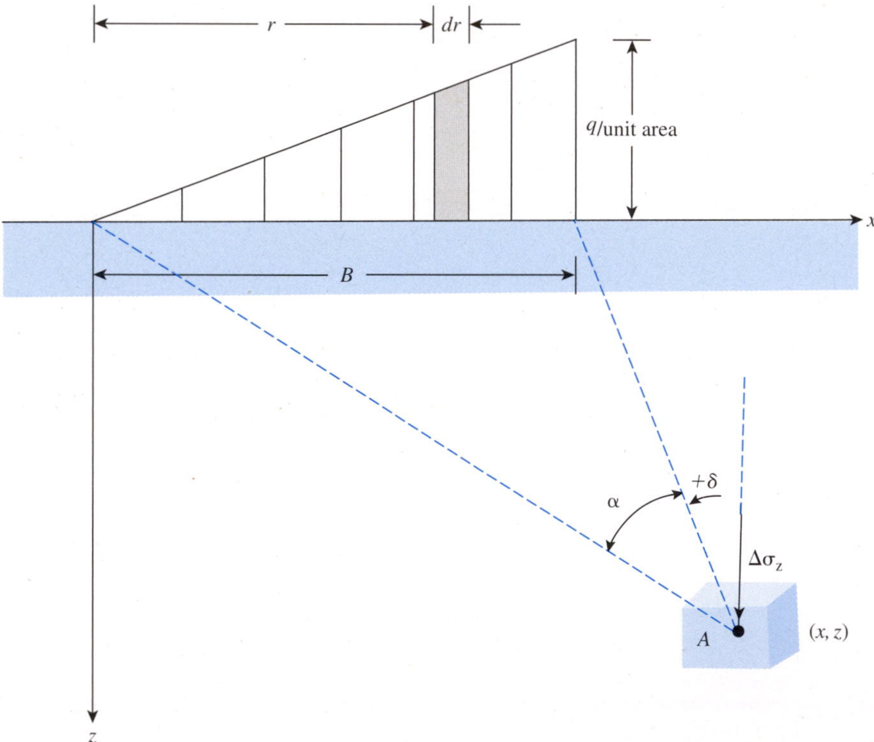

Figure 10.16 Linearly increasing vertical loading on an infinite strip

Table 10.6 Variation of $\Delta\sigma_z/q$ with $2x/B$ and $2z/B$ [Eq. (10.21)]

$\frac{2x}{B}$	2z/B: 0	0.5	1.0	1.5	2.0	2.5	3.0	4.0	5.0
−3	0	0.0003	0.0018	0.00054	0.0107	0.0170	0.0235	0.0347	0.0422
−2	0	0.0008	0.0053	0.0140	0.0249	0.0356	0.0448	0.0567	0.0616
−1	0	0.0041	0.0217	0.0447	0.0643	0.0777	0.0854	0.0894	0.0858
0	0	0.0748	0.1273	0.1528	0.1592	0.1553	0.1469	0.1273	0.1098
1	0.5	0.4797	0.4092	0.3341	0.2749	0.2309	0.1979	0.1735	0.1241
2	0.5	0.4220	0.3524	0.2952	0.2500	0.2148	0.1872	0.1476	0.1211
3	0	0.0152	0.0622	0.1010	0.1206	0.1268	0.1258	0.1154	0.1026
4	0	0.0019	0.0119	0.0285	0.0457	0.0596	0.0691	0.0775	0.0776
5	0	0.0005	0.0035	0.0097	0.0182	0.0274	0.0358	0.0482	0.0546

Example 10.10

Refer to Figure 10.17. For a linearly increasing vertical loading on an infinite strip, given: $B = 2$ m; $q = 100$ kN/m². Determine the vertical stress $\Delta\sigma_z$ at A (−1 m, 1.5 m).

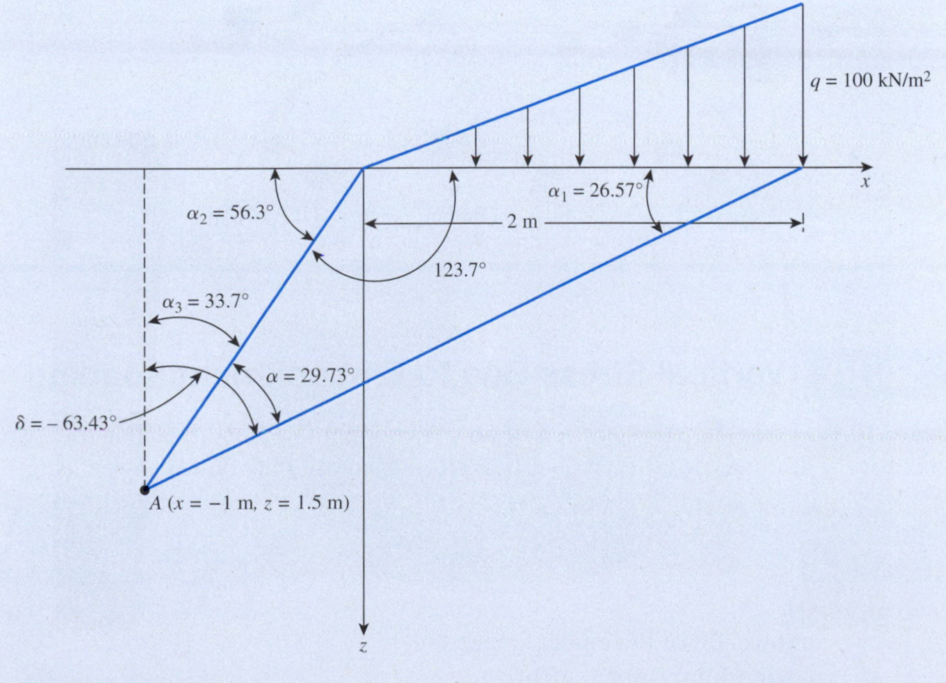

Figure 10.17

Solution

Referring to Figure 10.17,

$$\alpha_1 = \tan^{-1}\left(\frac{1.5}{3}\right) = 26.57°$$

$$\alpha_2 = \tan^{-1}\left(\frac{1.5}{1}\right) = 56.3°$$

$$\alpha = \alpha_2 - \alpha_1 = 56.3 - 26.57 = 29.73°$$

$$\alpha_3 = 90 - \alpha_2 = 90 - 56.3 = 33.7°$$

$$\delta = -(\alpha_3 + \alpha) = -(33.7 + 29.73) = -63.43°$$

$$2\delta = -126.86°$$

From Eq. (10.21),

$$\frac{\Delta\sigma_z}{q} = \frac{1}{2\pi}\left(\frac{2x}{B}\alpha - \sin 2\delta\right) = \frac{1}{2\pi}\left[\frac{2 \times (-1)}{2}\left(\frac{\pi}{180} \times 29.73\right) - \sin(-126.86)\right]$$

$$= \frac{1}{2\pi}[-0.519 - (-0.8)] = 0.0447$$

Compare this value of $\frac{\Delta\sigma_z}{q}$ with

$\frac{2x}{B} = \frac{(2)(-1)}{2} = -1$ and $\frac{2z}{B} = \frac{(2)(1.5)}{2} = 1.5$ given in Table 10.6. It matches, so

$$\Delta\sigma_z = (0.0447)(q) = (0.0447)(100) = \mathbf{4.47\ kN/m^2}$$

10.10 Vertical Stress Due to Embankment Loading

Figure 10.18 shows the cross section of an embankment of height H. For this two-dimensional loading condition the vertical stress increase may be expressed as

$$\Delta\sigma_z = \frac{q_o}{\pi}\left[\left(\frac{B_1 + B_2}{B_2}\right)(\alpha_1 + \alpha_2) - \frac{B_1}{B_2}(\alpha_2)\right] \tag{10.22}$$

where $q_o = \gamma H$
γ = unit weight of the embankment soil
H = height of the embankment

$$\alpha_1\ (\text{radians}) = \tan^{-1}\left(\frac{B_1 + B_2}{z}\right) - \tan^{-1}\left(\frac{B_1}{z}\right) \tag{10.23}$$

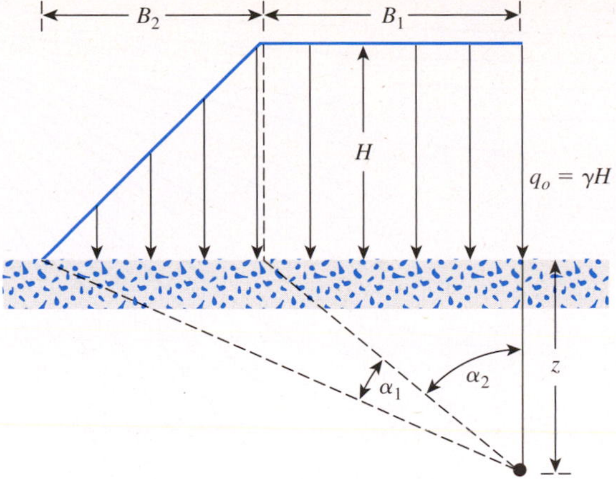

Figure 10.18 Embankment loading

$$\alpha_2 = \tan^{-1}\left(\frac{B_1}{z}\right) \quad (10.24)$$

For a detailed derivation of the equation, see Das (2014). A simplified form of Eq. (10.22) is

$$\Delta\sigma_z = q_o I_2 \quad (10.25)$$

where I_2 = a function of B_1/z and B_2/z.

The variation of I_2 with B_1/z and B_2/z is shown in Figure 10.19 (Osterberg, 1957).

Example 10.11

An embankment is shown in Figure 10.20a. Determine the stress increase under the embankment at points A_1 and A_2.

Solution

$$\gamma H = (17.5)(7) = 122.5 \text{ kN/m}^2$$

Stress Increase at A_1

The left side of Figure 10.20b indicates that $B_1 = 2.5$ m and $B_2 = 14$ m. So,

$$\frac{B_1}{z} = \frac{2.5}{5} = 0.5; \frac{B_2}{z} = \frac{14}{5} = 2.8$$

According to Figure 10.19, in this case, $I_2 = 0.445$. Because the two sides in Figure 10.20b are symmetrical, the value of I_2 for the right side will also be 0.445. So,

$$\Delta\sigma_z = \Delta\sigma_{z(1)} + \Delta\sigma_{z(2)} = q_o[I_{2(\text{Left})} + I_{2(\text{Right})}]$$
$$= 122.5[0.445 + 0.445] = \mathbf{109.03\ kN/m^2}$$

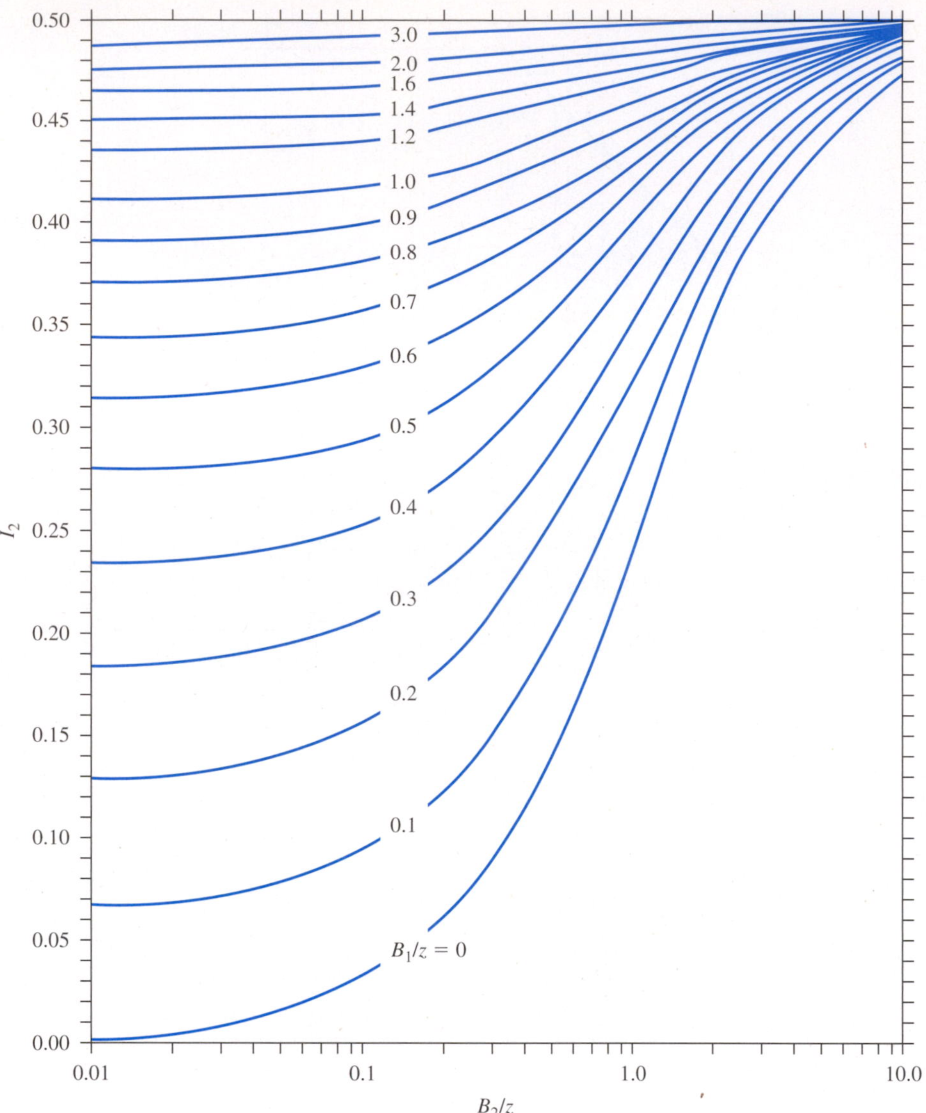

Figure 10.19 Osterberg's chart for determination of vertical stress due to embankment loading

Stress Increase at A_2

Refer to Figure 10.20c. For the left side, $B_2 = 5$ m and $B_1 = 0$. So,

$$\frac{B_2}{z} = \frac{5}{5} = 1; \frac{B_1}{z} = \frac{0}{5} = 0$$

According to Figure 10.19, for these values of B_2/z and B_1/z, $I_2 = 0.24$. So,

$$\Delta\sigma_{z(1)} = 43.75(0.24) = 10.5 \text{ kN/m}^2$$

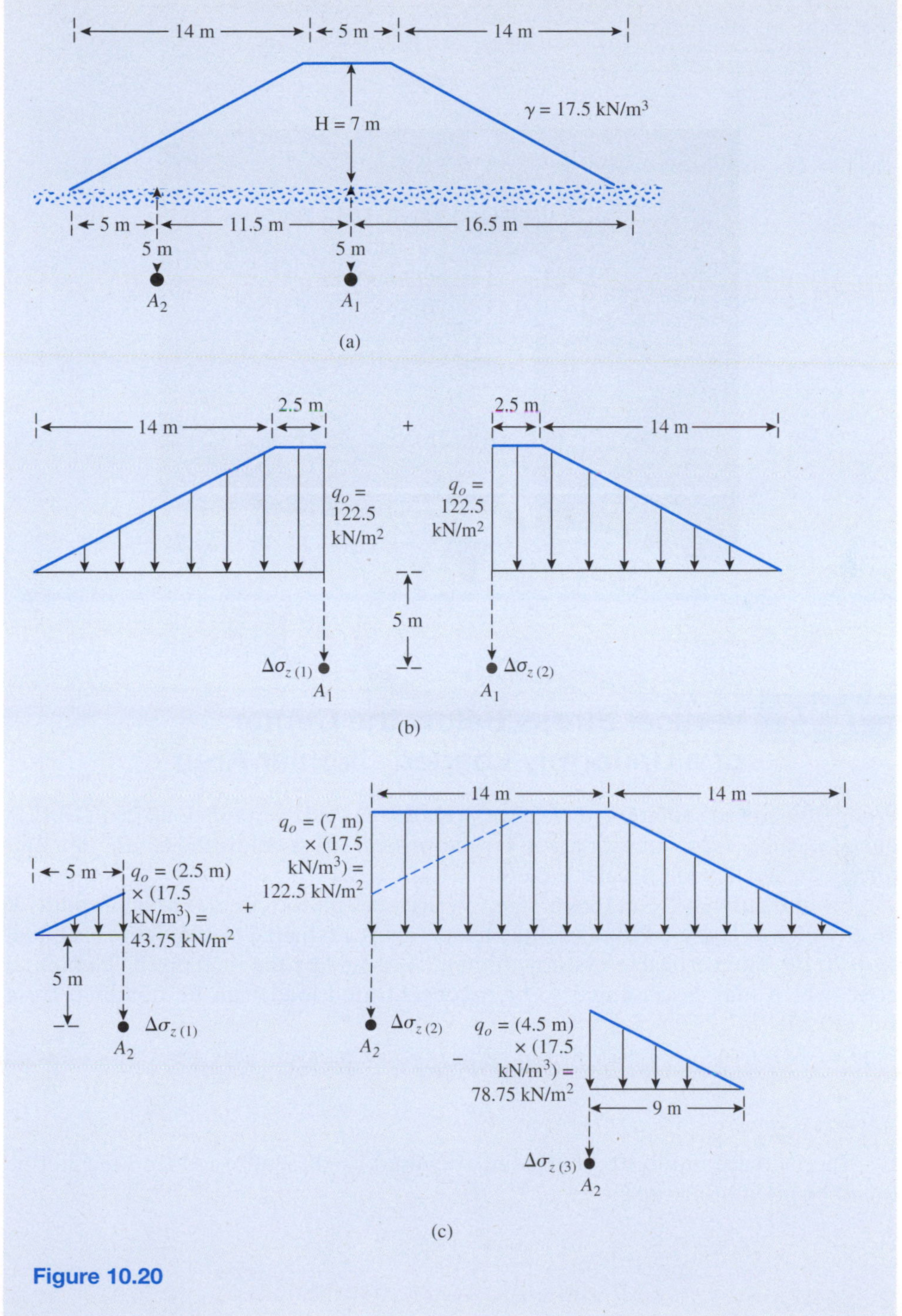
14 m
5 m
14 m
H = 7 m
γ = 17.5 kN/m³
5 m
11.5 m
16.5 m
5 m
5 m
A2
A1
(a)
14 m
2.5 m
+
2.5 m
14 m
qo = 122.5 kN/m²
qo = 122.5 kN/m²
5 m
Δσz (1)
A1
Δσz (2)
A1
(b)
14 m
14 m
qo = (7 m) × (17.5 kN/m³) = 122.5 kN/m²
5 m
qo = (2.5 m) × (17.5 kN/m³) = 43.75 kN/m²
+
5 m
Δσz (1)
A2
Δσz (2)
A2
qo = (4.5 m) × (17.5 kN/m³) = 78.75 kN/m²
–
9 m
Δσz (3)
A2
(c)

Figure 10.20

For the middle section,

$$\frac{B_2}{z} = \frac{14}{5} = 2.8; \frac{B_1}{z} = \frac{14}{5} = 2.8$$

Thus, $I_2 = 0.495$. So,

$$\Delta\sigma_{z(2)} = 0.495(122.5) = 60.64 \text{ kN/m}^2$$

For the right side,

$$\frac{B_2}{z} = \frac{9}{5} = 1.8; \frac{B_1}{z} = \frac{0}{5} = 0$$

and $I_2 = 0.335$. So,

$$\Delta\sigma_{z(3)} = (78.75)(0.335) = 26.38 \text{ kN/m}^2$$

Total stress increase at point A_2 is

$$\Delta\sigma_z = \Delta\sigma_{z(1)} + \Delta\sigma_{z(2)} - \Delta\sigma_{z(3)} = 10.5 + 60.64 - 26.38 = \mathbf{44.76\ kN/m^2}$$

10.11 Vertical Stress Below the Center of a Uniformly Loaded Circular Area

Using Boussinesq's solution for vertical stress $\Delta\sigma_z$ caused by a point load [Eq. (10.12)], one also can develop an expression for the vertical stress below the center of a uniformly loaded flexible circular area.

From Figure 10.21, let the intensity of pressure on the circular area of radius R be equal to q. The total load on the elemental area (shaded in the figure) is equal to $qr\,dr\,d\alpha$. The vertical stress, $d\sigma_z$, at point A caused by the load on the elemental area (which may be assumed to be a concentrated load) can be obtained from Eq. (10.12):

$$d\sigma_z = \frac{3(qr\,dr\,d\alpha)}{2\pi} \frac{z^3}{(r^2 + z^2)^{5/2}} \tag{10.26}$$

The increase in the stress at point A caused by the entire loaded area can be found by integrating Eq. (10.26):

$$\Delta\sigma_z = \int d\sigma_z = \int_{\alpha=0}^{\alpha=2\pi} \int_{r=0}^{r=R} \frac{3q}{2\pi} \frac{z^3 r}{(r^2 + z^2)^{5/2}}\, dr\, d\alpha$$

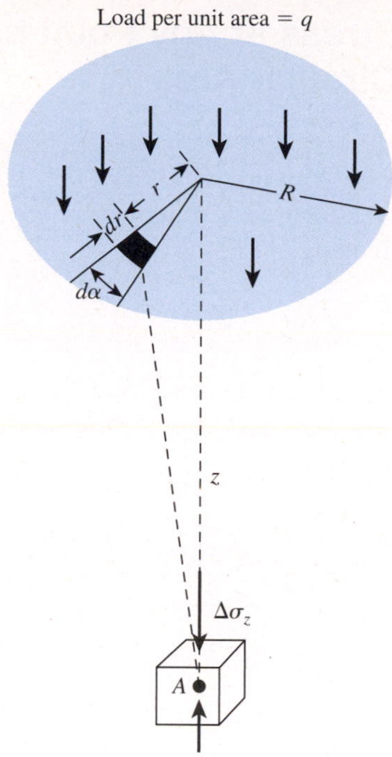

Figure 10.21 Vertical stress below the center of a uniformly loaded flexible circular area

So,

$$\Delta\sigma_z = q\left\{1 - \frac{1}{[(R/z)^2 + 1]^{3/2}}\right\} \tag{10.27}$$

The variation of $\Delta\sigma_z/q$ with z/R as obtained from Eq. (10.27) is given in Table 10.7. The value of $\Delta\sigma_z$ decreases rapidly with depth, and at $z = 5R$, it is about 6% of q, which is the intensity of pressure at the ground surface.

Table 10.7 Variation of $\Delta\sigma_z/q$ with z/R [Eq. (10.27)]

z/R	$\Delta\sigma_z/q$	z/R	$\Delta\sigma_z/q$
0	1	1.0	0.6465
0.02	0.9999	1.5	0.4240
0.05	0.9998	2.0	0.2845
0.10	0.9990	2.5	0.1996
0.2	0.9925	3.0	0.1436
0.4	0.9488	4.0	0.0869
0.5	0.9106	5.0	0.0571
0.8	0.7562		

10.12 Vertical Stress at Any Point below a Uniformly Loaded Circular Area

A detailed tabulation for calculation of vertical stress below a uniformly loaded flexible circular area was given by Ahlvin and Ulery (1962). Referring to Figure 10.22, we find that $\Delta\sigma_z$ at any point A located at a depth z at any distance r from the center of the loaded area can be given as

$$\Delta\sigma_z = q(A' + B') \tag{10.28}$$

where A' and B' are functions of z/R and r/R. (See Tables 10.8 and 10.9.)

Table 10.8 Variation of A' with z/R and r/R*

	r/R								
z/R	0	0.2	0.4	0.6	0.8	1	1.2	1.5	2
0	1.0	1.0	1.0	1.0	1.0	0.5	0	0	0
0.1	0.90050	0.89748	0.88679	0.86126	0.78797	0.43015	0.09645	0.02787	0.00856
0.2	0.80388	0.79824	0.77884	0.73483	0.63014	0.38269	0.15433	0.05251	0.01680
0.3	0.71265	0.70518	0.68316	0.62690	0.52081	0.34375	0.17964	0.07199	0.02440
0.4	0.62861	0.62015	0.59241	0.53767	0.44329	0.31048	0.18709	0.08593	0.03118
0.5	0.55279	0.54403	0.51622	0.46448	0.38390	0.28156	0.18556	0.09499	0.03701
0.6	0.48550	0.47691	0.45078	0.40427	0.33676	0.25588	0.17952	0.10010	
0.7	0.42654	0.41874	0.39491	0.35428	0.29833	0.21727	0.17124	0.10228	0.04558
0.8	0.37531	0.36832	0.34729	0.31243	0.26581	0.21297	0.16206	0.10236	
0.9	0.33104	0.32492	0.30669	0.27707	0.23832	0.19488	0.15253	0.10094	
1	0.29289	0.28763	0.27005	0.24697	0.21468	0.17868	0.14329	0.09849	0.05185
1.2	0.23178	0.22795	0.21662	0.19890	0.17626	0.15101	0.12570	0.09192	0.05260
1.5	0.16795	0.16552	0.15877	0.14804	0.13436	0.11892	0.10296	0.08048	0.05116
2	0.10557	0.10453	0.10140	0.09647	0.09011	0.08269	0.07471	0.06275	0.04496
2.5	0.07152	0.07098	0.06947	0.06698	0.06373	0.05974	0.05555	0.04880	0.03787
3	0.05132	0.05101	0.05022	0.04886	0.04707	0.04487	0.04241	0.03839	0.03150
4	0.02986	0.02976	0.02907	0.02802	0.02832	0.02749	0.02651	0.02490	0.02193
5	0.01942	0.01938				0.01835			0.01573
6	0.01361					0.01307			0.01168
7	0.01005					0.00976			0.00894
8	0.00772					0.00755			0.00703
9	0.00612					0.00600			0.00566
10								0.00477	0.00465

*Source: From Ahlvin, R. G., and H. H. Ulery. Tabulated Values for Determining the Complete Pattern of Stresses, Strains, and Deflections Beneath a Uniform Circular Load on a Homogeneous Half Space. In Highway Research Bulletin 342, Highway Research Board, National Research Council, Washington, D.C., 1962, Tables 1 and 2, p. 3. Reproduced with permission of the Transportation Research Board.

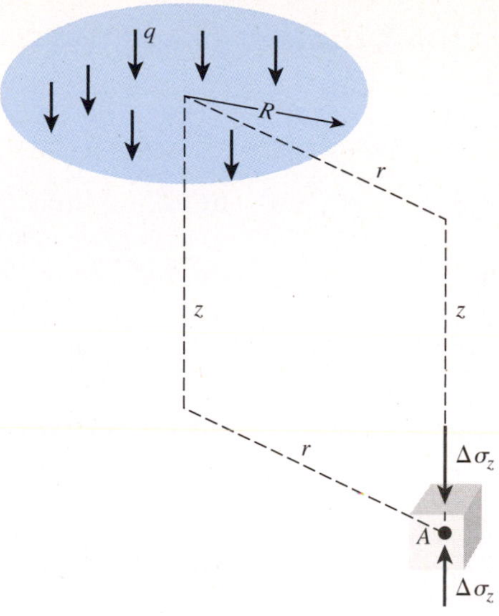

Figure 10.22 Vertical stress at any point below a uniformly loaded circular area

Table 10.8 (*continued*)

	r/R								
z/R	**3**	**4**	**5**	**6**	**7**	**8**	**10**	**12**	**14**
0	0	0	0	0	0	0	0	0	0
0.1	0.00211	0.00084	0.00042						
0.2	0.00419	0.00167	0.00083	0.00048	0.00030	0.00020			
0.3	0.00622	0.00250							
0.4									
0.5	0.01013	0.00407	0.00209	0.00118	0.00071	0.00053	0.00025	0.00014	0.00009
0.6									
0.7									
0.8									
0.9									
1	0.01742	0.00761	0.00393	0.00226	0.00143	0.00097	0.00050	0.00029	0.00018
1.2	0.01935	0.00871	0.00459	0.00269	0.00171	0.00115			
1.5	0.02142	0.01013	0.00548	0.00325	0.00210	0.00141	0.00073	0.00043	0.00027
2	0.02221	0.01160	0.00659	0.00399	0.00264	0.00180	0.00094	0.00056	0.00036
2.5	0.02143	0.01221	0.00732	0.00463	0.00308	0.00214	0.00115	0.00068	0.00043
3	0.01980	0.01220	0.00770	0.00505	0.00346	0.00242	0.00132	0.00079	0.00051
4	0.01592	0.01109	0.00768	0.00536	0.00384	0.00282	0.00160	0.00099	0.00065
5	0.01249	0.00949	0.00708	0.00527	0.00394	0.00298	0.00179	0.00113	0.00075
6	0.00983	0.00795	0.00628	0.00492	0.00384	0.00299	0.00188	0.00124	0.00084
7	0.00784	0.00661	0.00548	0.00445	0.00360	0.00291	0.00193	0.00130	0.00091
8	0.00635	0.00554	0.00472	0.00398	0.00332	0.00276	0.00189	0.00134	0.00094
9	0.00520	0.00466	0.00409	0.00353	0.00301	0.00256	0.00184	0.00133	0.00096
10	0.00438	0.00397	0.00352	0.00326	0.00273	0.00241			

Table 10.9 Variation of B' with z/R and r/R*

	r/R								
z/R	0	0.2	0.4	0.6	0.8	1	1.2	1.5	2
0	0	0	0	0	0	0	0	0	0
0.1	0.09852	0.10140	0.11138	0.13424	0.18796	0.05388	−0.07899	−0.02672	−0.00845
0.2	0.18857	0.19306	0.20772	0.23524	0.25983	0.08513	−0.07759	−0.04448	−0.01593
0.3	0.26362	0.26787	0.28018	0.29483	0.27257	0.10757	−0.04316	−0.04999	−0.02166
0.4	0.32016	0.32259	0.32748	0.32273	0.26925	0.12404	−0.00766	−0.04535	−0.02522
0.5	0.35777	0.35752	0.35323	0.33106	0.26236	0.13591	0.02165	−0.03455	−0.02651
0.6	0.37831	0.37531	0.36308	0.32822	0.25411	0.14440	0.04457	−0.02101	
0.7	0.38487	0.37962	0.36072	0.31929	0.24638	0.14986	0.06209	−0.00702	−0.02329
0.8	0.38091	0.37408	0.35133	0.30699	0.23779	0.15292	0.07530	0.00614	
0.9	0.36962	0.36275	0.33734	0.29299	0.22891	0.15404	0.08507	0.01795	
1	0.35355	0.34553	0.32075	0.27819	0.21978	0.15355	0.09210	0.02814	−0.01005
1.2	0.31485	0.30730	0.28481	0.24836	0.20113	0.14915	0.10002	0.04378	0.00023
1.5	0.25602	0.25025	0.23338	0.20694	0.17368	0.13732	0.10193	0.05745	0.01385
2	0.17889	0.18144	0.16644	0.15198	0.13375	0.11331	0.09254	0.06371	0.02836
2.5	0.12807	0.12633	0.12126	0.11327	0.10298	0.09130	0.07869	0.06022	0.03429
3	0.09487	0.09394	0.09099	0.08635	0.08033	0.07325	0.06551	0.05354	0.03511
4	0.05707	0.05666	0.05562	0.05383	0.05145	0.04773	0.04532	0.03995	0.03066
5	0.03772	0.03760				0.03384			0.02474
6	0.02666					0.02468			0.01968
7	0.01980					0.01868			0.01577
8	0.01526					0.01459			0.01279
9	0.01212					0.01170			0.01054
10								0.00924	0.00879

* *Source:* From Ahlvin, R. G., and H. H. Ulery. Tabulated Values for Determining the Complete Pattern of Stresses, Strains, and Deflections Beneath a Uniform Circular Load on a Homogeneous Half Space. In Highway Research Bulletin 342, Highway Research Board, National Research Council, Washington, D.C., 1962, Tables 1 and 2, p. 3. Reproduced with permission of the Transportation Research Board.

Example 10.12

Consider a uniformly loaded flexible circular area on the ground surface, as shown in Fig. 10.22. Given: $R = 3$ m and uniform load $q = 100$ kN/m^2.

Calculate the increase in vertical stress at depths of 1.5, 3, 4.5, 6, and 12 m below the ground surface for points at (a) $r = 0$ and (b) $r = 4.5$ m.

Solution

From Eq. (10.28),

$$\Delta\sigma_z = q\,(A' + B')$$

Given $R = 3$ m and $q = 100$ kN/m^2.

Table 10.9 (*continued*)

	r/R								
z/R	3	4	5	6	7	8	10	12	14
0	0	0	0	0	0	0	0	0	0
0.1	−0.00210	−0.00084	−0.00042						
0.2	−0.00412	−0.00166	−0.00083	−0.00024	−0.00015	−0.00010			
0.3	−0.00599	−0.00245							
0.4									
0.5	−0.00991	−0.00388	−0.00199	−0.00116	−0.00073	−0.00049	−0.00025	−0.00014	−0.00009
0.6									
0.7									
0.8									
0.9									
1	−0.01115	−0.00608	−0.00344	−0.00210	−0.00135	−0.00092	−0.00048	−0.00028	−0.00018
1.2	−0.00995	−0.00632	−0.00378	−0.00236	−0.00156	−0.00107			
1.5	−0.00669	−0.00600	−0.00401	−0.00265	−0.00181	−0.00126	−0.00068	−0.00040	−0.00026
2	0.00028	−0.00410	−0.00371	−0.00278	−0.00202	−0.00148	−0.00084	−0.00050	−0.00033
2.5	0.00661	−0.00130	−0.00271	−0.00250	−0.00201	−0.00156	−0.00094	−0.00059	−0.00039
3	0.01112	0.00157	−0.00134	−0.00192	−0.00179	−0.00151	−0.00099	−0.00065	−0.00046
4	0.01515	0.00595	0.00155	−0.00029	−0.00094	−0.00109	−0.00094	−0.00068	−0.00050
5	0.01522	0.00810	0.00371	0.00132	0.00013	−0.00043	−0.00070	−0.00061	−0.00049
6	0.01380	0.00867	0.00496	0.00254	0.00110	0.00028	−0.00037	−0.00047	−0.00045
7	0.01204	0.00842	0.00547	0.00332	0.00185	0.00093	−0.00002	−0.00029	−0.00037
8	0.01034	0.00779	0.00554	0.00372	0.00236	0.00141	0.00035	−0.00008	−0.00025
9	0.00888	0.00705	0.00533	0.00386	0.00265	0.00178	0.00066	0.00012	−0.00012
10	0.00764	0.00631	0.00501	0.00382	0.00281	0.00199			

Part a

We can prepare the following table. (*Note:* $r/R = 0$. A' and B' values are from Tables 10.8 and 10.9.)

Depth, z (m)	z/R	A'	B'	$\Delta\sigma_z$ (kN/m²)
1.5	0.5	0.553	0.358	**91.1**
3	1.0	0.293	0.354	**64.7**
4.5	1.5	0.168	0.256	**42.4**
6	2.0	0.106	0.179	**28.5**
12	4.0	0.03	0.057	**8.7**

Part b

$$r/R = 4.5/3 = 1.5$$

Depth, z (m)	z/R	A'	B'	$\Delta\sigma_z$ (kN/m²)
1.5	0.5	0.095	−0.035	**6.0**
3	1.0	0.098	0.028	**12.6**
4.5	1.5	0.08	0.057	**13.7**
6	2.0	0.063	0.064	**12.7**
12	4.0	0.025	0.04	**6.5**

10.13 Vertical Stress Caused by a Rectangularly Loaded Area

Boussinesq's solution also can be used to calculate the vertical stress increase below a flexible rectangular loaded area, as shown in Figure 10.23. The loaded area is located at the ground surface and has length L and width B. The uniformly distributed load per unit area is equal to q. To determine the increase in the vertical stress ($\Delta\sigma_z$) at point A, which is located at depth z below the corner of the rectangular area, we need to consider a small elemental area $dx\ dy$ of the rectangle. (This is shown in Figure 10.23.) The load on this elemental area can be given by

$$dq = q\ dx\ dy \tag{10.29}$$

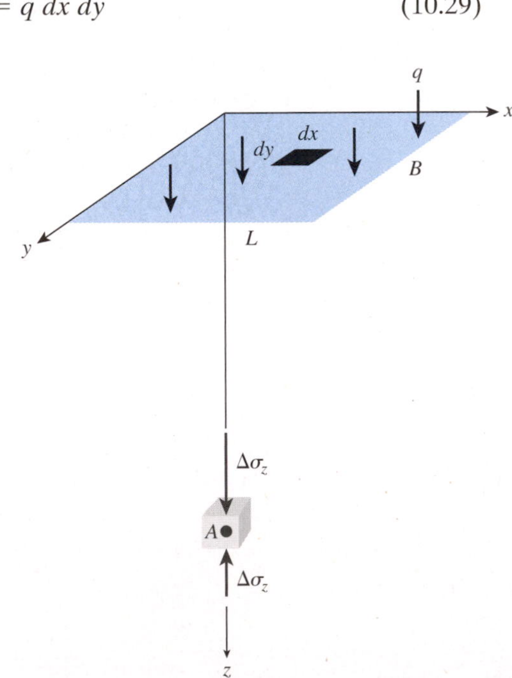

Figure 10.23 Vertical stress below the corner of a uniformly loaded flexible rectangular area

The increase in the stress ($d\sigma_z$) at point A caused by the load dq can be determined by using Eq. (10.12). However, we need to replace P with $dq = q\ dx\ dy$ and r^2 with $x^2 + y^2$. Thus,

$$d\sigma_z = \frac{3q\ dx\ dy\ z^3}{2\pi(x^2 + y^2 + z^2)^{5/2}} \tag{10.30}$$

The increase in the stress, at point A caused by the entire loaded area can now be determined by integrating the preceding equation. We obtain

$$\Delta\sigma_z = \int d\sigma_z = \int_{y=0}^{B}\int_{x=0}^{L} \frac{3qz^3(dx\,dy)}{2\pi(x^2 + y^2 + z^2)^{5/2}} = qI_3 \tag{10.31}$$

where

$$I_3 = \frac{1}{4\pi}\left[\frac{2mn\sqrt{m^2 + n^2 + 1}}{m^2 + n^2 + m^2n^2 + 1}\left(\frac{m^2 + n^2 + 2}{m^2 + n^2 + 1}\right) + \tan^{-1}\left(\frac{2mn\sqrt{m^2 + n^2 + 1}}{m^2 + n^2 - m^2n^2 + 1}\right)\right] \tag{10.32}$$

$$m = \frac{B}{z} \tag{10.33}$$

$$n = \frac{L}{z} \tag{10.34}$$

The arctangent term in Eq. (10.32) must be a positive angle in radians. When $m^2 + n^2 + 1 < m^2 n^2$, it becomes a negative angle. So a term π should be added to that angle.

The variation of I_3 with m and n is shown in Table 10.10 and Figure 10.24.

The increase in the stress at any point below a rectangularly loaded area can be found by using Eq. (10.31). This can be explained by reference to Figure 10.25. Let us determine the stress at a point below point A' at depth z. The loaded area can be divided into four rectangles as shown. The point A' is the corner common to all four rectangles. The increase in the stress at depth z below point A' due to each rectangular area can now be calculated by using Eq. (10.31). The total stress increase caused by the entire loaded area can be given by

$$\Delta\sigma_z = q[I_{3(1)} + I_{3(2)} + I_{3(3)} + I_{3(4)}] \tag{10.35}$$

where $I_{3(1)}$, $I_{3(2)}$, $I_{3(3)}$, and $I_{3(4)}$ = values of I_3 for rectangles 1, 2, 3, and 4, respectively.

In most cases the vertical stress increase below the center of a rectangular area (Figure 10.26) is important. This stress increase can be given by the relationship

$$\Delta\sigma_z = qI_4 \tag{10.36}$$

where

$$I_4 = \frac{2}{\pi}\left[\frac{m_1n_1}{\sqrt{1 + m_1^2 + n_1^2}}\,\frac{1 + m_1^2 + 2n_1^2}{(1 + n_1^2)(m_1^2 + n_1^2)} + \sin^{-1}\frac{m_1}{\sqrt{m_1^2 + n_1^2}\sqrt{1 + n_1^2}}\right] \tag{10.37}$$

Table 10.10 Variation of I_3 with m and n [Eq. (10.32)]

	m									
n	0.1	0.2	0.3	0.4	0.5	0.6	0.7	0.8	0.9	1.0
0.1	0.0047	0.0092	0.0132	0.0168	0.0198	0.0222	0.0242	0.0258	0.0270	0.0279
0.2	0.0092	0.0179	0.0259	0.0328	0.0387	0.0435	0.0474	0.0504	0.0528	0.0547
0.3	0.0132	0.0259	0.0374	0.0474	0.0559	0.0629	0.0686	0.0731	0.0766	0.0794
0.4	0.0168	0.0328	0.0474	0.0602	0.0711	0.0801	0.0873	0.0931	0.0977	0.1013
0.5	0.0198	0.0387	0.0559	0.0711	0.0840	0.0947	0.1034	0.1104	0.1158	0.1202
0.6	0.0222	0.0435	0.0629	0.0801	0.0947	0.1069	0.1168	0.1247	0.1311	0.1361
0.7	0.0242	0.0474	0.0686	0.0873	0.1034	0.1169	0.1277	0.1365	0.1436	0.1491
0.8	0.0258	0.0504	0.0731	0.0931	0.1104	0.1247	0.1365	0.1461	0.1537	0.1598
0.9	0.0270	0.0528	0.0766	0.0977	0.1158	0.1311	0.1436	0.1537	0.1619	0.1684
1.0	0.0279	0.0547	0.0794	0.1013	0.1202	0.1361	0.1491	0.1598	0.1684	0.1752
1.2	0.0293	0.0573	0.0832	0.1063	0.1263	0.1431	0.1570	0.1684	0.1777	0.1851
1.4	0.0301	0.0589	0.0856	0.1094	0.1300	0.1475	0.1620	0.1739	0.1836	0.1914
1.6	0.0306	0.0599	0.0871	0.1114	0.1324	0.1503	0.1652	0.1774	0.1874	0.1955
1.8	0.0309	0.0606	0.0880	0.1126	0.1340	0.1521	0.1672	0.1797	0.1899	0.1981
2.0	0.0311	0.0610	0.0887	0.1134	0.1350	0.1533	0.1686	0.1812	0.1915	0.1999
2.5	0.0314	0.0616	0.0895	0.1145	0.1363	0.1548	0.1704	0.1832	0.1938	0.2024
3.0	0.0315	0.0618	0.0898	0.1150	0.1368	0.1555	0.1711	0.1841	0.1947	0.2034
4.0	0.0316	0.0619	0.0901	0.1153	0.1372	0.1560	0.1717	0.1847	0.1954	0.2042
5.0	0.0316	0.0620	0.0901	0.1154	0.1374	0.1561	0.1719	0.1849	0.1956	0.2044
6.0	0.0316	0.0620	0.0902	0.1154	0.1374	0.1562	0.1719	0.1850	0.1957	0.2045

n	1.2	1.4	1.6	1.8	2.0	2.5	3.0	4.0	5.0	6.0
0.1	0.0293	0.0301	0.0306	0.0309	0.0311	0.0314	0.0315	0.0316	0.0316	0.0316
0.2	0.0573	0.0589	0.0599	0.0606	0.0610	0.0616	0.0618	0.0619	0.0620	0.0620
0.3	0.0832	0.0856	0.0871	0.0880	0.0887	0.0895	0.0898	0.0901	0.0901	0.0902
0.4	0.1063	0.1094	0.1114	0.1126	0.1134	0.1145	0.1150	0.1153	0.1154	0.1154
0.5	0.1263	0.1300	0.1324	0.1340	0.1350	0.1363	0.1368	0.1372	0.1374	0.1374
0.6	0.1431	0.1475	0.1503	0.1521	0.1533	0.1548	0.1555	0.1560	0.1561	0.1562
0.7	0.1570	0.1620	0.1652	0.1672	0.1686	0.1704	0.1711	0.1717	0.1719	0.1719
0.8	0.1684	0.1739	0.1774	0.1797	0.1812	0.1832	0.1841	0.1847	0.1849	0.1850
0.9	0.1777	0.1836	0.1874	0.1899	0.1915	0.1938	0.1947	0.1954	0.1956	0.1957
1.0	0.1851	0.1914	0.1955	0.1981	0.1999	0.2024	0.2034	0.2042	0.2044	0.2045
1.2	0.1958	0.2028	0.2073	0.2103	0.2124	0.2151	0.2163	0.2172	0.2175	0.2176
1.4	0.2028	0.2102	0.2151	0.2184	0.2206	0.2236	0.2250	0.2260	0.2263	0.2264
1.6	0.2073	0.2151	0.2203	0.2237	0.2261	0.2294	0.2309	0.2320	0.2323	0.2325
1.8	0.2103	0.2183	0.2237	0.2274	0.2299	0.2333	0.2350	0.2362	0.2366	0.2367
2.0	0.2124	0.2206	0.2261	0.2299	0.2325	0.2361	0.2378	0.2391	0.2395	0.2397
2.5	0.2151	0.2236	0.2294	0.2333	0.2361	0.2401	0.2420	0.2434	0.2439	0.2441
3.0	0.2163	0.2250	0.2309	0.2350	0.2378	0.2420	0.2439	0.2455	0.2461	0.2463
4.0	0.2172	0.2260	0.2320	0.2362	0.2391	0.2434	0.2455	0.2472	0.2479	0.2481
5.0	0.2175	0.2263	0.2324	0.2366	0.2395	0.2439	0.2460	0.2479	0.2486	0.2489
6.0	0.2176	0.2264	0.2325	0.2367	0.2397	0.2441	0.2463	0.2482	0.2489	0.2492

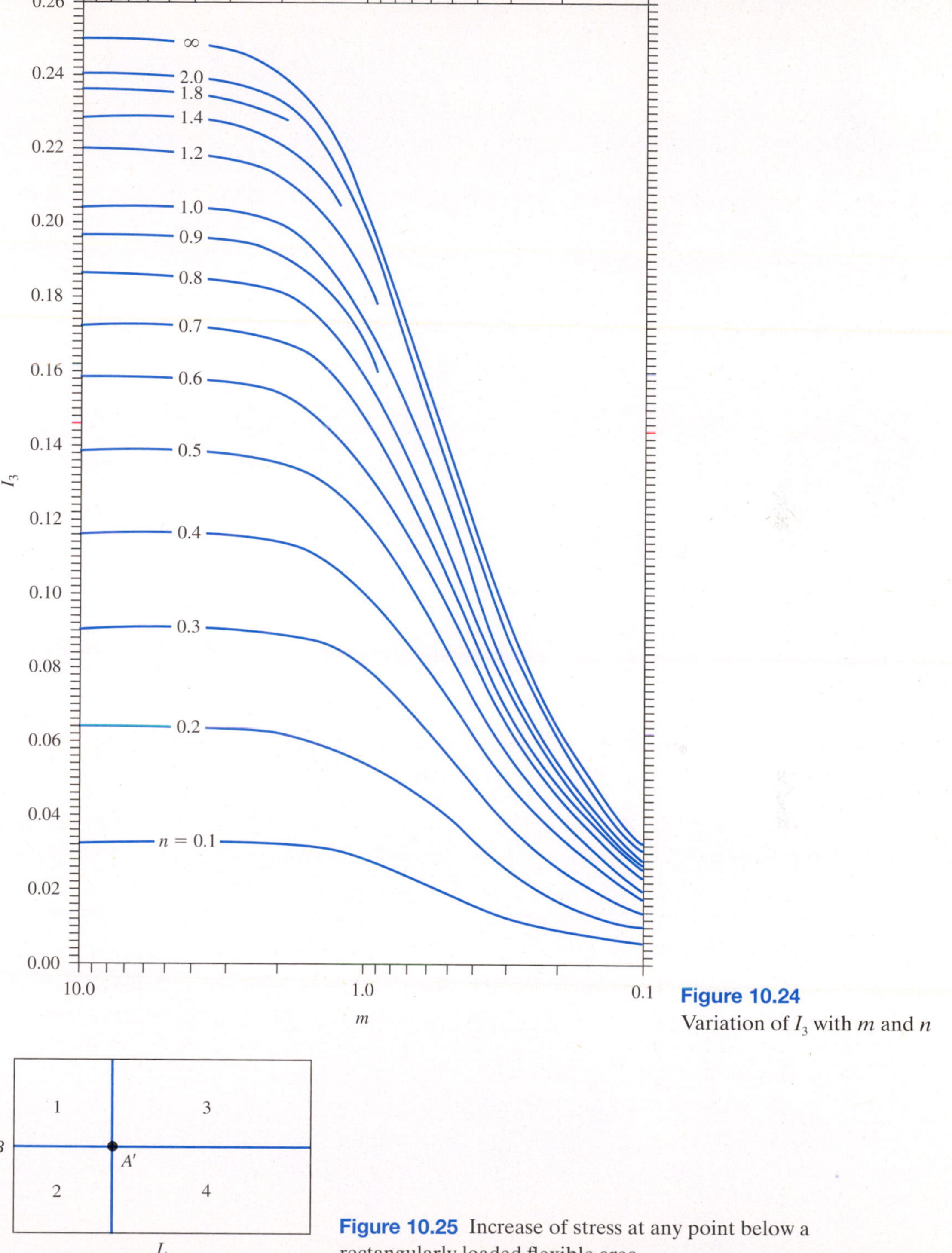

Figure 10.24 Variation of I_3 with m and n

Figure 10.25 Increase of stress at any point below a rectangularly loaded flexible area

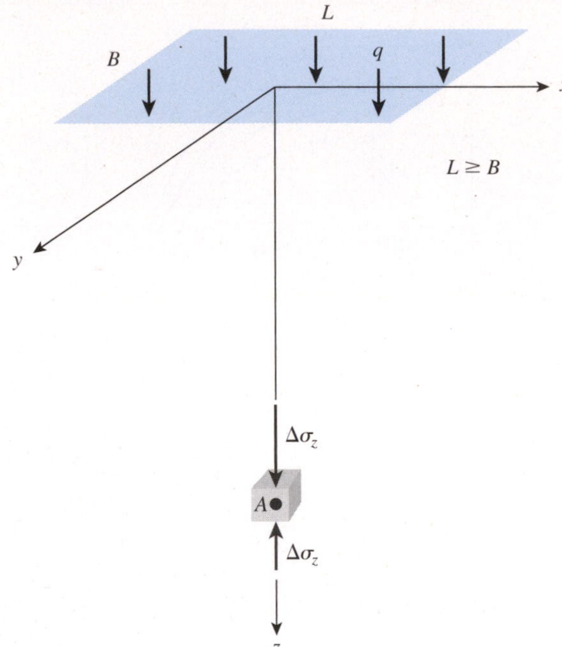

Figure 10.26 Vertical stress below the center of a uniformly loaded flexible rectangular area

$$m_1 = \frac{L}{B} \tag{10.38}$$

$$n_1 = \frac{z}{b} \tag{10.39}$$

$$b = \frac{B}{2} \tag{10.40}$$

The variation of I_4 with m_1 and n_1 is given in Table 10.11.

Example 10.13

The plan of a uniformly loaded rectangular area is shown in Figure 10.27a. Determine the vertical stress increase $\Delta\sigma_z$ below point A' at a depth of $z = 4$ m.

Solution

The stress increase $\Delta\sigma_z$ can be written as

$$\Delta\sigma_z = \Delta\sigma_{z(1)} - \Delta\sigma_{z(2)}$$

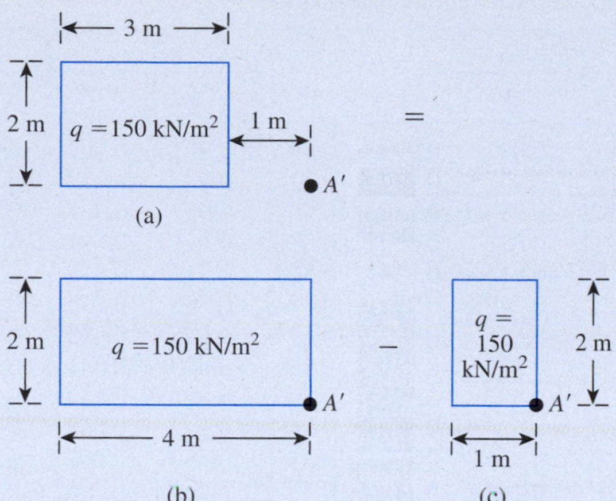

Figure 10.27

where

$\Delta\sigma_{z(1)}$ = stress increase due to the loaded area shown in Figure 10.27b
$\Delta\sigma_{z(2)}$ = stress increase due to the loaded area shown in Figure 10.27c

For the loaded area shown in Figure 10.27b:

$$m = \frac{B}{z} = \frac{2}{4} = 0.5$$

$$n = \frac{L}{z} = \frac{4}{4} = 1$$

From Figure 10.24 for $m = 0.5$ and $n = 1$, the value of $I_3 = 0.1225$. So

$$\Delta\sigma_{z(1)} = qI_3 = (150)(0.1225) = 18.38 \text{ kN/m}^2$$

Similarly, for the loaded area shown in Figure 10.27c:

$$m = \frac{B}{z} = \frac{1}{4} = 0.25$$

$$n = \frac{L}{z} = \frac{2}{4} = 0.5$$

Thus, $I_3 = 0.0473$. Hence,

$$\Delta\sigma_{z(2)} = (150)(0.0473) = 7.1 \text{ kN/m}^2$$

So

$$\Delta\sigma_z = \Delta\sigma_{z(1)} - \Delta\sigma_{z(2)} = 18.38 - 7.1 = \mathbf{11.28\ kN/m^2}$$

Table 10.11 Variation of I_4 with m_1 and n_1 [Eq. (10.37)]

	m_1									
n_1	1	2	3	4	5	6	7	8	9	10
0.20	0.994	0.997	0.997	0.997	0.997	0.997	0.997	0.997	0.997	0.997
0.40	0.960	0.976	0.977	0.977	0.977	0.977	0.977	0.977	0.977	0.977
0.60	0.892	0.932	0.936	0.936	0.937	0.937	0.937	0.937	0.937	0.937
0.80	0.800	0.870	0.878	0.880	0.881	0.881	0.881	0.881	0.881	0.881
1.00	0.701	0.800	0.814	0.817	0.818	0.818	0.818	0.818	0.818	0.818
1.20	0.606	0.727	0.748	0.753	0.754	0.755	0.755	0.755	0.755	0.755
1.40	0.522	0.658	0.685	0.692	0.694	0.695	0.695	0.696	0.696	0.696
1.60	0.449	0.593	0.627	0.636	0.639	0.640	0.641	0.641	0.641	0.642
1.80	0.388	0.534	0.573	0.585	0.590	0.591	0.592	0.592	0.593	0.593
2.00	0.336	0.481	0.525	0.540	0.545	0.547	0.548	0.549	0.549	0.549
3.00	0.179	0.293	0.348	0.373	0.384	0.389	0.392	0.393	0.394	0.395
4.00	0.108	0.190	0.241	0.269	0.285	0.293	0.298	0.301	0.302	0.303
5.00	0.072	0.131	0.174	0.202	0.219	0.229	0.236	0.240	0.242	0.244
6.00	0.051	0.095	0.130	0.155	0.172	0.184	0.192	0.197	0.200	0.202
7.00	0.038	0.072	0.100	0.122	0.139	0.150	0.158	0.164	0.168	0.171
8.00	0.029	0.056	0.079	0.098	0.113	0.125	0.133	0.139	0.144	0.147
9.00	0.023	0.045	0.064	0.081	0.094	0.105	0.113	0.119	0.124	0.128
10.00	0.019	0.037	0.053	0.067	0.079	0.089	0.097	0.103	0.108	0.112

10.14 Influence Chart for Vertical Pressure

Equation (10.27) can be rearranged and written in the form

$$\frac{R}{z} = \sqrt{\left(1 - \frac{\Delta\sigma_z}{q}\right)^{-2/3} - 1} \tag{10.41}$$

Note that R/z and $\Delta\sigma_z/q$ in this equation are nondimensional quantities. The values of R/z that correspond to various pressure ratios are given in Table 10.12.

Table 10.12 Values of R/z for Various Pressure Ratios [Eq. (10.41)]

$\Delta\sigma_z/q$	R/z	$\Delta\sigma_z/q$	R/z
0	0	0.55	0.8384
0.05	0.1865	0.60	0.9176
0.10	0.2698	0.65	1.0067
0.15	0.3383	0.70	1.1097
0.20	0.4005	0.75	1.2328
0.25	0.4598	0.80	1.3871
0.30	0.5181	0.85	1.5943
0.35	0.5768	0.90	1.9084
0.40	0.6370	0.95	2.5232
0.45	0.6997	1.00	∞
0.50	0.7664		

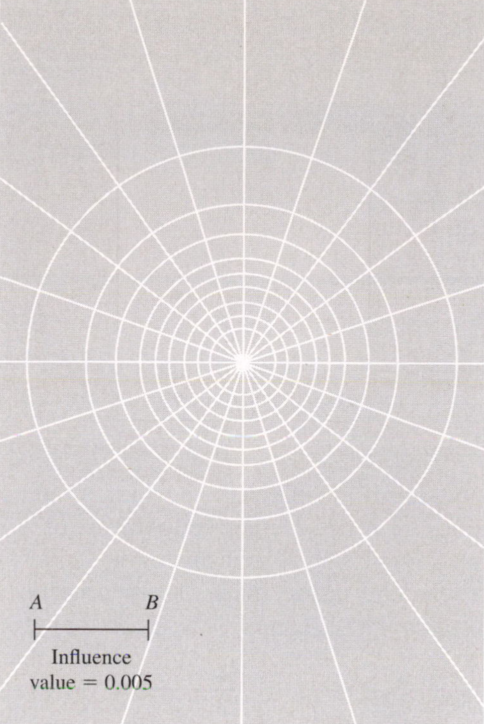

Figure 10.28 Influence chart for vertical pressure based on Boussinesq's theory (*Bulletin No. 338. Influence Charts for Computation of Stresses in Elastic. Foundations, by Nathan M. Newmark. University of Illinois, 1942.*)

Using the values of R/z obtained from Eq. (10.41) for various pressure ratios, Newmark (1942) presented an influence chart that can be used to determine the vertical pressure at any point below a uniformly loaded flexible area of any shape.

Figure 10.28 shows an influence chart that has been constructed by drawing concentric circles. The radii of the circles are equal to the R/z values corresponding to $\Delta\sigma_z/q = 0, 0.1, 0.2, \ldots, 1$. (*Note:* For $\Delta\sigma_z/q = 0$, $R/z = 0$, and for $\Delta\sigma_z/q = 1$, $R/z = \infty$, so nine circles are shown.) The unit length for plotting the circles is $\overline{AB}$. The circles are divided by several equally spaced radial lines. The influence value of the chart is given by $1/N$, where N is equal to the number of elements in the chart. In Figure 10.28, there are 200 elements; hence, the influence value is 0.005.

The procedure for obtaining vertical pressure at any point below a loaded area is as follows:

Step 1. Determine the depth z below the uniformly loaded area at which the stress increase is required.

Step 2. Plot the plan of the loaded area with a scale of z equal to the unit length of the chart ($\overline{AB}$).

Step 3. Place the plan (plotted in step 2) on the influence chart in such a way that the point below which the stress is to be determined is located at the center of the chart.

Step 4. Count the number of elements (M) of the chart enclosed by the plan of the loaded area.

The increase in the pressure at the point under consideration is given by

$$\Delta\sigma_z = (IV)qM \tag{10.42}$$

where IV = influence value
q = pressure on the loaded area

Example 10.14

The cross section and plan of a column foundation are shown in Figure 10.29a. Find the increase in vertical stress produced by the column footing at point A.

Solution

Point A is located at a depth 3 m below the bottom of the foundation. The plan of the square foundation has been replotted to a scale of $\overline{AB} = 3$ m and placed on the influence chart (Figure 10.29b) in such a way that point A on the plan falls directly over the center of the chart. The number of elements inside the outline of the plan is about 48.5. Hence,

$$\Delta\sigma_z = (IV)qM = 0.005\left(\frac{660}{3 \times 3}\right)48.5 = \mathbf{17.78\ kN/m^2}$$

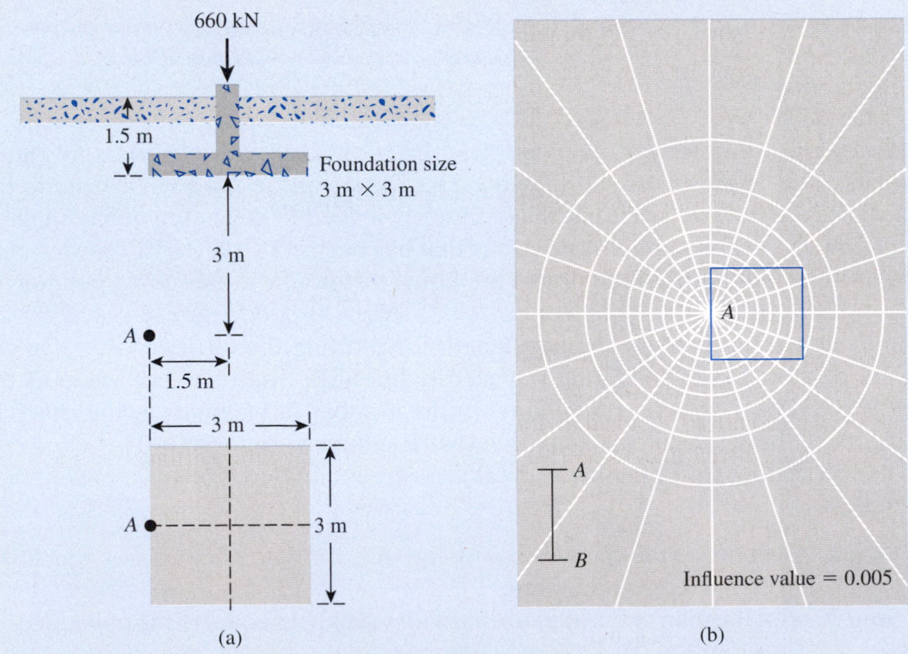

Figure 10.29 (a) Cross section and plan of a column foundation; (b) determination of stress at A by use of Newmark's influence chart.

10.15 Westergaard's Solution for Vertical Stress Due to a Point Load

Boussinesq's solution for stress distribution due to a point load was presented in Section 10.4. The stress distribution due to various types of loading discussed in Sections 10.4 through 10.14 is based on integration of Boussinesq's solution.

Westergaard (1938) has proposed a solution for the determination of the vertical stress due to a point load P in an elastic solid medium in which there exist alternating layers with thin rigid reinforcements (Figure 10.30a). This type of assumption may be an idealization of a clay layer with thin seams of sand. For such

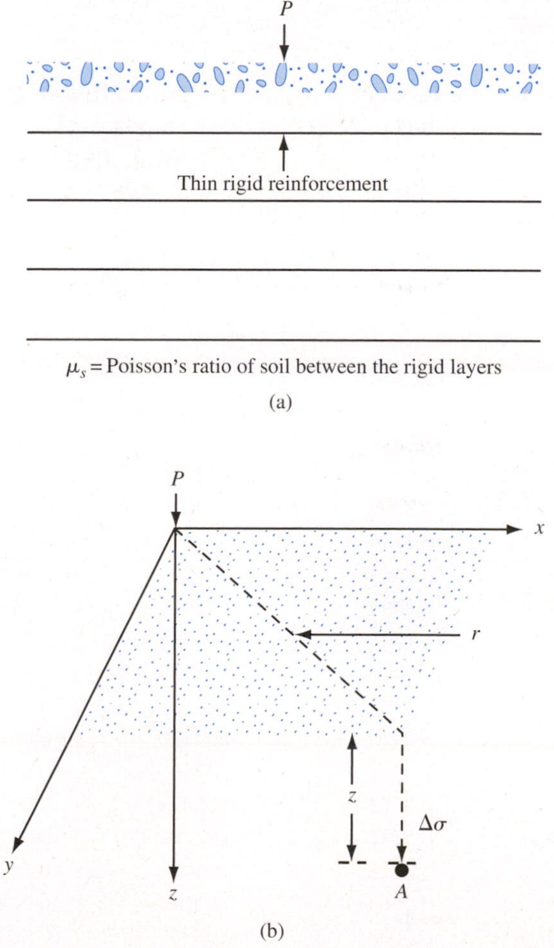

Figure 10.30 Westergaard's solution for vertical stress due to a point load

an assumption, the vertical stress increase at a point A (Figure 10.30b) can be given as

$$\Delta\sigma_z = \frac{P\eta}{2\pi z^2}\left[\frac{1}{\eta^2 + (r/z)^2}\right]^{3/2} \tag{10.43}$$

where

$$\eta = \sqrt{\frac{1 - 2\mu_s}{2 - 2\mu_s}} \tag{10.44}$$

μ_s = Poisson's ratio of the solid between the rigid reinforcements

$r = \sqrt{x^2 + y^2}$

Equation (10.43) can be rewritten as

$$\Delta\sigma_z = \left(\frac{P}{z^2}\right)I_5 \tag{10.45}$$

where

$$I_5 = \frac{1}{2\pi\eta^2}\left[\left(\frac{r}{\eta z}\right)^2 + 1\right]^{-3/2} \tag{10.46}$$

Table 10.13 gives the variation of I_5 with μ_s.

Table 10.13 Variation of I_5 [Eq. (10.46)]

	I_5		
r/z	$\mu_s = 0$	$\mu_s = 0.2$	$\mu_s = 0.4$
0	0.3183	0.4244	0.9550
0.1	0.3090	0.4080	0.8750
0.2	0.2836	0.3646	0.6916
0.3	0.2483	0.3074	0.4997
0.4	0.2099	0.2491	0.3480
0.5	0.1733	0.1973	0.2416
0.6	0.1411	0.1547	0.1700
0.7	0.1143	0.1212	0.1221
0.8	0.0925	0.0953	0.0897
0.9	0.0751	0.0756	0.0673
1.0	0.0613	0.0605	0.0516
1.5	0.0247	0.0229	0.0173
2.0	0.0118	0.0107	0.0076
2.5	0.0064	0.0057	0.0040
3.0	0.0038	0.0034	0.0023
4.0	0.0017	0.0015	0.0010
5.0	0.0009	0.0008	0.0005

Example 10.15

Solve Example 10.3 using Westergaard's solution. Use $\mu_s = 0.3$. Compare this solution with $\Delta\sigma_z$ versus z obtained based on Boussinesq's solution.

Solution

$r = 5$ m. From Eq. (10.44),

$$\eta = \sqrt{\frac{1 - 2\mu_s}{2 - 2\mu_s}} = \sqrt{\frac{1 - (2)(0.3)}{2 - (2)(0.3)}} = 0.535$$

Now the following table can be prepared.

r (m)	z (m)	$\dfrac{r}{\eta z}$	I_5 [Eq. (10.46)]	$\Delta\sigma_z$ [Eq. (10.45)] (kN/m^2)
5	0	∞	0	**0**
5	2	1.67	0.0051	**0.0064**
5	4	2.34	0.0337	**0.0105**
5	6	1.56	0.0874	**0.0121**
5	10	0.935	0.2167	**0.0108**
5	20	0.467	0.4136	**0.0052**

Figure 10.31 shows the comparison of the same problem between the Boussinesq solution and Westergaard solution.

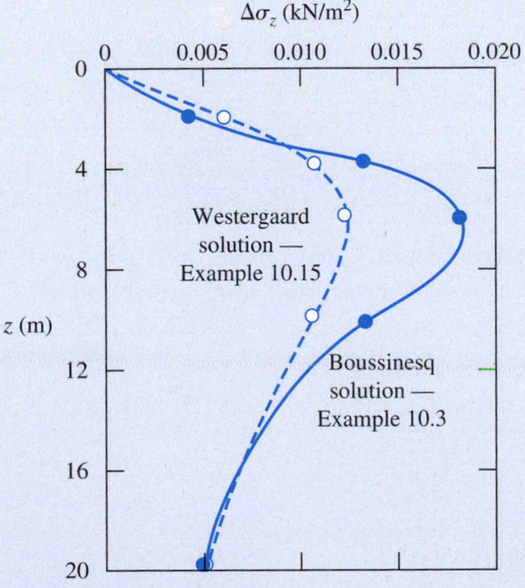

Figure 10.31

10.16 Stress Distribution for Westergaard Material

Stress due to a circularly loaded area

Referring to Figure 10.21, if the circular area is located on a Westergaard-type material, the increase in vertical stress, $\Delta\sigma_z$, at a point located at a depth z immediately below the center of the area can be given as

$$\Delta\sigma_z = q\left\{1 - \frac{\eta}{\left[\eta^2 + \left(\frac{R}{z}\right)^2\right]^{1/2}}\right\} \tag{10.47}$$

The term η has been defined in Eq. (10.44). The variations of $\Delta\sigma_z/q$ with R/z and $\mu_s = 0$ are given in Table 10.14.

Stress due to a uniformly loaded flexible rectangular area

Refer to Figure 10.23. If the flexible rectangular area is located on a Westergaard-type material, the stress increase at point A can be given as

$$\Delta\sigma_z = \frac{q}{2\pi}\left[\cot^{-1}\sqrt{\eta^2\left(\frac{1}{m^2} + \frac{1}{n^2}\right) + \eta^4\left(\frac{1}{m^2n^2}\right)}\right] \tag{10.48}$$

where

$$m = \frac{B}{z}$$

$$n = \frac{L}{z}$$

or

$$\frac{\Delta\sigma_z}{q} = \frac{1}{2\pi}\left[\cot^{-1}\sqrt{\eta^2\left(\frac{1}{m^2} + \frac{1}{n^2}\right) + \eta^4\left(\frac{1}{m^2n^2}\right)}\right] = I_w \tag{10.49}$$

Table 10.15 gives the variation of I_w with m and n (for $\mu_s = 0$). Figure 10.32 also provides a plot of I_w (for $\mu_s = 0$) for various values of m and n.

Table 10.14 Variation of $\Delta\sigma_z/q$ with R/z and $\mu_s = 0$ [Eq. (10.47)]

R/z	$\Delta\sigma_z/q$	R/z	$\Delta\sigma_z/q$	R/z	$\Delta\sigma_z/q$
0.00	0.0	1.50	0.5736	4.00	0.8259
0.25	0.0572	1.75	0.6254	5.00	0.8600
0.33	0.0938	2.00	0.6667	6.00	0.8830
0.50	0.1835	2.25	0.7002	7.00	0.8995
0.75	0.3140	2.50	0.7278	8.00	0.9120
1.00	0.4227	2.75	0.7510	9.00	0.9217
1.25	0.5076	3.00	0.7706	10.00	0.9295

Table 10.15 Variation of I_w with m and n ($\mu_s = 0$)

	n								
m	0.1	0.2	0.4	0.5	0.6	1.0	2.0	5.0	10.0
0.1	0.0031	0.0061	0.0110	0.0129	0.0144	0.0182	0.0211	0.0211	0.0223
0.2	0.0061	0.0118	0.0214	0.0251	0.0282	0.0357	0.0413	0.0434	0.0438
0.4	0.0110	0.0214	0.0390	0.0459	0.0516	0.0658	0.0768	0.0811	0.0847
0.5	0.0129	0.0251	0.0459	0.0541	0.0610	0.0781	0.0916	0.0969	0.0977
0.6	0.0144	0.0282	0.0516	0.0610	0.0687	0.0886	0.1044	0.1107	0.1117
1.0	0.0183	0.0357	0.0658	0.0781	0.0886	0.1161	0.1398	0.1491	0.1515
2.0	0.0211	0.0413	0.0768	0.0916	0.1044	0.1398	0.1743	0.1916	0.1948
5.0	0.0221	0.0435	0.0811	0.0969	0.1107	0.1499	0.1916	0.2184	0.2250
10.0	0.0223	0.0438	0.0817	0.0977	0.1117	0.1515	0.1948	0.2250	0.2341

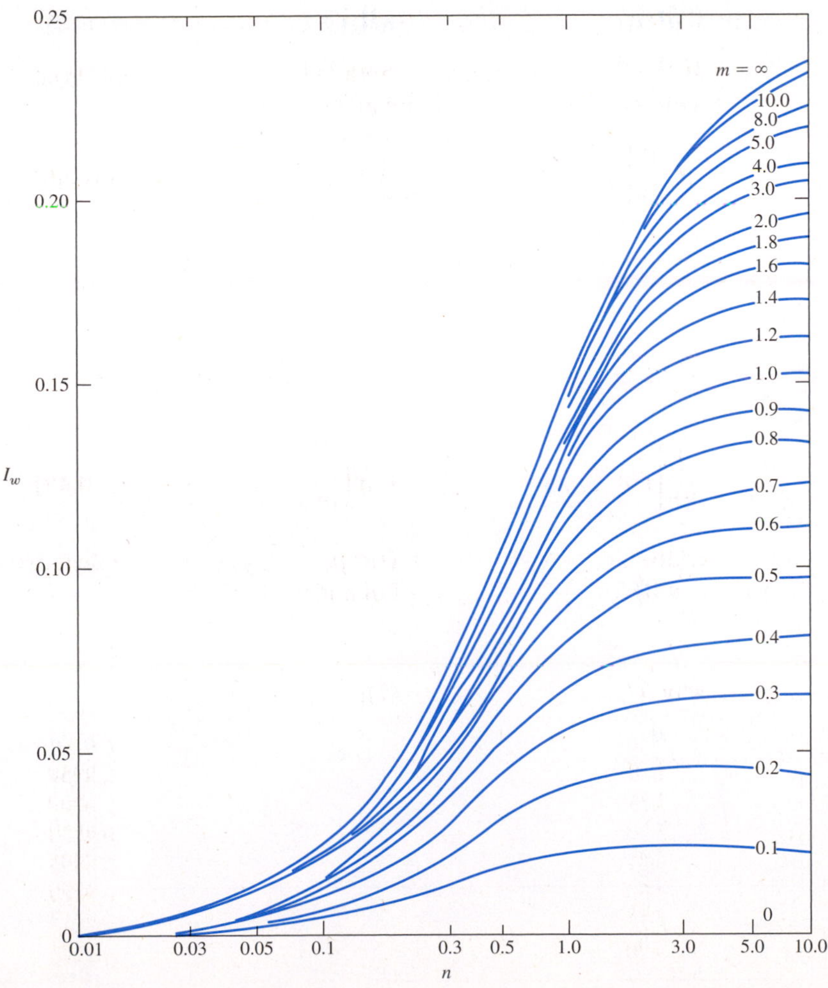

Figure 10.32 Variation of I_w ($\mu_s = 0$) [Eq. (10.49)] for various values of m and n

Example 10.16

Consider a flexible circular loaded area with $R = 4$ m. Let $q = 300$ kN/m^2. Calculate and compare the variation of $\Delta\sigma_z$ below the center of the circular area using Boussinesq's theory and Westergaard's theory (with $\mu_s = 0$) for $z = 0$ to 12 m.

Solution

Boussinesq's solution (see Table 10.7) with $R = 4$ m, $q = 300$ kN/m^2:

z (m)	$\frac{z}{R}$	$\frac{\Delta\sigma_z}{q}$	$\Delta\sigma_z$ (kN/m^2)
0	0	1	300
0.4	0.1	0.9990	299.7
2.0	0.5	0.9106	273.18
4.0	1.0	0.6465	193.95
6.0	1.5	0.4240	127.2
8.0	2.0	0.2845	85.35
10.0	2.5	0.1996	59.88
12.0	3.0	0.1436	43.08

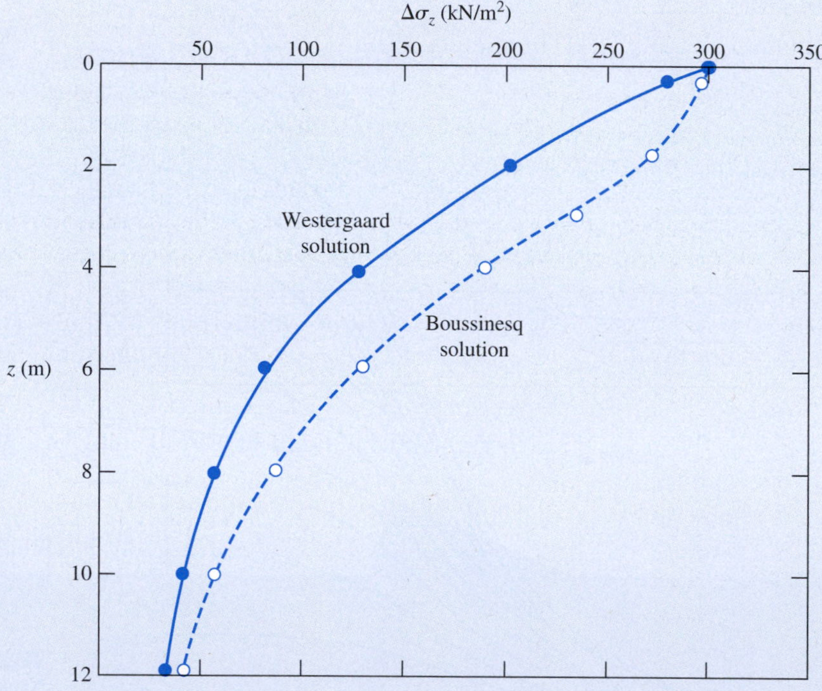

Figure 10.33

Westergaard's solution (see Table 10.14):

z (m)	$\frac{z}{R}$	$\frac{\Delta\sigma_z}{q}$	$\Delta\sigma_z$ (kN/m^2)
0	0	1	300
0.4	0.1	0.9295	278.85
2.0	0.5	0.6667	200.01
4.0	1.0	0.4227	126.81
6.0	1.5	0.275	82.5
8.0	2.0	0.1835	55.05
10.0	2.5	0.130	39.0
12.0	3.0	0.0938	28.14

The plot of $\Delta\sigma_z$ versus z is shown in Figure 10.33.

10.17 Summary

In this chapter, we have studied the following:

- The procedure to determine the normal and shear stresses on an inclined plane based on the stress conditions on a two-dimensional soil element [Eqs. (10.3) and (10.4)].
- The principles of Mohr's circle and the pole method to determine the stress along a plane have been provided in Sections 10.2 and 10.3, respectively.
- The vertical stress ($\Delta\sigma_z$) produced at any point in a homogeneous, elastic, and isotropic medium as a result of various types of load applied on the surface of an infinitely large half-space has been presented. The following table provides a list of the type of loading and the corresponding relationships to determine vertical stress.

Type of loading	Equation number to estimate $\Delta\sigma_z$
Point load	10.12
Vertical line load	10.15
Horizontal line load	10.17
Vertical strip load	10.19
Horizontal strip load	10.20
Linearly increasing vertical load on a strip	10.21
Embankment loading	10.22
Uniformly loaded circular area	10.27, 10.28
Uniformly loaded rectangular area	10.31, 10.36

- The concept of using an influence chart to determine the vertical pressure at any point below a loaded area is given in Section 10.14.
- Vertical stress calculations in a Westergaard material due to a point load, circularly loaded area, and flexible rectangular area are provided in Eqs. (10.43), (10.47) and (10.48), respectively.

The equations and graphs presented in this chapter are based entirely on the principles of the theory of elasticity; however, one must realize the limitations of these theories when they are applied to a soil medium. This is because soil deposits, in general, are not homogeneous, perfectly elastic, and isotropic. Hence, some deviations from the theoretical stress calculations can be expected in the field. Only a limited number of field observations are available in the literature for comparision purposes. On the basis of these results, it appears that one could expect a difference of ± 25 to 30% between theoretical estimates and actual field values.

Problems

10.1 A soil element is shown in Figure 10.34. Determine the following:
- **a.** Maximum and minimum principal stresses
- **b.** Normal and shear stresses on plane *AB*

Use Eqs. (10.3), (10.4), (10.6), and (10.7).

10.2 Repeat Problem 10.1 for the element shown in Figure 10.35.

10.3 Using the principles of Mohr's circles for the soil element shown in Figure 10.36, determine the following:
- **a.** Maximum and minimum principal stresses
- **b.** Normal and shear stresses on plane *AB*

10.4 Repeat Problem 10.3 for the element shown in Figure 10.37.

10.5 A soil element is shown in Figure 10.38. Using the pole method, determine:
- **a.** Maximum and minimum principal stresses
- **b.** Normal and shear stresses on plane *AB*

10.6 Repeat Problem 10.5 for the element shown in Figure 10.39.

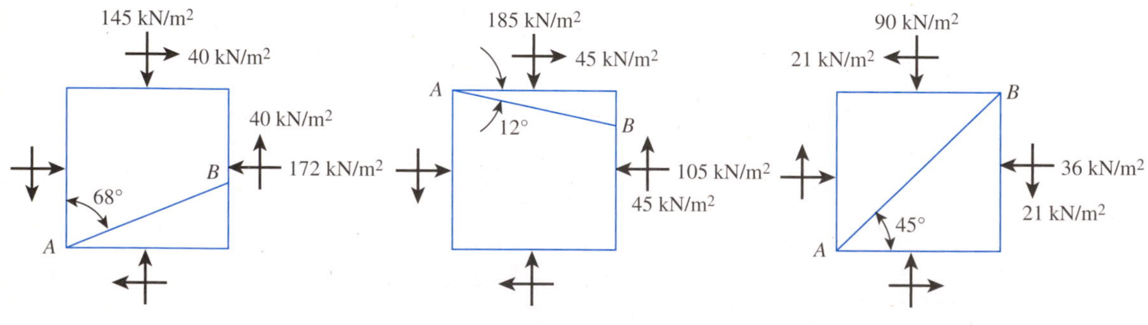

Figure 10.34

Figure 10.35

Figure 10.36

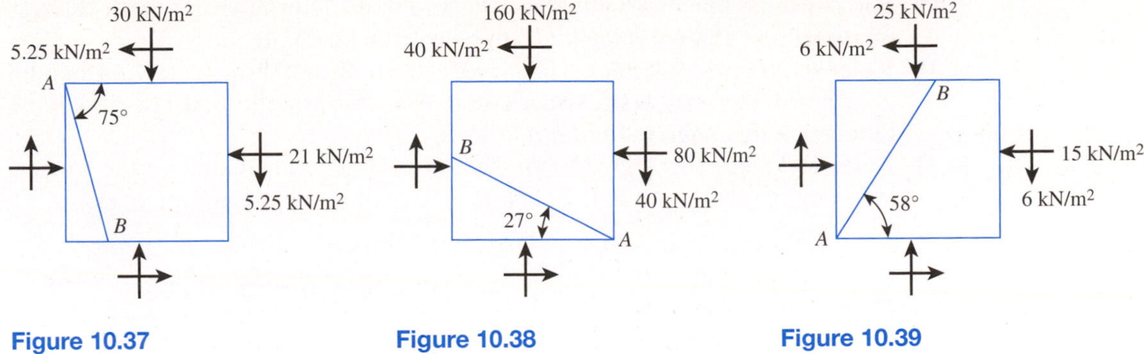

Figure 10.37

Figure 10.38

Figure 10.39

10.7 Point loads of magnitude 125, 250, and 500 kN act at B, C, and D, respectively (Figure 10.40). Determine the increase in vertical stress at a depth of 10 m below the point A. Use Boussinesq's equation.

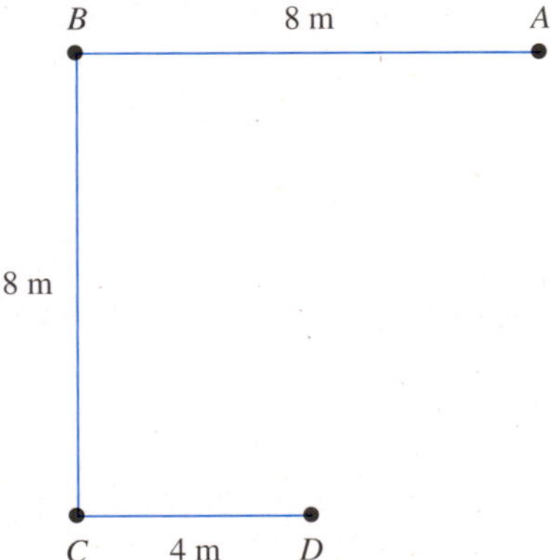

Figure 10.40

10.8 Refer to Figure 10.41. Determine the vertical stress increase, $\Delta\sigma_z$, at point A with the following values: $q_1 = 110$ kN/m, $q_2 = 440$ kN/m, $x_1 = 6$ m, $x_2 = 3$ m, and $z = 4$ m.

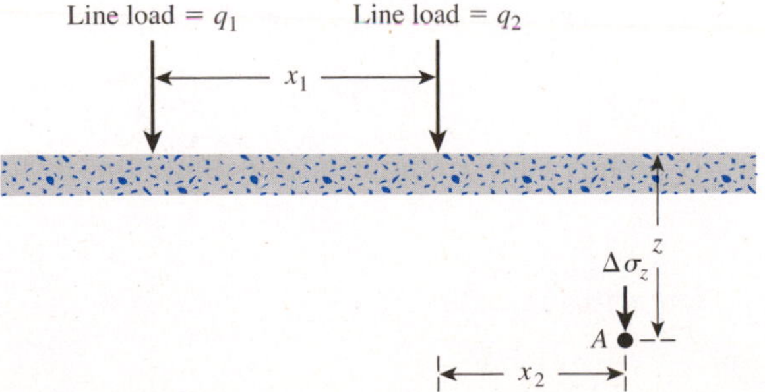

Figure 10.41

10.9 For the same line loads given in Problem 10.8, determine the vertical stress increase, $\Delta\sigma_z$, at a point located 4 m below the line load, q_2.

10.10 Refer to Figure 10.41. Given: $q_2 = 55$ kN/m, $x_1 = 5.5$ m, $x_2 = 2.44$ m, and $z = 2.1$ m. If the vertical stress increase at point A due to the loading is 3.7 kN/m², determine the magnitude of q_1.

10.11 Refer to Figure 10.42. Due to application of line loads q_1 and q_2, the vertical stress increase at point A is 58 kN/m². Determine the magnitude of q_2.

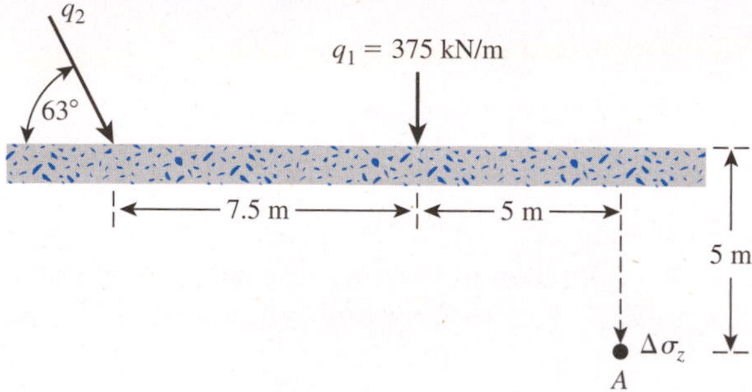

Figure 10.42

10.12 Refer to Figure 10.43. A strip load of $q = 70$ kN/m² is applied over a width with $B = 14.6$ m. Determine the increase in vertical stress at point A located $z = 6.4$ m below the surface. Given $x = 8.8$ m.

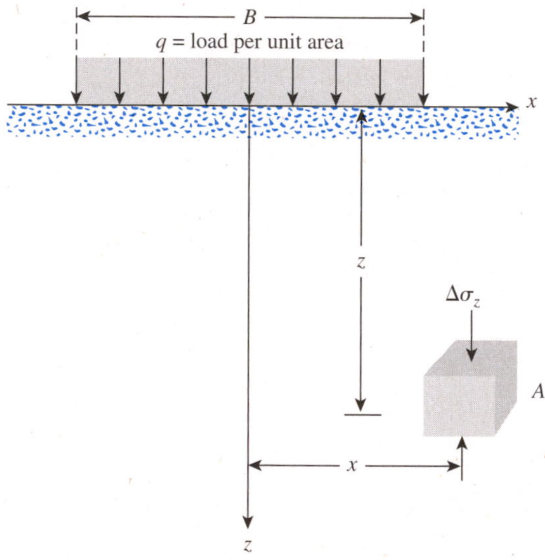

Figure 10.43

10.13 Repeat Problem 10.12 for $q = 700$ kN/m^2, $B = 8$ m, and $z = 4$ m. In this case, point A is located below the centerline under the strip load.

10.14 An earth embankment is shown in Figure 10.44. Determine the stress increase at point A due to the embankment load. Given: $\beta = 25°$, $\gamma = 18.7$ kN/m^3, $x = 16.75$ m, $y = 8.53$ m, and $z = 6.1$ m.

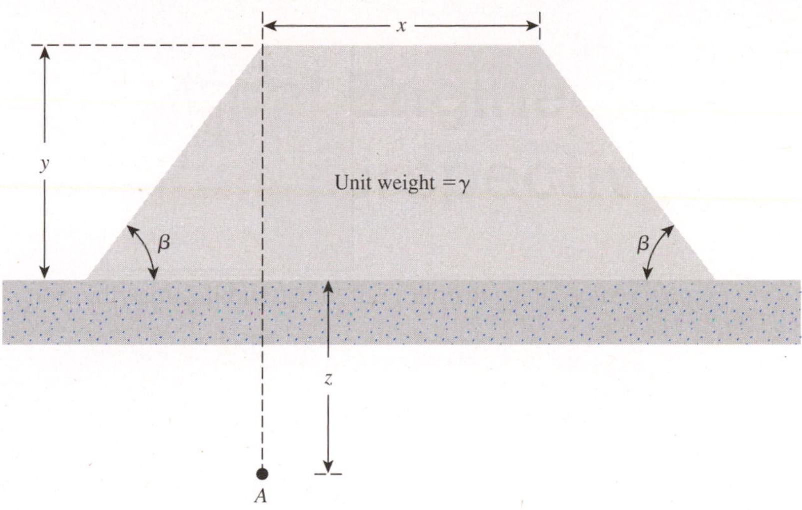

Figure 10.44

10.15 For the embankment shown in Figure 10.45, determine the vertical stress increases at points A, B, and C.

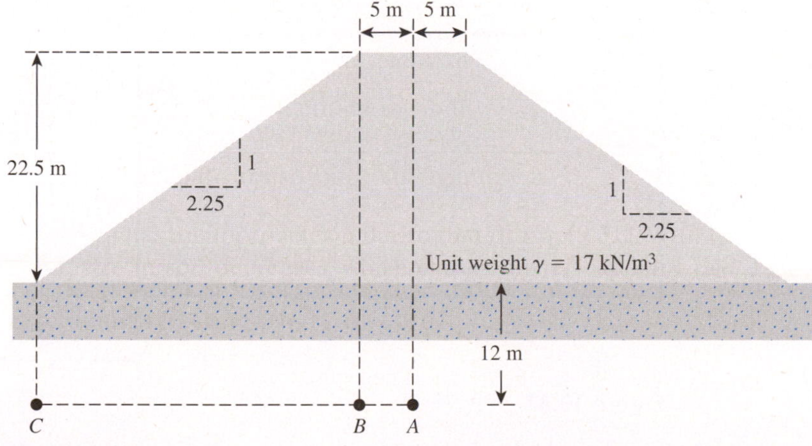

Figure 10.45

10.16 Refer to Figure 10.46. A flexible circular area of radius 6 m is uniformly loaded. Given: $q = 565$ kN/m^2. Using Newmark's chart, determine the increase in vertical stress, $\Delta\sigma_z$, at point A.

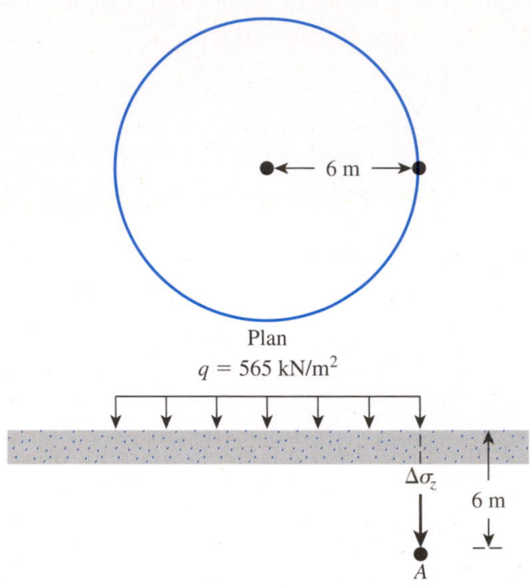

Figure 10.46

10.17 Refer to Figure 10.47. A flexible rectangular area is subjected to a uniformly distributed load of $q = 330$ kN/m^2. Determine the increase in vertical stress, $\Delta\sigma_z$, at a depth of $z = 6$ m under points A, B, and C.

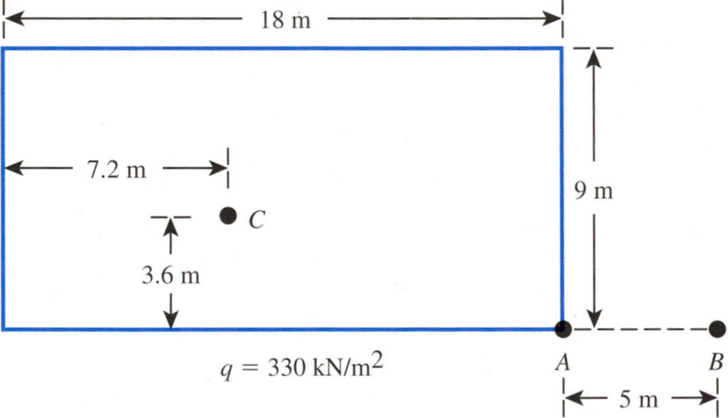

Figure 10.47

10.18 Refer to the flexible loaded rectangular area shown in Figure 10.47. Using Eq. (10.36), determine the vertical stress increase below the center of the loaded area at depths $z = 3, 6, 9, 12$, and 15 m.

10.19 Figure 10.48 shows the schematic of a circular water storage facility resting on the ground surface. The radius of the storage tank, $R = 2.5$ m, and the maximum height of water, $h_w = 4$ m. Determine the vertical stress increase, $\Delta\sigma_z$, at points 0, 2, 4, 8, and 10 m below the ground surface along the centerline of the tank. Use Boussinesq's theory [Eq. (10.27)].

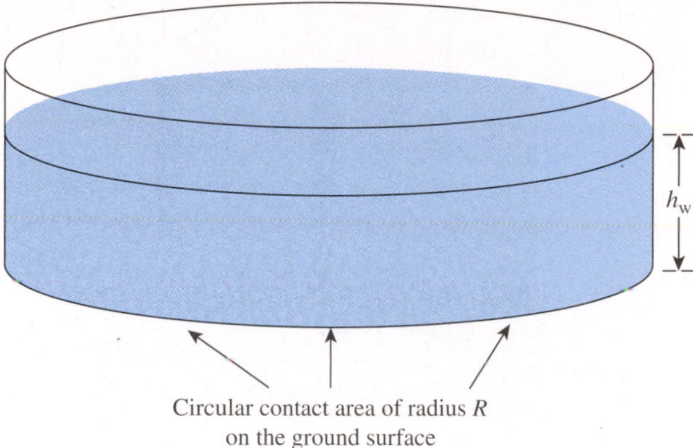

Figure 10.48

10.20 Redo Problem 10.19 using Westergaard's solution (Table 10.14) and compare with the solution by Boussinesq's theory. Assume $\mu_s = 0$.

10.21 Refer to Figure 10.48. If $R = 4$ m and h_w = height of water = 5 m, determine the vertical stress increases 2 m below the loaded area at radial distances where $r = 0, 2, 4, 6,$ and 8 m.

10.22 Refer to Figure 10.49. For the linearly increasing vertical loading on an infinite strip of width 5 m, determine the vertical stress increase, $\Delta\sigma_z$, at A.

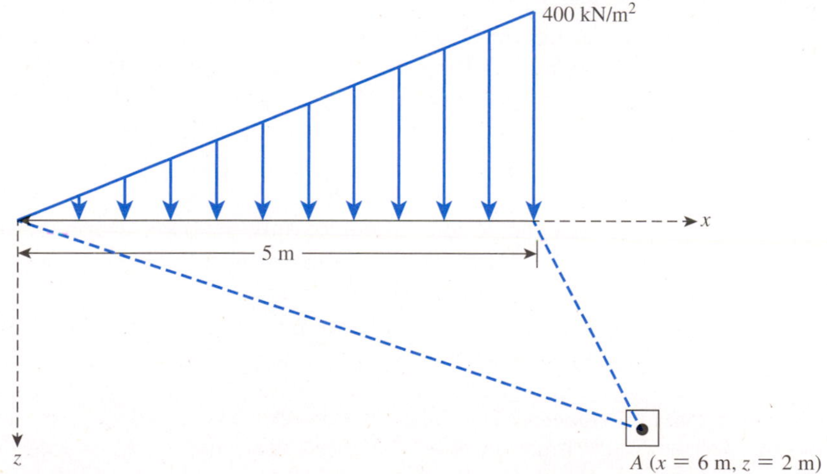

Figure 10.49

Critical Thinking Problems

10.C.1 *EB* and *FG* are two planes inside a soil element *ABCD* as shown in Figure 10.50.

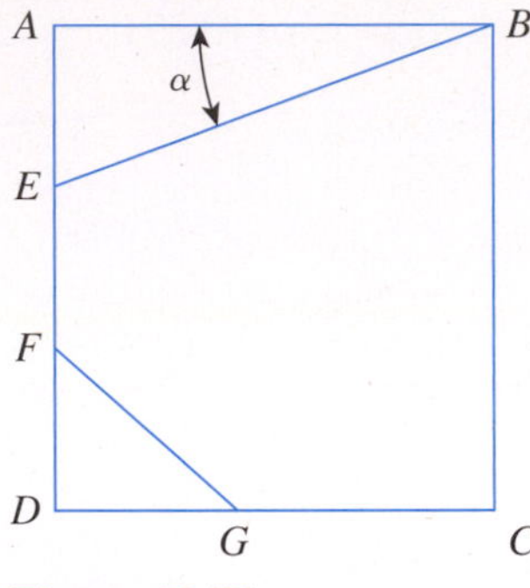

Figure 10.50

Stress conditions on the two planes are
Plane *EB*: $\sigma_{EB} = 25$ kN/m^2; $\tau_{EB} = +10$ kN/m^2
Plane *FG*: $\sigma_{FG} = 10$ kN/m^2; $\tau_{FG} = -5$ kN/m^2
(*Note:* Mohr's circle sign conventions for stresses are used above)

Given $\alpha = 25°$, determine:

a. The maximum and minimum principal stresses
b. The angle between the planes *EB* and *FG*
c. The external stresses on planes *AB* and *BC* that would cause the above internal stresses on planes *EB* and *FG*

10.C.2 A soil element beneath a pavement experiences principal stress rotations when the wheel load, *W*, passes over it and moves away, as shown in Figure 10.51. In this case, the wheel load has passed over points *A* and *B* and is now over point *C*. The general state of stress at these points is similar to the one shown by a stress block at point *D*. The phenomenon of principal stress rotation influences the permanent deformation behavior of the pavement layers.

Investigate how the magnitude and the orientations of the principal stresses vary with distance from the point of application of the wheel load. Consider the case shown in Figure 10.51. An unpaved aggregate road with a thickness of 610 mm and unit weight of 19.4 kN/m^3 is placed over a soil subgrade. A typical single-axle wheel load, $W = 40$ kN, is applied uniformly over a circular contact area with a radius of $R = 150$ mm (tire pressure of 565 kN/m^2). The horizontal and shear stresses at each point are calculated from a linear elastic finite element analysis for a two-layer pavement and are presented in the following table.

Element at	Radial distance, r (m)	Horizontal stress, σ_x (kN/m^2)	Shear stress, τ (kN/m^2)	Vertical stress, σ_y (kN/m^2)	σ_1 (kN/m^2)	σ_3 (kN/m^2)	α_i (deg)
A	0.457	25	17				
B	0.267	32	45				
C	0	7	0				

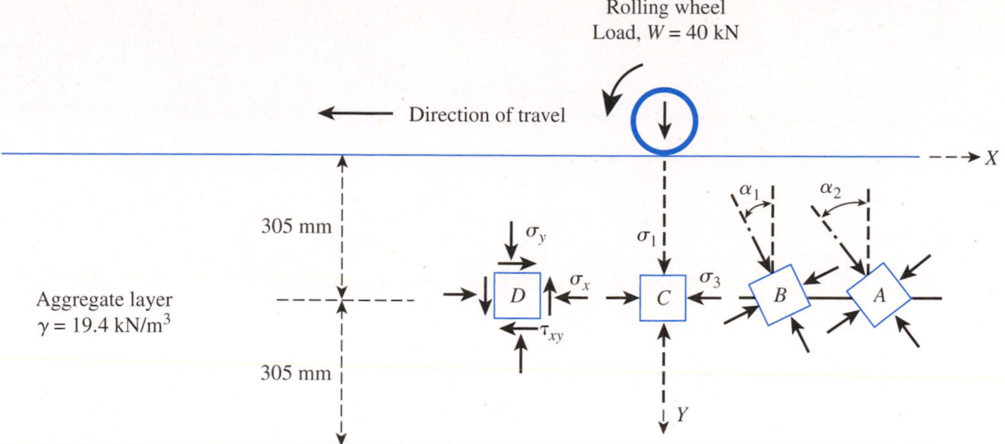

Figure 10.51

a. Use Eq. (10.28) to calculate the vertical stress increases at soil elements *A, B,* and *C* that are located at radial distances 0.457, 0.267, and 0 m, respectively, from the center of the load. Determine the total vertical stress (σ_y) due to wheel load, the overburden pressure at each point, and enter these values in the table.

b. Use the pole method to determine the maximum and minimum principal stresses (σ_1 and σ_3) for elements *A*, *B*, and *C*. Also determine the orientation (α_i) of the principal stress with respect to the vertical. Enter these values in the table.

c. Plot the variations of σ_1 and α_i with normalized radial distance, *r/R,* from the center of loading.

References

Ahlvin, R. G., and Ulery, H. H. (1962). "Tabulated Values for Determining the Complete Pattern of Stresses, Strains, and Deflections Beneath a Uniform Circular Load on a Homogeneous Half Space," in *Highway Research Bulletin 342*, Transportation Research Board, National Research Council, Washington, D.C., 1–13.

Boussinesq, J. (1883). *Application des Potentials à L'Etude de L'Equilibre et du Mouvement des Solides Elastiques*, Gauthier-Villars, Paris.

Das, B. (2014). *Advanced Soil Mechanics*, 4th ed., Taylor and Francis, London.

Newmark, N. M. (1942). "Influence Charts for Computation of Stresses in Elastic Soil," University of Illinois Engineering Experiment Station, *Bulletin No. 338*.

Osterberg, J. O. (1957). "Influence Values for Vertical Stresses in Semi-Infinite Mass Due to Embankment Loading," *Proceedings*, Fourth International Conference on Soil Mechanics and Foundation Engineering, London, Vol. 1, 393–396.

Westergaard, H. M. (1938). "A Problem of Elasticity Suggested by a Problem in Soil Mechanics: Soft Material Reinforced by Numerous Strong Horizontal Sheets," in *Contribution to the Mechanics of Solids*, Stephen Timoshenko 60th Anniversary Vol., Macmillan, New York.

CHAPTER 11

Compressibility of Soil

11.1 Introduction

A stress increase caused by the construction of foundations or other loads compresses soil layers. The compression is caused by (a) deformation of soil particles, (b) relocations of soil particles, and (c) expulsion of water or air from the void spaces. In general, the soil settlement caused by loads may be divided into three broad categories:

1. *Elastic settlement* (or *immediate settlement*), which is caused by the elastic deformation of dry soil and of moist and saturated soils without any change in the moisture content. Elastic settlement calculations generally are based on equations derived from the theory of elasticity.
2. *Primary consolidation settlement*, which is the result of a volume change in saturated cohesive soils because of expulsion of the water that occupies the void spaces.
3. *Secondary consolidation settlement*, which is observed in saturated cohesive soils and organic soil and is the result of the plastic adjustment of soil fabrics. It is an additional form of compression that occurs at constant effective stress.

This chapter presents the fundamental principles for estimating the elastic and consolidation settlements of soil layers under superimposed loadings.

The total settlement of a foundation can then be given as

$$S_T = S_c + S_s + S_e$$

where S_T = total settlement
S_c = primary consolidation settlement
S_s = secondary consolidation settlement
S_e = elastic settlement

When foundations are constructed on very compressible clays, the consolidation settlement can be several times greater than the elastic settlement.

This chapter will cover the following:

- Procedure for calculating elastic settlement
- Consolidation test procedure in the laboratory
- Estimation of consolidation settlement (primary and secondary)
- Time rate of primary consolidation settlement
- Methods to accelerate consolidation settlement
- Methods to reduce postconstruction settlement of structures

ELASTIC SETTLEMENT

11.2 Contact Pressure and Settlement Profile

Elastic, or immediate, settlement of foundations (S_e) occurs directly after the application of a load without a change in the moisture content of the soil. The magnitude of the contact settlement will depend on the flexibility of the foundation and the type of material on which it is resting.

In Chapter 10, the relationships for determining the increase in stress (which causes elastic settlement) due to the application of line load, strip load, embankment load, circular load, and rectangular load were based on the following assumptions:

- The load is applied at the ground surface.
- The loaded area is *flexible*.
- The soil medium is homogeneous, elastic, isotropic, and extends to a great depth.

In general, foundations are not perfectly flexible and are embedded at a certain depth below the ground surface. It is instructive, however, to evaluate the distribution of the contact pressure under a foundation along with the settlement profile under idealized conditions. Figure 11.1a shows a *perfectly flexible* foundation resting on an elastic material such as saturated clay. If the foundation is subjected to a uniformly distributed load, the contact pressure will be uniform and the foundation will experience a sagging profile. On the other hand, if we consider a *perfectly rigid* foundation resting on the ground surface subjected to a uniformly distributed load, the contact pressure and foundation settlement profile will be as shown in Figure 11.1b. The foundation will undergo a uniform settlement and the contact pressure will be redistributed.

The settlement profile and contact pressure distribution described are true for soils in which the modulus of elasticity is fairly constant with depth. In the case of cohesionless sand, the modulus of elasticity increases with depth. Additionally, there is a lack of lateral confinement on the edge of the foundation at the ground surface. The sand at the edge of a flexible foundation is pushed outward, and the deflection curve of the foundation takes a concave downward shape. The distributions of contact pressure and the settlement profiles of a flexible and a rigid foundation resting on sand and subjected to uniform loading are shown in Figures 11.2a and 11.2b, respectively.

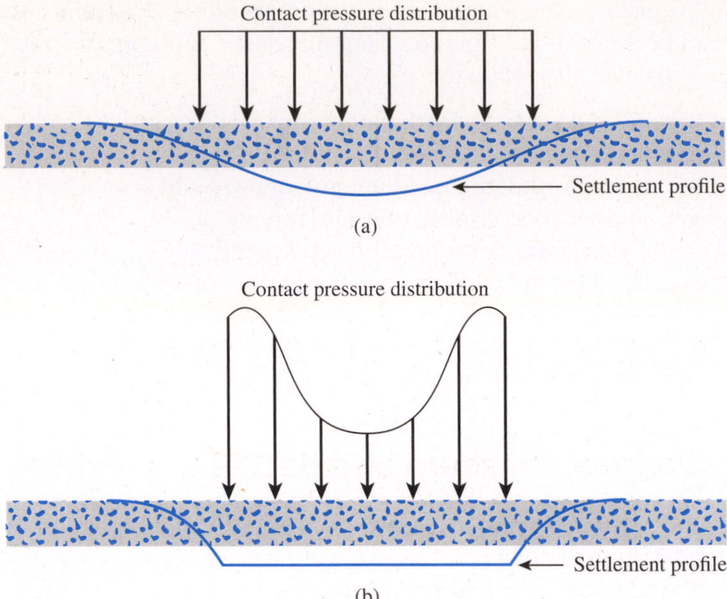

Figure 11.1 Elastic settlement profile and contact pressure in clay: (a) flexible foundation; (b) rigid foundation

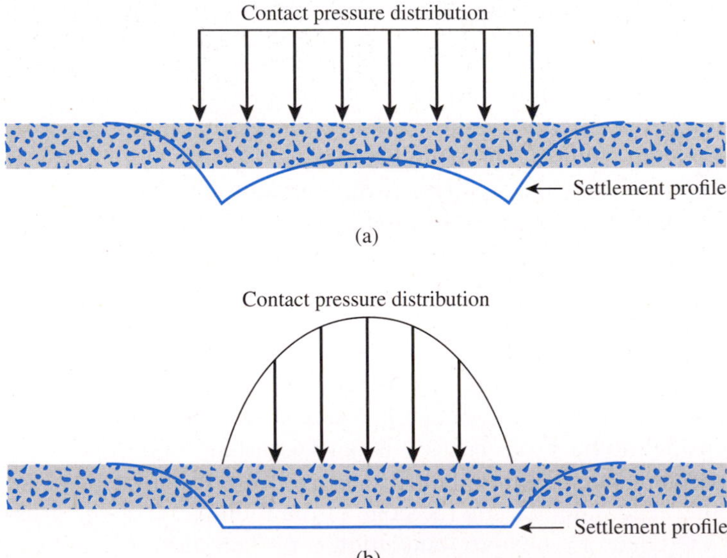

Figure 11.2 Elastic settlement profile and contact pressure in sand: (a) flexible foundation; (b) rigid foundation

11.3 Relations for Elastic Settlement Calculation

Figure 11.3 shows a shallow foundation subjected to a net force per unit area equal to $\Delta\sigma$. Let the Poisson's ratio and the modulus of elasticity of the soil supporting it be μ_s and E_s, respectively. Theoretically, if the foundation is perfectly flexible, the settlement may be expressed as

$$S_e = \Delta\sigma(\alpha B')\frac{1-\mu_s^2}{E_s}I_s I_f \tag{11.1}$$

where $\Delta\sigma$ = net applied pressure on the foundation

μ_s = Poisson's ratio of soil

E_s = average modulus of elasticity of the soil under the foundation measured from $z = 0$ to about $z = 5B$

B' = $B/2$ for center of foundation

= B for corner of foundation

I_s = shape factor (Steinbrenner, 1934)

$$= F_1 + \frac{1-2\mu_s}{1-\mu_s}F_2 \tag{11.2}$$

$$F_1 = \frac{1}{\pi}(A_0 + A_1) \tag{11.3}$$

$$F_2 = \frac{n'}{2\pi}\tan^{-1}A_2 \tag{11.4}$$

$$A_0 = m'\ln\frac{(1+\sqrt{m'^2+1})\sqrt{m'^2+n'^2}}{m'(1+\sqrt{m'^2+n'^2+1})} \tag{11.5}$$

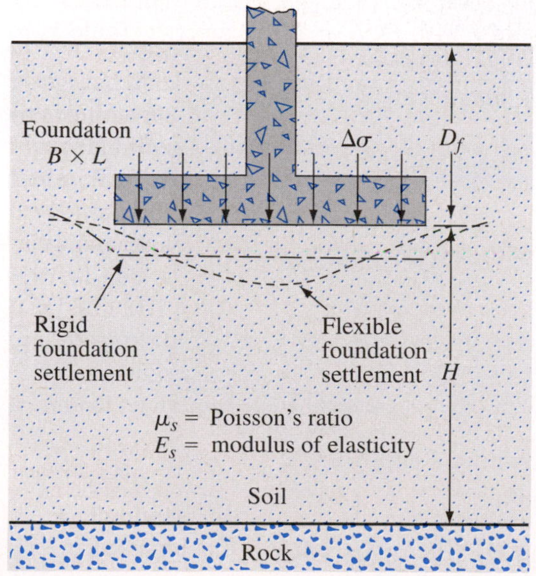

Figure 11.3 Elastic settlement of flexible and rigid foundations

$$A_1 = \ln \frac{(m' + \sqrt{m'^2 + 1})\sqrt{1 + n'^2}}{m' + \sqrt{m'^2 + n'^2 + 1}} \tag{11.6}$$

$$A_2 = \frac{m'}{n'\sqrt{m'^2 + n'^2 + 1}} \tag{11.7}$$

$$I_f = \text{depth factor (Fox, 1948)} = f\left(\frac{D_f}{B}, \mu_s, \text{and } \frac{L}{B}\right) \tag{11.8}$$

α = factor that depends on the location on the foundation where settlement is being calculated

- For calculation of settlement at the *center* of the foundation:

$$\alpha = 4$$

$$m' = \frac{L}{B}$$

$$n' = \frac{H}{\left(\frac{B}{2}\right)}$$

- For calculation of settlement at a *corner* of the foundation:

$$\alpha = 1$$

$$m' = \frac{L}{B}$$

$$n' = \frac{H}{B}$$

The variations of F_1 and F_2 [Eqs. (11.3) and (11.4)] with m' and n' are given in Tables 11.1 and 11.2 respectively. Also the variation of I_f with D_f/B and μ_s is given in Table 11.3. Note that *when* $D_f = 0$, *the value of* $I_f = 1$ *in all cases.*

The elastic settlement of a *rigid foundation* can be estimated as

$$S_{e(\text{rigid})} \approx 0.93 S_{e(\text{flexible, center})} \tag{11.9}$$

Due to the nonhomogeneous nature of soil deposits, the magnitude of E_s may vary with depth. For that reason, Bowles (1987) recommended using a weighted average value of E_s in Eq. (11.1) or

$$E_s = \frac{\sum E_{s(i)} \Delta z}{\bar{z}} \tag{11.10}$$

where $E_{s(i)}$ = soil modulus of elasticity within a depth Δz

$\bar{z}$ = H or $5B$, whichever is smaller

Representative values of the modulus of elasticity and Poisson's ratio for different types of soils are given in Tables 11.4 and 11.5, respectively.

Example 11.1

A rigid shallow foundation 1 m × 1 m in plan is shown in Figure 11.4. Calculate the elastic settlement at the center of the foundation.

Solution

Given: $B = 1$ m and $L = 1$ m. Note that $\bar{z} = 5\text{ m} = 5B$. From Eq. (11.10),

$$E_s = \frac{\sum E_{s(i)}\Delta z}{\bar{z}}$$

$$= \frac{(8000)(2) + (6000)(1) + (10{,}000)(2)}{5} = 8400\text{ kN/m}^2$$

For the *center of the foundation,*

$$\alpha = 4$$

$$m' = \frac{L}{B} = \frac{1}{1} = 1$$

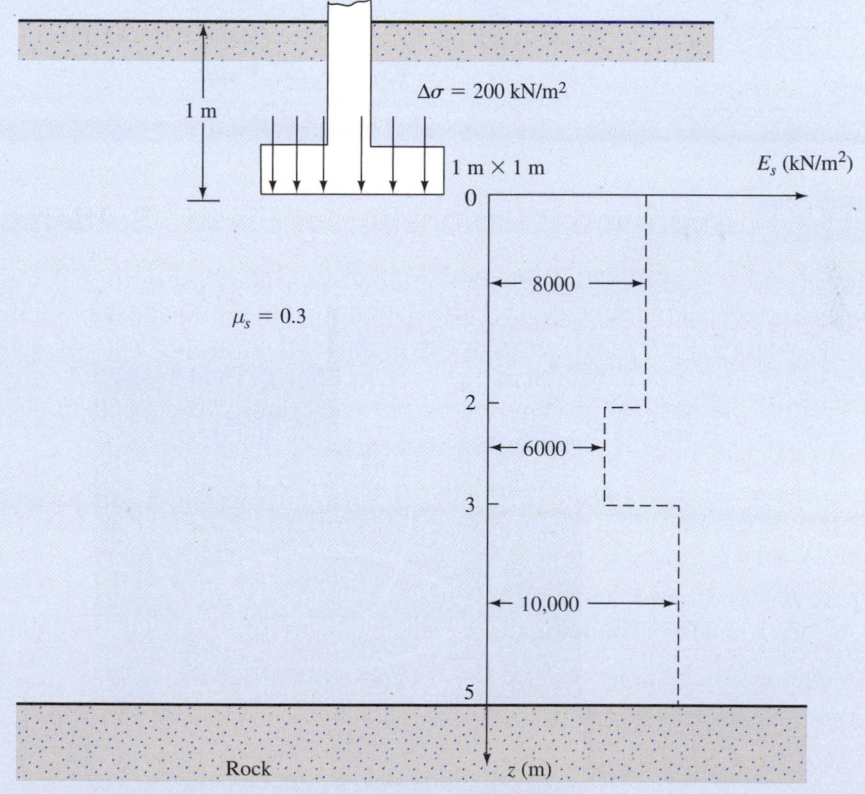

Figure 11.4

$$n' = \frac{H}{\left(\frac{B}{2}\right)} = \frac{5}{\left(\frac{1}{2}\right)} = 10$$

From Tables 11.1 and 11.2, $F_1 = 0.498$ and $F_2 = 0.016$. From Eq. (11.2),

$$I_s = F_1 + \frac{1 - 2\mu_s}{1 - \mu_s} F_2$$
$$= 0.498 + \frac{1 - 0.6}{1 - 0.3}(0.016) = 0.507$$

Again, $\frac{D_f}{B} = \frac{1}{1} = 1$, $\frac{L}{B} = 1$, $\mu_s = 0.3$. From Table 11.3, $I_f = 0.65$. Hence,

$$S_{e(\text{flexible})} = \Delta\sigma(\alpha B')\frac{1 - \mu_s^2}{E_s} I_s I_f$$
$$= (200)\left(4 \times \frac{1}{2}\right)\left(\frac{1 - 0.3^2}{8400}\right)(0.507)(0.65) = 0.0143 \text{ m} = 14.3 \text{ mm}$$

Since the foundation is rigid, from Eq. (11.9),

$$S_e(\text{rigid}) = (0.93)(14.3) = \mathbf{13.3\ mm}$$

11.4 Improved Relationship for Elastic Settlement

Mayne and Poulos (1999) presented an improved relationship for calculating the elastic settlement of foundations. This relationship takes into account the rigidity of the foundation, the depth of embedment of the foundation, the increase in the modulus of elasticity of soil with depth, and the location of rigid layers at limited depth. In order to use this relationship, one needs to determine the equivalent diameter of a rectangular foundation, which is

$$B_e = \sqrt{\frac{4BL}{\pi}} \tag{11.11}$$

where B = width of foundation
L = length of foundation

For circular foundations,

$$B_e = B \tag{11.12}$$

where B = diameter of foundation.

Figure 11.5 shows a foundation having an equivalent diameter of B_e located at a depth D_f below the ground surface. Let the thickness of the foundation be t and

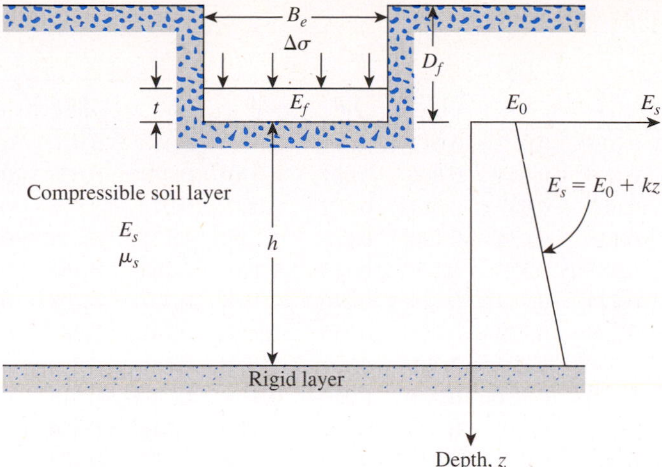

Figure 11.5 Improved relationship for elastic settlement

the modulus of elasticity of the foundation material be E_f. A rigid layer is located at a depth h below the bottom of the foundation. The modulus of elasticity of the compressible soil layer can be given as

$$E_s = E_o + kz \tag{11.13}$$

With the preceding parameters defined, the elastic settlement can be given as

$$S_e = \frac{\Delta\sigma B_e I_G I_F I_E}{E_o}(1 - \mu_s^2) \tag{11.14}$$

where I_G = influence factor for the variation of E_s with depth = $f(E_o, k, B_e, \text{and } h)$
I_F = foundation rigidity correction factor
I_E = foundation embedment correction factor

Figure 11.6 shows the variation of I_G with $\beta = E_o/kB_e$ and h/B_e. The foundation rigidity correction factor can be expressed as

$$I_F = \frac{\pi}{4} + \frac{1}{4.6 + 10\left(\dfrac{E_f}{E_0 + \dfrac{B_e}{2}k}\right)\left(\dfrac{2t}{B_e}\right)^3} \tag{11.15}$$

Similarly, the embedment correction factor is

$$I_E = 1 - \frac{1}{3.5\exp(1.22\mu_s - 0.4)\left(\dfrac{B_e}{D_f} + 1.6\right)} \tag{11.16}$$

Figures 11.7 and 11.8 show the variations of I_F and I_E expressed by Eqs. (11.15) and (11.16).

Table 11.1 Variation of F_1 with m' and n'

	m'									
n'	1.0	1.2	1.4	1.6	1.8	2.0	2.5	3.0	3.5	4.0
0.25	0.014	0.013	0.012	0.011	0.011	0.011	0.010	0.010	0.010	0.010
0.50	0.049	0.046	0.044	0.042	0.041	0.040	0.038	0.038	0.037	0.037
0.75	0.095	0.090	0.087	0.084	0.082	0.080	0.077	0.076	0.074	0.074
1.00	0.142	0.138	0.134	0.130	0.127	0.125	0.121	0.118	0.116	0.115
1.25	0.186	0.183	0.179	0.176	0.173	0.170	0.165	0.161	0.158	0.157
1.50	0.224	0.224	0.222	0.219	0.216	0.213	0.207	0.203	0.199	0.197
1.75	0.257	0.259	0.259	0.258	0.255	0.253	0.247	0.242	0.238	0.235
2.00	0.285	0.290	0.292	0.292	0.291	0.289	0.284	0.279	0.275	0.271
2.25	0.309	0.317	0.321	0.323	0.323	0.322	0.317	0.313	0.308	0.305
2.50	0.330	0.341	0.347	0.350	0.351	0.351	0.348	0.344	0.340	0.336
2.75	0.348	0.361	0.369	0.374	0.377	0.378	0.377	0.373	0.369	0.365
3.00	0.363	0.379	0.389	0.396	0.400	0.402	0.402	0.400	0.396	0.392
3.25	0.376	0.394	0.406	0.415	0.420	0.423	0.426	0.424	0.421	0.418
3.50	0.388	0.408	0.422	0.431	0.438	0.442	0.447	0.447	0.444	0.441
3.75	0.399	0.420	0.436	0.447	0.454	0.460	0.467	0.458	0.466	0.464
4.00	0.408	0.431	0.448	0.460	0.469	0.476	0.484	0.487	0.486	0.484
4.25	0.417	0.440	0.458	0.472	0.481	0.484	0.495	0.514	0.515	0.515
4.50	0.424	0.450	0.469	0.484	0.495	0.503	0.516	0.521	0.522	0.522
4.75	0.431	0.458	0.478	0.494	0.506	0.515	0.530	0.536	0.539	0.539
5.00	0.437	0.465	0.487	0.503	0.516	0.526	0.543	0.551	0.554	0.554
5.25	0.443	0.472	0.494	0.512	0.526	0.537	0.555	0.564	0.568	0.569
5.50	0.448	0.478	0.501	0.520	0.534	0.546	0.566	0.576	0.581	0.584
5.75	0.453	0.483	0.508	0.527	0.542	0.555	0.576	0.588	0.594	0.597
6.00	0.457	0.489	0.514	0.534	0.550	0.563	0.585	0.598	0.606	0.609
6.25	0.461	0.493	0.519	0.540	0.557	0.570	0.594	0.609	0.617	0.621
6.50	0.465	0.498	0.524	0.546	0.563	0.577	0.603	0.618	0.627	0.632
6.75	0.468	0.502	0.529	0.551	0.569	0.584	0.610	0.627	0.637	0.643
7.00	0.471	0.506	0.533	0.556	0.575	0.590	0.618	0.635	0.646	0.653
7.25	0.474	0.509	0.538	0.561	0.580	0.596	0.625	0.643	0.655	0.662
7.50	0.477	0.513	0.541	0.565	0.585	0.601	0.631	0.650	0.663	0.671
7.75	0.480	0.516	0.545	0.569	0.589	0.606	0.637	0.658	0.671	0.680
8.00	0.482	0.519	0.549	0.573	0.594	0.611	0.643	0.664	0.678	0.688
8.25	0.485	0.522	0.552	0.577	0.598	0.615	0.648	0.670	0.685	0.695
8.50	0.487	0.524	0.555	0.580	0.601	0.619	0.653	0.676	0.692	0.703
8.75	0.489	0.527	0.558	0.583	0.605	0.623	0.658	0.682	0.698	0.710
9.00	0.491	0.529	0.560	0.587	0.609	0.627	0.663	0.687	0.705	0.716
9.25	0.493	0.531	0.563	0.589	0.612	0.631	0.667	0.693	0.710	0.723
9.50	0.495	0.533	0.565	0.592	0.615	0.634	0.671	0.697	0.716	0.719
9.75	0.496	0.536	0.568	0.595	0.618	0.638	0.675	0.702	0.721	0.735
10.00	0.498	0.537	0.570	0.597	0.621	0.641	0.679	0.707	0.726	0.740
20.00	0.529	0.575	0.614	0.647	0.677	0.702	0.756	0.797	0.830	0.858
50.00	0.548	0.598	0.640	0.678	0.711	0.740	0.803	0.853	0.895	0.931
100.00	0.555	0.605	0.649	0.688	0.722	0.753	0.819	0.872	0.918	0.956

Table 11.1 (*continued*)

	m'									
n'	4.5	5.0	6.0	7.0	8.0	9.0	10.0	25.0	50.0	100.0
0.25	0.010	0.010	0.010	0.010	0.010	0.010	0.010	0.010	0.010	0.010
0.50	0.036	0.036	0.036	0.036	0.036	0.036	0.036	0.036	0.036	0.036
0.75	0.073	0.073	0.072	0.072	0.072	0.072	0.071	0.071	0.071	0.071
1.00	0.114	0.113	0.112	0.112	0.112	0.111	0.111	0.110	0.110	0.110
1.25	0.155	0.154	0.153	0.152	0.152	0.151	0.151	0.150	0.150	0.150
1.50	0.195	0.194	0.192	0.191	0.190	0.190	0.189	0.188	0.188	0.188
1.75	0.233	0.232	0.229	0.228	0.227	0.226	0.225	0.223	0.223	0.223
2.00	0.269	0.267	0.264	0.262	0.261	0.260	0.259	0.257	0.256	0.256
2.25	0.302	0.300	0.296	0.294	0.293	0.291	0.291	0.287	0.287	0.287
2.50	0.333	0.331	0.327	0.324	0.322	0.321	0.320	0.316	0.315	0.315
2.75	0.362	0.359	0.355	0.352	0.350	0.348	0.347	0.343	0.342	0.342
3.00	0.389	0.386	0.382	0.378	0.376	0.374	0.373	0.368	0.367	0.367
3.25	0.415	0.412	0.407	0.403	0.401	0.399	0.397	0.391	0.390	0.390
3.50	0.438	0.435	0.430	0.427	0.424	0.421	0.420	0.413	0.412	0.411
3.75	0.461	0.458	0.453	0.449	0.446	0.443	0.441	0.433	0.432	0.432
4.00	0.482	0.479	0.474	0.470	0.466	0.464	0.462	0.453	0.451	0.451
4.25	0.516	0.496	0.484	0.473	0.471	0.471	0.470	0.468	0.462	0.460
4.50	0.520	0.517	0.513	0.508	0.505	0.502	0.499	0.489	0.487	0.487
4.75	0.537	0.535	0.530	0.526	0.523	0.519	0.517	0.506	0.504	0.503
5.00	0.554	0.552	0.548	0.543	0.540	0.536	0.534	0.522	0.519	0.519
5.25	0.569	0.568	0.564	0.560	0.556	0.553	0.550	0.537	0.534	0.534
5.50	0.584	0.583	0.579	0.575	0.571	0.568	0.585	0.551	0.549	0.548
5.75	0.597	0.597	0.594	0.590	0.586	0.583	0.580	0.565	0.583	0.562
6.00	0.611	0.610	0.608	0.604	0.601	0.598	0.595	0.579	0.576	0.575
6.25	0.623	0.623	0.621	0.618	0.615	0.611	0.608	0.592	0.589	0.588
6.50	0.635	0.635	0.634	0.631	0.628	0.625	0.622	0.605	0.601	0.600
6.75	0.646	0.647	0.646	0.644	0.641	0.637	0.634	0.617	0.613	0.612
7.00	0.656	0.658	0.658	0.656	0.653	0.650	0.647	0.628	0.624	0.623
7.25	0.666	0.669	0.669	0.668	0.665	0.662	0.659	0.640	0.635	0.634
7.50	0.676	0.679	0.680	0.679	0.676	0.673	0.670	0.651	0.646	0.645
7.75	0.685	0.688	0.690	0.689	0.687	0.684	0.681	0.661	0.656	0.655
8.00	0.694	0.697	0.700	0.700	0.698	0.695	0.692	0.672	0.666	0.665
8.25	0.702	0.706	0.710	0.710	0.708	0.705	0.703	0.682	0.676	0.675
8.50	0.710	0.714	0.719	0.719	0.718	0.715	0.713	0.692	0.686	0.684
8.75	0.717	0.722	0.727	0.728	0.727	0.725	0.723	0.701	0.695	0.693
9.00	0.725	0.730	0.736	0.737	0.736	0.735	0.732	0.710	0.704	0.702
9.25	0.731	0.737	0.744	0.746	0.745	0.744	0.742	0.719	0.713	0.711
9.50	0.738	0.744	0.752	0.754	0.754	0.753	0.751	0.728	0.721	0.719
9.75	0.744	0.751	0.759	0.762	0.762	0.761	0.759	0.737	0.729	0.727
10.00	0.750	0.758	0.766	0.770	0.770	0.770	0.768	0.745	0.738	0.735
20.00	0.878	0.896	0.925	0.945	0.959	0.969	0.977	0.982	0.965	0.957
50.00	0.962	0.989	1.034	1.070	1.100	1.125	1.146	1.265	1.279	1.261
100.00	0.990	1.020	1.072	1.114	1.150	1.182	1.209	1.408	1.489	1.499

Table 11.2 Variation of F_2 with m' and n'

	m'									
n'	**1.0**	**1.2**	**1.4**	**1.6**	**1.8**	**2.0**	**2.5**	**3.0**	**3.5**	**4.0**
0.25	0.049	0.050	0.051	0.051	0.051	0.052	0.052	0.052	0.052	0.052
0.50	0.074	0.077	0.080	0.081	0.083	0.084	0.086	0.086	0.0878	0.087
0.75	0.083	0.089	0.093	0.097	0.099	0.101	0.104	0.106	0.107	0.108
1.00	0.083	0.091	0.098	0.102	0.106	0.109	0.114	0.117	0.119	0.120
1.25	0.080	0.089	0.096	0.102	0.107	0.111	0.118	0.122	0.125	0.127
1.50	0.075	0.084	0.093	0.099	0.105	0.110	0.118	0.124	0.128	0.130
1.75	0.069	0.079	0.088	0.095	0.101	0.107	0.117	0.123	0.128	0.131
2.00	0.064	0.074	0.083	0.090	0.097	0.102	0.114	0.121	0.127	0.131
2.25	0.059	0.069	0.077	0.085	0.092	0.098	0.110	0.119	0.125	0.130
2.50	0.055	0.064	0.073	0.080	0.087	0.093	0.106	0.115	0.122	0.127
2.75	0.051	0.060	0.068	0.076	0.082	0.089	0.102	0.111	0.119	0.125
3.00	0.048	0.056	0.064	0.071	0.078	0.084	0.097	0.108	0.116	0.122
3.25	0.045	0.053	0.060	0.067	0.074	0.080	0.093	0.104	0.112	0.119
3.50	0.042	0.050	0.057	0.064	0.070	0.076	0.089	0.100	0.109	0.116
3.75	0.040	0.047	0.054	0.060	0.067	0.073	0.086	0.096	0.105	0.113
4.00	0.037	0.044	0.051	0.057	0.063	0.069	0.082	0.093	0.102	0.110
4.25	0.036	0.042	0.049	0.055	0.061	0.066	0.079	0.090	0.099	0.107
4.50	0.034	0.040	0.046	0.052	0.058	0.063	0.076	0.086	0.096	0.104
4.75	0.032	0.038	0.044	0.050	0.055	0.061	0.073	0.083	0.093	0.101
5.00	0.031	0.036	0.042	0.048	0.053	0.058	0.070	0.080	0.090	0.098
5.25	0.029	0.035	0.040	0.046	0.051	0.056	0.067	0.078	0.087	0.095
5.50	0.028	0.033	0.039	0.044	0.049	0.054	0.065	0.075	0.084	0.092
5.75	0.027	0.032	0.037	0.042	0.047	0.052	0.063	0.073	0.082	0.090
6.00	0.026	0.031	0.036	0.040	0.045	0.050	0.060	0.070	0.079	0.087
6.25	0.025	0.030	0.034	0.039	0.044	0.048	0.058	0.068	0.077	0.085
6.50	0.024	0.029	0.033	0.038	0.042	0.046	0.056	0.066	0.075	0.083
6.75	0.023	0.028	0.032	0.036	0.041	0.045	0.055	0.064	0.073	0.080
7.00	0.022	0.027	0.031	0.035	0.039	0.043	0.053	0.062	0.071	0.078
7.25	0.022	0.026	0.030	0.034	0.038	0.042	0.051	0.060	0.069	0.076
7.50	0.021	0.025	0.029	0.033	0.037	0.041	0.050	0.059	0.067	0.074
7.75	0.020	0.024	0.028	0.032	0.036	0.039	0.048	0.057	0.065	0.072
8.00	0.020	0.023	0.027	0.031	0.035	0.038	0.047	0.055	0.063	0.071
8.25	0.019	0.023	0.026	0.030	0.034	0.037	0.046	0.054	0.062	0.069
8.50	0.018	0.022	0.026	0.029	0.033	0.036	0.045	0.053	0.060	0.067
8.75	0.018	0.021	0.025	0.028	0.032	0.035	0.043	0.051	0.059	0.066
9.00	0.017	0.021	0.024	0.028	0.031	0.034	0.042	0.050	0.057	0.064
9.25	0.017	0.020	0.024	0.027	0.030	0.033	0.041	0.049	0.056	0.063
9.50	0.017	0.020	0.023	0.026	0.029	0.033	0.040	0.048	0.055	0.061
9.75	0.016	0.019	0.023	0.026	0.029	0.032	0.039	0.047	0.054	0.060
10.00	0.016	0.019	0.022	0.025	0.028	0.031	0.038	0.046	0.052	0.059
20.00	0.008	0.010	0.011	0.013	0.014	0.016	0.020	0.024	0.027	0.031
50.00	0.003	0.004	0.004	0.005	0.006	0.006	0.008	0.010	0.011	0.013
100.00	0.002	0.002	0.002	0.003	0.003	0.003	0.004	0.005	0.006	0.006

Table 11.2 (*continued*)

	m'									
n'	4.5	5.0	6.0	7.0	8.0	9.0	10.0	25.0	50.0	100.0
0.25	0.053	0.053	0.053	0.053	0.053	0.053	0.053	0.053	0.053	0.053
0.50	0.087	0.087	0.088	0.088	0.088	0.088	0.088	0.088	0.088	0.088
0.75	0.109	0.109	0.109	0.110	0.110	0.110	0.110	0.111	0.111	0.111
1.00	0.121	0.122	0.123	0.123	0.124	0.124	0.124	0.125	0.125	0.125
1.25	0.128	0.130	0.131	0.132	0.132	0.133	0.133	0.134	0.134	0.134
1.50	0.132	0.134	0.136	0.137	0.138	0.138	0.139	0.140	0.140	0.140
1.75	0.134	0.136	0.138	0.140	0.141	0.142	0.142	0.144	0.144	0.145
2.00	0.134	0.136	0.139	0.141	0.143	0.144	0.145	0.147	0.147	0.148
2.25	0.133	0.136	0.140	0.142	0.144	0.145	0.146	0.149	0.150	0.150
2.50	0.132	0.135	0.139	0.142	0.144	0.146	0.147	0.151	0.151	0.151
2.75	0.130	0.133	0.138	0.142	0.144	0.146	0.147	0.152	0.152	0.153
3.00	0.127	0.131	0.137	0.141	0.144	0.145	0.147	0.152	0.153	0.154
3.25	0.125	0.129	0.135	0.140	0.143	0.145	0.147	0.153	0.154	0.154
3.50	0.122	0.126	0.133	0.138	0.142	0.144	0.146	0.153	0.155	0.155
3.75	0.119	0.124	0.131	0.137	0.141	0.143	0.145	0.154	0.155	0.155
4.00	0.116	0.121	0.129	0.135	0.139	0.142	0.145	0.154	0.155	0.156
4.25	0.113	0.119	0.127	0.133	0.138	0.141	0.144	0.154	0.156	0.156
4.50	0.110	0.116	0.125	0.131	0.136	0.140	0.143	0.154	0.156	0.156
4.75	0.107	0.113	0.123	0.130	0.135	0.139	0.142	0.154	0.156	0.157
5.00	0.105	0.111	0.120	0.128	0.133	0.137	0.140	0.154	0.156	0.157
5.25	0.102	0.108	0.118	0.126	0.131	0.136	0.139	0.154	0.156	0.157
5.50	0.099	0.106	0.116	0.124	0.130	0.134	0.138	0.154	0.156	0.157
5.75	0.097	0.103	0.113	0.122	0.128	0.133	0.136	0.154	0.157	0.157
6.00	0.094	0.101	0.111	0.120	0.126	0.131	0.135	0.153	0.157	0.157
6.25	0.092	0.098	0.109	0.118	0.124	0.129	0.134	0.153	0.157	0.158
6.50	0.090	0.096	0.107	0.116	0.122	0.128	0.132	0.153	0.157	0.158
6.75	0.087	0.094	0.105	0.114	0.121	0.126	0.131	0.153	0.157	0.158
7.00	0.085	0.092	0.103	0.112	0.119	0.125	0.129	0.152	0.157	0.158
7.25	0.083	0.090	0.101	0.110	0.117	0.123	0.128	0.152	0.157	0.158
7.50	0.081	0.088	0.099	0.108	0.115	0.121	0.126	0.152	0.156	0.158
7.75	0.079	0.086	0.097	0.106	0.114	0.120	0.125	0.151	0.156	0.158
8.00	0.077	0.084	0.095	0.104	0.112	0.118	0.124	0.151	0.156	0.158
8.25	0.076	0.082	0.093	0.102	0.110	0.117	0.122	0.150	0.156	0.158
8.50	0.074	0.080	0.091	0.101	0.108	0.115	0.121	0.150	0.156	0.158
8.75	0.072	0.078	0.089	0.099	0.107	0.114	0.119	0.150	0.156	0.158
9.00	0.071	0.077	0.088	0.097	0.105	0.112	0.118	0.149	0.156	0.158
9.25	0.069	0.075	0.086	0.096	0.104	0.110	0.116	0.149	0.156	0.158
9.50	0.068	0.074	0.085	0.094	0.102	0.109	0.115	0.148	0.156	0.158
9.75	0.066	0.072	0.083	0.092	0.100	0.107	0.113	0.148	0.156	0.158
10.00	0.065	0.071	0.082	0.091	0.099	0.106	0.112	0.147	0.156	0.158
20.00	0.035	0.039	0.046	0.053	0.059	0.065	0.071	0.124	0.148	0.156
50.00	0.014	0.016	0.019	0.022	0.025	0.028	0.031	0.071	0.113	0.142
100.00	0.007	0.008	0.010	0.011	0.013	0.014	0.016	0.039	0.071	0.113

Table 11.3 Variation of I_f with L/B and D_f/B

L/B	D_f/B	I_f: $\mu_s = 0.3$	$\mu_s = 0.4$	$\mu_s = 0.5$
1	0.5	0.77	0.82	0.85
	0.75	0.69	0.74	0.77
	1	0.65	0.69	0.72
2	0.5	0.82	0.86	0.89
	0.75	0.75	0.79	0.83
	1	0.71	0.75	0.79
5	0.5	0.87	0.91	0.93
	0.75	0.81	0.86	0.89
	1	0.78	0.82	0.85

Table 11.4 Representative Values of the Modulus of Elasticity of Soil

Soil type	E_s kN/m²
Soft clay	1800–3500
Hard clay	6000–14,000
Loose sand	10,000–28,000
Dense sand	35,000–70,000

Table 11.5 Representative Values of Poisson's Ratio

Type of soil	Poisson's ratio, μ_s
Loose sand	0.2–0.4
Medium sand	0.25–0.4
Dense sand	0.3–0.45
Silty sand	0.2–0.4
Soft clay	0.15–0.25
Medium clay	0.2–0.5

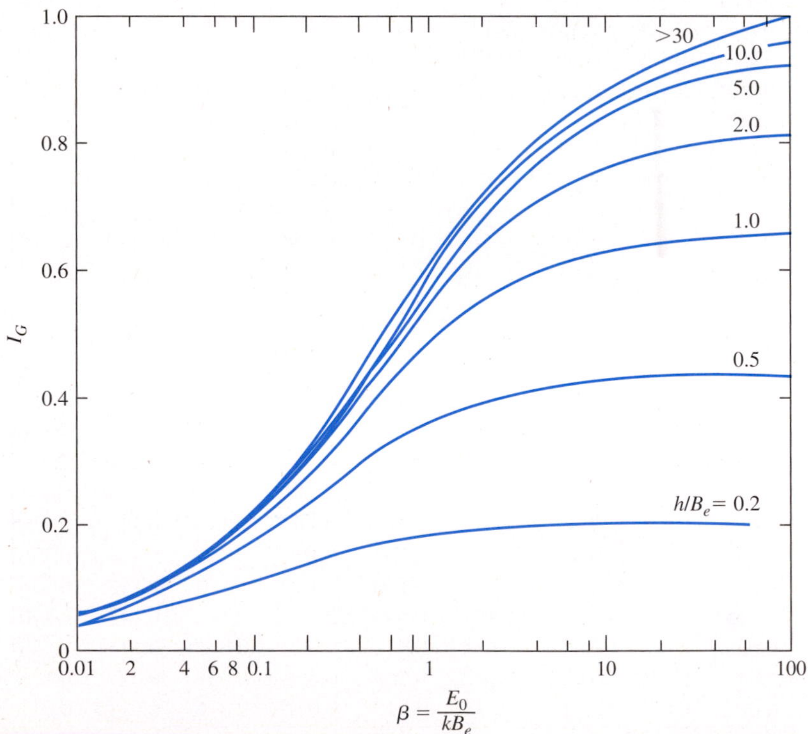

Figure 11.6 Variation of I_G with β

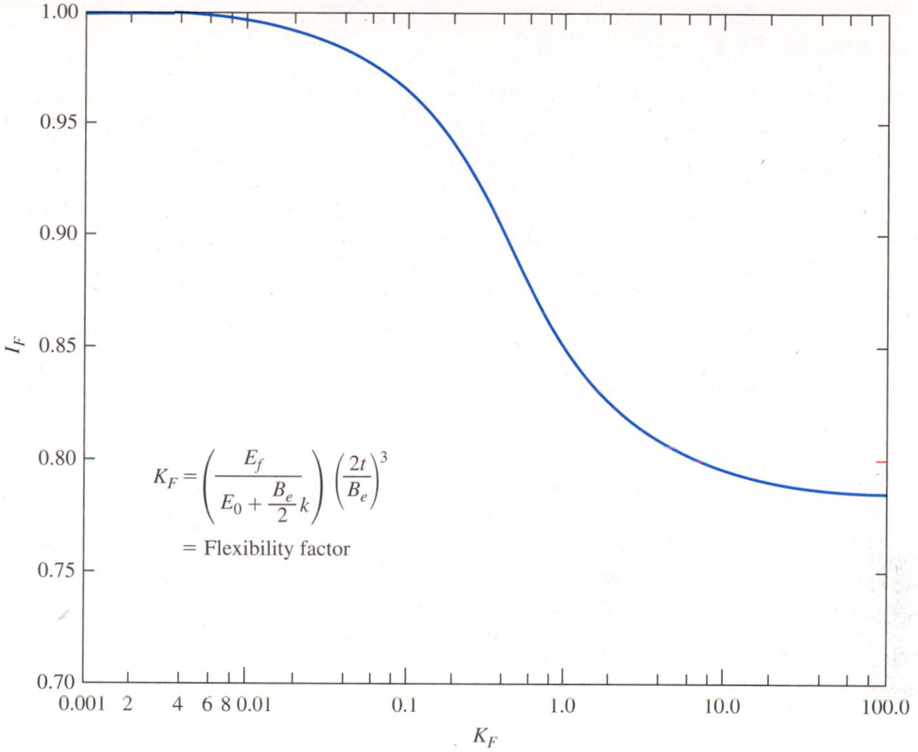

Figure 11.7 Variation of rigidity correction factor, I_F, with flexibility factor, K_F. [Eq. (11.15)]

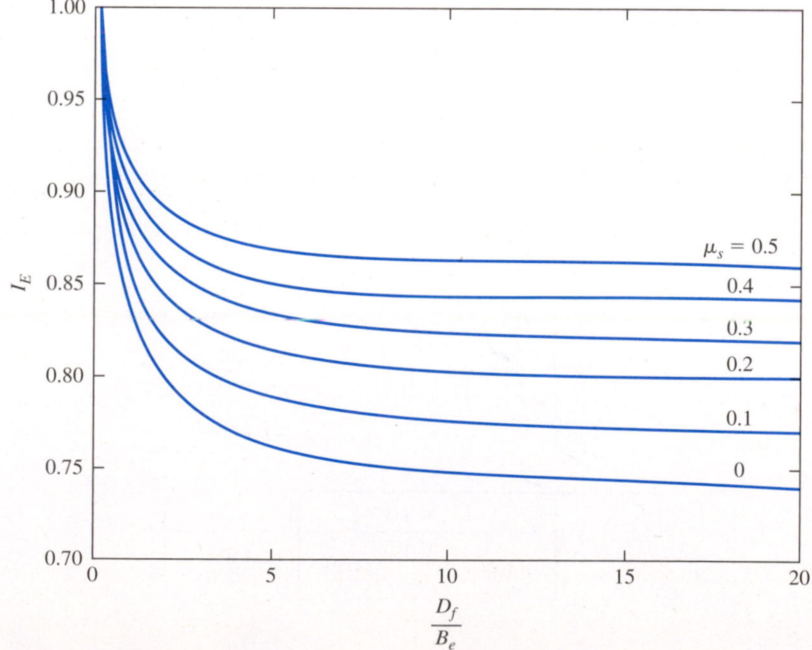

Figure 11.8 Variation of embedment correction factor, I_E [Eq. (11.16)]

Example 11.2

Refer to Figure 11.5. For a shallow foundation supported by a silty clay, the following are given:

Length, $L = 1.5$ m
Width, $B = 1$ m
Depth of foundation, $D_f = 1$ m
Thickness of foundation, $t = 0.23$ m
Load per unit area, $\Delta\sigma = 190$ kN/m^2
$E_f = 15 \times 10^6$ kN/m^2

The silty clay soil had the following properties:

$h = 2$ m
$\mu_s = 0.3$
$E_o = 9000$ kN/m^2
$k = 500$ kN/m^2/m

Estimate the elastic settlement of the foundation.

Solution

From Eq. (11.11), the equivalent diameter is

$$B_e = \sqrt{\frac{4BL}{\pi}} = \sqrt{\frac{(4)(1.5)(1)}{\pi}} = 1.38 \text{ m}$$

$$\Delta\sigma = 190 \text{ kN/m}^2$$

$$\beta = \frac{E_o}{kB_e} = \frac{9000}{(500)(1.38)} = 13.04$$

$$\frac{h}{B_e} = \frac{2}{1.38} = 1.45$$

From Figure 11.6, for $\beta = 13.04$ and $h/B_e = 1.45$, the value of $I_G \approx 0.74$. Thus, from Eq. (11.15),

$$I_F = \frac{\pi}{4} + \frac{1}{4.6 + 10\left(\dfrac{E_f}{E_o + \dfrac{B_e}{2}k}\right)\left(\dfrac{2t}{B_e}\right)^3}$$

$$= \frac{\pi}{4} + \frac{1}{4.6 + 10\left[\dfrac{15 \times 10^6}{9000 + \left(\dfrac{1.38}{2}\right)(500)}\right]\left[\dfrac{(2)(0.23)}{1.38}\right]^3} = 0.787$$

From Eq. (11.16),

$$I_E = 1 - \frac{1}{3.5\exp(1.22\mu_s - 0.4)\left(\dfrac{B_e}{D_f} + 1.6\right)}$$

$$= 1 - \frac{1}{3.5\exp[(1.22)(0.3) - 0.4]\left(\dfrac{1.38}{1} + 1.6\right)} = 0.907$$

From Eq. (11.14),

$$S_e = \frac{\Delta\sigma B_e I_G I_F I_E}{E_o}(1 - \mu_s^2) = \frac{(190)(1.38)(0.74)(0.787)(0.907)}{9000}(1 - 0.3^2)$$

$$= 0.014 \text{ m} \approx \mathbf{14\ mm}$$

CONSOLIDATION SETTLEMENT

11.5 Fundamentals of Consolidation

When a saturated soil layer is subjected to a stress increase, the pore water pressure is increased suddenly. In sandy soils that are highly permeable, the drainage caused by the increase in the pore water pressure is completed immediately. Pore water drainage is accompanied by a reduction in the volume of the soil mass, which results in settlement. Because of rapid drainage of the pore water in sandy soils, elastic settlement and consolidation occur simultaneously.

When a saturated compressible clay layer is subjected to a stress increase, elastic settlement occurs immediately. Because the hydraulic conductivity of clay is significantly smaller than that of sand, the excess pore water pressure generated by loading gradually dissipates over a long period. Thus, the associated volume change (that is, the consolidation) in the clay may continue long after the elastic settlement. The settlement caused by consolidation in clay may be several times greater than the elastic settlement.

The time-dependent deformation of saturated clayey soil can be best understood by considering a simple model that consists of a cylinder with a spring at its center. Let the inside area of the cross section of the cylinder be equal to A. The cylinder is filled with water and has a frictionless watertight piston and valve as shown in Figure 11.9a. At this time, if we place a load P on the piston (Figure 11.9b) and keep the valve closed, the entire load will be taken by the water in the cylinder because water is *incompressible*. The spring will not go through any deformation. The excess hydrostatic pressure at this time can be given as

$$\Delta u = \frac{P}{A} \tag{11.17}$$

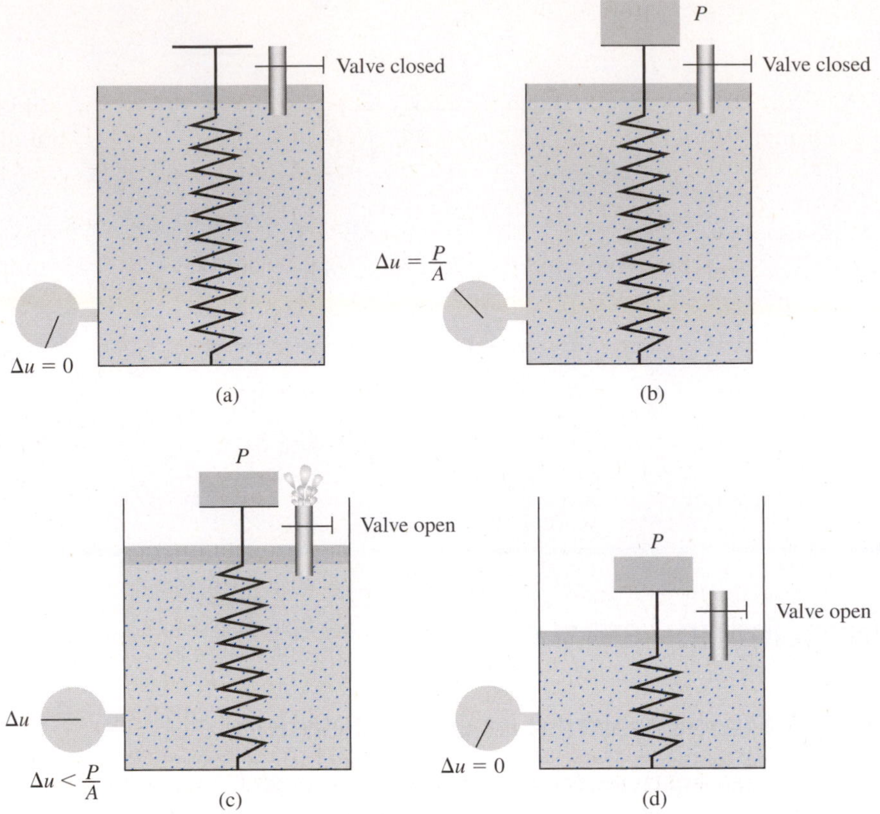

Figure 11.9 Spring-cylinder model for consolidation in saturated clay

This value can be observed in the pressure gauge attached to the cylinder.

In general, we can write

$$P = P_s + P_w \tag{11.18}$$

where P_s = load carried by the spring and P_w = load carried by the water.

From the preceding discussion, we can see that when the valve is closed after the placement of the load P,

$$P_s = 0 \quad \text{and} \quad P_w = P$$

Now, if the valve is opened, the water will flow outward (Figure 11.9c). This flow will be accompanied by a reduction of the excess hydrostatic pressure and an increase in the compression of the spring. So, at this time, Eq. (11.18) will hold. However,

$$P_s > 0 \quad \text{and} \quad P_w < P \quad (\text{that is, } \Delta u < P/A)$$

After some time, the excess hydrostatic pressure will become zero and the system will reach a state of equilibrium, as shown in Figure 11.9d. Now we can write

$$P_s = P \quad \text{and} \quad P_w = 0$$

and

$$P = P_s + P_w$$

With this in mind, we can analyze the strain of a saturated clay layer subjected to a stress increase (Figure 11.10a). Consider the case where a layer of saturated clay of thickness H that is confined between two layers of sand is being subjected to an instantaneous increase of *total stress* of $\Delta\sigma$.

As soon as $\Delta\sigma$ is applied on the ground surface, the level of water in the standpipes will rise. The curve that represents the locus of the water level in the standpipes at any given time represents an *isocrone*.

- At time $t = 0$ (Isocrone I_1)

  $$\Delta h = \Delta h_1 (\text{for } z = 0 \text{ to } z = H)$$

  At this time, the increase in pore water pressure from $z = 0$ to $z = H$ is (due to low hydraulic conductivity of clay)

  $$\Delta u = (\Delta h_1)(\gamma_w) = \Delta\sigma$$

  where γ_w = unit weight of water.
  From the principle of effective stress,

  $$\Delta\sigma = \Delta\sigma' + \Delta u \tag{11.19}$$

  where $\Delta\sigma'$ = increase in effective stress.
  Hence, at $t = 0$ ($z = 0$ to $z = H$)
  $\Delta u = \Delta\sigma$ (i.e., the entire incremental stress is carried by water)
  and

  $$\Delta\sigma' = 0$$

  This is similar to what is shown in Figure 11.9b. The variation of $\Delta\sigma$, Δu, and $\Delta\sigma'$ for $z = 0$ to $z = H$ is shown in Figure 11.10b.

- At time $t > 0$ (Isocrone I_2)

  The water in the void spaces will start to be squeezed out and will drain in both directions into the sand layer. By this process, the excess pore water pressure at any depth z will gradually decrease. Isocrone I_2 shows the variation of Δh in standpipes,
  or

  $$\Delta h = \Delta h_2 = f(z)$$

  Hence, the pore water pressure increase

  $$\Delta u = (\Delta h_2)(\gamma_w) < \Delta\sigma$$

  and

  $$\Delta\sigma' = \Delta\sigma - \Delta u$$

  This is similar to the situation shown in Figure 11.9c. The variation of $\Delta\sigma$, Δu, and $\Delta\sigma'$ at time $t > 0$ is shown in Figure 11.10c.

- At time $t = \infty$ (Isocrone I_3)

Theoretically, at time $t = \infty$ the entire pore water pressure would be dissipated by drainage from all points of the clay layer. This is shown by Isocrone I_3, or

$$\Delta h = \Delta h_3 = 0 \text{ (for } z = 0 \text{ to } z = H)$$

Thus

$$\Delta u = 0$$

and

$$\Delta\sigma' = \Delta\sigma$$

The total stress increase $\Delta\sigma$ is now carried by the soil structure. The variation of $\Delta\sigma$, Δu, and $\Delta\sigma'$ is shown in Figure 11.10d. This is similar to the case shown in Figure 11.9d.

This gradual process of drainage under an additional load application and the associated transfer of excess pore water pressure to effective stress cause the time-dependent settlement in the clay soil layer. This is called consolidation.

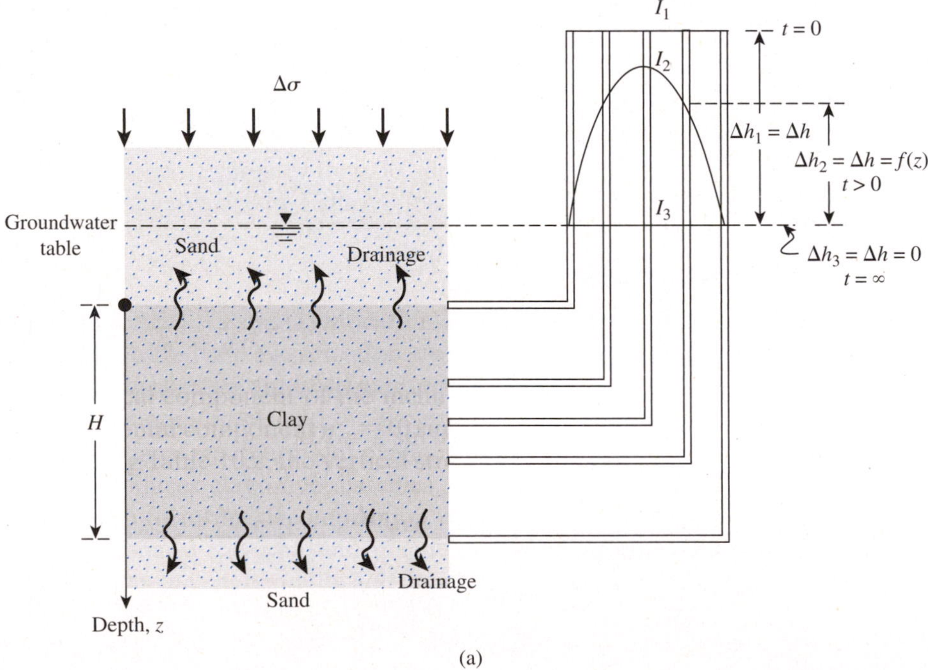

(a)

Figure 11.10 Variation of total stress, pore water pressure, and effective stress in a clay layer drained at top and bottom as the result of an added stress, $\Delta\sigma$ *(Continued)*

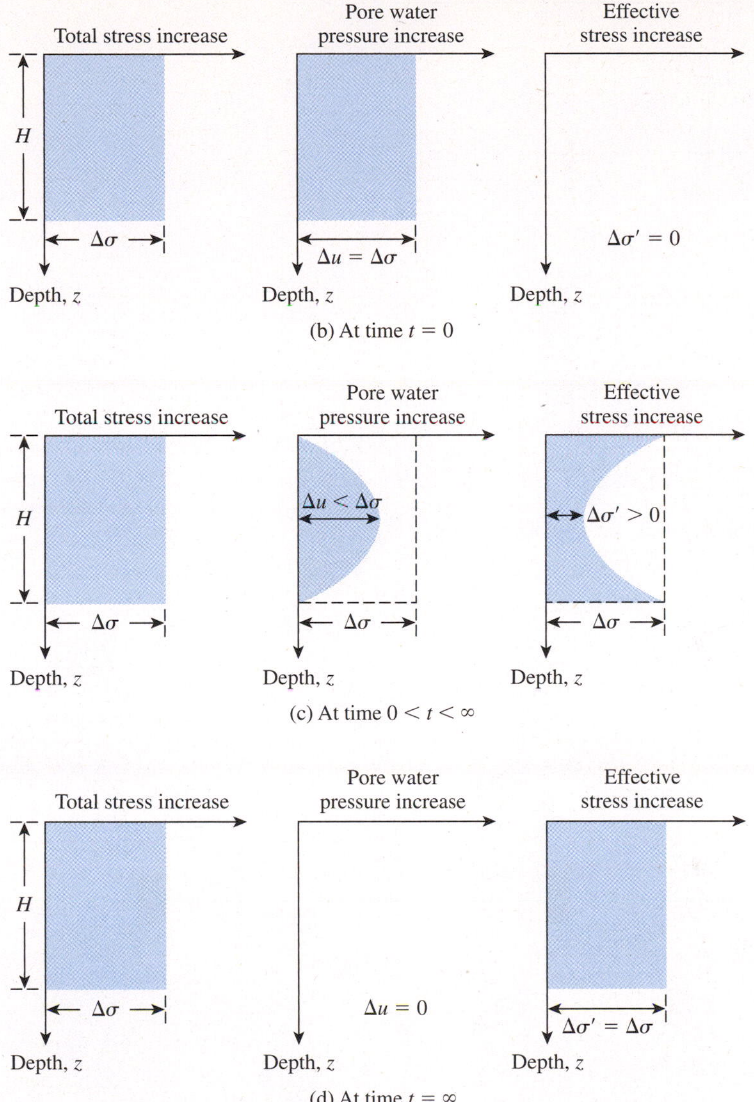

Figure 11.10 *(Continued)*

11.6 One-Dimensional Laboratory Consolidation Test

The one-dimensional consolidation testing procedure was first suggested by Terzaghi. This test is performed in a *consolidometer* (sometimes referred to as an *oedometer*). The schematic diagram of a consolidometer is shown in Figure 11.11a. Figure 11.11b shows a photograph of a consolidometer. The soil specimen is placed

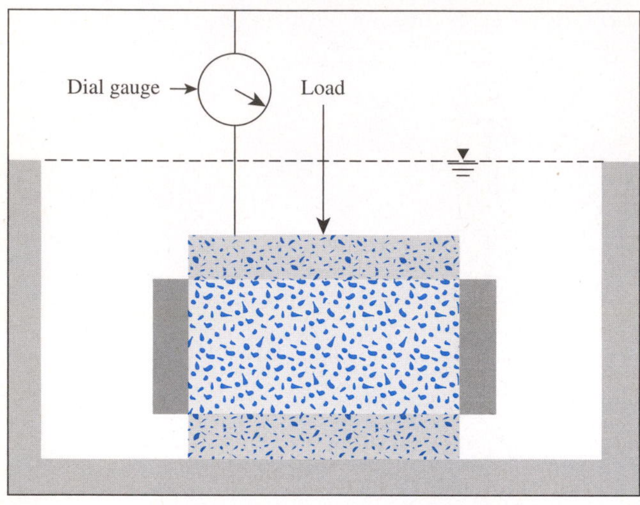

(a)

(b)

(c)

Figure 11.11 (a) Schematic diagram of a consolidometer; (b) photograph of a consolidometer; (c) a consolidation test in progress (right-hand side) (*Courtesy of Braja M. Das, Henderson, Nevada*)

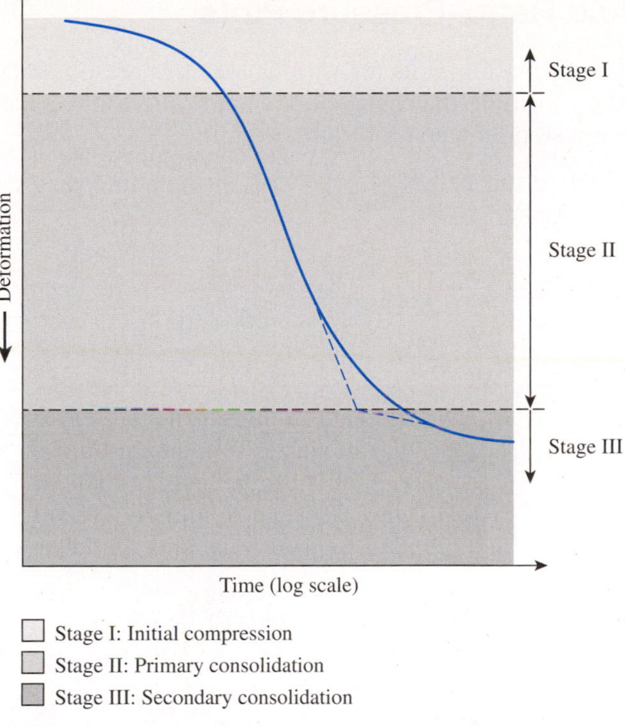

Figure 11.12 Time–deformation plot during consolidation for a given load increment

inside a metal ring with two porous stones, one at the top of the specimen and another at the bottom. The specimens are usually 64 mm ($\approx$ 2.5 in.) in diameter and 25 mm ($\approx$ 1 in.) thick. The load on the specimen is applied through a lever arm, and compression is measured by a micrometer dial gauge. The specimen is kept under water during the test. Each load usually is kept for 24 hours. After that, the load usually is doubled, which doubles the pressure on the specimen, and the compression measurement is continued. At the end of the test, the dry weight of the test specimen is determined. Figure 11.11c shows a consolidation test in progress (right-hand side).

The general shape of the plot of deformation of the specimen against time for a given load increment is shown in Figure 11.12. From the plot, we can observe three distinct stages, which may be described as follows:

Stage 1. Initial compression, which is caused mostly by preloading

Stage 2. Primary consolidation, during which excess pore water pressure gradually is transferred into effective stress because of the expulsion of pore water

Stage 3. Secondary consolidation, which occurs after complete dissipation of the excess pore water pressure, when some deformation of the specimen takes place because of the plastic readjustment of soil fabric

11.7 Void Ratio–Pressure Plots

After the time–deformation plots for various loadings are obtained in the laboratory, it is necessary to study the change in the void ratio of the specimen with pressure. Following is a step-by-step procedure for doing so:

Step 1. Calculate the height of solids, H_s, in the soil specimen (Figure 11.13) using the equation

$$H_s = \frac{W_s}{AG_s\gamma_w} = \frac{M_s}{AG_s\rho_w} \tag{11.20}$$

where W_s = dry weight of the specimen
M_s = dry mass of the specimen
A = area of the specimen
G_s = specific gravity of soil solids
γ_w = unit weight of water
ρ_w = density of water

Step 2. Calculate the initial height of voids as

$$H_v = H - H_s \tag{11.21}$$

where H = initial height of the specimen.

Step 3. Calculate the initial void ratio, e_o, of the specimen, using the equation

$$e_o = \frac{V_v}{V_s} = \frac{H_v\,A}{H_s\,A} = \frac{H_v}{H_s} \tag{11.22}$$

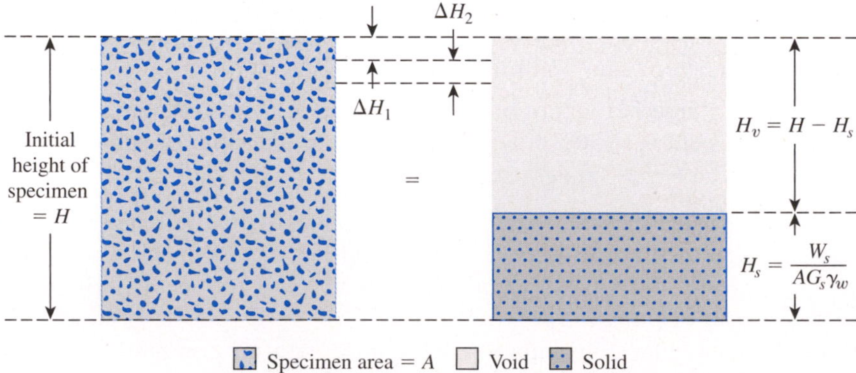

Figure 11.13 Change of height of specimen in one-dimensional consolidation test

Step 4. For the first incremental loading, σ_1 (total load/unit area of specimen), which causes a deformation ΔH_1, calculate the change in the void ratio as

$$\Delta e_1 = \frac{\Delta H_1}{H_s} \tag{11.23}$$

(ΔH_1 is obtained from the initial and the final dial readings for the loading).

It is important to note that, at the end of consolidation, total stress σ_1 is equal to effective stress σ'_1.

Step 5. Calculate the new void ratio after consolidation caused by the pressure increment as

$$e_1 = e_o - \Delta e_1 \tag{11.24}$$

For the next loading, σ_2 (*note:* σ_2 equals the cumulative load per unit area of specimen), which causes additional deformation ΔH_2, the void ratio at the end of consolidation can be calculated as

$$e_2 = e_1 - \frac{\Delta H_2}{H_s} \tag{11.25}$$

At this time, σ_2 = effective stress, σ'_2. Proceeding in a similar manner, one can obtain the void ratios at the end of the consolidation for all load increments.

The effective stress σ' and the corresponding void ratios (e) at the end of consolidation are plotted on semilogarithmic graph paper. The typical shape of such a plot is shown in Figure 11.14.

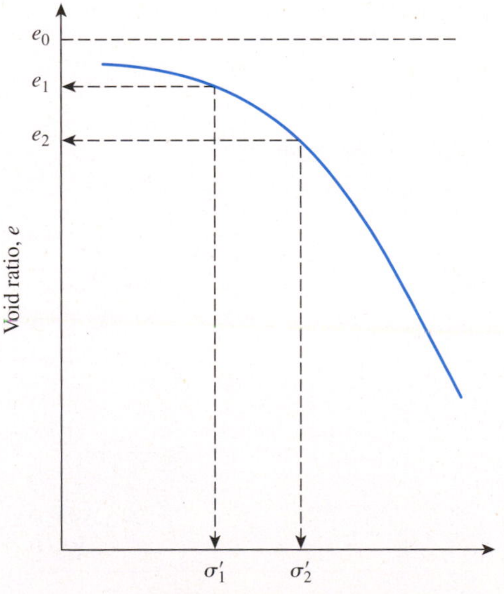

Figure 11.14 Typical plot of e against log σ'

Example 11.3

Following are the results of a laboratory consolidation test on a soil specimen obtained from the field: Dry mass of specimen = 128 g, height of specimen at the beginning of the test = 2.54 cm, G_s = 2.75, and area of the specimen = 30.68 cm^2.

Effective pressure, σ' (kN/m^2)	Final height of specimen at the end of consolidation (cm)
0	2.540
50	2.488
100	2.465
200	2.431
400	2.389
800	2.324
1600	2.225
3200	2.115

Make necessary calculations and draw an e versus log σ' curve.

Solution

From Eq. (11.20),

$$H_s = \frac{W_s}{AG_s\gamma_w} = \frac{M_s}{AG_s\rho_w} = \frac{128 \text{ g}}{(30.68 \text{ cm}^2)(2.75)(1\text{g/cm}^3)} = 1.52 \text{ cm}$$

Now the following table can be prepared.

Effective pressure, σ' (kN/m^2)	Height at the end of consolidation, H (cm)	$H_v = H - H_s$ (cm)	$e = H_v/H_s$
0	2.540	1.02	0.671
50	2.488	0.968	0.637
100	2.465	0.945	0.622
200	2.431	0.911	0.599
400	2.389	0.869	0.572
800	2.324	0.804	0.529
1600	2.225	0.705	0.464
3200	2.115	0.595	0.390

The e versus log σ' plot is shown in Figure 11.15.

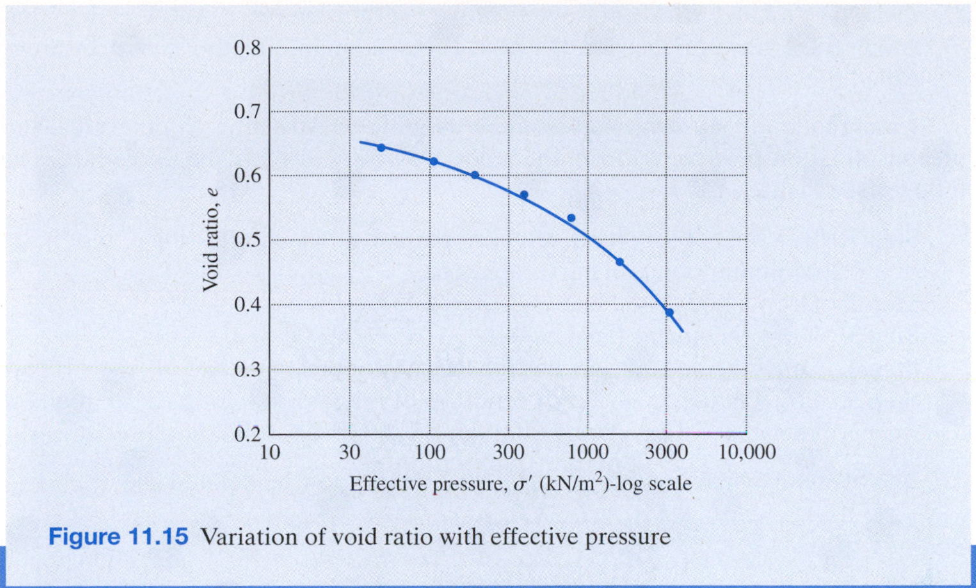

Figure 11.15 Variation of void ratio with effective pressure

11.8 Normally Consolidated and Overconsolidated Clays

Figure 11.14 shows that the upper part of the e-log σ' plot is somewhat curved with a flat slope, followed by a linear relationship for the void ratio with log σ' having a steeper slope. This phenomenon can be explained in the following manner:

A soil in the field at some depth has been subjected to a certain maximum effective past pressure in its geologic history. This maximum effective past pressure may be equal to or less than the existing effective overburden pressure at the time of sampling. The reduction of effective pressure in the field may be caused by natural geologic processes or human processes. During the soil sampling, the existing effective overburden pressure is also released, which results in some expansion. When this specimen is subjected to a consolidation test, a small amount of compression (that is, a small change in void ratio) will occur when the effective pressure applied is less than the maximum effective overburden pressure in the field to which the soil has been subjected in the past. When the effective pressure on the specimen becomes greater than the maximum effective past pressure, the change in the void ratio is much larger, and the e-log σ' relationship is practically linear with a steeper slope.

This relationship can be verified in the laboratory by loading the specimen to exceed the maximum effective overburden pressure, and then unloading and reloading again. The e-log σ' plot for such cases is shown in Figure 11.16, in which *cd* represents unloading and *dfg* represents the reloading process.

This leads us to the two basic definitions of clay based on stress history:

1. *Normally consolidated,* whose present effective overburden pressure is the maximum pressure that the soil was subjected to in the past.

2. *Overconsolidated,* whose present effective overburden pressure is less than that which the soil experienced in the past. The maximum effective past pressure is called the *preconsolidation pressure.*

Casagrande (1936) suggested a simple graphic construction to determine the preconsolidation pressure σ'_c from the laboratory e-log σ' plot. The procedure is as follows (see Figure 11.17):

Step 1. By visual observation, establish point a, at which the e-log σ' plot has a minimum radius of curvature.
Step 2. Draw a horizontal line ab.
Step 3. Draw the line ac tangent at a.
Step 4. Draw the line ad, which is the bisector of the angle bac.
Step 5. Project the straight-line portion gh of the e-log σ' plot back to intersect line ad at f. The abscissa of point f is the preconsolidation pressure, σ'_c.

The overconsolidation ratio (OCR) for a soil can now be defined as

$$OCR = \frac{\sigma'_c}{\sigma'} \tag{11.26}$$

where σ'_c = preconsolidation pressure of a specimen
σ' = present effective vertical pressure

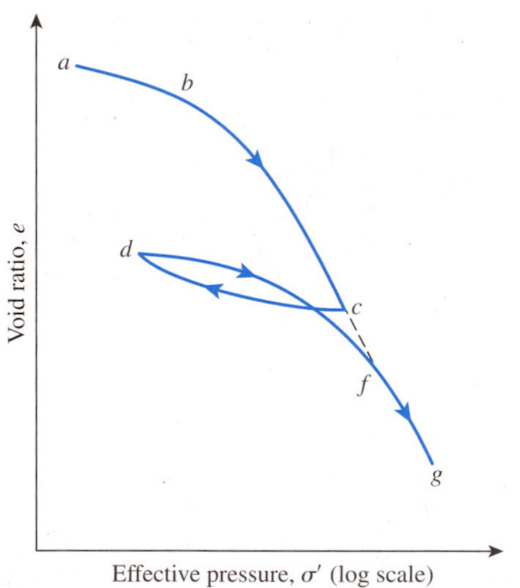

Figure 11.16 Plot of e against log σ' showing loading, unloading, and reloading branches

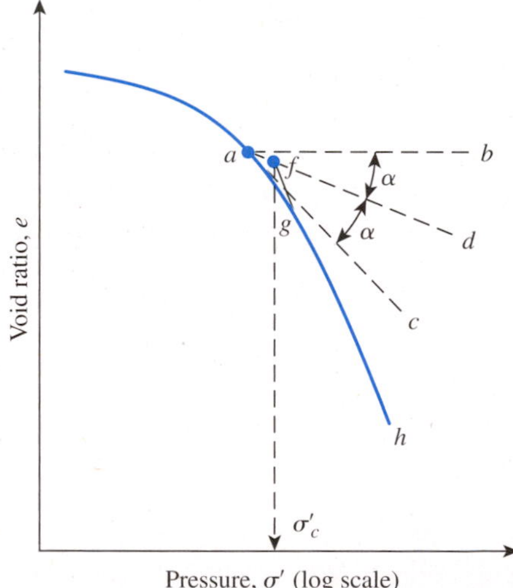

Figure 11.17 Graphic procedure for determining preconsolidation pressure

In the literature, some empirical relationships are available to predict the preconsolidation pressure. Some examples are given next.

- Stas and Kulhawy (1984):

$$\frac{\sigma'_c}{p_a} = 10^{[1.11 - 1.62(LI)]} \tag{11.27}$$

where p_a = atmospheric pressure ($\approx$100 kN/m^2)
LI = liquidity index

- Hansbo (1957)

$$\sigma'_c = \alpha_{(\text{VST})}\, c_{u(\text{VST})} \tag{11.28}$$

where $\alpha_{(\text{VST})}$ = an empirical coefficient $= \dfrac{222}{LL(\%)}$

$c_{u(\text{VST})}$ = undrained shear strength obtained from vane shear test (Chapter 12)

In any case, these above relationships may change from soil to soil. They may be taken as an initial approximation.

Example 11.4

Following are the results of a laboratory consolidation test.

Pressure, σ' (kN/m^2)	Void ratio, e
50	0.840
100	0.826
200	0.774
400	0.696
800	0.612
1000	0.528

Using Casagrande's procedure, determine the preconsolidation pressure σ'_c.

Solution

Figure 11.18 shows the e-log σ' plot. In this plot, a is the point where the radius of curvature is minimum. The preconsolidation pressure is determined using the procedure shown in Figure 11.17. From the plot, $\boldsymbol{\sigma'_c}$ **= 160 kN/m^2**.

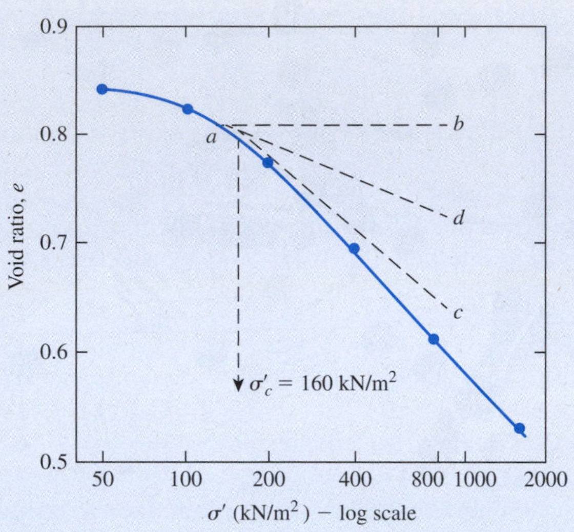

Figure 11.18

Example 11.5

A soil profile is shown in Figure 11.19. Using Eq. (11.27), estimate the preconsolidation pressure σ'_c and overconsolidation ratio OCR at point A.

Solution

Liquidity index:

$$LI = \frac{w - PL}{LL - PL} = \frac{32 - 23}{51 - 23} = 0.32$$

From Eq. (11.27),

$$\sigma'_c = (p_a)10^{[1.11-1.62(LI)]} = (100)10^{[1.11-(1.62)(0.32)]} = \mathbf{390.48\ kN/m^2}$$

At A, the present effective pressure is

$$\sigma' = (2.5)(15.6) + (5)(19 - 9.81) = 84.95\ \text{kN/m}^2$$

So,

$$OCR = \frac{390.48}{84.95} \approx \mathbf{4.6}$$

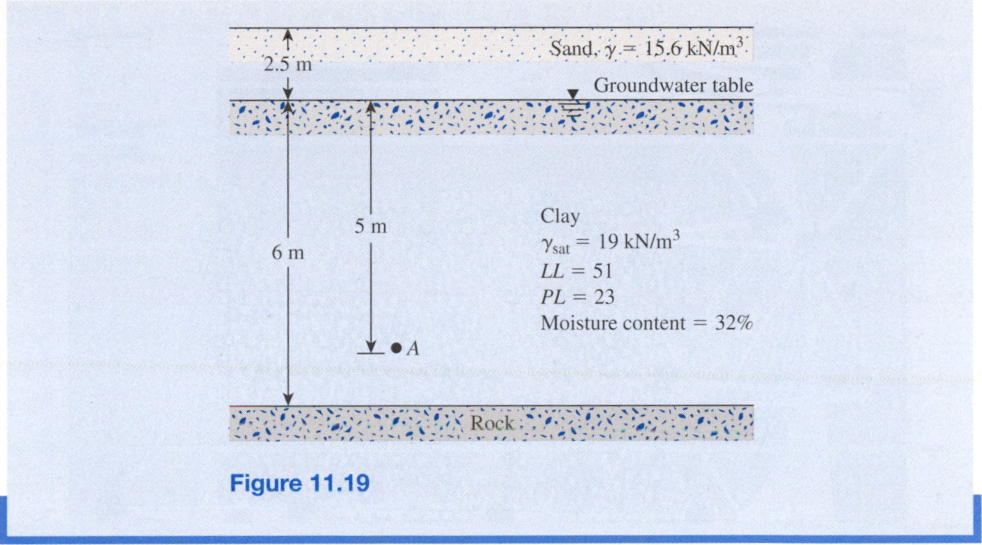

Figure 11.19

11.9 Effect of Disturbance on Void Ratio–Pressure Relationship

A soil specimen will be remolded when it is subjected to some degree of disturbance. This remolding will result in some deviation of the e-log σ' plot as observed in the laboratory from the actual behavior in the field. The field e-log σ' plot can be reconstructed from the laboratory test results in the manner described in this section (Terzaghi and Peck, 1967).

Normally consolidated clay of low to medium plasticity (Figure 11.20)

Step 1. In Figure 11.20, curve 2 is the laboratory e-log σ' plot. From this plot, determine the preconsolidation pressure, $\sigma'_c = \sigma'_o$ (that is, the present effective overburden pressure). Knowing where $\sigma'_c = \sigma'_o$, draw vertical line ab.

Step 2. Calculate the void ratio in the field, e_o. Draw horizontal line cd.

Step 3. Calculate $0.4e_o$ and draw line ef. (*Note:* f is the point of intersection of the line with curve 2.)

Step 4. Join points f and g. Note that g is the point of intersection of lines ab and cd. This is the *virgin compression curve.*

It is important to point out that if a soil is remolded completely, the general position of the e-log σ' plot will be as represented by curve 3.

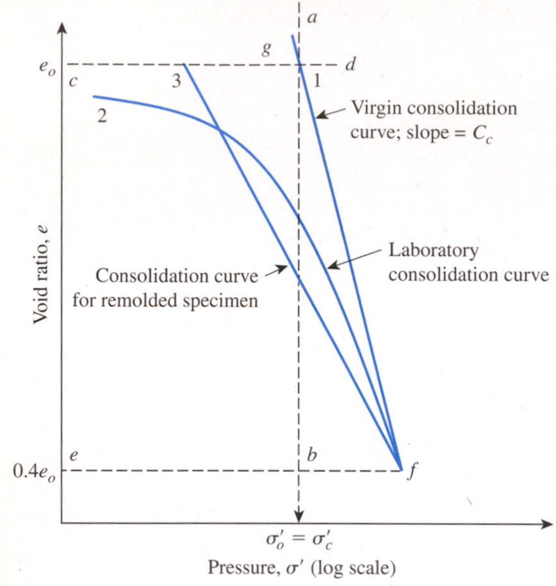

Figure 11.20 Consolidation characteristics of normally consolidated clay of low to medium sensitivity

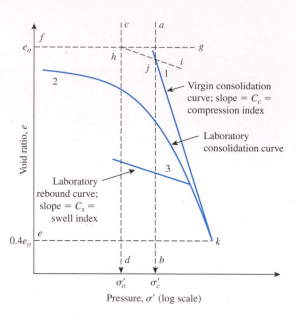

Figure 11.21 Consolidation characteristics of overconsolidated clay of low to medium sensitivity

Overconsolidated clay of low to medium plasticity (Figure 11.21)

Step 1. In Figure 11.21, curve 2 is the laboratory e-log σ' plot (loading), and curve 3 is the laboratory unloading, or rebound, curve. From curve 2, determine the preconsolidation pressure σ'_c. Draw the vertical line *ab*.

Step 2. Determine the field effective overburden pressure σ'_o. Draw vertical line *cd*.

Step 3. Determine the void ratio in the field, e_o. Draw the horizontal line *fg*. The point of intersection of lines *fg* and *cd* is *h*.

Step 4. Draw a line *hi*, which is parallel to curve 3 (which is practically a straight line). The point of intersection of lines *hi* and *ab* is *j*.

Step 5. Join points *j* and *k*. Point *k* is on curve 2, and its ordinate is $0.4e_o$.

The field consolidation plot will take a path *hjk*. The recompression path in the field is *hj* and is parallel to the laboratory rebound curve (Schmertmann, 1953).

11.10 Calculation of Settlement from One-Dimensional Primary Consolidation

With the knowledge gained from the analysis of consolidation test results, we can now proceed to calculate the probable settlement caused by primary consolidation in the field, assuming one-dimensional consolidation.

Let us consider a saturated clay layer of thickness H and cross-sectional area A under an existing average effective overburden pressure, σ_o'. Because of an increase of effective pressure, $\Delta\sigma'$, let the primary settlement be S_c. Thus, the change in volume (Figure 11.22) can be given by

$$\Delta V = V_0 - V_1 = HA - (H - S_c)A = S_c A \tag{11.29}$$

where V_0 and V_1 are the initial and final volumes, respectively. However, the change in the total volume is equal to the change in the volume of voids, ΔV_v. Hence,

$$\Delta V = S_c A = V_{v_0} - V_{v_1} = \Delta V_v \tag{11.30}$$

where V_{v_0} and V_{v_1} are the initial and final void volumes, respectively. From the definition of void ratio, it follows that

$$\Delta V_v = \Delta e V_s \tag{11.31}$$

where Δe = change of void ratio. But

$$V_s = \frac{V_0}{1 + e_o} = \frac{AH}{1 + e_o} \tag{11.32}$$

where e_o = initial void ratio at volume V_0. Thus, from Eqs. (11.29) through (11.32),

$$\Delta V = S_c A = \Delta e V_s = \frac{AH}{1 + e_o}\Delta e$$

or

$$S_c = H\frac{\Delta e}{1 + e_o} \tag{11.33}$$

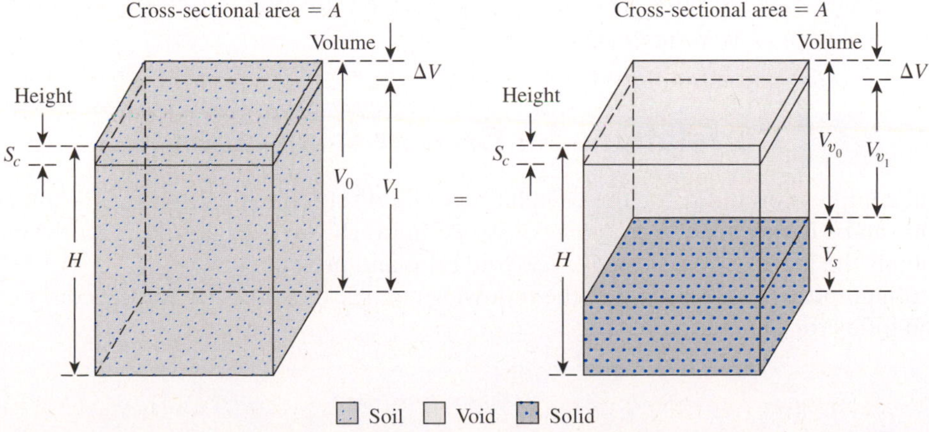

Figure 11.22 Settlement caused by one-dimensional consolidation

For normally consolidated clays that exhibit a linear e-log σ' relationship (see Figure 11.20),

$$\Delta e = C_c[\log(\sigma'_o + \Delta\sigma') - \log \sigma'_o] \tag{11.34}$$

where C_c = slope of the e-log σ' plot and is defined as the compression index. Substitution of Eq. (11.34) into Eq. (11.33) gives

$$S_c = \frac{C_c H}{1 + e_o} \log\left(\frac{\sigma'_o + \Delta\sigma'}{\sigma'_o}\right) \tag{11.35}$$

In overconsolidated clays (see Figure 11.21), for $\sigma'_o + \Delta\sigma' \leq \sigma'_c$, field e-log σ' variation will be along the line hj, the slope of which will be approximately equal to that for the laboratory rebound curve. The slope of the rebound curve C_s is referred to as the *swell index;* so

$$\Delta e = C_s[\log(\sigma'_o + \Delta\sigma') - \log \sigma'_o] \tag{11.36}$$

From Eqs. (11.33) and (11.36), we obtain

$$S_c = \frac{C_s H}{1 + e_o} \log\left(\frac{\sigma'_o + \Delta\sigma'}{\sigma'_o}\right) \tag{11.37}$$

If $\sigma'_o + \Delta\sigma' > \sigma'_c$, then

$$S_c = \frac{C_s H}{1 + e_o} \log\frac{\sigma'_c}{\sigma'_o} + \frac{C_c H}{1 + e_o} \log\left(\frac{\sigma'_o + \Delta\sigma'}{\sigma'_c}\right) \tag{11.38}$$

However, if the e-log σ' curve is given, one can simply pick Δe off the plot for the appropriate range of pressures. This number may be substituted into Eq. (11.33) for the calculation of settlement, S_c.

11.11 Correlations for Compression Index (C_c)

The compression index for the calculation of field settlement caused by consolidation can be determined by graphic construction (as shown in Figure 11.20) after one obtains the laboratory test results for void ratio and pressure.

Skempton (1944) suggested the following empirical expression for the compression index for undisturbed clays:

$$C_c = 0.009(LL - 10) \tag{11.39}$$

where LL = liquid limit.

Several other correlations for the compression index are also available. They have been developed by tests on various clays. Some of these correlations are given in Table 11.6.

On the basis of observations on several natural clays, Rendon-Herrero (1983) gave the relationship for the compression index in the form

$$C_c = 0.141 G_s^{1.2}\left(\frac{1 + e_o}{G_s}\right)^{2.38} \tag{11.40}$$

Nagaraj and Murty (1985) expressed the compression index as

$$C_c = 0.2343\left[\frac{LL(\%)}{100}\right]G_s \tag{11.41}$$

Based on the modified Cam clay model, Wroth and Wood (1978) have shown that

$$C_c \approx 0.5 G_s \frac{[PI(\%)]}{100} \tag{11.42}$$

where PI = plasticity index.

If an average value of G_s is taken to be about 2.7 (Kulhawy and Mayne, 1990)

$$C_c \approx \frac{PI}{74} \tag{11.43}$$

More recently, Park and Koumoto (2004) expressed the compression index by the following relationship:

$$C_c = \frac{n_o}{371.747 - 4.275 n_o} \tag{11.44}$$

where n_o = *in situ* porosity of the soil(%).

Table 11.6 Correlations for Compression Index, C_c*

Equation	Reference	Region of applicability
$C_c = 0.007(LL - 7)$	Skempton (1944)	Remolded clays
$C_c = 0.01 w_N$		Chicago clays
$C_c = 1.15(e_o - 0.27)$	Nishida (1956)	All clays
$C_c = 0.30(e_o - 0.27)$	Hough (1957)	Inorganic cohesive soil: silt, silty clay, clay
$C_c = 0.0115 w_N$		Organic soils, peats, organic silt, and clay
$C_c = 0.0046(LL - 9)$		Brazilian clays
$C_c = 0.75(e_o - 0.5)$		Soils with low plasticity
$C_c = 0.208 e_o + 0.0083$		Chicago clays
$C_c = 0.156 e_o + 0.0107$		All clays

*After Rendon-Herrero, 1980. With permission from ASCE.
Note: e_o = in situ void ratio; w_N = in situ water content.

11.12 Correlations for Swell Index (C_s)

The swell index is appreciably smaller in magnitude than the compression index and generally can be determined from laboratory tests. In most cases,

$$C_s \simeq \frac{1}{5} \text{ to } \frac{1}{10} C_c$$

The swell index was expressed by Nagaraj and Murty (1985) as

$$C_s = 0.0463\left[\frac{LL(\%)}{100}\right]G_s \tag{11.45}$$

Based on the modified Cam clay model, Kulhawy and Mayne (1990) have shown that

$$C_s \approx \frac{PI}{370} \tag{11.46}$$

Example 11.6

The following are the results of a laboratory consolidation test:

Pressure, σ' (kN/m^2)	Void ratio, e	Remarks	Pressure, σ' (kN/m^2)	Void ratio, e	Remarks
25	0.93	Loading	800	0.61	Loading
50	0.92		1600	0.52	
100	0.88		800	0.535	Unloading
200	0.81		400	0.555	
400	0.69		200	0.57	

a. Calculate the compression index and the ratio of C_s/C_c.
b. On the basis of the average e-log σ' plot, calculate the void ratio at $\sigma'_o = 1000$ kN/m^2.

Solution

Part a

The e versus log σ' plot is shown in Figure 11.23. From the average e-log σ' plot, for the loading and unloading branches, the following values can be determined:

Branch	e	σ'_o (kN/m^2)
Loading	0.8	200
	0.7	400
Unloading	0.57	200
	0.555	400

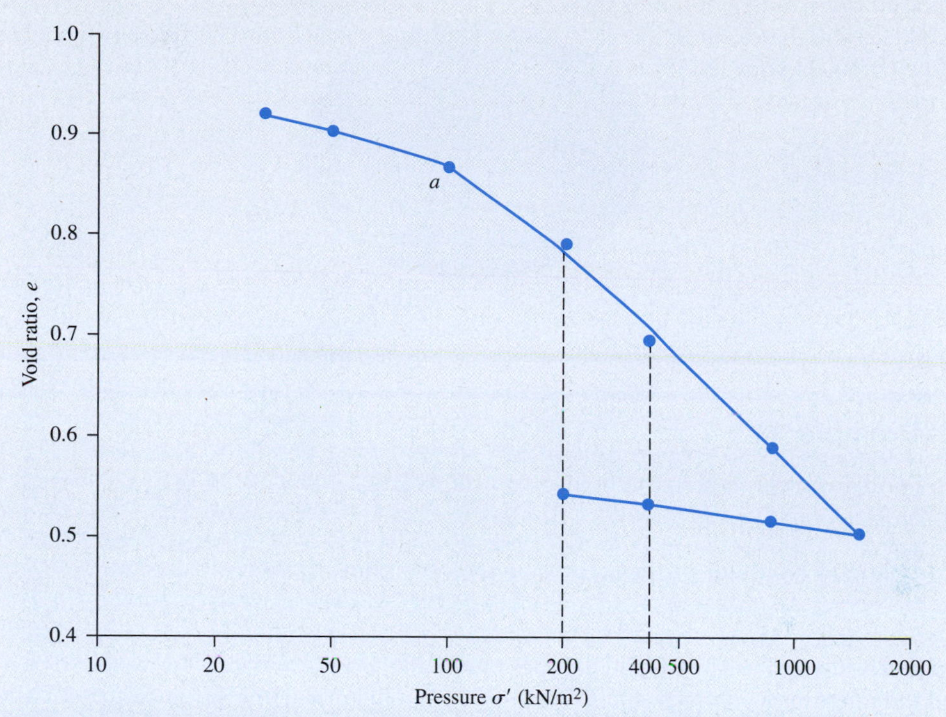

Figure 11.23 Plot of e versus log σ'

From the loading branch,

$$C_c = \frac{e_1 - e_2}{\log \dfrac{\sigma'_2}{\sigma'_1}} = \frac{0.8 - 0.7}{\log\left(\dfrac{400}{200}\right)} = \mathbf{0.33}$$

From the unloading branch,

$$C_s = \frac{e_1 - e_2}{\log \dfrac{\sigma'_2}{\sigma'_1}} = \frac{0.57 - 0.555}{\log\left(\dfrac{400}{200}\right)} = 0.0498 \approx 0.05$$

$$\frac{C_s}{C_c} = \frac{0.05}{0.33} = \mathbf{0.15}$$

Part b

$$C_c = \frac{e_1 - e_3}{\log \dfrac{\sigma'_3}{\sigma'_1}}$$

We know that $e_1 = 0.8$ at $\sigma'_1 = 200$ kN/m² and that $C_c = 0.33$ [part (a)]. Let $\sigma'_3 = 1000$ kN/m². So,

$$0.33 = \frac{0.8 - e_3}{\log\left(\frac{1000}{200}\right)}$$

$$e_3 = 0.8 - 0.33 \log\left(\frac{1000}{200}\right) \approx \mathbf{0.57}$$

Example 11.7

For a given clay soil in the field, given: $G_s = 2.68, e_o = 0.75$. Estimate C_c based on Eqs. (11.40) and (11.44).

Solution

From Eq. (11.40),

$$C_c = 0.141 G_s^{1.2}\left(\frac{1 + e_o}{G_s}\right)^{2.38} = (0.141)(2.68)^{1.2}\left(\frac{1 + 0.75}{2.68}\right)^{2.38} \approx \mathbf{0.167}$$

From Eq. (11.44),

$$C_c = \frac{n_o}{371.747 - 4.275 n_o}$$

$$n_o = \frac{e_o}{1 + e_o} = \frac{0.75}{1 + 0.75} = 0.429$$

$$C_c = \frac{(0.429)(100)}{371.747 - (4.275)(0.429 \times 100)} = \mathbf{0.228}$$

Note: It is important to know that the empirical correlations are approximations only and may deviate from one soil to another.

Example 11.8

A soil profile is shown in Figure 11.24. If a uniformly distributed load, $\Delta\sigma$, is applied at the ground surface, what is the settlement of the clay layer caused by primary consolidation if

a. The clay is normally consolidated

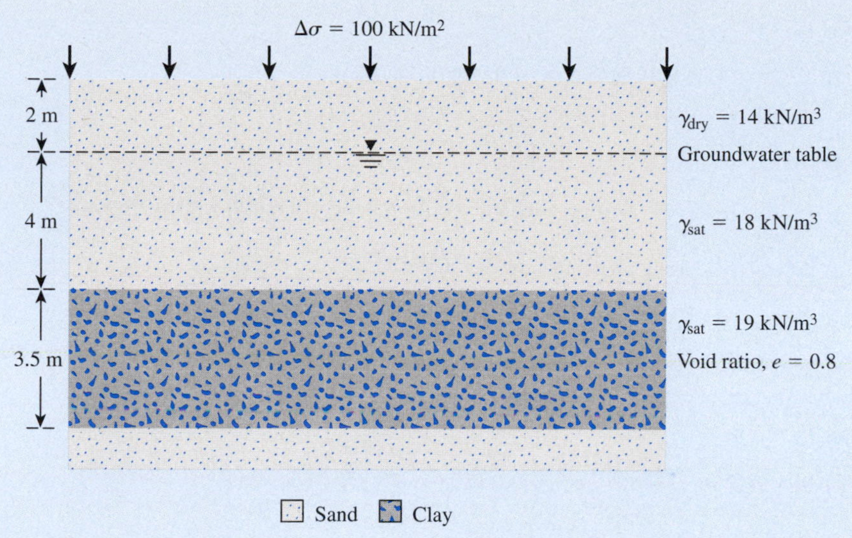

Figure 11.24

b. The preconsolidation pressure, $\sigma'_c = 200$ kN/m²
c. $\sigma'_c = 150$ kN/m²

Use $C_s \approx \frac{1}{5} C_c$; and Eq. (11.40).

Solution

Part a

The average effective stress at the middle of the clay layer is

$$\sigma'_o = 2\gamma_{dry} + 4[\gamma_{sat(sand)} - \gamma_w] + \frac{3.5}{2}[\gamma_{sat(clay)} - \gamma_w]$$

$$\sigma'_o = (2)(14) + 4(18 - 9.81) + 1.75(19 - 9.81) = 76.08 \text{ kN/m}^2$$

$$\gamma_{sat(clay)} = 19 \text{ kN/m}^3 = \frac{(G_s + e)\gamma_w}{1 + e} = \frac{(G_s + 0.8)(9.81)}{1 + 0.8}; \; G_s = 2.686$$

From Eq. (11.40),

$$C_c = 0.141 G_s^{1.2}\left(\frac{1 + e_o}{G_s}\right)^{2.38} = (0.141)(2.686)^{1.2}\left(\frac{1 + 0.8}{2.686}\right)^{2.38} = 0.178$$

From Eq. (11.35),

$$S_c = \frac{C_c H}{1 + e_o} \log\left(\frac{\sigma'_o + \Delta\sigma'}{\sigma'_o}\right)$$

So,

$$S_c = \frac{(0.178)(3.5)}{1 + 0.8} \log\left(\frac{76.08 + 100}{76.08}\right) = 0.126 \text{ m} = \mathbf{126 \text{ mm}}$$

Part b

$$\sigma'_o + \Delta\sigma' = 76.08 + 100 = 176.08 \text{ kN/m}^2$$

$$\sigma'_c = 200 \text{ kN/m}^2$$

Because $\sigma'_o + \Delta\sigma' < \sigma'_c$, use Eq. (11.37):

$$S_c = \frac{C_s H}{1 + e_o} \log\left(\frac{\sigma'_o + \Delta\sigma'}{\sigma'_o}\right)$$

$$C_s = \frac{C_c}{5} = \frac{0.178}{5} = 0.0356$$

$$S_c = \frac{(0.0356)(3.5)}{1 + 0.8} \log\left(\frac{76.08 + 100}{76.08}\right) = 0.025 \text{ m} = \mathbf{25 \text{ mm}}$$

Part c

$$\sigma'_o = 76.08 \text{ kN/m}^2$$

$$\sigma'_o + \Delta\sigma' = 176.08 \text{ kN/m}^2$$

$$\sigma'_c = 150 \text{ kN/m}^2$$

Because $\sigma'_o < \sigma'_c < \sigma'_o + \Delta\sigma'$, use Eq. (11.38):

$$S_c = \frac{C_s H}{1 + e_o} \log \frac{\sigma'_c}{\sigma'_o} + \frac{C_c H}{1 + e_o} \log\left(\frac{\sigma'_o + \Delta\sigma'}{\sigma'_c}\right)$$

$$= \frac{(0.0356)(3.5)}{1.8} \log\left(\frac{150}{76.08}\right) + \frac{(0.178)(3.5)}{1.8} \log\left(\frac{176.08}{150}\right)$$

$$\approx 0.0445 \text{ m} = \mathbf{44.5 \text{ mm}}$$

Example 11.9

Refer to Example 11.8. For each part, calculate and plot a graph of e vs. σ' at the beginning and end of consolidation.

Solution

For each part, $\boldsymbol{e} = \mathbf{0.8}$ at the beginning of consolidation. For e at the end of consolidation, the following calculations can be made.

Part a

$$e = 0.8 - C_c \log\left(\frac{\sigma'_o + \Delta\sigma'}{\sigma'_o}\right) = 0.8 - 0.178 \log\left(\frac{176.08}{76.08}\right) = \mathbf{0.735}$$

Part b

$$e = 0.8 - C_s \log\left(\frac{\sigma'_o + \Delta\sigma'}{\sigma'_o}\right) = 0.8 - 0.0356 \log\left(\frac{176.08}{76.08}\right) = \mathbf{0.787}$$

Part c

$$e = 0.8 - \left[C_s \log\left(\frac{\sigma'_c}{\sigma'_o}\right) + C_c \log\left(\frac{\sigma'_o + \Delta\sigma'}{\sigma'_c}\right)\right]$$

$$= 0.8 - \left[0.0356 \log\left(\frac{150}{76.08}\right) + 0.178 \log\left(\frac{176.08}{150}\right)\right]$$

$$= 0.8 - 0.0105 - 0.0124 = \mathbf{0.771}$$

A plot of e versus $\log \sigma'$ is shown in Figure 11.25.

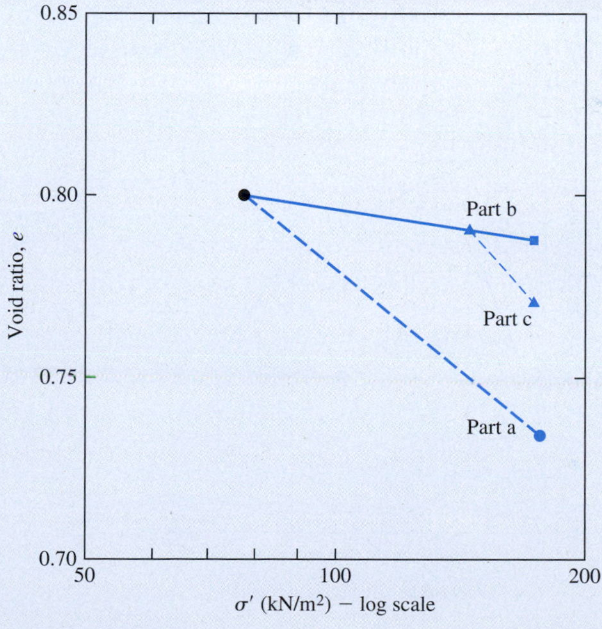

Figure 11.25

Example 11.10

A soil profile is shown in Figure 11.26a. Laboratory consolidation tests were conducted on a specimen collected from the middle of the clay layer. The field consolidation curve interpolated from the laboratory test results is shown in Figure 11.26b. Calculate the settlement in the field caused by primary consolidation for a surcharge of 60 kN/m² applied at the ground surface.

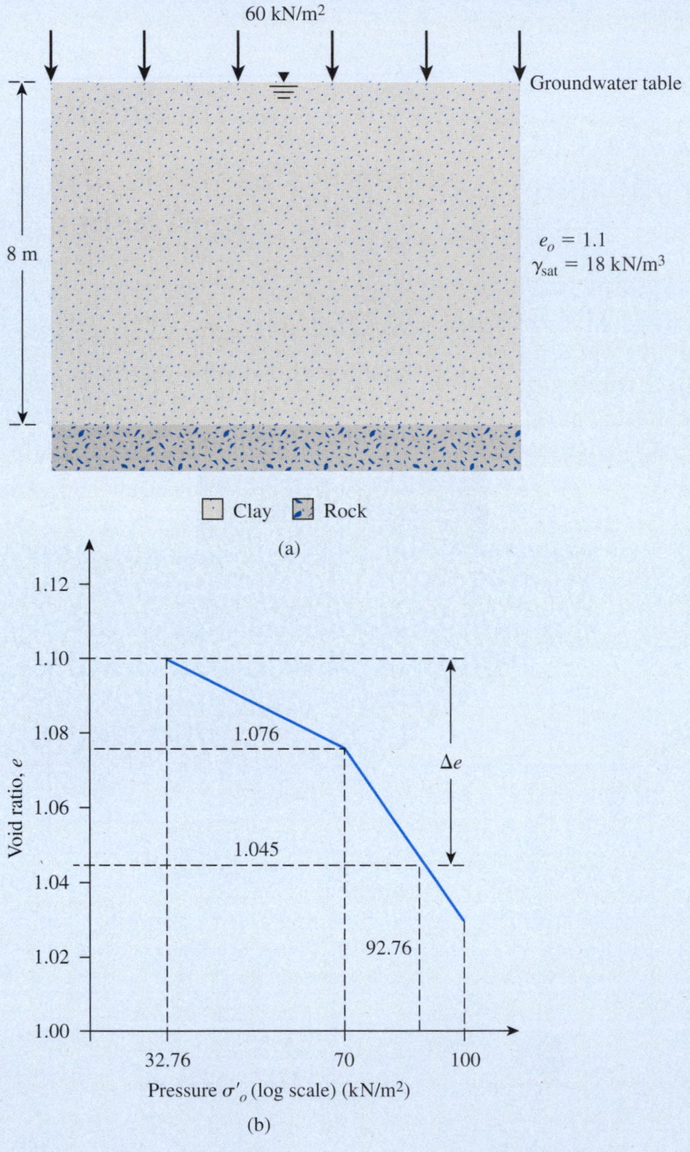

Figure 11.26 (a) Soil profile (b) field consolidation curve

Solution

$$\sigma'_o = (4)(\gamma_{sat} - \gamma_w) = 4(18.0 - 9.81)$$
$$= 32.76 \text{ kN/m}^2$$
$$e_o = 1.1$$
$$\Delta\sigma' = 60 \text{ kN/m}^2$$
$$\sigma_O' + \Delta\sigma' = 32.76 + 60 = 92.76 \text{ kN/m}^2$$

The void ratio corresponding to 92.76 kN/m^2 (see Figure 11.26b) is 1.045. Hence, $\Delta e = 1.1 - 1.045 = 0.055$. We have

$$\text{Settlement, } S_c = H\frac{\Delta e}{1 + e_o} \quad \text{[Eq. (11.33)]}$$

So,

$$S_c = 8\frac{(0.055)}{1 + 1.1} = 0.21 \text{ m} = \mathbf{210\ mm}$$

11.13 Secondary Consolidation Settlement

Section 11.6 showed that at the end of primary consolidation (that is, after complete dissipation of excess pore water pressure) some settlement is observed because of the plastic adjustment of soil fabrics. This stage of consolidation is called *secondary consolidation*. During secondary consolidation the plot of deformation against the log of time is practically linear (see Figure 11.12). The variation of the void ratio, e, with time t for a given load increment will be similar to that shown in Figure 11.12. This variation is shown in Figure 11.27. From Figure 11.27, the secondary compression index can be defined as

$$C_\alpha = \frac{\Delta e}{\log t_2 - \log t_1} = \frac{\Delta e}{\log (t_2/t_1)} \quad (11.47)$$

where C_α = secondary compression index
Δe = change of void ratio
t_1, t_2 = time

The magnitude of the secondary consolidation can be calculated as

$$S_s = C'_\alpha H \log\left(\frac{t_2}{t_1}\right) \quad (11.48)$$

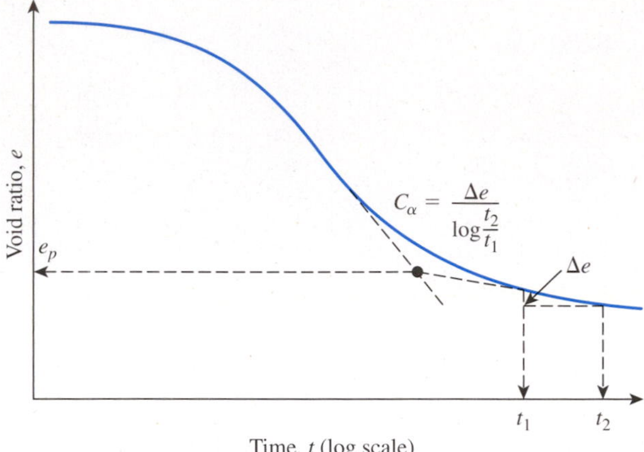

Figure 11.27 Variation of e with log t under a given load increment and definition of secondary consolidation index

and

$$C'_\alpha = \frac{C_\alpha}{1 + e_p} \tag{11.49}$$

where e_p = void ratio at the end of primary consolidation (see Figure 11.27)
H = thickness of clay layer

The general magnitudes of C'_α as observed in various natural deposits are as follows:

- Overconsolidated clays = 0.001 or less
- Normally consolidated clays = 0.005 to 0.03
- Organic soil = 0.04 or more

Mersri and Godlewski (1977) compiled the ratio of C_α/C_c for a number of natural clays. From this study, it appears that C_α/C_c for

- Inorganic clays and silts $\approx 0.04 \pm 0.01$
- Organic clays and silts $\approx 0.05 \pm 0.01$
- Peats $\approx 0.075 \pm 0.01$

Secondary consolidation settlement is more important than primary consolidation in organic and highly compressible inorganic soils. In overconsolidated inorganic clays, the secondary compression index is very small and of less practical significance.

Example 11.11

For a normally consolidated clay layer in the field, the following values are given:

Thickness of clay layer = 2.6 m
Void ratio, e_o = 0.8

Compression index, $C_c = 0.28$
Average effective pressure on the clay layer, $\sigma'_o = 127\ \text{kN/m}^2$
$\Delta\sigma' = 46.5\ \text{kN/m}^2$
Secondary compression index, $C_\alpha = 0.02$

What is the total consolidation settlement of the clay layer five years after the completion of primary consolidation settlement? (*Note:* Time for completion of primary settlement = 1.5 years.)

Solution

From Eq. (11.49),

$$C'_\alpha = \frac{C_\alpha}{1 + e_p}$$

The value of e_p can be calculated as

$$e_p = e_o - \Delta e_{\text{primary}}$$

Combining Eqs. (11.33) and (11.34), we find that

$$\Delta e = C_c \log\left(\frac{\sigma'_o + \Delta\sigma'}{\sigma'_o}\right) = 0.28 \log\left(\frac{127 + 46.5}{127}\right)$$
$$= 0.038$$

Primary consolidation:

$$S_c = \frac{\Delta eH}{1 + e_o} = \frac{(0.038)(2.6 \times 1000)}{1 + 0.8} = 54.9\ \text{mm.}$$

It is given that $e_o = 0.8$, and thus,

$$e_p = 0.8 - 0.038 = 0.762$$

Hence,

$$C'_\alpha = \frac{0.02}{1 + 0.762} = 0.011$$

From Eq. (11.48),

$$S_s = C'_\alpha H \log\left(\frac{t_2}{t_1}\right) = (0.011)(2.6 \times 1000) \log\left(\frac{5}{1.5}\right) \approx 14.95\ \text{mm}$$

Total consolidation settlement = primary consolidation (S_c) + secondary settlement (S_s). So

Total consolidation settlement = 54.9 + 14.95 = **69.85 mm.**

11.14 Time Rate of Consolidation

The total settlement caused by primary consolidation resulting from an increase in the stress on a soil layer can be calculated by the use of one of the three equations—(11.35), (11.37), or (11.38)—given in Section 11.10. However, they do not provide any information regarding the rate of primary consolidation. Terzaghi (1925) proposed the first theory to consider the rate of one-dimensional consolidation for saturated clay soils. The mathematical derivations are based on the following six assumptions (also see Taylor, 1948):

1. The clay–water system is homogeneous.
2. Saturation is complete.
3. Compressibility of water is negligible.
4. Compressibility of soil grains is negligible (but soil grains rearrange).
5. The flow of water is in one direction only (that is, in the direction of compression).
6. Darcy's law is valid.

Figure 11.28a shows a layer of clay of thickness $2H_{dr}$ (*Note:* H_{dr} = length of maximum drainage path) that is located between two highly permeable sand layers. If the clay layer is subjected to an increased pressure of $\Delta\sigma$, the pore water pressure at any point A in the clay layer will increase. For one-dimensional consolidation, water will be squeezed out in the vertical direction toward the sand layers.

Figure 11.28b shows the flow of water through a prismatic element at A. For the soil element shown,

Rate of outflow of water − Rate of inflow of water = Rate of volume change

Thus,

$$\left(v_z + \frac{\partial v_z}{\partial z}\,dz\right)dx\,dy - v_z\,dx\,dy = \frac{\partial V}{\partial t}$$

where V = volume of the soil element
v_z = velocity of flow in z direction

or

$$\frac{\partial v_z}{\partial z}\,dx\,dy\,dz = \frac{\partial V}{\partial t} \tag{11.50}$$

Using Darcy's law, we have

$$v_z = ki = -k\frac{\partial h}{\partial z} = -\frac{k}{\gamma_w}\frac{\partial u}{\partial z} \tag{11.51}$$

where u = excess pore water pressure caused by the increase of stress.

From Eqs. (11.50) and (11.51),

$$-\frac{k}{\gamma_w}\frac{\partial^2 u}{\partial z^2} = \frac{1}{dx\,dy\,dz}\frac{\partial V}{\partial t} \tag{11.52}$$

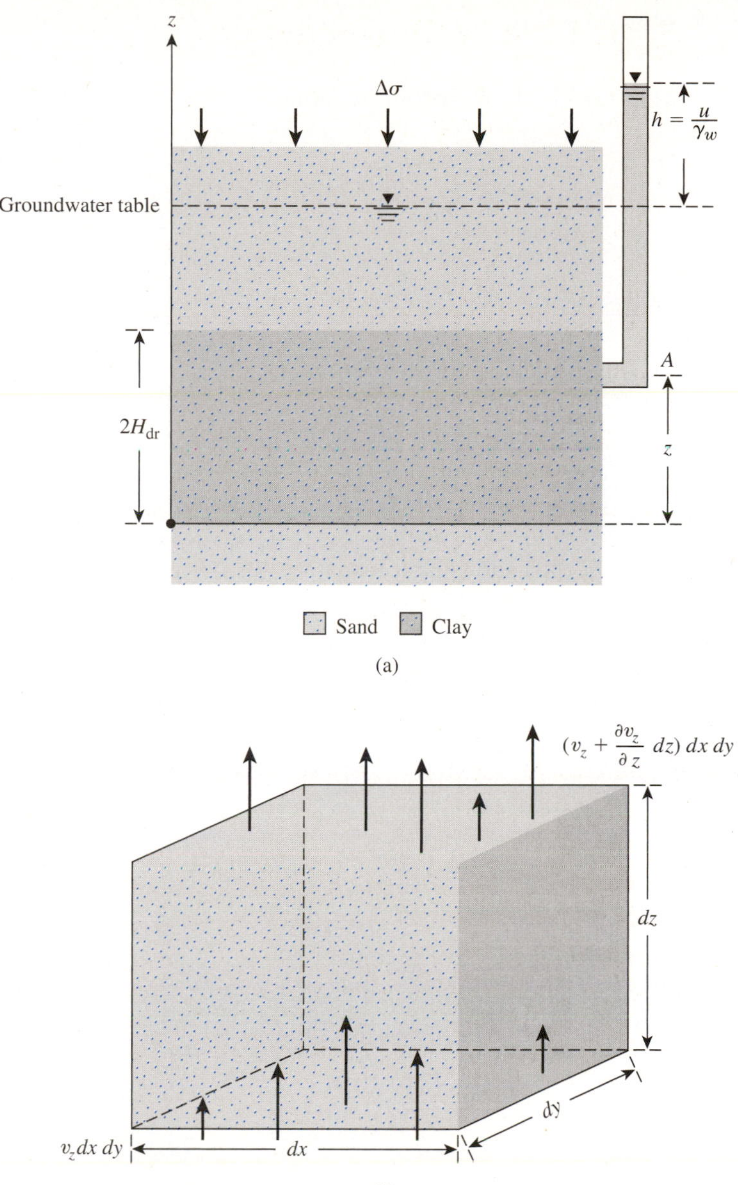

Figure 11.28 (a) Clay layer undergoing consolidation; (b) flow of water at A during consolidation

During consolidation, the rate of change in the volume of the soil element is equal to the rate of change in the volume of voids. Thus,

$$\frac{\partial V}{\partial t} = \frac{\partial V_v}{\partial t} = \frac{\partial(V_s + eV_s)}{\partial t} = \frac{\partial V_s}{\partial t} + V_s\frac{\partial e}{\partial t} + e\frac{\partial V_s}{\partial t} \tag{11.53}$$

where V_s = volume of soil solids
V_v = volume of voids

But (assuming that soil solids are incompressible)

$$\frac{\partial V_s}{\partial t} = 0$$

and

$$V_s = \frac{V}{1 + e_o} = \frac{dx\,dy\,dz}{1 + e_o}$$

Substitution for $\partial V_s/\partial t$ and V_s in Eq. (11.53) yields

$$\frac{\partial V}{\partial t} = \frac{dx\,dy\,dz}{1 + e_o}\frac{\partial e}{\partial t} \tag{11.54}$$

where e_o = initial void ratio.

Combining Eqs. (11.52) and (11.54) gives

$$-\frac{k}{\gamma_w}\frac{\partial^2 u}{\partial z^2} = \frac{1}{1 + e_o}\frac{\partial e}{\partial t} \tag{11.55}$$

The change in the void ratio is caused by the increase of effective stress (i.e., a decrease of excess pore water pressure). Assuming that they are related linearly, we have

$$\partial e = a_v \partial(\Delta\sigma') = -a_v \partial u \tag{11.56}$$

where $\partial(\Delta\sigma')$ = change in effective pressure
a_v = coefficient of compressibility (a_v can be considered constant for a narrow range of pressure increase) $= \frac{\Delta e}{\Delta\sigma'}$

Combining Eqs. (11.55) and (11.56) gives

$$-\frac{k}{\gamma_w}\frac{\partial^2 u}{\partial z^2} = -\frac{a_v}{1 + e_o}\frac{\partial u}{\partial t} = -m_v\frac{\partial u}{\partial t}$$

where

$$m_v = \text{coefficient of volume compressibility} = a_v/(1 + e_o) \tag{11.57}$$

or,

$$\frac{\partial u}{\partial t} = c_v\frac{\partial^2 u}{\partial z^2} \tag{11.58}$$

where

$$c_v = \text{coefficient of consolidation} = k/(\gamma_w m_v) \tag{11.59}$$

Thus,

$$c_v = \frac{k}{\gamma_w m_v} = \frac{k}{\gamma_w\left(\dfrac{a_v}{1 + e_o}\right)} \tag{11.60}$$

Equation (11.58) is the basic differential equation of Terzaghi's consolidation theory and can be solved with the following boundary conditions:

$$z = 0, \quad u = 0$$

$$z = 2H_{\text{dr}}, \quad u = 0$$

$$t = 0, \quad u = u_o$$

The solution yields

$$u = \sum_{m=0}^{m=\infty}\left[\frac{2u_o}{M}\sin\left(\frac{Mz}{H_{\text{dr}}}\right)\right]e^{-M^2T_v} \tag{11.61}$$

where m = an integer
$M = (\pi/2)(2m + 1)$
u_o = initial excess pore water pressure

$$T_v = \frac{c_v t}{H_{\text{dr}}^2} = \text{time factor} \tag{11.62}$$

The time factor is a nondimensional number.

Because consolidation progresses by the dissipation of excess pore water pressure, the degree of consolidation at a distance z at any time t is

$$U_z = \frac{u_o - u_z}{u_o} = 1 - \frac{u_z}{u_o} \tag{11.63}$$

where u_z = excess pore water pressure at time t.

Equations (11.61) and (11.63) can be combined to obtain the degree of consolidation at any depth z. This is shown in Figure 11.29.

The average degree of consolidation for the entire depth of the clay layer at any time t can be written from Eq. (11.63) as

$$U = \frac{S_{c(t)}}{S_c} = 1 - \frac{\left(\dfrac{1}{2H_{\text{dr}}}\right)\displaystyle\int_0^{2H_{\text{dr}}} u_z\,dz}{u_o} \tag{11.64}$$

where U = average degree of consolidation
$S_{c(t)}$ = settlement of the layer at time t
S_c = ultimate settlement of the layer from primary consolidation

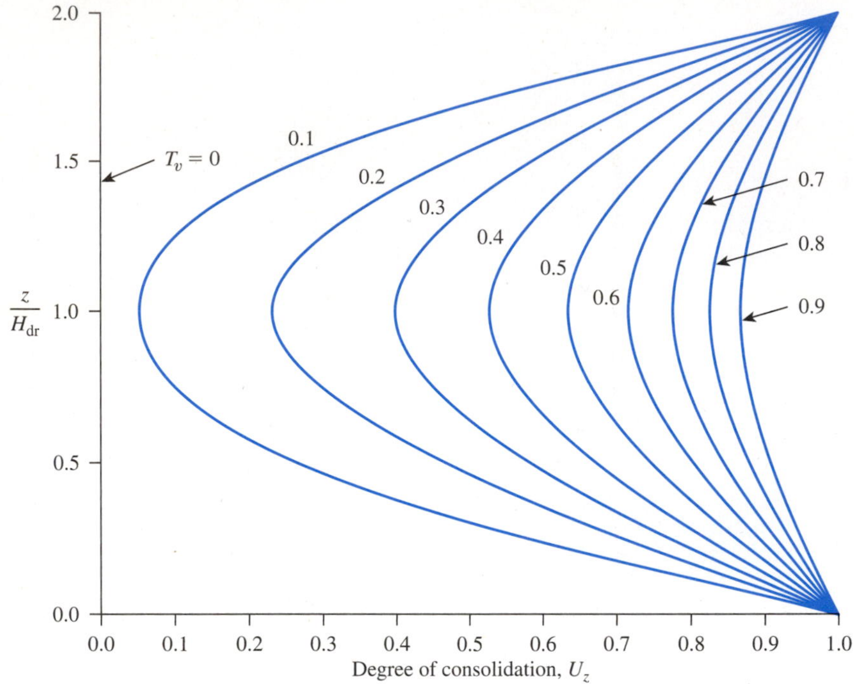

Figure 11.29 Variation of U_z with T_v and z/H_{dr}

Substitution of the expression for excess pore water pressure u_z given in Eq. (11.61) into Eq. (11.64) gives

$$U = 1 - \sum_{m=0}^{m=\infty} \frac{2}{M^2} e^{-M^2 T_v} \tag{11.65}$$

The variation in the average degree of consolidation with the nondimensional time factor, T_v, is given in Figure 11.30, which represents the case where u_o is the same for the entire depth of the consolidating layer.

The values of the time factor and their corresponding average degrees of consolidation for the case presented in Figure 11.30 may also be approximated by the following simple relationship:

$$\text{For } U = 0 \text{ to } 60\%, \quad T_v = \frac{\pi}{4}\left(\frac{U\%}{100}\right)^2 \tag{11.66}$$

$$\text{For } U > 60\%, \ T_v = 1.781 - 0.933 \log (100 - U\%) \tag{11.67}$$

Table 11.7 gives the variation of T_v with U on the basis of Eqs. (11.66) and (11.67).

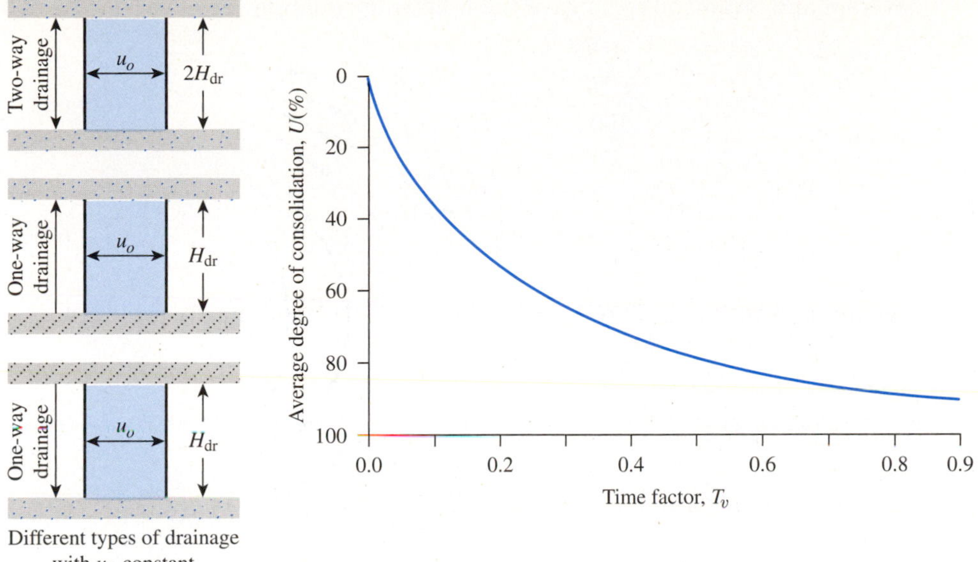

Figure 11.30 Variation of average degree of consolidation with time factor, T_v (u_o constant with depth)

Table 11.7 Variation of T_v with U

U (%)	T_v	U (%)	T_v	U (%)	T_v	U (%)	T_v
0	0	26	0.0531	52	0.212	78	0.529
1	0.00008	27	0.0572	53	0.221	79	0.547
2	0.0003	28	0.0615	54	0.230	80	0.567
3	0.00071	29	0.0660	55	0.239	81	0.588
4	0.00126	30	0.0707	56	0.248	82	0.610
5	0.00196	31	0.0754	57	0.257	83	0.633
6	0.00283	32	0.0803	58	0.267	84	0.658
7	0.00385	33	0.0855	59	0.276	85	0.684
8	0.00502	34	0.0907	60	0.286	86	0.712
9	0.00636	35	0.0962	61	0.297	87	0.742
10	0.00785	36	0.102	62	0.307	88	0.774
11	0.0095	37	0.107	63	0.318	89	0.809
12	0.0113	38	0.113	64	0.329	90	0.848
13	0.0133	39	0.119	65	0.340	91	0.891
14	0.0154	40	0.126	66	0.352	92	0.938
15	0.0177	41	0.132	67	0.364	93	0.993
16	0.0201	42	0.138	68	0.377	94	1.055
17	0.0227	43	0.145	69	0.390	95	1.129
18	0.0254	44	0.152	70	0.403	96	1.219
19	0.0283	45	0.159	71	0.417	97	1.336
20	0.0314	46	0.166	72	0.431	98	1.500
21	0.0346	47	0.173	73	0.446	99	1.781
22	0.0380	48	0.181	74	0.461	100	∞
23	0.0415	49	0.188	75	0.477		
24	0.0452	50	0.197	76	0.493		
25	0.0491	51	0.204	77	0.511		

Sivaram and Swamee (1977) gave the following equation for U varying from 0 to 100%:

$$\frac{U\%}{100} = \frac{(4T_v/\pi)^{0.5}}{[1 + (4T_v/\pi)^{2.8}]^{0.179}} \tag{11.68}$$

or

$$T_v = \frac{(\pi/4)(U\%/100)^2}{[1 - (U\%/100)^{5.6}]^{0.357}} \tag{11.69}$$

Equations (11.68) and (11.69) give an error in T_v of less than 1% for $0\% < U < 90\%$ and less than 3% for $90\% < U < 100\%$.

Example 11.12

Refer to Figure 11.28a and Eqs. (11.61) and (11.63). From Eqs. (11.61) and (11.63),

$$U_z = 1 - \sum_{m=0}^{m=\infty} \frac{2}{M} \sin\left(\frac{Mz}{H_{dr}}\right) e^{-M^2 T_v}$$

Determine the degree of consolidation U_z at $z = H_{dr}/3$ and $T_v = 0.3$.

Solution

Here $H_{dr}/3$ and $M = (\pi/2)(2m + 1)$. Now the following table can be prepared.

Step 1. $\frac{z}{H_{dr}}$	$\frac{1}{3}$	$\frac{1}{3}$	$\frac{1}{3}$		
Step 2. T_v	0.3	0.3	0.3		
Step 3. m	0	1	2		
Step 4. M	$\frac{\pi}{2}$	$\frac{3\pi}{2}$	$\frac{5\pi}{2}$		
Step 5. $\frac{Mz}{H_{dr}}$	$\frac{\pi}{6}$	$\frac{\pi}{2}$	$\frac{5\pi}{6}$		
Step 6. $\frac{2}{M}$	1.273	0.4244	0.2546		
Step 7. $e^{-M^2T_v}$	0.4770	0.00128	≈ 0		
Step 8. $\sin\left(\frac{Mz}{H_{dr}}\right)$	0.5	1.0	0.5		
Step 9. $\left(\frac{2}{M}\right)\left(e^{-M^2T_v}\right) \times \left[\sin\left(\frac{Mz}{H_{dr}}\right)\right]$	0.3036	0.0005	≈ 0		$\Sigma 0.304$

So,

$$U_z = 1 - 0.304 = 0.6959 = \mathbf{69.59\%}$$

Note: In the previous table, we need not go beyond $m = 2$ since the expression in Step 9 is negligible for $m \geq 3$.

Comment: Compare $U_z = 69.59\%$ just calculated to that shown in Figure 11.29.

Example 11.13

The time required for 50% consolidation of a 25-mm-thick clay layer (drained at both top and bottom) in the laboratory is 3 min 15 sec. How long (in days) will it take for a 3-m-thick clay layer of the same clay in the field under the same pressure increment to reach 50% consolidation? In the field, the clay layer is drained at the top only.

Solution

$$T_{50} = \frac{c_v t_{\text{lab}}}{H^2_{\text{dr(lab)}}} = \frac{c_v t_{\text{field}}}{H^2_{\text{dr(field)}}}$$

or

$$\frac{t_{\text{lab}}}{H^2_{\text{dr(lab)}}} = \frac{t_{\text{field}}}{H^2_{\text{dr(field)}}}$$

$$\frac{195 \text{ sec}}{\left(\frac{0.025 \text{ m}}{2}\right)^2} = \frac{t_{\text{field}}}{(3 \text{ m})^2}$$

$$t_{\text{field}} = 11{,}232{,}000 \text{ sec} = \mathbf{130\ days}$$

Example 11.14

Refer to Example 11.13. How long (in days) will it take in the field for 70% primary consolidation to occur?

Solution

$$T_v = \frac{c_v t}{H^2_{\text{dr}}}$$

From Table 11.7, for $U = 50\%$, $T_v = 0.197$ and for $U = 70\%$, $T_v = 0.403$. So,

$$\frac{T_{50}}{T_{70}} = \frac{\left(\frac{c_v t_{50}}{H^2_{\text{dr}}}\right)}{\left(\frac{c_v t_{70}}{H^2_{\text{dr}}}\right)}$$

Hence,

$$\frac{0.197}{0.403} = \frac{t_{50}}{t_{70}} = \frac{130 \text{ days}}{t_{70}}; t_{70} \approx \mathbf{266 \text{ days}}$$

Example 11.15

A 3-m-thick layer (double drainage) of saturated clay under a surcharge loading underwent 90% primary consolidation in 75 days. Find the coefficient of consolidation of clay for the pressure range.

Solution

$$T_{90} = \frac{c_v t_{90}}{H_{dr}^2}$$

Because the clay layer has two-way drainage, $H_{dr} = 3 \text{ m}/2 = 1.5$ m. Also, $T_{90} = 0.848$ (see Table 11.7). So,

$$0.848 = \frac{c_v(75 \times 24 \times 60 \times 60)}{(1.5 \times 100)^2}$$

$$c_v = \frac{0.848 \times 2.25 \times 10^4}{75 \times 24 \times 60 \times 60} = \mathbf{0.00294 \text{ cm}^2\text{/sec}}$$

Example 11.16

For a normally consolidated clay,

- $\sigma'_o = 200 \text{ kN/m}^2$
- $\sigma'_o + \Delta\sigma = 400 \text{ kN/m}^2$
- $e = e_o = 1.22$
- $e = 0.98$

The hydraulic conductivity, k, of the clay for the loading range is 0.61×10^{-4} m/day. How long (in days) will it take for a 4-m-thick clay layer (drained on one side) in the field to reach 60% consolidation?

Solution

The coefficient of volume compressibility is

$$m_v = \frac{a_v}{1 + e_{av}} = \frac{\left(\dfrac{\Delta e}{\Delta\sigma'}\right)}{1 + e_{av}}$$

$$\Delta e = 1.22 - 0.98 = 0.24$$

$$\Delta\sigma' = 400 - 200 = 200 \text{ kN/m}^2$$

$$e_{av} = \frac{1.22 + 0.98}{2} = 1.1$$

So,

$$m_v = \frac{\dfrac{0.24}{200}}{1 + 1.1} = 0.00057 \text{ m}^2/\text{kN}$$

$$c_v = \frac{k}{m_v \gamma_w} = \frac{0.61 \times 10^{-4} \text{ m/day}}{(0.00057 \text{ m}^2/\text{kN})(9.81 \text{ kN/m}^3)} = 0.0109 \text{ m}^2/\text{day}$$

$$T_{60} = \frac{c_v t_{60}}{H_{dr}^2}$$

$$t_{60} = \frac{T_{60} H_{dr}^2}{c_v}$$

From Table 11.7, for $U = 60\%$, $T_{60} = 0.286$, so

$$t_{60} = \frac{(0.286)(4)^2}{0.0109} = \mathbf{419.8 \text{ days}}$$

Example 11.17

For a laboratory consolidation test on a soil specimen (drained on both sides), the following results were obtained.

- Thickness of the clay specimen = 25 mm
- $\sigma'_1 = 50 \text{ kN/m}^2$
- $\sigma'_2 = 120 \text{ kN/m}^2$
- $e_1 = 0.92$
- $e_2 = 0.78$
- Time for 50% consolidation = 2.5 min

Determine the hydraulic conductivity of the clay for the loading range.

Solution

$$m_v = \frac{a_v}{1 + e_{av}} = \frac{\left(\dfrac{\Delta e}{\Delta\sigma'}\right)}{1 + e_{av}}$$

$$= \frac{\dfrac{0.92 - 0.78}{120 - 50}}{1 + \dfrac{0.92 + 0.78}{2}} = 0.00108 \text{ m}^2/\text{kN}$$

$$c_v = \frac{T_{50} H_{dr}^2}{t_{50}}$$

From Table 11.7, for $U = 50\%$, $T_v = 0.197$, so

$$c_v = \frac{(0.197)\left(\dfrac{0.025 \text{ m}}{2}\right)^2}{2.5 \text{ min}} = 1.23 \times 10^{-5} \text{ m}^2/\text{min}$$

$$k = c_v m_v \gamma_w = (1.23 \times 10^{-5})(0.00108)(9.81)$$

$$= \mathbf{1.303 \times 10^{-7}\ m/min}$$

11.15 Construction Time Correction of Consolidation Settlement

Until now, we have assumed that the entire load causing consolidation was applied *instantaneously*. In reality, the buildings or embankments are constructed over several months, as shown in Figure 11.31. Therefore the clay layer is subjected to ramp loading where the pressure q is applied over the construction period t_0. From Terzaghi's one-dimensional consolidation theory, it can be shown that,

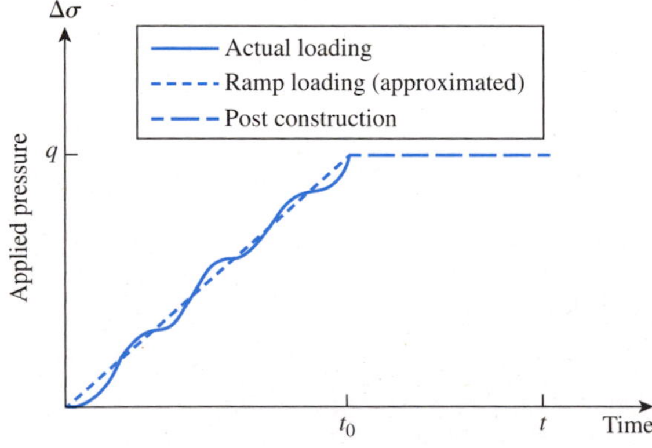

Figure 11.31 Ramp loading

during ramp loading, the average degree of consolidation U and the time factor T_v are related by

$$U = 1 - \frac{1}{T_v}\left[\sum_{m=0}^{\infty}\left(\frac{2}{M^4}\right)(1 - e^{-M^2T_v})\right] \tag{11.70}$$

The variation of U with T_v is shown in Figure 11.32 (Hanna et al., 2013; Sivakugan et al. 2014). Also shown in the figure are the U–T_v variations for the instantaneous loading [Eq. (11.65)], which is quite different.

Denoting the degree of consolidation at the end of construction as U_0, the remaining excess pore water pressure can be considered as being applied instantaneously at time t_0. The average degree of consolidation at time t ($> t_0$) then can be computed as

$$U_t = U_0 + (1 - U_0)U_{t-t_0} \tag{11.71}$$

where U_{t-t_0} is the average degree of consolidation for duration of $t - t_0$, assuming instantaneous loading. Therefore, appropriate values from Table 11.7 should be used for computing U_{t-t_0}. The procedure for using Eq. (11.71) and Figure 11.32 is shown in Example 11.18.

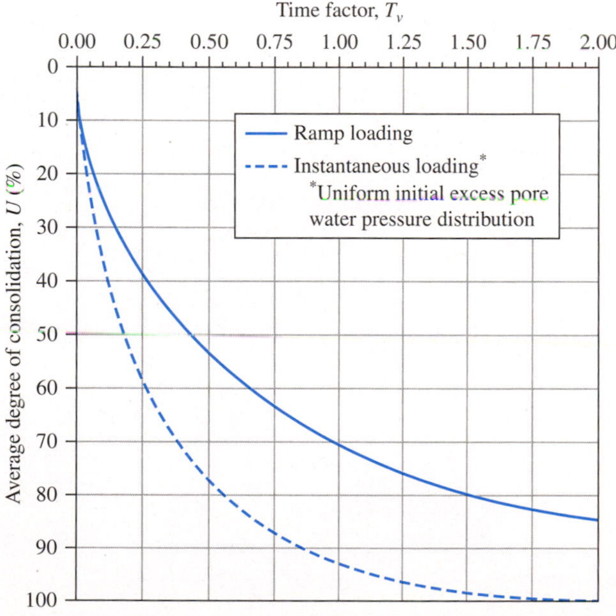

Figure 11.32 U–T_v relationships for ramp loading and instantaneous loading

Example 11.18

Consider a 3-m thick clay layer with one-way drainage. The laboratory-determined values for the clay are

Coefficient of volume compressibility, $m_v = 0.9\ \text{m}^2/\text{MN}$
Coefficient of consolidation, $c_v = 1.6 \times 10^{-2}\ \text{m}^2/\text{day}$

The clay layer is subjected to a ramp load as shown in Figure 11.33. Estimate the consolidation settlement at

a. Time $t = 100$ days
b. Time $t = 500$ days

Solution

Part a

At time $t = 100$ days, $\Delta\sigma = 100\ \text{kN/m}^2$. With this load, the final consolidation settlement at $t = \infty$ can be calculated as follows. We know

$$m_v = \frac{a_v}{1 + e_o} = \frac{\left(\dfrac{\Delta e}{\Delta\sigma'}\right)}{1 + e_o}$$

From Eq. (11.33),

$$S_c = H\left(\frac{\Delta e}{1 + e_o}\right) = m_v \Delta\sigma H$$

So,

$$S_c = (0.9\ \text{m}^2/\text{MN})(0.1\ \text{MN/m}^2)(3\ \text{m}) = 0.27\ \text{m} = 270\ \text{mm}$$

Again, at $t = 100$ days, the time factor is

$$T_v = \frac{c_v t}{H_{\text{dr}}^2} = \frac{(1.6 \times 10^{-2}\ \text{m}^2/\text{day})(100\ \text{days})}{(3\ \text{m})^2} = 0.178$$

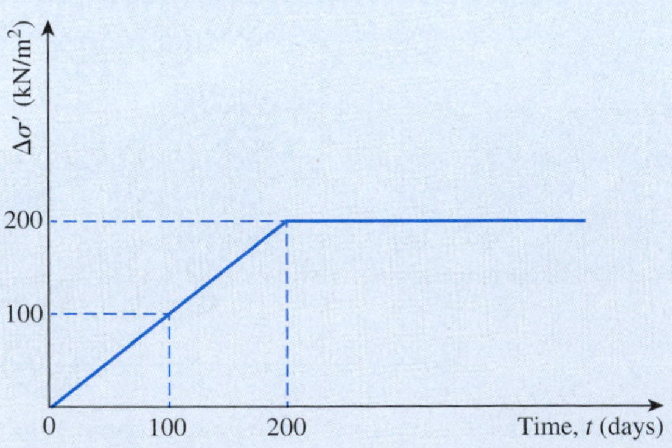

Figure 11.33

From Figure 11.32, for $T_v = 0.178$, the magnitude of U (for *ramp loading*) is about 32%. So,

$$S_{c(t=100\text{ days})} = 270 \times 0.32 = \mathbf{86.4\ mm}$$

Part b

The *ramp loading* (construction) ends at $t = t_0 = 200$ days. The final consolidation settlement at $t = \infty$ with $\Delta\sigma = 200\ \text{kN/m}^2$ will be

$$S_c = m_v \Delta\sigma H = (0.9\ \text{m}^2/\text{MN})(0.2)(3) = 0.54\ \text{m} = 540\ \text{mm}$$

The magnitude of T_v at the end of construction (ramp loading) is

$$T_v = \frac{c_v t_0}{H_{\text{dr}}^2} = \frac{(1.6 \times 10^{-2})(200)}{(3)^2} = 0.356$$

The magnitude of U_0 at the end of ramp loading (Figure 11.32) is about 44%. At $t = 500$ days, $t - t_0 = 500 - 200 = 300$ days. So,

$$T_{v(t-t_0)} = \frac{c_v t_{(t-t_0)}}{H_{\text{dr}}^2} = \frac{(1.6 \times 10^{-2})(300)}{(3)^2} = 0.533$$

For *instantaneous loading*, Figure 11.32 (or Table 11.7) gives U_{t-t_0} to be about 78%. Thus,

$$U_{t=500\text{ days}} = U_0 + (1 - U_0)U_{t-t_0} = 0.44 + (1 - 0.44)(0.78) \approx 0.877$$

Hence, the consolidation settlement at $t = 500$ days is

$$S_{c(t=500\text{ days})} = S_{c(t=\infty)}(0.877) = (540\ \text{mm})(0.877) = 473.58\ \text{mm} \approx \mathbf{474\ mm}$$

11.16 Determination of Coefficient of Consolidation

The coefficient of consolidation c_v generally decreases as the liquid limit of soil increases. The range of variation of c_v for a given liquid limit of soil is wide.

For a given load increment on a specimen, two graphical methods commonly are used for determining c_v from laboratory one-dimensional consolidation tests. The first is the *logarithm-of-time method* proposed by Casagrande and Fadum (1940), and the other is the *square-root-of-time method* given by Taylor (1942). More recently, at least two other methods were proposed. They are the *hyperbola method* (Sridharan and Prakash, 1985) and the *early stage log-t method* (Robinson and Allam, 1996). The general procedures for obtaining c_v by these methods are described in this section.

Logarithm-of-time method

For a given incremental loading of the laboratory test, the specimen deformation against log-of-time plot is shown in Figure 11.34. The following constructions are needed to determine c_v.

Step 1. Extend the straight-line portions of primary and secondary consolidations to intersect at A. The ordinate of A is represented by d_{100}—that is, the deformation at the end of 100% primary consolidation.

Step 2. The initial curved portion of the plot of deformation versus log t is approximated to be a parabola on the natural scale. Select times t_1 and t_2 on the curved portion such that $t_2 = 4t_1$. Let the difference of specimen deformation during time $(t_2 - t_1)$ be equal to x.

Step 3. Draw a horizontal line DE such that the vertical distance BD is equal to x. The deformation corresponding to the line DE is d_0 (that is, deformation at 0% consolidation).

Step 4. The ordinate of point F on the consolidation curve represents the deformation at 50% primary consolidation, and its abscissa represents the corresponding time (t_{50}).

Step 5. For 50% average degree of consolidation, $T_v = 0.197$ (see Table 11.7), so,

$$T_{50} = \frac{c_v t_{50}}{H_{dr}^2}$$

or

$$c_v = \frac{0.197 H_{dr}^2}{t_{50}} \tag{11.72}$$

where H_{dr} = average longest drainage path during consolidation.

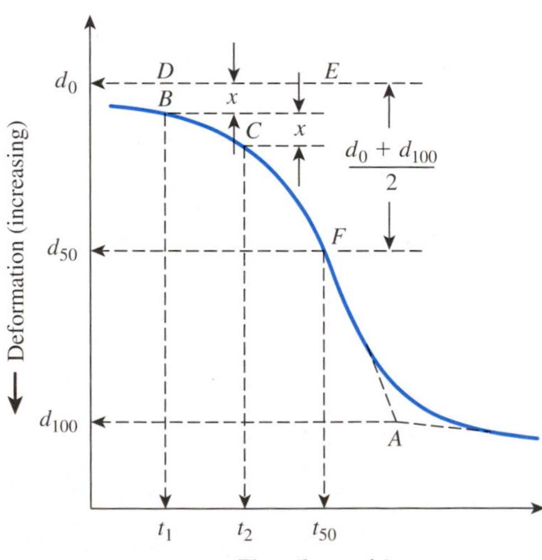

Figure 11.34 Logarithm-of-time method for determining coefficient of consolidation

For specimens drained at both top and bottom, H_{dr} equals one-half the average height of the specimen during consolidation. For specimens drained on only one side, H_{dr} equals the average height of the specimen during consolidation.

Square-root-of-time method

In the square-root-of-time method, a plot of deformation against the square root of time is made for the incremental loading (Figure 11.35). Other graphic constructions required are as follows:

Step 1. Draw a line AB through the early portion of the curve.

Step 2. Draw a line AC such that $\overline{OC} = 1.15\overline{OB}$. The abscissa of point D, which is the intersection of AC and the consolidation curve, gives the square root of time for 90% consolidation ($\sqrt{t_{90}}$).

Step 3. For 90% consolidation, $T_{90} = 0.848$ (see Table 11.7), so

$$T_{90} = 0.848 = \frac{c_v t_{90}}{H_{dr}^2}$$

or

$$c_v = \frac{0.848 H_{dr}^2}{t_{90}} \tag{11.73}$$

H_{dr} in Eq. (11.73) is determined in a manner similar to that in the logarithm-of-time method.

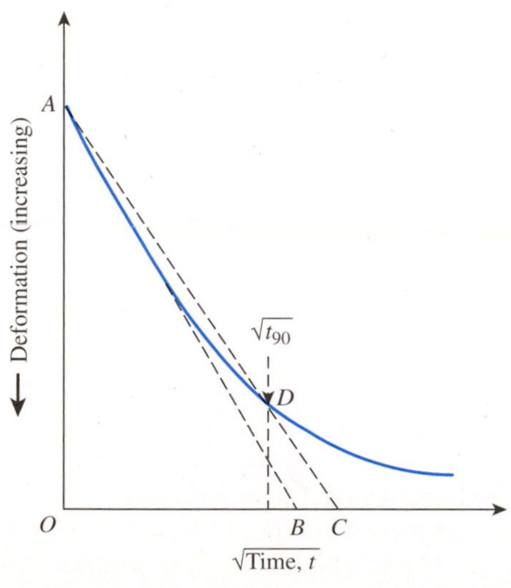

Figure 11.35 Square-root-of-time fitting method

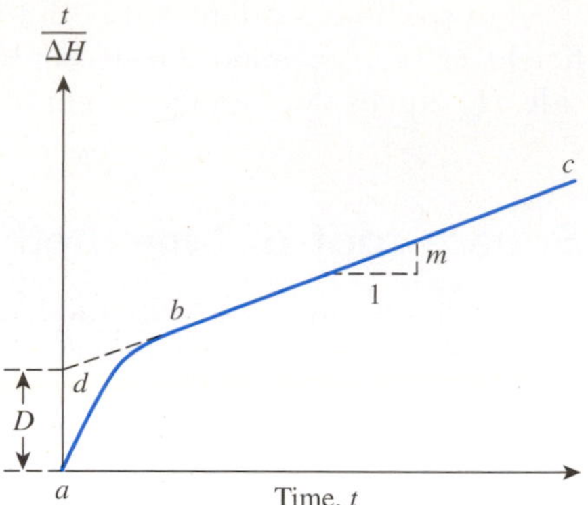

Figure 11.36 Hyperbola method for determination of c_v

Hyperbola method

In the hyperbola method, the following procedure is recommended for the determination of c_v.

Step 1. Obtain the time t and the specimen deformation (ΔH) from the laboratory consolidation test.
Step 2. Plot the graph of $t/\Delta H$ against t as shown in Figure 11.36.
Step 3. Identify the straight-line portion bc and project it back to point d. Determine the intercept D.
Step 4. Determine the slope m of the line bc.
Step 5. Calculate c_v as

$$c_v = 0.3\left(\frac{mH_{\mathrm{dr}}^2}{D}\right) \tag{11.74}$$

Note that because the unit of D is time/length and the unit of m is (time/length)/time = 1/length, the unit of c_v is

$$\frac{\left(\dfrac{1}{\text{Length}}\right)(\text{Length})^2}{\left(\dfrac{\text{Time}}{\text{Length}}\right)} = \frac{(\text{Length})^2}{\text{Time}}$$

The hyperbola method is fairly simple to use, and it gives good results for $U = 60$ to 90%.

Early stage log-*t* method

The early stage log-t method, an extension of the logarithm-of-time method, is based on specimen deformation against log-of-time plot as shown in Figure 11.37. According to this method, follow steps 2 and 3 described for the logarithm-of-time method to determine d_0. Draw a horizontal line DE through d_0. Then draw a tangent through

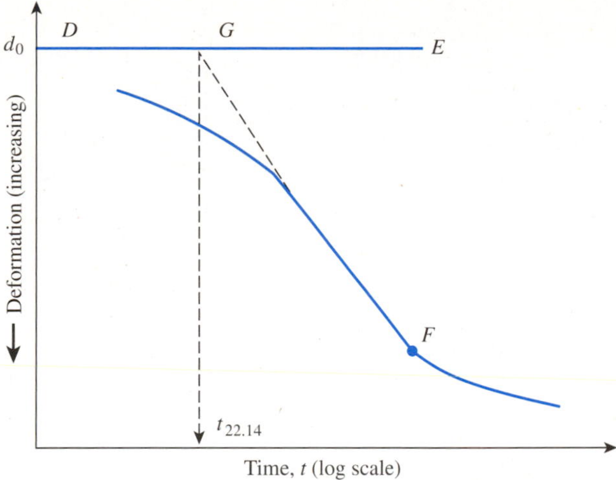

Figure 11.37 Early stage log-*t* method

the point of inflection, *F*. The tangent intersects line *DE* at point *G*. Determine the time *t* corresponding to *G*, which is the time at $U = 22.14\%$. So

$$c_v = \frac{0.0385H_{\text{dr}}^2}{t_{22.14}} \tag{11.75}$$

In most cases, for a given soil and pressure range, the magnitude of c_v determined by using the *logarithm-of-time method* provides the *lowest value*. The *highest value* is obtained from the *early stage log-t method*. The primary reason is because the early stage log-*t* method uses the earlier part of the consolidation curve, whereas the logarithm-of-time method uses the lower portion of the consolidation curve. When the lower portion of the consolidation curve is taken into account, the effect of secondary consolidation plays a role in the magnitude of c_v. This fact has been demonstrated for several soils by Robinson and Allam (1996).

Several investigators also have reported that the c_v value obtained from the field is substantially higher than that obtained from laboratory tests conducted by using conventional testing methods (that is, logarithm-of-time and square-root-of-time methods). These have been summarized in a paper by Leroueil (1988).

Example 11.19

During a laboratory consolidation test, the time and dial gauge readings obtained from an increase of pressure on the specimen from 50 kN/m^2 to 100 kN/m^2 are given here.

Time (min)	Dial gauge reading (cm × 10⁴)	Time (min)	Dial gauge reading (cm × 10⁴)
0	3975	16.0	4572
0.1	4082	30.0	4737
0.25	4102	60.0	4923
0.5	4128	120.0	5080
1.0	4166	240.0	5207
2.0	4224	480.0	5283
4.0	4298	960.0	5334
8.0	4420	1440.0	5364

Using the logarithm-of-time method, determine c_v. The average height of the specimen during consolidation was 2.24 cm, and it was drained at the top and bottom.

Solution

The semilogarithmic plot of dial reading versus time is shown in Figure 11.38. For this, $t_1 = 0.1$ min, $t_2 = 0.4$ min have been used to determine d_0. Following the procedure outlined in Figure 11.34, $t_{50} \approx 19$ min. From Eq. (11.72),

$$c_v = \frac{0.197H_{dr}^2}{t_{50}} = \frac{0.197\left(\frac{2.24}{2}\right)^2}{19} = 0.013 \text{ cm}^2/\text{min} = \mathbf{2.17 \times 10^{-4}\ cm^2/sec}$$

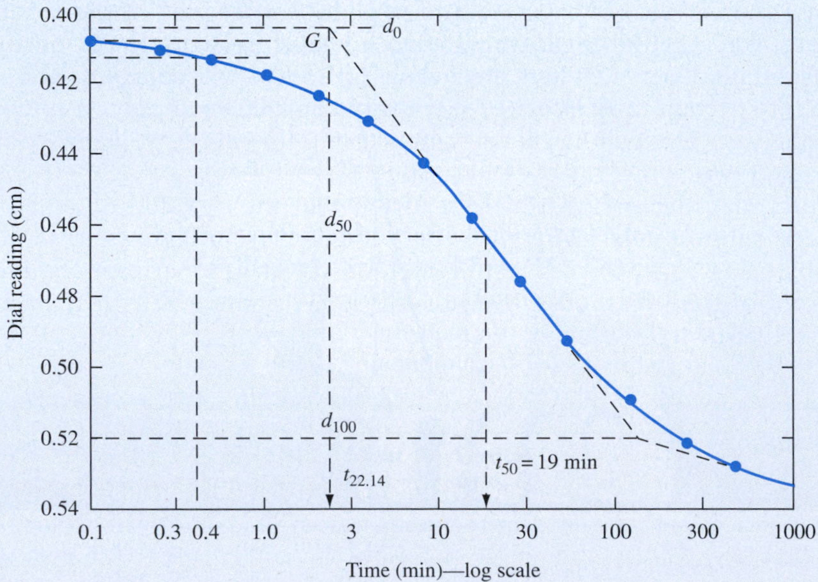

Figure 11.38

Example 11.20

Refer to the laboratory test results of a consolidation test given in Example 11.19. Using the hyperbola method, determine c_v.

Solution

The following table can now be prepared.

Time, t (min)	Dial reading (cm)	ΔH (cm)	$\frac{t}{\Delta H}$
0	0.3975	0	0
0.10	0.4082	0.0107	9.346
0.25	0.4102	0.0127	19.89
0.50	0.4128	0.0153	32.68
1.00	0.4166	0.0191	52.36
2.00	0.4224	0.0249	80.32
4.00	0.4298	0.0323	123.84
8.00	0.4420	0.0445	179.78
16.00	0.4572	0.0597	268.00
30.00	0.4737	0.0762	393.70
60.00	0.4923	0.0948	623.91
120.00	0.5080	0.1105	1085.97

The plot of $t/\Delta H$ versus time (t) is shown in Figure 11.39. From this plot,

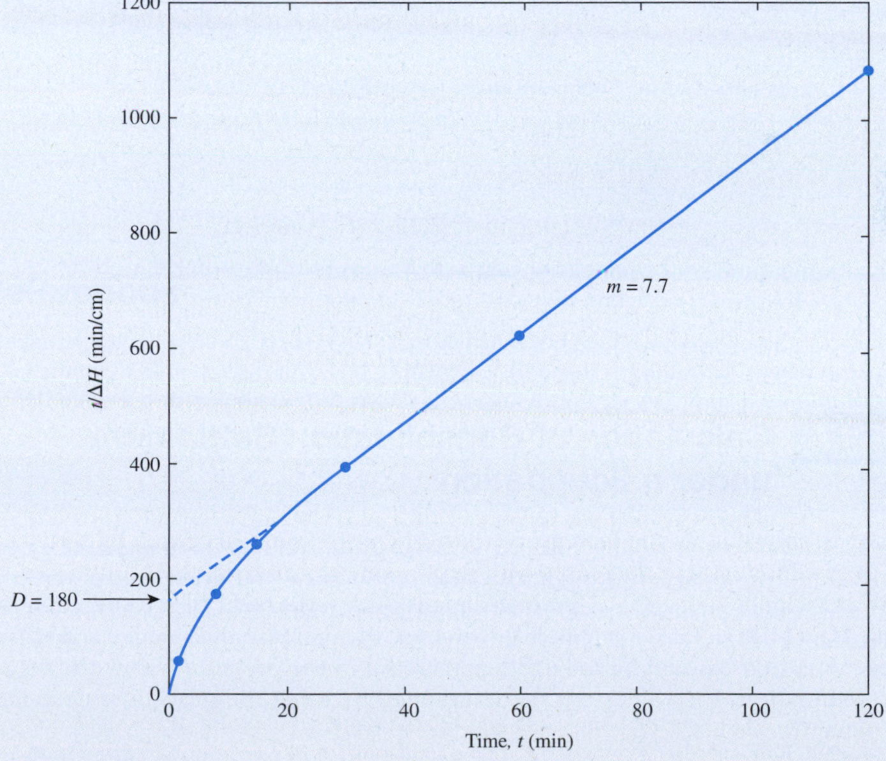

Figure 11.39

$$D \approx 180$$

$$m = \frac{1085.97 - 623.91}{60} \approx 7.7$$

So, from Eq. (11.74),

$$c_v = \frac{0.3mH_{dr}^2}{D} = \frac{(0.3)(7.7)\left(\frac{2.24}{2}\right)^2}{180} = 0.0161 \text{ cm}^2/\text{min} = \mathbf{2.68 \times 10^{-4}\ cm^2/sec}$$

Example 11.21

Refer to the laboratory test results of a consolidation test given in Example 11.19. Using the early stage log-t method, determine c_v.

Solution

Refer to Figure 11.38. A tangent is drawn through the point of inflection. It intersects the d_o line at G. The time corresponding to point G is 2.57 min. From Eq. (11.75),

$$c_v = \frac{0.0385(H_{dr})^2}{t_{22.14}} = \frac{(0.0385)\left(\frac{2.24}{2}\right)^2}{2.57}$$

$$= 0.01879 \text{ cm/min} = \mathbf{3.13 \times 10^{-4}\ cm^2/sec}$$

11.17 Calculation of Consolidation Settlement under a Foundation

Chapter 10 showed that the increase in the vertical stress in soil caused by a load applied over a limited area decreases with depth z measured from the ground surface downward. Hence to estimate the one-dimensional settlement of a foundation, we can use Eq. (11.35), (11.37), or (11.38). However, the increase of effective stress, $\Delta\sigma'$, in these equations should be the average increase in the pressure below the center of the foundation. The values can be determined by using the procedure described in Chapter 10.

Assuming that the pressure increase varies parabolically, using Simpson's rule, we can estimate the value of $\Delta\sigma'_{av}$ as

$$\Delta\sigma'_{av} = \frac{\Delta\sigma'_t + 4\Delta\sigma'_m + \Delta\sigma'_b}{6} \tag{11.76}$$

where $\Delta\sigma'_t$, $\Delta\sigma'_m$, and $\Delta\sigma'_b$ represent the increase in the effective pressure at the top, middle, and bottom of the layer, respectively.

Example 11.22

Calculate the settlement of the 3-m-thick clay layer (Figure 11.40) that will result from the load carried by a 1.5-m-square foundation. The clay is normally consolidated. Use the weighted average method [Eq. (11.76)] to calculate the average increase of effective pressure in the clay layer.

Solution

For normally consolidated clay, from Eq. (11.35),

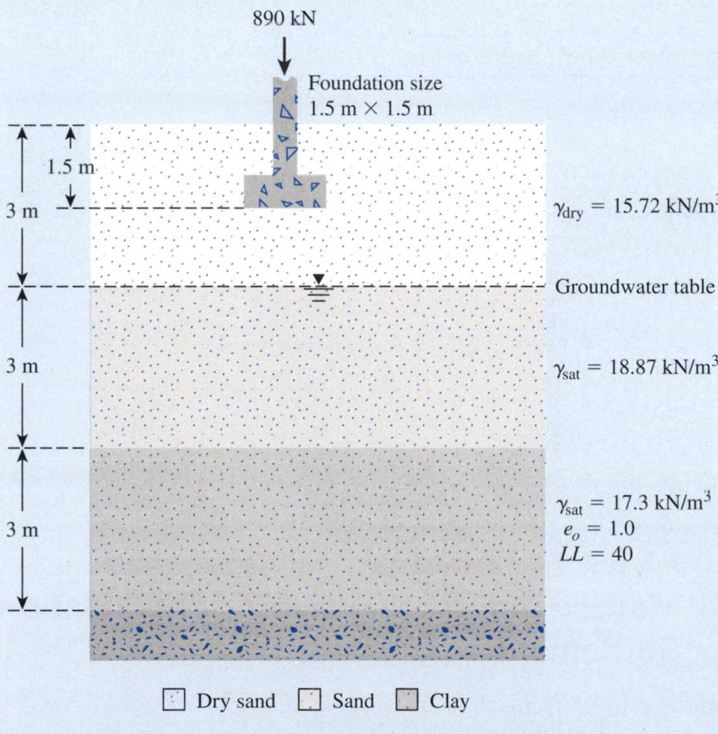

Figure 11.40

$$S_c = \frac{C_c H}{1 + e_o} \log \frac{\sigma'_o + \Delta\sigma'_{av}}{\sigma'_o}$$

where

$$C_c = 0.009(LL - 10) = 0.009(40 - 10) = 0.27$$
$$H = 3 \times 1000 = 3000 \text{ mm.}$$
$$e_o = 1.0$$

$$\sigma'_o = 3 \text{ m} \times \gamma_{\text{dry(sand)}} + 3 \text{ m}[\gamma_{\text{sat(sand)}} - 9.81] + \frac{3}{2}[\gamma_{\text{sat(clay)}} - 9.81]$$
$$= 3 \times 15.72 + 3(18.87 - 9.81) + 1.5(17.3 - 9.81)$$
$$= 85.58 \text{ kN/m}^2$$

From Eq. (11.76),

$$\Delta\sigma'_{av} = \frac{\Delta\sigma'_t + 4\Delta\sigma'_m + \Delta\sigma'_b}{6}$$

$\Delta\sigma'_t$, $\Delta\sigma'_m$, and $\Delta\sigma'_b$ below the center of the foundation can be obtained from Eq. (10.36).

Now we can prepare the following table (*Note:* $L/B = 5/5 = 1$):

m_1	z (m)	$b = B/2$ (m)	$n_1 = z/b$	q (kN/m²)	I_4	$\Delta\sigma' = qI_4$ (kN/m²)
1	4.5	0.75	6	$\frac{890}{1.5 \times 1.5} = 395.6$	0.051	$20.18 = \Delta\sigma'_t$
1	6.0	0.75	8	395.6	0.029	$11.47 = \Delta\sigma'_m$
1	7.5	0.75	10	395.6	0.019	$7.57 = \Delta\sigma'_b$

So,

$$\Delta\sigma'_{av} = \frac{20.18 + (4)(11.47) + 7.52}{6} = 12.26 \text{ kN/m}^2$$

Hence,

$$S_c = \frac{(0.27)(3000)}{1 + 1} \log \frac{85.58 + 12.26}{85.58} \approx \mathbf{23.6 \text{ mm}}.$$

11.18 Methods for Accelerating Consolidation Settlement

In many instances, *sand drains* and *prefabricated vertical drains* are used in the field to accelerate consolidation settlement in soft, normally consolidated clay layers and to achieve precompression before the construction of a desired foundation. Sand drains are constructed by drilling holes through the clay layer(s) in the field at regular intervals. The holes then are backfilled with sand. This can be achieved by several means, such as (a) rotary drilling and then backfilling with sand; (b) drilling by continuous flight auger with a hollow stem and backfilling with sand (through the hollow stem); and (c) driving hollow steel piles. The soil inside the pile is then jetted out, and the hole is backfilled with sand. Figure 11.41 shows a schematic diagram of sand drains. After backfilling the drill holes with sand, a surcharge is applied at the ground surface. This surcharge will increase the pore water pressure in the clay. The excess pore water pressure in the clay will be dissipated by drainage—both vertically and radially to the sand drains—which accelerates settlement of the clay layer. In Figure 11.41a, note that the radius of the sand drains is r_w. Figure 11.41b shows the plan of the layout of the sand drains. The effective zone from which the radial drainage will be directed toward a given sand drain is approximately cylindrical, with a diameter of d_e. The surcharge that needs to be applied at the ground surface and the length of time it has to be maintained to achieve the desired degree of consolidation will be a function of r_w, d_e, and other soil parameters. Figure 11.42 shows a sand drain installation in progress.

Prefabricated vertical drains (PVDs), which also are referred to as *wick* or *strip drains,* originally were developed as a substitute for the commonly used sand drain. With the advent of materials science, these drains are manufactured from synthetic polymers such as polypropylene and high-density polyethylene. PVDs normally are manufactured with a corrugated or channeled synthetic core enclosed by a geotextile filter, as shown schematically in Figure 11.43. Installation rates reported in the literature are on the order of 0.1 to 0.3 m/sec, excluding equipment mobilization and setup time. PVDs have been used extensively in the past for expedient consolidation of low permeability soils under surface surcharge. The main advantage of PVDs over sand drains is that they do not require drilling and, thus, installation is much faster. Figure 11.44 shows the installation of PVDs in the field.

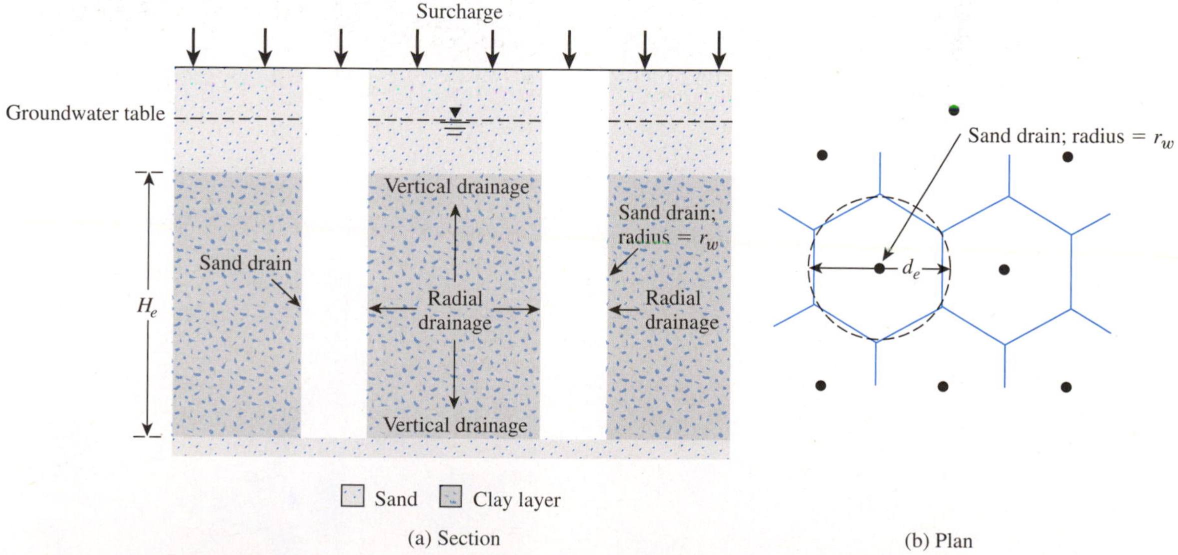

Figure 11.41 Sand drains

Figure 11.42 Sand drain installation in progress (*Courtesy of E.C. Shin, University of Incheon, South Korea*)

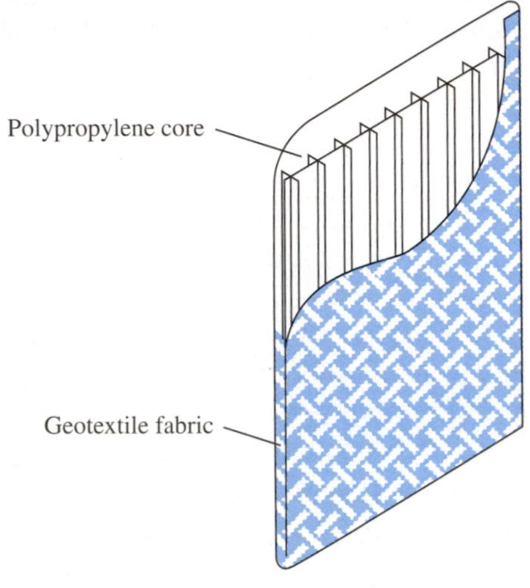

Figure 11.43 Prefabricated vertical drain (PVD)

Figure 11.44 Installation of PVDs in progress (*Courtesy of E.C. Shin, University of Incheon, South Korea*)

11.19 Summary

This chapter discussed theories relating to the settlement of foundations. Following is a summary of the topics covered.

- Total settlement of a foundation is the sum of elastic settlement and consolidation settlement. Consolidation settlement has two components—primary and secondary.
- Elastic settlement (Sections 11.3 and 11.4) is primarily a function of the size (length and width) and rigidity of the foundation, the modulus of elasticity and Poisson's ratio of the soil supporting the foundation, and the intensity of the load applied.
- Consolidation is a time-dependent process of settlement of the saturated clay layer(s) located below the groundwater table.
- Primary consolidation settlement can be calculated using Eqs. (11.35), (11.37), and (11.38).
- Empirical relationships for compression index and swell index needed to estimate primary consolidation settlement are given in Sections 11.11 and 11.12.
- Secondary consolidation settlement can be estimated using Eq. (11.48).
- The degree of consolidation at any time after load application is a function of the nondimensional time factor T_v [see Table 11.7 and Eqs. (11.66)–(11.69)].
- Construction time correction of consolidation settlement due to ramp loading has been discussed in Section 11.15.
- The coefficient of consolidation for a given loading range can be obtained by using logarithm-of-time, square-root-of-time, hyperbola, and early stage log-t methods (Section 11.16).
- Sand drains and prefabricated vertical drains may be used to accelerate the consolidation process in the field (Section 11.18).

There are several case histories in the literature for which the fundamental principles of soil compressibility have been used to predict and compare the actual total settlement and the time rate of settlement of soil profiles under superimposed loading. In some cases, the actual and predicted maximum settlements agree remarkably well; in many others, the predicted settlements deviate to a large extent from the actual settlements observed. The disagreement in the latter cases may have several causes:

1. Improper evaluation of soil properties
2. Nonhomogeneity and irregularity of soil profiles
3. Error in the evaluation of the net stress increase with depth, which induces settlement

The variation between the predicted and observed time rate of settlement may also be due to

- Improper evaluation of c_v (see Section 11.16)
- Presence of irregular sandy seams within the clay layer, which reduces the length of the maximum drainage path, H_{dr}

Problems

11.1 A vertical column load, $P = 600$ kN, is applied to a rigid concrete foundation with dimensions $B = 1$ m and $L = 2$ m, as shown in Figure 11.45. The foundation rests at a depth $D_f = 0.75$ m on a uniform dense sand with the following properties: average modulus of elasticity, $E_s = 20{,}600$ kN/m^2, and Poisson's ratio, $\mu_s = 0.3$. Estimate the elastic settlement due to the net applied pressure, $\Delta\sigma$, on the foundation. Given: $H = 5$ m.

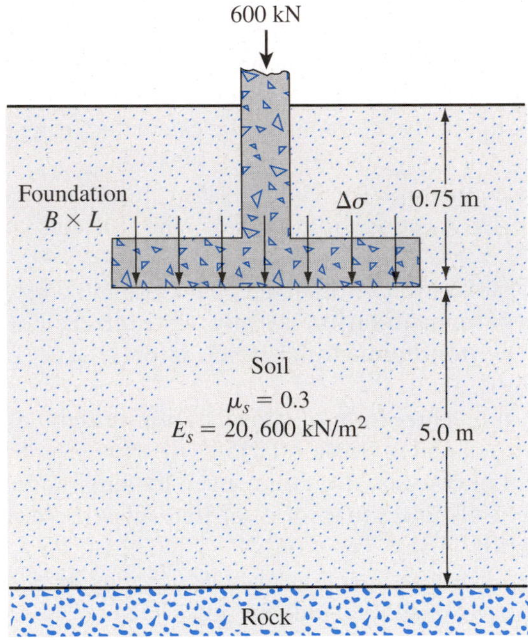

Figure 11.45

11.2 A rigid reinforced concrete foundation is subjected to a column load of 388 kN. The foundation measures 2.4 m × 2.4 m in plan and rests on 6.5 m of layered soil underlain by rock. The soil layers have the following characteristics:

Layer type	Thickness (m)	E_s (kN/m^2)	γ_d (kN/m^3)
Loose sand	1.5	15,167	16.5
Medium clay	1.5–4.0	6756	17.1
Dense sand	4.0–6.5	62,735	19.2

If the foundation depth $D_f = 1.2$ m and $\mu_s = 0.4$ for all layers, estimate the elastic settlement of the foundation. Use Eq. (11.1)

11.3 Following are the results of a laboratory consolidation test on a sample of undisturbed clay obtained from the field.

Pressure, σ' (kg/cm^2)	Final height of specimen (cm)
0	1.9
0.25	1.763
0.5	1.742
1.0	1.702
2.0	1.656
4.0	1.604
8.0	1.550

The height of the specimen at the beginning of the test was 1.9 cm, and the diameter was 6.35 cm. The mass of the dry specimen was 91 g. Estimate the compression index and the preconsolidation pressure from the e – log σ' curve. Given: $G_s = 2.72$.

11.4 The following are the results of a consolidation test on a sample of a clayey soil.

e	Pressure, σ' (kN/m^2)
1.113	25
1.106	50
1.066	100
0.982	200
0.855	400
0.735	800
0.63	1600
0.66	800
0.675	400
0.685	200

a. Plot the e-log σ' curve.

b. Using Casagrande's method, determine the preconsolidation pressure.

c. Calculate the compression index, C_c and the ratio of C_s/C_c.

11.5 Organic soils are typically characterized by high void ratio, low specific gravity, and high compressibility. Following are the results of a consolidation test on a sample of organic soil obtained in southwest Florida.

Pressure, σ' (kN/m^2)	Change in dial reading (mm.)
4.8	0.279
9.6	0.302
24	0.411
48	0.579
96	0.902
192	1.996
384	3.921

Given are the initial height of the specimen = 19 mm., weight of dry specimen = 18 g, area of specimen = 31.67 cm.^{2}, and $G_s = 2.55$.

a. Plot the e-log σ' curve.
b. Determine the preconsolidation pressure.
c. Calculate the compression index, C_c.

11.6 The coordinates of two points on the virgin compression curve are as follows:

e	σ' (kN/m^2)
1.72	96
1.36	192

Determine the void ratio that corresponds to a pressure of 6300 lb/ft^2.

11.7 Figure 11.46 shows a foundation with dimensions $B \times L$ supporting a column load, P. The foundation rests on a sandy soil underlain by a clay layer. Estimate the primary consolidation settlement of the clay due to the foundation load. Given: $P = 350$ kN, $B = L = 2.5$ m, $D_f = 2$ m, $H_1 = 3$ m, and $H_2 = 4.5$ m. The soil properties are as follows:

Sand: $\gamma_d = 16.4$ kN/m^3; $\gamma_{sat} = 18.8$ kN/m^3

Clay: $\gamma_{sat} = 17.6$ kN/m^3; $e_0 = 0.82$; $LL = 48$; normally consolidated

(Use $\Delta\sigma' = \dfrac{P}{BL}$ at the middle of the clay layer.)

11.8 Redo Problem 11.7 using the weighted average method [Eq. (11.76)] to calculate the stress increase in the clay layer.

11.9 Refer to Figure 11.46. If the consolidation properties of the clay are represented by the test results given in Problem 11.4, determine the primary

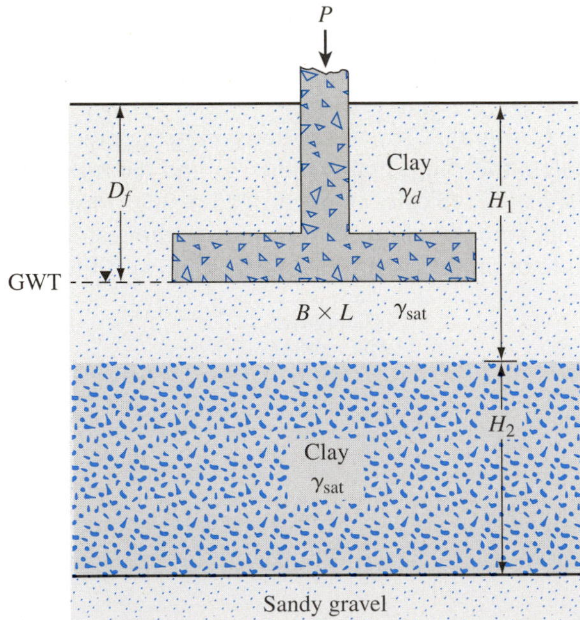

Figure 11.46

consolidation settlement of the clay layer. Given: $P = 675$ kN, $B = L = 3$ m, $D_f = 2$ m, $H_1 = 4$ m, and $H_2 = 6$ m. The soil properties are

Sand: $\gamma_d = 15$ kN/m^3; $\gamma_{sat} = 18$ kN/m^3

Clay: $\gamma_{sat} = 18.9$ kN/m^3; $e_0 = 1.18$

(Use $\Delta\sigma' = \dfrac{P}{BL}$ at the middle of the clay layer.)

11.10 Consider the soil profile shown in Figure 11.47 subjected to the uniformly distributed load, $\Delta\sigma$, on the ground surface. Given: $\Delta\sigma = 45$ kN/m^2, $H_1 = 1.4$ m, $H_2 = 2.8$ m, and $H_3 = 4.9$ m. Soil characteristics are

Sand: $\gamma_d = 17.6$ kN/m^3; $\gamma_{sat} = 19.3$ kN/m^3
Clay: $\gamma_{sat} = 18.7$ kN/m^3; $LL = 46$; $e = 0.71$; $C_s = \dfrac{1}{5}C_c$

Estimate the primary consolidation settlement of the clay if

a. The clay is normally consolidated
b. The preconsolidation pressure, $\sigma_c' = 100$ kN/m^2

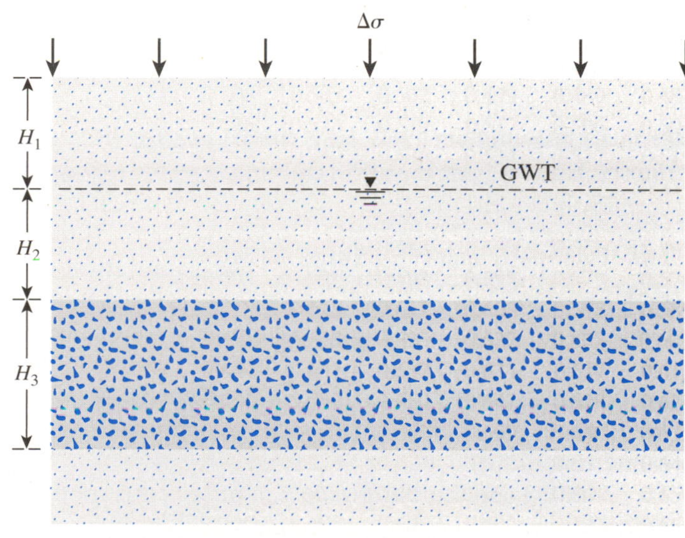

Figure 11.47

11.11 Refer to Figure 11.47. Estimate the primary consolidation settlement in the clay layer. Given: $\Delta\sigma = 105$ kN/m^2, $H_1 = 2.2$ m, $H_2 = 4.4$ m, and $H_3 = 7.5$ m. Soil characteristics are

Sand: $e = 0.58$; $G_s = 2.67$
Clay: $LL = 49$; $e = 1.08$; $G_s = 2.71$; $\sigma_c' = 210$ kN/m^2; $C_s = \dfrac{1}{6}C_c$

11.12 An undisturbed clay has a preconsolidation pressure of 182 kN/m^2 at a void ratio of 1.1. Laboratory consolidation test yields the following data:

e	σ' (kN/m²)
0.98	200
0.84	400

a. Determine the void ratio that corresponds to a pressure 575 kN/m².

b. What would be the void ratio at a pressure of 144 kN/m²? Given: $C_s = \frac{1}{5}C_c$.

11.13 A 4-m clay layer in the field has a current effective stress of $\sigma_0{}' = 58$ kN/m². There is a net stress increase of $\Delta\sigma = 195$ kN/m² due to a foundation load. Only four data points are available from a consolidation test on the clay, as shown. Estimate the primary consolidation settlement of the clay layer in the field.

e	σ' (kN/m²)
1.08	20
1.05	80
0.94	200
0.79	400

11.14 Refer to Problem 11.11. How long will it take for 85% consolidation to be over in the field? Given: $c_v = 0.18$ cm²/min.

11.15 Refer to Problem 11.11.

a. What would be the degree of consolidation in the middle of the clay layer 180 days after the application of the foundation load? Given: $c_v = 0.18$ cm²/min.

b. What would be the remaining excess pore water pressure in the middle of the clay layer 180 days after construction?

11.16 For the consolidation test data given in Problem 11.12,

a. Determine the coefficient of volume compressibility for the pressure range stated.

b. If $c_v = 0.013$ cm²/sec, determine k in m/sec corresponding to the average void ratio within the pressure range.

11.17 The time for 75% consolidation of a 25-mm clay specimen (drained at top and bottom) in the laboratory is 22 minutes. How long will it take for a 6-m thick clay layer in the field to undergo 55% consolidation under the same pressure increment? In the field, there is a rock layer at the bottom of the clay.

11.18 A 5.5-m thick clay layer in the field (drained on one side) is normally consolidated. When the pressure is increased from 72 kN/m² to 144 kN/m², the void ratio decreases from 1.12 to 0.98. The hydraulic conductivity, k, of the clay during the above loading range was found to be 4.3×10^{-7} cm/sec.

a. How long (in days) will it take for the clay layer to reach 70% consolidation?

b. What is the settlement at that time (that is, at 70% consolidation)?

11.19 For a laboratory consolidation test on a 25 mm thick clay specimen (drained on both ends), the following data were obtained:

e	σ' (kN/m^2)
0.92	150
0.77	300

If the time for 65% consolidation is 5.25 min, determine the hydraulic conductivity of the clay for the loading range.

11.20 A 6.4-m thick saturated clay layer (two-way drainage) subjected to surcharge loading underwent 87% primary consolidation in 300 days.

a. Find the coefficient of consolidation of the clay for the pressure range.

b. How long will it take for a 19-mm. thick specimen of the same undisturbed clay to undergo 75% consolidation in a laboratory test?

11.21 Refer to Figure 11.48. A square foundation, 2.5 × 2.5 m in size, supports a column load of 478 kN. The soil characteristics are given in the figure. Field monitoring indicated that the foundation settlement was 46 mm at the end of 2 years.

a. Estimate the average stress increase in the clay layer due to the applied load.

b. Estimate the primary consolidation settlement.

c. What is the degree of consolidation after 2 years?

d. Estimate the coefficient of consolidation for the pressure range.

e. Estimate the settlement in 3 years.

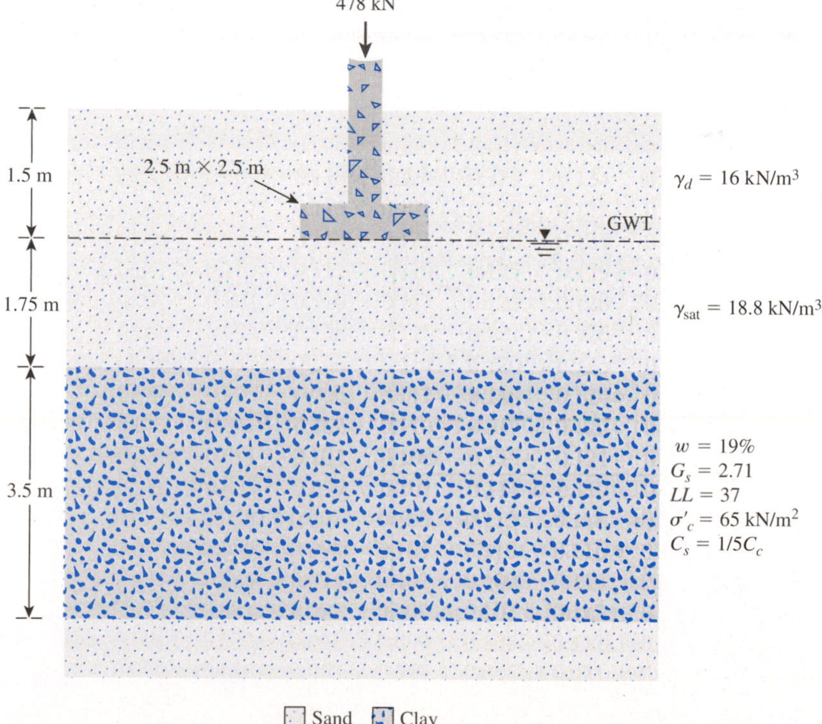

Figure 11.48

Critical Thinking Problem

11.C.1 Foundation engineers are often challenged by the existence of soft compressible soils at the construction site. Figure 11.49 shows a soil profile with a silty sand (γ = 17 kN/m^3; γ_{sat} = 19.2 kN/m^3) underlain by high plasticity clay (γ_{sat} = 18.8 kN/m^3) and a peat layer (γ_{sat} = 15 kN/m^3), followed by dense sand. To expedite consolidation and minimize future settlement, an additional 1.75-m thick fill material, compacted to a unit weight of 20.1 kN/m^3, will be placed on top of the silty sand layer. The plan area of the fill is 8 m × 8 m. The fill load will be left in place for 18 months, after which construction will begin with the fill becoming part of the permanent foundation. Undisturbed samples collected from the clay and organic layers had the following properties:

Layer	C_c	C_α	c_v (cm^2/sec)	e_0
Clay	0.31	0.048	0.006	1.08
Peat	7.2	0.273	0.029	6.4

a. Estimate the total consolidation settlement under the action of the fill load. Consider both the clay and peat layers to be normally consolidated.
b. Estimate the time for 99% primary consolidation in each layer. Are the layers singly or doubly drained? Explain.
c. Estimate the secondary compression in each layer up to end of 18 months.
d. What will be the total settlement after 18 months?
e. What is the remaining excess pore water pressure at point A two months after the application of the fill load?

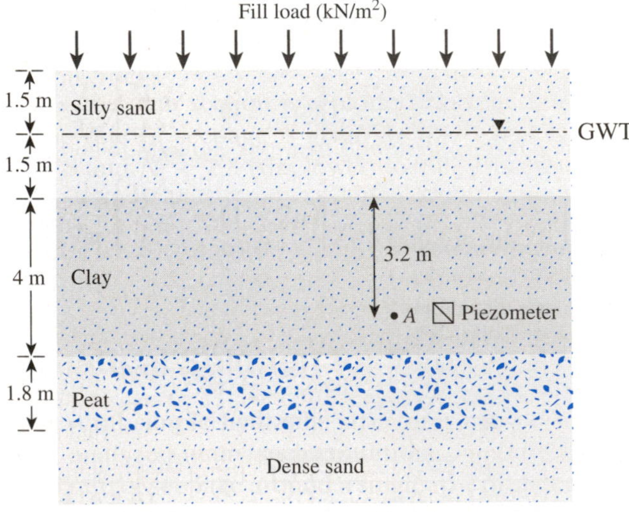

Figure 11.49

f. Determine the effective stress at point A two months after the application of the fill load.

g. A piezometer was installed at point A to monitor the pore water pressure. What should be the piezometer reading ($u_{piezometer}$) two months after the fill load was applied?

References

Bowles, J. E. (1987). "Elastic Foundation Settlement on Sand Deposits," *Journal of Geotechnical Engineering*, ASCE, Vol. 113, No. 8, 846–860.

Casagrande, A. (1936). "Determination of the Preconsolidation Load and Its Practical Significance," *Proceedings,* 1st International Conference on Soil Mechanics and Foundation Engineering, Cambridge, Mass., Vol. 3, 60–64.

Casagrande, A., and Fadum, R. E. (1940). "Notes on Soil Testing for Engineering Purposes," Harvard University Graduate School of Engineering Publication No. 8.

Fox, E. N. (1948). "The Mean Elastic Settlement of a Uniformly Loaded Area at a Depth Below the Ground Surface," *Proceedings*, 2nd International Conference on Soil Mechanics and Foundation Engineering, Rotterdam, Vol. 1, pp. 129–132.

Hanna, D., Sivakugan, N., and Lovisa, J. (2013). "Simple Approach to Consolidation Due to Constant Rate Loading in Clays," *International Journal of Geomechanics*, ASCE, Vol. 13, No. 2, 193–196.

Hansbo, S. (1957). "A New Approach to the Determination of Shear Strength of Clay by the Fall Cone Test," *Report 14*, Swedish Geotechnical Institute, Stockholm.

Hough, B. K. (1957). *Basic Soils Engineering,* Ronald Press, New York.

Kulhawy, F. H., and Mayne, P. W. (1990). *Manual of Estimating Soil Properties for Foundation Design*, Final Report (*EL-6800*), Electric Power Research Institute (EPRI), Palo Alto, CA.

Leroueil, S. (1988). "Tenth Canadian Geotechnical Colloquium: Recent Developments in Consolidation of Natural Clays," *Canadian Geotechnical Journal*, Vol. 25, No. 1, 85–107.

Mayne, P. W., and Poulos, H. G. (1999). "Approximate Displacement Influence Factors for Elastic Shallow Foundations," *Journal of Geotechnical and Geoenvironmental Engineering*, ASCE, Vol. 125, No. 6, 453–460.

Mesri, G., and Godlewski, P. M. (1977). "Time and Stress—Compressiblity Interrelationship," *Journal of the Geotechnical Engineering Division*, ASCE, Vol. 103, No. GT5, 417–430.

Nagaraj, T., and Murty, B. R. S. (1985). "Prediction of the Preconsolidation Pressure and Recompression Index of Soils," *Geotechnical Testing Journal,* Vol. 8, No. 4, 199–202.

Nishida, Y. (1956). "A Brief Note on Compression Index of Soils," *Journal of the Soil Mechanics and Foundations Division,* ASCE, Vol. 82, No. SM3, 1027-1–1027-14.

Park, J. H., and Koumoto, T. (2004). "New Compression Index Equation," *Journal of Geotechnical and Geoenvironmental Engineering,* ASCE, Vol. 130, No. 2, 223–226.

Rendon-Herrero, O. (1983). "Universal Compression Index Equation," *Discussion, Journal of Geotechnical Engineering,* ASCE, Vol. 109, No. 10, 1349.

Rendon-Herrero, O. (1980). "Universal Compression Index Equation," *Journal of the Geotechnical Engineering Division,* ASCE, Vol. 106, No. GT11, 1179–1200.

Robinson, R. G., and Allam, M. M. (1996). "Determination of Coefficient of Consolidation from Early Stage of log *t* Plot," *Geotechnical Testing Journal,* ASTM, Vol. 19, No. 3, 316–320.

Schmertmann, J. H. (1953). "Undisturbed Consolidation Behavior of Clay," *Transactions,* ASCE, Vol. 120, 1201.

Sivakugan, N., Lovisa, J., Ameratunga, J., and Das, B.M. (2014). "Consolidation Settlement Due to Ramp Loading," *International Journal of Geotechnical Engineering*, Vol. 8, No. 2, 191–196.

Sivaram, B., and Swamee, A. (1977). "A Computational Method for Consolidation Coefficient," *Soils and Foundations,* Vol. 17, No. 2, 48–52.

Skempton, A. W. (1944). "Notes on the Compressibility of Clays," *Quarterly Journal of the Geological Society of London,* Vol. 100, 119–135.

Sridharan, A., and Prakash, K. (1985). "Improved Rectangular Hyperbola Method for the Determination of Coefficient of Consolidation," *Geotechnical Testing Journal,* ASTM, Vol. 8, No. 1, 37–40.

Stas, C.V., and Kulhawy, F. H. (1984). "Critical Evaluation of Design Methods for Foundations under Axial Uplift and Compression Loading," *Report EL-3771*, Electric Power Research Institute (EPRI), Palo Alto, CA.

Steinbrenner, W. (1934). "Tafeln zur Setzungsberechnung," *Die Strasse,* Vol. 1, 121–124.

Taylor, D. W. (1942). "Research on Consolidation of Clays," *Serial No. 82,* Department of Civil and Sanitary Engineering, Massachusetts Institute of Technology, Cambridge, Mass.

Taylor, D. W. (1948). *Fundamentals of Soil Mechanics,* Wiley, New York.

Terzaghi, K. (1925). *Erdbaumechanik auf Bodenphysikalischer Grundlager,* Deuticke, Vienna.

Terzaghi, K., and Peck, R. B. (1967). *Soil Mechanics in Engineering Practice,* 2nd ed., Wiley, New York.

Wroth, C. P., and Wood, D. M. (1978). "The Correlation of Index Properties with Some Basic Engineering Properties of Soils," *Canadian Geotechnical Journal,* Vol. 15, No. 2, 137–145.

CHAPTER 12

Shear Strength of Soil

12.1 Introduction

The *shear strength* of a soil mass is the internal resistance per unit area that the soil mass can offer to resist failure and sliding along any plane inside it. One must understand the nature of shearing resistance in order to analyze soil stability problems, such as bearing capacity, slope stability, and lateral pressure on earth-retaining structures. The following will be introduced in this chapter:

- Shear strength parameters of soil
- Laboratory testing of soil under various drainage conditions to estimate the shear strength parameters
- Effect of remolding on shear strength of cohesive soils
- Effect of variation of shear strength depending on the direction of load application
- Use of vane shear to obtain shear strength of saturated cohesive soils

12.2 Mohr–Coulomb Failure Criterion

Mohr (1900) presented a theory for rupture in materials that contended that a material fails because of a critical combination of normal stress and shearing stress and not from either maximum normal or shear stress alone. Thus, the functional relationship between normal stress and shear stress on a failure plane can be expressed in the following form:

$$\tau_f = f(\sigma) \tag{12.1}$$

The failure envelope defined by Eq. (12.1) is a curved line. For most soil mechanics problems, it is sufficient to approximate the shear stress on the failure plane as a linear function of the normal stress (Coulomb, 1776). This linear function can be written as

$$\tau_f = c + \sigma \tan \phi \tag{12.2}$$

where c = cohesion
ϕ = angle of internal friction
σ = normal stress on the failure plane
τ_f = shear strength

The preceding equation is called the *Mohr–Coulomb failure criterion.*

In saturated soil, the total normal stress at a point is the sum of the effective stress (σ') and pore water pressure (u), or

$$\sigma = \sigma' + u$$

The effective stress σ' is carried by the soil solids. The Mohr–Coulomb failure criterion, expressed in terms of effective stress, will be of the form

$$\tau_f = c' + \sigma' \tan \phi' \tag{12.3}$$

where c' = cohesion and ϕ' = friction angle, based on effective stress.

Thus, Eqs. (12.2) and (12.3) are expressions of shear strength based on total stress and effective stress respectively. The value of c' for sand and inorganic silt is 0. For normally consolidated clays, c' can be approximated at 0. Overconsolidated clays have values of c' that are greater than 0. The angle of friction, ϕ', is sometimes referred to as the *drained angle of friction*. Typical values of ϕ' for some granular soils are given in Table 12.1.

Table 12.1 Typical Values of Drained Angle of Friction for Sands and Silts

Soil type	**ϕ' (deg)**
Sand: Rounded grains	
Loose	27–30
Medium	30–35
Dense	35–38
Sand: Angular grains	
Loose	30–35
Medium	35–40
Dense	40–45
Gravel with some sand	34–48
Silts	26–35

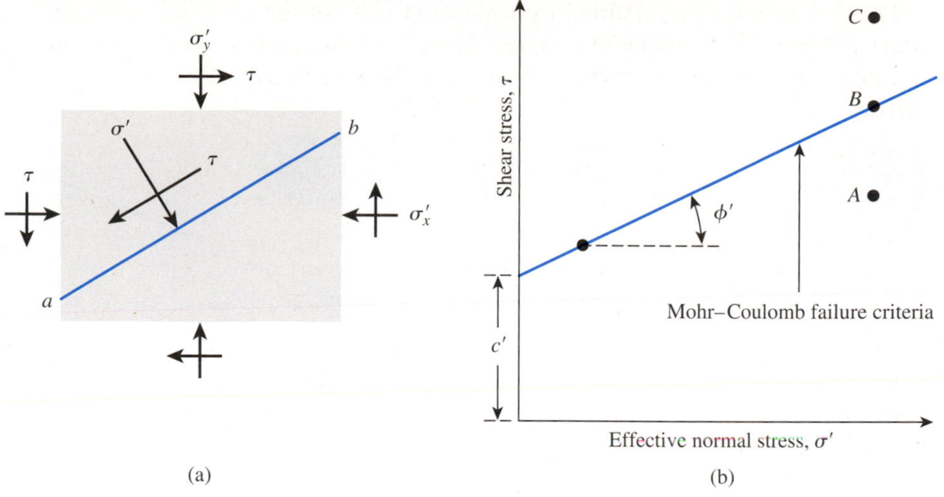

Figure 12.1 Mohr–Coulomb failure criterion

The significance of Eq. (12.3) can be explained by referring to Figure 12.1, which shows an elemental soil mass. Let the effective normal stress and the shear stress on the plane *ab* be σ' and τ, respectively. Figure 12.1b shows the plot of the failure envelope defined by Eq. (12.3). If the magnitudes of σ' and τ on plane *ab* are such that they plot as point *A* in Figure 12.1b, shear failure will not occur along the plane. If the effective normal stress and the shear stress on plane *ab* plot as point *B* (which falls on the failure envelope), shear failure will occur along that plane. A state of stress on a plane represented by point *C* cannot exist, because it plots above the failure envelope, and shear failure in a soil would have occurred already.

12.3 Inclination of the Plane of Failure Caused by Shear

As stated by the Mohr–Coulomb failure criterion, failure from shear will occur when the shear stress on a plane reaches a value given by Eq. (12.3). To determine the inclination of the failure plane with the major principal plane, refer to Figure 12.2, where σ_1' and σ_3' are, respectively, the major and minor effective principal stresses. The failure plane *EF* makes an angle θ with the major principal plane. To determine the angle θ and the relationship between σ_1' and σ_3', refer to Figure 12.3, which is a plot of the Mohr's circle for the state of stress shown in Figure 12.2 (see Chapter 10). In Figure 12.3, *fgh* is the failure envelope defined by the relationship $\tau_f = c' + \sigma' \tan \phi'$. The radial line *ab* defines the major principal plane (*CD* in Figure 12.2), and the radial line *ad* defines the failure plane (*EF* in Figure 12.2). It can be shown that $\angle bad = 2\theta = 90 + \phi'$, or

$$\theta = 45 + \frac{\phi'}{2} \tag{12.4}$$

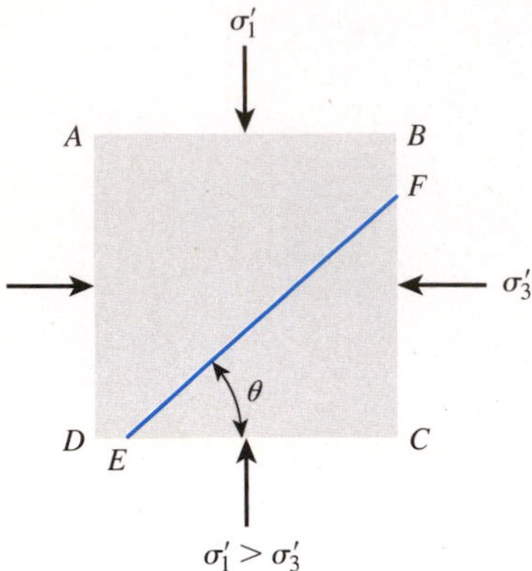

Figure 12.2 Inclination of failure plane in soil with major principal plane

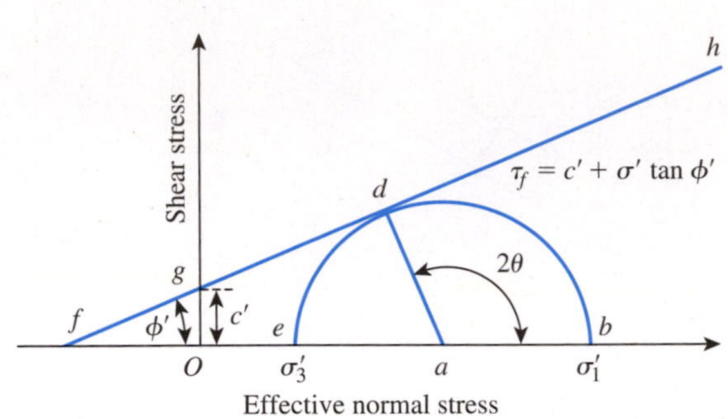

Figure 12.3 Mohr's circle and failure envelope

Again, from Figure 12.3,

$$\frac{\overline{ad}}{\overline{fa}} = \sin \phi' \tag{12.5}$$

$$\overline{fa} = fO + Oa = c' \cot \phi' + \frac{\sigma'_1 + \sigma'_3}{2} \tag{12.6a}$$

Also,

$$\overline{ad} = \frac{\sigma'_1 - \sigma'_3}{2} \tag{12.6b}$$

Substituting Eqs. (12.6a) and (12.6b) into Eq. (12.5), we obtain

$$\sin \phi' = \frac{\dfrac{\sigma'_1 - \sigma'_3}{2}}{c' \cot \phi' + \dfrac{\sigma'_1 + \sigma'_3}{2}}$$

or

$$\sigma'_1 = \sigma'_3 \left(\frac{1 + \sin \phi'}{1 - \sin \phi'}\right) + 2c'\left(\frac{\cos \phi'}{1 - \sin \phi'}\right) \tag{12.7}$$

However,

$$\frac{1 + \sin \phi'}{1 - \sin \phi'} = \tan^2\left(45 + \frac{\phi'}{2}\right)$$

and

$$\frac{\cos \phi'}{1 - \sin \phi'} = \tan\left(45 + \frac{\phi'}{2}\right)$$

Thus,

$$\sigma_1' = \sigma_3' \tan^2\left(45 + \frac{\phi'}{2}\right) + 2c' \tan\left(45 + \frac{\phi'}{2}\right) \quad (12.8)$$

An expression similar to Eq. (12.8) could also be derived using Eq. (12.2) (that is, total stress parameters c and ϕ), or

$$\sigma_1 = \sigma_3 \tan^2\left(45 + \frac{\phi}{2}\right) + 2c \tan\left(45 + \frac{\phi}{2}\right) \quad (12.9)$$

12.4 Laboratory Test for Determination of Shear Strength Parameters

There are several laboratory methods available to determine the shear strength parameters (i.e., c, ϕ, c', ϕ') of various soil specimens in the laboratory. They are as follows:

- Direct shear test
- Triaxial test
- Direct simple shear test
- Plane strain triaxial test
- Torsional ring shear test

The direct shear test and the triaxial test are the two commonly used techniques for determining the shear strength parameters. These two tests will be described in detail in the sections that follow.

12.5 Direct Shear Test

The direct shear test is the oldest and simplest form of shear test arrangement. A diagram of the direct shear test apparatus is shown in Figure 12.4. The test equipment consists of a metal shear box in which the soil specimen is placed. The soil specimens may be square or circular in plan. The size of the specimens generally used is about 51 mm × 51 mm or 102 mm × 102 mm across and about 25 mm high. The box is split horizontally into halves. Normal force on the specimen is applied from the top of the shear box. The normal stress on the specimens can be as great as 1050 kN/m^2. Shear force is applied by moving one-half of the box relative to the other to cause failure in the soil specimen.

Depending on the equipment, the shear test can be either stress controlled or strain controlled. In stress-controlled tests, the shear force is applied in equal increments until the specimen fails. The failure occurs along the plane of split of the shear box. After the application of each incremental load, the shear displacement of the top half of the box is measured by a horizontal dial gauge. The change in the height

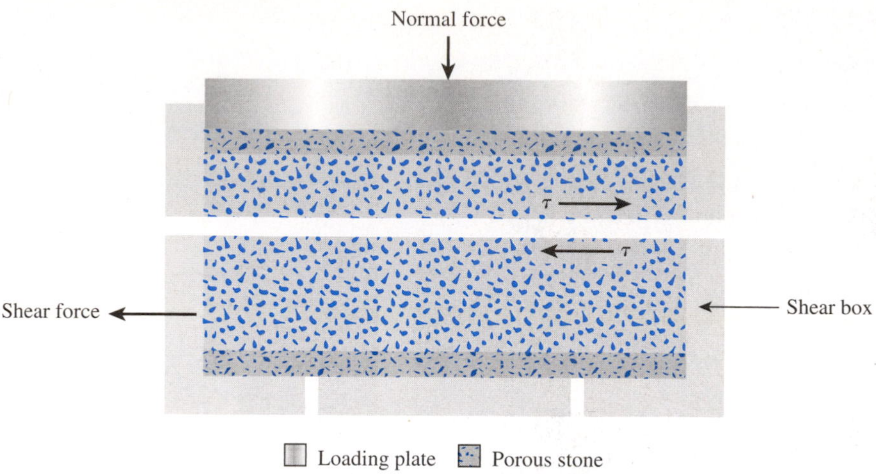

Figure 12.4 Diagram of direct shear test arrangement

of the specimen (and thus the volume change of the specimen) during the test can be obtained from the readings of a dial gauge that measures the vertical movement of the upper loading plate.

In strain-controlled tests, a constant rate of shear displacement is applied to one-half of the box by a motor that acts through gears. The constant rate of shear displacement is measured by a horizontal dial gauge. The resisting shear force of the soil corresponding to any shear displacement can be measured by a horizontal proving ring or load cell. The volume change of the specimen during the test is obtained in a manner similar to that in the stress-controlled tests. Figure 12.5 shows a photograph of strain-controlled direct shear test equipment.

The advantage of the strain-controlled tests is that in the case of dense sand, peak shear resistance (that is, at failure) as well as lesser shear resistance (that is, at a point after failure called *ultimate strength*) can be observed and plotted. In stress-controlled tests, only the peak shear resistance can be observed and plotted. Note that the peak shear resistance in stress-controlled tests can be only approximated because failure occurs at a stress level somewhere between the prefailure load increment and the failure load increment. Nevertheless, compared with strain-controlled tests, stress-controlled tests probably model real field situations better.

For a given test, the normal stress can be calculated as

$$\sigma = \text{Normal stress} = \frac{\text{Normal force}}{\text{Cross-sectional area of the specimen}} \tag{12.10}$$

The resisting shear stress for any shear displacement can be calculated as

$$\tau = \text{Shear stress} = \frac{\text{Resisting shear force}}{\text{Cross-sectional area of the specimen}} \tag{12.11}$$

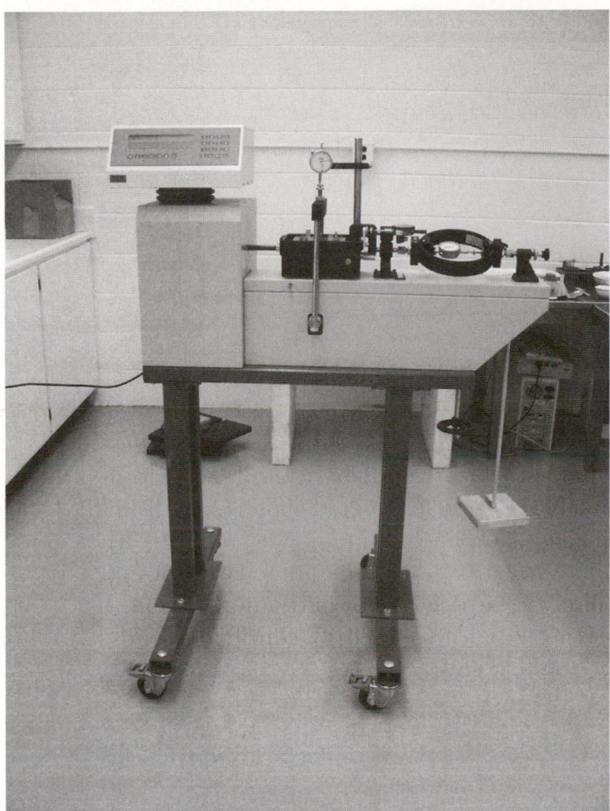

Figure 12.5 Strain-controlled direct shear equipment (*Courtesy of Braja M. Das, Henderson, Nevada*)

Figure 12.6 shows a typical plot of shear stress and change in the height of the specimen against shear displacement for dry loose and dense sands. These observations were obtained from a strain-controlled test. The following generalizations can be developed from Figure 12.6 regarding the variation of resisting shear stress with shear displacement:

1. In loose sand, the resisting shear stress increases with shear displacement until a failure shear stress of τ_f is reached. After that, the shear resistance remains approximately constant for any further increase in the shear displacement.
2. In dense sand, the resisting shear stress increases with shear displacement until it reaches a failure stress of τ_f. This τ_f is called the *peak shear strength*. After failure stress is attained, the resisting shear stress gradually decreases as shear displacement increases until it finally reaches a constant value called the *ultimate shear strength*.

Since the height of the specimen changes during the application of the shear force (as shown in Figure 12.6), it is obvious that the void ratio of the sand changes (at least in the vicinity of the split of the shear box). Figure 12.7 shows the nature of variation of the void ratio for loose and dense sands with shear displacement. At

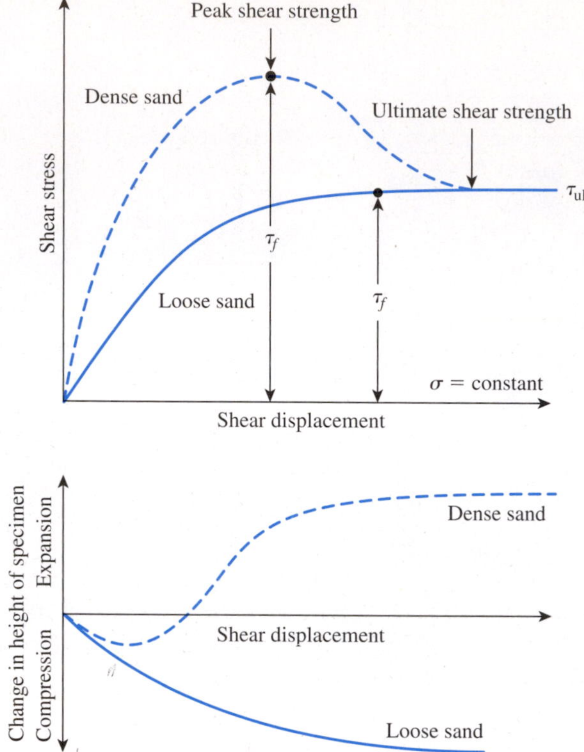

Figure 12.6 Plot of shear stress and change in height of specimen against shear displacement for loose and dense dry sand (direct shear test)

large shear displacement, the void ratios of loose and dense sands become practically the same, and this is termed the *critical void ratio*. It is important to note that, in dry sand,

$$\sigma = \sigma'$$

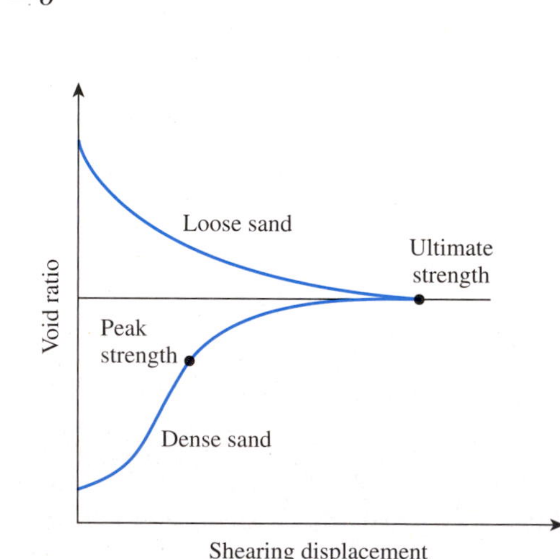

Figure 12.7 Nature of variation of void ratio with shearing displacement

and

$$c' = 0$$

Direct shear tests are repeated on similar specimens at various normal stresses. The normal stresses and the corresponding values of τ_f obtained from a number of tests are plotted on a graph from which the shear strength parameters are determined. Figure 12.8 shows such a plot for tests on a dry sand. The equation for the average line obtained from experimental results is

$$\tau_f = \sigma' \tan \phi' \tag{12.12}$$

So, the friction angle can be determined as follows:

$$\phi' = \tan^{-1}\left(\frac{\tau_f}{\sigma'}\right) \tag{12.13}$$

It is important to note that *in situ* cemented sands may show a c' intercept.

If the variation of the ultimate shear strength (τ_{ult}) with normal stress is known, it can be plotted as shown in Figure 12.8. The average plot can be expressed as

$$\tau_{ult} = \sigma' \tan \phi'_{ult} \tag{12.14}$$

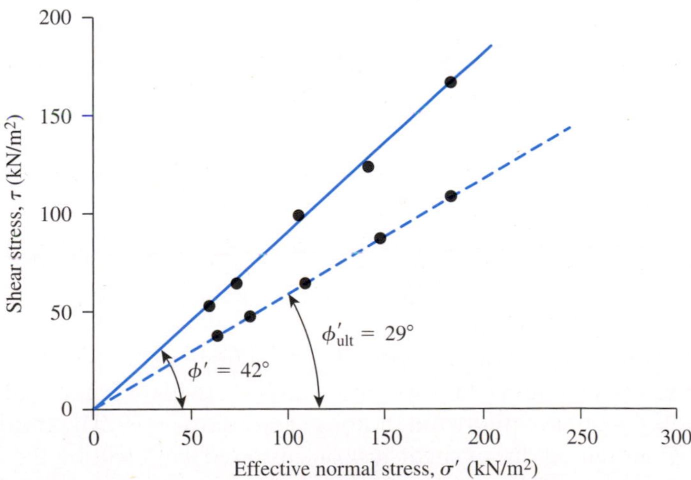

Figure 12.8 Determination of shear strength parameters for a dry sand using the results of direct shear tests

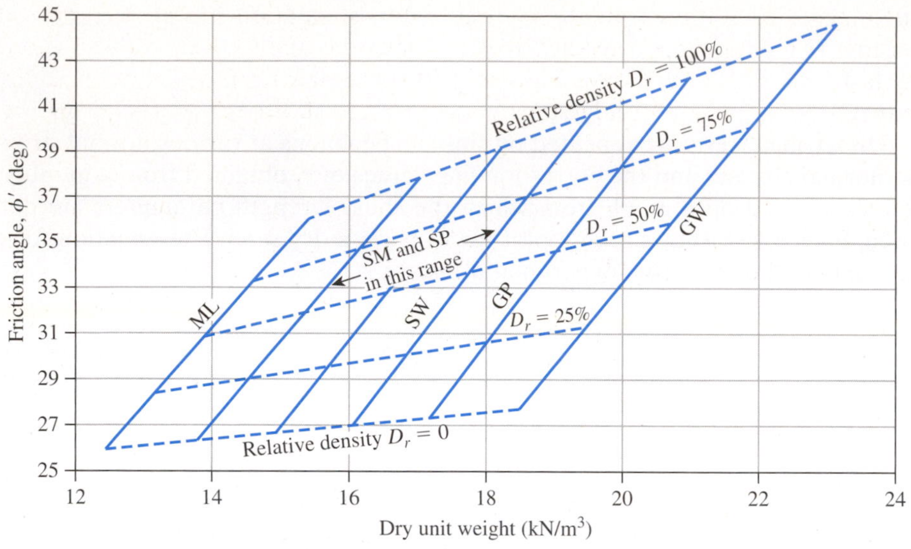

Figure 12.9 Friction angles of granular soils (*Based on U.S. Navy, 1986*)

or

$$\phi'_{\text{ult}} = \tan^{-1}\left(\frac{\tau_{\text{ult}}}{\sigma'}\right) \tag{12.15}$$

Typical values of peak friction angle for granular soils suggested by U.S. Navy (1986) are shown in Figure 12.9.

12.6 Drained Direct Shear Test on Saturated Sand and Clay

In the direct shear test arrangement, the shear box that contains the soil specimen is generally kept inside a container that can be filled with water to saturate the specimen. A *drained test* is made on a saturated soil specimen by keeping the rate of loading slow enough so that the excess pore water pressure generated in the soil is dissipated completely by drainage. Pore water from the specimen is drained through two porous stones. (See Figure 12.4.)

Because the hydraulic conductivity of sand is high, the excess pore water pressure generated due to loading (normal and shear) is dissipated quickly. Hence, for an ordinary loading rate, essentially full drainage conditions exist. The friction angle, ϕ', obtained from a drained direct shear test of saturated sand will be the same as that for a similar specimen of dry sand.

The hydraulic conductivity of clay is very small compared with that of sand. When a normal load is applied to a clay soil specimen, a sufficient length of time must elapse for full consolidation—that is, for dissipation of excess pore water pressure. For this reason, the shearing load must be applied very slowly. The test may last from two to five days. Figure 12.10 shows the results of a drained direct shear test on an overconsolidated clay. Figure 12.11 shows the plot of τ_f against σ'

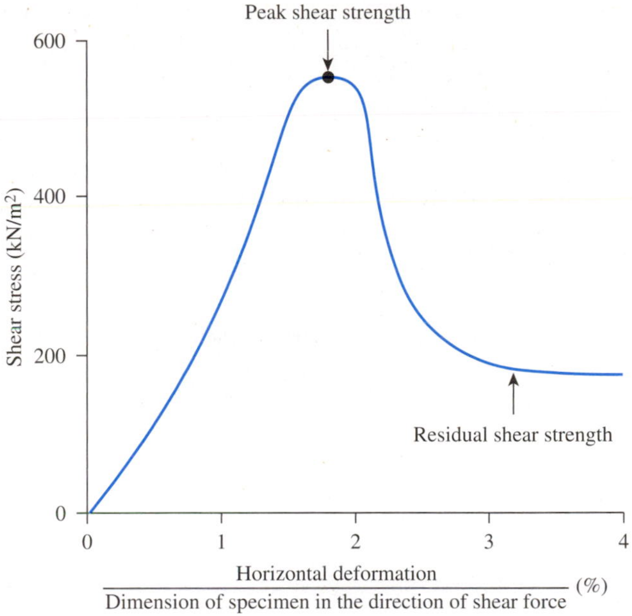

Figure 12.10 Results of a drained direct shear test on an overconsolidated clay [*Note:* Residual shear strength in clay is similar to ultimate shear strength in sand (see Figures 12.6 and 12.7)]

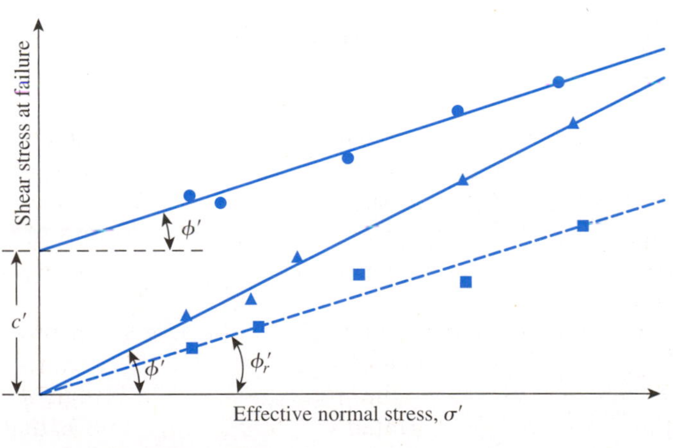

Figure 12.11 Failure envelope for clay obtained from drained direct shear tests

obtained from a number of drained direct shear tests on a normally consolidated clay and an overconsolidated clay. Note that the value of $c' \simeq 0$ for a normally consolidated clay.

Similar to the ultimate shear strength in the case of sand (Figure 12.7), at large shearing displacements, we can obtain the *residual shear strength* of clay (τ_r) in a drained test. This is shown in Figure 12.10. Figure 12.11 shows the plot of τ_r versus σ'. The average plot *will pass through the origin* and can be expressed as

$$\tau_r = \sigma \tan \phi'_r$$

or

$$\phi'_r = \tan^{-1}\left(\frac{\tau_r}{\sigma'}\right) \quad (12.16)$$

The drained angle of friction, ϕ', of normally consolidated clays generally decreases with the plasticity index of soil. This fact is illustrated in Figure 12.12 for a number of clays from data compiled by Sorensen and Okkels (2013).

From this plot,

$$\phi' = 43 - 10 \log (PI) \text{ (mean)} \quad (12.17)$$

and

$$\phi' = 39 - 11 \log (PI) \text{ (lower bound)} \quad (12.18)$$

Skempton (1964) provided the results of the variation of the residual angle of friction, ϕ'_r, of a number of clayey soils with the clay-size fraction ($\leq 2\ \mu$m) present. The following table shows a summary of these results.

Soil	**Clay-size fraction (%)**	**Residual friction angle, ϕ'_r (deg)**
Selset	17.7	29.8
Wiener Tegel	22.8	25.1
Jackfield	35.4	19.1
Oxford clay	41.9	16.3
Jari	46.5	18.6
London clay	54.9	16.3
Walton's Wood	67	13.2
Weser-Elbe	63.2	9.3
Little Belt	77.2	11.2
Biotite	100	7.5

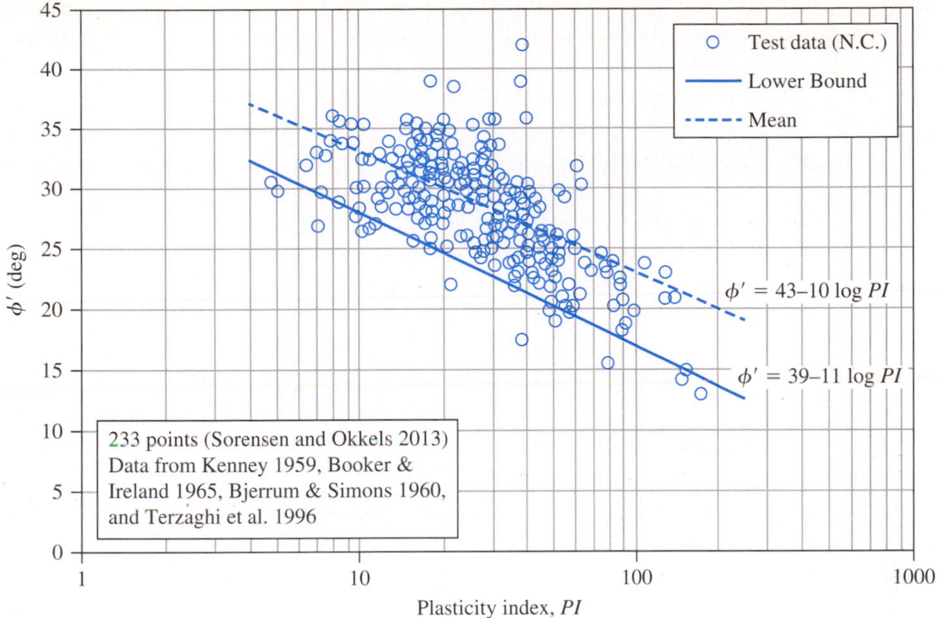

Figure 12.12 Variation of ϕ' with plasticity index (PI) for several normally consolidated clays (*Adapted from Sorensen and Okkels, 2013*)

12.7 General Comments on Direct Shear Test

The direct shear test is simple to perform, but it has some inherent shortcomings. The reliability of the results may be questioned because the soil is not allowed to fail along the weakest plane but is forced to fail along the plane of split of the shear box. Also, the shear stress distribution over the shear surface of the specimen is not uniform. Despite these shortcomings, the direct shear test is the simplest and most economical test for a dry or saturated sandy soil.

In many foundation design problems, one must determine the angle of friction between the soil and the material in which the foundation is constructed (Figure 12.13).

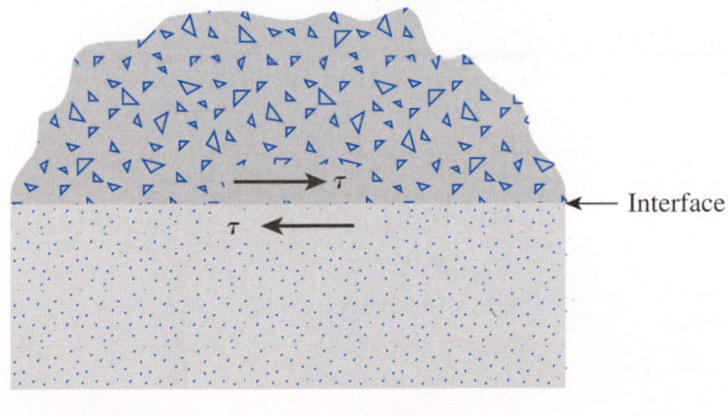

Figure 12.13 Interface of a foundation material and soil

The foundation material may be concrete, steel, or wood. The shear strength along the surface of contact of the soil and the foundation can be given as

$$\tau_f = c'_a + \sigma' \tan \delta' \tag{12.19}$$

where c'_a = adhesion

δ' = effective angle of friction between the soil and the foundation meterial

Note that the preceding equation is similar in form to Eq. (12.3). The shear strength parameters between a soil and a foundation material can be conveniently determined by a direct shear test. This is a great advantage of the direct shear test. The foundation material can be placed in the bottom part of the direct shear test box and then the soil can be placed above it (that is, in the top part of the box), as shown in Figure 12.14, and the test can be conducted in the usual manner.

Figure 12.15 shows the results of direct shear tests conducted in this manner with a quartz sand and concrete, wood, and steel as foundation materials, with $\sigma' = 100 \text{ kN/m}^2$.

It was mentioned briefly in Section 12.1 [related to Eq. (12.1)] that Mohr's failure envelope is curvilinear in nature, and Eq. (12.2) is only an approximation. This fact should be kept in mind when considering problems at higher confining pressures. Figure 12.16 shows the decrease of ϕ' and δ' with the increase of normal stress (σ') for the same materials discussed in Figure 12.15. This can be explained by referring to Figure 12.17, which shows a curved Mohr's failure envelope. If a direct shear test is conducted with $\sigma' = \sigma'_{(1)}$, the shear strength will be $\tau_{f(1)}$. So,

$$\delta'_1 = \tan^{-1}\left[\frac{\tau_{f(1)}}{\sigma'_{(1)}}\right]$$

This is shown in Figure 12.17. In a similar manner, if the test is conducted with $\sigma' = \sigma'_{(2)}$, then

$$\delta' = \delta'_2 = \tan^{-1}\left[\frac{\tau_{f(2)}}{\sigma'_{(2)}}\right]$$

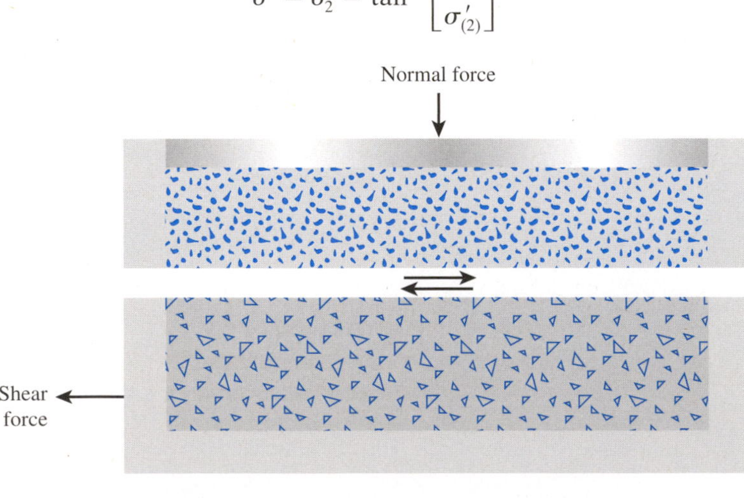

Figure 12.14 Direct shear test to determine interface friction angle

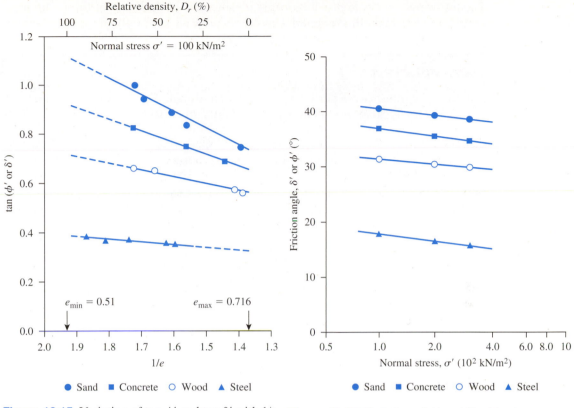

Figure 12.15 Variation of tan ϕ' and tan δ' with $1/e$ [*Note:* e = void ratio, $\sigma' = 100$ kN/m², quartz sand] (*After Acar, Durgunoglu, and Tumay, 1982. With permission from ASCE.*)

Figure 12.16 Variation of ϕ' and δ' with σ' (*Note:* Relative density = 45%; quartz sand) (*After Acar, Durgunoglu, and Tumay, 1982. With permission from ASCE.*)

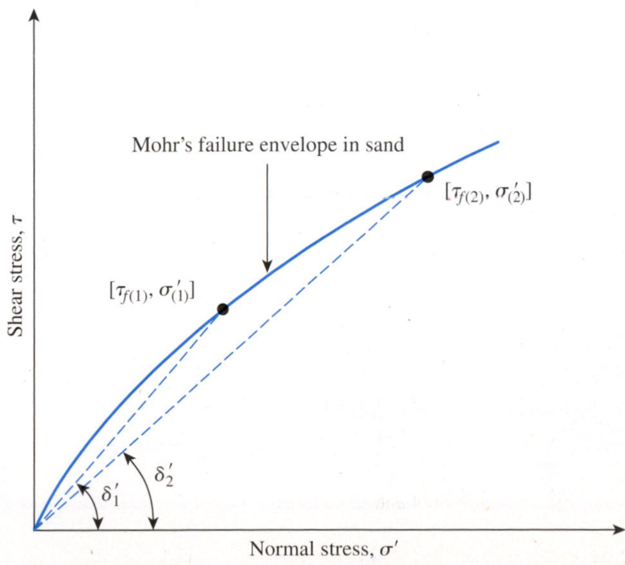

Figure 12.17 Curvilinear nature of Mohr's failure envelope in sand

As can be seen from Figure 12.17, $\delta'_2 < \delta'_1$ since $\sigma'_2 > \sigma'_{(1)}$. Keeping this in mind, it must be realized that the values of ϕ' given in Table 12.1 are only the average values.

Example 12.1

Direct shear tests were performed on a dry, sandy soil. The size of the specimen was 50 mm. × 50 mm. × 19 mm in. Test results are as follows:

Test no.	Normal force (N)	Normal[a] stress $\sigma = \sigma'$ (kN/m²)	Shear force at failure (N)	Shear stress[b] at failure τ_f (kN/m²)
1	89	35.6	53.4	21.4
2	133	53.2	81.4	32.6
3	311	124.4	187.3	74.9
4	445	178.0	267.3	106.9

$$^{a}\sigma'(\text{kN/m}^2) = \frac{\text{normal force}}{\text{area of specimen}} = \frac{(\text{normal force})}{(1000)(0.05\text{m})(0.05\text{m})}$$

$$^{b}\tau_f(\text{kN/m}^2) = \frac{\text{shear force}}{\text{area of specimen}} = \frac{(\text{shear force})}{(1000)(0.05\text{m})(0.05\text{m})}$$

Find the shear stress parameters.

Solution

The shear stresses, τ_f, obtained from the tests are plotted against the normal stresses in Figure 12.18, from which $\boldsymbol{c' = 0, \phi' = 32°}$.

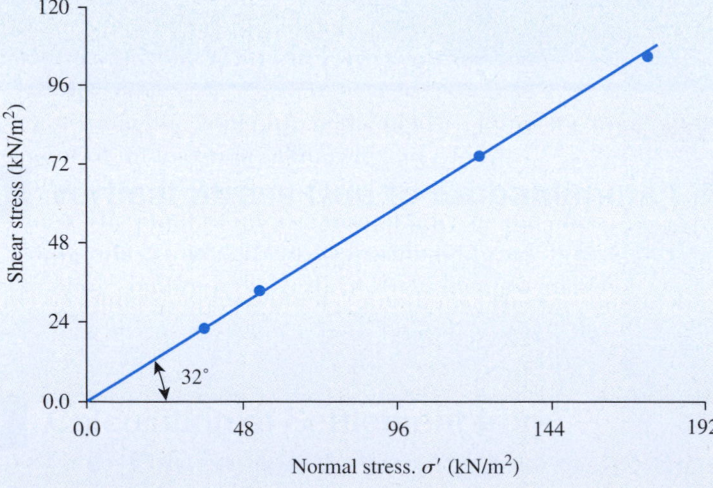

Figure 12.18

Example 12.2

Following are the results of four drained direct shear tests on an *overconsolidated clay*:

Diameter of specimen = 50 mm
Height of specimen = 25 mm

Test no.	Normal force, N (N)	Shear force at failure, S_{peak} (N)	Residual shear force, $S_{residual}$ (N)
1	150	157.5	44.2
2	250	199.9	56.6
3	350	257.6	102.9
4	550	363.4	144.5

Determine the relationships for *peak shear strength* (τ_f) and *residual shear strength* (τ_r).

Solution

Area of the specimen $(A) = (\pi/4)\left(\frac{50}{1000}\right)^2 = 0.0019634 \text{ m}^2$. Now the following table can be prepared.

Test no.	Normal force, N (N)	Normal stress, σ' (kN/m²)	Peak shear force, S_{peak} (N)	$\tau_f = \frac{S_{peak}}{A}$ (kN/m²)	Residual shear force, $S_{residual}$ (N)	$\tau_r = \frac{S_{residual}}{A}$ (kN/m²)
1	150	76.4	157.5	80.2	44.2	22.5
2	250	127.3	199.9	101.8	56.6	28.8
3	350	178.3	257.6	131.2	102.9	52.4
4	550	280.1	363.4	185.1	144.5	73.6

The variations of τ_f and τ_r with σ' are plotted in Figure 12.19. From the plots, we find that

Peak strength: $\tau_f(\text{kN/m}^2) = \mathbf{40 + \sigma' \tan 27}$
Residual strength: $\tau_r(\text{kN/m}^2) = \mathbf{\sigma' \tan 14.6}$

(*Note:* For all *overconsolidated clays*, the residual shear strength can be expressed as

$$\tau_r = \sigma' \tan \phi'_r$$

where ϕ'_r = effective residual friction angle.)

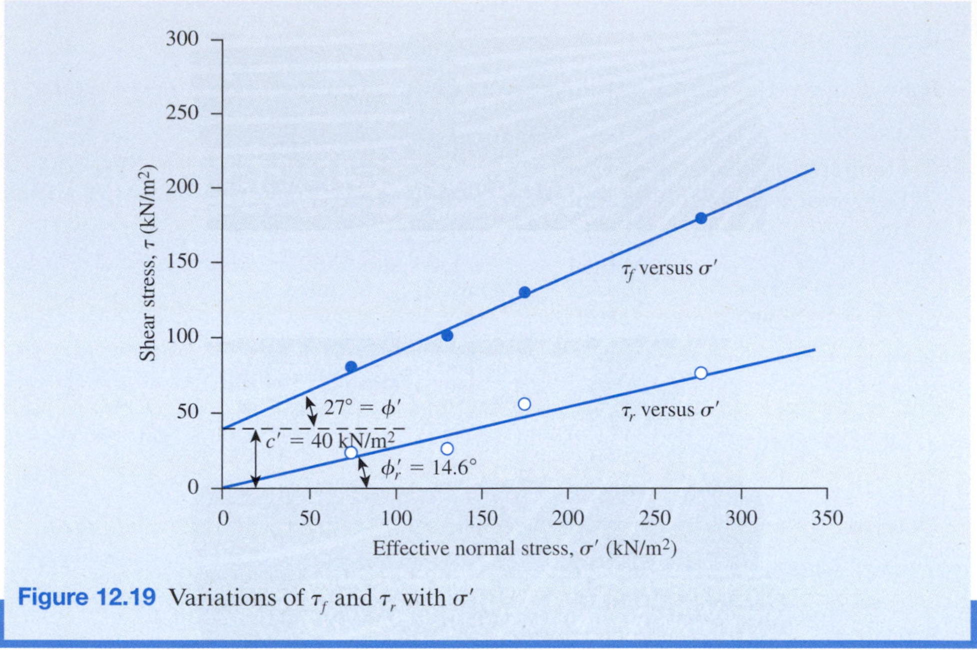

Figure 12.19 Variations of τ_f and τ_r with σ'

12.8 Triaxial Shear Test-General

The triaxial shear test is one of the most reliable methods available for determining shear strength parameters. It is used widely for research and conventional testing. A diagram of the triaxial test layout is shown in Figure 12.20. Figure 12.21 shows a triaxial test in progress in the laboratory.

In this test, a soil specimen about 36 mm in diameter and 76 mm long generally is used. The specimen is encased by a thin rubber membrane and placed inside a plastic cylindrical chamber that usually is filled with water or glycerine. The specimen is subjected to a confining pressure by compression of the fluid in the chamber. (*Note:* Air is sometimes used as a compression medium.) To cause shear failure in the specimen, one must apply axial stress (sometimes called *deviator stress*) through a vertical loading ram. This stress can be applied in one of two ways:

1. Application of dead weights or hydraulic pressure in equal increments until the specimen fails. (Axial deformation of the specimen resulting from the load applied through the ram is measured by a dial gauge.)
2. Application of axial deformation at a constant rate by means of a geared or hydraulic loading press. This is a strain-controlled test.

The axial load applied by the loading ram corresponding to a given axial deformation is measured by a proving ring or load cell attached to the ram.

Connections to measure drainage into or out of the specimen, or to measure pressure in the pore water (as per the test conditions), also are provided. The following three standard types of triaxial tests generally are conducted:

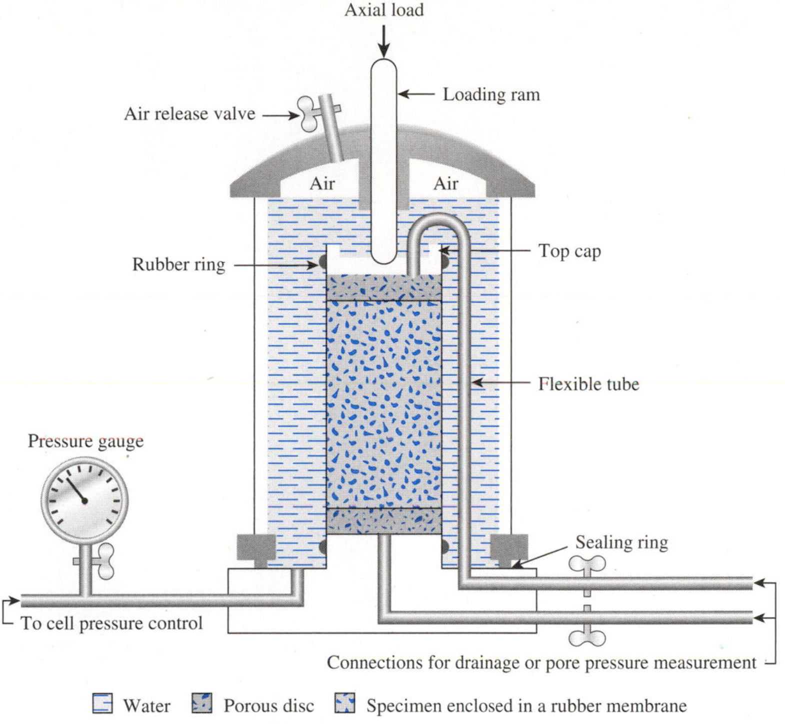

Figure 12.20 Diagram of triaxial test equipment (*After Bishop and Bjerrum, 1960, with permission from ASCE*)

1. Consolidated-drained test or drained test (CD test)
2. Consolidated-undrained test (CU test)
3. Unconsolidated-undrained test or undrained test (UU test)

The general procedures and implications for each of the tests in *saturated soils* are described in the following sections.

12.9 Consolidated-Drained Triaxial Test

In the CD test, the saturated specimen first is subjected to an all around confining pressure, σ_3, by compression of the chamber fluid (Figure 12.22a). As confining pressure is applied, the pore water pressure of the specimen increases by u_c (if drainage is prevented). This increase in the pore water pressure can be expressed as a nondimensional parameter in the form

$$B = \frac{u_c}{\sigma_3} \tag{12.20}$$

where B = Skempton's pore pressure parameter (Skempton, 1954).

Figure 12.21 A triaxial test in progress in the laboratory (*Courtesy of S. Vanapalli, University of Ottawa, Canada*)

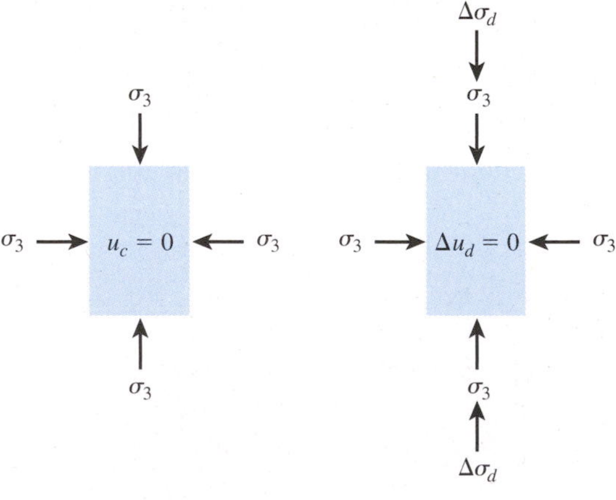

Figure 12.22 Consolidated-drained triaxial test: (a) specimen under chamber-confining pressure; (b) deviator stress application

For saturated soft soils, B is approximately equal to 1; however, for saturated stiff soils, the magnitude of B can be less than 1. Black and Lee (1973) gave the theoretical values of B for various soils at complete saturation. These values are listed in Table 12.2.

Now, if the connection to drainage is opened, dissipation of the excess pore water pressure, and thus consolidation, will occur. With time, u_c will become equal to 0. In saturated soil, the change in the volume of the specimen (ΔV_c) that takes place during consolidation can be obtained from the volume of pore water drained (Figure 12.23a). Next, the deviator stress, $\Delta\sigma_d$, on the specimen is increased very slowly (Figure 12.22b). The drainage connection is kept open, and the slow rate of deviator stress application allows complete dissipation of any pore water pressure that developed as a result ($\Delta u_d = 0$).

A typical plot of the variation of deviator stress against strain in loose sand and normally consolidated clay is shown in Figure 12.23b. Figure 12.23c shows a similar plot for dense sand and overconsolidated clay. The volume change, ΔV_d, of specimens that occurs because of the application of deviator stress in various soils is also shown in Figures 12.23d and 12.23e.

Because the pore water pressure developed during the test is completely dissipated, we have

$$\text{Total and effective confining stress} = \sigma_3 = \sigma_3'$$

and

$$\text{Total and effective axial stress at failure} = \sigma_3 + (\Delta\sigma_d)_f = \sigma_1 = \sigma_1'$$

In a triaxial test, σ_1' is the major principal effective stress at failure and σ_3' is the minor principal effective stress at failure.

Several tests on similar specimens can be conducted by varying the confining pressure. With the major and minor principal stresses at failure for each test the Mohr's circles can be drawn and the failure envelopes can be obtained. Figure 12.24 shows the type of effective stress failure envelope obtained for tests on sand and normally consolidated clay. The coordinates of the point of tangency of the failure envelope with a Mohr's circle (that is, point A) give the stresses (normal and shear) on the failure plane of that test specimen.

For normally consolidated clay, referring to Figure 12.24

$$\sin\phi' = \frac{AO'}{OO'}$$

Table 12.2 Theoretical Values of B at Complete Saturation

Type of soil	Theoretical value
Normally consolidated soft clay	0.9998
Lightly overconsolidated soft clays and silts	0.9988
Overconsolidated stiff clays and sands	0.9877
Very dense sands and very stiff clays at high confining pressures	0.9130

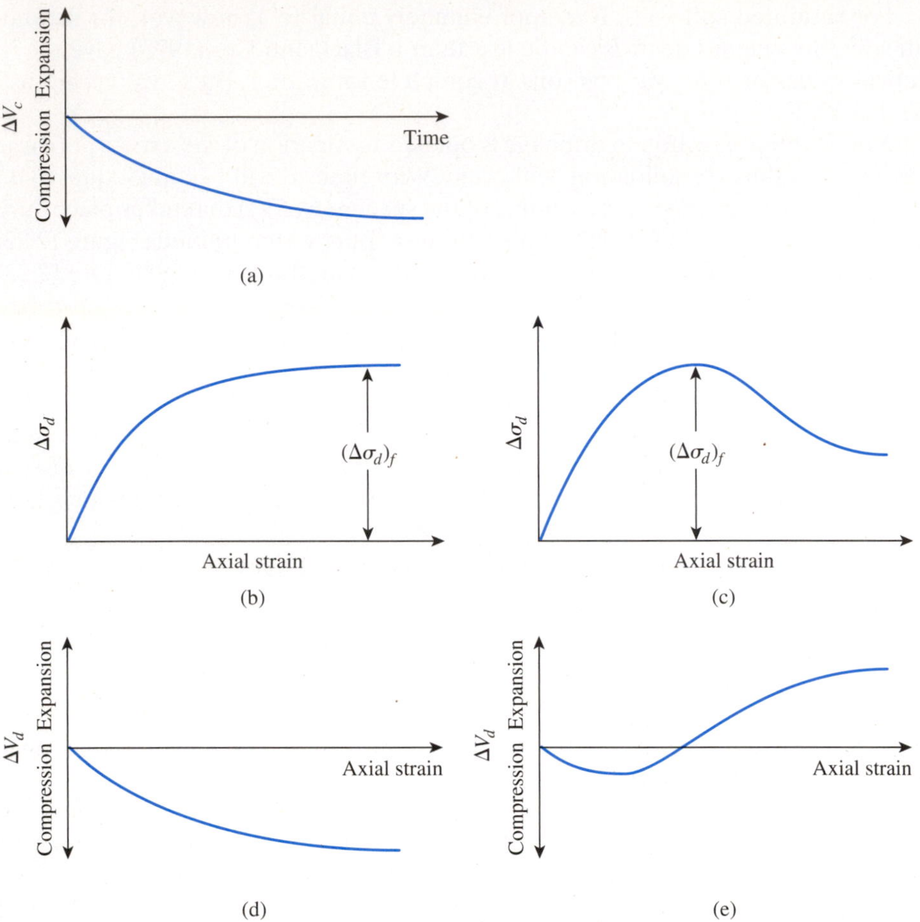

Figure 12.23 Consolidated-drained triaxial test: (a) volume change of specimen caused by chamber-confining pressure; (b) plot of deviator stress against axial strain in the vertical direction for loose sand and normally consolidated clay; (c) plot of deviator stress against axial strain in the vertical direction for dense sand and overconsolidated clay; (d) volume change in loose sand and normally consolidated clay during deviator stress application; (e) volume change in dense sand and overconsolidated clay during deviator stress application

or

$$\sin \phi' = \frac{\left(\dfrac{\sigma_1' - \sigma_3'}{2}\right)}{\left(\dfrac{\sigma_1' + \sigma_3'}{2}\right)}$$

$$\phi' = \sin^{-1}\left(\frac{\sigma_1' - \sigma_3'}{\sigma_1' + \sigma_3'}\right) \tag{12.21}$$

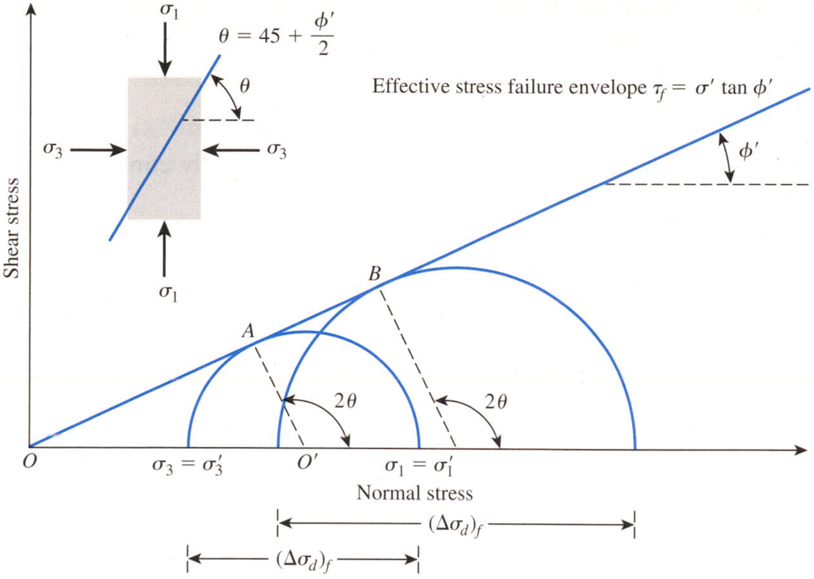

Figure 12.24 Effective stress failure envelope from drained tests on sand and normally consolidated clay

Also, the failure plane will be inclined at an angle of $\theta = 45 + \phi'/2$ to the major principal plane, as shown in Figure 12.24.

Overconsolidation results when a clay initially is consolidated under an all-around chamber pressure of σ_c ($= \sigma_c'$) and is allowed to swell by reducing the chamber pressure to σ_3 ($= \sigma_3'$). The failure envelope obtained from drained triaxial tests of such overconsolidated clay specimens shows two distinct branches (*ab* and *bc* in Figure 12.25). The

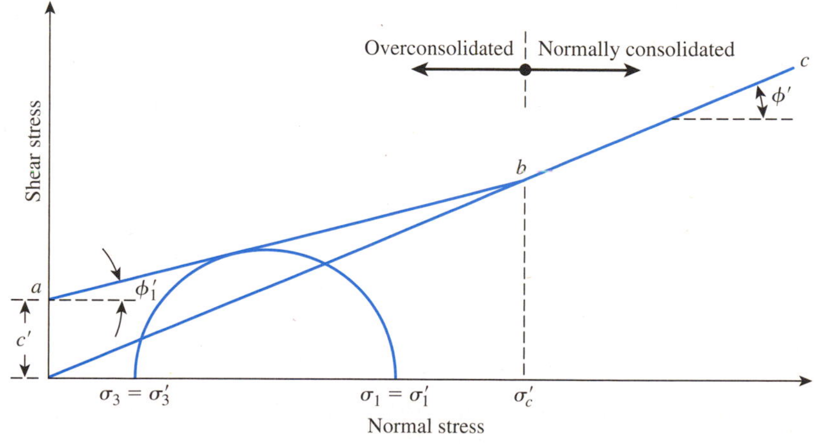

Figure 12.25 Effective stress failure envelope for overconsolidated clay

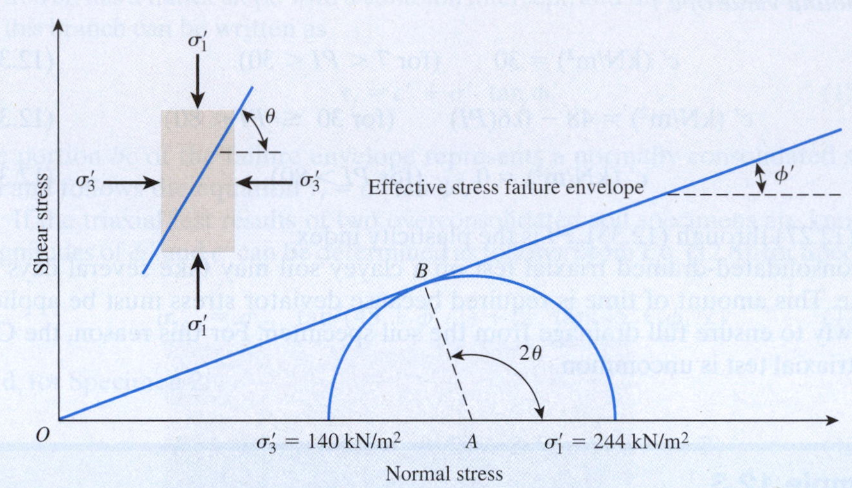

Figure 12.26

Part b

From Eq. (12.4),

$$\theta = 45 + \frac{\phi'}{2} = 45° + \frac{17.46°}{2} = \mathbf{53.73°}$$

Example 12.4

Refer to Example 12.3.

a. Find the normal stress σ' and the shear stress τ_f on the failure plane.
b. Determine the effective normal stress on the plane of maximum shear stress.

Solution

Part a

From Eqs. (10.8) and (10.9),

$$\sigma'(\text{on the failure plane}) = \frac{\sigma_1' + \sigma_3'}{2} + \frac{\sigma_1' - \sigma_3'}{2} \cos 2\theta$$

and

$$\tau_f = \frac{\sigma_1' - \sigma_3'}{2} \sin 2\theta$$

Substituting the values of $\sigma_1' = 244 \text{ kN/m}^2$, $\sigma_3' = 140 \text{ kN/m}^2$, and $\theta = 53.36°$ into the preceding equations, we get

$$\sigma' = \frac{244 + 140}{2} + \frac{244 - 140}{2} \cos(2 \times 53.73) = \mathbf{176.36 \ kN/m^2}$$

and

$$\tau_f = \frac{244 - 140}{2} \sin(2 \times 53.73) = \mathbf{49.59 \ kN/m^2}$$

Part b

From Eq. (10.9), it can be seen that the maximum shear stress will occur on the plane with $\theta = 45°$. From Eq. (10.8),

$$\sigma' = \frac{\sigma_1' + \sigma_3'}{2} + \frac{\sigma_1' - \sigma_3'}{2} \cos 2\theta$$

Substituting $\theta = 45°$ into the preceding equation gives

$$\sigma' = \frac{244 + 140}{2} + \frac{244 - 140}{2} \cos 90 = \mathbf{192 \ kN/m^2}$$

Example 12.5

The equation of the effective stress failure envelope for normally consolidated clayey soil is $\tau_f = \sigma' \tan 28°$. A drained triaxial test was conducted with the same soil at a chamber-confining pressure of 100 kN/m^2. Calculate the deviator stress at failure.

Solution

For normally consolidated clay, $c' = 0$. Thus, from Eq. (12.8),

$$\sigma_1' = \sigma_3' \tan^2\left(45 + \frac{\phi'}{2}\right)$$

$$\phi' = 28°$$

$$\sigma_1' = 100 \tan^2\left(45 + \frac{28}{2}\right) \approx 277 \text{ kN/m}^2$$

So,

$$(\Delta\sigma_d)_f = \sigma_1' - \sigma_3' = 277 - 100 = \mathbf{177 \ kN/m^2}$$

Example 12.6

The results of two drained triaxial tests on a saturated clay follow:

Specimen I:

$$\sigma_3 = 70 \text{ kN/m}^2$$
$$(\Delta\sigma_d)_f = 130 \text{ kN/m}^2$$

Specimen II:

$$\sigma_3 = 160 \text{ kN/m}^2$$
$$(\Delta\sigma_d)_f = 223.5 \text{ kN/m}^2$$

Determine the shear strength parameters.

Solution

Refer to Figure 12.27. For Specimen I, the principal stresses at failure are

$$\sigma_3' = \sigma_3 = 70 \text{ kN/m}^2$$

and

$$\sigma_1' = \sigma_1 = \sigma_3 + (\Delta\sigma_d)_f = 70 + 130 = 200 \text{ kN/m}^2$$

Similarly, the principal stresses at failure for Specimen II are

$$\sigma_3' = \sigma_3 = 160 \text{ kN/m}^2$$

and

$$\sigma_1' = \sigma_1 = \sigma_3 + (\Delta\sigma_d)_f = 160 + 223.5 = 383.5 \text{ kN/m}^2$$

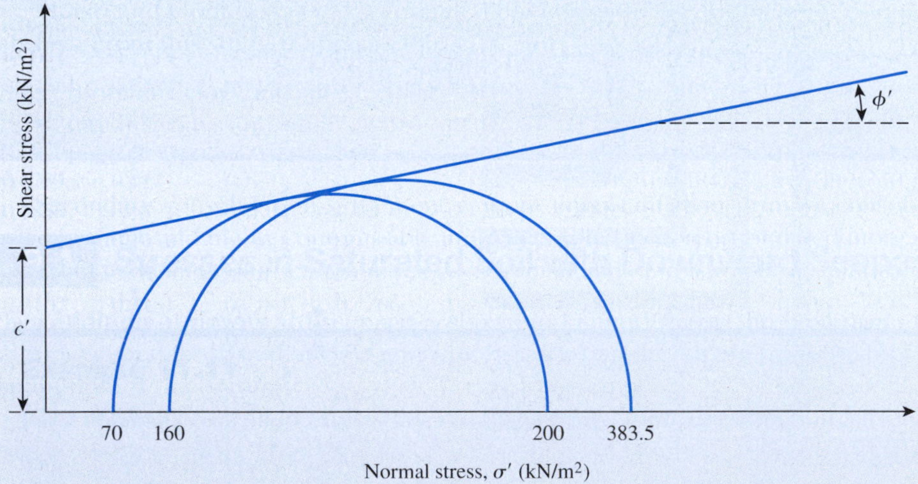

Figure 12.27

Now, from Eq. (12.25),

$$\phi'_1 = 2\left\{\tan^{-1}\left[\frac{\sigma'_{1(\mathrm{I})} - \sigma'_{1(\mathrm{II})}}{\sigma'_{3(\mathrm{I})} - \sigma'_{3(\mathrm{II})}}\right]^{0.5} - 45°\right\}$$

$$= 2\left\{\tan^{-1}\left[\frac{200 - 383.5}{70 - 160}\right]^{0.5} - 45°\right\} = \mathbf{20°}$$

Again, from Eq. (12.26),

$$c' = \frac{\sigma'_{1(\mathrm{I})} - \sigma'_{3(\mathrm{I})}\tan^2\left(45 + \frac{\phi'_1}{2}\right)}{2\tan\left(45 + \frac{\phi'_1}{2}\right)} = \frac{200 - 70\tan^2\left(45 + \frac{20}{2}\right)}{2\tan\left(45 + \frac{20}{2}\right)} = \mathbf{20\ kN/m^2}$$

Example 12.7

An undisturbed, normally consolidated clay soil specimen will be subjected to a consolidated-drained triaxial test with $\sigma'_3 = 65$ kN/m^2. The clay has a liquid limit (LL) of 52 and a plastic limit of 24. Using Eq. (12.17), estimate the approximate magnitude of σ'_1.

Solution

From Eq. (12.17),

$$\phi' = 43 - 10\log(PI) = 43 - 10\log(54 - 24) = 28.53°$$

For normally consolidated clay, $c' = 0$. So [Eq. (12.8)],

$$\sigma'_1 = \sigma'_3\tan^2\left(45 + \frac{\phi'}{2}\right) = 65\tan^2\left(45 + \frac{28.53}{2}\right) = 183.86\ \mathrm{kN/m^2} \approx \mathbf{184\ kN/m^2}$$

12.10 Consolidated-Undrained Triaxial Test

The consolidated-undrained test is the most common type of triaxial test. In this test, the saturated soil specimen is first consolidated by an all-around chamber fluid pressure, σ_3, that results in drainage (Figures 12.28a and 12.28b). After the pore water pressure generated by the application of confining pressure is dissipated, the deviator stress, $\Delta\sigma_d$, on the specimen is increased to cause shear failure (Figure 12.28c). During this phase of the test, the drainage line from the specimen is kept closed. Because drainage is not permitted, the pore water pressure, Δu_d, will increase. During

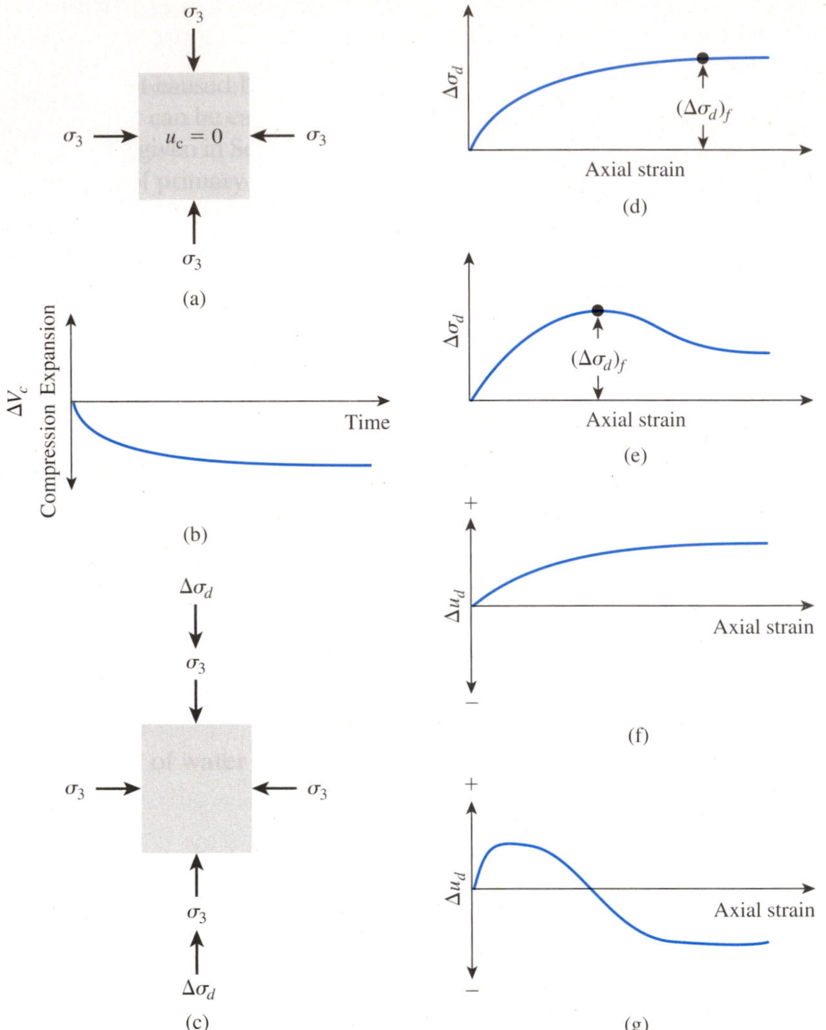

Figure 12.28 Consolidated-undrained test: (a) specimen under chamber-confining pressure; (b) volume change in specimen caused by confining pressure; (c) deviator stress application; (d) deviator stress against axial strain for loose sand and normally consolidated clay; (e) deviator stress against axial strain for dense sand and overconsolidated clay; (f) variation of pore water pressure with axial strain for loose sand and normally consolidated clay; (g) variation of pore water pressure with axial strain for dense sand and overconsolidated clay

the test, simultaneous measurements of $\Delta\sigma_d$ and Δu_d are made. The increase in the pore water pressure, Δu_d, can be expressed in a nondimensional form as

$$\overline{A} = \frac{\Delta u_d}{\Delta\sigma_d} \tag{12.34}$$

where $\overline{A}$ = Skempton's pore pressure parameter (Skempton, 1954).

The general patterns of variation of $\Delta\sigma_d$ and Δu_d with axial strain for sand and clay soils are shown in Figures 12.28d through 12.28g. In loose sand and normally consolidated clay, the pore water pressure increases with strain. In dense sand and overconsolidated clay, the pore water pressure increases with strain to a certain limit, beyond which it decreases and becomes negative (with respect to the atmospheric pressure). This decrease is because of a tendency of the soil to dilate. Figure 12.29 shows a triaxial soil specimen at failure during a consolidated-undrained test.

Unlike the consolidated-drained test, the total and effective principal stresses are not the same in the consolidated-undrained test. Because the pore water pressure at failure is measured in this test, the principal stresses may be analyzed as follows:

- Major principal stress at failure (total): $\sigma_3 + (\Delta\sigma_d)_f = \sigma_1$
- Major principal stress at failure (effective): $\sigma_1 - (\Delta u_d)_f = \sigma_1'$
- Minor principal stress at failure (total): σ_3
- Minor principal stress at failure (effective): $\sigma_3 - (\Delta u_d)_f = \sigma_3'$

In these equations, $(\Delta u_d)_f$ = pore water pressure at failure. The preceding derivations show that

$$\sigma_1 - \sigma_3 = \sigma_1' - \sigma_3'$$

Tests on several similar specimens with varying confining pressures may be conducted to determine the shear strength parameters. Figure 12.30 shows the total and

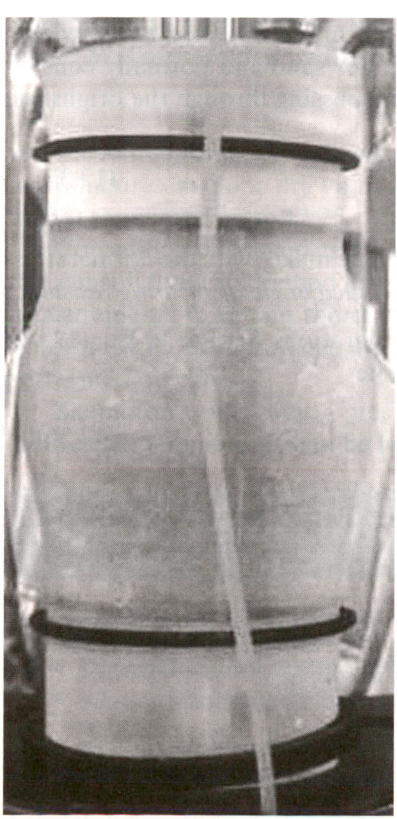

Figure 12.29 Triaxial soil specimen at failure during a consolidated-undrained test *(Courtesy of S. Varapalli, University of Ottawa, Canada)*

The reason for obtaining the same added axial stress $(\Delta\sigma_d)_f$ regardless of the confining pressure can be explained as follows. If a clay specimen (No. I) is consolidated at a chamber pressure σ_3 and then sheared to failure without drainage, the total stress conditions at failure can be represented by the Mohr's circle P in Figure 12.35. The pore pressure developed in the specimen at failure is equal to $(\Delta u_d)_f$. Thus, the major and minor principal effective stresses at failure are, respectively,

$$\sigma_1' = [\sigma_3 + (\Delta\sigma_d)_f] - (\Delta u_d)_f = \sigma_1 - (\Delta u_d)_f$$

and

$$\sigma_3' = \sigma_3 - (\Delta u_d)_f$$

Q is the effective stress Mohr's circle drawn with the preceding principal stresses. Note that the diameters of circles P and Q are the same.

Now let us consider another similar clay specimen (No. II) that has been consolidated under a chamber pressure σ_3 with initial pore pressure equal to zero. If the chamber pressure is increased by $\Delta\sigma_3$ without drainage, the pore water pressure will increase by an amount Δu_c. For saturated soils under isotropic stresses, the pore water pressure increase is equal to the total stress increase, so $\Delta u_c = \Delta\sigma_3$ $(B = 1)$. At this time, the effective confining pressure is equal to $\sigma_3 + \Delta\sigma_3 - \Delta u_c = \sigma_3 + \Delta\sigma_3 - \Delta\sigma_3 = \sigma_3$. This is the same as the effective confining pressure of Specimen I before the application of deviator stress. Hence, if Specimen II is sheared to failure by increasing the axial stress, it should fail at the same deviator stress $(\Delta\sigma_d)_f$ that was obtained for Specimen I. The total stress Mohr's circle at failure will be R (see Figure 12.35). The added pore pressure increase caused by the application of $(\Delta\sigma_d)_f$ will be $(\Delta u_d)_f$.

At failure, the minor principal effective stress is

$$[(\sigma_3 + \Delta\sigma_3)] - [\Delta u_c + (\Delta u_d)_f] = \sigma_3 - (\Delta u_d)_f = \sigma_3'$$

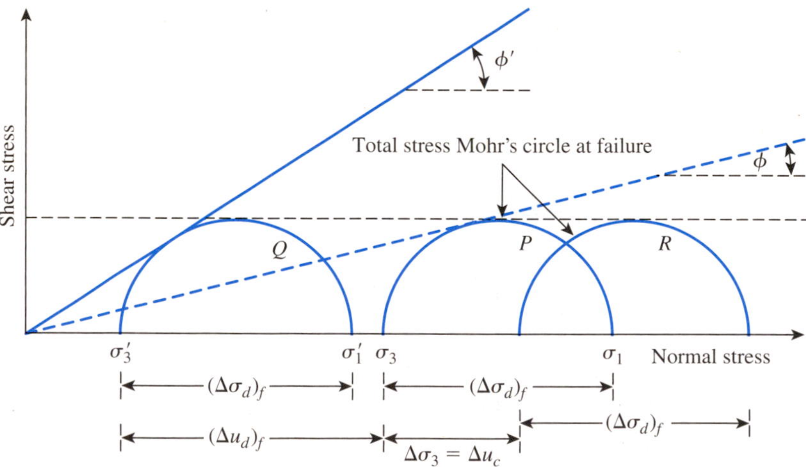

Figure 12.35 The $\phi = 0$ concept

Table 12.5 Correlations for c_{ur} (kN/m^2)

Investigator	Relationship
Leroueil et al. (1983)	$c_{ur} = \dfrac{1}{[(LI) - 0.21]^2}$
Hirata et al. (1990)	$c_{ur} = \exp[-3.36(LI) + 0.376]$
Terzaghi et al. (1996)	$c_{ur} = 2(LI)^{-2.8}$
Yang et al. (2006)	$c_{ur} = 159.6 \exp[-3.97(LI)]$

Note: LI = liquidity index

and the major principal effective stress is

$$[\sigma_3 + \Delta\sigma_3 + (\Delta\sigma_d)_f] - [\Delta u_c + (\Delta u_d)_f] = [\sigma_3 + (\Delta\sigma_d)_f] - (\Delta u_d)_f$$
$$= \sigma_1 - (\Delta u_d)_f = \sigma_1'$$

Thus, the effective stress Mohr's circle will still be Q because strength is a function of effective stress. Note that the diameters of circles P, Q, and R are all the same.

Any value of $\Delta\sigma_3$ could have been chosen for testing Specimen II. In any case, the deviator stress $(\Delta\sigma_d)_f$ to cause failure would have been the same as long as the soil was fully saturated and fully undrained during both stages of the test.

Several correlations have been suggested for the *undrained shear strength of remolded clay* (c_{ur}) in the past, and some are given in Table 12.5. It is important to point out that these relationships should be used as an approximation only. O'Kelly (2013) has also shown that, at a moisture content w, c_{ur} can be estimated as

$$\log c_{ur} = (1 - W_{LN})\left[\log\left(\frac{c_{ur(A)}}{c_{ur(B)}}\right)\right] + \log c_{ur(B)} \tag{12.43}$$

where $c_{ur(A)}$ = undrained shear strength at moisture content w_A
$c_{ur(B)}$ = undrained shear strength at moisture content w_B

$$W_{LN} = \frac{\log w - \log w_A}{\log w_B - \log w_A} \tag{12.44}$$

Example 12.10

Consider a saturated remolded clay soil. Given:

Liquid limit = 48
Plastic limit = 23
Moisture content of soil = 43%

Estimate the undrained shear strength c_{ur} using the equations of Leroueil et al. (1983) and Terzaghi et al. (1996) given in Table 12.5.

Solution

The liquidity index is

$$LI = \frac{w - PL}{LL - PL} = \frac{43 - 23}{48 - 23} = 0.8$$

From Leroueil et al. (1983),

$$c_{ur} = \frac{1}{[(LI) - 0.21]^2} = \frac{1}{(0.8 - 0.21)^2} = \mathbf{2.87\ kN/m^2}$$

From Terzaghi et al. (1996),

$$c_{ur} = 2(LI)^{-2.8} = (1)(0.8)^{-2.8} = \mathbf{3.74\ kN/m^2}$$

Example 12.11

Consider a remolded saturated clay. Given:

Moisture content, w (%)	Undrained shear strength, c_{ur} (kN/m²)
68	4.68
54	10.68

Estimate the undrained shear strength c_{ur} when the moisture content is 40%. Use Eq. (12.43).

Solution

Given:
At $w_A = 68\%$, the value of $c_{ur(A)} = 4.86\ \text{kN/m}^2$
At $w_B = 54\%$, the value of $c_{ur(B)} = 10.68\ \text{kN/m}^2$
$w = 40\%$

From Eq. (12.44),

$$W_{LN} = \frac{\log w - \log w_A}{\log w_B - \log w_A} = \frac{\log(40) - \log(68)}{\log(54) - \log(68)} = \frac{-0.231}{-0.101} = 2.287$$

From Eq. (12.43),

$$\log c_{ur} = (1 - 2.287)\left[\log\left(\frac{4.86}{10.68}\right)\right] + \log(10.68) = \mathbf{29.44\ kN/m^2}$$

12.12 Unconfined Compression Test on Saturated Clay

The unconfined compression test is a special type of unconsolidated-undrained test that is commonly used for clay specimens. In this test, the confining pressure σ_3 is 0. An axial load is rapidly applied to the specimen to cause failure. At failure, the total minor principal stress is zero and the total major principal stress is σ_1 (Figure 12.36). Because the undrained shear strength is independent of the confining pressure as long as the soil is fully saturated and fully undrained, we have

$$\tau_f = \frac{\sigma_1}{2} = \frac{q_u}{2} = c_u \qquad (12.45)$$

where q_u is the *unconfined compression strength*. Table 12.6 gives the approximate consistencies of clays on the basis of their unconfined compression strength. A photograph of unconfined compression test equipment is shown in Figure 12.37. Figures 12.38 and 12.39 show the failure in two specimens—one by shear and one by bulging—at the end of unconfined compression tests.

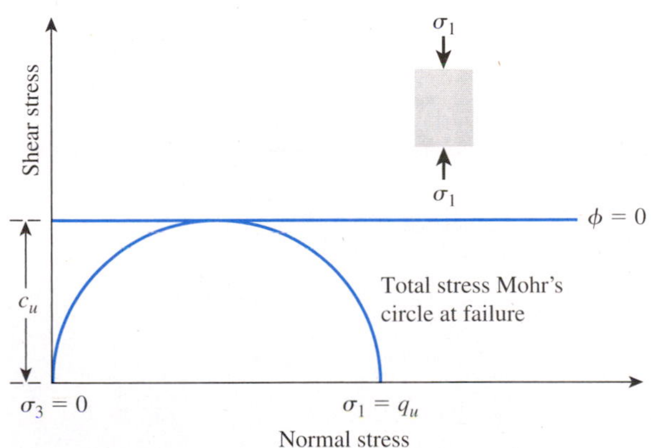

Figure 12.36 Unconfined compression test

Table 12.6 General Relationship of Consistency and Unconfined Compression Strength of Clays

Consistency	q_u kN/m²
Very soft	0–25
Soft	25–50
Medium	50–100
Stiff	100–200
Very stiff	200–400
Hard	>400

Figure 12.37 Unconfined compression test equipment (*Courtesy of ELE International*)

Figure 12.38 Failure by shear of an unconfined compression test specimen (*Courtesy of Braja M. Das, Henderson, Nevada*)

Figure 12.39 Failure by bulging of an unconfined compression test specimen (*Courtesy of Braja M. Das, Henderson, Nevada*)

Theoretically, for similar saturated clay specimens, the unconfined compression tests and the unconsolidated-undrained triaxial tests should yield the same values of c_u. In practice, however, unconfined compression tests on saturated clays yield slightly lower values of c_u than those obtained from unconsolidated-undrained tests.

12.13 Empirical Relationships between Undrained Cohesion (c_u) and Effective Overburden Pressure (σ'_o)

Several empirical relationships have been proposed between c_u and the effective overburden pressure σ'_o. The most commonly cited relationship is that given by Skempton (1957), which can be expressed as

$$\frac{c_{u(\text{VST})}}{\sigma'_o} = 0.11 + 0.0037(PI) \quad \text{(for normally consolidated clay)} \tag{12.46}$$

where $c_{u(\text{VST})}$ = undrained shear strength from vane shear test (see Section 12.16)
PI = plasticity index (%)

Chandler (1988) suggested that the preceding relationship will hold good for overconsolidated soil with an accuracy of ± 25%. This does not include sensitive and fissured clays. Ladd et al. (1977) proposed that

$$\frac{\left(\dfrac{c_u}{\sigma'_o}\right)_{\text{overconsolidated}}}{\left(\dfrac{c_u}{\sigma'_o}\right)_{\text{normally consolidated}}} = (OCR)^{0.8} \tag{12.47}$$

where OCR = overconsolidation ratio.

Jamiolkowski et al. (1985) suggested that

$$\frac{c_u}{\sigma'_o} = (0.23 \pm 0.04)(OCR)^{0.8} \tag{12.48}$$

Example 12.12

An overconsolidated clay deposit located below the groundwater table has the following:

Average present effective overburden pressure = 180 kN/m^2
Overconsolidation ratio = 2.6
Plasticity index = 19

Estimate the average undrained shear strength of the clay (that is, c_u). Use Eq. (12.48).

Solution

From Eq. (12.48),

$$\frac{c_u}{\sigma'_o} = (0.23 \pm 0.04)(OCR)^{0.8}$$

$$\frac{c_u}{180} = (0.23 \pm 0.04)(2.6)^{0.8}$$

$c_u = 104.38\ \text{kN/m}^2$ to $73.45\ \text{kN/m}^2$. So average $c_u \approx$ **88.9 kN/m²**

Example 12.13

Repeat Example 12.12 using Eqs. (12.46) and (12.47).

Solution

From Eq. (12.46) for normally consolidated clay,

$$\frac{c_{u(\text{VST})}}{\sigma'_o} = 0.11 + 0.0037PI = 0.11 + 0.0037(19) = 0.1803$$

From Eq. (12.47),

$$\left(\frac{c_u}{\sigma'_o}\right)_{\text{overconsolidated}} = (\text{OCR})^{0.8}\left(\frac{c_u}{\sigma'_o}\right)_{\text{normally consolidated}} = (2.6)^{0.8}(0.1803)$$

$$c_{u(\text{overconsolidated})} = (2.6)^{0.8}(0.1803)(180) = \mathbf{69.7\ kN/m^2}$$

12.14 Sensitivity and Thixotropy of Clay

For many naturally deposited clay soils, the unconfined compression strength is reduced greatly when the soils are tested after remolding without any change in the moisture content, as shown in Figure 12.40. This property of clay soils is called *sensitivity*. The degree of sensitivity may be defined as the ratio of the unconfined compression strength in an undisturbed state to that in a remolded state, or

$$S_t = \frac{c_{u\,(\text{undisturbed})}}{c_{u\,(\text{remolded})}} = \frac{\tau_{f\,(\text{undisturbed})}}{\tau_{f\,(\text{remolded})}} \tag{12.49}$$

The sensitivity ratio of most clays ranges from about 1 to 8; however, highly flocculent marine clay deposits may have sensitivity ratios ranging from about

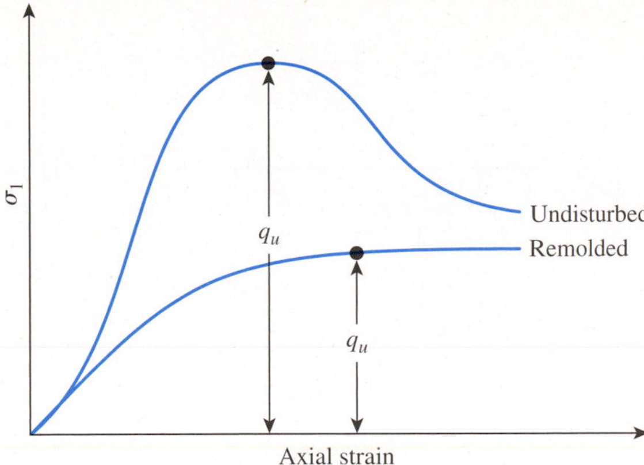

Figure 12.40 Unconfined compression strength for undisturbed and remolded clay

10 to 80. Some clays turn to viscous fluids upon remolding. These clays are found mostly in the previously glaciated areas of North America and Scandinavia. Such clays are referred to as *quick* clays. Rosenqvist (1953) classified clays on the basis of their sensitivity as follows:

Sensitivity	Classification
1	Insensitive
1–2	Slightly sensitive
2–4	Medium sensitive
4–8	Very sensitive
8–16	Slightly quick
16–32	Medium quick
32–64	Very quick
>64	Extra quick

The loss of strength of clay soils from remolding is caused primarily by the destruction of the clay particle structure that was developed during the original process of sedimentation.

If, however, after remolding, a soil specimen is kept in an undisturbed state (that is, without any change in the moisture content), it will continue to gain strength with time. This phenomenon is referred to as *thixotropy*. Thixotropy is a time-dependent, reversible process in which materials under constant composition and volume soften when remolded. This loss of strength is gradually regained with time when the materials are allowed to rest. This phenomenon is illustrated in Figure 12.41a.

Most soils, however, are partially thixotropic—that is, part of the strength loss caused by remolding is never regained with time. The nature of the strength-time variation for partially thixotropic materials is shown in Figure 12.41b. For soils, the difference between the undisturbed strength and the strength after thixotropic

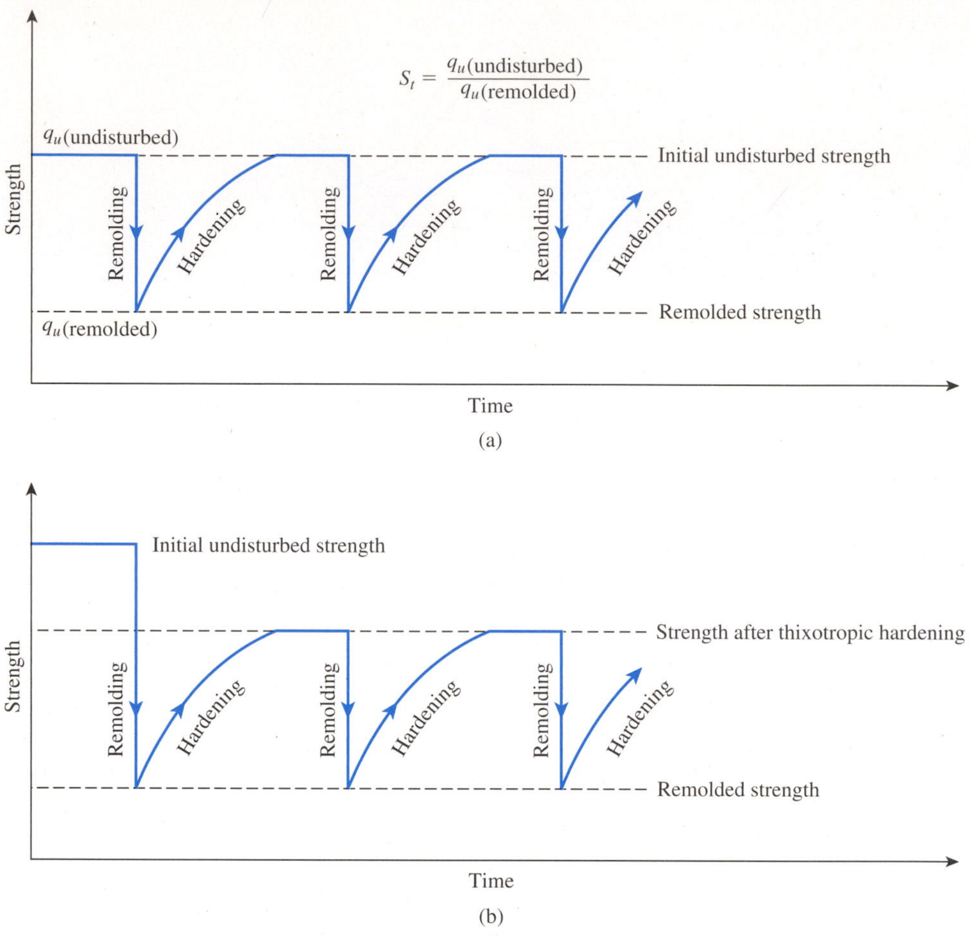

Figure 12.41 Behavior of (a) thixotropic material; (b) partially thixotropic material

hardening can be attributed to the destruction of the clay-particle structure that was developed during the original process of sedimentation.

12.15 Strength Anisotropy in Clay

The unconsolidated-undrained shear strength of some saturated clays can vary, depending on the direction of load application; this variation is referred to as *anisotropy with respect to strength*. Anisotropy is caused primarily by the nature of the deposition of the cohesive soils, and subsequent consolidation makes the clay particles orient perpendicular to the direction of the major principal stress. Parallel orientation of the clay particles can cause the strength of clay to vary with direction. Figure 12.42 shows an element of saturated clay in a deposit with the major principal stress making an angle α with respect to the horizontal. For anisotropic clays, the

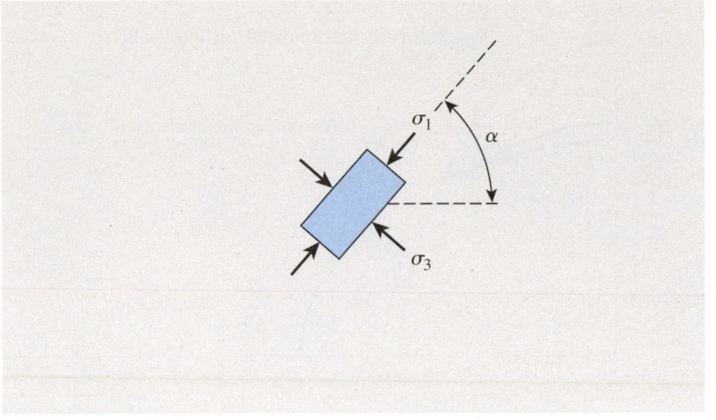

Figure 12.42 Strength anisotropy in clay

magnitude of c_u is a function of α. An example of the variation of c_u with α for undisturbed specimens of Winnipeg upper brown clay (Loh and Holt, 1974) is shown in Figure 12.43.

Based on several laboratory test results, Casagrande and Carrillo (1944) proposed the following relationship for the directional variation of undrained shear strength:

$$c_{u(\alpha)} = c_{u(\alpha = 0°)} + [c_{u(\alpha = 90°)} - c_{u(\alpha = 0°)}] \sin^2 \alpha \tag{12.50}$$

For normally consolidated clays, $c_{u(\alpha = 90°)} > c_{u(\alpha = 0°)}$; for overconsolidated clays, $c_{u(\alpha = 90°)} < c_{u(\alpha = 0°)}$. Figure 12.44 shows the directional variation for $c_{u(\alpha)}$ based on Eq. (12.50). The anisotropy with respect to strength for clays can have an important effect on various stability calculations.

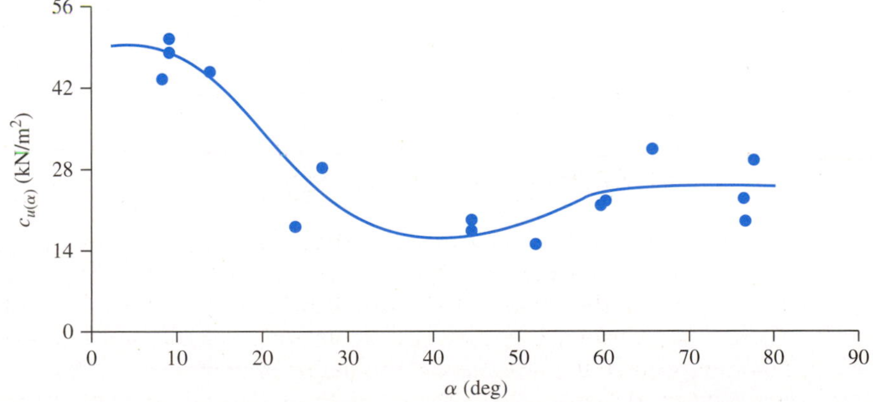

Figure 12.43 Directional variation of c_u for undisturbed Winnipeg upper brown clay *(Based on Loh and Holt, 1974)*

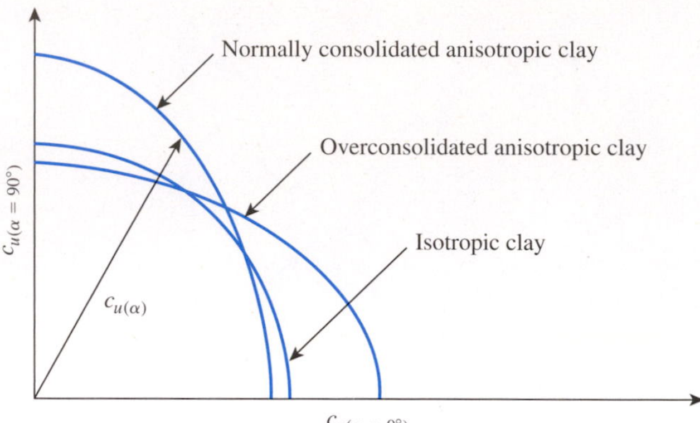

Figure 12.44 Graphical representation of Eq. (12.50)

12.16 Vane Shear Test

Fairly reliable results for the undrained shear strength, c_u ($\phi = 0$ concept), of very soft to medium cohesive soils may be obtained directly from vane shear tests. The shear vane usually consists of four thin, equal-sized steel plates welded to a steel torque rod (Figure 12.45). First, the vane is pushed into the soil. Then torque is applied at the top of the torque rod to rotate the vane at a uniform speed. A cylinder of soil of height h and diameter d will resist the torque until the soil fails. The undrained shear strength of the soil can be calculated as follows.

If T is the maximum torque applied at the head of the torque rod to cause failure, it should be equal to the sum of the resisting moment of the shear force along the side surface of the soil cylinder (M_s) and the resisting moment of the shear force at each end (M_e) (Figure 12.46):

$$T = M_s + \underbrace{M_e + M_e}_{\text{Two ends}} \tag{12.51}$$

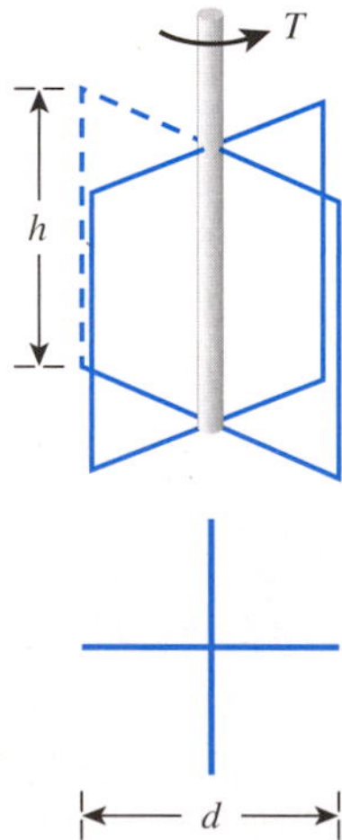

Figure 12.45 Diagram of vane shear test equipment

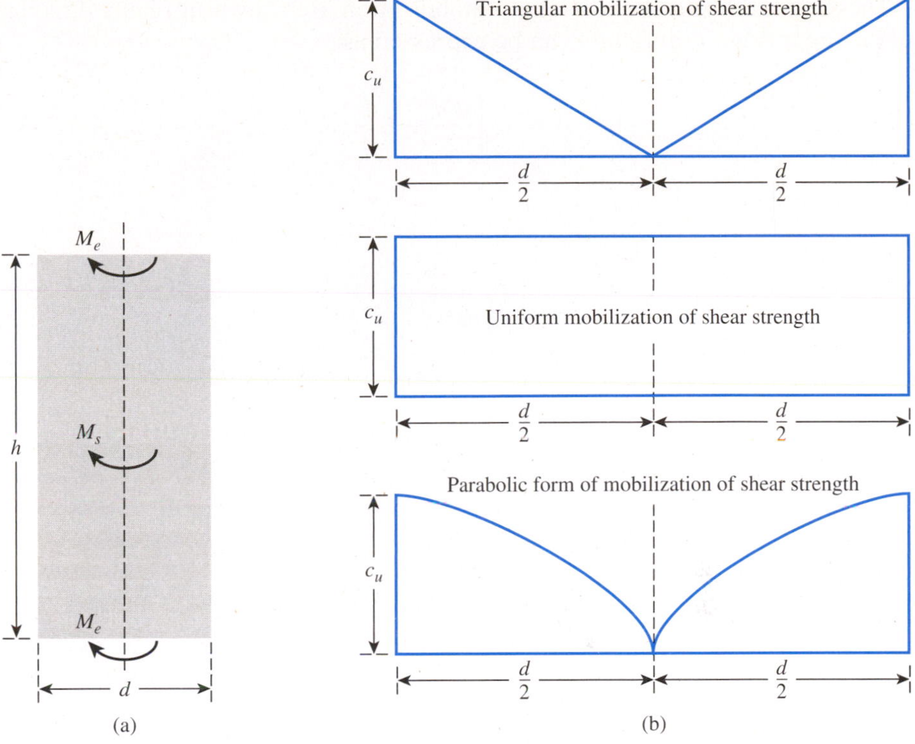

Figure 12.46 Derivation of Eq. (12.54): (a) resisting moment of shear force; (b) variations in shear strength-mobilization

The resisting moment, M_s, can be given as

$$M_s = \underbrace{(\pi dh)c_u}_{\text{Surface area}} \underbrace{(d/2)}_{\text{moment arm}} \tag{12.52}$$

where d = diameter of the shear vane
h = height of the shear vane

For the calculation of M_e, investigators have assumed several types of distribution of shear strength mobilization at the ends of the soil cylinder:

1. *Triangular.* Shear strength mobilization is c_u at the periphery of the soil cylinder and decreases linearly to zero at the center.
2. *Uniform.* Shear strength mobilization is constant (that is, c_u) from the periphery to the center of the soil cylinder.
3. *Parabolic.* Shear strength mobilization is c_u at the periphery of the soil cylinder and decreases parabolically to zero at the center.

These variations in shear strength mobilization are shown in Figure 12.46b. In general, the torque, T, at failure can be expressed as

$$T = \pi c_u \left[\frac{d^2 h}{2} + \beta \frac{d^3}{4} \right] \tag{12.53}$$

or

$$c_u = \frac{T}{\pi \left[\frac{d^2 h}{2} + \beta \frac{d^3}{4} \right]} \tag{12.54}$$

where $\beta = \frac{1}{2}$ for triangular mobilization of undrained shear strength
$\beta = \frac{2}{3}$ for uniform mobilization of undrained shear strength
$\beta = \frac{3}{5}$ for parabolic mobilization of undrained shear strength

Note that Eq. (12.54) usually is referred to as *Calding's equation.*

Vane shear tests can be conducted in the laboratory and in the field during soil exploration. The laboratory shear vane has dimensions of about 13 mm in diameter and 25 mm in height. Figure 12.47 shows a photograph of laboratory vane shear test equipment. Figure 12.48 shows the field vanes recommended by ASTM (2004). Table 12.7 gives the ASTM recommended dimensions of field vanes. The standard rate of torque application is 0.1°/sec. The maximum torque, T, applied to cause failure can be given as

$$T = f(c_u, h, d) \tag{12.55}$$

or

$$c_u = \frac{T}{K} \tag{12.56}$$

Table 12.7 Recommended Dimensions of Field Vanes*[a]

Casing size	Diameter, (mm)	Height, (mm)	Thickness of blade, (mm)	Diameter of rod, (mm)
AX	38.1	76.2	1.6	12.7
BX	50.8	101.6	1.6	12.7
NX	63.5	127.0	3.2	12.7
101.6 mm[b]	92.1	184.1	3.2	12.7

*_After ASTM, 2004. Copyright ASTM INTERNATIONAL. Reprinted with permission._

[a]Selection of vane size is directly related to the consistency of the soil being tested; that is, the softer the soil, the larger the vane diameter should be

[b]Inside diameter

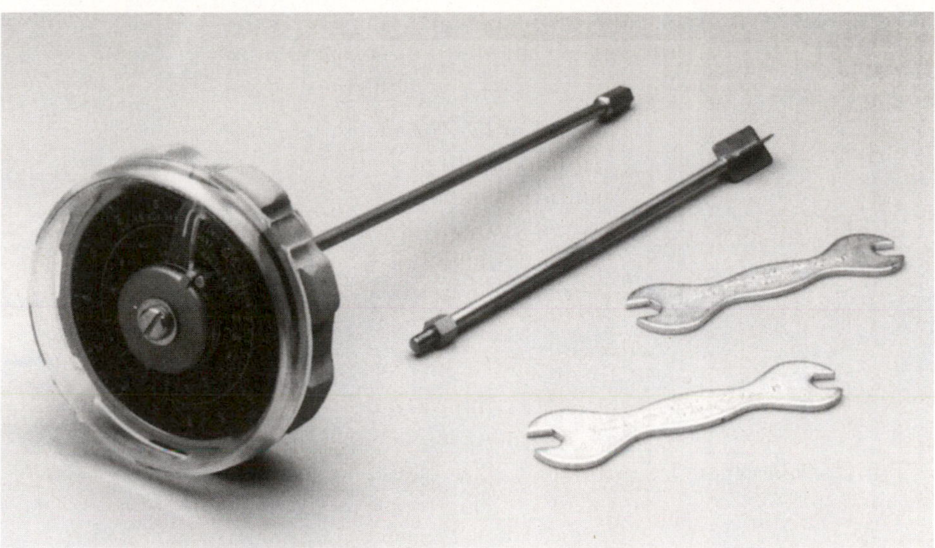

Figure 12.47 Laboratory vane shear test device (*Courtesy of ELE International*)

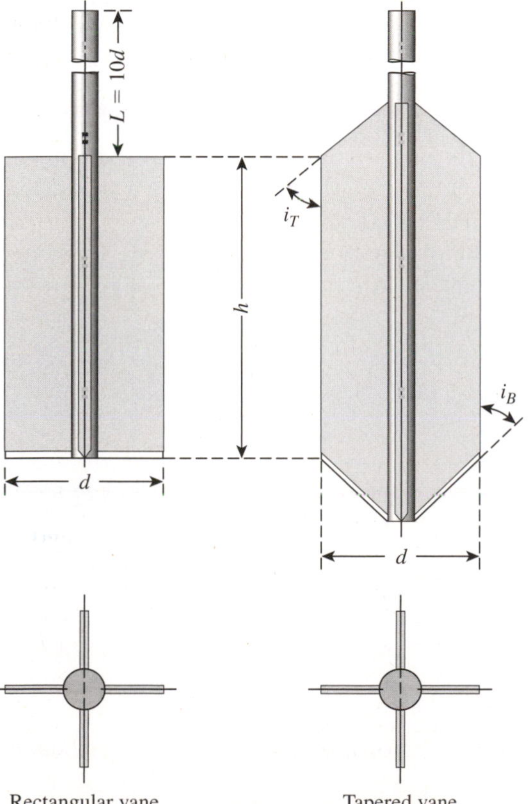

Figure 12.48 Geometry of field vanes (*Note:* i_T and i_B are usually 45°) (*From Annual Book of ASTM Standard (2004), 04.08, p. 346. Copyright ASTM INTERNATIONAL. Reprinted with permission.*)

According to ASTM (2010) for rectangular vanes,

$$K = \frac{\pi d^2}{2}\left(h + \frac{d}{3}\right) \tag{12.57}$$

If $h/d = 2$,

$$K = \frac{7\pi d^3}{6} \tag{12.58}$$

Thus,

$$c_u = \frac{6T}{7\pi d^3} \tag{12.59}$$

For tapered vanes,

$$K = \frac{\pi d^2}{12}\left(\frac{d}{\cos i_T} + \frac{d}{\cos i_B} + 6h\right) \tag{12.60}$$

The angles i_T and i_B are defined in Figure 12.48.

In the field, where considerable variation in the undrained shear strength can be found with depth, vane shear tests are extremely useful. In a short period, one can establish a reasonable pattern of the change of c_u with depth. However, if the clay deposit at a given site is more or less uniform, a few unconsolidated-undrained triaxial tests on undisturbed specimens will allow a reasonable estimation of soil parameters for design work. Vane shear tests also are limited by the strength of soils in which they can be used. The undrained shear strength obtained from a vane shear test also depends on the rate of application of torque T.

Bjerrum (1974) also showed that, as the plasticity of soils increases, c_u obtained from vane shear tests may give results that are unsafe for foundation design. For this reason, he suggested the correction

$$c_{u(\text{design})} = \lambda c_{u(\text{vane shear})} \tag{12.61}$$

where

$$\lambda = \text{correction factor} = 1.7 - 0.54 \log (PI) \tag{12.62}$$

$$PI = \text{plasticity index}$$

Morris and Williams (1994) gave the correlations of λ as

$$\lambda = 1.18e^{-0.08(PI)} + 0.57 \quad (\text{for } PI > 5) \tag{12.63}$$

and

$$\lambda = 7.01e^{-0.08(LL)} + 0.57 \quad (\text{for } LL > 20) \tag{12.64}$$

where LL = liquid limit (%).

Example 12.14

A soil profile is shown in Figure 12.49. The clay is normally consolidated. Its liquid limit is 60 and its plastic limit is 25. Estimate the unconfined compression strength of the clay at a depth of 10 m measured from the ground surface. Use Skempton's relationship from Eq. (12.46) and Eqs. (12.61) and (12.62).

Solution

For the saturated clay layer, the void ratio is

$$e = wG_s = (2.68)(0.3) = 0.8$$

The effective unit weight is

$$\gamma'_{\text{clay}} = \left(\frac{G_s - 1}{1 + e}\right)\gamma_w = \frac{(2.68 - 1)(9.81)}{1 + 0.8} = 9.16 \text{ kN/m}^3$$

The effective stress at a depth of 10 m from the ground surface is

$$\sigma'_o = 3\gamma_{\text{sand}} + 7\gamma'_{\text{clay}} = (3)(15.5) + (7)(9.16)$$
$$= 110.62 \text{ kN/m}^2$$

From Eq. (12.46),

$$\frac{c_{u(\text{VST})}}{\sigma'_o} = 0.11 + 0.0037(PI)$$

$$\frac{c_{u(\text{VST})}}{110.62} = 0.11 + 0.0037(60 - 25)$$

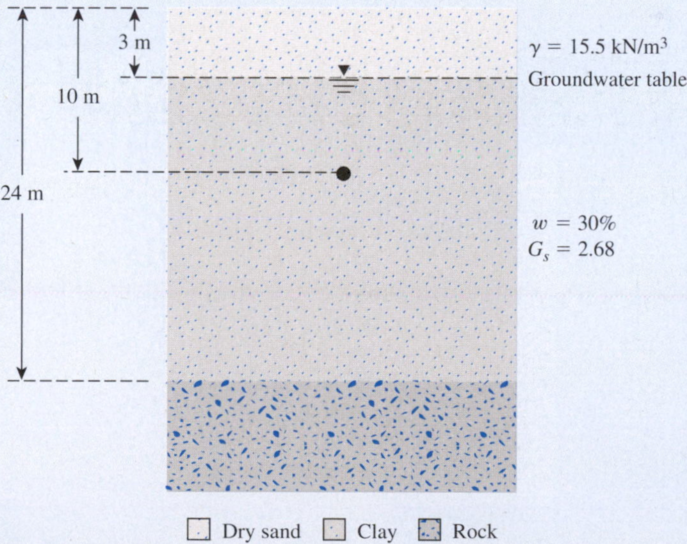

Figure 12.49

and

$$c_{u(\text{VST})} = 26.49 \text{ kN/m}^2$$

From Eqs. (12.61) and (12.62), we get

$$\begin{aligned} c_u &= \lambda c_{u(\text{VST})} \\ &= [1.7 - 0.54 \log (PI)]c_{u(\text{VST})} \\ &= [1.7 - 0.54 \log (60 - 25)]26.49 = 22.95 \text{ kN/m}^2 \end{aligned}$$

So the unconfined compression strength is

$$q_u = 2c_u = (2)(22.95) = \mathbf{45.9\ kN/m^2}$$

Example 12.15

Refer to Figure 12.48. Vane shear tests (tapered vane) were conducted in the clay layer. The vane dimensions were 63.5 mm (d) × 127 m (h), and $i_T = i_B = 45°$. For a test at a certain depth in the clay, the torque required to cause failure was 20 N · m. For the clay, liquid limit was 50 and plastic limit was 18. Estimate the undrained cohesion of the clay for use in the design by using each equation:

a. Bjerrum's λ relationship (Eq. 12.62)
b. Morris and Williams' λ and *PI* relationship (Eq. 12.63)
c. Morris and Williams' λ and *LL* relationship (Eq. 12.64)

Solution

Given $h/d = 127/63.5 = 2$.

Part a

From Eq. (12.60),

$$\begin{aligned} K &= \frac{\pi d^2}{12}\left(\frac{d}{\cos i_T} + \frac{d}{\cos i_B} + 6h\right) \\ &= \frac{\pi(0.0635)^2}{12}\left[\frac{0.0635}{\cos 45} + \frac{0.0635}{\cos 45} + 6(0.127)\right] \\ &= (0.001056)(0.0898 + 0.0898 + 0.762) \\ &= 0.000994 \end{aligned}$$

From Eq. (12.56),

$$\begin{aligned} c_{u(\text{VST})} &= \frac{T}{K} = \frac{20}{0.000994} \\ &= 20{,}121 \text{ N/m}^2 \approx 20.12 \text{ kN/m}^2 \end{aligned}$$

From Eqs. (12.61) and (12.62),

$$c_{u(\text{corrected})} = [1.7 - 0.54 \log (PI\%)]c_{u(\text{VST})}$$
$$= [1.7 - 0.54 \log(50 - 18)](20.12)$$
$$= \mathbf{17.85\ kN/m^2}$$

Part b
From Eqs. (12.63) and (12.61),

$$c_{u(\text{corrected})} = [1.18e^{-0.08(PI)} + 0.57]c_{u(\text{VST})}$$
$$= [1.18e^{-0.08(50-18)} + 0.57](20.12)$$
$$= \mathbf{13.3\ kN/m^2}$$

Part c
From Eqs. (12.64) and (12.61),

$$c_{u(\text{corrected})} = [7.01e^{-0.08(LL)} + 0.57]c_{u(\text{VST})}$$
$$= [7.01e^{-0.08(50)} + 0.57](20.12)$$
$$= \mathbf{14.05\ kN/m^2}$$

12.17 Other Methods for Determining Undrained Shear Strength

A modified form of the vane shear test apparatus is the *Torvane* (Figure 12.50), which is a handheld device with a calibrated spring. This instrument can be used for determining c_u for tube specimens collected from the field during soil exploration, and it can be used in the field. The Torvane is pushed into the soil and then rotated until the soil fails. The undrained shear strength can be read at the top of the calibrated dial.

Figure 12.51 shows a *pocket penetrometer,* which is pushed directly into the soil. The unconfined compression strength (q_u) is measured by a calibrated spring. This device can be used both in the laboratory and in the field.

12.18 Shear Strength of Unsaturated Cohesive Soils

The equation relating total stress, effective stress, and pore water pressure for unsaturated soils can be expressed as

$$\sigma' = \sigma - u_a + \chi(u_a - u_w) \tag{12.65}$$

where σ' = effective stress
σ = total stress
u_a = pore air pressure
u_w = pore water pressure

Figure 12.50
Torvane (*Courtesy of ELE International*)

Figure 12.51
Pocket penetrometer (*Courtesy of ELE International*)

When the expression for σ' is substituted into the shear strength equation [Eq. (12.3)], which is based on effective stress parameters, we get

$$\tau_f = c' + [\sigma - u_a + \chi(u_a - u_w)] \tan \phi' \tag{12.66}$$

The values of χ depend primarily on the degree of saturation. With ordinary triaxial equipment used for laboratory testing, it is not possible to determine accurately the effective stresses in unsaturated soil specimens, so the common practice is to conduct undrained triaxial tests on unsaturated specimens and measure only the total stress. Figure 12.52 shows a total stress failure envelope obtained from a number of undrained triaxial tests conducted with a given initial degree of saturation. The failure envelope is generally curved. Higher confining pressure causes higher compression of the air in void spaces; thus, the solubility of void air in void water is increased. For design purposes, the curved envelope is sometimes approximated as a straight line, as shown in Figure 12.52, with an equation as follows:

$$\tau_f = c + \sigma \tan \phi \tag{12.67}$$

(*Note:* c and ϕ in the preceding equation are empirical constants.)

Figure 12.53 shows the variation of the total stress envelopes with change of the initial degree of saturation obtained from undrained tests on an inorganic clay. Note that for these tests the specimens were prepared with approximately the same initial dry unit weight of about 16.7 kN/m³. For a given total normal stress, the shear stress needed to cause failure decreases as the degree of saturation increases. When the degree of saturation reaches 100%, the total stress failure envelope becomes a horizontal line that is the same as with the $\phi = 0$ concept.

In practical cases where a cohesive soil deposit may become saturated because of rainfall or a rise in the groundwater table, the strength of partially saturated clay should not be used for design considerations. Instead, the unsaturated soil specimens collected from the field must be saturated in the laboratory and the undrained strength determined.

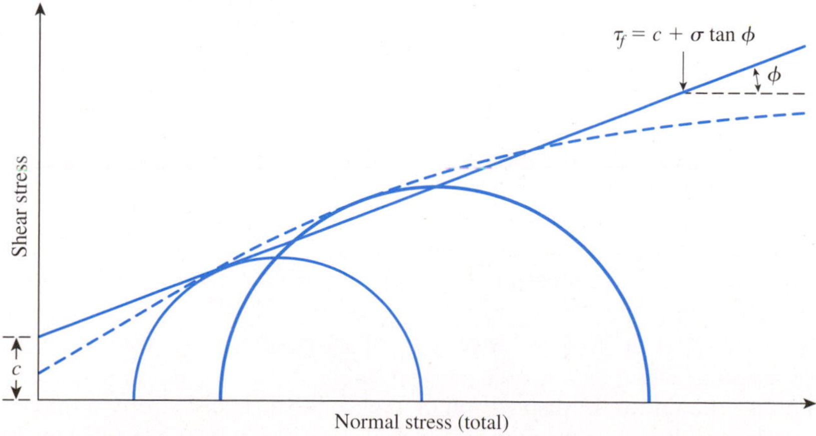

Figure 12.52 Total stress failure envelope for unsaturated cohesive soils

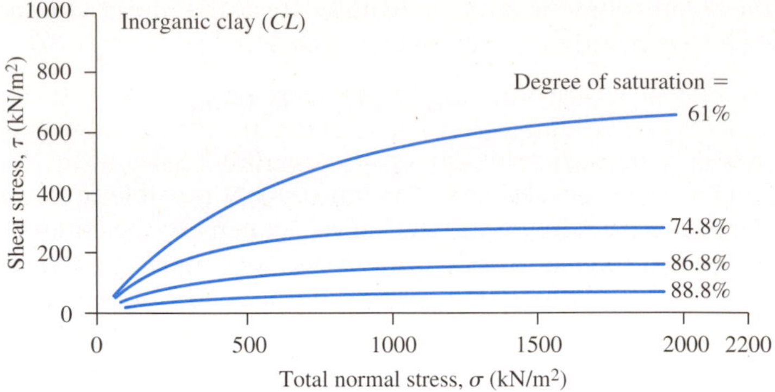

Figure 12.53 Variation of the total stress failure envelope with change of initial degree of saturation obtained from undrained tests of an inorganic clay (*After Casagrande and Hirschfeld, 1960. With permission from ASCE.*)

12.19 Summary

In this chapter, the shear strengths of granular and cohesive soils were examined. Laboratory procedures for determining the shear strength parameters were described.

A summary of subjects covered in this chapter is as follows:

- According to the Mohr–Coulomb failure criteria, the shear strength of soil can be expressed as

$$\tau_f = c' + \sigma' \tan \phi'$$

- Direct shear and triaxial are two commonly used laboratory test methods to determine the shear strength parameters of soil.
- Shear strength of soil is dependent on the drainage conditions. Triaxial tests can be conducted under three different drainage conditions:
  - Consolidated-drained (Section 12.9)
  - Consolidated-undrained (Section 12.10)
  - Unconsolidated-undrained (Section 12.11)
- The unconfined compression test is a special type of unconsolidated-undrained test (Section 12.12).
- Sensitivity is a loss of strength of cohesive soils due to remolding (Section 12.14).
- Due to the nature of disposition of clay soils, the shear strength may vary depending on the direction of load application (Section 12.15). This is referred to as strength anisotropy of clay.

- The vane shear test is another method to determine the undrained shear strength of clay soils in the laboratory and field (Section 12.16).

In textbooks, determination of the shear strength parameters of cohesive soils appears to be fairly simple. However, in practice, the proper choice of these parameters for design and stability checks of various earth, earth-retaining, and earth-supported structures is very difficult and requires experience and an appropriate theoretical background in geotechnical engineering. In this chapter, three types of strength parameters (*consolidated-drained, consolidated-undrained,* and *unconsolidated-undrained*) were introduced. Their use depends on drainage conditions.

Consolidated-drained strength parameters can be used to determine the long-term stability of structures such as earth embankments and cut slopes. Consolidated-undrained shear strength parameters can be used to study stability problems relating to cases where the soil initially is fully consolidated and then there is rapid loading. An excellent example of this is the stability of slopes of earth dams after rapid drawdown. The unconsolidated-undrained shear strength of clays can be used to evaluate the end-of-construction stability of saturated cohesive soils with the assumption that the load caused by construction has been applied rapidly and there has been little time for drainage to take place. The bearing capacity of foundations on soft saturated clays and the stability of the base of embankments on soft clays are examples of this condition.

Problems

12.1. Following data are given for a direct shear test conducted on dry silty sand:
Specimen dimensions: diameter = 71 mm; height = 25 mm
Normal stress: 150 kN/m^2
Shear force at failure: 276 N
a. Determine the angle of friction, ϕ'.
b. For a normal stress of 200 kN/m^2, what shear force is required to cause failure?

12.2 Consider the specimen in Problem 12.1.
a. If a direct shear test is conducted with a normal force of 675 N, what would be the principal stresses at failure?
b. What would be the inclination of the major principal plane with the horizontal?

12.3 For a dry sand specimen in a direct shear test box, the following are given:
Size of specimen: 71 mm × 71 mm × 31.7 mm (height)
Angle of friction: 41°
Normal stress: 152 kN/m^2
Determine the shear force required to cause failure

12.4 During a subsoil exploration program, undisturbed normally consolidated silty clay samples were collected in Shelby tubes from location *A* as shown in Figure 12.54.

a. Determine the angle of friction, ϕ'.
b. Determine the angle θ that the failure plane makes with the major principal plane.
c. Calculate the normal stress, σ'_f, and the shear stress, τ_f, on the failure plane.

12.16 The results of two consolidated-drained triaxial tests on a clayey sand are given here.

Specimen	Chamber pressure, σ_3 (kN/m^2)	Deviator stress, $(\sigma_1 - \sigma_3)_f$ (kN/m^2)
I	70	155
II	140	265

Calculate the shear strength parameters of the soil.

12.17 Consider the triaxial tests in Problem 12.16.
a. What are the normal and shear stresses on a plane inclined at 33° to the major principal plane for specimen I?
b. What are the normal and shear stresses on the failure plane at failure for specimen II?

12.18 A clay sample was consolidated in a triaxial test under an all-around confining pressure of 103 kN/m^2. The sample was then loaded to failure in undrained condition by applying an additional axial stress of 151.7 kN/m^2. A pore water pressure sensor recorded an excess pore pressure of $(\Delta u_d)_f = -62$ kN/m^2 at failure. Determine the undrained and drained friction angles for the soil.

12.19 The shear strength of a normally consolidated clay can be given by the equation $\tau_f = \sigma' \tan 21°$. The results of a consolidated-undrained test on the clay are
Chamber confining pressure = 225 kN/m^2
Deviator stress at failure = 112 kN/m^2
Determine:
a. The consolidated-undrained (total stress) friction angle
b. Pore water pressure developed in the specimen at failure

12.20 If a consolidated-drained test is conducted on the clay specimen of Problem 12.19 with the same chamber-confining pressure of 225 kN/m^2, what would be the deviator stress at failure?

12.21 A consolidated-undrained triaxial test was conducted on a dense sand with a chamber-confining pressure of 96 kN/m^2. Results showed that $\phi' = 27°$ and $\phi = 33°$. Determine the deviator stress and the pore water pressure at failure. If it were a loose sand, what would have been the expected behavior? Explain.

12.22 Undisturbed samples from a normally consolidated clay layer were collected during a subsoil exploration. Drained triaxial tests showed that the effective friction angle was $\phi' = 25°$. The unconfined compressive strength, q_u, of a similar specimen was found to be 133 kN/m^2. Find the pore pressure at failure for the unconfined compression test.

12.23 A 15-m thick normally consolidated clay layer is shown in Figure 12.55. The liquid limit and plastic limit of the soil are 39 and 20, respectively.

a. Using Eq. (12.46) given by Skempton (1957), estimate the undrained cohesion at a depth of 11 m below the ground surface as would be obtained by conducting a vane shear test.

b. Using Bjerrum's (1974) correction factor [Eqs. (12.61) and (12.62)], estimate the design value of the undrained shear strength determined in Part a.

c. If the clay has the potential of becoming overconsolidated up to OCR = 3.0 due to future ground improvement activities, what would be the new undrained cohesion? Use Eq. (12.47) given by Ladd et al. (1977).

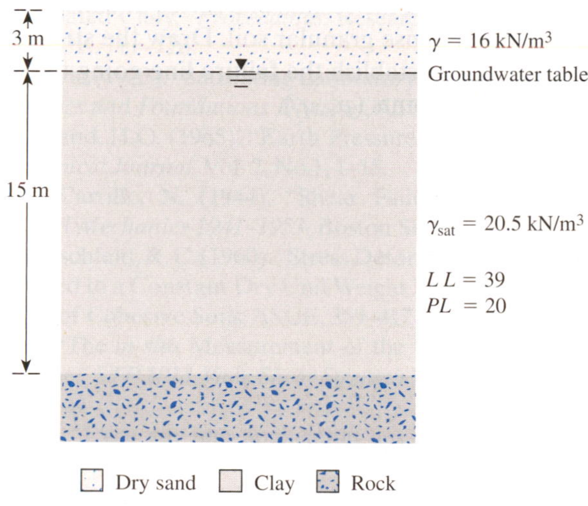

Figure 12.55

12.24 Refer to the clay soil in Figure 12.55. If the natural moisture content is 28%, estimate the undrained shear strength of remolded clay using the relationships given in Table 12.5 by

a. Leroueil et al. (1983)

b. Terzaghi et al. (1996)

Critical Thinking Problem

12.C.1 In this chapter, you learned different types of triaxial tests. During a triaxial test, the successive stress states experienced by the soil specimen can be represented by diagrams called stress paths. Schofield and Wroth (1968) defined the stress states in terms of p' (mean normal effective stress) and q (deviator stress) as follows:

$$p' = \frac{1}{3}(\sigma_1' + \sigma_2' + \sigma_3') = \frac{1}{3}(\sigma_1' + 2\sigma_3') \quad (\textit{Note}: \sigma_2' = \sigma_3' \text{ for triaxial tests})$$

$$q = \sigma_1 - \sigma_3$$

Rosenqvist, I. Th. (1953). "Considerations on the Sensitivity of Norwegian Quick Clays," *Geotechnique*, Vol. 3, No. 5, 195–200.

Schofield, A., and Wroth, P. (1968). *The Critical State Soil Mechanics*, McGraw Hill, Inc., New York.

Simons, N. E. (1960). "The Effect of Overconsolidation on the Shear Strength Characteristics of an Undisturbed Oslo Clay," *Proceedings*, Research Conference on Shear Strength of Cohesive Soils, ASCE, 747–763.

Skempton, A. W. (1954). "The Pore Water Coefficients *A* and *B*," *Geotechnique*, Vol. 4, 143–147.

Skempton, A. W. (1957). "Discussion: The Planning and Design of New Hong Kong Airport," *Proceedings*, Institute of Civil Engineers, London, Vol. 7, 305–307.

Skempton, A. W. (1964). "Long-Term Stability of Clay Slopes," *Geotechnique*, Vol. 14, 77.

Sorensen, K. K., and Okkels, N. (2013). "Correlation between Drained Shear Strength and Plasticity Index of Undisturbed Overconsolidated Clays," *Proceedings*, *18th International Conference on Soil Mechanics and Geotechnical Engineering*, Paris, Presses des Ponts, Vol. 1, 423–428.

Terzaghi, K., Peck, R. B., and Mesri, G. (1996). *Soil Mechanics in Engineering Practice*, 3rd Ed., Wiley, NY.

US Navy (1986). *Soil Mechanics—Design Manual 7.1*, Department of the Navy, Naval Facilities Engineering Command, U.S. Government Printing Office, Washington, D.C.

Yang, S. L., Kvalstad, T., Solheim, A., and Forsberg, C. F. (2006). "Parameter Studies of Sediments in the Storegga Slide Region," *Geo-Marine Letters*, Vol. 26, No. 4, 213–224.

CHAPTER 13

Lateral Earth Pressure: At-Rest, Rankine, and Coulomb

13.1 Introduction

Retaining structures such as retaining walls, basement walls, and bulkheads commonly are encountered in foundation engineering as they support slopes of earth masses. Proper design and construction of these structures require a thorough knowledge of the lateral forces that act between the retaining structures and the soil masses being retained. These lateral forces are caused by lateral earth pressure. The magnitude and distribution of lateral earth pressure depends on many factors, such as the shear strength parameters of the soil being retained, the inclination of the surface of the backfill, the height and inclination of the retaining wall at the wall–backfill interface, the nature of wall movement under lateral pressure, and the adhesion and friction angle at the wall–backfill interface. This chapter is devoted to the study of the various earth pressure theories and the influence of the above parameters on the magnitude of lateral earth pressure.

13.2 At-Rest, Active, and Passive Pressures

Consider a mass of soil shown in Figure. 13.1a. The mass is bounded by a *frictionless wall* of height AB. A soil element located at a depth z is subjected to a vertical effective pressure, σ'_o, and a horizontal effective pressure, σ'_h. There are no shear stresses

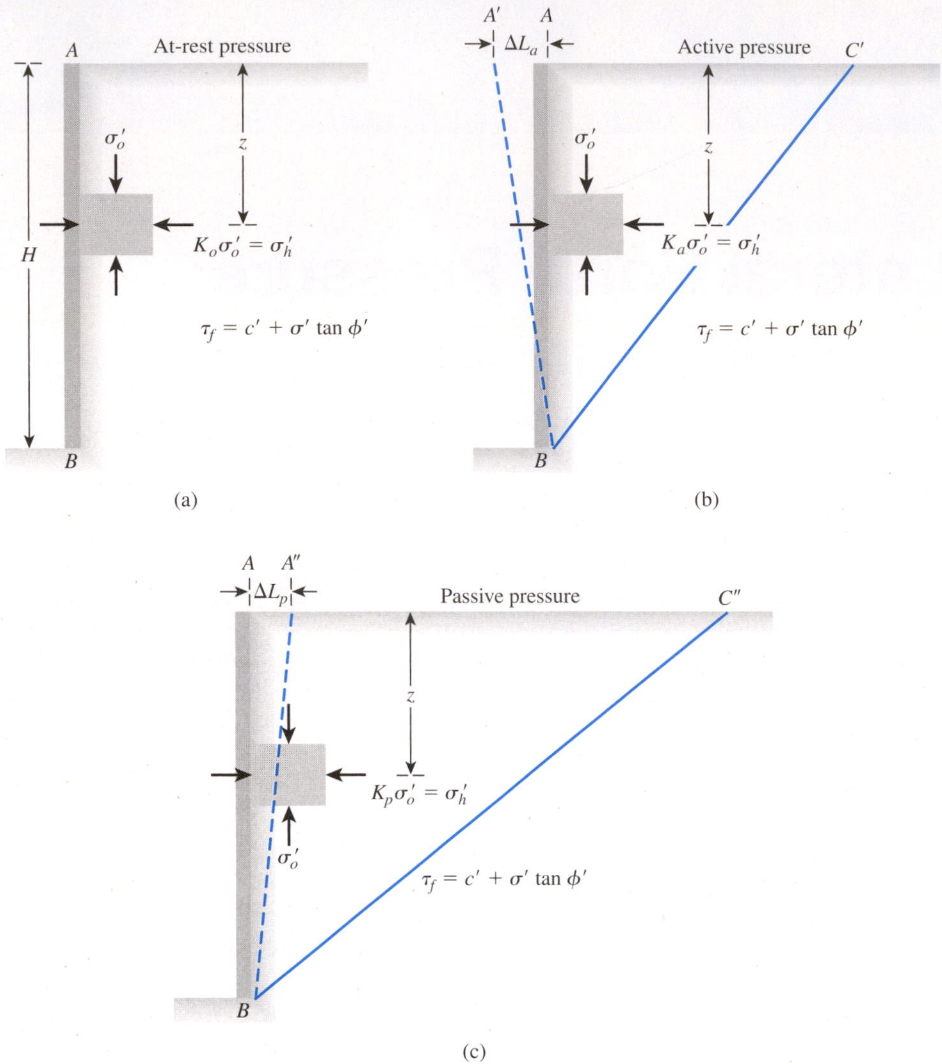

Figure 13.1 Definition of at-rest, active, and passive pressures (*Note:* Wall *AB* is frictionless)

on the vertical and horizontal planes of the soil element. Let us define the ratio of σ'_h to σ'_o as a nondimensional quantity *K*, or

$$K = \frac{\sigma'_h}{\sigma'_o} \tag{13.1}$$

Now, three possible cases may arise concerning the retaining wall; they are described as follows:

Case 1 If the wall AB is static—that is, if it does not move either to the right or to the left of its initial position—the soil mass will be in a state of *static equilibrium*. In that case, σ'_h is referred to as the *at-rest earth pressure,* or

$$K = K_o = \frac{\sigma'_h}{\sigma'_o} \tag{13.2}$$

where K_o = at-rest earth pressure coefficient.

Case 2 If the frictionless wall rotates sufficiently about its bottom to a position of $A'B$ (Figure 13.1b), then a triangular soil mass ABC' adjacent to the wall will reach a state of *plastic equilibrium* and will fail sliding down the plane BC'. At this time, the horizontal effective stress, $\sigma'_h = \sigma'_a$, will be referred to as *active pressure*. Now,

$$K = K_a = \frac{\sigma'_h}{\sigma'_o} = \frac{\sigma'_a}{\sigma'_o} \tag{13.3}$$

where K_a = active earth pressure coefficient.

Case 3 If the frictionless wall rotates sufficiently about its bottom to a position $A''B$ (Figure 13.1c), then a triangular soil mass ABC'' will reach a state of *plastic equilibrium* and will fail sliding upward along the plane BC''. The horizontal effective stress at this time will be $\sigma'_h = \sigma'_p$, the so-called *passive pressure*. In this case,

$$K = K_p = \frac{\sigma'_h}{\sigma'_o} = \frac{\sigma'_p}{\sigma'_o} \tag{13.4}$$

where K_p = passive earth pressure coefficient

Figure 13.2 shows the nature of variation of lateral earth pressure with the wall tilt. Typical values of $\Delta L_a/H$ ($\Delta L_a = A'A$ in Figure 13.1b) and $\Delta L_p/H$ ($\Delta L_p = A''A$ in Figure 13.1c) for attaining the active and passive states in various soils are given in Table 13.1.

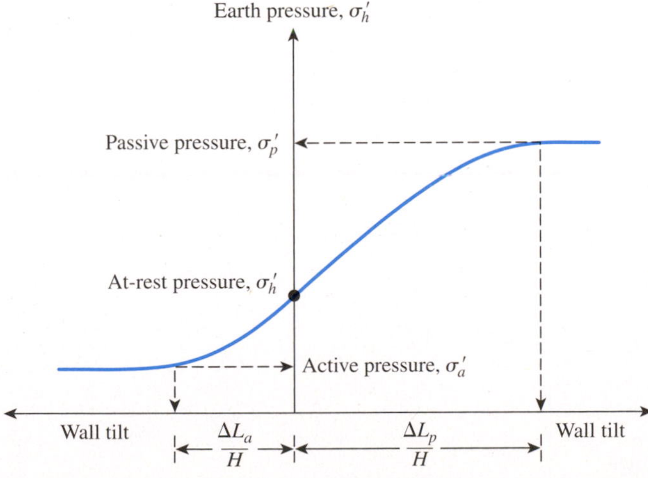

Figure 13.2 Variation of the magnitude of lateral earth pressure with wall tilt

Table 13.1 Typical Values of $\Delta L_a/H$ and $\Delta L_p/H$

Soil type	$\Delta L_a/H$	$\Delta L_p/H$
Loose sand	0.001–0.002	0.01
Dense sand	0.0005–0.001	0.005
Soft clay	0.02	0.04
Stiff clay	0.01	0.02

AT-REST LATERAL EARTH PRESSURE

13.3 Earth Pressure At-Rest

The fundamental concept of earth pressure at rest was discussed in the preceding section. In order to define the earth pressure coefficient K_o at rest, we refer to Figure 13.3, which shows a wall AB retaining a dry soil with a unit weight of γ. The wall is static. At a depth z,

$$\text{Vertical effective stress} = \sigma'_o = \gamma z$$

$$\text{Horizontal effective stress} = \sigma'_h = K_o \gamma z$$

So,

$$K_o = \frac{\sigma'_h}{\sigma'_o} = \text{at-rest earth pressure coefficient}$$

For coarse-grained soils, the coefficient of earth pressure at rest can be estimated by using the empirical relationship (Jaky, 1944)

$$K_o = 1 - \sin \phi' \tag{13.5}$$

where ϕ' = drained friction angle.

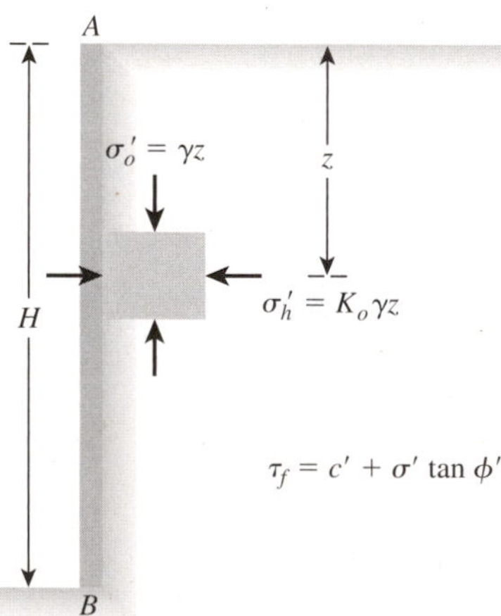

Figure 13.3 Earth pressure at rest

While designing a wall that may be subjected to lateral earth pressure at rest, one must take care in evaluating the value of K_o. Sherif, Fang, and Sherif (1984), on the basis of their laboratory tests, showed that Jaky's equation for K_o [Eq. (13.5)] gives good results when the backfill is loose sand. However, for a dense, compacted sand backfill, Eq. (13.5) may grossly underestimate the lateral earth pressure at rest. This underestimation results because of the process of compaction of backfill. For this reason, they recommended the design relationship

$$K_o = (1 - \sin\phi) + \left[\frac{\gamma_d}{\gamma_{d(\min)}} - 1\right]5.5 \tag{13.6}$$

where γ_d = actual compacted dry unit weight of the sand behind the wall
$\gamma_{d(\min)}$ = dry unit weight of the sand in the loosest state (Chapter 3)

The increase of K_o observed from Eq. (13.6) compared to Eq. (13.5) is due to over consolidation. For that reason, Mayne and Kulhawy (1982), after evaluating 171 soils, recommended a modification to Eq. (13.5). Or

$$K_o = (1 - \sin\phi')(OCR)^{\sin\phi'} \tag{13.7}$$

where

$$OCR = \text{overconsolidation ratio} = \frac{\text{preconsolidation pressure, } \sigma'_c}{\text{present effective overburden pressure, } \sigma'_o}$$

Equation (13.7) is valid for soils ranging from clay to gravel.

For fine-grained, normally consolidated soils, Massarsch (1979) suggested the following equation for K_o:

$$K_o = 0.44 + 0.42\left[\frac{PI\,(\%)}{100}\right] \tag{13.8}$$

For overconsolidated clays, the coefficient of earth pressure at rest can be approximated as

$$K_{o(\text{overconsolidated})} = K_{o(\text{normally consolidated})}\sqrt{OCR} \tag{13.9}$$

Figure 13.4 shows the distribution of lateral earth pressure at rest on a wall of height H retaining a dry soil having a unit weight of γ. The total force per unit length of the wall, P_o, is equal to the area of the pressure diagram, so

$$P_o = \frac{1}{2}K_o\gamma H^2 \tag{13.10}$$

Table 13.3 Passive Earth Pressure Coefficient, K_p [Eq. (13.40)]

↓ α (deg)	ϕ' (deg) →						
	28	30	32	34	36	38	40
0	2.770	3.000	3.255	3.537	3.852	4.204	4.599
5	2.715	2.943	3.196	3.476	3.788	4.136	4.527
10	2.551	2.775	3.022	3.295	3.598	3.937	4.316
15	2.284	2.502	2.740	3.003	3.293	3.615	3.977
20	1.918	2.132	2.362	2.612	2.886	3.189	3.526
25	1.434	1.664	1.894	2.135	2.394	2.676	2.987

13.10 A Generalized Case for Rankine Active and Passive Pressure—Granular Backfill

In Sections 13.6, 13.7, 13.8, and 13.9, we discussed the Rankine active and passive pressure cases for a frictionless wall with a vertical back and horizontal and inclined backfill of granular soil. This can be extended to general cases of frictionless wall with inclined backfill (granular soil), as shown in Figure 13.14 (Chu, 1991).

Rankine active case

For the Rankine active case, the lateral earth pressure (σ'_a) at a depth z can be given as

$$\sigma'_a = \frac{\gamma z \cos\alpha\sqrt{1 + \sin^2\phi' - 2\sin\phi'\cos\psi_a}}{\cos\alpha + \sqrt{\sin^2\phi' - \sin^2\alpha}} \tag{13.41}$$

where

$$\psi_a = \sin^{-1}\left(\frac{\sin\alpha}{\sin\phi'}\right) - \alpha + 2\theta \tag{13.42}$$

The pressure σ'_a will be inclined at an angle β with the plane drawn at right angle to the backface of the wall, and

$$\beta = \tan^{-1}\left(\frac{\sin\phi'\sin\psi_a}{1 - \sin\phi'\cos\psi_a}\right) \tag{13.43}$$

The active force P_a for unit length of the wall then can be calculated as

$$P_a = \frac{1}{2}\gamma H^2 K_{a(R)} \tag{13.44}$$

where

$$K_{a(R)} = \frac{\cos(\alpha - \theta)\sqrt{1 + \sin^2\phi' - 2\sin\phi'\cos\psi_a}}{\cos^2\theta(\cos\alpha + \sqrt{\sin^2\phi' - \sin^2\alpha})}$$

= Rankine active earth-pressure coefficient for generalized case (13.45)

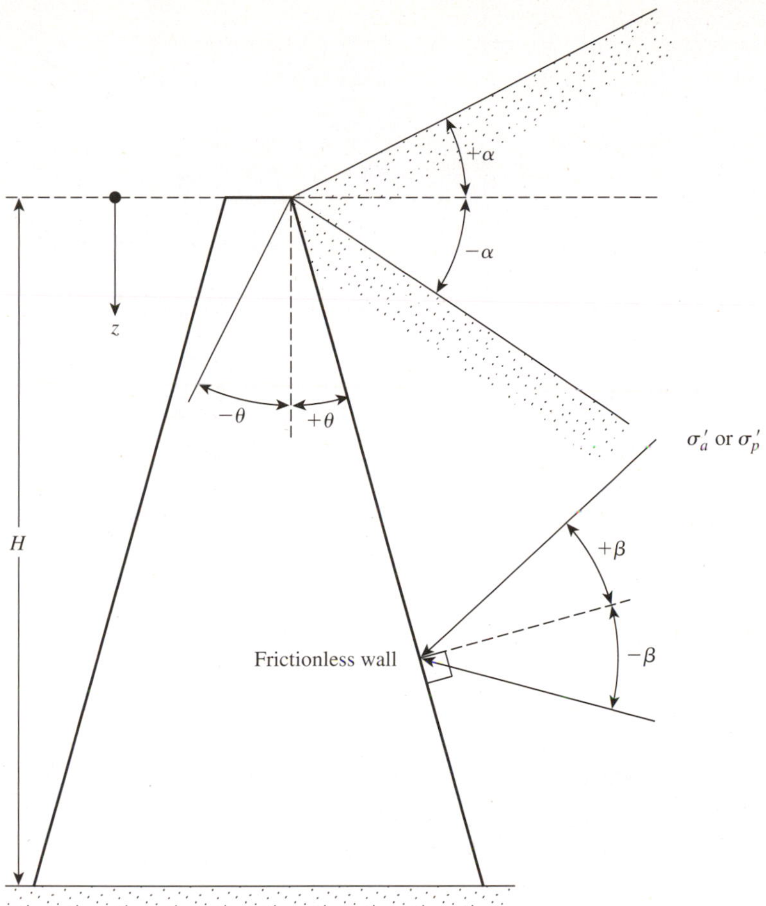

Figure 13.14 General case for Rankine active and passive pressures

The location and direction of the resultant force P_a is shown in Figure 13.15a. Also shown in this figure is the failure wedge, ABC. Note that BC will be inclined at an angle η, or

$$\eta = \frac{\pi}{4} + \frac{\phi'}{2} + \frac{\alpha}{2} - \frac{1}{2}\sin^{-1}\left(\frac{\sin\alpha}{\sin\phi'}\right) \tag{13.46}$$

As a special case, for a vertical backface of the wall (that is, $\theta = 0$) as shown in Figure 13.13, Eqs. (13.44) and (13.45) simplify to the following, which is same as Eqs. (13.37) and (13.38)

$$P_a = \frac{1}{2}K_{a(R)}\gamma H^2$$

where

$$K_{a(R)} = \cos\alpha \frac{\cos\alpha - \sqrt{\cos^2\alpha - \cos^2\phi'}}{\cos\alpha + \sqrt{\cos^2\alpha - \cos^2\phi'}}$$

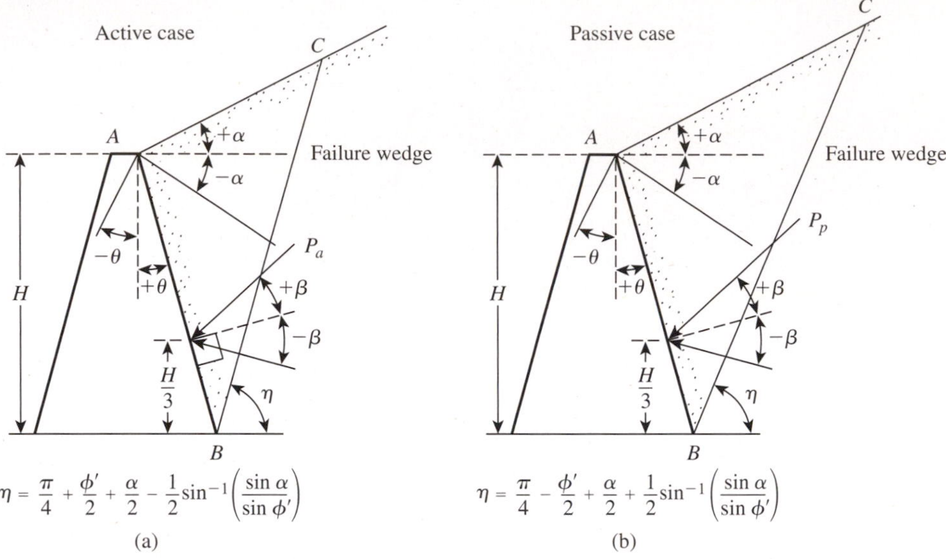

Figure 13.15 Location and direction of resultant Rankine force

Rankine passive case

Similar to the active case, for the Rankine passive case, we can obtain the following relationships.

$$\sigma'_p = \frac{\gamma z \cos\alpha\sqrt{1 + \sin^2\phi' + 2\sin\phi\cos\psi_p}}{\cos\alpha - \sqrt{\sin^2\phi' - \sin^2\alpha}} \tag{13.47}$$

where

$$\psi_p = \sin^{-1}\left(\frac{\sin\alpha}{\sin\phi'}\right) + \alpha - 2\theta \tag{13.48}$$

The inclination β of σ'_p, as shown in Figure 13.14, is

$$\beta = \tan^{-1}\left(\frac{\sin\phi'\sin\psi_p}{1 + \sin\phi'\cos\psi_p}\right) \tag{13.49}$$

The passive force per unit length of the wall is

$$P_p = \frac{1}{2}\gamma H^2 K_{p(R)}$$

where

$$K_{p(R)} = \frac{\cos(\alpha - \theta)\sqrt{1 + \sin^2\phi' + 2\sin\phi'\cos\psi_p}}{\cos^2\theta(\cos\alpha - \sqrt{\sin^2\phi' - \sin^2\alpha})} \tag{13.50}$$

The location and direction of P_p along with the failure wedge is shown in Figure 13.15b. For walls with vertical backface, $\theta = 0$,

$$P_p = \frac{1}{2}K_{p(R)}\gamma H^2$$

where

$$K_{p(R)} = \cos\alpha \frac{\cos\alpha + \sqrt{\cos^2\alpha - \cos^2\phi'}}{\cos\alpha - \sqrt{\cos^2\alpha - \cos^2\phi'}}$$

The preceding two equations are the same as Eqs. (13.39) and (13.40).

13.11 Diagrams for Lateral Earth-Pressure Distribution against Retaining Walls with Vertical Back

Backfill—cohesionless soil with horizontal ground surface

Active case Figure 13.16a shows a retaining wall with cohesionless soil backfill that has a horizontal ground surface. The unit weight and the angle of friction of the soil are γ and ϕ', respectively.

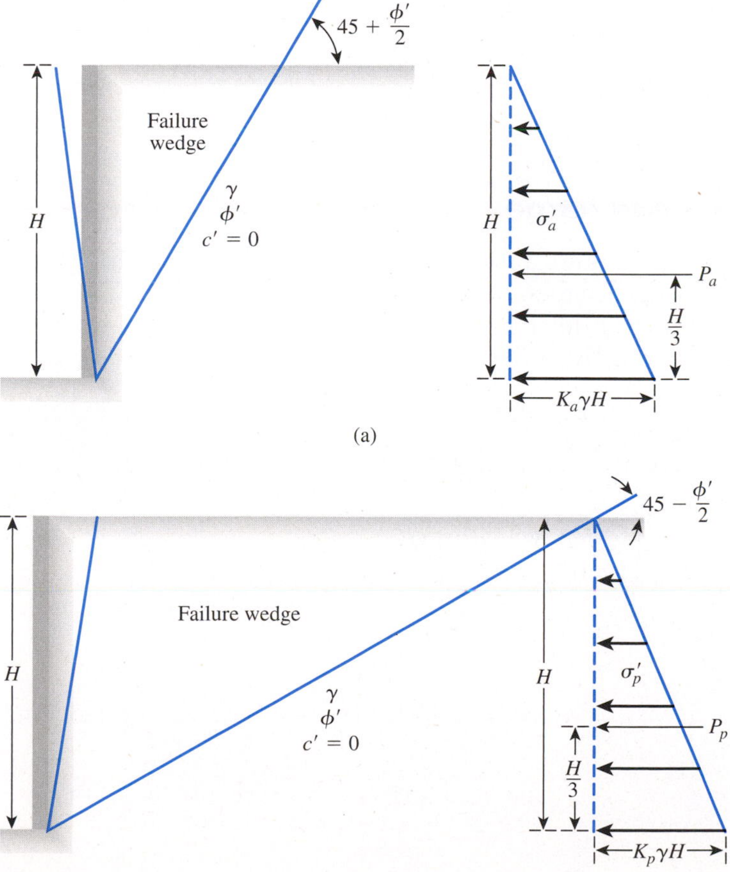

Figure 13.16 Pressure distribution against a retaining wall for cohesionless soil backfill with horizontal ground surface: (a) Rankine's active state; (b) Rankine's passive state

For Rankine's active state, the earth pressure at any depth against the retaining wall can be given by Eq. (13.31):

$$\sigma'_a = K_a \gamma z \quad (\textit{Note: } c' = 0)$$

Note that σ'_a increases linearly with depth, and at the bottom of the wall, it is

$$\sigma'_a = K_a \gamma H \tag{13.51}$$

The total force per unit length of the wall is equal to the area of the pressure diagram, so

$$P_a = \frac{1}{2} K_a \gamma H^2 \tag{13.52}$$

Passive case The lateral pressure distribution against a retaining wall of height H for Rankine's passive state is shown in Figure 13.16b. The lateral earth pressure at any depth z [Eq. (13.36), $c' = 0$] is

$$\sigma'_p = K_p \gamma H \tag{13.53}$$

The total force per unit length of the wall is

$$P_p = \frac{1}{2} K_p \gamma H^2 \tag{13.54}$$

Backfill—partially submerged cohesionless soil supporting a surcharge

Active case Figure 13.17a shows a frictionless retaining wall of height H and a backfill of cohesionless soil. The groundwater table is located at a depth of H_1 below the ground surface, and the backfill is supporting a surcharge pressure of q per unit area. From Eq. (13.33), the effective active earth pressure at any depth can be given by

$$\sigma'_a = K_a \sigma'_o \tag{13.55}$$

where σ'_o and σ'_a = the effective vertical pressure and lateral pressure, respectively. At $z = 0$,

$$\sigma_o = \sigma'_o = q \tag{13.56}$$

and

$$\sigma'_a = K_a q \tag{13.57}$$

At depth $z = H_1$,

$$\sigma'_o = (q + \gamma H_1) \tag{13.58}$$

and

$$\sigma'_a = K_a(q + \gamma H_1) \tag{13.59}$$

At depth $z = H$,

$$\sigma'_o = (q + \gamma H_1 + \gamma' H_2) \tag{13.60}$$

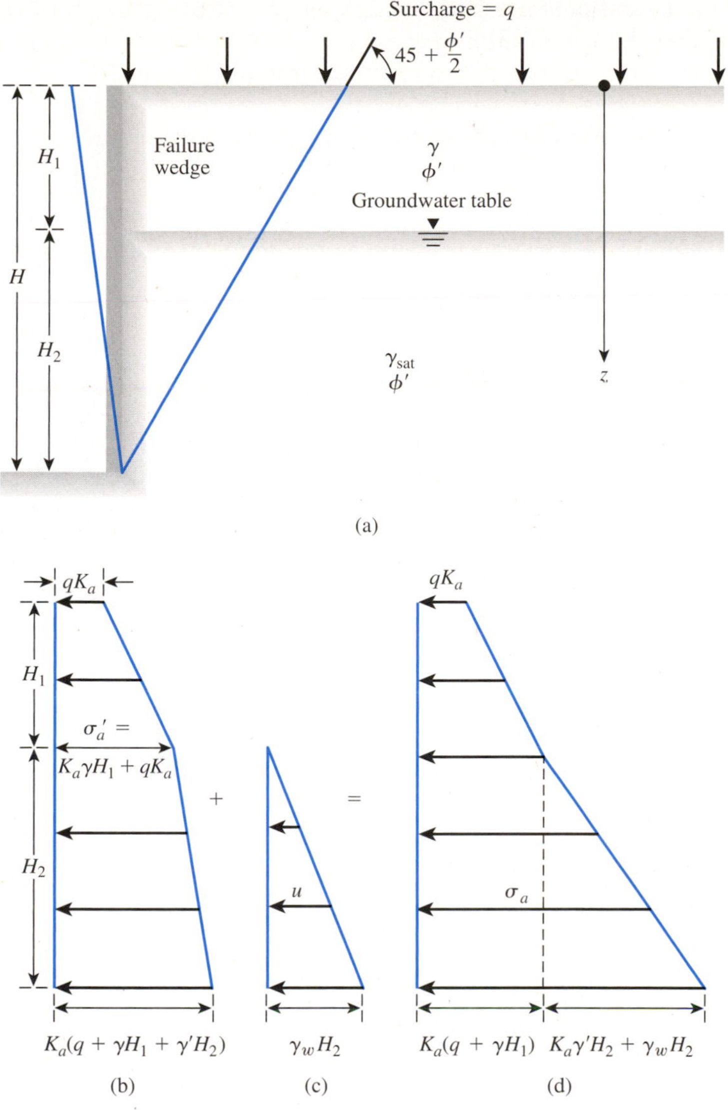

Figure 13.17 Rankine's active earth-pressure distribution against a retaining wall with partially submerged cohesionless soil backfill supporting a surcharge

and

$$\sigma'_a = K_a(q + \gamma H_1 + \gamma' H_2) \tag{13.61}$$

where $\gamma' = \gamma_{sat} - \gamma_w$. The variation of σ'_a with depth is shown in Figure 13.17b.

The lateral pressure on the wall from the pore water between $z = 0$ and H_1 is 0, and for $z > H_1$, it increases linearly with depth (Figure 13.17c). At $z = H$,

$$u = \gamma_w H_2$$

The total lateral-pressure diagram (Figure 13.17d) is the sum of the pressure diagrams shown in Figures 13.17b and 13.17c. The total active force per unit length of the wall is the area of the total pressure diagram. Thus,

$$P_a = K_a qH + \frac{1}{2}K_a\gamma H_1^2 + K_a\gamma H_1 H_2 + \frac{1}{2}(K_a\gamma' + \gamma_w)H_2^2 \qquad (13.62)$$

Passive case Figure 13.18a shows the same retaining wall as was shown in Figure 13.17a. Rankine's passive pressure at any depth against the wall can be given by Eq. (13.36):

$$\sigma'_p = K_p\sigma'_o$$

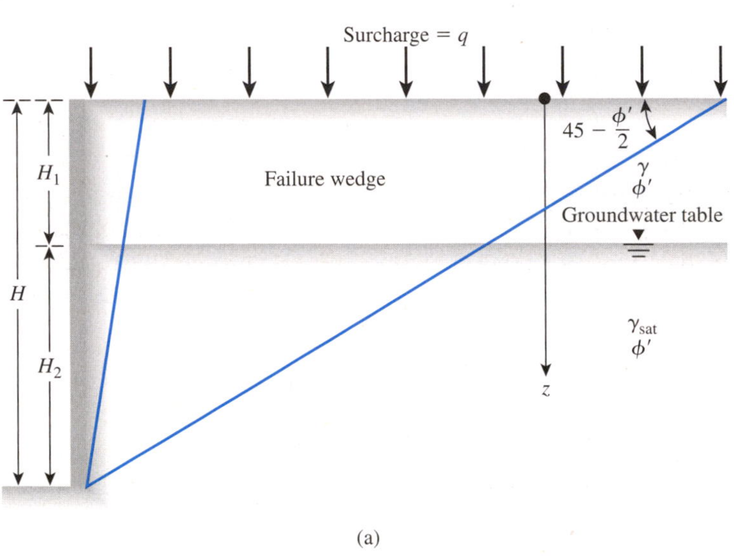

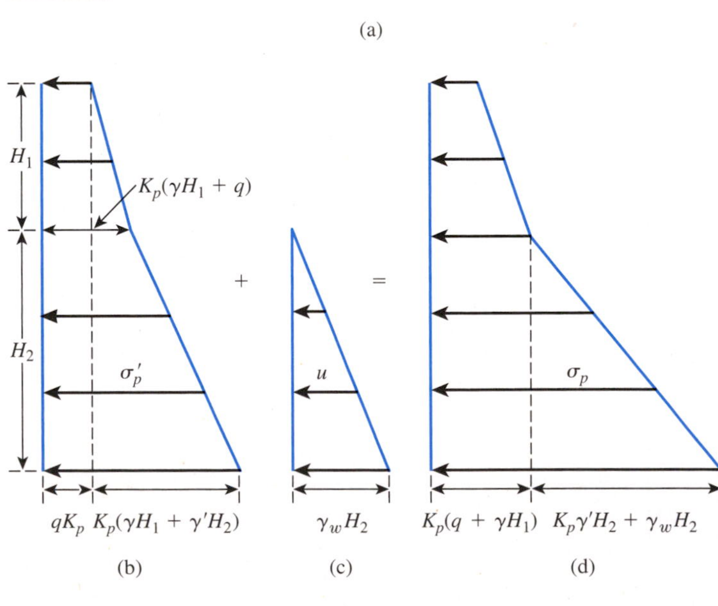

Figure 13.18 Rankine's passive earth-pressure distribution against a retaining wall with partially submerged cohesionless soil backfill supporting a surcharge

Using the preceding equation, we can determine the variation of σ'_p with depth, as shown in Figure 13.18b. The variation of the pressure on the wall from water with depth is shown in Figure 13.18c. Figure 13.18d shows the distribution of the total pressure σ_p with depth. The total lateral passive force per unit length of the wall is the area of the diagram given in Figure 13.18d, or

$$P_p = K_p q H + \frac{1}{2} K_p \gamma H_1^2 + K_p \gamma H_1 H_2 + \frac{1}{2}(K_p \gamma' + \gamma_w) H_2^2 \tag{13.63}$$

Backfill—cohesive soil with horizontal backfill

Active case Figure 13.19a shows a frictionless retaining wall with a cohesive soil backfill. The active pressure against the wall at any depth below the ground surface can be expressed as [Eq. (13.31)]

$$\sigma'_a = K_a \gamma z - 2\sqrt{K_a} c'$$

The variation of $K_a \gamma z$ with depth is shown in Figure 13.19b, and the variation of $2\sqrt{K_a} c'$ with depth is shown in Figure 13.19c. Note that $2\sqrt{K_a} c'$ is not a function of z; hence, Figure 13.19c is a rectangle. The variation of the net value of σ'_a with depth is plotted in Figure 13.19d. Also note that, because of the effect of cohesion, σ'_a is negative in the upper part of the retaining wall. The depth z_o at which the active pressure becomes equal to 0 can be found from Eq. (13.31) as

$$K_a \gamma z_o - 2\sqrt{K_a} c' = 0$$

or

$$z_o = \frac{2c'}{\gamma \sqrt{K_a}} \tag{13.64}$$

For the undrained condition—that is, $\phi = 0$, $K_a = \tan^2 45 = 1$, and $c = c_u$ (undrained cohesion)—from Eq. (13.34),

$$z_o = \frac{2c_u}{\gamma} \tag{13.65}$$

So, with time, tensile cracks at the soil–wall interface will develop up to a depth z_o.

The total active force per unit length of the wall can be found from the area of the total pressure diagram (Figure 13.19d), or

$$P_a = \frac{1}{2} K_a \gamma H^2 - 2\sqrt{K_a} c' H \tag{13.66}$$

For the $\phi = 0$ condition,

$$P_a = \frac{1}{2} \gamma H^2 - 2 c_u H \tag{13.67}$$

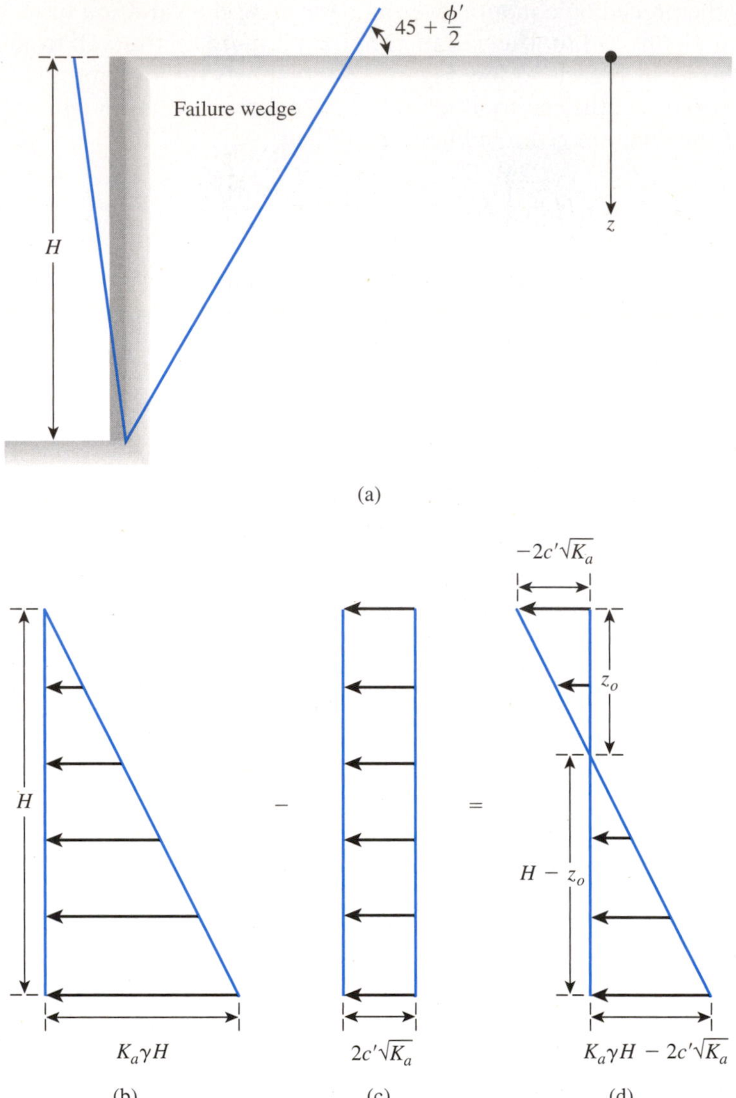

Figure 13.19 Rankine's active earth-pressure distribution against a retaining wall with cohesive soil backfill

For calculation of the total active force, common practice is to take the tensile cracks into account. Because no contact exists between the soil and the wall up to a depth of z_o after the development of tensile cracks, only the active pressure distribution against the wall between $z = 2c'/(\gamma\sqrt{K_a})$ and H (Figure 13.19d) is considered. In this case,

$$P_a = \frac{1}{2}(K_a\gamma H - 2\sqrt{K_a}c')\left(H - \frac{2c'}{\gamma\sqrt{K_a}}\right)$$

$$= \frac{1}{2}K_a\gamma H^2 - 2\sqrt{K_a}c'H + 2\frac{c'^2}{\gamma} \quad (13.68)$$

For the $\phi = 0$ condition,

$$P_a = \frac{1}{2}\gamma H^2 - 2c_u H + 2\frac{c_u^2}{\gamma} \quad (13.69)$$

Passive case Figure 13.20a shows the same retaining wall with backfill similar to that considered in Figure 13.19a. Rankine's passive pressure against the wall at depth z can be given by [Eq. (13.35)]

$$\sigma'_p = K_p\gamma z + 2\sqrt{K_p}c'$$

At $z = 0$,

$$\sigma'_p = 2\sqrt{K_p}c' \quad (13.70)$$

and at $z = H$,

$$\sigma'_p = K_p\gamma H + 2\sqrt{K_p}c' \quad (13.71)$$

The variation of σ'_p with depth is shown in Figure 13.20b. The passive force per unit length of the wall can be found from the area of the pressure diagrams as

$$P_p = \frac{1}{2}K_p\gamma H^2 + 2\sqrt{K_p}c'H \quad (13.72a)$$

For the $\phi = 0$ condition, $K_p = 1$ and

$$P_p = \frac{1}{2}\gamma H^2 + 2c_u H \quad (13.72b)$$

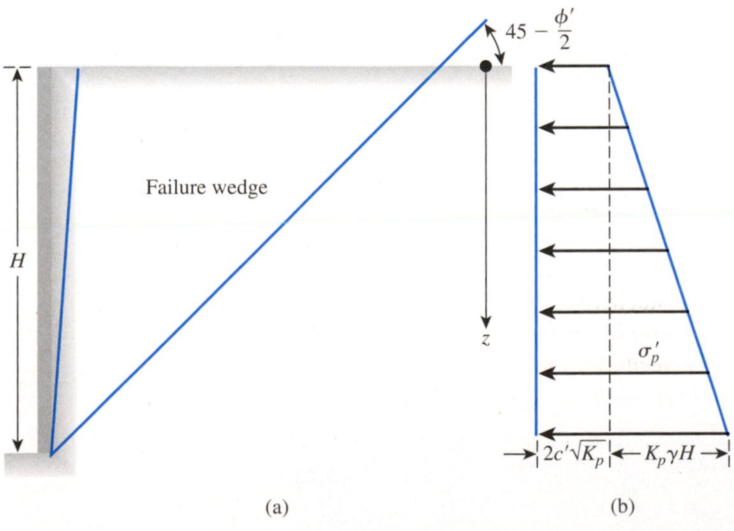

Figure 13.20 Rankine's passive earth-pressure distribution against a retaining wall with cohesive soil backfill

Example 13.6

An 6-m-high retaining wall is shown in Figure 13.21a. Determine:

a. Rankine active force per unit length of the wall and the location of the resultant
b. Rankine passive force per unit length of the wall and the location of the resultant

Solution

Part a

Because $c' = 0$, to determine the active force, we can use Eq. (13.33).

$$\sigma'_a = K_a\sigma'_o = K_a\gamma z$$

$$K_a = \frac{1 - \sin\phi'}{1 + \sin\phi'} = \frac{1 - \sin 36}{1 + \sin 36} = 0.26$$

At $z = 0, \sigma'_a = 0$; at $z = 6$ m,

$$\sigma'_a = (0.26)(16)(6) = 24.96 \text{ kN/m}^2$$

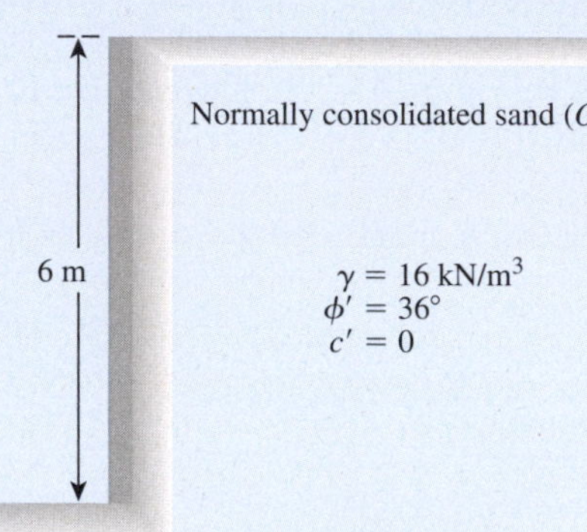

(a)

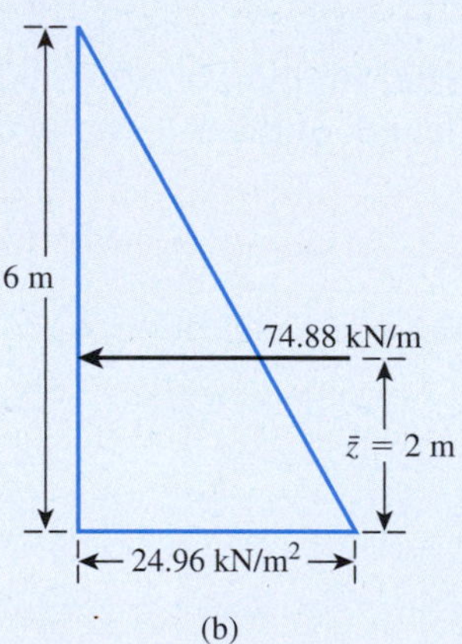

(b)

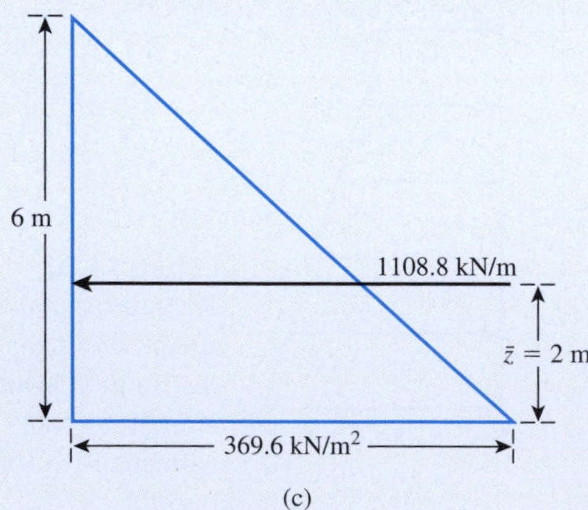

(c)

Figure 13.21

The pressure-distribution diagram is shown in Figure 13.21b. The active force per unit length of the wall is

$$P_a = \frac{1}{2}(6)(24.96) = \mathbf{74.88\ kN/m}$$

Also,

$$\bar{z} = \mathbf{2\ m}$$

Part b

To determine the passive force, we are given that $c' = 0$. So, from Eq. (13.36),

$$\sigma'_p = K_p\sigma'_o = K_p\gamma z$$

$$K_p = \frac{1 + \sin\phi'}{1 - \sin\phi'} = \frac{1 + \sin 36}{1 - \sin 36} = 3.85$$

At $z = 0, \sigma'_p = 0$; at $z = 6$ m,

$$\sigma'_p = (3.85)(16)(6) = 369.6\ \text{kN/m}^2$$

The pressure-distribution diagram is shown in Figure 13.21c. The passive force per unit length of the wall is

$$P_p = \frac{1}{2}(6)(369.6) = \mathbf{1108.8\ kN/m}$$

Also,

$$\bar{z} = \frac{6}{3} = \mathbf{2\ m}$$

Example 13.7

For the retaining wall shown in Figure 13.22a, determine the force per unit length of the wall for Rankine's active state. Also find the location of the resultant.

Solution

Given that $c' = 0$, we know that $\sigma'_a = K_a\sigma'_o$. For the upper layer of the soil, Rankine's active earth-pressure coefficient is

$$K_a = K_{a(1)} = \frac{1 - \sin 30°}{1 + \sin 30°} = \frac{1}{3}$$

For the lower layer,

$$K_a = K_{a(2)} = \frac{1 - \sin 35°}{1 + \sin 35°} = 0.271$$

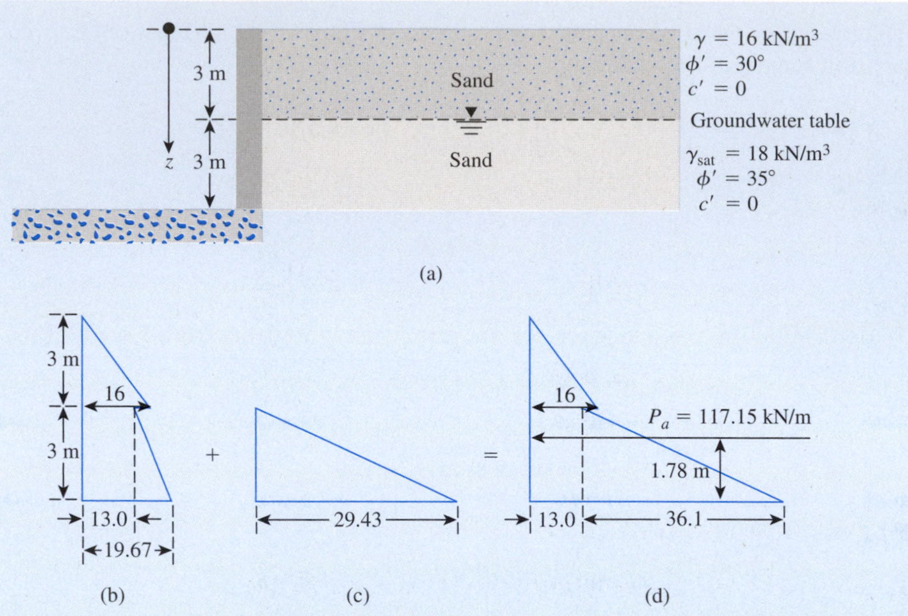

Figure 13.22 Retaining wall and pressure diagrams for determining Rankine's active earth pressure. (*Note:* The units of pressure in (b), (c), and (d) are kN/m²)

At $z = 0$, $\sigma'_o = 0$. At $z = 3$ m (just inside the bottom of the upper layer), $\sigma'_o = 3 \times 16 = 48$ kN/m². So

$$\sigma'_a = K_{a(1)}\sigma'_o = \frac{1}{3} \times 48 = 16 \text{ kN/m}^2$$

Again, at $z = 3$ m (in the lower layer), $\sigma'_o = 3 \times 16 = 48$ kN/m², and

$$\sigma'_a = K_{a(2)}\sigma'_o = (0.271)(48) = 13.0 \text{ kN/m}^2$$

At $z = 6$ m,

$$\sigma'_o = 3 \times 16 + 3(18 - \underset{\substack{\uparrow \\ \gamma_w}}{9.81}) = 72.57 \text{ kN/m}^2$$

and

$$\sigma'_a = K_{a(2)}\sigma'_o = (0.271)(72.57) = 19.67 \text{ kN/m}^2$$

The variation of σ'_a with depth is shown in Figure 13.22b.

The lateral pressures due to the pore water are as follows.

At $z = 0$: $u = 0$
At $z = 3$ m: $u = 0$
At $z = 6$ m: $u = 3 \times \gamma_w = 3 \times 9.81 = 29.43$ kN/m²

The variation of u with depth is shown in Figure 13.22c, and that for σ_a (total active pressure) is shown in Figure 13.22d. Thus,

$$P_a = \left(\tfrac{1}{2}\right)(3)(16) + 3(13.0) + \left(\tfrac{1}{2}\right)(3)(36.1) = 24 + 39.0 + 54.15 = \mathbf{117.15\ kN/m}$$

The location of the resultant can be found by taking the moment about the bottom of the wall:

$$\bar{z} = \frac{24\left(3 + \frac{3}{3}\right) + 39.0\left(\frac{3}{2}\right) + 54.15\left(\frac{3}{3}\right)}{117.15}$$

$$= \mathbf{1.78\ m}$$

Example 13.8

A frictionless retaining wall is shown in Figure 13.23a. Determine:

a. The active force P_a after the tensile crack occurs
b. The passive force P_p

Solution

Part a

Given $\phi' = 26°$, we have

$$K_a = \frac{1 - \sin \phi'}{1 + \sin \phi'} = \frac{1 - \sin 26°}{1 + \sin 26°} = 0.39$$

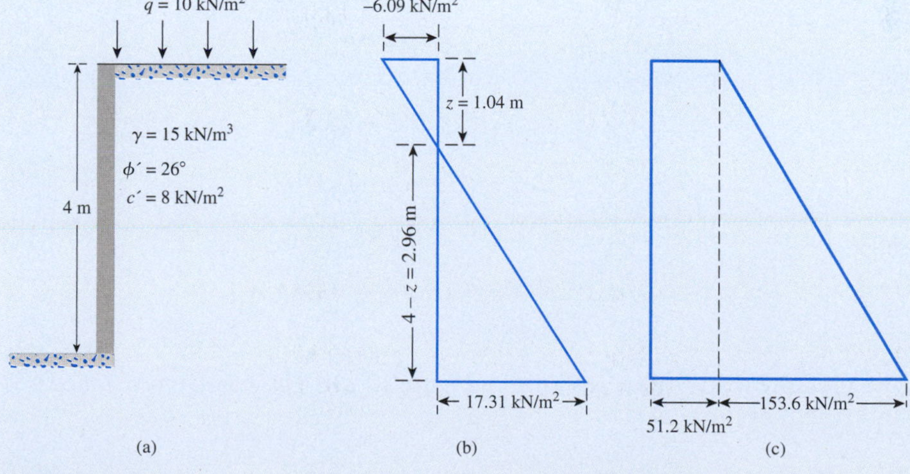

Figure 13.23

From Eq. (13.31),

$$\sigma'_a = K_a\sigma'_o - 2c'\sqrt{K_a}$$

At $z = 0$,

$$\sigma'_a = (0.39)(10) - (2)(8)\sqrt{0.39} = 3.9 - 9.99 = -6.09 \text{ kN/m}^2$$

At $z = 4$ m,

$$\sigma'_a = (0.39)[10 + (4)(15)] - (2)(8)\sqrt{0.39} = 27.3 - 9.99$$
$$= 17.31 \text{ kN/m}^2$$

The pressure distribution is shown in Figure 13.23b. From this diagram,

$$\frac{6.09}{z_o} = \frac{17.31}{4 - z_o}$$

or

$$z_o = 1.04 \text{ m}$$

After the tensile crack occurs,

$$P_a = \frac{1}{2}(4 - z_o)(17.31) = \left(\frac{1}{2}\right)(2.96)(17.31) = \mathbf{25.62\ kN/m}$$

Part b

Given $\phi' = 26°$, we have

$$K_p = \frac{1 + \sin\phi'}{1 - \sin\phi'} = \frac{1 + \sin 26°}{1 - \sin 26°} = \frac{1.4384}{0.5616} = 2.56$$

From Eq. (13.35),

$$\sigma'_p = K_p\sigma'_o + 2\sqrt{K_p}c'$$

At $z = 0, \sigma'_o = 10 \text{ kN/m}^2$ and

$$\sigma_p = (2.56)(10) + 2\sqrt{2.56}(8)$$
$$= 25.6 + 25.6 = 51.2 \text{ kN/m}^2$$

Again, at $z = 4 \text{ m}, \sigma'_o = (10 + 4 \times 15) = 70 \text{ kN/m}^2$ and

$$\sigma'_p = (2.56)(70) + 2\sqrt{2.56}(8)$$
$$= 204.8 \text{ kN/m}^2$$

The pressure distribution is shown in Figure 13.23c. The passive resistance per unit length of the wall is

$$P_p = (51.2)(4) + \frac{1}{2}(4)(153.6)$$
$$= 204.8 + 307.2 = \mathbf{512\ kN/m}$$

Example 13.9

A retaining wall is shown in Figure 13.24a. Determine P_a after the occurrence of the tensile crack.

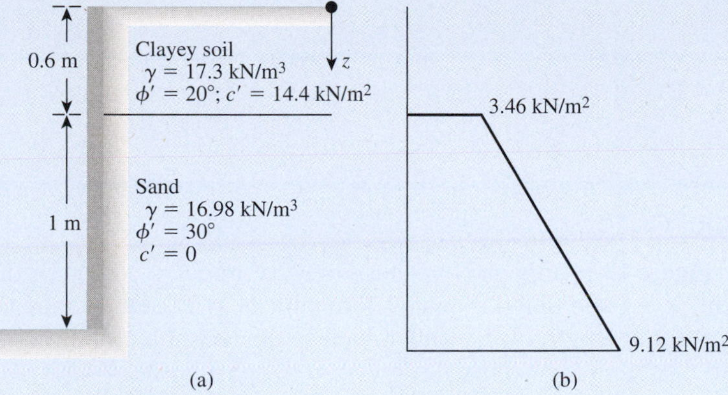

Figure 13.24

Solution

For the upper layer,

$$K_a = K_{a(1)} = \tan^2\left(45 - \frac{20}{2}\right) = 0.49$$

From Eq (13.64),

$$z_o = \frac{2c'}{\gamma\sqrt{K_a}} = \frac{(2)(14.4)}{(17.3)\sqrt{0.49}} = 2.38 \text{ m}$$

Since the depth of the clayey soil layer is 0.6 m (which is less than z_o), the tensile crack will develop up to $z = 0.6$ m. Now

$$K_a = K_{a(2)} = \tan^2\left(45 - \frac{30}{2}\right) = \frac{1}{3}$$

At $z = 0.6$ m,

$$\sigma_o = \sigma'_o = (0.6)(17.3) = 10.38 \text{ kN/m}^2$$

So,

$$\sigma'_a = \sigma'_o K_{a(2)} = (10.38)\left(\frac{1}{3}\right) = 3.46 \text{ kN/m}^2$$

At $z = 1.6$ m,

$$\sigma'_o = (0.6)(17.3) + (1)(16.98) = 10.38 + 16.98 = 27.36 \text{ kN/m}^2$$

$$\sigma'_a = \sigma'_o K_{a(2)} = (27.36)\left(\frac{1}{3}\right) = 9.12 \text{ kN/m}^2$$

The pressure distribution diagram after the occurrence of the tensile crack is shown in Figure 13.24b. From this

$$P_a = \left(\frac{1}{2}\right)(3.46 + 9.12)(1) = \mathbf{6.29\ kN/m}$$

Example 13.10

Refer to Figure 13.14. Given: $H = 4$ m, $\alpha = +20°$, and $\theta = +20°$. For the granular backfill, it is given that $\gamma = 18.08$ kN/m^3 and $\phi' = 30°$. Determine the active force P_a per unit length of the wall as well as the location and direction of the resultant.

Solution

From Eq. (13.42),

$$\psi_a = \sin^{-1}\left(\frac{\sin\alpha}{\sin\phi'}\right) - \alpha + 2\theta = \sin^{-1}\left(\frac{\sin 20}{\sin 30}\right) - 20° + (2)(20°)$$

$$= 43.16° - 20° + 40° = 63.16°$$

From Eq. (13.45),

$$K_{a(R)} = \frac{\cos(\alpha - \theta)\sqrt{1 + \sin^2\phi' - 2\sin\phi'\cos\psi_a}}{\cos^2\theta(\cos\alpha + \sqrt{\sin^2\phi' - \sin^2\alpha})}$$

$$= \frac{\cos(20 - 20)\sqrt{1 + \sin^2 30 - (2)(\sin 30)(\cos 63.16)}}{\cos^2 20(\cos 20 + \sqrt{\sin^2 30 - \sin^2 20})} = 0.776$$

From Eq. (13.44),

$$P_a = \frac{1}{2}\gamma H^2 K_{a(R)} = \frac{1}{2}(18.08)(4)^2(0.776) = \mathbf{112.24\ kN/m}$$

From Eq. (13.43),

$$\beta = \tan^{-1}\left(\frac{\sin\phi'\sin\psi_a}{1 - \sin\phi'\cos\psi_a}\right)$$

$$= \tan^{-1}\left[\frac{(\sin 30)\,(\sin 63.16)}{1 - (\sin 30)\,(\cos 63.16)}\right] = 29.95° \approx 30°$$

The resultant will act a distance of 12/3 = **1.33 m** above the bottom of the wall with $\beta =$ **30°**.

COULOMB'S EARTH PRESSURE THEORY

More than 200 years ago, Coulomb (1776) presented a theory for active and passive earth pressures against retaining walls. In this theory, Coulomb assumed that the failure surface is a plane. The *wall friction* was taken into consideration. The following sections discuss the general principles of the derivation of Coulomb's earth-pressure theory for a cohesionless backfill (shear strength defined by the equation $\tau_f = \sigma' \tan \phi'$).

13.12 Coulomb's Active Pressure

Let AB (Figure 13.25a) be the back face of a retaining wall supporting a granular soil, the surface of which is constantly sloping at an angle α with the horizontal. BC is a trial failure surface. In the stability consideration of the probable failure wedge ABC, the following forces are involved (per unit length of the wall):

1. W—the weight of the soil wedge.
2. F—the resultant of the shear and normal forces on the surface of failure, BC. This is inclined at an angle of ϕ' to the normal drawn to the plane BC.
3. P_a—the active force per unit length of the wall. The direction of P_a is inclined at an angle δ' to the normal drawn to the face of the wall that supports the soil. δ' is the angle of friction between the soil and the wall.

The force triangle for the wedge is shown in Figure 13.19b. From the law of sines, we have

$$\frac{W}{\sin(90 + \theta + \delta' - \beta + \phi')} = \frac{P_a}{\sin(\beta - \phi')} \tag{13.73}$$

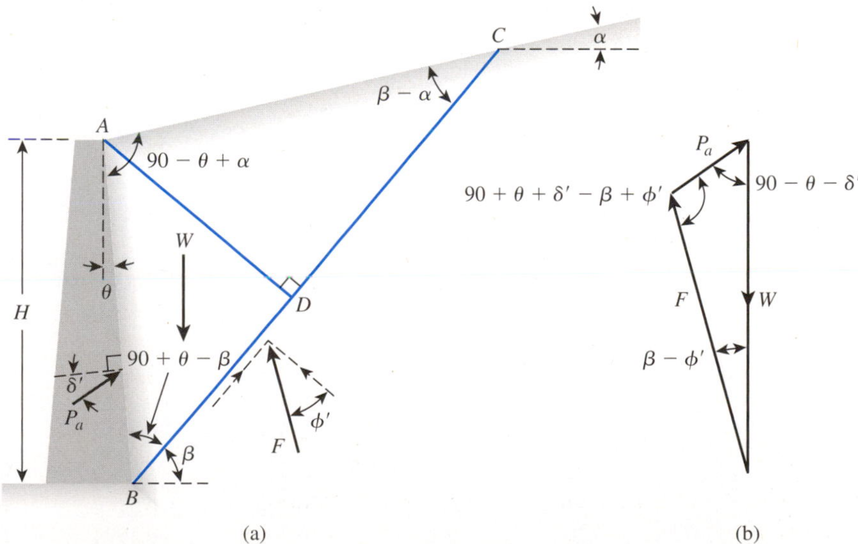

Figure 13.25 Coulomb's active pressure: (a) trial failure wedge; (b) force polygon

or

$$P_a = \frac{\sin(\beta - \phi')}{\sin(90 + \theta + \delta' - \beta + \phi')} W \tag{13.74}$$

The preceding equation can be written in the form

$$P_a = \frac{1}{2}\gamma H^2 \left[\frac{\cos(\theta - \beta)\cos(\theta - \alpha)\sin(\beta - \phi')}{\cos^2\theta \sin(\beta - \alpha)\sin(90 + \theta + \delta' - \beta + \phi')} \right] \tag{13.75}$$

where γ = unit weight of the backfill. The values of $\gamma, H, \theta, \alpha, \phi'$, and δ' are constants, and β is the only variable. To determine the critical value of β for maximum P_a, we have

$$\frac{dP_a}{d\beta} = 0 \tag{13.76}$$

After solving Eq. (13.76), when the relationship of β is substituted into Eq. (13.75), we obtain Coulomb's active earth pressure as

$$P_a = \frac{1}{2} K_a \gamma H^2 \tag{13.77}$$

where K_a is Coulomb's active earth-pressure coefficient and is given by

$$K_a = \frac{\cos^2(\phi' - \theta)}{\cos^2\theta \cos(\delta' + \theta)\left[1 + \sqrt{\frac{\sin(\delta' + \phi')\sin(\phi' - \alpha)}{\cos(\delta' + \theta)\cos(\theta - \alpha)}}\right]^2} \tag{13.78}$$

Note that when $\alpha = 0°$, $\theta = 0°$, and $\delta' = 0°$, Coulomb's active earth-pressure coefficient becomes equal to $(1 - \sin\phi')/(1 + \sin\phi')$, which is the same as Rankine's earth-pressure coefficient given earlier in this chapter.

The variation of the values of K_a for retaining walls with a vertical back ($\theta = 0°$) and horizontal backfill ($\alpha = 0°$) is given in Table 13.4. From this table, note that for a given value of ϕ', the effect of wall friction is to reduce somewhat the active earth-pressure coefficient.

Tables 13.5 and 13.6 also give the variation of K_a [Eq. (13.78)] for various values of α, ϕ', θ, and δ' ($\delta' = \frac{2}{3}\phi'$ in Table 13.5 and $\delta' = \frac{1}{2}\phi'$ in Table 13.6).

Table 13.4 Values of K_a [Eq. (13.78)] for $\theta = 0°$, $\alpha = 0°$

	δ' (deg) →					
↓ ϕ' (deg)	0	5	10	15	20	25
28	0.3610	0.3448	0.3330	0.3251	0.3203	0.3186
30	0.3333	0.3189	0.3085	0.3014	0.2973	0.2956
32	0.3073	0.2945	0.2853	0.2791	0.2755	0.2745
34	0.2827	0.2714	0.2633	0.2579	0.2549	0.2542
36	0.2596	0.2497	0.2426	0.2379	0.2354	0.2350
38	0.2379	0.2292	0.2230	0.2190	0.2169	0.2167
40	0.2174	0.2089	0.2045	0.2011	0.1994	0.1995
42	0.1982	0.1916	0.1870	0.1841	0.1828	0.1831

Table 13.5 Values of K_a [Eq. (13.78)] (*Note:* $\delta' = \frac{2}{3}\phi'$)

α (deg)	ϕ' (deg)	θ (deg) 0	5	10	15	20	25
0	28	0.3213	0.3588	0.4007	0.4481	0.5026	0.5662
	29	0.3091	0.3467	0.3886	0.4362	0.4908	0.5547
	30	0.2973	0.3349	0.3769	0.4245	0.4794	0.5435
	31	0.2860	0.3235	0.3655	0.4133	0.4682	0.5326
	32	0.2750	0.3125	0.3545	0.4023	0.4574	0.5220
	33	0.2645	0.3019	0.3439	0.3917	0.4469	0.5117
	34	0.2543	0.2916	0.3335	0.3813	0.4367	0.5017
	35	0.2444	0.2816	0.3235	0.3713	0.4267	0.4919
	36	0.2349	0.2719	0.3137	0.3615	0.4170	0.4824
	37	0.2257	0.2626	0.3042	0.3520	0.4075	0.4732
	38	0.2168	0.2535	0.2950	0.3427	0.3983	0.4641
	39	0.2082	0.2447	0.2861	0.3337	0.3894	0.4553
	40	0.1998	0.2361	0.2774	0.3249	0.3806	0.4468
	41	0.1918	0.2278	0.2689	0.3164	0.3721	0.4384
	42	0.1840	0.2197	0.2606	0.3080	0.3637	0.4302
5	28	0.3431	0.3845	0.4311	0.4843	0.5461	0.6190
	29	0.3295	0.3709	0.4175	0.4707	0.5325	0.6056
	30	0.3165	0.3578	0.4043	0.4575	0.5194	0.5926
	31	0.3039	0.3451	0.3916	0.4447	0.5067	0.5800
	32	0.2919	0.3329	0.3792	0.4324	0.4943	0.5677
	33	0.2803	0.3211	0.3673	0.4204	0.4823	0.5558
	34	0.2691	0.3097	0.3558	0.4088	0.4707	0.5443
	35	0.2583	0.2987	0.3446	0.3975	0.4594	0.5330
	36	0.2479	0.2881	0.3338	0.3866	0.4484	0.5221
	37	0.2379	0.2778	0.3233	0.3759	0.4377	0.5115
	38	0.2282	0.2679	0.3131	0.3656	0.4273	0.5012
	39	0.2188	0.2582	0.3033	0.3556	0.4172	0.4911
	40	0.2098	0.2489	0.2937	0.3458	0.4074	0.4813
	41	0.2011	0.2398	0.2844	0.3363	0.3978	0.4718
	42	0.1927	0.2311	0.2753	0.3271	0.3884	0.4625
10	28	0.3702	0.4164	0.4686	0.5287	0.5992	0.6834
	29	0.3548	0.4007	0.4528	0.5128	0.5831	0.6672
	30	0.3400	0.3857	0.4376	0.4974	0.5676	0.6516
	31	0.3259	0.3713	0.4230	0.4826	0.5526	0.6365
	32	0.3123	0.3575	0.4089	0.4683	0.5382	0.6219
	33	0.2993	0.3442	0.3953	0.4545	0.5242	0.6078
	34	0.2868	0.3314	0.3822	0.4412	0.5107	0.5942
	35	0.2748	0.3190	0.3696	0.4283	0.4976	0.5810
	36	0.2633	0.3072	0.3574	0.4158	0.4849	0.5682
	37	0.2522	0.2957	0.3456	0.4037	0.4726	0.5558
	38	0.2415	0.2846	0.3342	0.3920	0.4607	0.5437
	39	0.2313	0.2740	0.3231	0.3807	0.4491	0.5321
	40	0.2214	0.2636	0.3125	0.3697	0.4379	0.5207
	41	0.2119	0.2537	0.3021	0.3590	0.4270	0.5097
	42	0.2027	0.2441	0.2921	0.3487	0.4164	0.4990

(*continued*)

Table 13.5 *(continued)*

α (deg)	ϕ' (deg)	θ (deg) 0	5	10	15	20	25
15	28	0.4065	0.4585	0.5179	0.5868	0.6685	0.7670
	29	0.3881	0.4397	0.4987	0.5672	0.6483	0.7463
	30	0.3707	0.4219	0.4804	0.5484	0.6291	0.7265
	31	0.3541	0.4049	0.4629	0.5305	0.6106	0.7076
	32	0.3384	0.3887	0.4462	0.5133	0.5930	0.6895
	33	0.3234	0.3732	0.4303	0.4969	0.5761	0.6721
	34	0.3091	0.3583	0.4150	0.4811	0.5598	0.6554
	35	0.2954	0.3442	0.4003	0.4659	0.5442	0.6393
	36	0.2823	0.3306	0.3862	0.4513	0.5291	0.6238
	37	0.2698	0.3175	0.3726	0.4373	0.5146	0.6089
	38	0.2578	0.3050	0.3595	0.4237	0.5006	0.5945
	39	0.2463	0.2929	0.3470	0.4106	0.4871	0.5805
	40	0.2353	0.2813	0.3348	0.3980	0.4740	0.5671
	41	0.2247	0.2702	0.3231	0.3858	0.4613	0.5541
	42	0.2146	0.2594	0.3118	0.3740	0.4491	0.5415
20	28	0.4602	0.5205	0.5900	0.6714	0.7689	0.8880
	29	0.4364	0.4958	0.5642	0.6445	0.7406	0.8581
	30	0.4142	0.4728	0.5403	0.6195	0.7144	0.8303
	31	0.3935	0.4513	0.5179	0.5961	0.6898	0.8043
	32	0.3742	0.4311	0.4968	0.5741	0.6666	0.7799
	33	0.3559	0.4121	0.4769	0.5532	0.6448	0.7569
	34	0.3388	0.3941	0.4581	0.5335	0.6241	0.7351
	35	0.3225	0.3771	0.4402	0.5148	0.6044	0.7144
	36	0.3071	0.3609	0.4233	0.4969	0.5856	0.6947
	37	0.2925	0.3455	0.4071	0.4799	0.5677	0.6759
	38	0.2787	0.3308	0.3916	0.4636	0.5506	0.6579
	39	0.2654	0.3168	0.3768	0.4480	0.5342	0.6407
	40	0.2529	0.3034	0.3626	0.4331	0.5185	0.6242
	41	0.2408	0.2906	0.3490	0.4187	0.5033	0.6083
	42	0.2294	0.2784	0.3360	0.4049	0.4888	0.5930

Example 13.11

Refer to Figure 13.25. Given: $\alpha = 10°$; $\theta = 5°$; $H = 4$ m; unit weight of soil, $\gamma = 15$ kN/m^3; soil friction angle, $\phi' = 30°$; and $\delta' = 15°$. Estimate the active force, P_a, per unit length of the wall. Also, state the direction and location of the resultant force, P_a.

Solution

From Eq. (13.77),

$$P_a = \frac{1}{2}\gamma H^2 K_a$$

For $\phi' = 30°; \delta' = 15°$—that is, $\frac{\delta'}{\phi'} = \frac{15}{30} = \frac{1}{2}$; $\alpha = 10°$; and $\theta = 5°$, the magnitude of K_a is 0.3872 (Table 13.6). So,

$$P_a = \frac{1}{2}(15)(4)^2(0.3872) = \mathbf{46.46\ kN/m}$$

The resultant will act at a vertical distance equal to $H/3 = 4/3 = 1.33$ m above the bottom of the wall and will be inclined at an angle of 15° (= δ') to the back face of the wall.

Table 13.6 Values of K_a [Eq. (13.78)] (*Note:* $\delta' = \phi'/2$)

α (deg)	ϕ' (deg)	θ (deg) 0	5	10	15	20	25
0	28	0.3264	0.3629	0.4034	0.4490	0.5011	0.5616
	29	0.3137	0.3502	0.3907	0.4363	0.4886	0.5492
	30	0.3014	0.3379	0.3784	0.4241	0.4764	0.5371
	31	0.2896	0.3260	0.3665	0.4121	0.4645	0.5253
	32	0.2782	0.3145	0.3549	0.4005	0.4529	0.5137
	33	0.2671	0.3033	0.3436	0.3892	0.4415	0.5025
	34	0.2564	0.2925	0.3327	0.3782	0.4305	0.4915
	35	0.2461	0.2820	0.3221	0.3675	0.4197	0.4807
	36	0.2362	0.2718	0.3118	0.3571	0.4092	0.4702
	37	0.2265	0.2620	0.3017	0.3469	0.3990	0.4599
	38	0.2172	0.2524	0.2920	0.3370	0.3890	0.4498
	39	0.2081	0.2431	0.2825	0.3273	0.3792	0.4400
	40	0.1994	0.2341	0.2732	0.3179	0.3696	0.4304
	41	0.1909	0.2253	0.2642	0.3087	0.3602	0.4209
	42	0.1828	0.2168	0.2554	0.2997	0.3511	0.4117
5	28	0.3477	0.3879	0.4327	0.4837	0.5425	0.6115
	29	0.3337	0.3737	0.4185	0.4694	0.5282	0.5972
	30	0.3202	0.3601	0.4048	0.4556	0.5144	0.5833
	31	0.3072	0.3470	0.3915	0.4422	0.5009	0.5698
	32	0.2946	0.3342	0.3787	0.4292	0.4878	0.5566
	33	0.2825	0.3219	0.3662	0.4166	0.4750	0.5437
	34	0.2709	0.3101	0.3541	0.4043	0.4626	0.5312
	35	0.2596	0.2986	0.3424	0.3924	0.4505	0.5190
	36	0.2488	0.2874	0.3310	0.3808	0.4387	0.5070
	37	0.2383	0.2767	0.3199	0.3695	0.4272	0.4954
	38	0.2282	0.2662	0.3092	0.3585	0.4160	0.4840
	39	0.2185	0.2561	0.2988	0.3478	0.4050	0.4729
	40	0.2090	0.2463	0.2887	0.3374	0.3944	0.4620
	41	0.1999	0.2368	0.2788	0.3273	0.3840	0.4514
	42	0.1911	0.2276	0.2693	0.3174	0.3738	0.4410

(*continued*)

Table 13.9 Values of K_a'' [Eq. (13.84)] with $\theta = 5°$ and $k_v = 0$

			ϕ' (deg)				
k_h	δ' (deg)	α (deg)	28	30	35	40	45
0.1	0	0	0.462	0.432	0.363	0.303	0.250
0.2			0.543	0.509	0.432	0.365	0.306
0.3			0.647	0.606	0.516	0.439	0.372
0.4			0.792	0.736	0.622	0.530	0.451
0.5			1.054	0.936	0.762	0.642	0.546
0.1	0	5	0.496	0.462	0.386	0.320	0.263
0.2			0.594	0.554	0.465	0.389	0.324
0.3			0.733	0.678	0.566	0.475	0.398
0.4			0.996	0.875	0.702	0.583	0.489
0.5			—	—	0.914	0.728	0.604
0.1	0	10	0.540	0.500	0.413	0.339	0.277
0.2			0.669	0.616	0.507	0.419	0.345
0.3			0.912	0.801	0.635	0.521	0.430
0.4			—	—	0.842	0.660	0.540
0.5			—	—	—	0.879	0.687
0.1	$\phi'/2$	0	0.434	0.407	0.344	0.291	0.245
0.2			0.526	0.494	0.423	0.362	0.329
0.3			0.651	0.610	0.523	0.451	0.389
0.4			0.836	0.775	0.657	0.566	0.541
0.5			1.202	1.050	0.847	0.719	0.623
0.1	$\phi'/2$	5	0.472	0.440	0.369	0.309	0.258
0.2			0.585	0.546	0.460	0.389	0.330
0.3			0.756	0.698	0.583	0.494	0.421
0.4			1.111	0.959	0.761	0.636	0.542
0.5			—	—	1.063	0.841	0.705
0.1	$\phi'/2$	10	0.521	0.482	0.399	0.331	0.274
0.2			0.674	0.619	0.509	0.424	0.355
0.3			0.990	0.855	0.670	0.551	0.462
0.4			—	—	0.952	0.739	0.609
0.5			—	—	—	1.067	0.827
0.1	$\frac{2}{3}\phi'$	0	0.433	0.407	0.347	0.296	0.253
0.2			0.530	0.499	0.431	0.373	0.323
0.3			0.664	0.624	0.540	0.471	0.413
0.4			0.868	0.806	0.689	0.601	0.531
0.5			1.284	1.119	0.907	0.781	0.689
0.1	$\frac{2}{3}\phi'$	5	0.472	0.441	0.373	0.316	0.267
0.2			0.593	0.554	0.471	0.403	0.346
0.3			0.779	0.719	0.605	0.519	0.449
0.4			1.175	1.012	0.805	0.681	0.589
0.5			—	—	1.159	0.923	0.787
0.1	$\frac{2}{3}\phi'$	10	0.524	0.486	0.405	0.339	0.284
0.2			0.688	0.632	0.523	0.440	0.373
0.3			1.036	0.892	0.701	0.583	0.495
0.4			—	—	1.024	0.800	0.668
0.5			—	—	—	1.195	0.936

Table 13.10 Values of K_a'' [Eq. (13.84)] with $\theta = 10°$ and $k_v = 0$

			ϕ' (deg)				
k_h	δ' (deg)	α (deg)	28	30	35	40	45
0.1	0	0	0.500	0.470	0.402	0.342	0.288
0.2			0.582	0.549	0.473	0.406	0.346
0.3			0.688	0.648	0.559	0.483	0.415
0.4			0.838	0.782	0.669	0.577	0.498
0.5			1.115	0.993	0.816	0.695	0.599
0.1	0	5	0.539	0.505	0.428	0.362	0.303
0.2			0.639	0.599	0.510	0.434	0.368
0.3			0.782	0.728	0.615	0.524	0.446
0.4			1.060	0.935	0.758	0.638	0.543
0.5			—	—	0.984	0.793	0.665
0.1	0	10	0.588	0.548	0.460	0.385	0.320
0.2			0.721	0.668	0.558	0.469	0.393
0.3			0.979	0.863	0.693	0.577	0.484
0.4			—	—	0.914	0.725	0.601
0.5			—	—	—	0.962	0.760
0.1	$\phi'/2$	0	0.477	0.450	0.388	0.334	0.287
0.2			0.573	0.542	0.471	0.410	0.412
0.3			0.705	0.665	0.579	0.507	0.446
0.4			0.904	0.843	0.724	0.634	0.700
0.5			1.311	1.147	0.935	0.805	0.709
0.1	$\phi'/2$	5	0.520	0.488	0.417	0.357	0.304
0.2			0.641	0.601	0.515	0.444	0.383
0.3			0.825	0.765	0.649	0.559	0.485
0.4			1.216	1.053	0.845	0.717	0.621
0.5			—	—	1.188	0.950	0.808
0.1	$\phi'/2$	10	0.577	0.538	0.454	0.384	0.324
0.2			0.742	0.685	0.573	0.486	0.414
0.3			1.089	0.944	0.750	0.628	0.535
0.4			—	—	1.068	0.840	0.704
0.5			—	—	—	1.221	0.958
0.1	$\frac{2}{3}\phi'$	0	0.479	0.452	0.394	0.343	0.299
0.2			0.581	0.551	0.484	0.427	0.378
0.3			0.724	0.685	0.603	0.536	0.479
0.4			0.948	0.885	0.768	0.682	0.614
0.5			1.419	1.239	1.017	0.889	0.800
0.1	$\frac{2}{3}\phi'$	5	0.524	0.493	0.425	0.367	0.318
0.2			0.653	0.614	0.531	0.464	0.406
0.3			0.856	0.796	0.680	0.594	0.524
0.4			1.302	1.124	0.906	0.779	0.687
0.5			—	—	1.319	1.064	0.923
0.1	$\frac{2}{3}\phi'$	10	0.584	0.545	0.463	0.396	0.340
0.2			0.762	0.705	0.594	0.510	0.441
0.3			1.151	0.995	0.794	0.672	0.582
0.4			—	—	1.167	0.924	0.786
0.5			—	—	—	1.400	1.112

Table 13.11 Values of K''_a [Eq. (13.84)] with $\theta = 15°$ and $k_v = 0$

			ϕ' (deg)				
k_h	δ' (deg)	α (deg)	28	30	35	40	45
0.1	0	0	0.543	0.514	0.446	0.385	0.329
0.2			0.626	0.594	0.519	0.451	0.391
0.3			0.735	0.696	0.608	0.532	0.464
0.4			0.892	0.836	0.723	0.631	0.552
0.5			1.190	1.061	0.879	0.756	0.659
0.1	0	5	0.587	0.554	0.477	0.409	0.348
0.2			0.691	0.651	0.562	0.485	0.417
0.3			0.841	0.785	0.672	0.579	0.500
0.4			1.139	1.007	0.824	0.701	0.604
0.5			—	—	1.069	0.868	0.736
0.1	0	10	0.643	0.603	0.514	0.437	0.369
0.2			0.783	0.729	0.617	0.525	0.447
0.3			1.059	0.937	0.761	0.641	0.545
0.4			—	—	1.000	0.802	0.672
0.5			—	—	—	1.062	0.846
0.1	$\phi'/2$	0	0.526	0.499	0.438	0.384	0.336
0.2			0.627	0.596	0.527	0.467	0.526
0.3			0.769	0.729	0.644	0.573	0.512
0.4			0.987	0.925	0.805	0.714	0.963
0.5			1.450	1.270	1.044	0.911	0.814
0.1	$\phi'/2$	5	0.576	0.544	0.473	0.412	0.358
0.2			0.705	0.665	0.580	0.508	0.446
0.3			0.906	0.845	0.727	0.636	0.561
0.4			1.349	1.169	0.948	0.815	0.717
0.5			—	—	1.348	1.088	0.938
0.1	$\phi'/2$	10	0.642	0.602	0.517	0.445	0.384
0.2			0.822	0.763	0.648	0.559	0.485
0.3			1.211	1.053	0.847	0.719	0.623
0.4			—	—	1.214	0.965	0.820
0.5			—	—	—	1.420	1.125
0.1	$\frac{2}{3}\phi'$	0	0.530	0.505	0.447	0.397	0.353
0.2			0.640	0.611	0.545	0.490	0.442
0.3			0.796	0.758	0.677	0.613	0.559
0.4			1.046	0.983	0.866	0.782	0.718
0.5			1.597	1.395	1.157	1.029	0.944
0.1	$\frac{2}{3}\phi'$	5	0.583	0.553	0.485	0.427	0.378
0.2			0.724	0.685	0.603	0.536	0.479
0.3			0.949	0.887	0.771	0.684	0.616
0.4			1.465	1.266	1.033	0.902	0.811
0.5			—	—	1.531	1.250	1.105
0.1	$\frac{2}{3}\phi'$	10	0.654	0.615	0.532	0.464	0.406
0.2			0.851	0.792	0.679	0.593	0.523
0.3			1.296	1.123	0.908	0.781	0.689
0.4			—	—	1.351	1.083	0.938
0.5			—	—	—	1.681	1.353

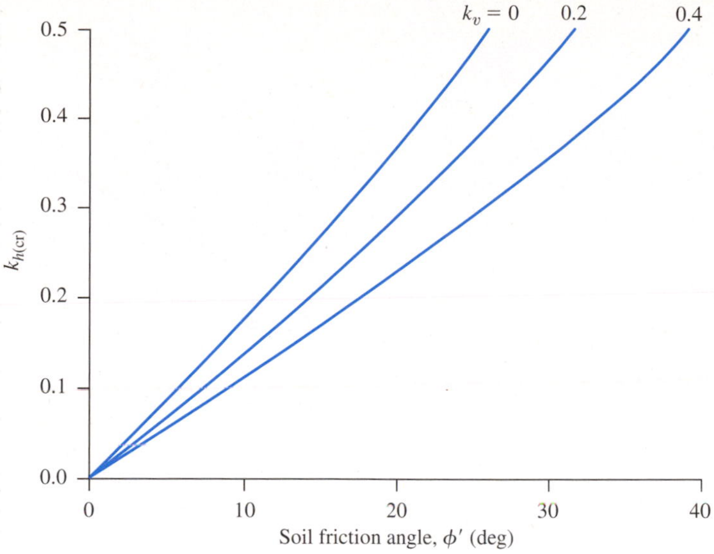

Figure 13.29 Critical values of horizontal acceleration [Eq. (13.90)]

Location of line of action of resultant force, P_{ae}

Seed and Whitman (1970) proposed a simple procedure to determine the location of the line of action of the resultant, P_{ae}. Their method is as follows:

1. Let

$$P_{ae} = P_a + \Delta P_{ae} \tag{13.91}$$

where P_a = Coulomb's active force as determined from Eq. (13.77)
ΔP_{ae} = additional active force caused by the earthquake effect

2. Calculate P_a [Eq. (13.77)].
3. Calculate P_{ae} [Eq. (13.83)].
4. Calculate $\Delta P_{ae} = P_{ae} - P_a$.
5. According to Figure 13.30, P_a will act at a distance of $H/3$ from the base of the wall. Also, ΔP_{ae} will act at a distance of $0.6H$ from the base of the wall.
6. Calculate the location of P_{ae} as

$$\bar{z} = \frac{P_a\left(\frac{H}{3}\right) + \Delta P_{ae}(0.6H)}{P_{ae}} \tag{13.92}$$

where $\bar{z}$ = distance of the line of action of P_{ae} from the base of the wall.

Note that the line of action of P_{ae} will be inclined at an angle of δ' to the normal drawn to the back face of the retaining wall. It is very important to realize that this method of determining P_{ae} is approximate and does not actually model the soil dynamics.

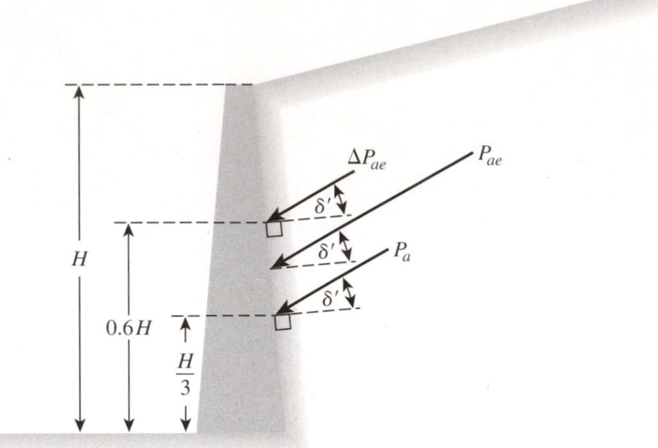

Figure 13.30 Location of the line of action of P_{ae}

Active case ($c' - \phi'$ backfill)

Shukla et al. (2009) developed a procedure for estimation of P_{ae} for a retaining wall with a vertical back face and horizontal backfill with a $c' - \phi'$ soil (Figure 13.31a). In Figure 13.31a, ABC is the trial failure wedge. The following assumptions have been made in the analysis:

1. The effect of tensile crack is not taken into account.
2. The friction and adhesion between the back face of the wall and the backfill are neglected.

Figure 13.31b shows the polygon for all the forces acting on the wedge ABC. The notations are similar to those shown in Figure 13.27. According to this analysis, the critical wedge angle $\eta = \eta_c$ for maximum value of P_{ae} can be given as

$$\tan \eta_c = \frac{\sin\phi' \sin(\phi' - \overline{\beta}) + m \sin 2\phi' + \left[\begin{array}{l} \sin\phi' \sin(\phi' - \overline{\beta})\cos\overline{\beta} \\ + 4m^2 \cos^2\phi' + 2m\cos\phi' \\ \{\sin\phi' \cos\overline{\beta} + \sin(\phi' - \overline{\beta})\} \end{array}\right]^{0.5}}{\sin\phi' \cos(\phi' - \overline{\beta}) + 2m\cos^2\phi'} \tag{13.93}$$

where

$$m = \frac{c' \cos\overline{\beta}}{\gamma H(1 - k_v)} \tag{13.94}$$

For definition of $\overline{\beta}$, see Eq. (13.85).

Thus, the magnitude of P_{ae} can be expressed as

$$\frac{P_{ae}}{\gamma H^2} = P^*_{ae} = \frac{1}{2}(1 - k_v)K_{ae\gamma} - c^* K_{aec} \tag{13.95}$$

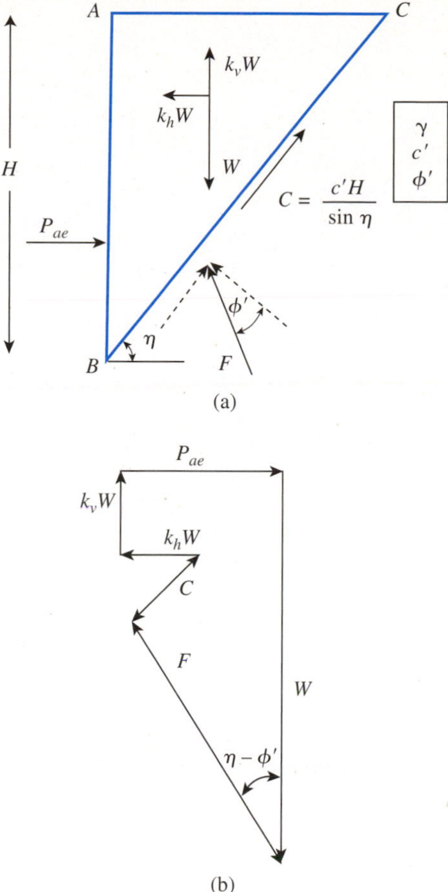

Figure 13.31 Estimation of P_{ae} with $c' - \phi'$ backfill: (a) trial failure wedge, (b) force polygon

where

$$c^* = \frac{c'}{\gamma H} \tag{13.96}$$

$$K_{ae\gamma} = \frac{\cos(\phi' - \overline{\beta}) - \dfrac{\sin(\phi' - \overline{\beta})}{\tan \eta_c}}{\cos\overline{\beta}(\cos\phi' + \tan\eta_c \sin\phi')} \tag{13.97}$$

$$K_{aec} = \frac{\cos\phi'(1 + \tan^2\eta_c)}{\tan\eta_c(\cos\phi' + \tan\eta_c \sin\phi')} \tag{13.98}$$

Figure 13.32 gives plots of P^*_{ae} against ϕ' for various values of c^* and k_h ($k_v = 0$).

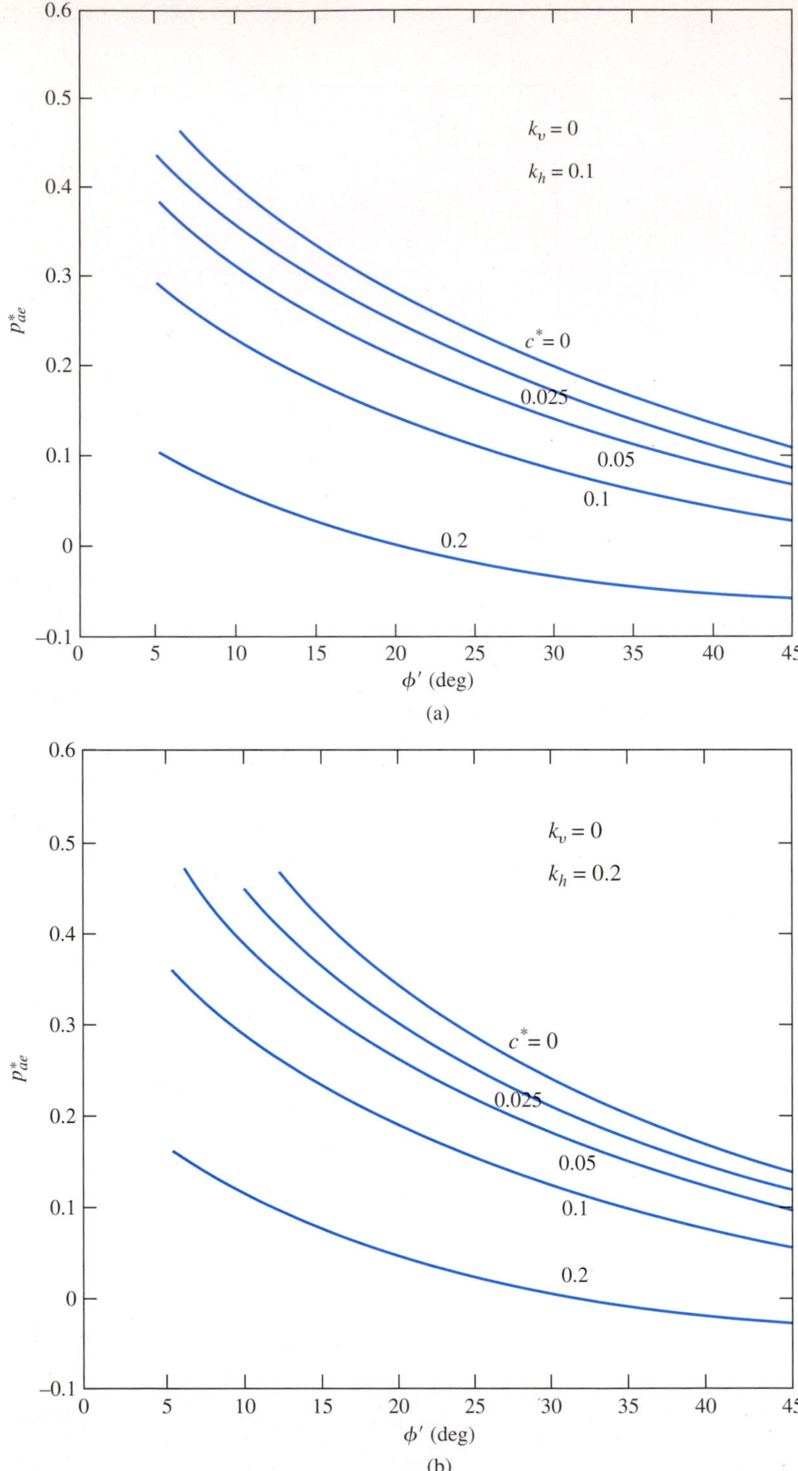

Figure 13.32 Plot of P^*_{ae} vs. ϕ' for various values of c^*: (a) $k_h = 0.1$, (b) $k_h = 0.2$, (c) $k_h = 0.3$, (d) $k_h = 0.4$ (*Note:* $k_v = 0$)

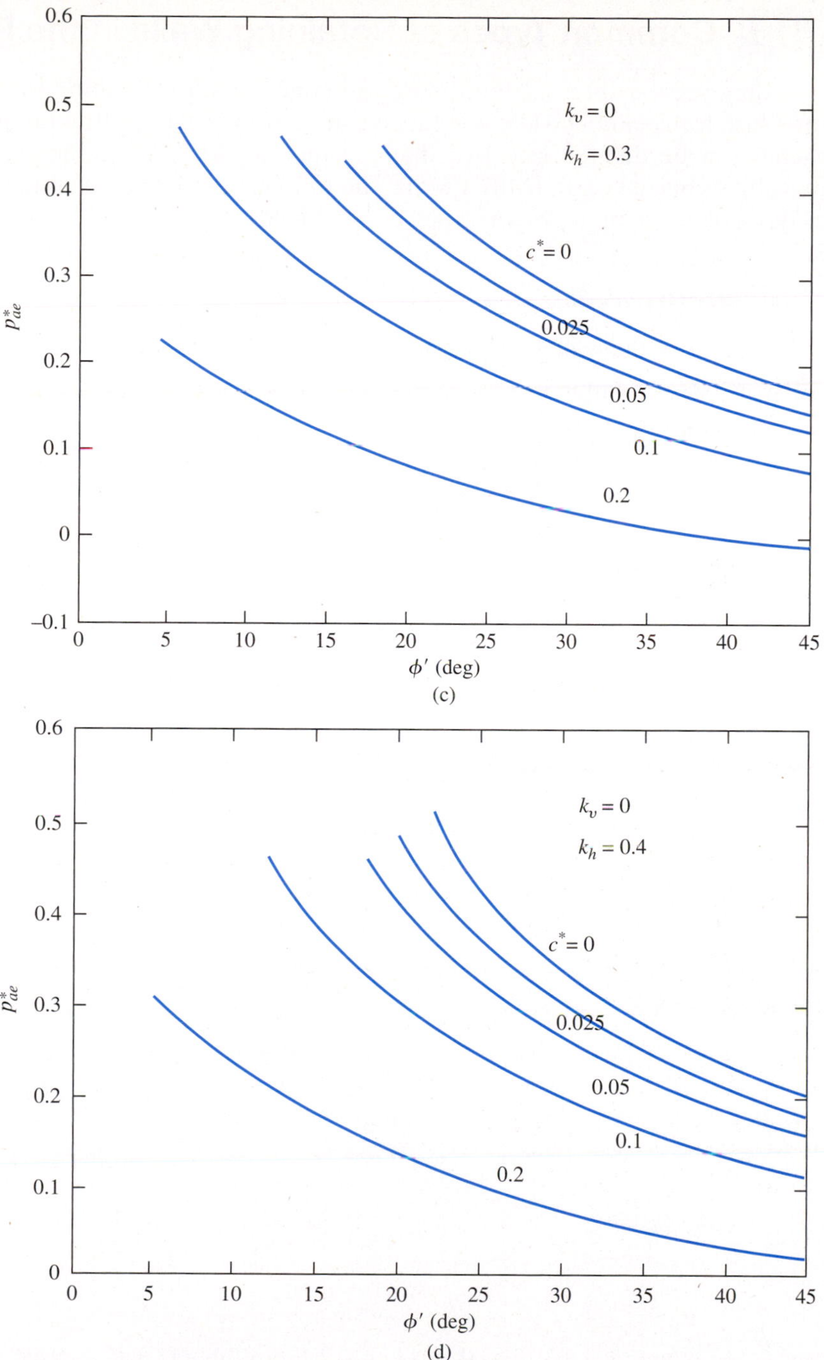

Figure 13.32 *(Continued)*

13.15 Common Types of Retaining Walls in the Field

The preceding sections present the theoretical concepts for estimating the lateral earth pressure for retaining walls. In practice, the common types of retaining walls constructed can be divided into two major categories: rigid retaining walls and mechanically stabilized earth (MSE) walls. The following is a brief overview of the various types of retaining walls constructed in the field.

Rigid retaining walls

Under this category, the wall may be subdivided to four categories. They are:

1. Gravity retaining walls
2. Semigravity retaining walls
3. Cantilever retaining walls
4. Counterfort retaining walls

Gravity retaining walls (Figure 13.33a) are constructed with plain concrete or stone masonry. They depend on their own weight and any soil resting on the masonry for stability. This type of construction is not economical for high walls.

In many cases, a small amount of steel may be used for the construction of gravity walls, thereby minimizing the size of wall sections. Such walls generally are referred to as *semigravity walls* (Figure 13.33b).

Cantilever retaining walls (Figure 13.33c) are made of reinforced concrete that consists of a thin stem and a base slab. This type of wall is economical to a height of about 8 m. Figure 13.34 shows a cantilever retaining wall under construction.

Counterfort retaining walls (Figure 13.33d) are similar to cantilever walls. At regular intervals, however, they have thin, vertical concrete slabs known as *counterforts* that tie the wall and the base slab together. The purpose of the counterforts is to reduce the shear and the bending moments.

Example 13.12

For a retaining wall with a cohesionless soil backfill, $\gamma = 15.5$ kN/m^3, $\phi' = 30°$, $\delta' = 15°$, $\theta = 0°$, $\alpha = 0°$, $H = 4$ m, $k_v = 0$, and $k_h = 0.2$. Determine P_{ae}. Also determine the location of the resultant line of action of P_{ae}—that is, $\bar{z}$.

Solution

From Eq. (13.83),

$$P_{ae} = \frac{1}{2}\gamma H^2(1 - k_v)K''_a$$

Given: $\dfrac{\delta'}{\phi'} = \dfrac{15}{30} = 0.5$, $\alpha = 0$, $k_v = 0$, and $k_h = 0.2$. From Table 13.8, $K''_a = 0.452$. Thus,

$$P_{ae} = \left(\frac{1}{2}\right)(15.5)(4)^2(1 - 0)(0.452) = \mathbf{56.05\ kN/m}$$

We now locate the resultant line of action. From Eq. (13.77),

$$P_a = \tfrac{1}{2}K_a\gamma H^2$$

For $\phi' = 30°$ and $\delta' = 15°$, $K_a = 0.3014$ (Table 13.6), so

$$P_a = \tfrac{1}{2}(0.3014)(15.5)(4)^2 = 37.37 \text{ kN/m}$$

Hence, $\Delta P_{ae} = 56.05 - 37.37 = 18.68$ kN/m. From Eq. (13.92)

$$\bar{z} = \frac{P_a\left(\frac{H}{3}\right) + \Delta P_{ae}(0.6H)}{P_{ae}} = \frac{(37.37)\left(\frac{4}{3}\right) + (18.68)(2.4)}{56.05} = \mathbf{1.69\ m}$$

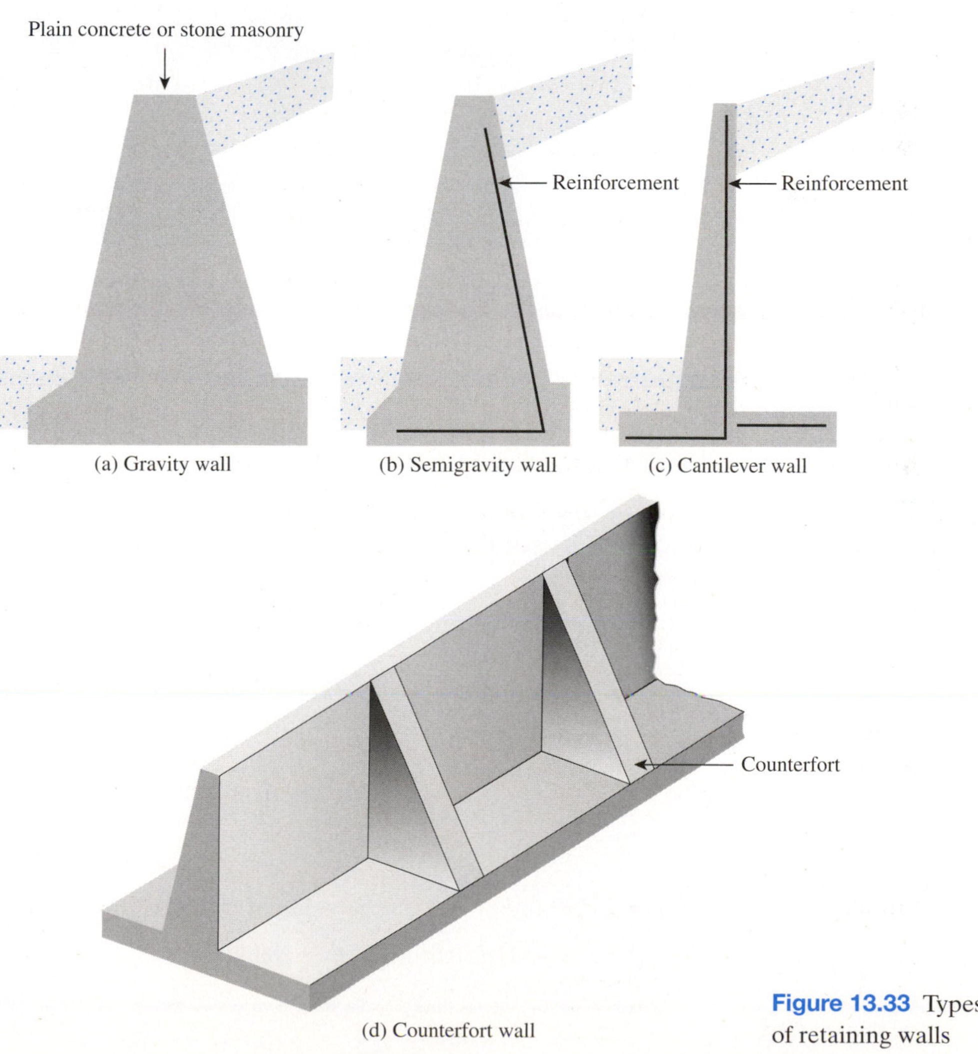

Figure 13.33 Types of retaining walls

Figure 13.34 A cantilever retaining wall under construction (*Courtesy of Dharma Shakya, Geotechnical Solutions, Inc., Irvine, California*)

Example 13.13

For a retaining wall with a vertical backfill, the following are given.

- $H = 8.54$ m
- $\phi' = 20°$
- $c' = 7.9$ kN/m^2
- $\gamma = 18.55$ kN/m^3
- $k_h = 0.1$

Determine the magnitude of the active force, P_{ae}.

Solution

From Eq. (13.96),

$$c^* = \frac{c'}{\gamma H} = \frac{7.9}{(18.55)(8.54)} = 0.0499 \approx 0.05$$

$$\phi' = 20°$$

From Figure 13.32a, for $\phi' = 20°$ and $c^* = 0.05$, the value of $P^*_{ae} \approx 0.207$. Hence

$$P_{ae} = P^*_{ae}\gamma H^2 = (0.207)(18.55)(8.54)^2 = \mathbf{280\ kN/m}$$

Mechanically stabilized earth (MSE) walls

Mechanically stabilized earth walls are flexible walls, and they are becoming more common nowadays. The main components of these types of walls are

- *Backfill*—which is granular soil
- *Reinforcement* in the backfill
- A *cover* (or *skin*) on the front face

The reinforcement can be thin galvanized steel strips, geogrid, or geotextile. In most cases, precast concrete slabs are used as skin. The slabs are grooved to fit into each other so that soil cannot flow between the joints. Thin galvanized steel also can be used as skin when the reinforcements are metallic strips. When metal skins are used, they are bolted together, and reinforcing strips are placed between the skins.

Figure 13.35 shows an MSE wall with metallic strips as reinforcement along with a metal skin. Figure 13.36 shows some typical MSE walls with geogrid reinforcement in the backfill. Figures 13.37 and 13.38 show MSE wall construction with geotextile and geogrid reinforcement, respectively.

The retaining walls are designed using various earth-pressure theories described in this chapter. For actual wall design, refer to any foundation engineering book.

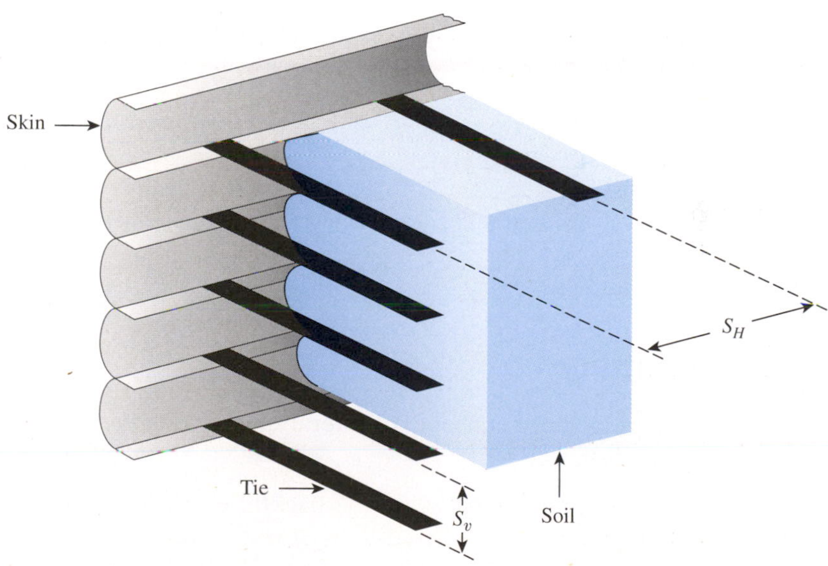

Figure 13.35 *MSE* wall with metallic strip reinforcement and metallic skin

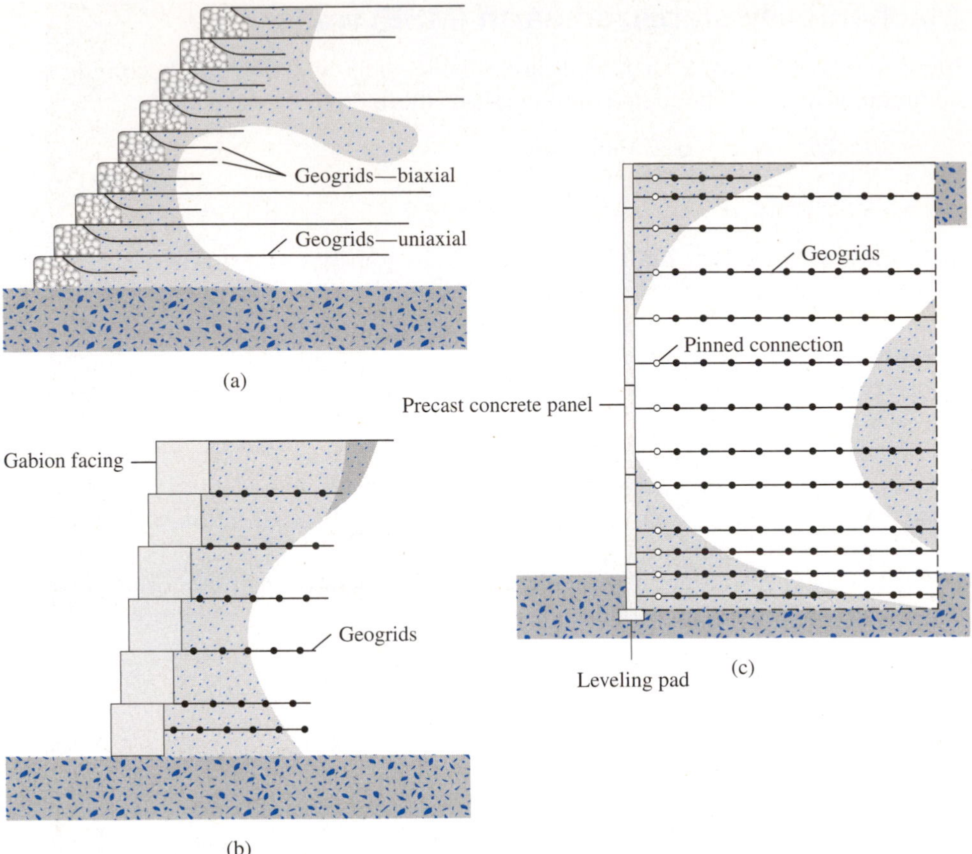

Figure 13.36 Typical schematic diagrams of retaining walls with geogrid reinforcement: (a) geogrid wraparound wall; (b) wall with gabion facing; (c) concrete-panel-faced wall

13.16 Summary

This chapter covers the general concepts of lateral earth pressure. Following is a summary of the topics discussed:

- When the wall does not yield toward the backfill or away from the backfill, the lateral earth pressure is referred to as *at-rest earth pressure*. The at-rest earth pressure coefficients are given in Eqs. (13.5), (13.7), (13.8), and (13.9).
- The Rankine active earth pressure (frictionless wall—Section 13.6) can be given by Eq. (13.34). The Rankine active earth-pressure coefficient is given by Eq. (13.33), or

$$K_a = \tan^2\left(45 - \frac{\phi'}{2}\right)$$

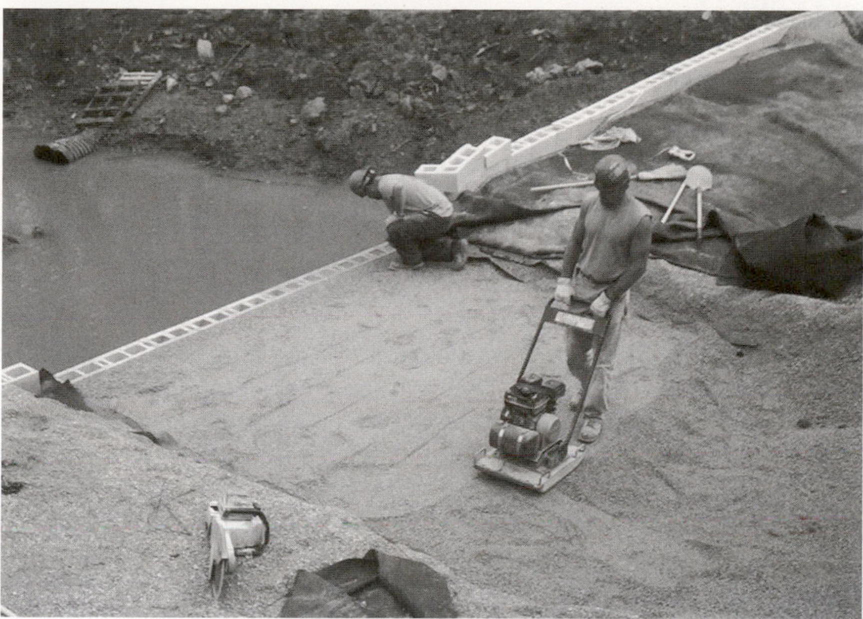

Figure 13.37 Construction of a retaining wall with geotextile reinforcement (*Courtesy of Jonathan T. H. Wu, University of Colorado at Denver, Denver, Colorado*)

Figure 13.38 HDPE (high density polyethylene) geogrid-reinforced wall with precast concrete panel facing under construction (*Courtesy of Tensar International Corporation, Atlanta, Georgia*)

- The Rankine passive earth-pressure (frictionless wall—Section 13.7) can be given by Eq. (13.35). The Rankine passive earth-pressure coefficient [Eq. (13.36)] is

$$K_p = \tan^2\left(45 + \frac{\phi'}{2}\right)$$

- The lateral earth pressure with wall friction and granular backfill can be obtained from Coulomb's analysis. Coulomb's active and passive earth pressure coefficients are given in Eqs. (13.78) and (13.80), respectively.
- The analysis of active earth pressure with granular and c'–ϕ' soil backfill subjected to earthquake forces was discussed in Section 13.14. The active earth pressure coefficient with granular backfill is given in Eq. (13.84). Similarly, for a c'–ϕ' soil backfill, the lateral earth pressure coefficients can be given by Eqs. (13.97) and (13.98).

Problems

13.1 through 13.4 Figure 13.39 shows a retaining wall that is restrained from yielding. For each problem, determine the magnitude of the lateral earth force per unit length of the wall. Also, find the location of the resultant, $\bar{z}$, measured from the bottom of the wall.

Problem	H	ϕ' (deg)	γ	Over-consolidation ratio, OCR
13.1	7.5 m	31	19.2 kN/m^3	2.2
13.2	8.25 m	38	18.9 kN/m^3	1.9
13.3	11 m	28	17.9 kN/m^3	1.3
13.4	4.85 m	41	18.5 kN/m^3	1.0

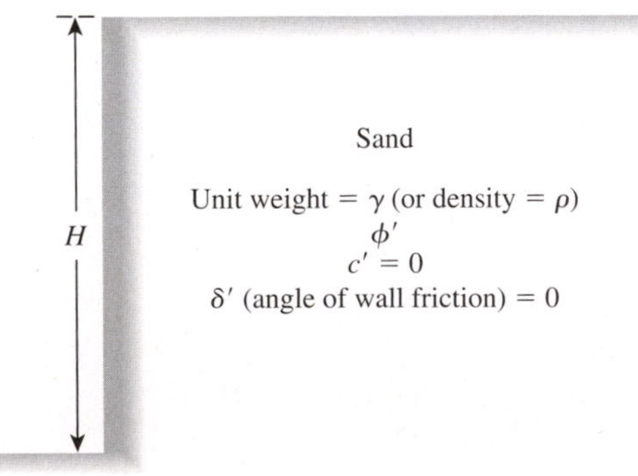

Figure 13.39

13.5 A 5-m high unyielding retaining wall is subjected to a surcharge point load, $Q = 95$ kN, on the ground surface (Figure 13.40a). Determine the increase in lateral pressure, σ'_h, on the wall (on the plane of the section shown) at $z = 1, 2,$ 3, 4, and 5 m due to the point load. (*Note*: z = distance measured downwards from the top of the wall.)

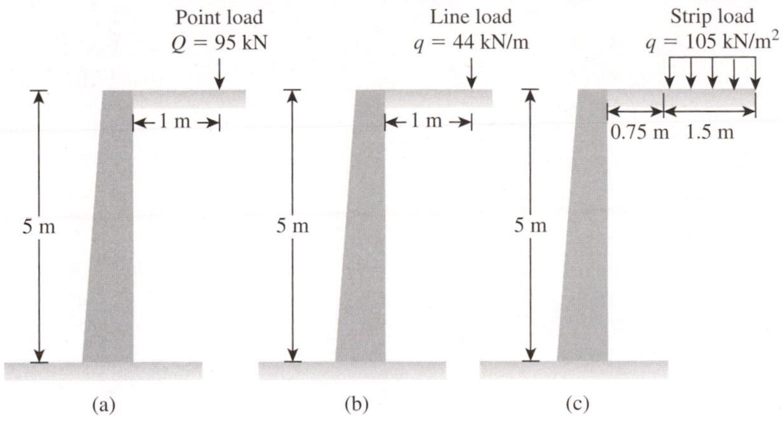

Figure 13.40

13.6 The 5-m high retaining wall in Figure 13.40b is subjected to a surcharge line load of $q = 44$ kN/m on the ground surface. Determine the increase in lateral force per unit length of the wall due to the line load.

13.7 The 5-m high retaining wall in Figure 13.40c is subjected to a surcharge strip load of $q = 105$ kN/m^2 on the ground surface. Determine the total force per unit length of the wall due to the strip loading only.

13.8 Refer to Problem 13.7. Determine the location of the resultant total force $\bar{z}$. Use Eq. (13.27).

13.9 through 13.12 Assume that the retaining wall shown in Figure 13.39 is frictionless. For each problem, determine the Rankine active force per unit length of the wall, the variation of active earth pressure with depth, and the location of the resultant.

Problem	H	ϕ' (deg)	γ
13.9	8.85 m	34	18.7 kN/m^3
13.10	12.2 m	42	19.6 kN/m^3
13.11	9 m	28	16.7 kN/m^3
13.12	12 m	38	19.5 kN/m^3

13.13 through 13.16 Assume that the retaining wall shown in Figure 13.39 is frictionless. For each problem, determine the Rankine passive force per unit length of the wall, the variation of active earth pressure with depth, and the location of the resultant.

Problem	H	ϕ' (deg)	γ
13.13	5.5 m	23	15.5 kN/m^3
13.14	7.6 m	33	18.7 kN/m^3
13.15	9.6 m	39	19 kN/m^3
13.16	13.5 m	29	18.3 kN/m^3

13.17 through 13.19 A retaining wall is shown in Figure 13.41. For each problem, determine the Rankine active force, P_a, per unit length of the wall and the location of the resultant.

Problem	H	H_1	γ_1	γ_2	ϕ'_1 (deg)	ϕ'_2 (deg)	q
13.17	6.4 m	1.2 m	17.1 kN/m^3	19 kN/m^3	26	26	0
13.18	8 m	3 m	13 kN/m^3	18.8 kN/m^3	30	30	16 kN/m^2
13.19	12 m	4 m	17 kN/m^3	23.2 kN/m^3	36	42	25 kN/m^2

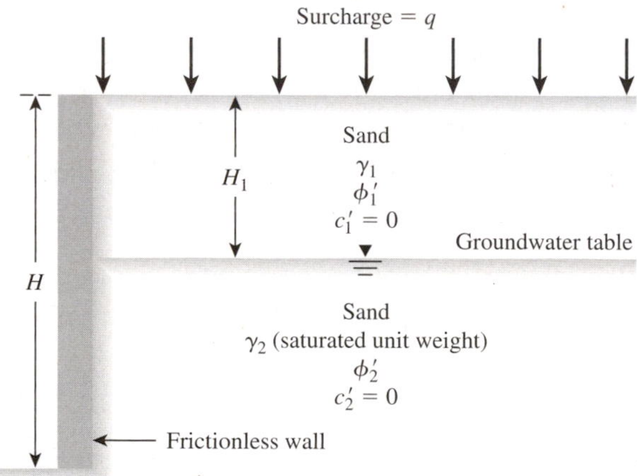

Figure 13.41

13.20 For the partially submerged backfill in Problem 13.17 (Figure 13.41), determine the Rankine's passive force per unit length of the wall and the location of the resultant.

13.21 Figure 13.13 shows a frictionless wall with a sloping granular backfill. Given: $H = 7$ m, $\alpha = 12°$, $\phi' = 28°$, and $\gamma = 18.6$ kN/m^3.

a. Determine the magnitude of active pressure, σ'_a, at the bottom of the wall.

b. Determine Rankine active force, P_a, per unit length of the wall and its location and direction.

13.22 For the data given in Problem 13.21, determine the Rankine passive force, P_p, per unit length of the wall, and its location and direction.

13.23 An 8.5-m high retaining wall with a vertical back face retains a homogeneous saturated soft clay. The saturated unit weight of the clay is 19.6 kN/m^3. Laboratory tests showed that the undrained shear strength, c_u, of the clay is 22 kN/m^2.

a. Make the necessary calculations and draw the variation of Rankine's active pressure on the wall with depth.

b. Find the depth up to which tensile crack can occur.

c. Determine the total active force per unit length of the wall before the tensile crack occurs.

d. Determine the total active force per unit length of the wall after the tensile crack occurs. Also find the location of the resultant.

13.24 Redo Problem 13.23 assuming that a surcharge pressure of 20 kN/m^2 is applied on top of thc backfill.

13.25 A 17.9-m high retaining wall with a vertical back face has a $c' - \phi'$soil for backfill material. Properties of the backfill material are as follows: $\gamma = 18.6$ kN/m^3, $c' = 53$ kN/m^2, and $\phi' = 24°$. Considering the existence of the tensile crack, determine the Rankine active force, P_a, per unit length of the wall.

13.26 Consider the retaining wall shown in Figure 13.42. The height of the wall is 8 m, and the unit weight of the sand backfill is 21.1 kN/m^3. Using Coulomb's equation, calculate the active force, P_a, on the wall for the following values of the angle of wall friction. Also, comment on the direction and location of the resultant. Consider the following two cases of wall friction:

a. $\delta' = 25°$

b. $\delta' = 19°$

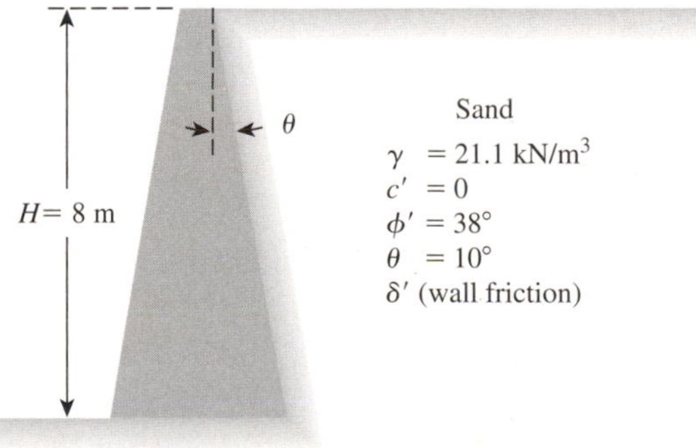

Figure 13.42

13.27 Refer to Figure 13.27. Given: $H = 7.5$ m, $\theta = 10°$, $\alpha = 5°$, $\gamma = 17.9$ kN/m^3, $\phi' = 28°$, $\delta' = \frac{1}{2}\phi'$, $k_h = 0.4$, $k_v = 0$. Determine the active force, P_{ae}, per unit length of the retaining wall. Also find the location of the resultant line of action of P_{ae}.

Critical Thinking Problem

13.C.1 Figure 13.14 provides a generalized case for Rankine active pressure on a frictionless retaining wall with inclined back and a sloping granular backfill. You are required to develop some compaction guidelines for the backfill soil when $\theta = 10°$ and $\alpha = 0°, 10°$, and $20°$. Laboratory direct shear tests on the granular soil revealed that the effective friction angle varies with the dry unit weight as follows:

Dry unit weight, γ (kN/m^3)	Friction angle, ϕ' (deg)
16.5	28
18.7	32
19.5	36

The data show that the soil friction angle increases as the compacted unit weight increases. You already know from Chapter 12 that higher friction angle means better shear strength and stability. However, according to Eq. (13.38), higher unit weight also means higher active force P_a on the wall, which is not desirable. To further investigate if higher friction angle indeed has a beneficial effect, prepare a design chart by plotting the variations of $P_a/0.5H^2$ (which is equal to $K_{a(R)}\gamma$) for various values of the backfill slope α and the friction angle ϕ'. Explain how this chart may aid a geotechnical engineer in developing guidelines for backfill construction for a given height (H) of the retaining wall.

References

Chu, S. C. (1991). "Rankine Analysis of Active and Passive Pressures on Dry Sand," *Soils and Foundations*, Vol. 31, No. 4, 115–120.

Coulomb, C. A. (1776). "Essai sur une Application des Règles de Maximis et Minimis a quelques Problèmes de Statique, relatifs a l'Architecture," *Mem. Roy. des Sciences,* Paris, Vol. 3, 38.

Gerber, E. (1929). *Untersuchungen über die Druckverteilung im Örlich belasteten Sand*, Technische Hochschule, Zurich.

Jaky, J. (1944). "The Coefficient of Earth Pressure at Rest," *Journal of the Society of Hungarian Architects and Engineers,* Vol. 7, 355–358.

Jarquio, R. (1981). "Total Lateral Surcharge Pressure Due to Strip Load," *Journal of the Geotechnical Engineering Division*, American Society of Civil Engineers, Vol. 107, No. GT10, 1424–1428.

Massarsch, K. R. (1979). "Lateral Earth Pressure in Normally Consolidated Clay," *Proceedings of the Seventh European Conference on Soil Mechanics and Foundation Engineering*, Brighton, England, Vol. 2, 245–250.

Mayne, P. W., and Kulhawy, F. H. (1982). "K_o–OCR Relationships in Soil," *Journal of the Geotechnical Division,* ASCE, Vol. 108, No. 6, 851–872.

Mononobe, N. (1929). "Earthquake-Proof Construction of Masonry Dams," *Proceeding* World Engineering Conference, Vol. 9, pp. 274–280.

Okabe, S. (1926). "General Theory of Earth Pressure," *Journal of the Japanese Society of Civil Engineers,* Tokyo, Vol. 12, No. 1.

Rankine, W. M. J. (1857). "On Stability on Loose Earth," *Philosophic Transactions of Royal Society,* London, Part I, 9–27.

Seed, H. B., and Whitman, R. V. (1970). "Design of Earth Retaining Structures for Dynamic Loads," *Proceedings,* Specialty Conference on Lateral Stresses in the Ground and Design of Earth Retaining Structures, ASCE, 103–147.

Sherif, M. A., Fang, Y. S., and Sherif, R. I. (1984). "K_A and K_O Behind Rotating and Non-Yielding Walls," *Journal of Geotechnical Engineering,* ASCE, Vol. 110, No. GT1, 41–56.

Shukla, S. K., Gupta, S. K., and Sivakugan, N. (2009). "Active Earth Pressure on Retaining Wall for c–ϕ Soil Backfill Under Seismic Loading Condition," *Journal of Geotechnical and Geoenvironmental Engineering,* ASCE, Vol. 135, No. 5, 690–696.

Spangler, M. G. (1938). "Horizontal Pressures on Retaining Walls Due to Concentrated Surface Loads," Iowa State University Engineering Experiment Station, *Bulletin*, No. 140.

CHAPTER 14

Lateral Earth Pressure: Curved Failure Surface

14.1 Introduction

In Chapter 13, we considered Coulomb's earth pressure theory, in which the retaining wall was considered to be rough. The potential failure surfaces in the backfill were considered to be planes. In reality, most failure surfaces in soil are curved. There are several instances where the assumption of plane failure surfaces in soil may provide unsafe results. Examples of these cases are the estimation of passive pressure and braced cuts. This chapter describes procedures by which passive earth pressure and lateral earth pressure on braced cuts can be estimated using curved failure surfaces in the soil.

14.2 Retaining Walls with Friction

In reality, retaining walls are rough, and shear forces develop between the face of the wall and the backfill. To understand the effect of wall friction on the failure surface, let us consider a rough retaining wall AB with a horizontal granular backfill as shown in Figure 14.1.

In the active case (Figure 14.1a), when the wall AB moves to a position $A'B$, the soil mass in the active zone will be stretched outward. This will cause a downward motion of the soil relative to the wall. This motion causes a downward shear on the wall (Figure 14.1b), and it is called a *positive wall friction in the active case*. If δ' is the angle of friction between the wall and the backfill, then the resultant active force P_a will be inclined at an angle δ' to the normal drawn to the back face

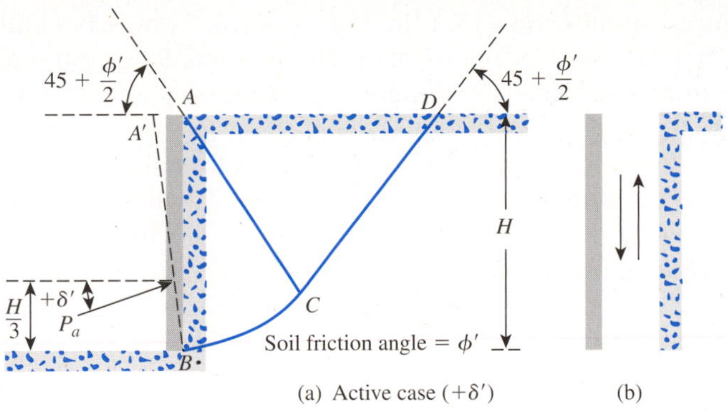

(a) Active case (+δ′)

(b)

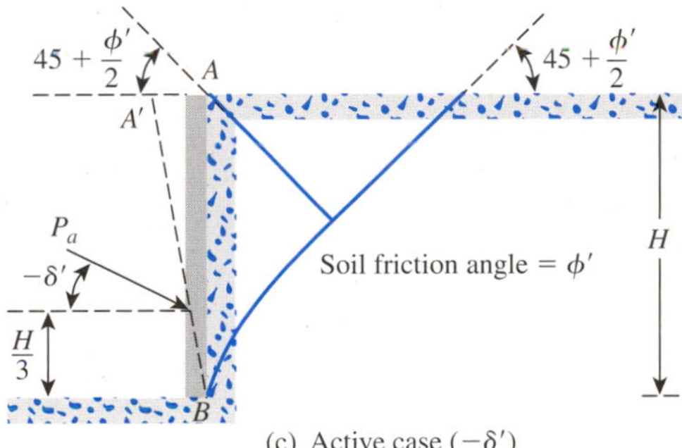

(c) Active case (−δ′)

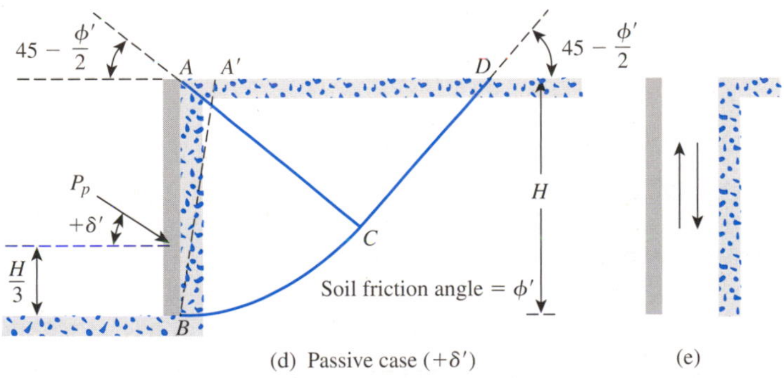

(d) Passive case (+δ′)

(e)

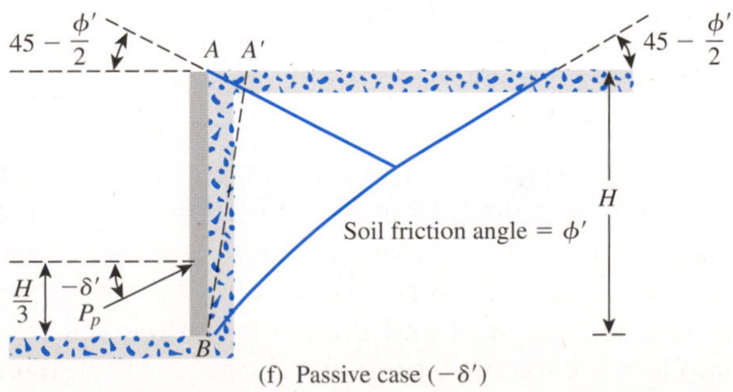

(f) Passive case (−δ′)

Figure 14.1 Effect of wall friction on failure surface

of the retaining wall. Advanced studies show that the failure surface in the backfill can be represented by *BCD*, as shown in Figure 14.1a. The portion *BC* is curved, and the portion *CD* of the failure surface is a straight line. Rankine's active state exists in the zone *ACD*.

Under certain conditions, if the wall shown in Figure 14.1a is forced downward with reference to the backfill, the direction of the active force, P_a, will change as shown in Figure 14.1c. This is a situation of negative wall friction $(-\delta')$ in the active case. Figure 14.1c also shows the nature of the failure surface in the backfill.

The effect of wall friction for the passive state is shown in Figures 14.1d and e. When the wall *AB* is pushed to a position *A′B* (Figure 14.1d), the soil in the passive zone will be compressed. The result is an upward motion relative to the wall. The upward motion of the soil will cause an upward shear on the retaining wall (Figure 14.1e). This is referred to as *positive wall friction in the passive case*. The resultant passive force, P_p, will be inclined at an angle δ' to the normal drawn to the back face of the wall. The failure surface in the soil has a curved lower portion *BC* and a straight upper portion *CD*. Rankine's passive state exists in the zone *ACD*.

If the wall shown in Figure 14.1d is forced upward relative to the backfill by a force, then the direction of the passive force P_p will change as shown in Figure 14.1f. This is *negative wall friction in the passive case* $(-\delta')$. Figure 14.1f also shows the nature of the failure surface in the backfill under such a condition.

For practical considerations, in the case of loose granular backfill, the angle of wall friction δ' is taken to be equal to the angle of friction of soil, ϕ'. For dense granular backfills, δ' is smaller than ϕ' and is in the range of $\phi'/2 \le \delta' \le (2/3)\,\phi'$.

The assumption of plane failure surface gives reasonably good results while calculating active earth pressure. However, the assumption that the failure surface is a plane in Coulomb's theory grossly overestimates the passive resistance of walls, particularly for $\delta' > \phi'/2$.

14.3 Properties of a Logarithmic Spiral

The case of passive pressure shown in Figure 14.1d (case of $+\delta'$) is the most common one encountered in design and construction. Also, the curved failure surface represented by *BC* in Figure 14.1d is assumed most commonly to be the arc of a logarithmic spiral. In a similar manner, the failure surface in soil in the case of braced cuts (Sections 14.9 to 14.10) also is assumed to be the arc of a logarithmic spiral. Hence, some useful ideas concerning the properties of a logarithmic spiral are described in this section.

The equation of the logarithmic spiral generally used in solving problems in soil mechanics is of the form

$$r = r_o e^{\beta \tan\phi'} \tag{14.1}$$

where r = radius of the spiral
r_o = starting radius at $\beta = 0$
ϕ' = angle of friction of soil
β = angle between r and r_o

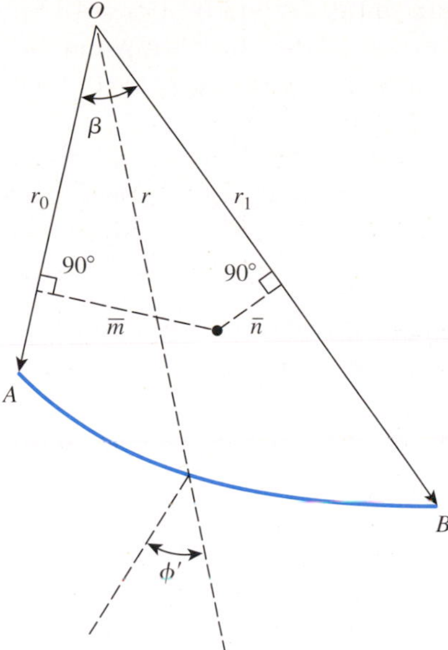

Figure 14.2 General parameters of a logarithmic spiral

The basic parameters of a logarithmic spiral are shown in Figure 14.2, in which O is the center of the spiral. The area of the sector OAB is given by

$$A = \int_0^{\theta} \frac{1}{2} r(r\, d\beta) \tag{14.2}$$

Substituting the values of r from Eq. (14.1) into Eq. (14.2), we get

$$\begin{aligned} A &= \int_0^{\beta_1} \frac{1}{2} r_o^2 e^{2\beta \tan\phi'}\, d\beta \\ &= \frac{r_1^2 - r_o^2}{4\tan\phi'} \end{aligned} \tag{14.3}$$

The location of the centroid can be defined by the distances $\overline{m}$ and $\overline{n}$ (Figure 14.2), measured from OA and OB, respectively, and can be given by the following equations (Hijab, 1956):

$$\overline{m} = \frac{4}{3} r_o \frac{\tan\phi'}{(9\tan^2\phi' + 1)} \left[\frac{\left(\dfrac{r_1}{r_o}\right)^3 (3\tan\phi' \sin\beta - \cos\beta) + 1}{\left(\dfrac{r_1}{r_o}\right)^2 - 1} \right] \tag{14.4}$$

$$\overline{n} = \frac{4}{3} r_o \frac{\tan\phi'}{(9\tan^2\phi' + 1)} \left[\frac{\left(\dfrac{r_1}{r_o}\right)^3 - 3\tan\phi' \sin\beta - \cos\beta}{\left(\dfrac{r_1}{r_o}\right)^2 - 1} \right] \tag{14.5}$$

Another important property of the logarithmic spiral defined by Eq. (14.1) is that any radial line makes an angle ϕ' with the normal to the curve drawn at the point where the radial line and the spiral intersect. This basic property is useful particularly in solving problems related to lateral earth pressure.

PASSIVE EARTH PRESSURE

14.4 Procedure for Determination of Passive Earth Pressure (P_p)—Cohesionless Backfill

Figure 14.1d shows the curved failure surface in the granular backfill of a retaining wall of height H. The shear strength of the granular backfill is expressed as

$$\tau_f = \sigma' \tan \phi' \tag{14.6}$$

The curved lower portion BC of the failure wedge is an arc of a logarithmic spiral defined by Eq. (14.1). The center of the log spiral lies on the line CA (not necessarily within the limits of points C and A). The upper portion CD is a straight line that makes an angle of $(45 - \phi'/2)$ degrees with the horizontal. The soil in the zone ACD is in *Rankine's passive state*.

Figure 14.3 shows the procedure for evaluating the passive resistance by trial wedges (Terzaghi and Peck, 1967). The retaining wall is first drawn to scale as shown in Figure 14.3a. The line C_1A is drawn in such a way that it makes an angle of $(45 - \phi'/2)$ degrees with the surface of the backfill. BC_1D_1 is a trial wedge in which BC_1 is the arc of a logarithmic spiral. According to the equation $r_1 = r_o e^{\beta \tan \phi'}$, O_1 is the center of the spiral. (*Note:* $\overline{O_1B} = r_o$ and $\overline{O_1C_1} = r_1$ and $\angle BO_1C_1 = \beta$; refer to Figure 14.2.)

Now let us consider the stability of the soil mass ABC_1C_1' (Figure 14.3b). For equilibrium, the following forces per unit length of the wall are to be considered:

1. Weight of the soil in zone $ABC_1C_1' = W_1 = (\gamma)(\text{Area of } ABC_1C_1')(1)$
2. The vertical face, C_1C_1' is in the zone of Rankine's passive state; hence, the force acting on this face is

$$P_{d(1)} = \frac{1}{2}\gamma(d_1)^2 \tan^2\left(45 + \frac{\phi'}{2}\right) \tag{14.7}$$

   where $d_1 = \overline{C_1C_1'}$. $P_{d(1)}$ acts horizontally at a distance of $d_1/3$ measured vertically upward from C_1.
3. F_1 is the resultant of the shear and normal forces that act along the surface of sliding, BC_1. At any point on the curve, according to the property of the logarithmic spiral, a radial line makes an angle ϕ' with the normal. Because the resultant, F_1, makes an angle ϕ' with the normal to the spiral at its point of application, its line of application will coincide with a radial line and will pass through the point O_1.

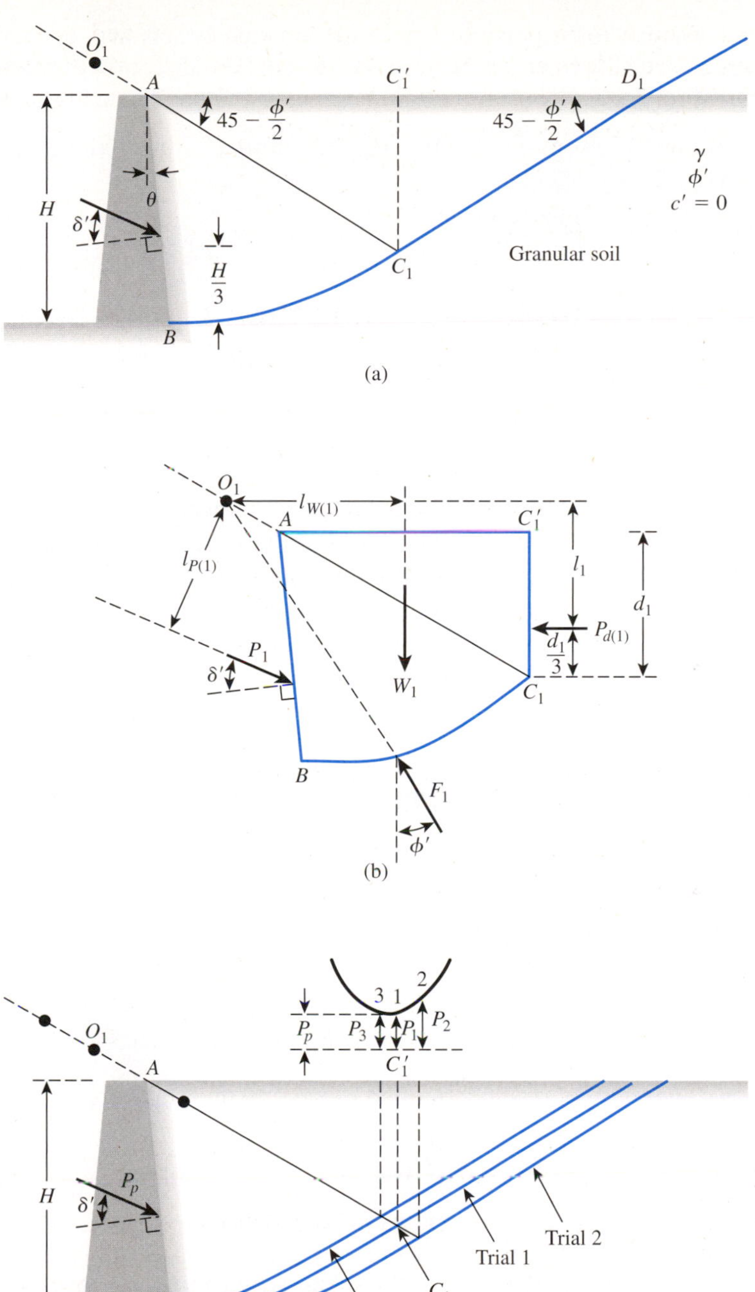

Figure 14.3 Passive earth pressure against retaining wall with curved failure surface

4. P_1 is the passive force per unit length of the wall. It acts at a distance of $H/3$ measured vertically from the bottom of the wall. The direction of the force P_1 is inclined at an angle δ' with the normal drawn to the back face of the wall.

Now, taking the moments of W_1, $P_{d(1)}$, F_1, and P_1 about the point O_1, for equilibrium, we have

$$W_1[l_{w(1)}] + P_{d(1)}[l_1] + F_1[0] = P_1[l_{p(1)}] \tag{14.8}$$

or

$$P_1 = \frac{1}{l_{p(1)}}[W_1 l_{w(1)} + P_{d(1)} l_1] \tag{14.9}$$

where $l_{w(1)}$, l_1, and $l_{P(1)}$ are moment arms for the forces W_1, $P_{d(1)}$, and P_1, respectively.

The preceding procedure for finding the trial passive force per unit length of the wall is repeated for several trial wedges such as those shown in Figure 14.3c. Let $P_1, P_2, P_3, \ldots, P_n$ be the forces that correspond to trial wedges $1, 2, 3, \ldots, n$, respectively. The forces are plotted to some scale as shown in the upper part of the figure. A smooth curve is plotted through the points $1, 2, 3, \ldots, n$. The lowest point of the smooth curve defines the actual passive force, P_p, per unit length of the wall.

14.5 Coefficient of Passive Earth Pressure (K_p)

Referring to the retaining wall with a *granular backfill* ($c' = 0$) shown in Figure 14.3, the passive earth pressure K_p can be expressed as

$$P_p = \frac{1}{2} K_p \gamma H^2 \tag{14.10}$$

or

$$K_p = \frac{P_p}{0.5\gamma H^2} \tag{14.11}$$

Following is a summary of results obtained by several investigators.

Procedure of Terzaghi and Peck

Using the procedure of Terzaghi and Peck (1967) described in Section 14.4, the passive earth-pressure coefficient can be evaluated for various combinations of θ, δ', and ϕ'. Figure 14.4 shows the variation of K_p for $\phi' = 30°$ and $40°$ (for $\theta = 0$) with δ'.

Solution by the method of slices

Shields and Tolunay (1973) improved the trail wedge solutions described in Section 14.4 using the *method of slices* to consider the stability of the trial soil wedge such as $ABC_1C'_1$ in Figure 14.3a. The details of the analysis are beyond the scope of this text. However, the values of K_p (passive earth-pressure coefficient) obtained by this method are given

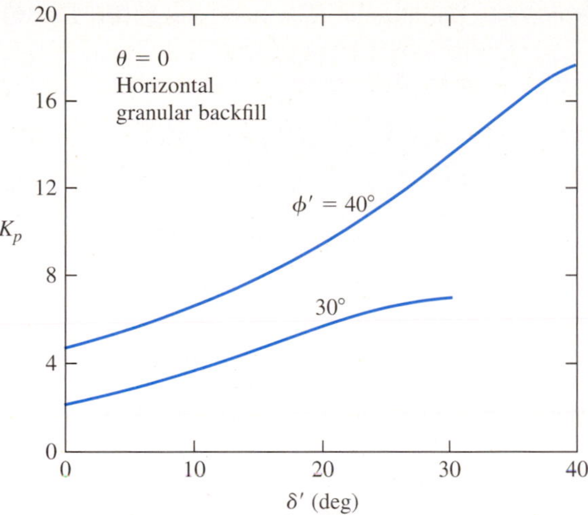

Figure 14.4 Variation of K_p with ϕ' and δ' based on the procedure of Terzaghi and Peck (1967) (*Note:* $\theta = 0$) (*Based on Terzaghi and Peck, 1967*)

in Table 14.1. It is important to note that the values of K_p given in Table 14.1 are for a retaining wall with a vertical back face and horizontal backfill of granular soil. Shields and Tolunay (1973) actually determined the *horizontal component of* P_p [i.e., $P_{p(H)}$] or

$$P_{p(H)} = \frac{1}{2}\gamma H^2 K'_p \tag{14.12}$$

However,

$$P_{p(H)} = P_p \cos\delta' = \frac{1}{2}\gamma H^2 K_p \cos\delta' \tag{14.13}$$

or

$$K_p = \frac{K'_p}{\cos\delta'} \tag{14.14}$$

The values of K_p given in Table 14.1 are based on Eq. (14.14.).

Table 14.1 Shields and Tolunay's Values of K_p Based on the Method of Slices

ϕ' (deg)	δ' (deg)									
	0	**5**	**10**	**15**	**20**	**25**	**30**	**35**	**40**	**45**
20	2.04	2.27	2.47	2.64	2.87					
25	2.46	2.78	3.08	3.34	3.61	4.01				
30	3.00	3.44	3.86	4.28	4.68	5.12	5.81			
35	3.69	4.31	4.91	5.53	6.17	6.85	7.61	8.85		
40	4.69	5.46	3.36	7.30	8.30	9.39	10.60	12.00	14.40	
45	5.83	7.09	8.43	9.89	11.49	13.28	15.31	17.65	20.36	25.46

Table 14.2 Variation of K_p with ϕ' and δ' [Eq. (14.16)]

ϕ' (deg)	δ' (deg)	K_p
20	10	2.52
	20	2.92
30	15	4.44
	30	5.81
40	20	8.91
	40	14.4

Solution by the lower-bound theorem of plasticity

Lancellotta (2002) analyzed the passive problem by using the lower-bound theorem of plasticity for retaining walls with a vertical back face ($\theta = 0$) and a horizontal granular backfill. Lancellotta (2002) provided the expression for K'_p [as defined by Eq. (14.12)] as

$$K'_p = \left[\frac{\cos\delta'}{1 - \sin\phi'}\left(\cos\delta' + \sqrt{\sin^2\phi' - \sin^2\delta'}\right)\right]e^{2\eta\tan\phi'} \quad (14.15)$$

Using Eq. (14.14),

$$K_p = \left[\left(\frac{1}{1 - \sin\phi'}\right)\left(\cos\delta' + \sqrt{\sin^2\phi' - \sin^2\delta'}\right)\right]e^{2\eta\tan\phi'} \quad (14.16)$$

where

$$2\eta \text{ (in radians)} = \sin^{-1}\left(\frac{\sin\delta'}{\sin\phi'}\right) + \delta' \quad (14.17)$$

The variation of K_p [Eq. (14.16)] with ϕ' and δ' is given in Table 14.2 and Figure 14.5.

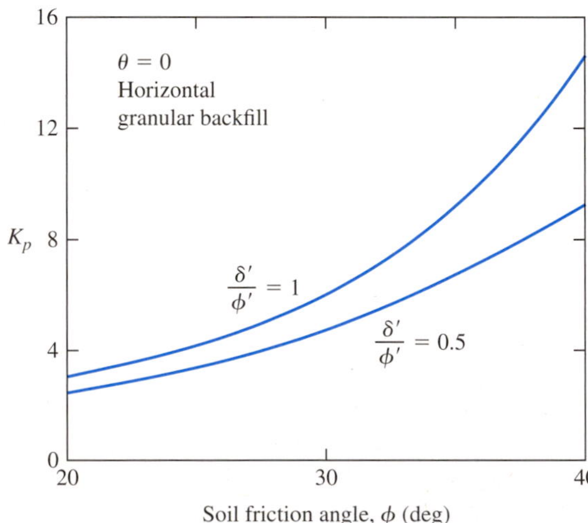

Figure 14.5 Lancellotta's solution for K_p—variation with ϕ' and $\frac{\delta'}{\phi'}$

Table 14.3 Variation of K_p with ϕ' and δ' using Sokolowskiĭ's Method of Characteristics ($\theta = 0$, Horizontal Granular Backfill)

ϕ' (deg)	δ' (deg)	K_p
20	10	2.55
	20	3.04
30	15	4.62
	30	6.55
40	20	9.68
	40	18.27

Solution by method of characteristics

Sokolowskiĭ (1965) developed the procedure for numerical solution based on the method of characteristics. Using this procedure, the variation of K_p is shown in Table 14.3 ($\theta = 0$, horizontal granular backfill).

Solution by the method of triangular slices

Zhu and Qian (2000) used the method of triangular slices (such as in the zone of ABC_1 in Figure 14.3a) to obtain the variation of K_p. According to this analysis,

$$K_p = K_{p(\delta'=0)}R \tag{14.18}$$

where K_p = passive earth pressure coefficient for a given value of θ, δ', and ϕ'

$K_{p(\delta'=0)}$ = K_p for a given value of θ, ϕ', with $\delta' = 0$

R = modification factor which is a function of ϕ', θ, δ'/ϕ'

The variations of $K_{p(\delta'=0)}$ are given in Table 14.4. The interpolated values of R are given in Table 14.5.

Table 14.4 Variation of $K_{p(\delta'=0)}$ [see Eq. (14.18) and Figure 14.3a]*

	θ (deg)						
ϕ' (deg)	30	25	20	15	10	5	0
20	1.70	1.69	1.72	1.77	1.83	1.92	2.04
21	1.74	1.73	1.76	1.81	1.89	1.99	2.12
22	1.77	1.77	1.80	1.87	1.95	2.06	2.20
23	1.81	1.81	1.85	1.92	2.01	2.13	2.28
24	1.84	1.85	1.90	1.97	2.07	2.21	2.37
25	1.88	1.89	1.95	2.03	2.14	2.28	2.46
26	1.91	1.93	1.99	2.09	2.21	2.36	2.56
27	1.95	1.98	2.05	2.15	2.28	2.45	2.66
28	1.99	2.02	2.10	2.21	2.35	2.54	2.77
29	2.03	2.07	2.15	2.27	2.43	2.63	2.88
30	2.07	2.11	2.21	2.34	2.51	2.73	3.00

(continued)

Table 14.4 *(continued)*

	θ (deg)						
ϕ' (deg)	30	25	20	15	10	5	0
31	2.11	2.16	2.27	2.41	2.60	2.83	3.12
32	2.15	2.21	2.33	2.48	2.68	2.93	3.25
33	2.20	2.26	2.39	2.56	2.77	3.04	3.39
34	2.24	2.32	2.45	2.64	2.87	3.16	3.53
35	2.29	2.37	2.52	2.72	2.97	3.28	3.68
36	2.33	2.43	2.59	2.80	3.07	3.41	3.84
37	2.38	2.49	2.66	2.89	3.18	3.55	4.01
38	2.43	2.55	2.73	2.98	3.29	3.69	4.19
39	2.48	2.61	2.81	3.07	3.41	3.84	4.38
40	2.53	2.67	2.89	3.17	3.53	4.00	4.59
41	2.59	2.74	2.97	3.27	3.66	4.16	4.80
42	2.64	2.80	3.05	3.38	3.80	4.34	5.03
43	2.70	2.88	3.14	3.49	3.94	4.52	5.27
44	2.76	2.94	3.23	3.61	4.09	4.72	5.53
45	2.82	3.02	3.32	3.73	4.25	4.92	5.80

*Based on Zhu and Qian, 2000

Table 14.5 Variation of R [Eq. (14.18)]

		R for ϕ' (deg)			
θ (deg)	δ'/ϕ'	30	35	40	45
0	0.2	1.2	1.28	1.35	1.45
	0.4	1.4	1.6	1.8	2.2
	0.6	1.65	1.95	2.4	3.2
	0.8	1.95	2.4	3.15	4.45
	1.0	2.2	2.85	3.95	6.1
5	0.2	1.2	1.25	1.32	1.4
	0.4	1.4	1.6	1.8	2.1
	0.6	1.6	1.9	2.35	3.0
	0.8	1.9	2.35	3.05	4.3
	1.0	2.15	2.8	3.8	5.7
10	0.2	1.15	1.2	1.3	1.4
	0.4	1.35	1.5	1.7	2.0
	0.6	1.6	1.85	2.25	2.9
	0.8	1.8	2.25	2.9	4.0
	1.0	2.05	2.65	3.6	5.3
15	0.2	1.15	1.2	1.3	1.35
	0.4	1.35	1.5	1.65	1.95
	0.6	1.55	1.8	2.2	2.7
	0.8	1.8	2.2	2.8	3.8
	1.0	2.0	2.6	3.4	4.95
20	0.2	1.15	1.2	1.3	1.35
	0.4	1.35	1.45	1.65	1.9
	0.6	1.5	1.8	2.1	2.6
	0.8	1.8	2.1	2.6	3.55
	1.0	1.9	2.4	3.2	4.8

14.6 Caquot and Kerisel Solution for Passive Earth Pressure (Granular Backfill)

Figure 14.6 shows a retaining wall with an inclined back and a horizontal backfill. For this case, the passive pressure per unit length of the wall can be calculated as

$$P_p = \frac{1}{2}\gamma H_1^2 K_p \tag{14.19}$$

where K_p = the passive pressure coefficient

For the definition of H_1, refer to Figure 14.6. The variation of K_p determined by Caquot and Kerisel (1948) also is shown in Figure 14.6. It is important to note that the K_p values shown are for $\delta'/\phi' = 1$. If $\delta'/\phi' \neq 1$, the following procedure must be used to determine K_p.

1. Assume δ' and ϕ'.
2. Calculate δ'/ϕ'.
3. Using the ratio of δ'/ϕ' (step 2), determine the reduction factor, R', from Table 14.6.
4. Determine K_p from Figure 14.6 for $\delta'/\phi' = 1$
5. Calculate K_p for the required δ'/ϕ' as

$$K_p = (R')[K_{p(\delta'/\phi' = 1)}] \tag{14.20}$$

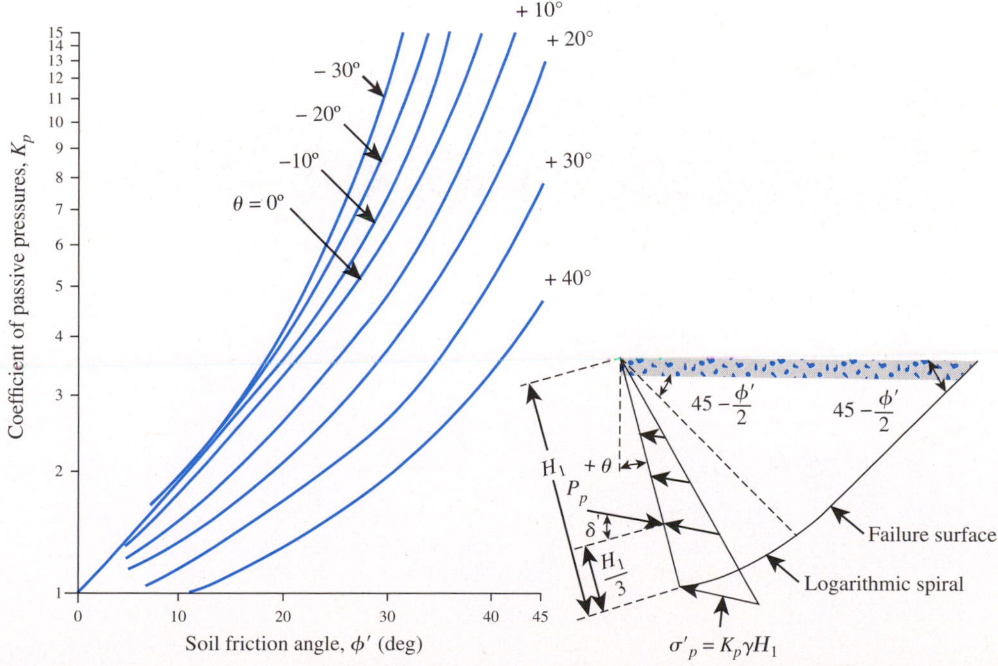

Figure 14.6 Caquot and Kerisel's solution for K_p [Eq. (14.19)]

Figure 14.7 shows a vertical retaining wall with an inclined granular backfill. For this case,

$$P_p = \frac{1}{2}\gamma H^2 K_p \tag{14.21}$$

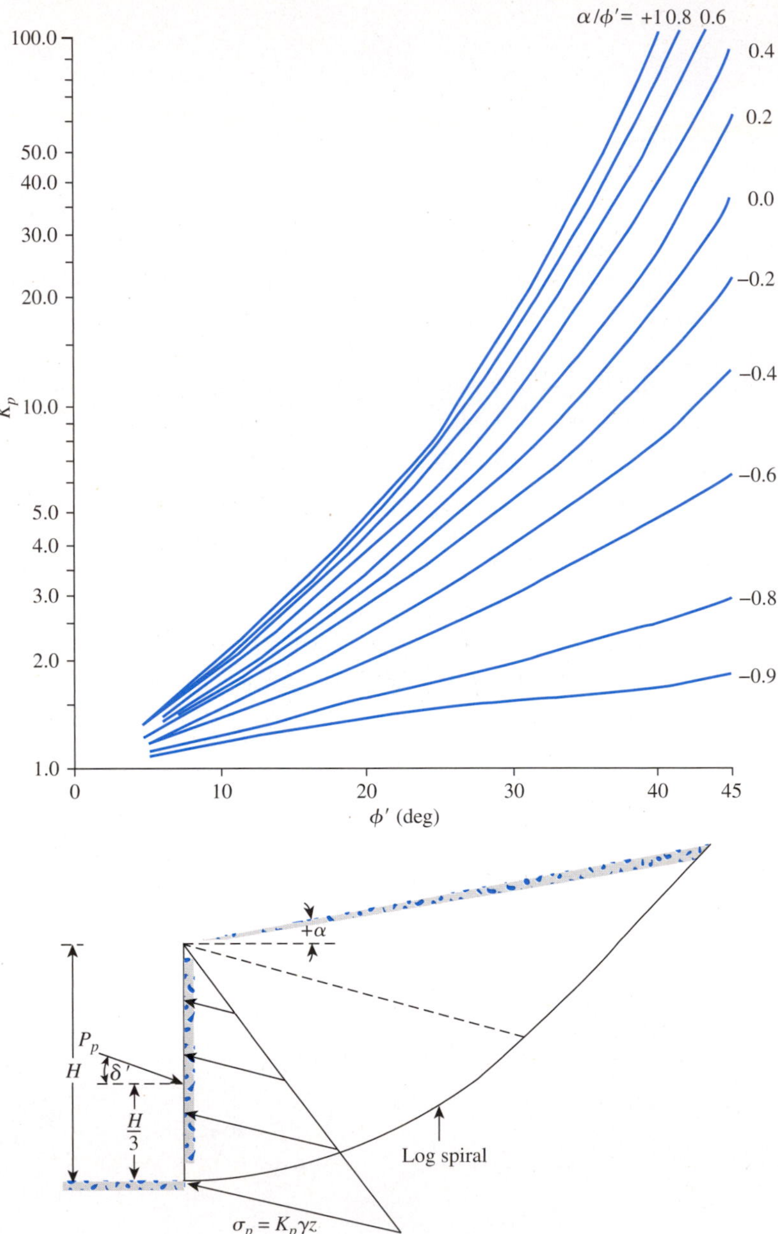

Figure 14.7 Caquot and Kerisel's solution for K_p [Eq. (14.21)]

Table 14.6 Caquot and Kerisel's Reduction Factor, R', for Passive Pressure Calculation

	δ'/ϕ'							
ϕ'	**0.7**	**0.6**	**0.5**	**0.4**	**0.3**	**0.2**	**0.1**	**0.0**
10	0.978	0.962	0.946	0.929	0.912	0.898	0.881	0.864
15	0.961	0.934	0.907	0.881	0.854	0.830	0.803	0.775
20	0.939	0.901	0.862	0.824	0.787	0.752	0.716	0.678
25	0.912	0.860	0.808	0.759	0.711	0.666	0.620	0.574
30	0.878	0.811	0.746	0.686	0.627	0.574	0.520	0.467
35	0.836	0.752	0.674	0.603	0.536	0.475	0.417	0.362
40	0.783	0.682	0.592	0.512	0.439	0.375	0.316	0.262
45	0.718	0.600	0.500	0.414	0.339	0.276	0.221	0.174

Caquot and Kerisel's solution (1948) for K_p to use in Eq. (14.21) is given in Figure 14.7 for $\delta'/\phi' = 1$. In order to determine K_p via Figure 14.7, the following steps are necessary:

Step 1. Determine α/ϕ' (note the sign of α).
Step 2. Knowing ϕ' and α/ϕ', use Figure 14.7 to determine K_p for $\delta'/\phi' = 1$.
Step 3. Calculate δ'/ϕ'.
Step 4. Go to Table 14.6 to determine the reduction factor, R'.
Step 5. $K_p = (R')[K_{p(\delta'/\phi' = 1)}]$. (14.22)

Example 14.1

Consider a 3-m-high (H) retaining wall with a vertical back ($\theta = 0°$) and a horizontal granular backfill. Given: $\gamma = 15.7$ kN/m^3, $\delta' = 15°$, and $\phi' = 30°$. Estimate the passive force, P_p, by using

a. Coulomb's theory
b. Terzaghi and Peck's wedge theory
c. Shields and Tolunay's solution (method of slices)
d. Zhu and Qian's solution (method of triangular slices)
e. Caquot and Kerisel's theory

Solution

Part a

From Eq. (13.79),

$$P_p = \frac{1}{2} K_p \gamma H^2$$

From Table 13.7, for $\phi' = 30°$ and $\delta' = 15°$, the value of K_p is 4.977. Thus,

$$P_p = \left(\frac{1}{2}\right)(4.977)(15.7)(3)^2 = \mathbf{351.6\ kN/m}$$

Part b

From Figure 14.4, for $\phi' = 30°$ and $\delta' = 15°$, the value of K_p is about 4.6. Thus,

$$P_p = \left(\frac{1}{2}\right)(4.6)(15.7)(3)^2 \approx \mathbf{325\ kN/m}$$

Part c

$$P_p = \frac{1}{2} K_p \gamma H^2$$

From Table 14.1, for $\phi' = 30°$ and $\delta' = 15°$ (i.e, $\frac{\delta'}{\phi'} = 0.5$) the value of K_p is 4.28. Hence,

$$P_p = \left(\frac{1}{2}\right)(4.28)(15.7)(3)^2 \approx \mathbf{302\ kN/m}$$

Part d

From Eq. (14.18),

$$K_p = K_{p(\delta'=0)} R$$

For $\phi' = 30°$ and $\theta = 0$, $K_{p(\delta'=0)}$ is equal to 3.0 (Table 14.4). Again, for $\theta = 0$ and $\delta'/\phi' = 0.5$, the value of R is about 1.52 (Table 14.5). Thus, $K_p = (3)(1.52) = 4.56$.

$$P_p = \left(\frac{1}{2}\right)(4.56)(15.7)(3)^2 = \mathbf{322\ kN/m}$$

Part e

From Eq. (14.19), with $\theta = 0$, $H_1 = H$,

$$P_p = \frac{1}{2} \gamma H^2 K_p$$

From Figure 14.6, for $\phi' = 30°$ and $\delta'/\phi' = 1$, the value of $K_{p(\delta'/\phi'=1)}$ is about 5.9. Also, from Table 14.6, with $\phi' = 30°$ and $\delta'/\phi' = 0.5$, the value of R' is 0.746.

Hence,

$$P_p = \frac{1}{2} \gamma H^2 K_p = \frac{1}{2}(15.7)(3)^2(0.746 \times 5.9) \approx \mathbf{311\ kN/m}$$

Example 14.2

Refer to Example 14.1. Estimate the passive force P_p using

a. Lancellotta's analysis (2002)
b. The method of characteristics of Sokolowskiĭ (1965)

Solution

Part a

From Table 14.2, for $\phi' = 30°$ and $\delta' = 15°$, the value of K_p is 4.44. So,

$$P_p = \frac{1}{2}K_p\gamma H^2 = \left(\frac{1}{2}\right)(4.44)(15.7)(3)^2 \approx \mathbf{314\ kN/m}$$

Part b

From Table 14.3, for $\phi' = 30°$ and $\delta' = 15°$, the value of $K_p = 4.62$. Hence

$$P_p = \frac{1}{2}K_p\gamma H^2 = \left(\frac{1}{2}\right)(4.62)(15.7)(3)^2 \approx \mathbf{326\ kN/m}$$

14.7 Passive Force on Walls with Earthquake Forces

The relationship for passive earth pressure on a retaining wall with a *horizontal backfill* and vertical back face under earthquake conditions (Figure 14.8) was evaluated by Subba Rao and Choudhury (2005) using the pseudo-static approach to the method of limit equilibrium. The failure surface in soil assumed in the analysis was similar to that shown in Figure 14.3 (with $\theta = 0$; that is, vertical back face) and in Figure 14.8. The notations used in the analysis were

H = Height of retaining wall

P_{pe} = Passive force per unit length of the wall

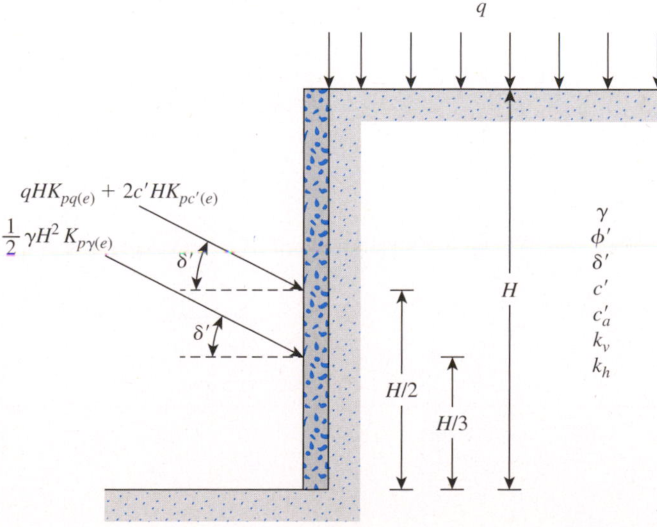

Figure 14.8 Passive pressure on wall (vertical back and horizontal backfill) with earthquake forces

ϕ' = Soil friction angle

δ' = Angle of wall friction

c' = Cohesion

c'_a = Soil-wall interfall adhesion

q = Surcharge

$$k_h = \frac{\text{Horizontal component of earthquake acceleration}}{\text{Acceleration due to gravity, } g}$$

$$k_v = \frac{\text{Vertical component of earthquake acceleration}}{\text{Acceleration due to gravity, } g}$$

Based on this analysis, the passive force P_{pe} can be expressed as

$$P_{pe} = \left[\frac{1}{2}\gamma H^2 K_{p\gamma(e)} + qHK_{pq(e)} + 2c'HK_{pc'(e)}\right]\frac{1}{\cos\delta'} \tag{14.23}$$

where $K_{p\gamma(e)}$, $K_{pq(e)}$, and $K_{pc'(e)}$ = passive earth-pressure coefficients in the normal direction to the wall.

The variations of $K_{p\gamma(e)}$ and $K_{pq(e)}$ for δ'/ϕ' = 0.5 and 1 are shown in Figures 14.9 and 14.10.

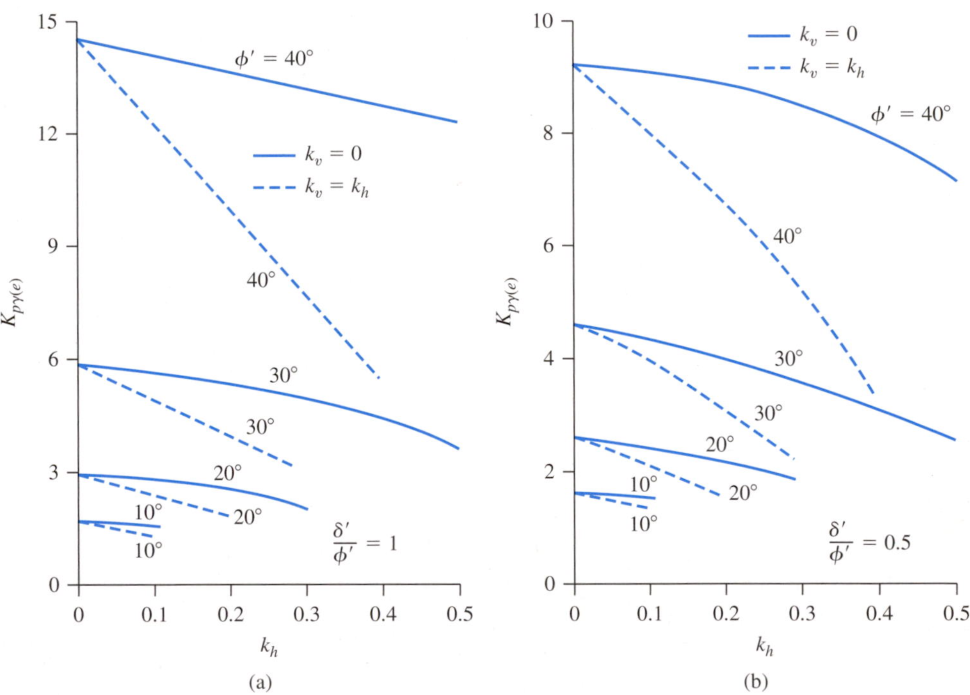

Figure 14.9 Variation of $K_{p\gamma(e)}$: (a) δ'/ϕ' = 1; (b) $\dfrac{\delta'}{\phi'}$ = 0.5

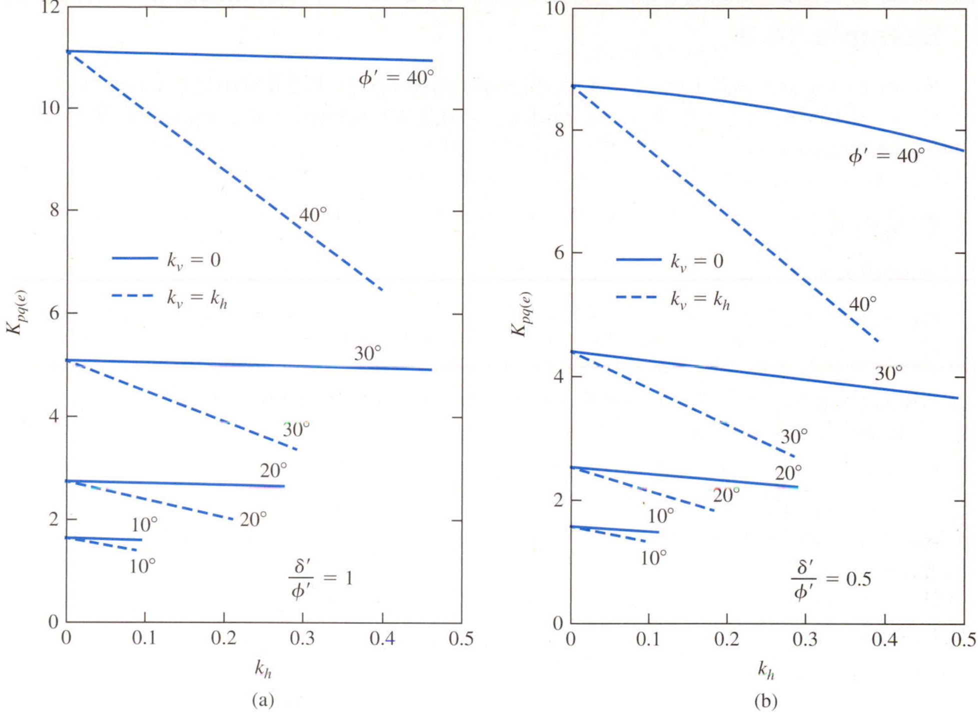

Figure 14.10 Variation of $K_{pq(e)}$: (a) $\dfrac{\delta'}{\phi'} = 1$, (b) $\dfrac{\delta'}{\phi'} = 0.5$

The locations of the components of P_{pe},

$$\left[\frac{\frac{1}{2}\gamma H^2 K_{p\gamma(e)}}{\cos\delta'}\right] \text{ and } \left[\frac{qHK_{pq(e)} + 2c'HK_{pc'(e)}}{\cos\delta'}\right]$$

are shown in Figure 14.8.

The variation of $K_{pc'(e)}$ is given in Table 14.7.

Table 14.7 Variation of $K_{pc'(e)}$

	$c'_a/c' = 0$		$c'_a/c' = \tan\delta'/\tan\phi'$	
ϕ' (deg)	$\delta'/\phi' = 0.5$	$\delta'/\phi' = 1.0$	$\delta'/\phi' = 0.5$	$\delta'/\phi' = 1.0$
10	1.32	1.45	1.55	1.69
20	1.81	2.18	2.07	2.33
30	2.66	3.42	2.94	3.49
40	4.33	5.95	4.65	5.97

Figure 14.12 Braced cut for Chicago subway construction (*Courtesy of Ralph B. Peck*)

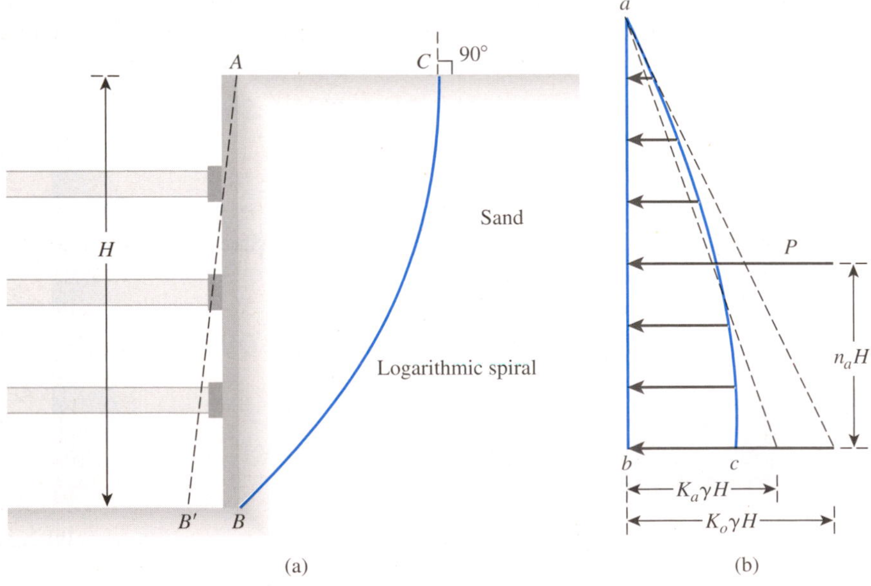

Figure 14.13 Earth pressure distribution against a wall with rotation about the top

Because the upper portion of the soil mass in the zone *ABC* does not undergo sufficient deformation, it does not pass into Rankine's active state. The sliding surface *BC* intersects the ground surface almost at 90°. The corresponding earth pressure will be somewhat parabolic, like *acb* shown in Figure 14.13b. With this type of pressure distribution, the point of application of the resultant active thrust, P_a, will be at a height of n_aH measured from the bottom of the wall, with $n_a > \frac{1}{3}$ (for triangular pressure distribution $n_a = \frac{1}{3}$). Theoretical evaluation and field measurements have shown that n_a could be as high as 0.55.

14.9 Determination of Active Thrust on Bracing Systems of Open Cuts—Granular Soil

The active thrust on the bracing system of open cuts can be estimated theoretically by using trial wedges and Terzaghi's general wedge theory (1941). The basic procedure for determination of the active thrust are described in this section.

Figure 14.14a shows a braced wall *AB* of height *H* that deforms by rotating about its top. The wall is assumed to be rough, with the angle of wall friction equal to δ'. The point of application of the active thrust (that is, n_aH) is assumed to be known. The curve of sliding is assumed to be an arc of a logarithmic spiral. As we discussed in the preceding section, the curve of sliding intersects the horizontal ground surface at 90°. To proceed with the trial wedge solution, let us select a point b_1. From b_1, a line $b_1 b'_1$ that makes an angle ϕ' with the ground surface is drawn. (Note that ϕ' = effective angle of friction of the soil.). The arc of the logarithmic spiral, b_1B, which defines the curve of sliding for this trial, can now be drawn, with the center of the spiral (point O_1) located on the line $b_1b'_1$. Note that the equation for the logarithmic spiral is given by $r_1 = r_o e^{\beta_1 \tan\phi'}$ and, in this case, $\overline{O_1b_1} = r_o$ and $\overline{O_1B} = r_1$. Also, it is interesting to see that the horizontal line that represents the ground surface is the normal to the curve of sliding at the point b_1, and that O_1b_1 is a radial line. The angle between them is equal to ϕ', which agrees with the property of the spiral.

To look at the equilibrium of the failure wedge, let us consider the following forces per unit length of the braced wall:

- W_1 = the weight of the wedge ABb_1 = (Area of ABb_1) × (γ) × (1).
- P_1 = the active thrust acting at a point n_aH measured vertically upward from the bottom of the cut and inclined at an angle δ' with the horizontal.
- F_1 = the resultant of the shear and normal forces that act along the trial failure surface. The line of action of the force F_1 will pass through the point O_1.

Now, taking the moments of these forces about O_1, we have

$$W_1[l_{w(1)}] + F_1(0) - P_1[l_{P(1)}] = 0$$

or

$$P_1 = \frac{W_1 l_{w(1)}}{l_{P(1)}} \tag{14.24}$$

where $l_{W(1)}$ and $l_{P(1)}$ are the moment arms for the forces W_1 and P_1, respectively.

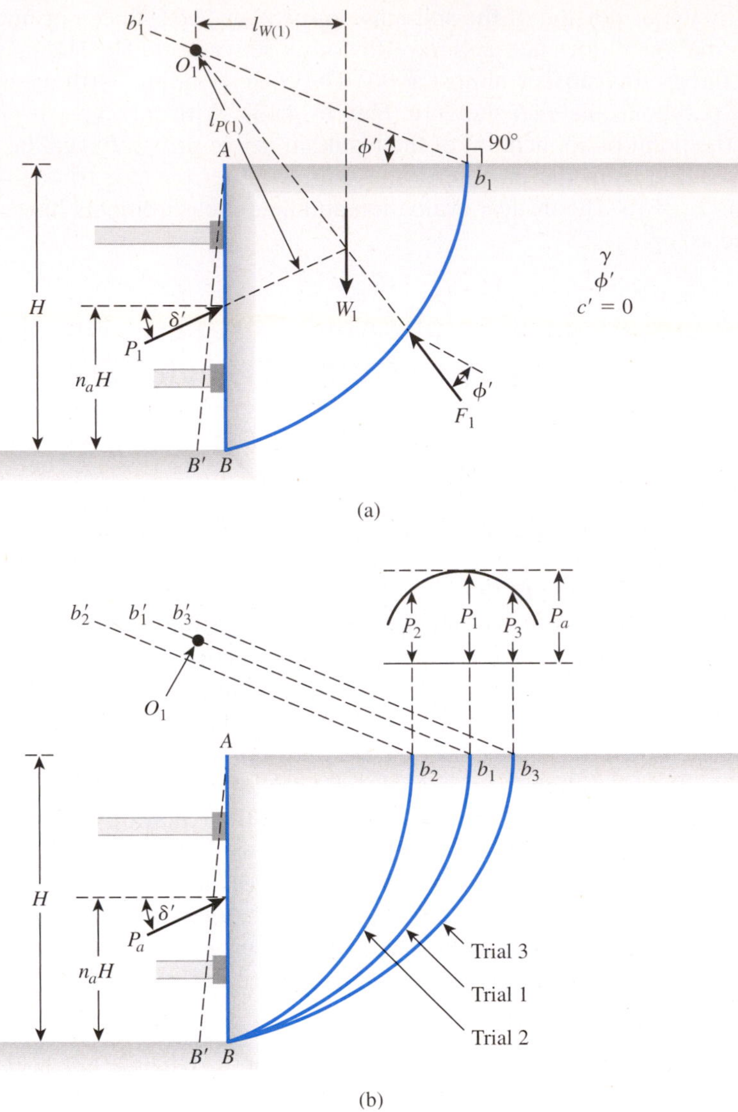

Figure 14.14 Determination of active force on bracing system of open cut in cohesionless soil

This procedure of finding the active thrust can now be repeated for several wedges such as ABb_2, ABb_3, . . ., ABb_n (Figure 14.14b). Note that the centers of the logarithmic-spiral arcs will lie on lines b_2b_2', b_3b_3', . . ., b_nb_n' respectively. The active thrusts $P_1, P_2, P_3, \ldots, P_n$ derived from the trial wedges are plotted to some scale in the upper portion of Figure 14.14b. The maximum point of the smooth curve drawn through these points will yield the desired maximum active thrust, P_a, on the braced wall.

Table 14.8 $P_a/0.5\gamma H^2$ Against ϕ', δ', and n_a ($c' = 0$) for Braced Cuts*

		$P_a/0.5\gamma H^2$						$P_a/0.5\gamma H^2$			
ϕ' (deg)	δ' (deg)	$n_a = 0.3$	$n_a = 0.4$	$n_a = 0.5$	$n_a = 0.6$	ϕ' (deg)	δ' (deg)	$n_a = 0.3$	$n_a = 0.4$	$n_a = 0.5$	$n_a = 0.6$
10	0	0.653	0.734	0.840	0.983	35	0	0.247	0.267	0.290	0.318
	5	0.623	0.700	0.799	0.933		5	0.239	0.258	0.280	0.318
	10	0.610	0.685	0.783	0.916		10	0.234	0.252	0.273	0.300
15	0	0.542	0.602	0.679	0.778		15	0.231	0.249	0.270	0.296
	5	0.518	0.575	0.646	0.739		20	0.231	0.248	0.269	0.295
	10	0.505	0.559	0.629	0.719		25	0.232	0.250	0.271	0.297
	15	0.499	0.554	0.623	0.714		30	0.236	0.254	0.276	0.302
20	0	0.499	0.495	0.551	0.622		35	0.243	0.262	0.284	0.312
	5	0.430	0.473	0.526	0.593	40	0	0.198	0.213	0.230	0.252
	10	0.419	0.460	0.511	0.575		5	0.192	0.206	0.223	0.244
	15	0.413	0.454	0.504	0.568		10	0.189	0.202	0.219	0.238
	20	0.413	0.454	0.504	0.569		15	0.187	0.200	0.216	0.236
25	0	0.371	0.405	0.447	0.499		20	0.187	0.200	0.216	0.235
	5	0.356	0.389	0.428	0.477		25	0.188	0.202	0.218	0.237
	10	0.347	0.378	0.416	0.464		30	0.192	0.205	0.222	0.241
	15	0.342	0.373	0.410	0.457		35	0.197	0.211	0.228	0.248
	20	0.341	0.372	0.409	0.456		40	0.205	0.220	0.237	0.259
	25	0.344	0.375	0.413	0.461	45	0	0.156	0.167	0.180	0.196
30	0	0.304	0.330	0.361	0.400		5	0.152	0.163	0.175	0.190
	5	0.293	0.318	0.347	0.384		10	0.150	0.160	0.172	0.187
	10	0.286	0.310	0.339	0.374		15	0.148	0.159	0.171	0.185
	15	0.282	0.306	0.334	0.368		20	0.149	0.159	0.171	0.185
	20	0.281	0.305	0.332	0.367		25	0.150	0.160	0.173	0.187
	25	0.284	0.307	0.335	0.370		30	0.153	0.164	0.176	0.190
	30	0.289	0.313	0.341	0.377		35	0.158	0.168	0.181	0.196
							40	0.164	0.175	0.188	0.204
							45	0.173	0.184	0.198	0.213

After Kim and Preber, 1969. With permission from ASCE.

Kim and Preber (1969) determined the values of $P_a/0.5\gamma H^2$ for braced excavations for various values of ϕ', δ', and n_a. These values are given in Table 14.8. In general, the average magnitude of P_a is about 10% greater when the wall rotation is about the top as compared with the value obtained by Coulomb's active earth-pressure theory.

14.10 Determination of Active Thrust on Bracing Systems for Cuts—Cohesive Soil

Using the principles of the general wedge theory, we also can determine the active thrust on bracing systems for cuts made in $c'-\phi'$ soil. Table 14.9 gives the variation of P_a in a nondimensional form for various values of ϕ', δ', n_a, and $c'/\gamma H$.

Table 14.9 Values of $P_a/0.5\gamma H^2$ for Cuts in a c'-ϕ' Soil with the Assumption $c_a' = c'(\tan\delta'/\tan\phi')$*

δ' (deg)	$n_a = 0.3$ and $c'/\gamma H = 0.1$	$n_a = 0.4$ and $c'/\gamma H = 0.1$	$n_a = 0.5$ and $c'/\gamma H = 0.1$
$\phi' = 15°$			
0	0.254	0.285	0.322
5	0.214	0.240	0.270
10	0.187	0.210	0.238
15	0.169	0.191	0.218
$\phi' = 20°$			
0	0.191	0.210	0.236
5	0.160	0.179	0.200
10	0.140	0.156	0.173
15	0.122	0.127	0.154
20	0.113	0.124	0.140
$\phi' = 25°$			
0	0.138	0.150	0.167
5	0.116	0.128	0.141
10	0.099	0.110	0.122
15	0.085	0.095	0.106
20	0.074	0.083	0.093
25	0.065	0.074	0.083
$\phi' = 30°$			
0	0.093	0.103	0.113
5	0.078	0.086	0.094
10	0.066	0.073	0.080
15	0.056	0.060	0.067
20	0.047	0.051	0.056
25	0.036	0.042	0.047
30	0.029	0.033	0.038

After Kim and Preber, 1969. With permission from ASCE.

14.11 Pressure Variation for Design of Sheetings, Struts, and Wales

The active thrust against sheeting in a braced cut, calculated by using the general wedge theory, does not explain the variation of the earth pressure with depth that is necessary for design work. An important difference between bracings in open cuts and retaining walls is that retaining walls fail as single units, whereas bracings in an open cut undergo progressive failure where one or more struts fail at one time.

Empirical lateral pressure diagrams against sheetings for the design of bracing systems have been given by Peck (1969). These pressure diagrams for cuts in sand, soft to medium clay, and stiff clay are given in Figure 14.15. Strut loads may be determined by assuming that the vertical members are hinged at each strut level except the topmost and bottommost ones (Figure 14.16). Example 14.4 illustrates the procedure for the calculation of strut loads.

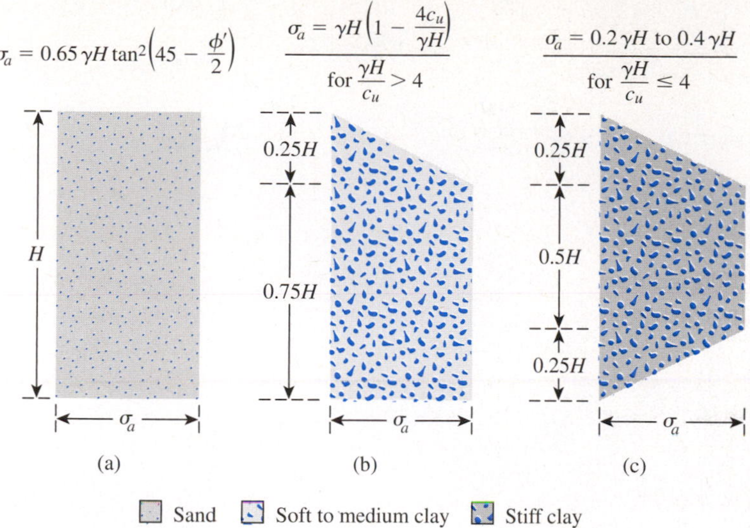

Figure 14.15 Peck's pressure diagrams for design of bracing systems

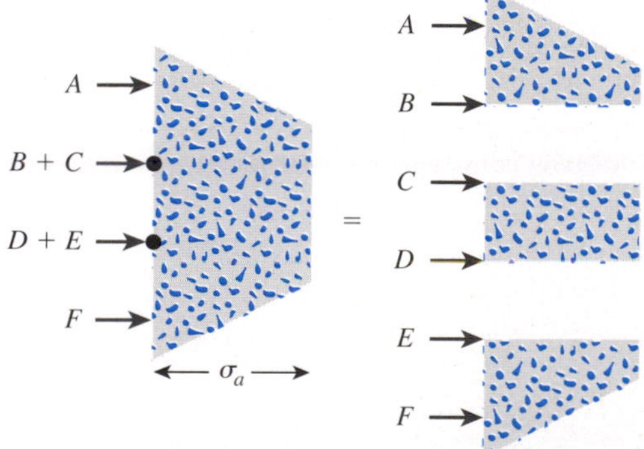

Figure 14.16 Determination of strut loads from empirical lateral pressure diagrams

Example 14.4

A 7.5-m-deep braced cut in sand is shown in Figure 14.17. In the plan, the struts are placed at spacings, $s = 2$ m center to center. Using Peck's empirical pressure diagram, calculate the design strut loads.

Solution

Refer to Figure 14.15a. For the lateral earth pressure diagram,

$$\sigma_a = 0.65\gamma H \tan^2\left(45 - \frac{\phi'}{2}\right) = (0.65)(16)(7.5)\tan^2\left(45 - \frac{30}{2}\right) = 26 \text{ kN/m}^2$$

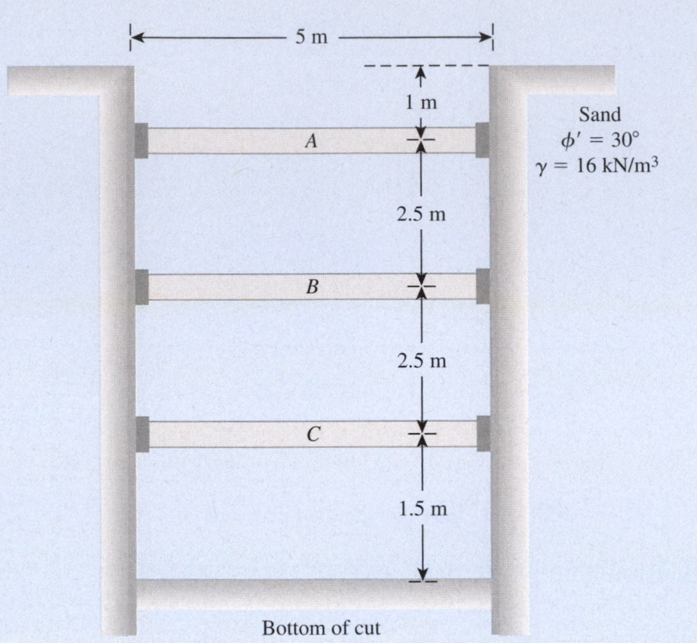

Figure 14.17 Braced cut in sand

Assume that the sheeting is hinged at strut level B. Now refer to the diagram in Figure 14.18. We need to find reactions at A, B_1, B_2, and C. Taking the moment about B_1, we have

$$2.5A = (26)(3.5)\left(\frac{3.5}{2}\right); \quad A = 63.7 \text{ kN/m}$$

Hence,

$$B_1 = (26)(3.5) - 63.7 = 27.3 \text{ kN/m}$$

Again, taking the moment about B_2, we have

$$2.5C = (26)(4)\left(\frac{4}{2}\right)$$

$$C = 83.2 \text{ kN/m}$$

So

$$B_2 = (26)(4) - 83.2 = 20.8 \text{ kN/m}$$

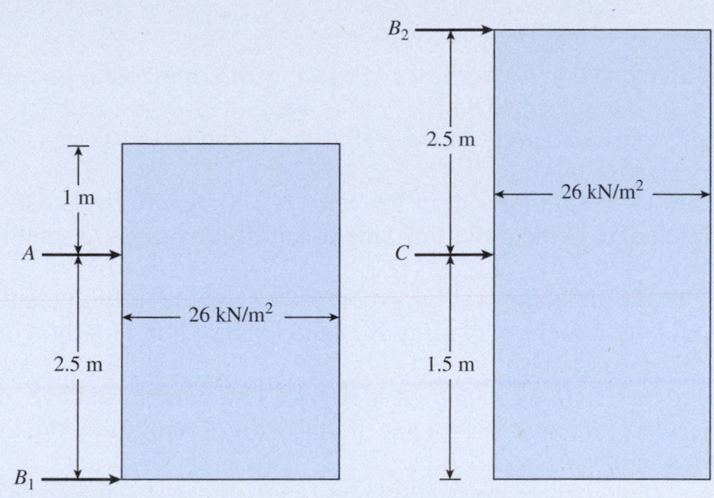

Figure 14.18 Calculation of strut loads from pressure envelope

The strut loads are:

At level A: $(A)(s) = (63.7)(2) = \mathbf{127.4\ kN}$

At level B: $(B_1 + B_2)(s) = (27.3 + 20.8)(2) = \mathbf{96.2\ kN}$

At level C: $(C)(s) = (83.2)(2) = \mathbf{166.4\ kN}$

14.12 Summary

This chapter covers two major topics:

- Estimation of passive pressure using curved failure surface in soil
- Lateral earth pressure on braced cuts using the general wedge theory and pressure envelopes for design of struts, wales, and sheet piles.

Passive pressure calculations using curved failure surface are essential for the case in which $\delta' > \phi'/2$, since plane-failure surface assumption provides results on the unsafe side for design. The passive pressure coefficient as obtained using the analyses of Terzaghi and Peck (1967), Shields and Tolunay (1973), Zhu and Qian (2000), Caquot and Kerisel (1948), Lancellotta (2002), and Sokolowskiĭ (1965) are given in Sections 14.5 and 14.6.

In the case of braced cuts, although the general wedge theory provides the force per unit length of the cut, it does not provide the nature of distribution of earth pressure with depth. For that reason, pressure envelopes are necessary for practical design. Section 14.11 presents the earth pressure envelopes recommended by Peck (1969) for cuts in sand, soft to medium clay, and stiff clay. It also provides the procedure for calculation of the strut loads in braced cuts.

Problems

14.1 A retaining wall has a vertical back face with a horizontal granular backfill. Given: $H = 6$ m, $\gamma = 18.5$ kN/m^3, $\phi' = 40°$, and $\delta' = \frac{1}{2}\phi'$. Estimate the passive force, P_p, per unit length of the wall using Terzaghi and Peck's (1967) wedge theory (Figure 14.4).

14.2 Refer to the retaining wall in Problem 14.1. Estimate the passive force, P_p, per unit length of the wall using Shields and Tolunay's (1973) method of slices (Table 14.1).

14.3 Refer to the retaining wall in Problem 14.1. Estimate the passive force, P_p, per unit length of the wall using Zhu and Qian's (2000) method of triangular slices. Use Eq. (14.18).

14.4 Refer to the retaining wall in Problem 14.1. Estimate the passive force, P_p, per unit length of the wall using Lancellotta's (2002) analysis by the lower bound theorem of plasticity. Use Table 14.2.

14.5 Refer to the retaining wall in Problem 14.1. Estimate the passive force, P_p, per unit length of the wall using Sokolowskiĭ (1965) solution by the method of characteristics (Table 14.3).

14.6 Refer to Figure 14.19. Given: $H = 4.9$ m, $\theta = 0, \alpha = 0, \gamma = 18.7$ kN/m^3, $\phi' = 30°$, and $\delta' = \frac{2}{3}\phi'$. Estimate the passive force, P_p, per unit length of the wall using the K_p values given by Shields and Tolunay's (1973) method of slices (Table 14.1).

14.7 Refer to the retaining wall shown in Figure 14.19. Given: $\theta = 10°$, $\alpha = 0$, $\gamma = 19.1$ kN/m^3, $\phi' = 35°$, $\delta' = 21°$, and $H = 5.5$ m. Estimate the passive force, P_p, per unit length of the wall using Zhu and Qian's (2000) method of triangular slices.

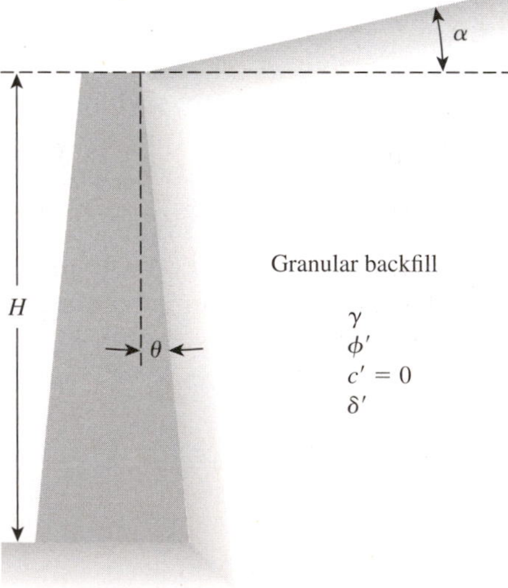

Figure 14.19

4. **Spread.** This is a form of slide (Figure 15.4) by translation. It occurs by "sudden movement of water-bearing seams of sands or silts overlain by clays or loaded by fills."
5. **Flow.** This is a downward movement of soil mass similar to a viscous fluid (Figure 15.5).

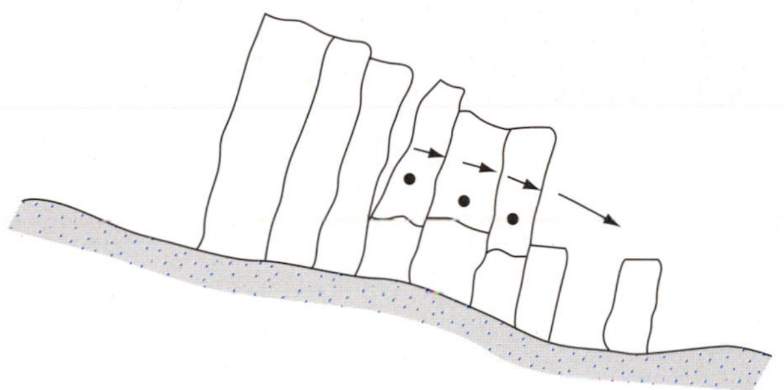

Figure 15.2 Slope failure by "toppling"

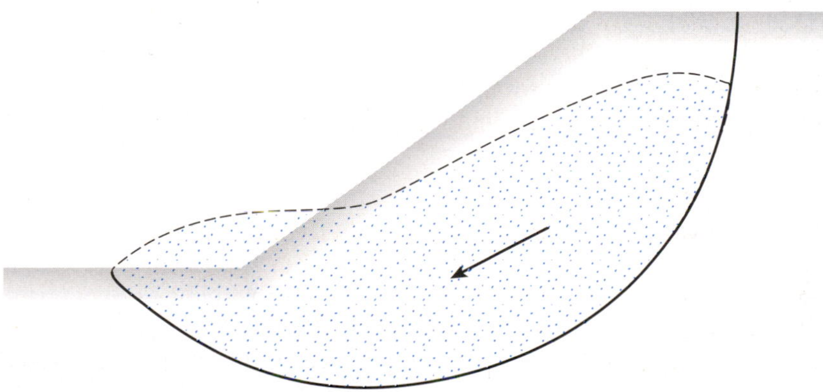

Figure 15.3 Slope failure by "sliding"

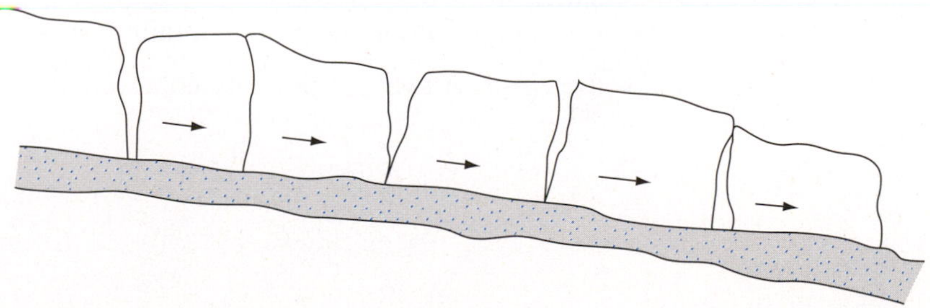

Figure 15.4 Slope failure by lateral "spreading"

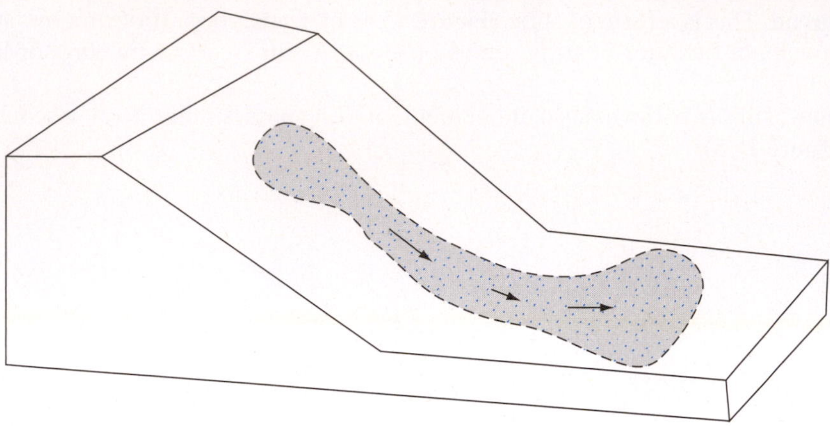

Figure 15.5 Slope failure by "flowing"

This chapter primarily relates to the quantitative analysis that falls under the category of *slide*. We will discuss in detail the following:

- Definition of factor of safety
- Stability of infinite slopes
- Stability of finite slopes with plane and circular failure surfaces
- Analysis of the stability of finite slopes with steady-state seepage condition

15.2 Factor of Safety

The task of the engineer charged with analyzing slope stability is to determine the factor of safety. Generally, the factor of safety is defined as

$$F_s = \frac{\tau_f}{\tau_d} \tag{15.1}$$

where F_s = factor of safety with respect to strength
τ_f = average shear strength of the soil
τ_d = average shear stress developed along the potential failure surface

The shear strength of a soil consists of two components, cohesion and friction, and may be written as

$$\tau_f = c' + \sigma' \tan \phi' \tag{15.2}$$

where c' = cohesion
ϕ' = angle of friction
σ' = normal stress on the potential failure surface

In a similar manner, we can write

$$\tau_d = c'_d + \sigma' \tan \phi'_d \tag{15.3}$$

where c'_d and ϕ'_d are, respectively, the cohesion and the angle of friction that develop along the potential failure surface. Substituting Eqs. (15.2) and (15.3) into Eq. (15.1), we get

$$F_s = \frac{c' + \sigma' \tan \phi'}{c'_d + \sigma' \tan \phi'_d} \tag{15.4}$$

Now we can introduce some other aspects of the factor of safety—that is, the factor of safety with respect to cohesion, $F_{c'}$, and the factor of safety with respect to friction, $F_{\phi'}$. They are defined as

$$F_{c'} = \frac{c'}{c'_d} \tag{15.5}$$

and

$$F_{\phi'} = \frac{\tan \phi'}{\tan \phi'_d} \tag{15.6}$$

When we compare Eqs. (15.4) through (15.6), we can see that when $F_{c'}$ becomes equal to $F_{\phi'}$, it gives the factor of safety with respect to strength. Or, if

$$\frac{c'}{c'_d} = \frac{\tan \phi'}{\tan \phi'_d}$$

then we can write

$$F_s = F_{c'} = F_{\phi'} \tag{15.7}$$

When F_s is equal to 1, the slope is in a state of impending failure. Generally, a value of 1.5 for the factor of safety with respect to strength is acceptable for the design of a stable slope.

15.3 Stability of Infinite Slopes

In considering the problem of slope stability, let us start with the case of an infinite slope as shown in Figure 15.6. The shear strength of the soil may be given by Eq. (15.2):

$$\tau_f = c' + \sigma' \tan \phi'$$

Assuming that the pore water pressure is zero, we will evaluate the factor of safety against a possible slope failure along a plane AB located at a depth H below the

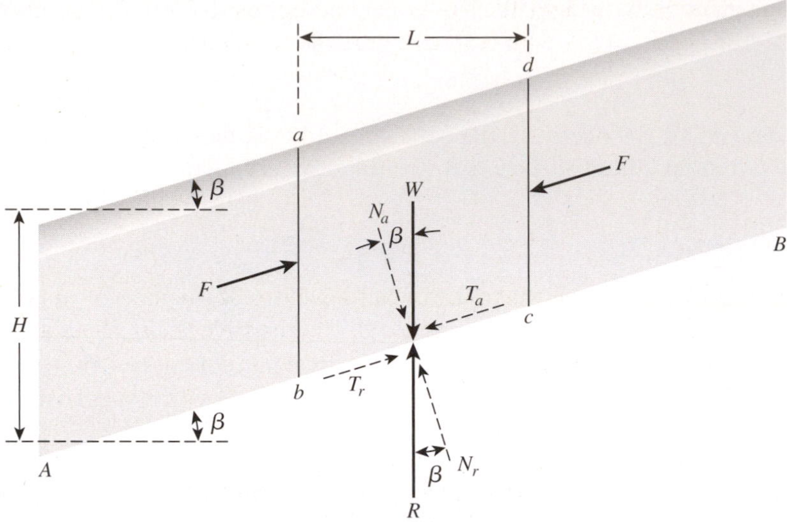

Figure 15.6 Analysis of infinite slope (without seepage)

ground surface. The slope failure can occur by the movement of soil above the plane AB from right to left.

Let us consider a slope element $abcd$ that has a unit length perpendicular to the plane of the section shown. The forces, F, that act on the faces ab and cd are equal and opposite and may be ignored. The weight of the soil element is

$$W = (\text{Volume of soil element}) \times (\text{Unit weight of soil}) = \gamma LH \quad (15.8)$$

The weight W can be resolved into two components:

1. Force perpendicular to the plane $AB = N_a = W\cos\beta = \gamma LH\cos\beta$.
2. Force parallel to the plane $AB = T_a = W\sin\beta = \gamma LH\sin\beta$. Note that this is the force that tends to cause the slip along the plane.

Thus, the effective normal stress and the shear stress at the base of the slope element can be given, respectively, as

$$\sigma' = \frac{N_a}{\text{Area of base}} = \frac{\gamma LH\cos\beta}{\left(\dfrac{L}{\cos\beta}\right)} = \gamma H\cos^2\beta \quad (15.9)$$

and

$$\tau = \frac{T_a}{\text{Area of base}} = \frac{\gamma LH\sin\beta}{\left(\dfrac{L}{\cos\beta}\right)} = \gamma H\cos\beta\sin\beta \quad (15.10)$$

The reaction to the weight W is an equal and opposite force R. The normal and tangential components of R with respect to the plane AB are

$$N_r = R \cos\beta = W \cos\beta \tag{15.11}$$

and

$$T_r = R \sin\beta = W \sin\beta \tag{15.12}$$

For equilibrium, the resistive shear stress that develops at the base of the element is equal to (T_r)/(Area of base) $= \gamma H \sin\beta \cos\beta$. The resistive shear stress also may be written in the same form as Eq. (15.3):

$$\tau_d = c'_d + \sigma' \tan\phi'_d$$

The value of the normal stress is given by Eq. (15.9). Substitution of Eq. (15.9) into Eq. (15.3) yields

$$\tau_d = c'_d + \gamma H \cos^2\beta \tan\phi'_d \tag{15.13}$$

Thus,

$$\gamma H \sin\beta \cos\beta = c'_d + \gamma H \cos^2\beta \tan\phi'_d$$

or

$$\frac{c'_d}{\gamma H} = \sin\beta \cos\beta - \cos^2\beta \tan\phi'_d$$

$$= \cos^2\beta(\tan\beta - \tan\phi'_d) \tag{15.14}$$

The factor of safety with respect to strength has been defined in Eq. (15.7), from which we get

$$\tan\phi'_d = \frac{\tan\phi'}{F_s} \quad \text{and} \quad c'_d = \frac{c'}{F_s}$$

Substituting the preceding relationships into Eq. (15.14), we obtain

$$F_s = \frac{c'}{\gamma H \cos^2\beta \tan\beta} + \frac{\tan\phi'}{\tan\beta} \tag{15.15}$$

For granular soils, $c' = 0$, and the factor of safety, F_s, becomes equal to $(\tan\phi')/(\tan\beta)$. This indicates that in an infinite slope in sand, the value of F_s is independent of the height H and the slope is stable as long as $\beta < \phi'$.

or

$$c'_d = \frac{1}{2}\gamma H\left[\frac{\sin(\beta - \theta)(\sin\theta - \cos\theta\tan\phi'_d)}{\sin\beta}\right] \tag{15.36}$$

The expression in Eq. (15.36) is derived for the trial failure plane AC. In an effort to determine the critical failure plane, we must use the principle of maxima and minima (for a given value of ϕ'_d) to find the angle θ where the developed cohesion would be maximum.

Thus, the first derivative of c'_d with respect to θ is set equal to zero, or

$$\frac{\partial c'_d}{\partial\theta} = 0 \tag{15.37}$$

Because γ, H, and β are constants in Eq. (15.36), we have

$$\frac{\partial}{\partial\theta}[\sin(\beta - \theta)(\sin\theta - \cos\theta\ \tan\phi'_d)] = 0 \tag{15.38}$$

Solving Eq. (15.38) gives the critical value of θ, or

$$\theta_{cr} = \frac{\beta + \phi'_d}{2} \tag{15.39}$$

Substitution of the value of $\theta = \theta_{cr}$ into Eq. (15.36) yields

$$c'_d = \frac{\gamma H}{4}\left[\frac{1 - \cos(\beta - \phi'_d)}{\sin\beta\ \cos\phi'_d}\right] \tag{15.40}$$

The preceding equation also can be written as

$$\frac{c'_d}{\gamma H} = m = \frac{1 - \cos(\beta - \phi'_d)}{4\ \sin\beta\ \cos\phi'_d} \tag{15.41}$$

where m = stability number.

The maximum height of the slope for which critical equilibrium occurs can be obtained by substituting $c'_d = c'$ and $\phi'_d = \phi'$ into Eq. (15.40). Thus,

$$H_{cr} = \frac{4c'}{\gamma}\left[\frac{\sin\beta\ \cos\phi'}{1 - \cos(\beta - \phi')}\right] \tag{15.42}$$

Example 15.2

A cut is to be made in a soil having $\gamma = 16.5\ \text{kN/m}^3$, $c' = 28.75\ \text{kN/m}^2$, and $\phi' = 15°$. The side of the cut slope will make an angle of 45° with the horizontal. What should be the depth of the cut slope that will have a factor of safety (F_s) of 3?

Solution

Given: $\phi' = 15°$; $c' = 28.75$ kN/m². If $F_s = 3$, then $F_{c'}$ and $F_{\phi'}$ should both be equal to 3.

$$F_{c'} = \frac{c'}{c'_d}$$

or

$$c'_d = \frac{c'}{F_{c'}} = \frac{c'}{F_s} = \frac{28.75}{3} = 9.58 \text{ kN/m}^2$$

Similarly,

$$F_{\phi'} = \frac{\tan \phi'}{\tan \phi'_d}$$

$$\tan \phi'_d = \frac{\tan \phi'}{F_{\phi'}} = \frac{\tan \phi'}{F_s} = \frac{\tan 15}{3}$$

or

$$\phi'_d = \tan^{-1}\left[\frac{\tan 15}{3}\right] = 5.1°$$

Substituting the preceding values of c'_d and ϕ'_d in Eq. (15.40),

$$H = \frac{4c'_d}{\gamma}\left[\frac{\sin \beta \cos \phi'_d}{1 - \cos(\beta - \phi'_d)}\right]$$

$$= \frac{4 \times 9.58}{16.5}\left[\frac{\sin 45 \cos 5.1}{1 - \cos(45 - 5.1)}\right]$$

$$= \mathbf{7.03\ m}$$

Example 15.3

Refer to Figure 15.9. For a trial failure surface *AC* in the slope, given: $H = 5$ m, $\beta = 55°$, $\theta = 35°$, and $\gamma = 17.5$ kN/m³. The shear strength parameters of the soil are $c' = 25$ kN/m² and $\phi' = 26°$. Determine the factor of safety F_s for the trial failure surface.

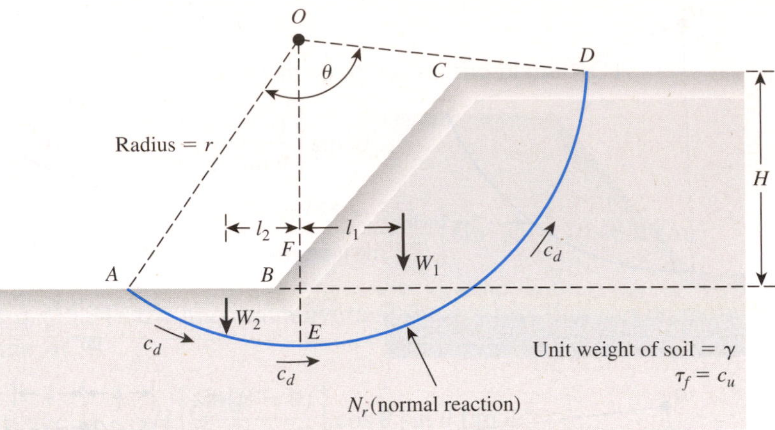

Figure 15.11 Stability analysis of slope in homogeneous saturated clay soil ($\phi = 0$)

arc of a circle that has a radius r. The center of the circle is located at O. Considering a unit length perpendicular to the section of the slope, we can give the weight of the soil above the curve AED as $W = W_1 + W_2$, where

$$W_1 = (\text{Area of } FCDEF)(\gamma)$$

and

$$W_2 = (\text{Area of } ABFEA)(\gamma)$$

Failure of the slope may occur by sliding of the soil mass. The moment of the driving force about O to cause slope instability is

$$M_d = W_1 l_1 - W_2 l_2 \tag{15.43}$$

where l_1 and l_2 are the moment arms.

The resistance to sliding is derived from the cohesion that acts along the potential surface of sliding. If c_d is the cohesion that needs to be developed, the moment of the resisting forces about O is

$$M_R = c_d(\widehat{AED})(1)(r) = c_d r^2 \theta \tag{15.44}$$

For equilibrium, $M_R = M_d$; thus,

$$c_d r^2 \theta = W_1 l_1 - W_2 l_2$$

or

$$c_d = \frac{W_1 l_1 - W_2 l_2}{r^2 \theta} \tag{15.45}$$

The factor of safety against sliding may now be found.

$$F_s = \frac{\tau_f}{c_d} = \frac{c_u}{c_d} \tag{15.46}$$

Note that the potential curve of sliding, *AED*, was chosen arbitrarily. The critical surface is that for which the ratio of c_u to c_d is a minimum. In other words, c_d is maximum. To find the critical surface for sliding, one must make a number of trials for different trial circles. The minimum value of the factor of safety thus obtained is the factor of safety against sliding for the slope, and the corresponding circle is the critical circle.

Stability problems of this type have been solved analytically by Fellenius (1927) and Taylor (1937). For the case of *critical circles*, the developed cohesion can be expressed by the relationship

$$c_d = \gamma H m$$

or

$$\frac{c_d}{\gamma H} = m \tag{15.47}$$

Note that the term m on the right-hand side of the preceding equation is nondimensional and is referred to as the *stability number*. The critical height (i.e., $F_s = 1$) of the slope can be evaluated by substituting $H = H_{cr}$ and $c_d = c_u$ (full mobilization of the undrained shear strength) into the preceding equation. Thus,

$$H_{cr} = \frac{c_u}{\gamma m} \tag{15.48}$$

Values of the stability number, m, for various slope angles, β, are given in Figure 15.12. Terzaghi used the term $\gamma H / c_d$, the reciprocal of m and called it the *stability factor*. Readers should be careful in using Figure 15.12 and note that it is valid for slopes of saturated clay and is applicable to only undrained conditions ($\phi = 0$).

In reference to Figure 15.12, the following must be pointed out:

1. For a slope angle β greater than 53°, the critical circle is always a toe circle. The location of the center of the critical toe circle may be found with the aid of Figure 15.13.
2. For $\beta < 53°$, the critical circle may be a toe, slope, or midpoint circle, depending on the location of the firm base under the slope. This is called the *depth function*, which is defined as

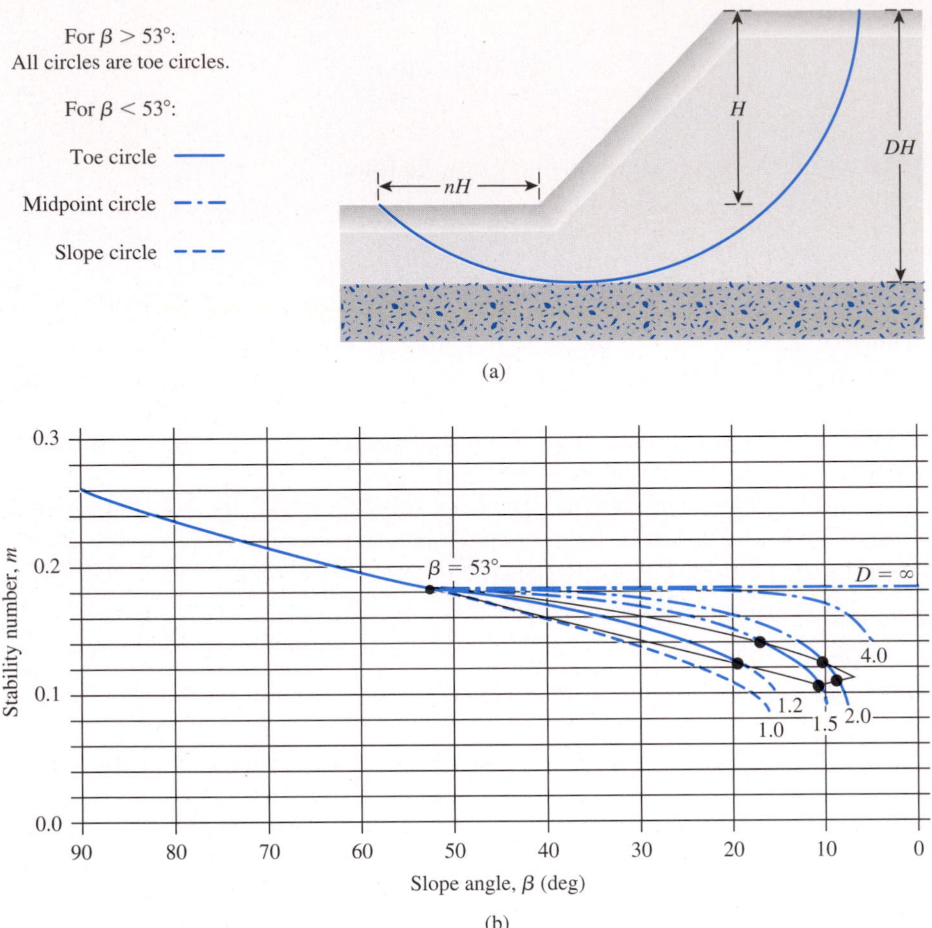

Figure 15.12 (a) Definition of parameters for midpoint circle type of failure; (b) plot of stability number against slope angle *(Adapted from Terzaghi and Peck, 1967. With permission of John Wiley & Sons, Inc.)*

$$D = \frac{\text{Vertical distance from top of slope to firm base}}{\text{Height of slope}} \tag{15.49}$$

3. When the critical circle is a midpoint circle (i.e., the failure surface is tangent to the firm base), its position can be determined with the aid of Figure 15.14.
4. The maximum possible value of the stability number for failure as a midpoint circle is 0.181.

Fellenius (1927) also investigated the case of critical toe circles for slopes with $\beta < 53°$. The location of these can be determined with the use of Figure 15.15 and Table 15.1. Note that these critical toe circles are not necessarily the most critical circles that exist.

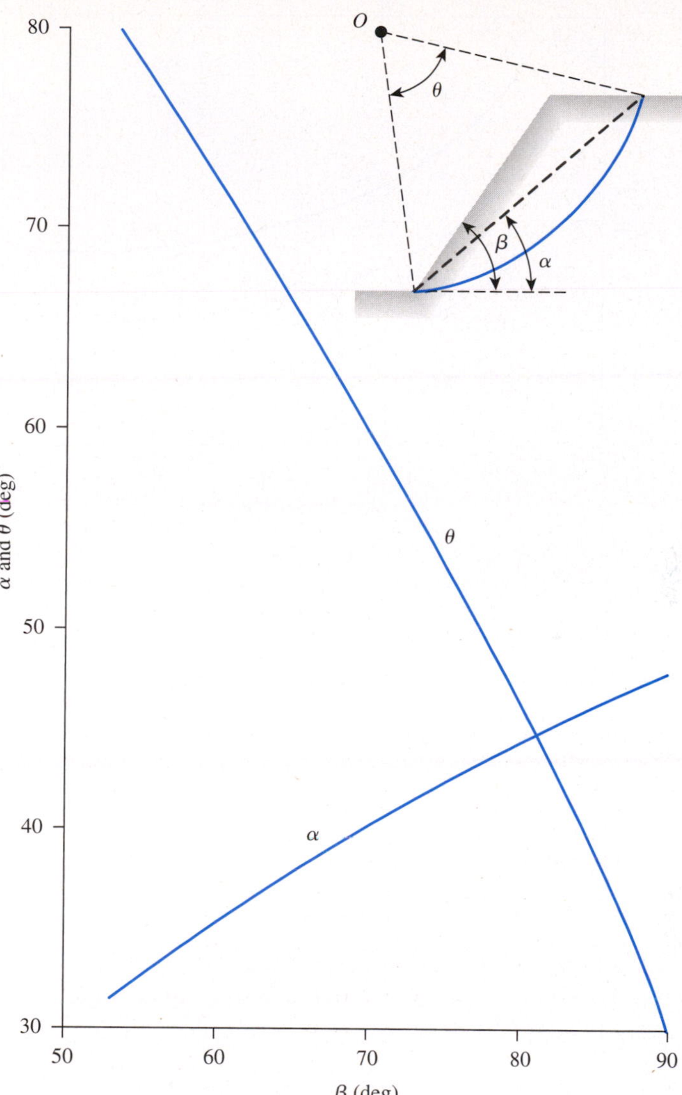

Figure 15.13 Location of the center of critical circles for $\beta > 53°$

Table 15.1 Location of the Center of Critical Toe Circles ($\beta < 53°$)

n'	β (deg)	α_1 (deg)	α_2 (deg)
1.0	45	28	37
1.5	33.68	26	35
2.0	26.57	25	35
3.0	18.43	25	35
5.0	11.32	25	37

Note: For notations of n', β, α_1, and α_2, see Figure 15.15.

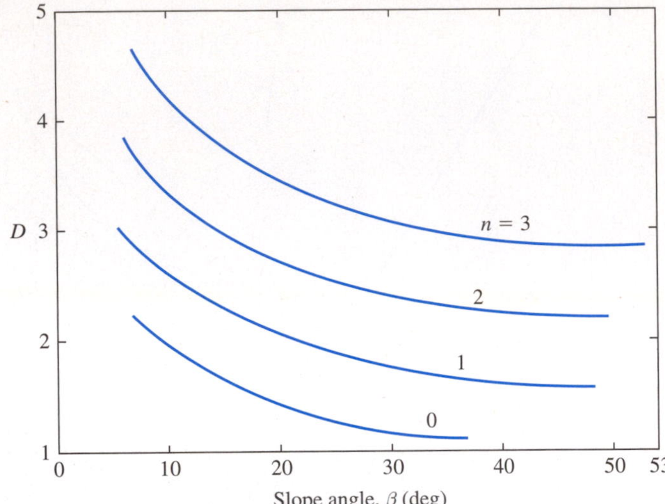

Figure 15.14 Location of midpoint circle (*Based on Fellenius, 1927; and Terzaghi and Peck, 1967*)

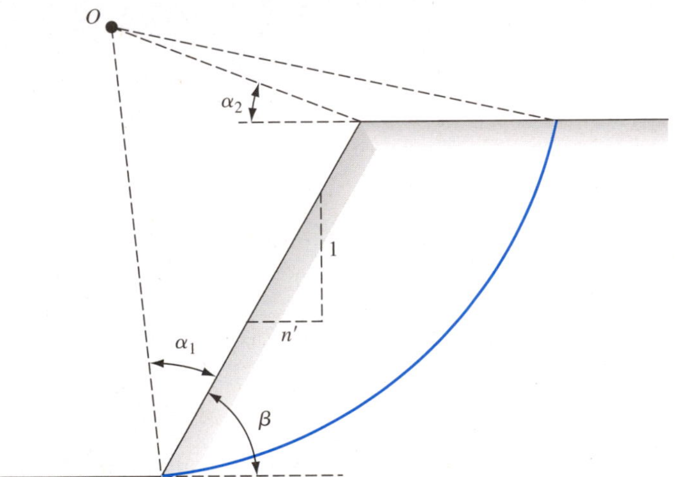

Figure 15.15 Location of the center of critical toe circles for $\beta < 53°$

Example 15.4

A cut slope in saturated clay (Figure 15.16) makes an angle 56° with the horizontal.

a. Determine the maximum depth up to which the cut could be made. Assume that the critical surface for sliding is circularly cylindrical. What will be the nature of the critical circle (i.e., toe, slope, or midpoint)?
b. How deep should the cut be made if a factor of safety of 2 against sliding is required?

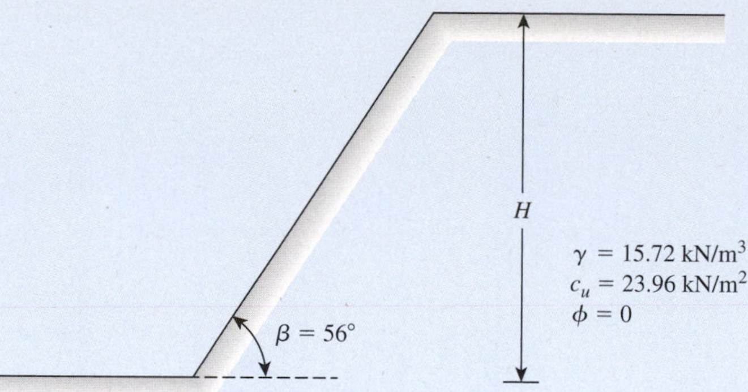

Figure 15.16

Solution

Part a

Since the slope angle $\beta = 56° > 53°$, the critical circle is a **toe circle**. From Figure 15.12 for $\beta = 56°, m = 0.185$. Using Eq. (15.48), we have

$$H_{cr} = \frac{c_u}{\gamma m} = \frac{23.96}{(15.72)(0.185)} = \mathbf{8.24\ m}$$

Part b

The developed cohesion is

$$c_d = \frac{c_u}{F_s} = \frac{23.96}{2} = 11.98\ \text{kN/m}^2$$

From Figure 15.12, for $\beta = 56°, m = 0.185$. Thus, we have

$$H = \frac{c_d}{\gamma m} = \frac{11.98}{(15.72)(0.185)} = \mathbf{4.12\ m}$$

Example 15.5

A cut slope is to be made in a soft saturated clay with its sides rising at an angle of 60° to the horizontal (Figure 15.17).
Given: $c_u = 40\ \text{kN/m}^2$ and $\gamma = 17.5\ \text{kN/m}^3$.

a. Determine the maximum depth up to which the excavation can be carried out.
b. Find the radius, r, of the critical circle when the factor of safety is equal to 1 (Part a).
c. Find the distance $\overline{BC}$.

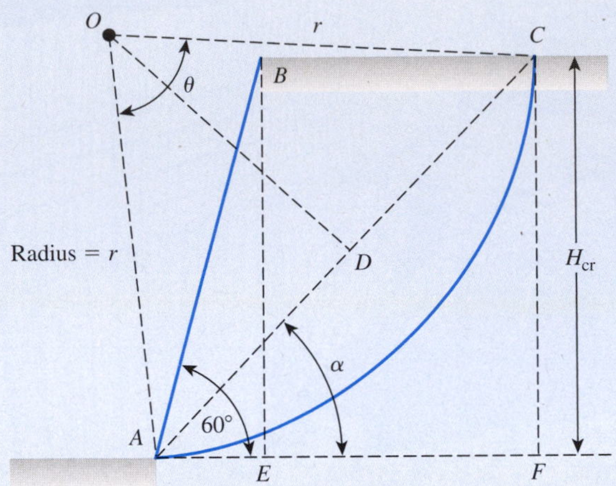

Figure 15.17

Solution

Part a

Since the slope angle $\beta = 60° > 53°$, the critical circle is a toe circle. From Figure 15.12, for $\beta = 60°$, the stability number = 0.195.

$$H_{cr} = \frac{c_u}{\gamma m} = \frac{40}{17.5 \times 0.195} = \mathbf{11.72\ m}$$

Part b

From Figure 15.17,

$$r = \frac{\overline{DC}}{\sin \frac{\theta}{2}}$$

But

$$\overline{DC} = \frac{\overline{AC}}{2} = \frac{\left(\frac{H_{cr}}{\sin \alpha}\right)}{2}$$

so,

$$r = \frac{H_{cr}}{2 \sin \alpha \ \sin \frac{\theta}{2}}$$

From Figure 15.13, for $\beta = 60°$, $\alpha = 35°$ and $\theta = 72.5°$. Substituting these values into the equation for r, we get

$$r = \frac{H_{cr}}{2 \sin \alpha \sin \dfrac{\theta}{2}}$$

$$= \frac{11.72}{2(\sin 35)(\sin 36.25)} = \mathbf{17.28\ m}$$

Part c

$$\overline{BC} = \overline{EF} = \overline{AF} - \overline{AE}$$

$$= H_{cr}(\cot \alpha - \cot 60°)$$

$$= 11.72(\cot 35 - \cot 60) = \mathbf{9.97\ m}$$

Example 15.6

A cut slope was excavated in a saturated clay. The slope made an angle of 40° with the horizontal. Slope failure occurred when the cut reached a depth of 7 m. Previous soil explorations showed that a rock layer was located at a depth of 10.5 m below the ground surface. Assuming an undrained condition and $\gamma_{sat} = 18$ kN/m^3, find the following.

a. Determine the undrained cohesion of the clay (use Figure 15.12).
b. What was the nature of the critical circle?
c. With reference to the toe of the slope, at what distance did the surface of sliding intersect the bottom of the excavation?

Solution

Part a

Referring to Figure 15.12,

$$D = \frac{10.5}{7} = 1.5$$

$$\gamma_{sat} = 18\ \text{kN/m}^3$$

$$H_{cr} = \frac{c_u}{\gamma m}$$

From Figure 15.12, for $\beta = 40°$ and $D = 1.5$, $m = 0.175$. So,

$$c_u = (H_{cr})(\gamma)(m) = (7)(18)(0.175) = \mathbf{22.05\ kN/m^2}$$

Part b
Midpoint circle.

Part c
Again, from Figure 15.14, for $D = 1.5, \beta = 40°; n = 0.9$. So,

$$\text{Distance} = (n)(H_{cr}) = (0.9)(7) = \mathbf{6.3\ m}$$

15.9 Slopes in Clay Soil with $\phi = 0$; and c_u Increasing with Depth

In many cases, the undrained cohesion, c_u, in normally consolidated clay increases with depth, as shown in Figure 15.18, or

$$c_{u(z)} = c_{u(z=0)} + a_0 z \tag{15.50}$$

where $c_{u(z)}$ = undrained shear strength at depth z
$c_{u(z=0)}$ = undrained shear strength at depth $z = 0$
a_0 = slope of the line of the plot of $c_{u(z)}$ vs. z

For such a condition, the critical circle will be a toe circle not a midpoint circle, since the strength increases with depth. Figure 15.19 shows a trial failure circle for this type of case. The moment of the driving force about O can be given as

$$M_d = \frac{\gamma H^3}{12}(1 - 2\cot^2\beta - 3\cot\alpha'\cot\beta + 3\cot\beta\cot\lambda + 3\cot\lambda\cot\alpha') \tag{15.51}$$

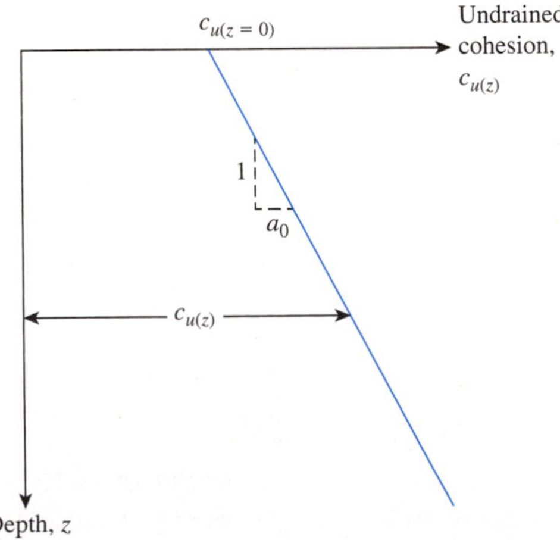

Figure 15.18 Increase of undrained cohesion with depth [Eq. (15.50)]

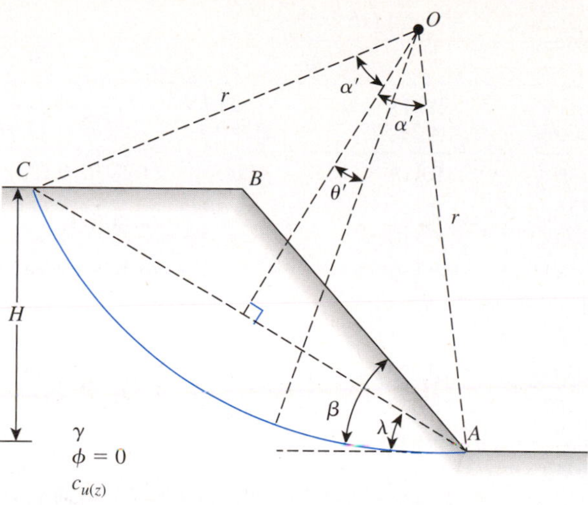

Figure 15.19 Analysis of slope in clay soil ($\phi = 0$ concept) with increasing undrained shear strength

In a similar manner, the moment of the resisting forces about O is

$$M_r = r\int_{-\alpha'}^{+\alpha'} c_{d(z)} r\, d\theta' \tag{15.52}$$

where $c_{d(z)} = c_{d(z=0)} + a_0 z$ (15.53)

The factor of safety against sliding is

$$F_s = \frac{M_r}{M_d} \tag{15.54}$$

Koppula (1984) has solved this problem in a slightly different form. His solution for obtaining the minimum factor of safety can be expressed as

$$m = \left[\frac{c_{u(z=0)}}{\gamma H}\right]\frac{1}{F_s} \tag{15.55}$$

where m = stability number, which is also a function of

$$c_R = \frac{a_0 H}{c_{u(z=0)}} \tag{15.56}$$

Table 15.2 gives the values of m for various values of c_R and β, which are slightly different from those expressed by Koppula (1984).

Table 15.2 Variation of m, c_R, and β [Eqs. (15.55) and (15.56)]

	m					
c_R	1H:1V $\beta = 45°$	1.5H:1V $\beta = 33.69°$	2H:1V $\beta = 26.57°$	3H:1V $\beta = 18.43°$	4H:1V $\beta = 14.04°$	5H:1V $\beta = 11.31°$
0.1	0.158	0.146	0.139	0.130	0.125	0.121
0.2	0.148	0.135	0.127	0.117	0.111	0.105
0.3	0.139	0.126	0.118	0.107	0.0995	0.0937
0.4	0. 131	0.118	0.110	0.0983	0.0907	0.0848
0.5	0.124	0.111	0.103	0.0912	0.0834	0.0775
1.0	0.0984	0.086	0.0778	0.0672	0.0600	0.0546
2.0	0.0697	0.0596	0.0529	0.0443	0.0388	0.0347
3.0	0.0541	0.0457	0.0402	0.0331	0.0288	0.0255
4.0	0.0442	0.0371	0.0325	0.0266	0.0229	0.0202
5.0	0.0374	0.0312	0.0272	0.0222	0.0190	0.0167
10.0	0.0211	0.0 175	0.0151	0.0121	0.0103	0.0090

Based on the analysis of Koppula (1984)

Example 15.7

A cut slope was excavated in saturated clay. The slope was made at an angle of 45° with the horizontal. Given:

- $c_{u(z)} = c_{u(z=0)} + a_o z = 9.58 \text{ kN/m}^2 + (8.65 \text{ kN/m}^3)z$
- $\gamma_{sat} = 19.18 \text{ kN/m}^3$
- Depth of cut $= H = 3.05$ m

Determine the factor of safety, F_s.

Solution

From Eq. (15.56)

$$c_R = \frac{a_0 H}{c_{u(z=0)}} = \frac{(8.65)(3.05)}{9.58} = 2.75$$

From Eq. (15.55)

$$m = \left[\frac{c_{u(z=0)}}{\gamma H}\right]\frac{1}{F_s}$$

Referring to Table 15.2, for $c_R = 2.75$ and $\beta = 45°$, we obtain $m = 0.058$. So,

$$0.058 = \left[\frac{9.58}{(19.18)(3.05)}\right]\frac{1}{F_s}$$

$$F_s = \mathbf{2.82}$$

15.10 Mass Procedure—Slopes in Homogeneous $c' - \phi'$ Soil

A slope in a homogeneous soil is shown in Figure 15.20a. The shear strength of the soil is given by

$$\tau_f = c' + \sigma' \tan \phi'$$

The pore water pressure is assumed to be zero. $\widehat{AC}$ is a trial circular arc that passes through the toe of the slope, and O is the center of the circle. Considering a unit length perpendicular to the section of the slope, we find

$$\text{Weight of soil wedge } ABC = W = (\text{Area of } ABC)(\gamma)$$

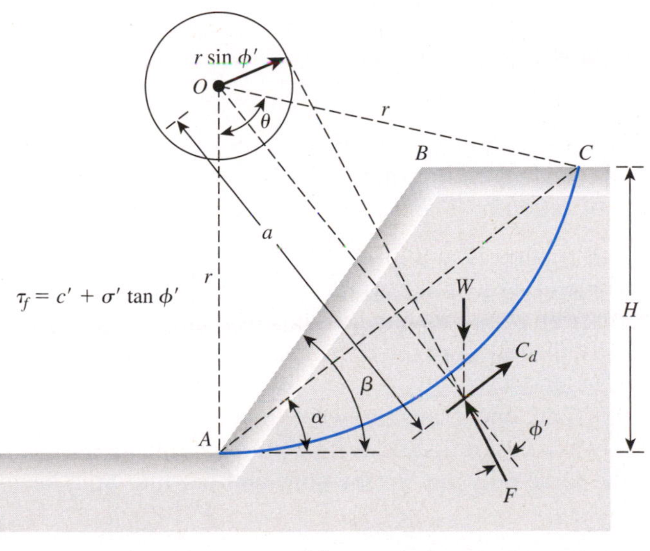

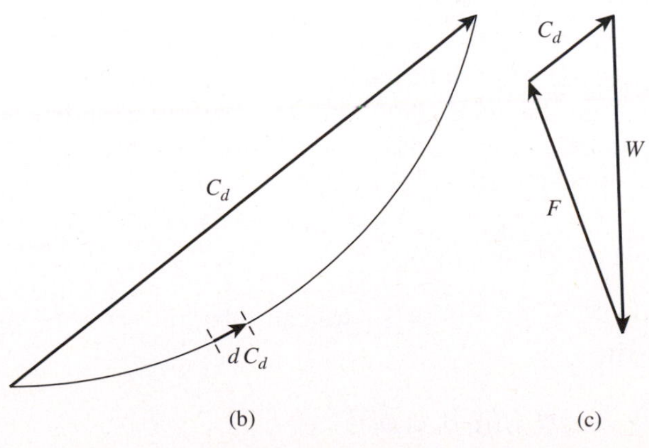

Figure 15.20 Stability analysis of slope in homogeneous $c' - \phi'$ soil

For equilibrium, the following other forces are acting on the wedge:

- C_d—resultant of the cohesive force that is equal to the cohesion per unit area developed times the length of the cord $\overline{AC}$. The magnitude of C_d is given by the following (Figure 15.20b).

$$C_d = c'_d\,(\overline{AC}) \tag{15.57}$$

C_d acts in a direction parallel to the cord $\overline{AC}$ (see Figure 15.20b) and at a distance a from the center of the circle O such that

$$C_d(a) = c'_d(\widehat{AC})r$$

or

$$a = \frac{c'_d(\widehat{AC})r}{C_d} = \frac{\widehat{AC}}{\overline{AC}}r \tag{15.58}$$

- F—the resultant of the normal and frictional forces along the surface of sliding. For equilibrium, the line of action of F will pass through the point of intersection of the line of action of W and C_d.

Now, if we assume that full friction is mobilized ($\phi'_d = \phi'$ or $F_{\phi'} = 1$), the line of action of F will make an angle of ϕ' with a normal to the arc and thus will be a tangent to a circle with its center at O and having a radius of $r \sin \phi'$. This circle is called the *friction circle*. Actually, the radius of the friction circle is a little larger than $r \sin \phi'$.

Because the directions of W, C_d, and F are known and the magnitude of W is known, a force polygon, as shown in Figure 15.20c, can be plotted. The magnitude of C_d can be determined from the force polygon. So the cohesion per unit area developed can be found.

$$c'_d = \frac{C_d}{\overline{AC}}$$

Determination of the magnitude of c'_d described previously is based on a trial surface of sliding. Several trials must be made to obtain the most critical sliding surface, along which the developed cohesion is a maximum. Thus, we can express the maximum cohesion developed along the critical surface as

$$c'_d = \gamma H[f(\alpha, \beta, \theta, \phi')] \tag{15.59}$$

For critical equilibrium—that is, $F_{c'} = F_{\phi'} = F_s = 1$—we can substitute $H = H_{cr}$ and $c'_d = c'$ into Eq. (15.59) and write

$$c' = \gamma H_{cr}[f(\alpha, \beta, \theta, \phi')]$$

or

$$\frac{c'}{\gamma H_{cr}} = f(\alpha, \beta, \theta, \phi') = m \tag{15.60}$$

where m = stability number. The values of m for various values of ϕ' and β are given in Figure 15.21, which is based on Taylor (1937). This can be used to determine the factor of safety, F_s, of the homogeneous slope. The procedure to do the analysis is given as

Step 1. Determine c', ϕ', γ, β and H.

Step 2. Assume several values of ϕ'_d (Note: $\phi'_d \leq \phi'$, such as $\phi'_{d(1)}$, $\phi'_{d(2)}$. . . . (Column 1 of Table 15.3)).

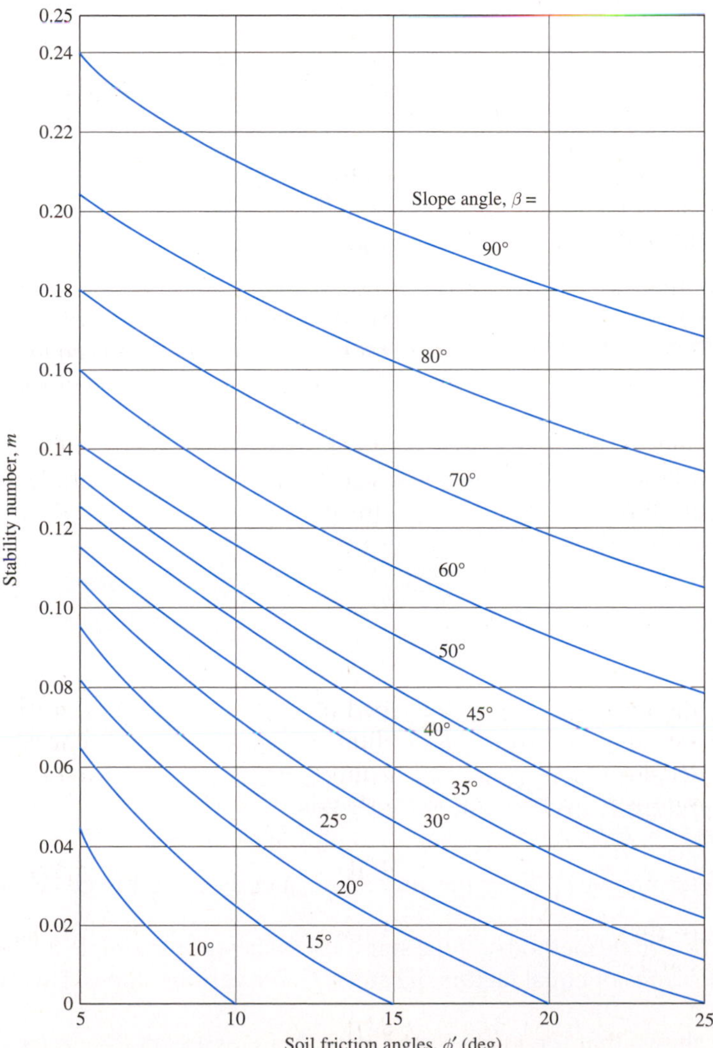

Figure 15.21 Taylor's stability number

Table 15.3 Determination of F_s by Friction Circle Method

ϕ'_d (1)	$F_{\phi'} = \dfrac{\tan \phi'}{\tan \phi'_d}$ (2)	m (3)	c'_d (4)	$F_{c'}$ (5)
$\phi'_{d(1)}$	$\dfrac{\tan \phi'}{\tan \phi'_{d(1)}}$	m_1	$m_1 \gamma H = c'_{d(1)}$	$\dfrac{c'}{c'_{d(1)}} = F_{c'(1)}$
$\phi'_{d(2)}$	$\dfrac{\tan \phi'}{\tan \phi'_{d(2)}}$	m_2	$m_2 \gamma H = c'_{d(2)}$	$\dfrac{c'}{c'_{d(2)}} = F_{c'(2)}$

Step 3. Determine $F_{\phi'}$ for each assumed value of ϕ'_d as (Column 2, Table 15.3)

$$F_{\phi'(1)} = \frac{\tan \phi'}{\tan \phi'_{d(1)}}$$

$$F_{\phi'(2)} = \frac{\tan \phi'}{\tan \phi'_{d(2)}}$$

Step 4. For each assumed value of ϕ'_d and β, determine m (that is, $m_1, m_2, m_3, \ldots$) from Figure 15.21 (Column 3, Table 15.3).

Step 5. Determine the developed cohesion for each value of m as (Column 4, Table 15.3)

$$c'_{d(1)} = m_1 \gamma H$$

$$c'_{d(2)} = m_2 \gamma H$$

Step 6. Calculate $F_{c'}$ for each value of c'_d (Column 5, Table 15.3), or

$$F_{c'(1)} = \frac{c'}{c'_{d(1)}}$$

$$F_{c'(2)} = \frac{c'}{c'_{d(2)}}$$

Step 7. Plot a graph of $F_{\phi'}$ versus the corresponding $F_{c'}$ (Figure 15.22) and determine $F_s = F_{\phi'} = F_{c'}$.

An example of determining F_s using the procedure just described is given in Example 15.8.

Using Taylor's friction circle method of slope stability (as shown in Example 15.8), Singh (1970) provided graphs of equal factors of safety, F_s, for various slopes. This is shown in Figure 15.23.

Calculations have shown that for $\phi > \sim 3°$, the critical circles are all *toe circles*.

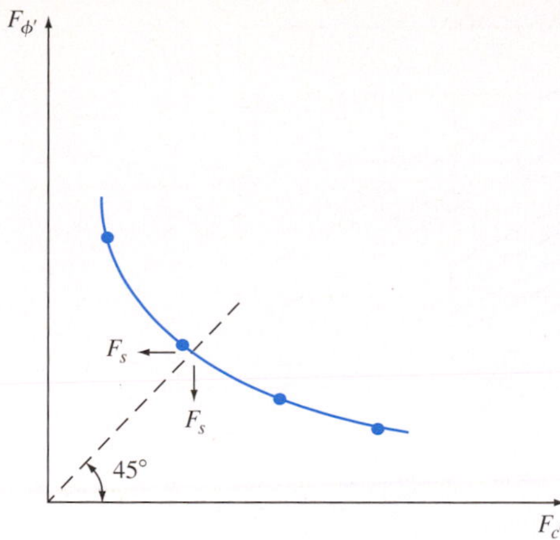

Figure 15.22 Plot of $F_{\phi'}$ versus $F_{c'}$ to determine F_s

Analysis of Michalowski (2002)

Michalowski (2002) made a stability analysis of simple slopes using the kinematic approach of limit analysis applied to a rigid rotational collapse mechanism. The failure surface in soil assumed in this study is an arc of a logarithmic spiral (Figure 15.24). The results of this study are summarized in Figure 15.25, from which F_s can be obtained directly (See Example 15.9). The interpolated values of $F_s/\tan \phi'$ for various values of $\dfrac{c'}{\gamma H \tan \phi'}$ are given in Table 15.4.

Table 15.4 Michalowski's stability numbers for simple slopes

	$\frac{F_s}{\tan \phi'}$					
$\frac{c'}{\gamma H \tan \phi'}$	$\beta = 15°$	$\beta = 30°$	$\beta = 45°$	$\beta = 60°$	$\beta = 75°$	$\beta = 90°$
0	3.85	1.82	1.10	0.64	0.35	0.14
0.05	4.78	2.56	1.74	1.27	0.94	0.65
0.1	5.30	3.09	2.19	1.67	1.29	0.96
0.2	6.43	3.98	2.96	2.37	1.88	1.47
0.3	7.30	4.79	3.66	2.96	2.41	1.91
0.4	8.21	5.46	4.33	3.57	2.91	2.35
0.5	9.15	6.20	4.96	4.13	3.41	2.74
1.0	13.10	9.68	8.07	6.79	5.72	4.71
1.5		13.12	11.08	9.51	8.07	6.71
2.0			14.00	12.16	10.39	8.62
2.5					12.72	10.57
3.0						12.46

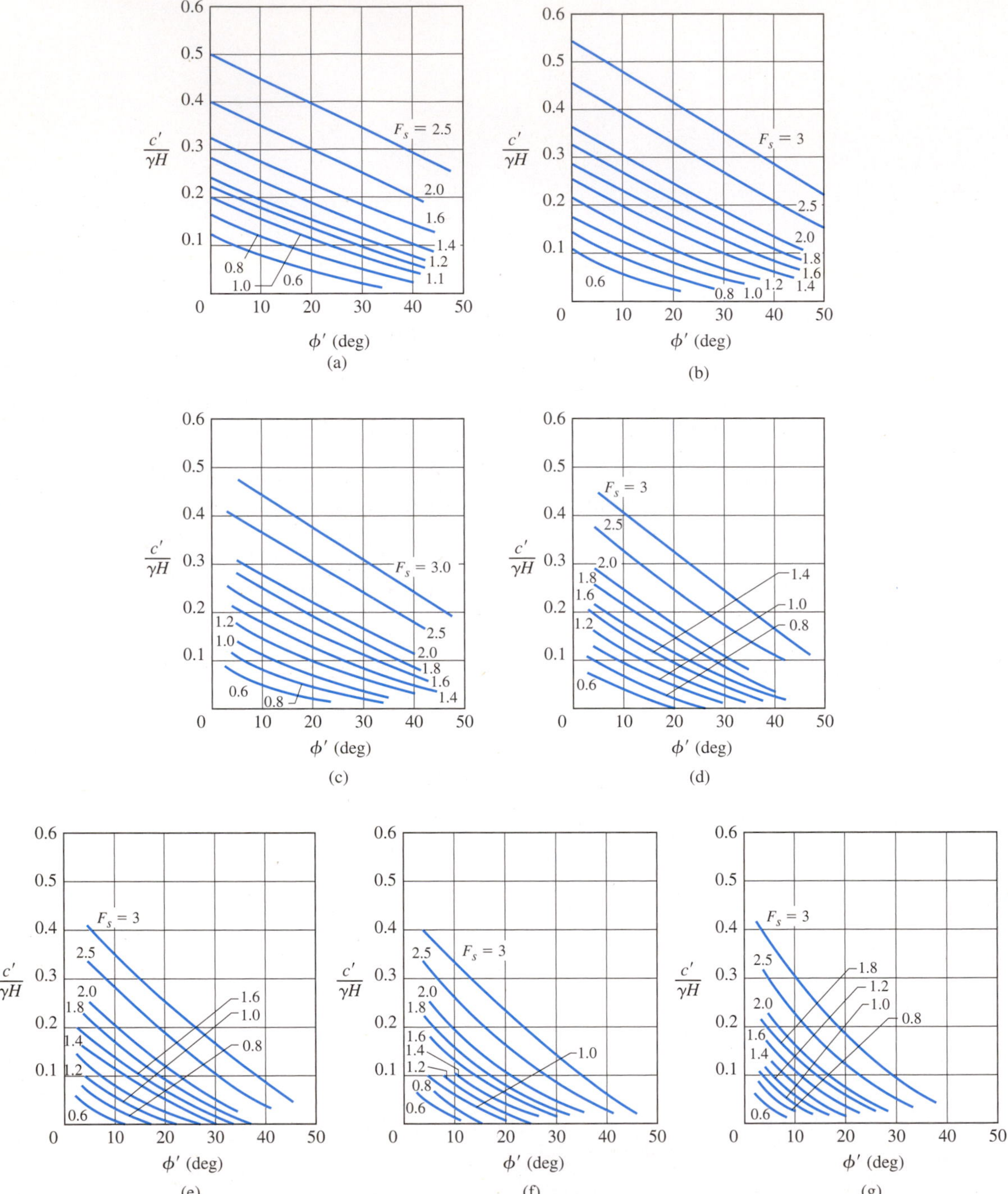

Figure 15.23 Contours of equal factors of safety: (a) slope − 1 vertical to 0.5 horizontal; (b) slope − 1 vertical to 0.75 horizontal; (c) slope − 1 vertical to 1 horizontal; (d) slope − 1 vertical to 1.5 horizontal; (e) slope − 1 vertical to 2 horizontal; (f) slope − 1 vertical to 2.5 horizontal; (g) slope − 1 vertical to 3 horizontal *(After Singh, 1970. With permission from ASCE.)*

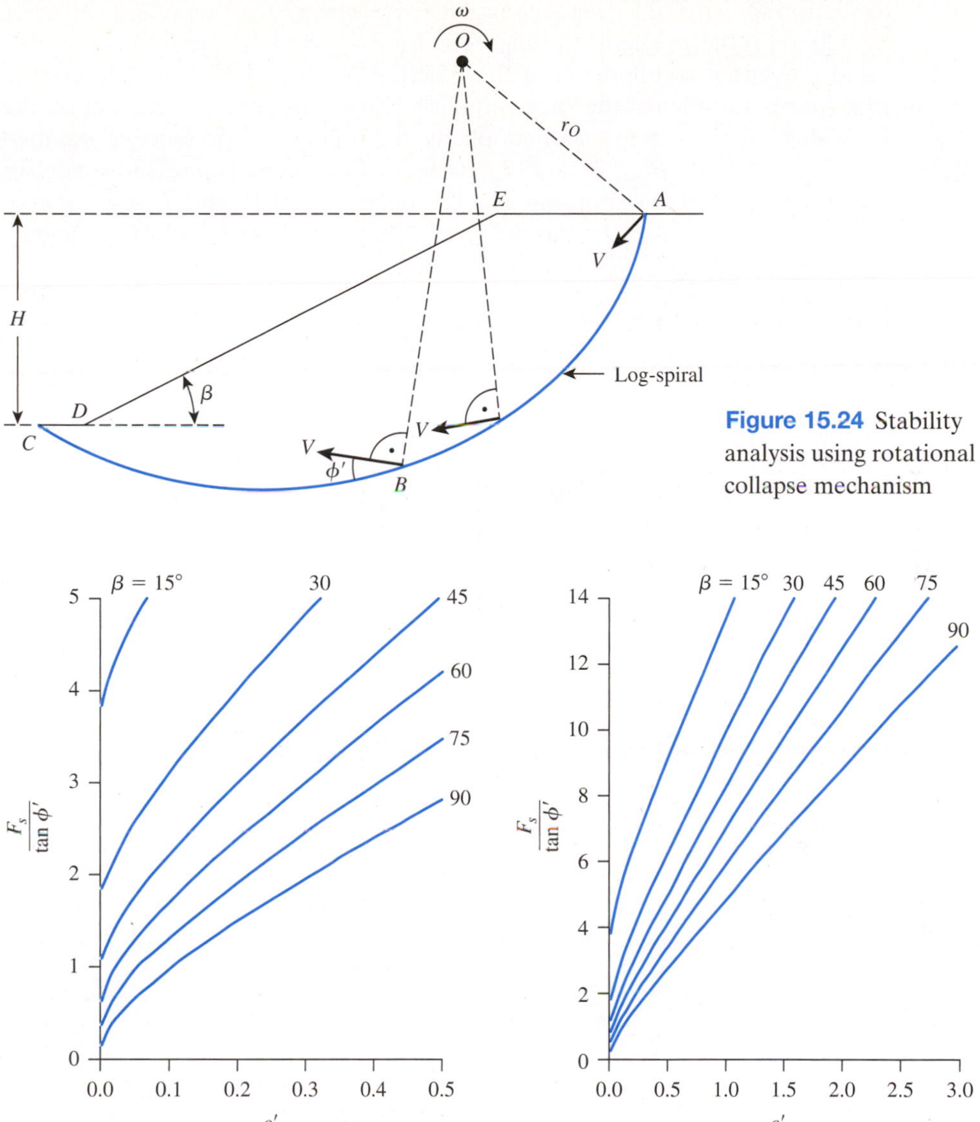

Figure 15.24 Stability analysis using rotational collapse mechanism

Figure 15.25 Michalowski's analysis for stability of simple slopes

15.11 Ordinary Method of Slices

Stability analysis by using the method of slices can be explained with the use of Figure 15.27a, in which *AC* is an arc of a circle representing the trial failure surface. The soil above the trial failure surface is divided into several vertical slices. The width of each slice need not be the same. Considering a unit length perpendicular to the cross section shown, the forces that act on a typical slice (nth slice) are shown in Figure 15.27b. W_n is the weight of the slice. The forces N_r and T_r, respectively, are the normal and tangential components of the reaction R. P_n and P_{n+1} are the normal forces that act on the sides of the slice. Similarly, the shearing forces that act on the sides of the slice are T_n and T_{n+1}. For simplicity, the pore water pressure is assumed to be zero. The forces P_n, P_{n+1}, T_n, and T_{n+1} are difficult to determine. However, we can make an approximate assumption that the resultants of P_n and T_n are equal in magnitude to the resultants of P_{n+1} and T_{n+1} and that their lines of action coincide.

Example 15.8

A slope with $\beta = 45°$ is to be constructed with a soil that has $\phi' = 20°$ and $c' = 24$ kN/m². The unit weight of the compacted soil will be 18.9 kN/m³.

a. Find the critical height of the slope.
b. If the height of the slope is 10 m, determine the factor of safety with respect to strength.

Solution

Part a

We have

$$m = \frac{c'}{\gamma H_{cr}}$$

From Figure 15.21, for $\beta = 45°$ and $\phi' = 20°$, $m = 0.06$. So

$$H_{cr} = \frac{c'}{\gamma m} = \frac{24}{(18.9)(0.06)} = \mathbf{21.1\ m}$$

Part b

If we assume that full friction is mobilized, then, referring to Figure 15.21 (for $\beta = 45°$ and $\phi'_d = \phi' = 20°$), we have

$$m = 0.06 = \frac{c'_d}{\gamma H}$$

or

$$c'_d = (0.06)(18.9)(10) = 11.34 \text{ kN/m}^2$$

Thus,

$$F_{\phi'} = \frac{\tan \phi'}{\tan \phi'_d} = \frac{\tan 20}{\tan 20} = 1$$

and

$$F_{c'} = \frac{c'}{c'_d} = \frac{24}{11.34} = 2.12$$

Since $F_{c'} \neq F_{\phi'}$, this is not the factor of safety with respect to strength.

Now we can make another trial. Let the developed angle of friction, ϕ'_d, be equal to 15°. For $\beta = 45°$ and the friction angle (ϕ'_d) equal to 15°, we find from Figure 15.21.

$$m = 0.083 = \frac{c'_d}{\gamma H}$$

or

$$c'_d = (0.083)(18.9)(10) = 15.69 \text{ kN/m}^2$$

For this trial,

$$F_{\phi'} = \frac{\tan \phi'}{\tan \phi'_d} = \frac{\tan 20}{\tan 15} = 1.36$$

and

$$F_{c'} = \frac{c'}{c'_d} = \frac{24}{15.69} = 1.53$$

Similar calculations of $F_{\phi'}$ and $F_{c'}$ for various assumed values of ϕ'_d are given in the following table.

ϕ'_d	$\tan \phi'_d$	$F_{\phi'}$	m	c'_d (kN/m^2)	$F_{c'}$
20	0.364	1.0	0.06	11.34	2.12
15	0.268	1.36	0.083	15.69	1.53
10	0.176	2.07	0.105	19.85	1.21
5	0.0875	4.16	0.136	25.70	0.93

The values of $F_{\phi'}$ are plotted against their corresponding values of $F_{c'}$ in Figure 15.26, from which we find

$$F_{c'} = F_{\phi'} = F_s = \mathbf{1.42}$$

Note: We could have found the value of F_s from Figure 15.23c. Since $\beta = 45°$, it is a slope of 1V:1H. For this slope

$$\frac{c'}{\gamma H} = \frac{24}{(18.9)(10)} = 0.127$$

From Figure 15.23c, for $c'/\gamma H = 0.127$, the value of $F_s \approx$ **1.4**.

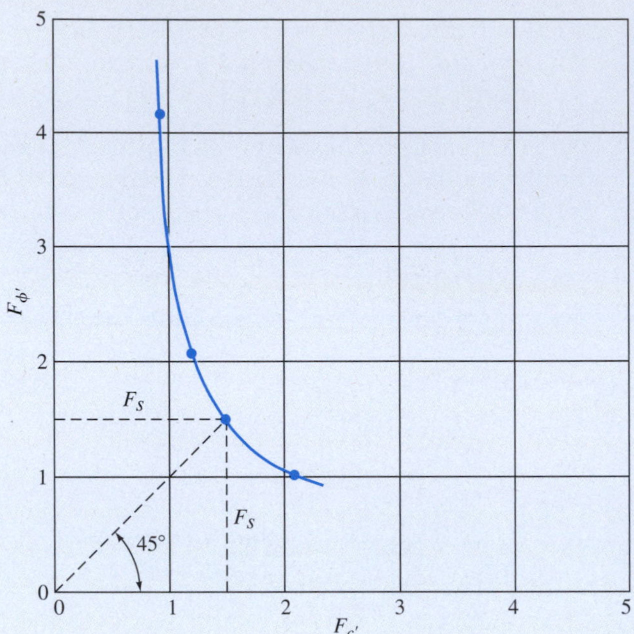

Figure 15.26

Example 15.9

Solve Example 15.8 using Michalowski's solution.

Solution

Part a

For critical height (H_{cr}), $F_s = 1$. Thus,

$$\frac{c'}{\gamma H \tan \phi'} = \frac{24}{(18.9)(H_{cr})(\tan 20)} = \frac{3.49}{H_{cr}}$$

$$\frac{F_s}{\tan \phi'} = \frac{1}{\tan 20} = 2.747$$

$$\beta = 45°$$

From Figure 15.25, for $\beta = 45°$ and $F_s/\tan \phi' = 2.747$, the value of $c'/\gamma H \tan \phi' \approx 0.17$. So

$$\frac{3.49}{H_{cr}} = 0.17; \quad H_{cr} = \mathbf{20.5\ m}$$

Part b

$$\frac{c'}{\gamma H \tan \phi'} = \frac{24}{(18.9)(10)(\tan 20)} = 0.349$$

$$\beta = 45°$$

From Figure 15.25, $F_s/\tan \phi' = 4$.

$$F_s = 4 \tan \phi' = (4)(\tan 20) = \mathbf{1.46}$$

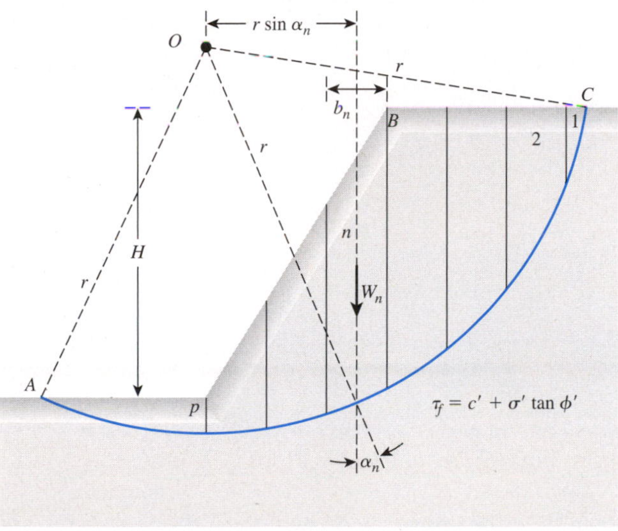

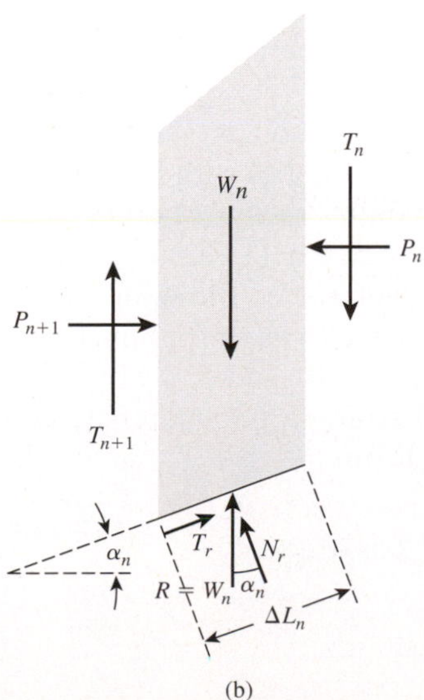

Figure 15.27 Stability analysis by ordinary method of slices: (a) trial failure surface; (b) forces acting on nth slice

For equilibrium consideration,

$$N_r = W_n \cos \alpha_n$$

The resisting shear force can be expressed as

$$T_r = \tau_d(\Delta L_n) = \frac{\tau_f(\Delta L_n)}{F_s} = \frac{1}{F_s}[c' + \sigma' \tan \phi']\Delta L_n \tag{15.61}$$

The normal stress, σ', in Eq. (15.61) is equal to

$$\frac{N_r}{\Delta L_n} = \frac{W_n \cos \alpha_n}{\Delta L_n}$$

For equilibrium of the trial wedge ABC, the moment of the driving force about O equals the moment of the resisting force about O, or

$$\sum_{n=1}^{n=p} W_n r \sin \alpha_n = \sum_{n=1}^{n=p} \frac{1}{F_s}\left(c' + \frac{W_n \cos \alpha_n}{\Delta L_n} \tan \phi'\right)(\Delta L_n)(r)$$

or

$$F_s = \frac{\sum_{n=1}^{n=p}(c'\Delta L_n + W_n \cos \alpha_n \tan \phi')}{\sum_{n=1}^{n=p} W_n \sin \alpha_n} \tag{15.62}$$

[*Note:* ΔL_n in Eq. (15.62) is approximately equal to $(b_n)/(\cos \alpha_n)$, where b_n = the width of the nth slice.]

Note that the value of α_n may be either positive or negative. The value of α_n is positive when the slope of the arc is in the same quadrant as the ground slope. To find the minimum factor of safety—that is, the factor of safety for the critical circle—one must make several trials by changing the center of the trial circle. This method generally is referred to as the *ordinary method of slices*.

For convenience, a slope in a homogeneous soil is shown in Figure 15.27. However, the method of slices can be extended to slopes with layered soil, as shown in Figure 15.28. The general procedure of stability analysis is the same. However, some minor points should be kept in mind. When Eq. (15.62) is used for the factor of safety calculation, the values of ϕ' and c' will not be the same for all slices. For example, for slice No. 3 (see Figure 15.28), we have to use a friction angle of $\phi' = \phi'_3$ and cohesion $c' = c'_3$; similarly, for slice No. 2, $\phi' = \phi'_2$ and $c' = c'_2$.

It is of interest to note that if total shear strength parameters (that is, $\tau_f = c + \tan \phi$) were used, Eq. (15.62) would take the form

$$F_s = \frac{\sum_{n=1}^{n=p}(c\Delta L_n + W_n \cos \alpha_n \tan \phi)}{\sum_{n=1}^{n=p} W_n \sin \alpha_n} \tag{15.63}$$

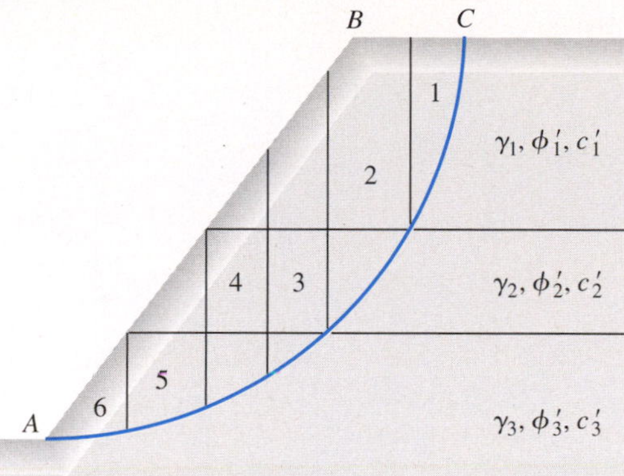

Figure 15.28 Stability analysis, by ordinary method of slices, for slope in layered soils

Example 15.10

Figure 15.29 shows a 10-m high slope in saturated clay. Given: the saturated unit weight of soil $\gamma = 19$ kN/m^3 and the undrained shear strength $c_u = 70$ kN/m^2. Determine the factor of safety F_s using the method of slices for the trial circle shown.

Solution

The trial wedge has been divided into nine slices. The following table gives the calculations for the driving moment M_d about O [also see Eq. (15.43)].

Slice no.	Weight (kN/m)	Moment arm (m)	M_d (kN-m/m)
1	$\frac{1}{2}(2.84)(0.6)(19) = 16.188$	$(6)(2.09) + (0.2) = 12.74$	**206.24**
2	$\frac{1}{2}(2.84 + 6.12)(2.09)(19) = 177.9$	$(5)(2.09) + \frac{2.09}{2} = 11.495$	**2044.96**
3	$\frac{1}{2}(6.12 + 8.21)(2.09)(19) = 284.52$	$(4)(2.09) + \frac{2.09}{2} = 9.405$	**2675.91**
4	$\frac{1}{2}(8.21 + 9.55)(2.09)(19) = 352.62$	$(3)(2.09) + \frac{2.09}{2} = 7.315$	**2579.41**
5	$\frac{1}{2}(9.55 + 7.16)(2.09)(19) = 331.78$	$(2)(2.09) + \frac{2.09}{2} = 5.225$	**1733.55**
6	$\frac{1}{2}(7.16 + 4.33)(2.09)(19) = 228.13$	$2.09 + \frac{2.09}{2} = 1.045$	**715.19**
7	$\frac{1}{2}(4.33 + 1.19)(2.09)(19) = 109.6$	$\frac{2.09}{2} = 1.045$	**114.53**
8	$\frac{1}{2}(1.19 + 0.9)(2.69)(19) = 53.41$	$-\frac{2.69}{2} = -1.345$	**−71.84**
9	$\frac{1}{2}(0.9 + 0)(2.69)(19) = 23.0$	$-\left(2.69 + \frac{2.69}{2}\right) = -4.035$	**−92.81**
			Σ9905.14 kN-m/m

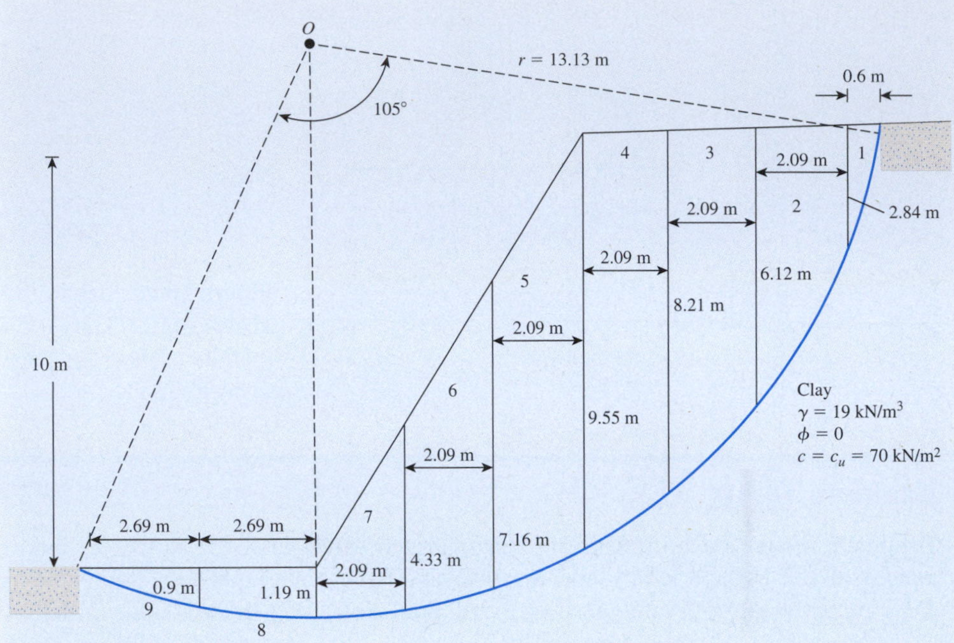

Figure 15.29

From Eq. (15.44),

$$\text{Resisting moment, } M_R = c_u r^2 \theta = (70)(13.13)^2 \left(\frac{105}{180}\right) \times \pi = 22{,}115.4 \text{ kN-m/m}$$

So,

$$F_s = \frac{M_R}{M_d} = \frac{22{,}115.4}{9{,}905.14} = \mathbf{2.23}$$

Example 15.11

Figure 15.30 shows a slope which has similar dimensions as in Figure 15.29 (Example 15.10). For the soil, given: $\gamma = 19$ kN/m³, $\phi' = 20°$, and $c' = 20$ kN/m². Determine F_s using the ordinary method of slices.

Solution

Since the magnitude of γ and the dimension slices are the same in Figures 15.29 and 15.30, the weight W_n for each slice will be the same as in Example 15.10. Now the following table can be prepared.

Slice no. (1)	W_n (kN/m) (2)	α_n (deg) (3)	$\sin \alpha_n$ (4)	$\cos \alpha_n$ (5)	ΔL_n (m) (6)	$W_n \sin \alpha_n$ (kN/m) (7)	$W_n \cos \alpha_n$ (kN/m) (8)
1	16.188	72	0.951	0.309	1.942	15.395	5.00
2	177.9	59	0.788	0.515	4.058	140.185	91.62
3	284.52	46	0.719	0.695	3.007	204.57	197.74
4	352.62	32	0.530	0.848	2.465	186.89	299.02
5	331.78	22	0.375	0.927	2.255	124.42	307.56
6	228.13	13	0.225	0.974	2.146	51.33	222.2
7	109.6	7	0.122	0.993	2.105	13.37	108.83
8	53.41	−6	−0.105	0.995	2.704	−5.61	53.14
9	23.0	−16	−0.276	0.961	2.799	−6.35	22.10
					$\Sigma \approx 23.48$ m	$\Sigma \approx 724.2$ kN/m	$\Sigma 1307.21$ kN/m

Note: $\Delta L_n = b_n/\cos\alpha_n$

$$F_s = \frac{(\Sigma \text{Col.6})(c') + (\Sigma \text{Col.8})(\tan \phi')}{(\Sigma \text{Col.7})}$$

$$= \frac{(23.48)(20) + (1307.21)(\tan 20)}{724.2}$$

$$= \mathbf{1.305}$$

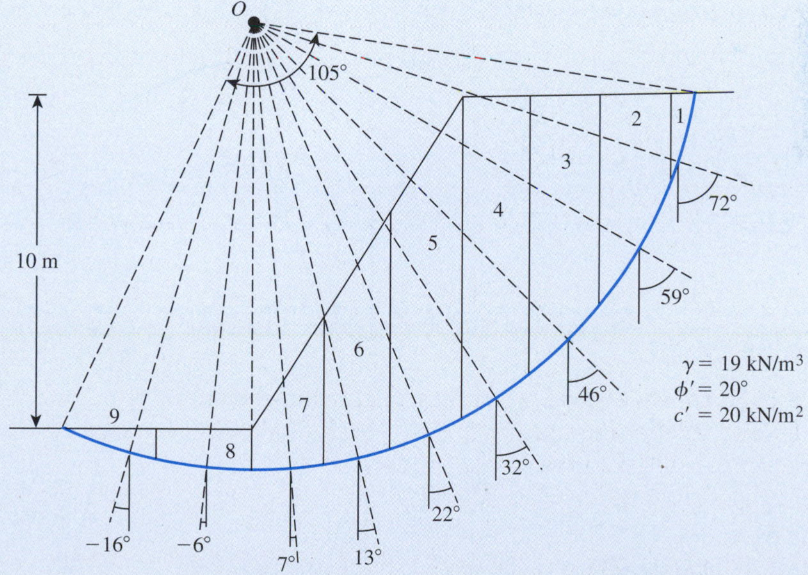

Figure 15.30 (*Note*: the width and height of each slice is same as in Figure 15.29)

15.12 Bishop's Simplified Method of Slices

In 1955, Bishop proposed a more refined solution to the ordinary method of slices. In this method, the effect of forces on the sides of each slice are accounted for to some degree. We can study this method by referring to the slope analysis presented in Figure 15.27. The forces that act on the nth slice shown in Figure 15.27b have been redrawn in Figure 15.31a. Now, let $P_n - P_{n+1} = \Delta P$ and $T_n - T_{n+1} = \Delta T$. Also, we can write

$$T_r = N_r(\tan \phi'_d) + c'_d \Delta L_n = N_r\left(\frac{\tan \phi'}{F_s}\right) + \frac{c' \Delta L_n}{F_s} \tag{15.64}$$

Figure 15.31b shows the force polygon for equilibrium of the nth slice. Summing the forces in the vertical direction gives

$$W_n + \Delta T = N_r \cos \alpha_n + \left[\frac{N_r \tan \phi'}{F_s} + \frac{c' \Delta L_n}{F_s}\right] \sin \alpha_n$$

or

$$N_r = \frac{W_n + \Delta T - \dfrac{c' \Delta L_n}{F_s} \sin \alpha_n}{\cos \alpha_n + \dfrac{\tan \phi' \sin \alpha_n}{F_s}} \tag{15.65}$$

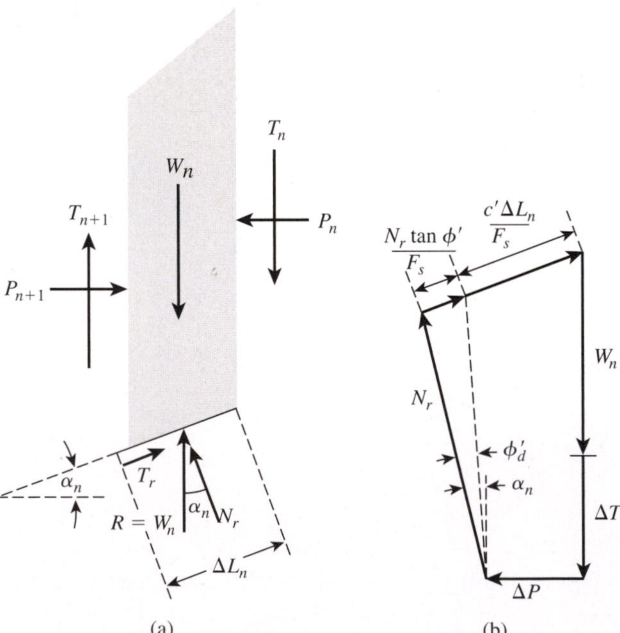

Figure 15.31 Bishop's simplified method of slices: (a) forces acting on the nth slice; (b) force polygon for equilibrium

For equilibrium of the wedge ABC (Figure 15.27a), taking the moment about O gives

$$\sum_{n=1}^{n=p} W_n r \sin \alpha_n = \sum_{n=1}^{n=p} T_r r \tag{15.66}$$

where

$$T_r = \frac{1}{F_s}(c' + \sigma' \tan \phi')\Delta L_n$$
$$= \frac{1}{F_s}(c'\Delta L_n + N_r \tan \phi') \tag{15.67}$$

Substitution of Eqs. (15.65) and (15.67) into Eq. (15.66) gives

$$F_s = \frac{\sum_{n=1}^{n=p} (c'b_n + W_n \tan \phi' + \Delta T \tan \phi')\frac{1}{m_{\alpha(n)}}}{\sum_{n=1}^{n=p} W_n \sin \alpha_n} \tag{15.68}$$

where

$$m_{\alpha(n)} = \cos \alpha_n + \frac{\tan \phi' \sin \alpha_n}{F_s} \tag{15.69}$$

Figure 15.32 shows the variation of $m_{\alpha(n)}$ with α_n and $\tan \phi'/F_s$.

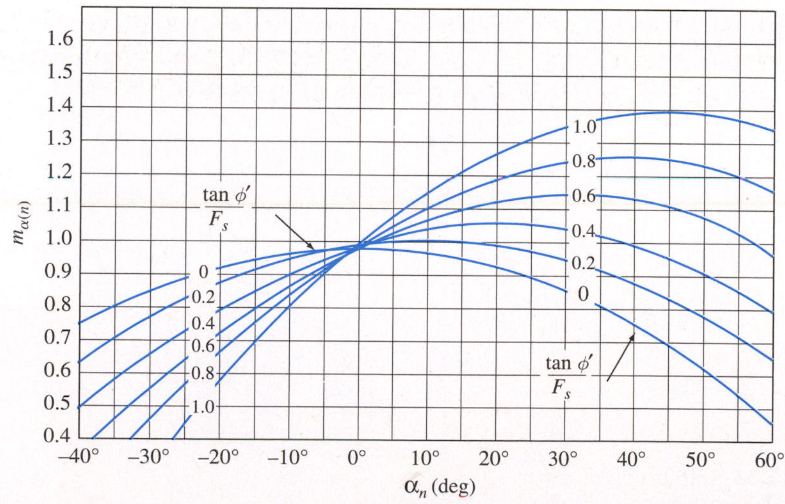

Figure 15.32 Variation of $m_{\alpha(n)}$ with α_n and $\tan \phi'/F_s$ [Eq. (15.69)]

For simplicity, if we let $\Delta T = 0$, Eq. (15.68) becomes

$$F_s = \frac{\sum_{n=1}^{n=p} (c'b_n + W_n \tan \phi')\frac{1}{m_{\alpha(n)}}}{\sum_{n=1}^{n=p} W_n \sin \alpha_n} \tag{15.70}$$

Note that the term F_s is present on both sides of Eq. (15.70). Hence, we must adopt a trial-and-error procedure to find the value of F_s. As in the method of ordinary slices, a number of failure surfaces must be investigated so that we can find the critical surface that provides the minimum factor of safety.

Bishop's simplified method is probably the most widely used. When incorporated into computer programs, it yields satisfactory results in most cases. The ordinary method of slices is presented in this chapter as a learning tool only. It is used rarely now because it is too conservative.

15.13 Stability Analysis by Method of Slices for Steady-State Seepage

The fundamentals of the ordinary method of slices and Bishop's simplified method of slices were presented in Sections 15.11 and 15.12, respectively, and we assumed the pore water pressure to be zero. However, for steady-state seepage through slopes, as is the situation in many practical cases, the pore water pressure must be considered when effective shear strength parameters are used. So we need to modify Eqs. (15.62) and (15.70) slightly.

Figure 15.33 shows a slope through which there is steady-state seepage. For the nth slice, the average pore water pressure at the bottom of the slice is equal to $u_n = h_n\gamma_w$. The total force caused by the pore water pressure at the bottom of the nth slice is equal to $u_n\Delta L_n$.

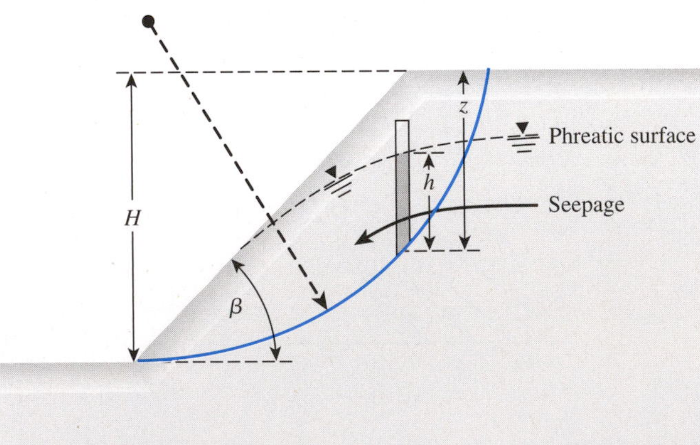

Figure 15.33 Stability analysis of slope with steady-state seepage

Thus, Eq. (15.62) for the ordinary method of slices will be modified to read as follows.

$$F_s = \frac{\sum_{n=1}^{n=p} c'\Delta L_n + [(W_n \cos\alpha_n - u_n \Delta L_n)] \tan\phi'}{\sum_{n=1}^{n=p} W_n \sin\alpha_n} \tag{15.71}$$

Similarly, Eq. (15.70) for Bishop's simplified method of slices will be modified to the form

$$F_s = \frac{\sum_{n=1}^{n=p} [c'b_n + (W_n - u_n b_n)\tan\phi'] \frac{1}{m_{(\alpha)n}}}{\sum_{n=1}^{n=p} W_n \sin\alpha_n} \tag{15.72}$$

Note that W_n in Eqs. (15.71) and (15.72) is the *total weight* of the slice.

Example 15.12

Consider the slope given in Example 15.11. However, there is steady-state seepage. The phreatic line is shown in Figure 15.34. Other parameters remaining the same, assume $\gamma_{sat} = 20.5$ kN/m³. Determine F_s using Eq. (15.71).

Solution

Following are the calculations for W_n and u_n.

Slice No. 1 $W_n = 16.188$ kN/m (same as in Example 15.11)

$u_n = 0$

Slice No. 2 $W_n = 177.9$ kN/m (same as in Example 15.11)

$u_n = 0$

Slice No. 3

$$W_n = \frac{1}{2}(6.12 + 6.27)(2.09)(19) + \frac{1}{2}(0 + 1.94)(2.09)(20.5)$$
$$= 287.56 \text{ kN/m}$$
$$u_n = \frac{1}{2}(0 + 1.94)(9.81) = 9.52 \text{ kN/m}^2$$

Slice No. 4

$$W_n = \frac{1}{2}(6.27 + 6.87)(2.09)(19) + \frac{1}{2}(1.94 + 2.68)(2.09)(20.5)$$
$$= 359.86 \text{ kN/m}$$
$$u_n = \frac{1}{2}(1.94 + 2.68)(9.81) = 22.66 \text{ kN/m}^2$$

Slice No. 5 $W_n = \frac{1}{2}(6.87 + 4.48)(2.09)(19) + \frac{1}{2}(2.68 + 2.68)(2.09)(20.5)$

$= 340.17$ kN/m

$u_n = \frac{1}{2}(2.68 + 2.68)(9.81) = 26.29 \text{ kN/m}^2$

Slice No. 6 $W_n = \frac{1}{2}(4.48 + 2.68)(2.09)(19) + \frac{1}{2}(2.68 + 1.65)(2.09)(20.5)$

$= 234.92$ kN/m

$u_n = \frac{1}{2}(2.68 + 1.65)(9.81) = 21.24 \text{ kN/m}^2$

Slice No. 7 $W_n = \frac{1}{2}(2.68 + 1.19)(2.09)(19) + \frac{1}{2}(1.65 + 0)(2.09)(20.5)$

$= 112.19$ kN/m

$u_n = \frac{1}{2}(1.65 + 0)(9.81) = 8.09 \text{ kN/m}^2$

Slice No. 8 $W_n = 53.41$ kN/m (same as in Example 15.11)

$u_n = 0$

Slice No. 9 $W_n = 23$ kN/m (same as in Example 15.11)

$u_n = 0$

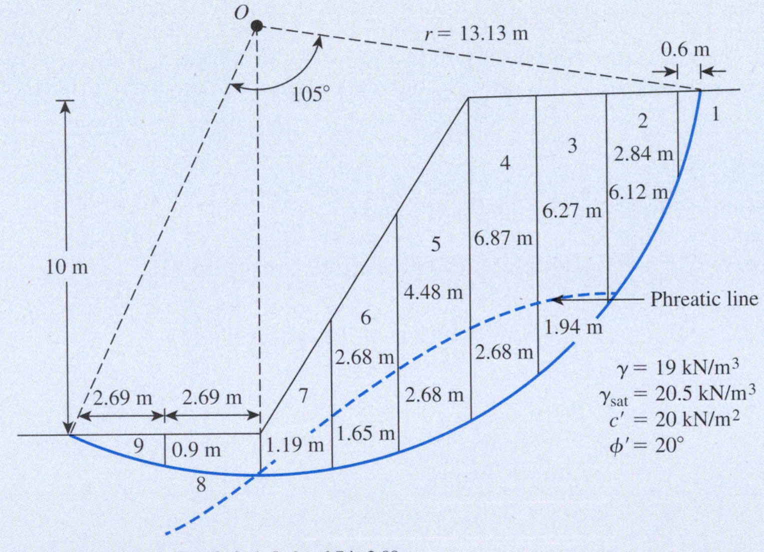

Figure 15.34 (*Note*: α_n value for each slice is same as in Figure 15.30)

Now the following table can be prepared.

Slice no. (1)	W_n (kN/m) (2)	$\sin\alpha_n$* (3)	$\cos\alpha_n$* (4)	u_n (kN/m²) (5)	ΔL_n* (m) (6)	$u_n\Delta L_n$ (kN/m) (7)	$W_n \cos\alpha_n$ (kN/m) (8)	$W_n \sin\alpha_n$ (kN/m) (9)	$W_n \cos\alpha_n - u_n\Delta L_n$ (kN/m) (10)
1	16.188	0.951	0.309	0	1.942	0	5.00	15.4	5.00
2	177.9	0.788	0.515	0	4.058	0	91.62	140.19	91.62
3	287.56	0.719	0.695	9.52	3.007	28.63	199.85	206.46	171.22
4	359.86	0.530	0.848	22.66	2.465	55.86	305.16	190.73	249.3
5	340.17	0.375	0.927	26.29	2.255	59.28	315.14	127.56	256.06
6	234.92	0.225	0.974	21.24	2.146	45.58	228.81	52.86	183.23
7	112.19	0.122	0.993	8.09	2.105	17.03	111.40	13.69	94.37
8	53.41	−0.105	0.995	0	2.704	0	53.14	−5.61	53.14
9	23.0	−0.276	0.961	0	2.799	0	22.10	−6.35	22.10
					Σ23.48 m			Σ735.23 kN/m	Σ1126.04 kN/m

*See table in Example 15.11

$$F_s = \frac{\sum c'\Delta L_n + \sum (W_n \cos\alpha_n - u_n\Delta L_n)\tan\phi'}{\sum W_n \sin\alpha_n}$$

$$= \frac{(20)(23.48) + (1126.04)(\tan 20)}{735.23} = 1.196 \approx \mathbf{1.2}$$

15.14 Solutions for Steady-State Seepage

Bishop and Morgenstern solution

Using Eq. (15.72), Bishop and Morgenstern (1960) developed tables for the calculation of F_s for simple slopes. The principles of these developments can be explained as follows. In Eq. (15.72),

$$W_n = \text{total weight of the } n\text{th slice} = \gamma b_n z_n \tag{15.73}$$

where z_n = average height of the nth slice. Also in Eq. (15.72),

$$u_n = h_n \gamma_w$$

So, we can let

$$r_{u(n)} = \frac{u_n}{\gamma z_n} = \frac{h_n \gamma_w}{\gamma z_n} \tag{15.74}$$

Note that $r_{u(n)}$ is a nondimensional quantity. Substituting Eqs. (15.73) and (15.74) into Eq. (15.72) and simplifying, we obtain

$$F_s = \left[\frac{1}{\sum_{n=1}^{n=p} \frac{b_n}{H}\frac{z_n}{H}\sin\alpha_n}\right] \times \sum_{n=1}^{n=p}\left\{\frac{\frac{c'}{\gamma H}\frac{b_n}{H} + \frac{b_n}{H}\frac{z_n}{H}[1 - r_{u(n)}]\tan\phi'}{m_{\alpha(n)}}\right\} \quad (15.75)$$

For a steady-state seepage condition, a weighted average value of $r_{u(n)}$ can be taken, which is a constant. Let the weighted averaged value of $r_{u(n)}$ be r_u. For most practical cases, the value of r_u may range up to 0.5. Thus,

$$F_s = \left[\frac{1}{\sum_{n=1}^{n=p} \frac{b_n}{H}\frac{z_n}{H}\sin\alpha_n}\right] \times \sum_{n=1}^{n=p}\left\{\frac{\left[\frac{c'}{\gamma H}\frac{b_n}{H} + \frac{b_n}{H}\frac{z_n}{H}(1 - r_u)\tan\phi'\right]}{m_{\alpha(n)}}\right\} \quad (15.76)$$

The factor of safety based on the preceding equation can be solved and expressed in the form

$$F_s = m' - n'r_u \quad (15.77)$$

where m' and n' = stability coefficients. Table 15.5 gives the values of m' and n' for various combinations of $c'/\gamma H, D, \phi'$, and β.

To determine F_s from Table 15.5, we must use the following step-by-step procedure:

Step 1. Obtain ϕ', β, and $c'/\gamma H$.
Step 2. Obtain r_u (weighted average value).
Step 3. From Table 15.5, obtain the values of m' and n' for D = 1, 1.25, and 1.5 (for the required parameters ϕ', β, r_u, and $c'/\gamma H$).
Step 4. Determine F_s, using the values of m' and n' for each value of D.
Step 5. The required value of F_s is the smallest one obtained in Step 4.

Spencer's solution

Bishop's simplified method of slices described in Sections 15.12, 15.13 and 15.14 satisfies the equations of equilibrium with respect to the moment but not with respect to the forces. Spencer (1967) has provided a method to determine the factor

Table 15.5 Values of m' and n' [Eq. (15.77)]

a. Stability coefficients m' and n' for $c'/\gamma H = 0$

	Stability coefficients for earth slopes							
	Slope 2:1		Slope 3:1		Slope 4:1		Slope 5:1	
ϕ'	m'	n'	m'	n'	m'	n'	m'	n'
10.0	0.353	0.441	0.529	0.588	0.705	0.749	0.882	0.917
12.5	0.443	0.554	0.665	0.739	0.887	0.943	1.109	1.153
15.0	0.536	0.670	0.804	0.893	1.072	1.139	1.340	1.393
17.5	0.631	0.789	0.946	1.051	1.261	1.340	1.577	1.639
20.0	0.728	0.910	1.092	1.213	1.456	1.547	1.820	1.892
22.5	0.828	1.035	1.243	1.381	1.657	1.761	2.071	2.153
25.0	0.933	1.166	1.399	1.554	1.865	1.982	2.332	2.424
27.5	1.041	1.301	1.562	1.736	2.082	2.213	2.603	2.706
30.0	1.155	1.444	1.732	1.924	2.309	2.454	2.887	3.001
32.5	1.274	1.593	1.911	2.123	2.548	2.708	3.185	3.311
35.0	1.400	1.750	2.101	2.334	2.801	2.977	3.501	3.639
37.5	1.535	1.919	2.302	2.558	3.069	3.261	3.837	3.989
40.0	1.678	2.098	2.517	2.797	3.356	3.566	4.196	4.362

b. Stability coefficients m' and n' for $c'/\gamma H = 0.025$ and $D = 1.00$

	Stability coefficients for earth slopes							
	Slope 2:1		Slope 3:1		Slope 4:1		Slope 5:1	
ϕ'	m'	n'	m'	n'	m'	n'	m'	n'
10.0	0.678	0.534	0.906	0.683	1.130	0.846	1.367	1.031
12.5	0.790	0.655	1.066	0.849	1.337	1.061	1.620	1.282
15.0	0.901	0.776	1.224	1.014	1.544	1.273	1.868	1.534
17.5	1.012	0.898	1.380	1.179	1.751	1.485	2.121	1.789
20.0	1.124	1.022	1.542	1.347	1.962	1.698	2.380	2.050
22.5	1.239	1.150	1.705	1.518	2.177	1.916	2.646	2.317
25.0	1.356	1.282	1.875	1.696	2.400	2.141	2.921	2.596
27.5	1.478	1.421	2.050	1.882	2.631	2.375	3.207	2.886
30.0	1.606	1.567	2.235	2.078	2.873	2.622	3.508	3.191
32.5	1.739	1.721	2.431	2.285	3.127	2.883	3.823	3.511
35.0	1.880	1.885	2.635	2.505	3.396	3.160	4.156	3.849
37.5	2.030	2.060	2.855	2.741	3.681	3.458	4.510	4.209
40.0	2.190	2.247	3.090	2.993	3.984	3.778	4.885	4.592

(continued)

Table 15.5 *(continued)*

c. Stability coefficients m' and n' for $c'/\gamma H = 0.025$ and $D = 1.25$

	Stability coefficients for earth slopes							
	Slope 2:1		Slope 3:1		Slope 4:1		Slope 5:1	
ϕ'	m'	n'	m'	n'	m'	n'	m'	n'
10.0	0.737	0.614	0.901	0.726	1.085	0.867	1.285	1.014
12.5	0.878	0.759	1.076	0.908	1.299	1.098	1.543	1.278
15.0	1.019	0.907	1.253	1.093	1.515	1.311	1.803	1.545
17.5	1.162	1.059	1.433	1.282	1.736	1.541	2.065	1.814
20.0	1.309	1.216	1.618	1.478	1.961	1.775	2.334	2.090
22.5	1.461	1.379	1.808	1.680	2.194	2.017	2.610	2.373
25.0	1.619	1.547	2.007	1.891	2.437	2.269	2.879	2.669
27.5	1.783	1.728	2.213	2.111	2.689	2.531	3.196	2.976
30.0	1.956	1.915	2.431	2.342	2.953	2.806	3.511	3.299
32.5	2.139	2.112	2.659	2.686	3.231	3.095	3.841	3.638
35.0	2.331	2.321	2.901	2.841	3.524	3.400	4.191	3.998
37.5	2.536	2.541	3.158	3.112	3.835	3.723	4.563	4.379
40.0	2.753	2.775	3.431	3.399	4.164	4.064	4.958	4.784

d. Stability coefficients m' and n' for $c'/\gamma H = 0.05$ and $D = 1.00$

	Stability coefficients for earth slopes							
	Slope 2:1		Slope 3:1		Slope 4:1		Slope 5:1	
ϕ'	m'	n'	m'	n'	m'	n'	m'	n'
10.0	0.913	0.563	1.181	0.717	1.469	0.910	1.733	1.069
12.5	1.030	0.690	1.343	0.878	1.688	1.136	1.995	1.316
15.0	1.145	0.816	1.506	1.043	1.904	1.353	2.256	1.567
17.5	1.262	0.942	1.671	1.212	2.117	1.565	2.517	1.825
20.0	1.380	1.071	1.840	1.387	2.333	1.776	2.783	2.091
22.5	1.500	1.202	2.014	1.568	2.551	1.989	3.055	2.365
25.0	1.624	1.338	2.193	1.757	2.778	2.211	3.336	2.651
27.5	1.753	1.480	1.380	1.952	3.013	2.444	3.628	2.948
30.0	1.888	1.630	2.574	2.157	3.261	2.693	3.934	3.259
32.5	2.029	1.789	2.777	2.370	3.523	2.961	4.256	3.585
35.0	2.178	1.958	2.990	2.592	3.803	3.253	4.597	3.927
37.5	2.336	2.138	3.215	2.826	4.103	3.574	4.959	4.288
40.0	2.505	2.332	3.451	3.071	4.425	3.926	5.344	4.668

Table 15.5 (*continued*)

e. Stability coefficients m' and n' for $c'/\gamma H = 0.05$ and $D = 1.25$

	Stability coefficients for earth slopes							
	Slope 2:1		Slope 3:1		Slope 4:1		Slope 5:1	
ϕ'	m'	n'	m'	n'	m'	n'	m'	n'
10.0	0.919	0.633	1.119	0.766	1.344	0.886	1.594	1.042
12.5	1.065	0.792	1.294	0.941	1.563	1.112	1.850	1.300
15.0	1.211	0.950	1.471	1.119	1.782	1.338	2.109	1.562
17.5	1.359	1.108	1.650	1.303	2.004	1.567	2.373	1.831
20.0	1.509	1.266	1.834	1.493	2.230	1.799	2.643	2.107
22.5	1.663	1.428	2.024	1.690	2.463	2.038	2.921	2.392
25.0	1.822	1.595	2.222	1.897	2.705	2.287	3.211	2.690
27.5	1.988	1.769	2.428	2.113	2.957	2.546	3.513	2.999
30.0	2.161	1.950	2.645	2.342	3.221	2.819	3.829	3.324
32.5	2.343	2.141	2.873	2.583	3.500	3.107	4.161	3.665
35.0	2.535	2.344	3.114	2.839	3.795	3.413	4.511	4.025
37.5	2.738	2.560	3.370	3.111	4.109	3.740	4.881	4.405
40.0	2.953	2.791	3.642	3.400	4.442	4.090	5.273	4.806

f. Stability coefficients m' and n' for $c'/\gamma H = 0.05$ and $D = 1.50$

	Stability coefficients for earth slopes							
	Slope 2:1		Slope 3:1		Slope 4:1		Slope 5:1	
ϕ'	m'	n'	m'	n'	m'	n'	m'	n'
10.0	1.022	0.751	1.170	0.828	1.343	0.974	1.547	1.108
12.5	1.202	0.936	1.376	1.043	1.589	1.227	1.829	1.399
15.0	1.383	1.122	1.583	1.260	1.835	1.480	2.112	1.690
17.5	1.565	1.309	1.795	1.480	2.084	1.734	2.398	1.983
20.0	1.752	1.501	2.011	1.705	2.337	1.993	2.690	2.280
22.5	1.943	1.698	2.234	1.937	2.597	2.258	2.990	2.585
25.0	2.143	1.903	2.467	2.179	2.867	2.534	3.302	2.902
27.5	2.350	2.117	2.709	2.431	3.148	2.820	3.626	3.231
30.0	2.568	2.342	2.964	2.696	3.443	3.120	3.967	3.577
32.5	2.798	2.580	3.232	2.975	3.753	3.436	4.326	3.940
35.0	3.041	2.832	3.515	3.269	4.082	3.771	4.707	4.325
37.5	3.299	3.102	3.817	3.583	4.431	4.128	5.112	4.735
40.0	3.574	3.389	4.136	3.915	4.803	4.507	5.543	5.171

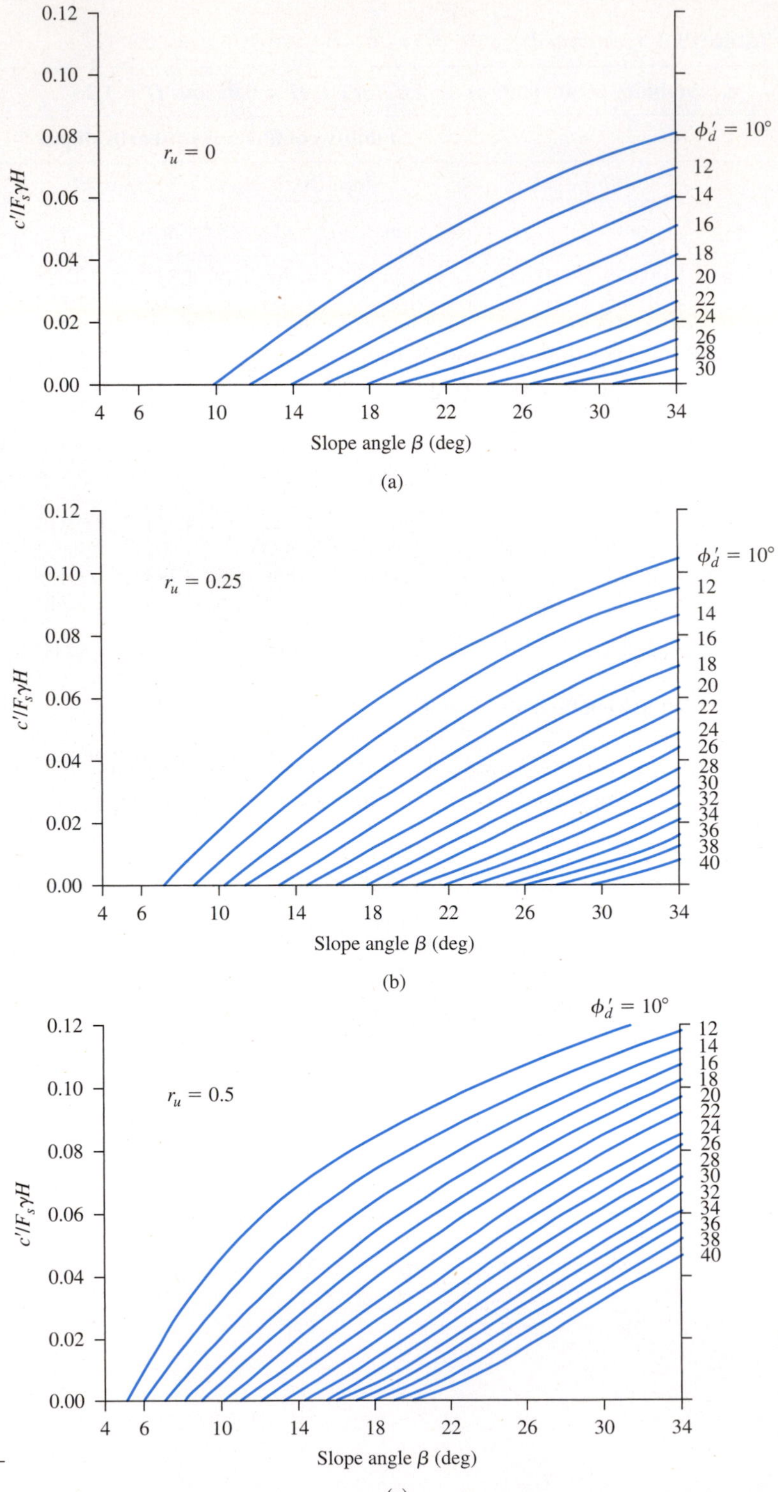

Figure 15.35 Spencer's solution—plot of $c'/F_s\gamma H$ versus β

of safety (F_s) by taking into account the interslice forces (P_n, T_n, P_{n+1}, T_{n+1}, as shown in Figure 15.31), which does satisfy the equations of equilibrium with respect to moment and forces. The details of this method of analysis are beyond the scope of this text; however, the final results of Spencer's work are summarized in this section in Figure 15.35. Note that r_u, as shown in Figure 15.35, is the same as that given in by Eq. (15.76).

In order to use the charts given in Figure 15.35 and to determine the required value of F_s, the following step-by-step procedure needs to be used.

Step 1. Determine c', γ, H, β, ϕ', and r_u for the given slope.

Step 2. Assume a value of F_s.

Step 3. Calculate $c'/[F_{s(\text{assumed})}\,\gamma H]$.

$$\uparrow$$
$$\text{Step 2}$$

Step 4. With the value of $c'/F_s\gamma H$ calculated in step 3 and the slope angle β, enter the proper chart in Figure 15.35 to obtain ϕ'_d. Note that Figures 15.35 a, b, and c, are, respectively, for r_u of 0, 0.25, and 0.5, respectively.

Step 5. Calculate $F_s = \tan\phi' / \tan\phi'_d$.

$$\uparrow$$
$$\text{Step 4}$$

Step 6. If the values of F_s as assumed in step 2 are not the same as those calculated in step 5, repeat steps 2, 3, 4, and 5 until they are the same.

Spencer (1967) also provided the necessary parameters to determine the location of the critical failure circle. These are given in Figures 15.36 through 15.40.

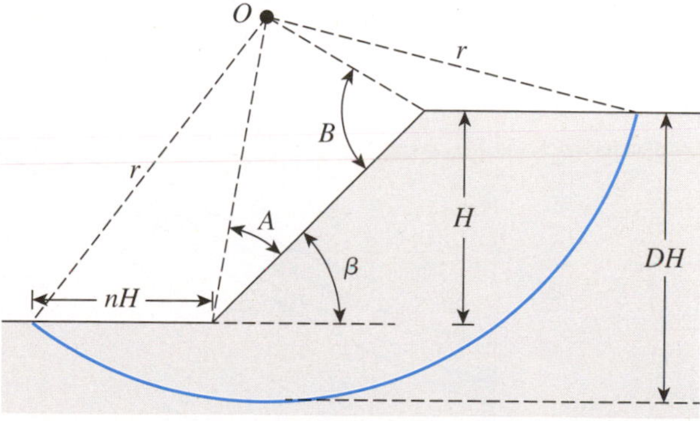

Figure 15.36 Parameters for determination of the critical circle—Spencer's method

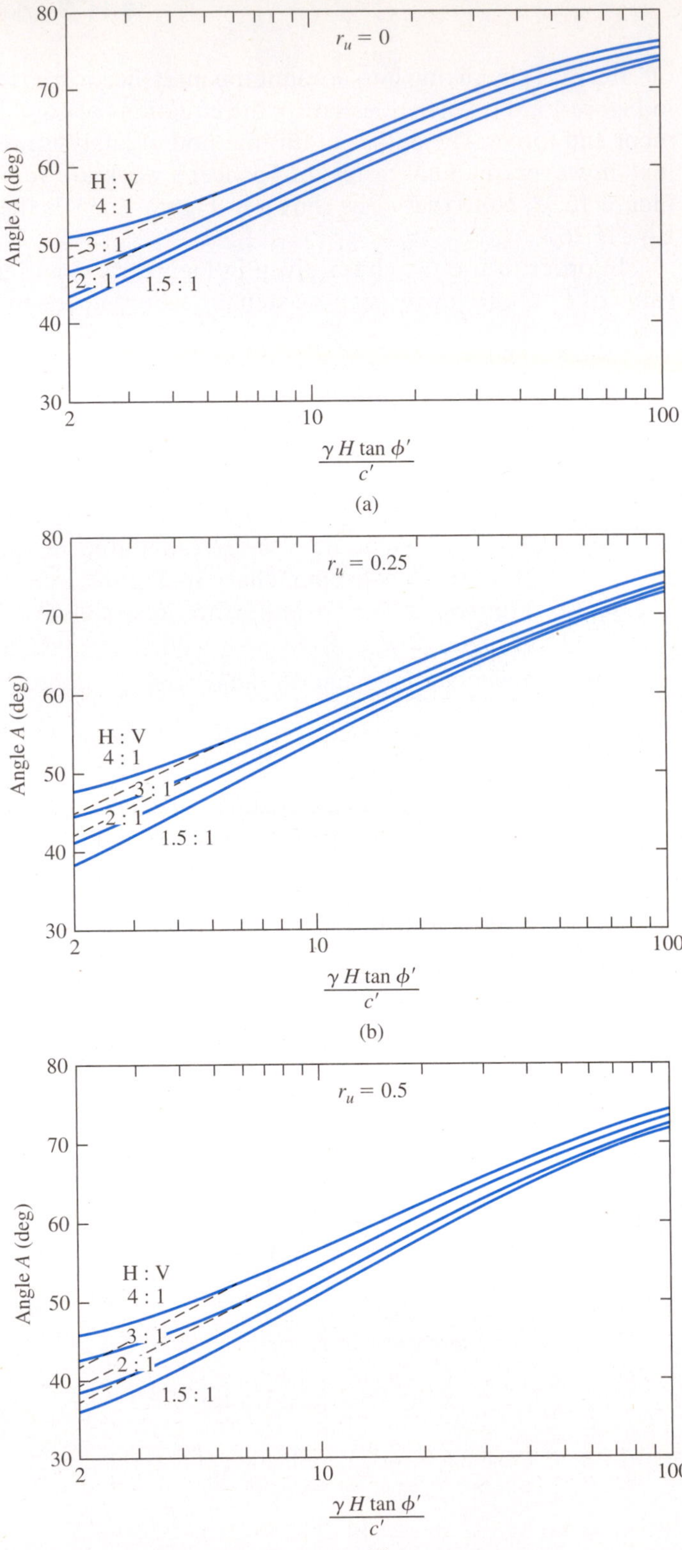

Figure 15.37 Variation of angle A for the critical circle based on Spencer's analysis (*Note*: the broken lines are for critical toe circle if the most critical circle is different than the critical toe circle)

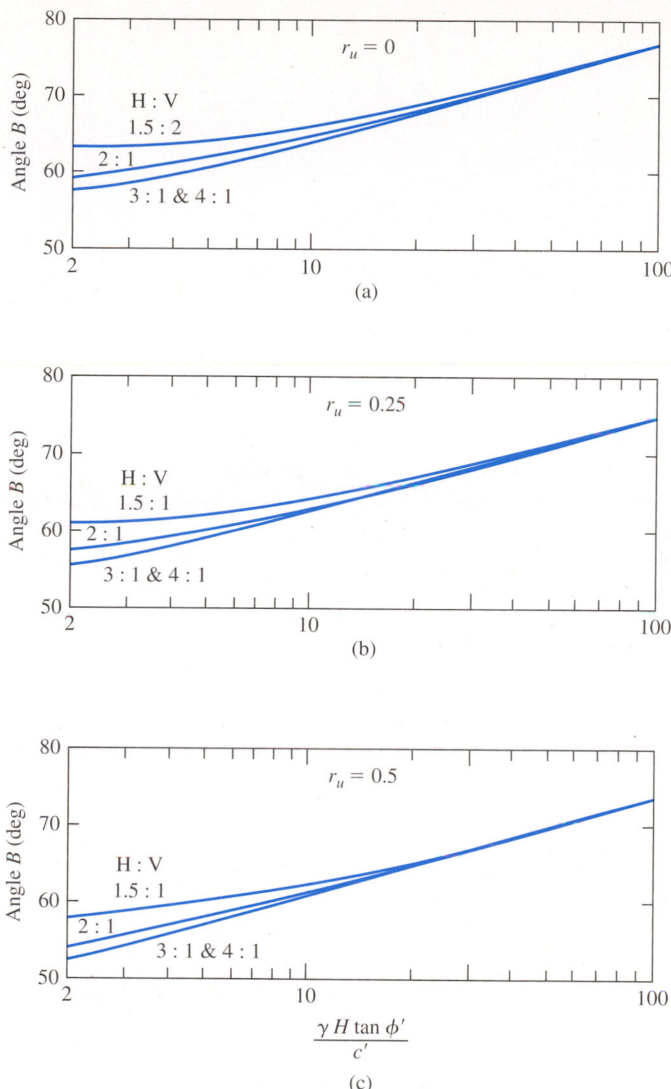

Figure 15.38 Variation of angle B for the critical circle based on Spencer's analysis

Michalowski's solution

Michalowski (2002) used the kinematic approach of limit analysis similar to that shown in Figures 15.24 and 15.25 to analyze slopes with steady-state seepage. The results of this analysis are summarized in Figure 15.41 for $r_u = 0.25$ and $r_u = 0.5$. Note that Figure 15.25 is applicable for the $r_u = 0$ condition. Tables 15.6 and 15.7 give the variation of $\frac{F_s}{\tan\phi'}$ for varying values of $\frac{c'}{\gamma H}\tan\phi'$.

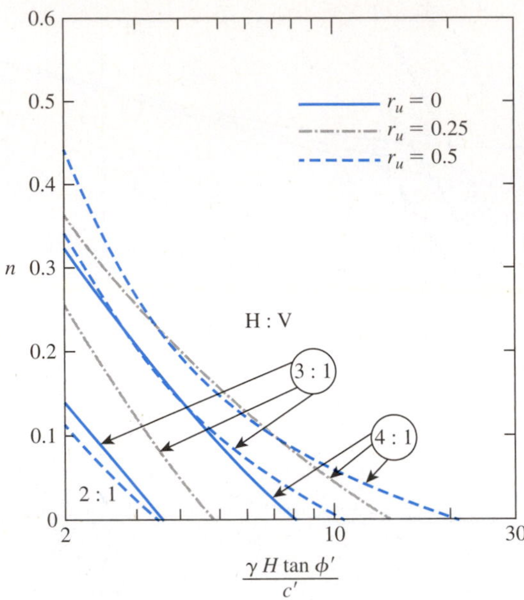

Figure 15.39 Variation of n for critical circles based on Spencer's analysis

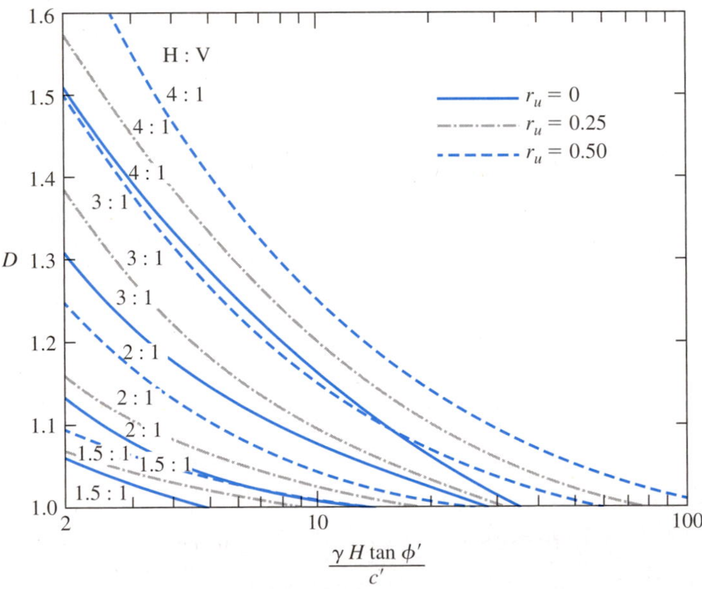

Figure 15.40 Variation of D for critical circles based on Spencer's analysis

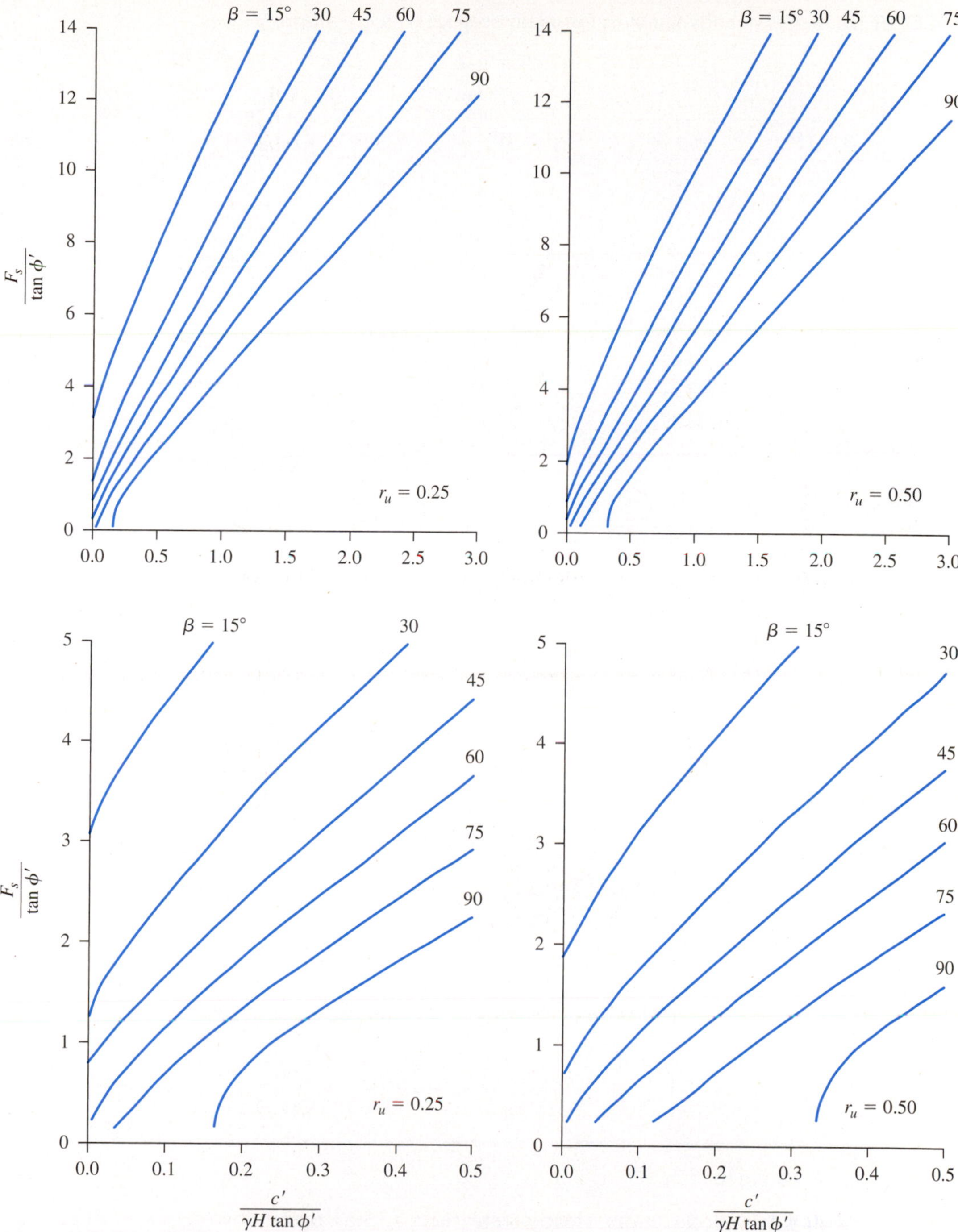

Figure 15.41 Michalowski's solution for steady-state seepage condition

Table 15.6 Michalowski's stability numbers for steady seepage with $r_u = 0.25$

$\frac{c'}{\gamma H \tan\phi'}$	$\frac{F_s}{\tan\phi'}$					
	$\beta = 15°$	$\beta = 30°$	$\beta = 45°$	$\beta = 60°$	$\beta = 75°$	$\beta = 90°$
0	3.09	1.21	0.82	0.21		
0.05	3.83	1.96	1.27	0.75	0.29	
0.1	4.38	2.43	1.66	1.14	0.68	
0.2	5.10	3.30	2.37	1.81	1.31	0.70
0.3	6.05	4.09	3.05	2.43	1.86	1.29
0.4	6.84	4.79	3.71	3.00	2.39	1.75
0.5	7.74	5.42	4.35	3.58	2.88	2.21
1.0	11.70	8.81	7.41	6.23	5.21	4.22
1.5		12.17	10.44	8.91	7.55	6.15
2.0			13.42	11.59	9.81	8.10
2.5				14.24	12.16	10.08
3.0						11.95

Table 15.7 Michalowski's stability numbers for steady seepage with $r_u = 0.50$

$\frac{c'}{\gamma H \tan\phi'}$	$\frac{F_s}{\tan\phi'}$					
	$\beta = 15°$	$\beta = 30°$	$\beta = 45°$	$\beta = 60°$	$\beta = 75°$	$\beta = 90°$
0	1.88	0.66	0.15			
0.05	2.57	1.32	0.74	0.30		
0.1	3.12	1.75	1.14	0.65		
0.2	4.04	2.54	1.83	1.28	0.71	
0.3	4.89	3.29	2.47	1.87	1.28	
0.4	5.55	3.97	3.10	2.43	1.82	1.06
0.5	6.45	4.65	3.71	3.00	2.31	1.58
1.0	10.07	8.00	6.73	5.63	4.64	3.69
1.5	13.32	11.03	9.62	8.32	6.96	5.64
2.0		14.11	12.65	10.93	9.28	7.55
2.5				13.62	11.56	9.49
3.0					13.88	11.38

Example 15.13

A given slope under steady-state seepage has the following: $H = 21.62$ m, $\phi' = 25°$, slope: 2H:1V, $c' = 20$ kN/m^2, $\gamma = 18.5$ kN/m^3, $r_u = 0.25$. Determine the factor of safety, F_s. Use Table 15.5.

Solution

$$\beta = \tan^{-1}\left(\frac{1}{2}\right) = 26.57°$$

$$\frac{c'}{\gamma H} = \frac{20}{(18.5)(21.62)} = 0.05$$

Now the following table can be prepared.

β (deg)	ϕ' (deg)	$c'/\gamma H$	D	m'^a	n'^b	$F_s = m' - n' r_u^c$
26.57	25	0.05	1.00	1.624[d]	1.338[d]	1.29
26.57	25	0.05	1.25	1.822[e]	1.595[e]	1.423
26.57	25	0.05	1.5	2.143[f]	1.903[f]	1.667

[a]From Table 15.5
[b]From Table 15.5
[c]Eq. (15.77); $r_u = 0.25$
[d]Table 15.5d
[e]Table 15.5e
[f]Table 15.5f

So,

$$F_s \simeq \mathbf{1.29}$$

Example 15.14

Solve Example 15.13 using Spencer's solution (Figure 15.35).

Solution

Given: $H = 21.62$ m, $\beta = 26.57°$, $c' = 20$ kN/m^2, $\gamma = 18.5$ kN/m^3, $\phi' = 25°$, and $r_u = 0.25$. Now the following table can be prepared.

β (deg)	$F_{s(\text{assumed})}$	$\frac{c'}{F_{s(\text{assumed})}\gamma H}$	$\phi_d'^a$ (deg)	$F_{s(\text{calculated})} = \frac{\tan\phi'}{\tan\phi_d'}$
26.57	1.1	0.0455	18	1.435
26.57	1.2	0.0417	19	1.354
26.57	1.3	0.0385	20	1.281
26.57	1.4	0.0357	21	1.215

[a]From Figure 15.35b

Figure 15.42 shows a plot of $F_{s(\text{assumed})}$ against $F_{s(\text{calculated})}$, from which $F_s \simeq$ **1.28.**

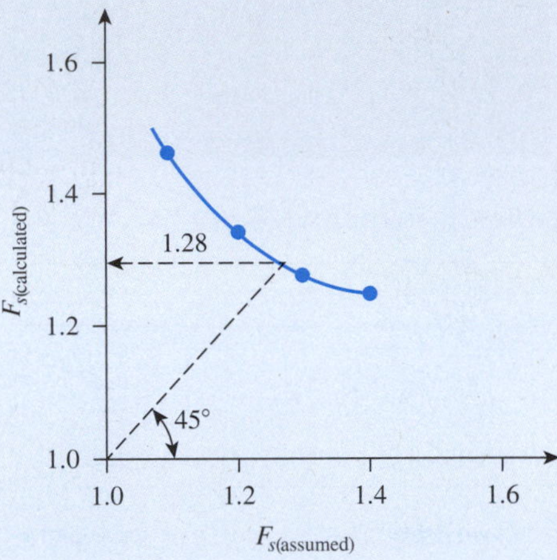

Figure 15.42

Example 15.15

For the slope described in Example 15.14, determine

a. The location of the center of the critical circle
b. The length of the radius of the critical circle

Solution

Part a

Refer to Figure 15.36 for the notation of various parameters. It is a 2H:1V slope, and $r_u = 0.25$. The critical circle is a toe circle (see Figure 15.37). For the slope,

$$\frac{\gamma H \tan \phi'}{c'} = \frac{(18.5)(21.62)(\tan 25)}{20} = 9.33$$

From Figures 15.37b and 15.38b, $A \approx 54°$; $B \approx 63°$. Figure 15.43 shows the location of the center of the critical circle.

Part b

In Figure 15.43, $\overline{PO}$ is the radius of the critical circle.

$$PQ = \sqrt{(21.62)^2 + (2 \times 21.62)^2} = 48.34 \text{ m}$$

$$\frac{\overline{PQ}}{\sin 63°} = \frac{r}{\sin 63}$$

$$r = \overline{PQ} = \mathbf{48.34 \text{ m}}$$

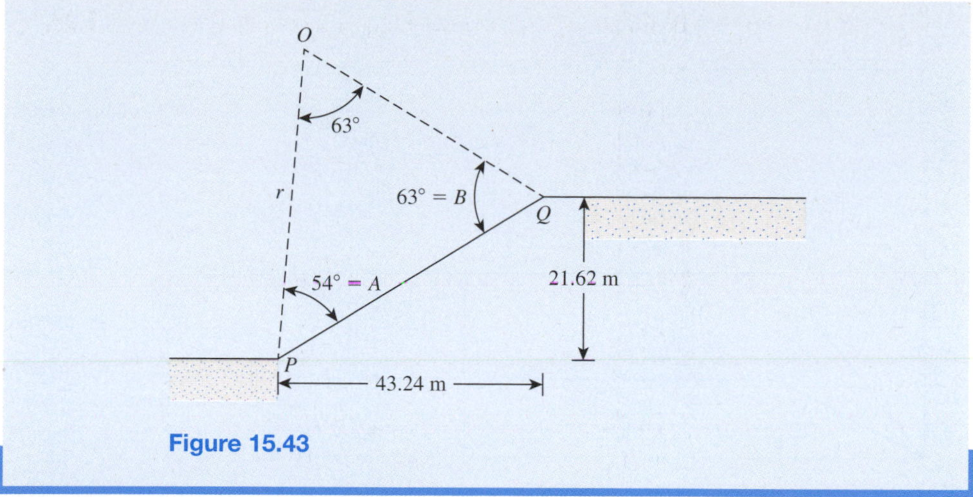

Figure 15.43

Example 15.16

Solve Example 15.13 using Michalowski's solution (Figure 15.41).

Solution

$$\frac{c'}{\gamma H \tan \phi'} = \frac{20}{(18.5)(21.62)(\tan 25)} = 0.107$$

For $r_u = 0.25$, from Figure 15.41, $\dfrac{F_s}{\tan \phi'} \approx 3.1$ So,

$$F_s = (3.1)(\tan 25) = \mathbf{1.45}$$

15.15 Fluctuation of Factor of Safety of Slopes in Clay Embankment on Saturated Clay

Figure 15.44a shows a clay embankment constructed on a *saturated soft clay*. Let P be a point on a potential failure surface APB that is an arc of a circle. Before construction of the embankment, the pore water pressure at P can be expressed as

$$u = h\gamma_w \tag{15.78}$$

Under ideal conditions, let us assume that the height of the fill needed for the construction of the embankment is placed uniformly, as shown in Figure 15.44b. At time $t = t_1$, the embankment height is equal to H, and it remains constant thereafter (that

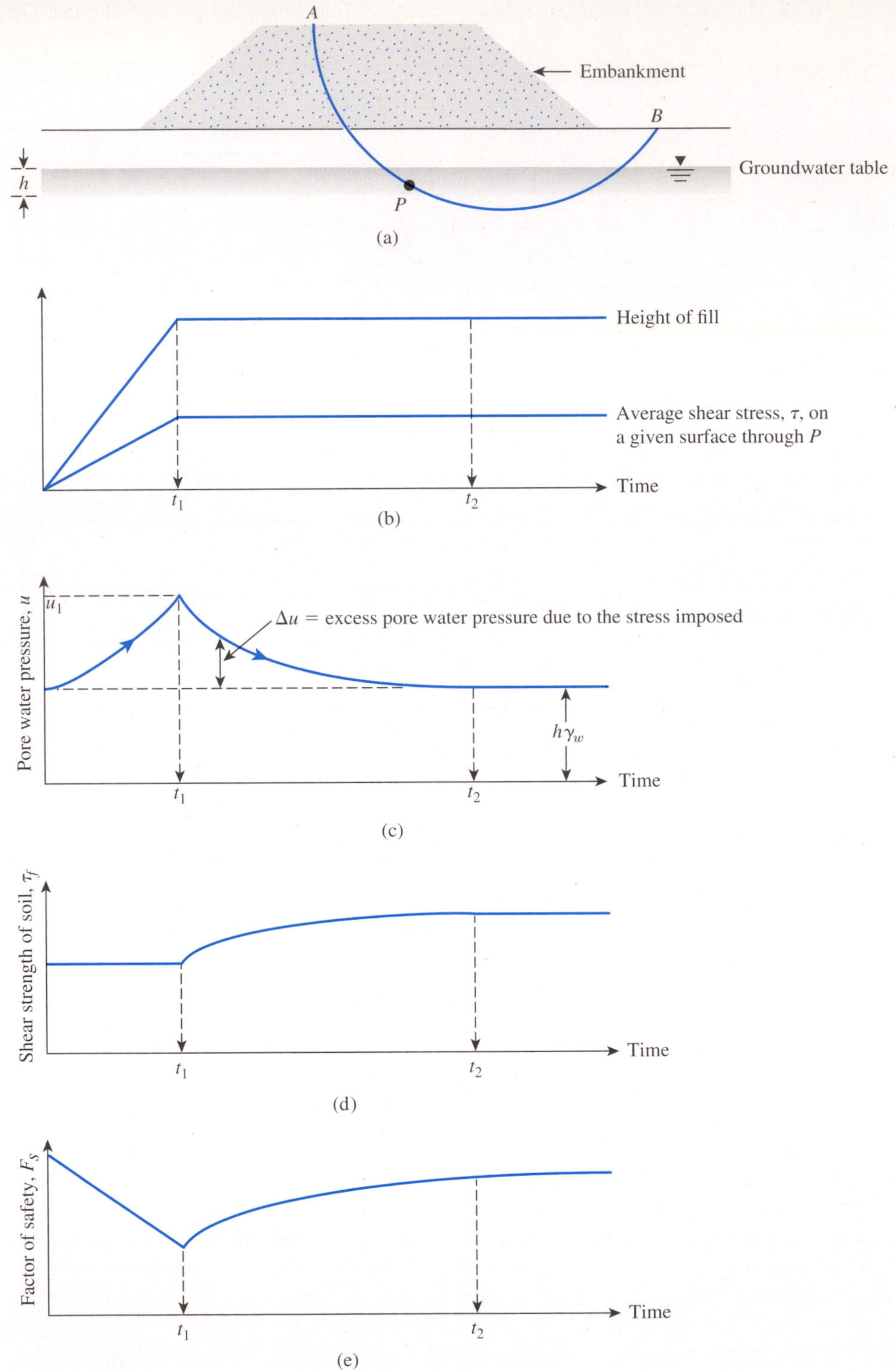

Figure 15.44 Factor of safety variation with time for embankment on soft clay (*Redrawn after Bishop and Bjerrum, 1960. With permission from ASCE.*)

is, $t > t_1$). The average shear stress increase, τ, on the potential failure surface caused by the construction of the embankment also is shown in Figure 15.44b. The value of τ will increase linearly with time up to time $t = t_1$ and remain constant thereafter.

The pore water pressure at point P (Figure 15.44a) will continue to increase as construction of the embankment progresses, as shown in Figure 15.44c. At time $t = t_1$, $u = u_1 > h\gamma_w$. This is because of the slow rate of drainage from the clay layer. However, after construction of the embankment is completed (that is, $t > t_1$), the pore water pressure gradually will decrease with time as the drainage (thus consolidation) progresses. At time $t \simeq t_2$,

$$u = h\gamma_w$$

For simplicity, if we assume that the embankment construction is rapid and that practically no drainage occurs during the construction period, the average *shear strength* of the clay will remain constant from $t = 0$ to $t = t_1$, or $\tau_f = c_u$ (undrained shear strength). This is shown in Figure 15.44d. For time $t > t_1$, as consolidation progresses, the magnitude of the shear strength, τ_f, will gradually increase. At time $t \geq t_2$—that is, after consolidation is completed—the average shear strength of the clay will be equal to $\tau_f = c' + \sigma' \tan \phi'$ (drained shear strength) (Figure 15.44d). The factor of safety of the embankment along the potential surface of sliding can be given as

$$F_s = \frac{\text{Average shear strength of clay, } \tau_f\text{, along sliding surface (Figure 14.44d)}}{\text{Average shear stress, } \tau\text{, along sliding surface (Figure 14.44b)}} \tag{15.79}$$

The general nature of the variation of the factor of safety, F_s, with time is shown in Figure 15.44e. As we can see from this figure, the magnitude of F_s initially decreases with time. At the end of construction (time $t = t_1$), the value of the factor of safety is a minimum. Beyond this point, the value of F_s continues to increase with drainage up to time $t = t_2$.

Cuts in saturated clay

Figure 15.45a shows a cut slope in a saturated soft clay in which APB is a circular potential failure surface. During advancement of the cut, the average shear stress, τ, on the potential failure surface passing through P will increase. The maximum value of the average shear stress, τ, will be attained at the end of construction—that is, at time $t = t_1$. This property is shown in Figure 15.45b.

Because of excavation of the soil, the effective overburden pressure at point P will decrease, which will induce a reduction in the pore water pressure. The variation of the net change of pore water pressure, Δu, is shown in Figure 15.45c. After excavation is complete (time $t > t_1$), the net negative excess pore water pressure will gradually dissipate. At time $t \geq t_2$, the magnitude of Δu will be equal to 0.

The variation of the average shear strength, τ_f, of the clay with time is shown in Figure 15.45d. Note that the shear strength of the soil after excavation gradually

A

Groundwater table

B

P

(a)

Shear stress, τ

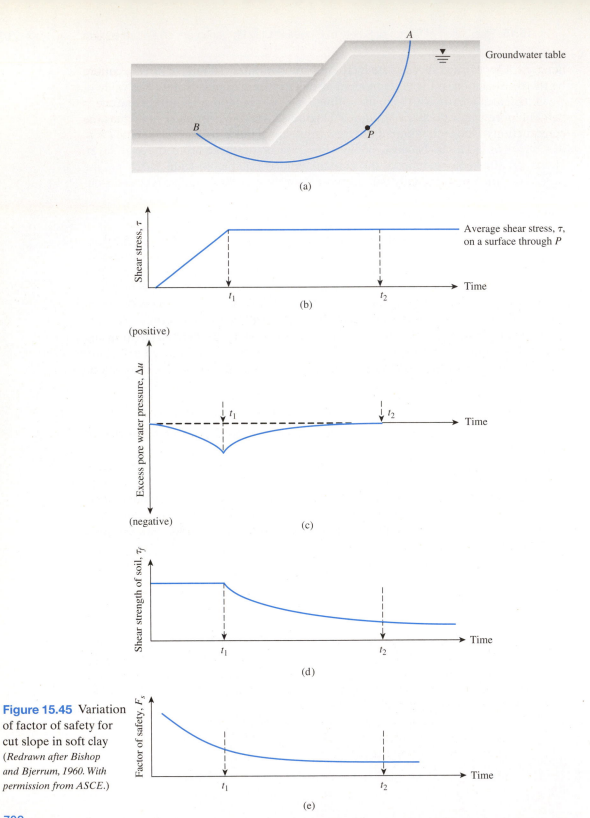

Average shear stress, τ, on a surface through P

Time

t_1 t_2

(b)

(positive)

Excess pore water pressure, Δu

t_1 t_2

Time

(negative)

(c)

Shear strength of soil, τ_f

Time

t_1 t_2

(d)

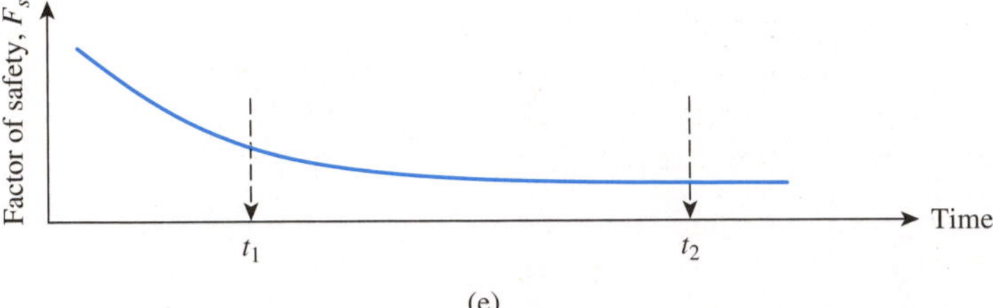

(e)

Figure 15.45 Variation of factor of safety for cut slope in soft clay (*Redrawn after Bishop and Bjerrum, 1960. With permission from ASCE.*)

decreases. This decrease occurs because of dissipation of the negative excess pore water pressure.

If the factor of safety of the cut slope, F_s, along the potential failure surface is defined by Eq. (15.79), its variation will be as shown in Figure 15.45e. Note that the magnitude of F_s decreases with time, and its minimum value is obtained at time $t \geq t_2$.

15.16 Summary

Following is a summary of the topics covered in this chapter:

- The factor of safety with respect to strength (F_s) occurs when [Eq. (15.7)]

$$F_s = F_{c'} = F_{\phi'}$$

- The factors of safety against sliding for infinite slopes for cases with and without seepage are given by Eqs. (15.15) and (15.28), respectively.
- The critical height of a finite slope with plane failure surface assumption can be given by Eq. (15.42).
- The modes of failure of finite slopes with circular failure surfaces can be categorized under (Section 15.7)
  - Slope failure
  - Base failure
- Stability analysis charts for clay slopes ($\phi = 0$ condition) are provided in Figures 15.12.
- Stability analysis charts for slopes with $c' - \phi'$ soil (pore water pressure equal to zero) are given in Figures 15.21, 15.23, and 15.25.
- Determination of factor of safety with respect to strength using the method of slices without seepage is described in Sections 15.11 and 15.12.
- Stability analysis of slopes with circular failure surface under steady-state seepage is presented in Sections 15.13 and 15.14.

Problems

15.1 Refer to the infinite slope shown in Figure 15.46. Given: $\beta = 19°$, $\gamma = 20$ kN/m^3, $\phi' = 33°$, and $c' = 47$ kN/m^2. Find the height, H, such that a factor of safety, F_s, of 3.1 is maintained against sliding along the soil-rock interface.

15.2 For the infinite slope shown in Figure 15.46, determine the height, H, for critical equilibrium. Given: $\beta = 27°$, $\gamma = 19.15$ kN/m^3, $\phi' = 18°$, and $c' = 35.9$ kN/m^3.

15.3 Determine the factor of safety, F_s, for the infinite slope shown in Figure 15.47, where seepage is occurring through the soil and the groundwater table coincides with the ground surface. Given: $H = 6$ m, $\beta = 22°$, $\gamma_{sat} = 20.8$ kN/m^3, $\phi' = 26°$, and $c' = 52$ kN/m^2.

Figure 15.46

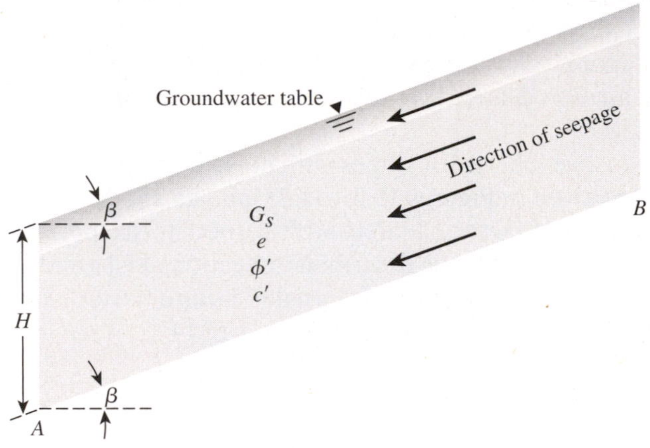

Figure 15.47

15.4 Figure 15.47 shows an infinite slope with $H = 9.14$ m and the groundwater table coinciding with the ground surface. If there is seepage through the soil, determine the factor of safety against sliding along the plane AB. The soil properties are $G_s = 2.7$, $e = 0.8$, $\beta = 16°$, $\phi' = 21°$, and $c' = 59.8$ kN/m^2.

15.5 An infinite slope is shown in Figure 15.48. The shear strength parameters at the interface of soil and rock are $\phi' = 23°$ and $c' = 34$ kN/m^2. Given $\rho = 2050$ kg/m^3. If $H = 7$ m and $\beta = 25°$, find the factor of safety against sliding on the rock surface.

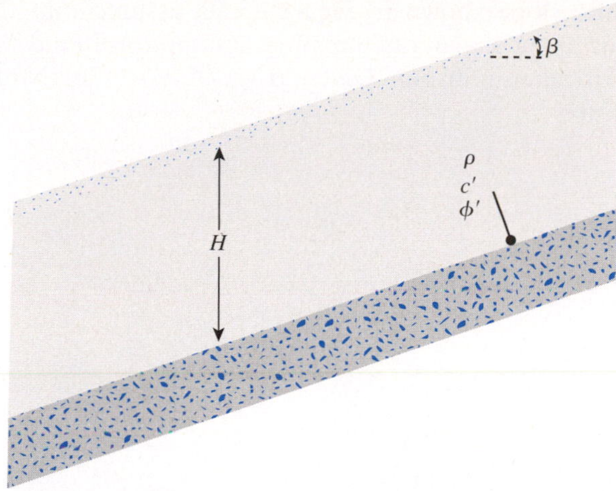

Figure 15.48

15.6 Refer to the infinite slope in Problem 15.5. If the factor of safety must be increased to 2.35, what would be the maximum height (H) of the slope?

15.7 Figure 15.49 shows a slope with an inclination of $\beta = 58°$. If AC represents a trial failure plane inclined at an angle $\theta = 32°$ with the horizontal, determine the factor of safety against sliding for the wedge ABC. Given: $H = 6$ m; $\gamma = 19$ kN/m^3, $\phi' = 21°$, and $c' = 38$ kN/m^2.

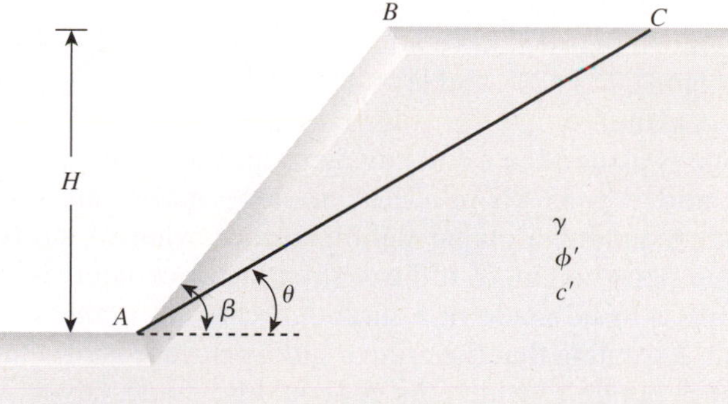

Figure 15.49

15.8 Refer to Problem 15.7. With all other conditions remaining the same, what would be the factor of safety against sliding for the trial wedge ABC if the height of the slope were 9 m?

15.9 Refer to the slope in Problem 15.7. Assume that the shear strength of the soil is improved by soil stabilization methods, and the new properties are as follows: $\gamma = 22$ kN/m^3, $\phi' = 32°$, and $c' = 75$ kN/m^2. What would be the improved factor of safety against sliding along the trial failure surface AC?

15.10 For the finite slope shown in Figure 15.50, assume that the slope failure would occur along a plane (Culmann's assumption). Find the height of the slope for critical equilibrium. Given: $\beta = 49°$, $\gamma = 17.6$ kN/m^3, $\phi' = 19°$, and $c' = 42$ kN/m^2.

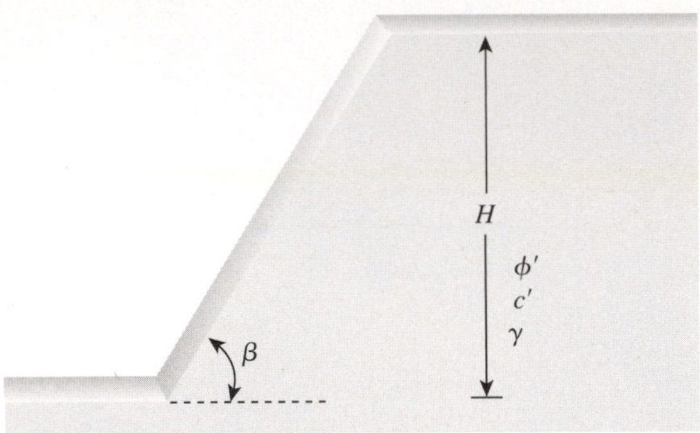

Figure 15.50

15.11 Refer to Figure 15.50. Using the soil parameters given in Problem 15.10, find the height of the slope, H, that will have a factor of safety of 2.95 against sliding. Assume that the critical sliding surface is a plane.

15.12 Refer to Figure 15.50. Given that $\beta = 55°$, $\gamma = 19$ kN/m^3, $\phi' = 17°$, $c' =$ 57.4 kN/m^2, and $H = 13.7$ m, determine the factor of safety with respect to sliding. Assume that the critical sliding surface is a plane.

15.13 The inclination of a finite slope is 1 vertical to 1 horizontal. Determine the slope height, H, that will have a factor of safety of 2.8 against sliding. Given: $\rho = 1950$ kg/m^3, $\phi' = 23°$, and $c' = 40$ kN/m^2. Assume that the critical sliding surface is a plane.

15.14 A cut slope is planned in a soil having properties as follows: $\gamma = 19.3$ kN/m^3, $\phi' = 19°$, and $c' = 55$ kN/m^2. The cut slope makes an angle 60° with the horizontal. Assuming a planar failure surface, what would be the maximum permissible slope height, H, if the desired factor of safety is 2.5?

15.15 A cut slope is to be made in a saturated clay at an angle $\beta = 60°$ with the horizontal. Assuming that the critical sliding surface is circular, determine the maximum depth up to which the cut could be made. Given: undrained shear strength, $c_u = 34$ kN/m^2 ($\phi = 0$ condition), and $\gamma = 17$ kN/m^3. What is the nature of the critical circle (toe, slope, or midpoint)?

15.16 For the cut slope described in Problem 15.15, how deep should the cut be made to ensure a factor of safety of 3.0 against sliding?

15.17 A cut slope in saturated clay is inclined at an angle $\beta = 26.57°$ (1 vertical to 2 horizontal) and has a height, $H = 7$ m. The undrained cohesion, c_u, increases with depth, z, as shown in Figure 15.51. Assuming the critical circle to be a toe circle, determine the factor of safety, F_s using the Koppula (1984) method. Given: $\gamma_{sat} = 18.6$ kN/m^3, $c_{u(z=0)} = 9$ kN/m^2, and $a_0 = 5$ kN/m^3.

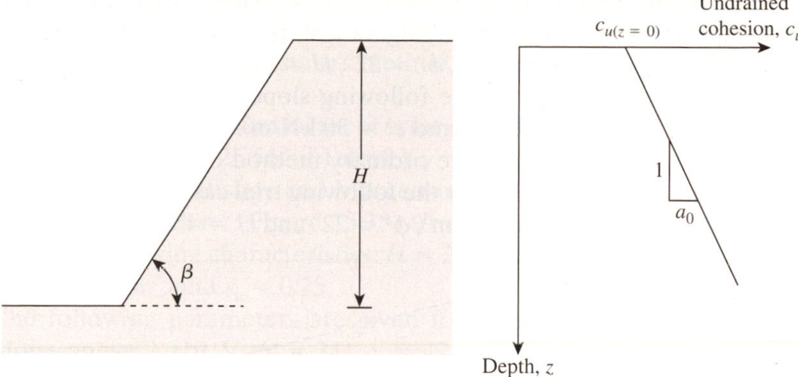

Figure 15.51

15.18 Refer to Figure 15.12. Determine the height of a slope (1 vertical to 3 horizontal) in saturated clay with an undrained shear strength of 59.8 kN/m^2 and a unit weight of 19.15 kN/m^3. The desired factor of safety against sliding is 2.75. Given is the depth function, $D = 2.0$.

15.19 Refer to Problem 15.18. What is the critical height of the slope? What is the nature of the critical circle?

15.20 A cut slope was excavated in a saturated clay with a slope angle $\beta = 57°$ with the horizontal. Slope failure occurred when the cut reached a depth of 8 m. Previous soil explorations showed that a rock layer was located at a depth of 12 m below the ground surface. Assuming an undrained condition and that $\gamma_{sat} = 19$ kN/m^3:

a. Determine the undrained cohesion of the clay (Figure 15.12).

b. What was the nature of the critical circle?

c. With reference to the top of the slope, at what distance did the surface of the sliding intersect the bottom of the excavation?

15.21 Refer to Figure 15.52. Using Michalowski's solution given in Figure 15.25 ($\phi' > 0$), determine the critical height of the slope for the following conditions.

a. $n' = 1.5$, $\phi' = 15°$, $c' = 51.4$ kN/m^2, and $\gamma = 19$ kN/m^3

b. $n' = 1$, $\phi' = 21°$, $c' = 43$ kN/m^2, and $\gamma = 18$ kN/m^3

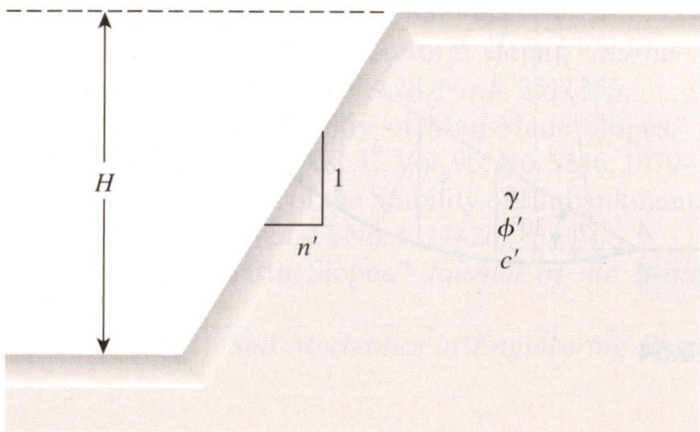

Figure 15.52

CHAPTER 16

Soil Bearing Capacity for Shallow Foundations

16.1 Introduction

The lowest part of a structure generally is referred to as the *foundation*. Its function is to transfer the load of the structure to the soil on which it is resting. A properly designed foundation transfers the load throughout the soil without overstressing the soil. Overstressing the soil can result in either excessive settlement or shear failure of the soil, both of which cause damage to the structure. Thus, geotechnical and structural engineers who design foundations must evaluate the bearing capacity of soils.

Depending on the structure and soil encountered, various types of foundations are used. Figure 16.1 shows the most common types of foundations. A *spread footing* is simply an enlargement of a load-bearing wall or column that makes it possible to spread the load of the structure over a larger area of the soil. In soil with low load-bearing capacity, the size of the spread footings required is impracticably large. In that case, it is more economical to construct the entire structure over a concrete pad. This is called a *mat foundation*.

Pile and *drilled shaft foundations* are used for heavier structures when larger depth is required for supporting the load. Piles are structural members made of timber, concrete, or steel that transmit the load of the superstructure to the lower layers of the soil. According to how they transmit their load into the subsoil, piles can be divided into two categories: friction piles and end-bearing piles. In the case of friction piles, the superstructure load is resisted by the shear stresses generated along the surface of the pile. In the end-bearing pile, the load carried by the pile is transmitted at its tip to a firm stratum.

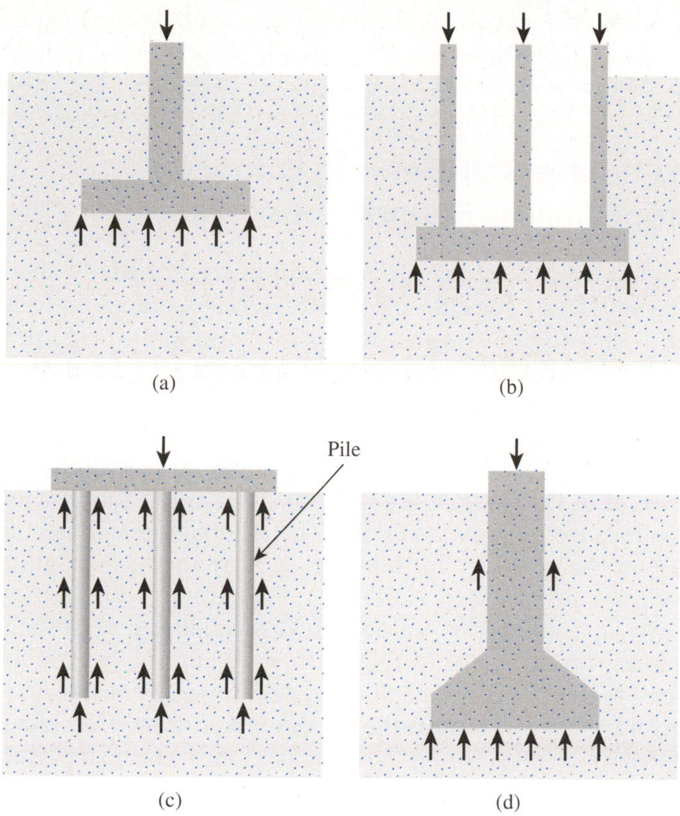

Figure 16.1 Common types of foundations: (a) spread footing; (b) mat foundation; (c) pile foundation; (d) drilled shaft foundation

In the case of drilled shafts, a shaft is drilled into the subsoil and then is filled with concrete. A metal casing may be used while the shaft is being drilled. The casing may be left in place or may be withdrawn during the placing of concrete. Generally, the diameter of a drilled shaft is much larger than that of a pile. The distinction between piles and drilled shafts becomes hazy at an approximate diameter of 1 m, and the definitions and nomenclature are inaccurate.

Spread footings and mat foundations generally are referred to as *shallow foundations*, whereas pile and drilled-shaft foundations are classified as *deep foundations*. In a more general sense, shallow foundations are foundations that have a depth-of-embedment-to-width ratio of approximately less than four. When the depth-of-embedment-to-width ratio of a foundation is greater than four, it may be classified as a deep foundation.

In this chapter, we discuss the soil-bearing capacity for shallow foundations. As mentioned before, for a foundation to function properly, (1) the settlement of soil caused by the load must be within the tolerable limit, and (2) shear failure of the soil supporting the foundation must not occur. Compressibility of soil—consolidation and elasticity theory—was introduced in Chapter 10. This chapter introduces the load-carrying capacity of shallow foundations based on the criteria of shear failure in soil.

16.2 Ultimate Soil-Bearing Capacity for Shallow Foundations

To understand the concept of the ultimate soil-bearing capacity and the mode of shear failure in soil, let us consider the case of a long rectangular footing of width B located at the surface of a dense sand layer (or stiff soil) shown in Figure 16.2a. When a uniformly distributed load of q per unit area is applied to the footing, it settles. If the uniformly distributed load (q) is increased, the settlement of the footing gradually increases. When the value of $q = q_u$ is reached (Figure 16.2b), bearing-capacity failure occurs; the footing undergoes a very large settlement without any further increase of q. The soil on one or both sides of the footing bulges, and the slip surface extends to the ground surface. The load-settlement relationship is like Curve I shown in Figure 16.2b. In this case, q_u is defined as the ultimate bearing capacity of soil.

The bearing-capacity failure just described is called a *general shear failure* and can be explained with reference to Figure 16.3a. When the foundation settles under the application of a load, a triangular wedge-shaped zone of soil (marked I) is pushed down, and, in turn, it presses the zones marked II and III sideways and then upward. At the ultimate pressure, q_u, the soil passes into a state of plastic equilibrium and failure occurs by sliding.

If the footing test is conducted instead in a loose-to-medium dense sand, the load-settlement relationship is like Curve II in Figure 16.2b. Beyond a certain value of $q = q'_u$, the load-settlement relationship becomes a steep, inclined straight line.

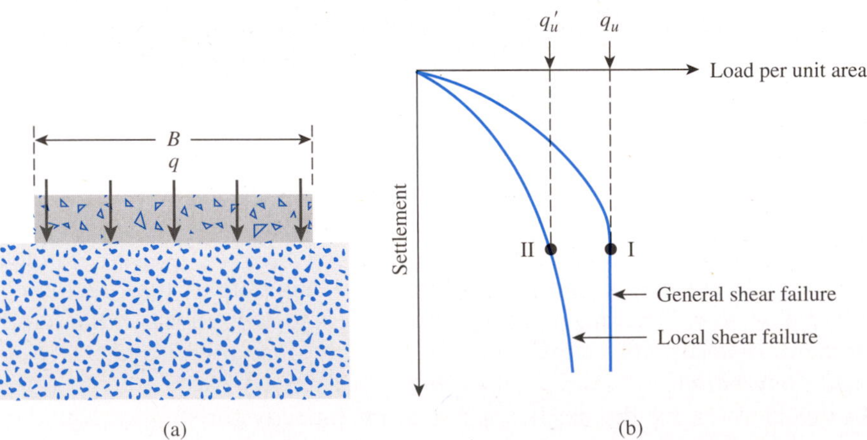

Figure 16.2 Ultimate soil-bearing capacity for shallow foundation: (a) model footing; (b) load-settlement relationship

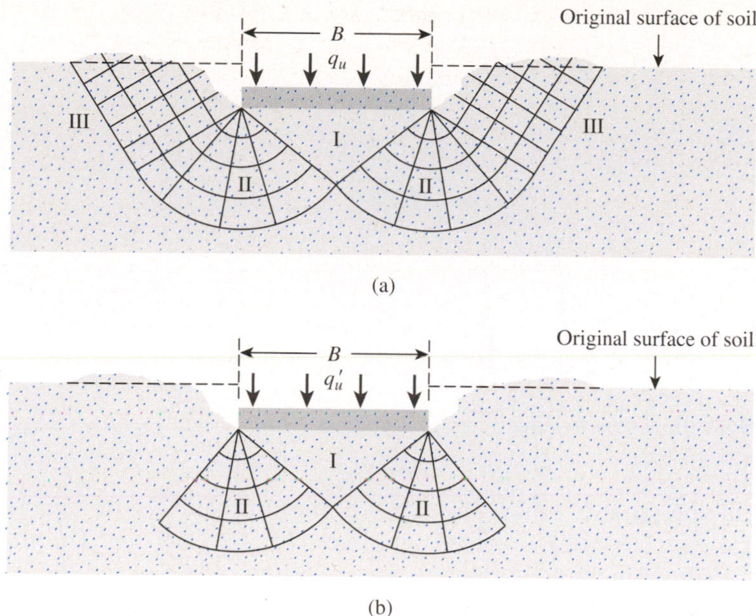

Figure 16.3 Modes of bearing-capacity failure in soil: (a) general shear failure of soil; (b) local shear failure of soil

In this case, q'_u is defined as the ultimate bearing capacity of soil. This type of soil failure is referred to as *local shear failure* and is shown in Figure 16.3b. The triangular wedge-shaped zone (marked I) below the footing moves downward, but unlike general shear failure, the slip surfaces end somewhere inside the soil. Some signs of soil bulging are seen, however.

16.3 Terzaghi's Ultimate Bearing Capacity Equation

In 1921, Prandtl published the results of his study on the penetration of hard bodies (such as metal punches) into a softer material. Terzaghi (1943) extended the plastic failure theory of Prandtl to evaluate the bearing capacity of soils for shallow strip footings. For practical considerations, a long wall footing (length-to-width ratio more than about five) may be called a *strip footing*. According to Terzaghi, a foundation may be defined as a shallow foundation if the depth D_f is less than or equal to its width B (Figure 16.4). He also assumed that, for ultimate soil-bearing capacity calculations, the weight of soil above the base of the footing may be replaced by a uniform surcharge, $q = \gamma D_f$.

The failure mechanism assumed by Terzaghi for determining the ultimate soil-bearing capacity (general shear failure) for a rough strip footing located at a depth D_f measured from the ground surface is shown in Figure 16.5a. The soil wedge ABJ (Zone I) is an elastic zone. Both AJ and BJ make an angle ϕ' with the horizontal. Zones marked II (AJE and BJD) are the radial shear zones, and zones marked III

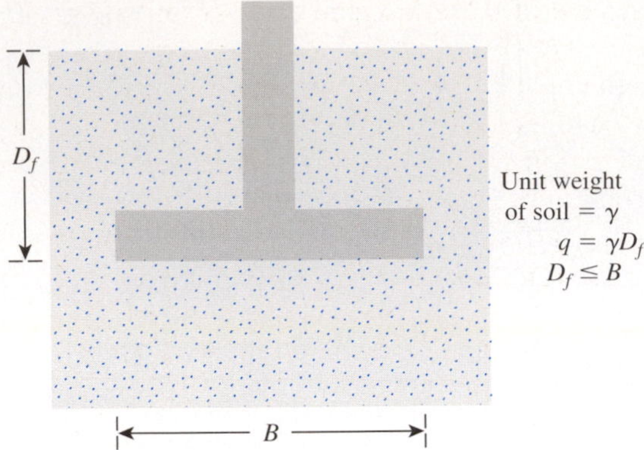

Figure 16.4 Shallow strip footing

are the Rankine passive zones. The rupture lines JD and JE are arcs of a logarithmic spiral, and DF and EG are straight lines. AE, BD, EG, and DF make angles of $45 - \phi'/2$ degrees with the horizontal. The equation of the arcs of the logarithmic spirals JD and JE may be given as (also see Section 14.3)

$$r = r_o e^{\theta \tan \phi'}$$

If the load per unit area, q_u, is applied to the footing and general shear failure occurs, the passive force P_p is acting on each of the faces of the soil wedge ABJ. This

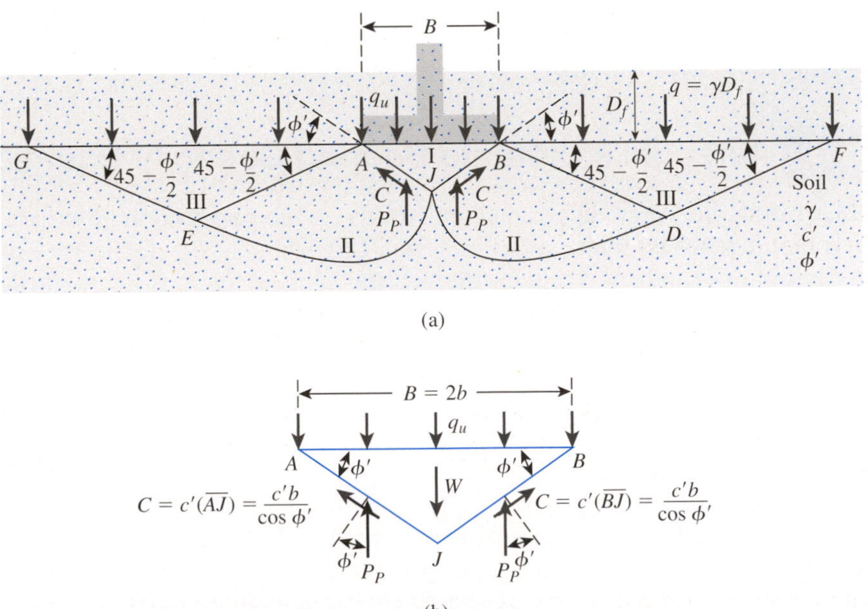

Figure 16.5 Terzaghi's bearing-capacity analysis

concept is easy to conceive of if we imagine that AJ and BJ are two walls that are pushing the soil wedges $AJEG$ and $BJDF$, respectively, to cause passive failure. P_p should be inclined at an angle δ' (which is the angle of wall friction) to the perpendicular drawn to the wedge faces (that is, AJ and BJ). In this case, δ' should be equal to the angle of friction of soil, ϕ'. Because AJ and BJ are inclined at an angle ϕ' to the horizontal, the direction of P_p should be vertical.

Now let us consider the free-body diagram of the wedge ABJ as shown in Figure 16.5b. Considering the unit length of the footing, we have, for equilibrium,

$$(q_u)(2b)(1) = -W + 2C\sin\phi' + 2P_p \tag{16.1}$$

where $b = B/2$
W = weight of soil wedge $ABJ = \gamma b^2 \tan\phi'$
C = cohesive force acting along each face, AJ and BJ, that is equal to the unit cohesion times the length of each face $= c'b/(\cos\phi')$

Thus,

$$2bq_u = 2P_p + 2bc'\tan\phi' - \gamma b^2 \tan\phi' \tag{16.2}$$

or

$$q_u = \frac{P_p}{b} + c'\tan\phi' - \frac{\gamma b}{2}\tan\phi' \tag{16.3}$$

The passive pressure in Eq. (16.2) and Eq.(16.3) is the sum of the contribution of the weight of soil γ, cohesion c', and surcharge q and can be expressed as

$$P_p = \frac{1}{2}\gamma(b\tan\phi')^2 K_\gamma + c'(b\tan\phi')K_c + q(b\tan\phi')K_q \tag{16.4}$$

where K_γ, K_c, and K_q are earth-pressure coefficients that are functions of the soil friction angle, ϕ'.

Combining Eqs. (16.3) and (16.4), we obtain

$$q_u = c'N_c + qN_q + \frac{1}{2}\gamma B N_\gamma$$

where

$$N_c = \tan\phi'(K_c + 1) \tag{16.5}$$

$$N_q = K_q \tan\phi' \tag{16.6}$$

$$N_\gamma = \frac{1}{2}\tan\phi'(K_\gamma \tan\phi' - 1) \tag{16.7}$$

The terms N_c, N_q, and N_γ are, respectively, the contributions of cohesion, surcharge, and unit weight of soil to the ultimate load-bearing capacity. It is extremely tedious to evaluate K_c, K_q, and K_γ. For this reason, Terzaghi used an approximate

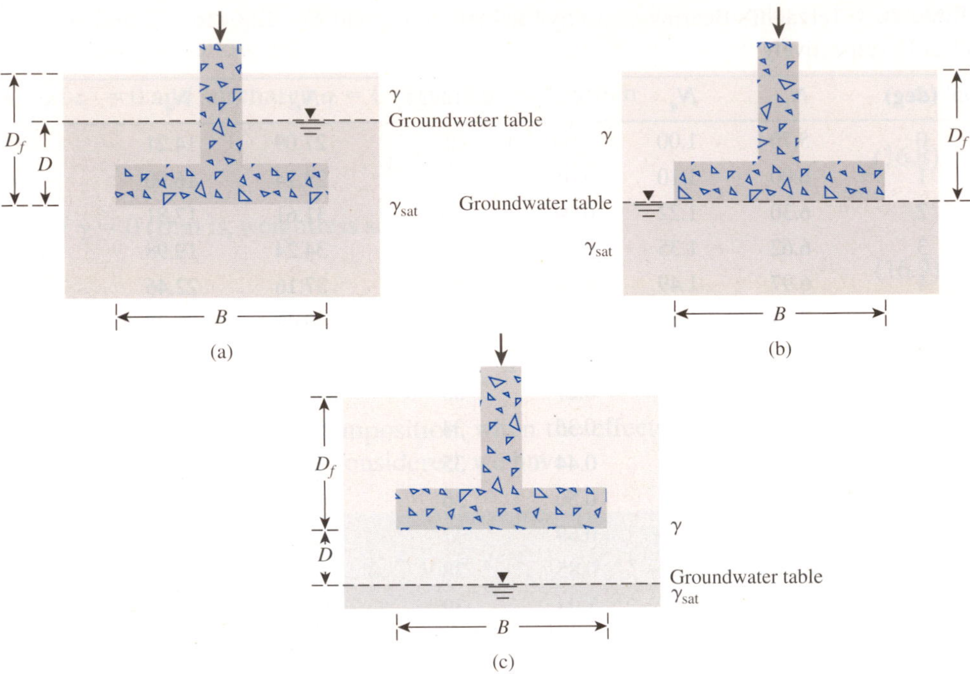

Figure 16.6 Effect of the location of groundwater table on the bearing capacity of shallow footing: (a) Case I; (b) Case II; (c) Case III

Case I (Figure 16.6a) If the groundwater table is located at a distance D above the bottom of the footing, the magnitude of q in the second term of the bearing-capacity equation should be calculated as

$$q = \gamma(D_f - D) + \gamma' D \tag{16.16}$$

where $\gamma' = \gamma_{sat} - \gamma_w$ = effective unit weight of soil. Also, the unit weight of soil, γ, that appears in the third term of the bearing-capacity equations should be replaced by γ'.

Case II (Figure 16.6b) If the groundwater table coincides with the bottom of the footing, the magnitude of q is equal to γD_f. However, the unit weight, γ, in the third term of the bearing-capacity equations should be replaced by γ'.

Case III (Figure 16.6c) When the groundwater table is at a depth D below the bottom of the footing, $q = \gamma D_f$. The magnitude of γ in the third term of the bearing-capacity equations should be replaced by γ_{av}.

$$\gamma_{av} = \frac{1}{B}[\gamma D + \gamma'(B - D)] \quad (\text{for } D \le B) \tag{16.17a}$$

$$\gamma_{av} = \gamma \quad (\text{for } D > B) \tag{16.17b}$$

16.5 Factor of Safety

Generally, a factor of safety, F_s, of about 3 or more is applied to the ultimate soil-bearing capacity to arrive at the value of the allowable bearing capacity. An F_s of 3 or more is not considered too conservative. In nature, soils are neither homogeneous nor isotropic. Much uncertainty is involved in evaluating the basic shear strength parameters of soil.

There are two basic definitions of the allowable bearing capacity of shallow footings. They are gross allowable bearing capacity, and net allowable bearing capacity.

The *gross allowable bearing capacity* can be calculated as

$$q_{\text{all}} = \frac{q_u}{F_s} \tag{16.18}$$

As defined by Eq. (16.18) q_{all} is the allowable load per unit area to which the soil under the footing should be subjected to avoid any chance of bearing capacity failure. It includes the contribution (Figure 16.7) of (a) the dead and live loads above the ground surface, $W_{(D+L)}$; (b) the self-weight of the footing, W_F; and (c) the weight of the soil located immediately above footing, W_S. Thus,

$$q_{\text{all}} = \frac{q_u}{F_s} = \left[\frac{W_{(D+L)} + W_F + W_S}{A}\right] \tag{16.19}$$

where A = area of the foundation.

The *net allowable bearing capacity* is the allowable load per unit area of the footing in excess of the existing vertical effective stress at the level of the footing. The vertical effective stress at the footing level is equal to $q = \gamma D_f$. So, the net ultimate load is

$$q_{u(\text{net})} = q_u - q \tag{16.20}$$

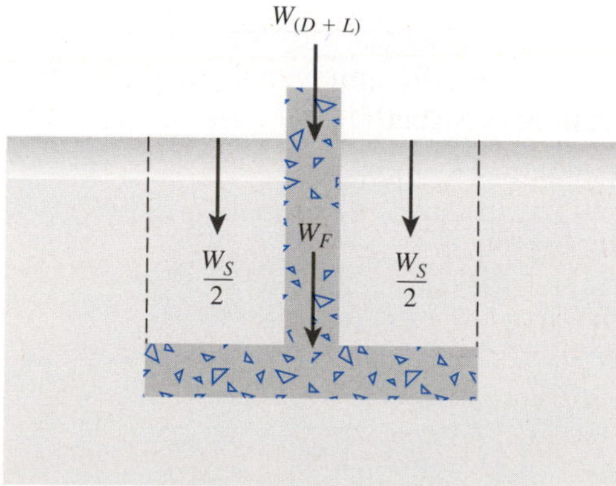

Figure 16.7 Contributions to q_{all}

Hence,

$$q_{\text{all(net)}} = \frac{q_{u(\text{net})}}{F_s} = \frac{q_u - q}{F_s} \tag{16.21}$$

If we assume that the weight of the soil and the weight of the concrete from which the footing is made are approximately the same, then

$$q = \gamma D_f \simeq \frac{W_S + W_F}{A}$$

Hence,

$$q_{\text{all(net)}} = \frac{W_{(D+L)}}{A} = \frac{q_u - q}{F_s} \tag{16.22}$$

Example 16.1

A square footing is 1.5 m × 1.5 m in plan. The soil supporting the foundation has a friction angle $\phi' = 20°$, and $c' = 15.2\ \text{kN/m}^2$. The unit weight of soil, γ, is 17.8 kN/m^3. Determine the allowable gross load on the footing with a factor of safety (F_s) of 4. Assume that the depth of the footing (D_f) is 1 meter and that general shear failure occurs in soil.

Solution

From Eq. (16.12),

$$q_u = 1.3c'N_c + qN_q + 0.4\gamma BN_\gamma$$

From Table 16.1, for $\phi' = 20°$,

$$N_c = 17.69$$

$$N_q = 7.44$$

$$N_\gamma = 3.64$$

Thus,

$$q_u = (1.3)(15.2)(17.69) + (1 \times 17.8)(7.44) + (0.4)(17.8)(1.5)(3.64)$$

$$= 349.55 + 132.43 + 38.87 = 520.85 \approx 521\ \text{kN/m}^2$$

So the allowable load per unit area of the footing is

$$q_{\text{all}} = \frac{q_u}{F_s} = \frac{521}{4} = 130.25\ \text{kN/m}^2 \approx 130\ \text{kN/m}^2$$

Thus, the total allowable gross load

$$Q = (130)B^2 = (130)(1.5 \times 1.5) = \mathbf{292.5\ kN}$$

Example 16.2

Refer to Example 16.1. Other quantities remaining the same, if the ground water table is located 1 m below the bottom of the footing, determine the allowable gross load per unit area with $F_s = 3$. Assume $\gamma_{sat} = 19 \text{ kN/m}^3$.

Solution

This is the case shown in Figure 16.6(c). $D = 1$ m. From Eq. (16.17a),

$$\gamma_{av} = \frac{1}{B}[\gamma D + \gamma'(B - D)]$$

$$= \frac{1}{1.5}[(17.8)(1) + (19 - 9.81))(1.5 - 1.0)] = 14.93 \text{ kN/m}^3$$

$$q_u = (1.3)(15.2)(17.69) + (1 \times 17.8)(7.44) + (0.4)(14.93)(1.5)(3.65)$$

$$= 349.55 + 132.43 + 32.61 = 514.59 \text{ kN/m}^2$$

$$q_{all} = \frac{q_u}{F_s} = \frac{515.59}{3} = \mathbf{171.53 \text{ kN/m}^2}$$

Example 16.3

A square footing is shown in Figure 16.8. The footing will carry a gross mass of 30,000 kg. Using a factor of safety of 3, determine the size of the footing—that is, the size of B. Use Eq. (16.12).

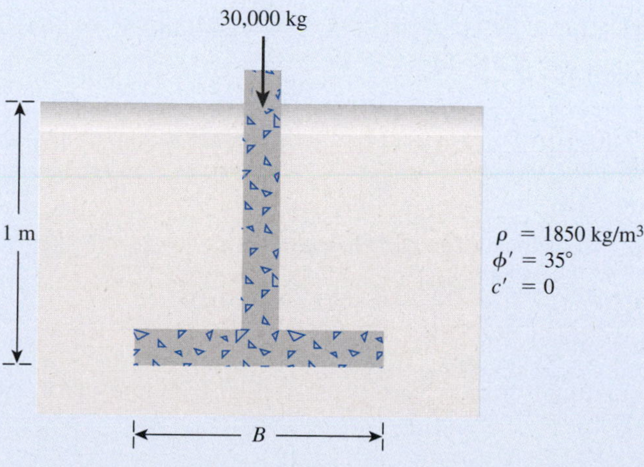

Figure 16.8

Solution

It is given that soil density = 1850 kg/m^3. So

$$\gamma = \frac{1850 \times 9.81}{1000} = 18.15 \text{ kN/m}^3$$

Total gross load to be supported by the footing is

$$\frac{(30{,}000)9.81}{1000} = 294.3 \text{ kN} = Q_{\text{all}}$$

From Eq. (16.12),

$$q_u = 1.3c'N_c + qN_q + 0.4\gamma BN_\gamma$$

With a factor of safety of 3,

$$q_{\text{all}} = \frac{q_u}{3} = \frac{1}{3}(1.3c'N_c + qN_q + 0.4\gamma BN_\gamma) \qquad \text{(a)}$$

Also,

$$q_{\text{all}} = \frac{Q_{\text{all}}}{B^2} = \frac{294.3}{B^2} \qquad \text{(b)}$$

From Eqs. (a) and (b),

$$\frac{294.3}{B^2} = \frac{1}{3}(1.3c'N_c + qN_q + 0.4\gamma BN_\gamma) \qquad \text{(c)}$$

From Table 16.1, for $\phi' = 35°$, $N_c = 57.75$, $N_q = 41.44$, and $N_\gamma = 45.41$. Substituting these values into Eq. (c) yields

$$\frac{294.3}{B^2} = \frac{1}{3}[(1.3)(0)(57.75) + (18.15 \times 1)(41.44) + 0.4(18.15)(B)(45.41)]$$

or

$$\frac{294.3}{B^2} = 250.7 + 109.9B$$

The preceding equation may now be solved by trial and error, and from that we get

$$B \simeq \mathbf{0.95 \text{ m}}$$

16.6 General Bearing Capacity Equation

After the development of Terzaghi's bearing-capacity equation, several investigators worked in this area and refined the solution (that is, Meyerhof, 1951 and 1963; Lundgren and Mortensen, 1953; Balla, 1962; Vesic, 1973; and Hansen, 1970). Different solutions show that the bearing-capacity factors N_c and N_q do not change much. However, for a given value of ϕ', the values of N_γ obtained by different investigators vary widely. This difference is because of the variation of the assumption of the wedge shape of soil located directly below the footing, as explained in the following paragraph.

While deriving the bearing-capacity equation for a strip footing, Terzaghi used the case of a rough footing and assumed that the sides *AJ* and *BJ* of the soil wedge *ABJ* (see Figure 16.5a) make an angle ϕ' with the horizontal. Later model tests (for example, DeBeer and Vesic, 1958) showed that Terzaghi's assumption of the general nature of the rupture surface in soil for bearing-capacity failure is correct. However, tests have shown that the sides *AJ* and *BJ* of the soil wedge *ABJ* make angles of about $45 + \phi'/2$ degrees (instead of ϕ') with the horizontal. This type of failure mechanism is shown in Figure 16.9. It consists of a Rankine active zone *ABJ* (Zone I), two radial shear zones (Zones II), and two Rankine passive zones (Zones III). The curves *JD* and *JE* are arcs of a logarithmic spiral.

On the basis of this type of failure mechanism, the ultimate bearing capacity of a strip footing may be evaluated by the approximate method of superimposition described in Section 16.3 as

$$q_u = q_c + q_q + q_\gamma \tag{16.23}$$

where q_c, q_q, and q_γ are the contributions of cohesion, surcharge, and unit weight of soil, respectively.

Reissner (1924) expressed q_q as

$$q_q = qN_q \tag{16.24}$$

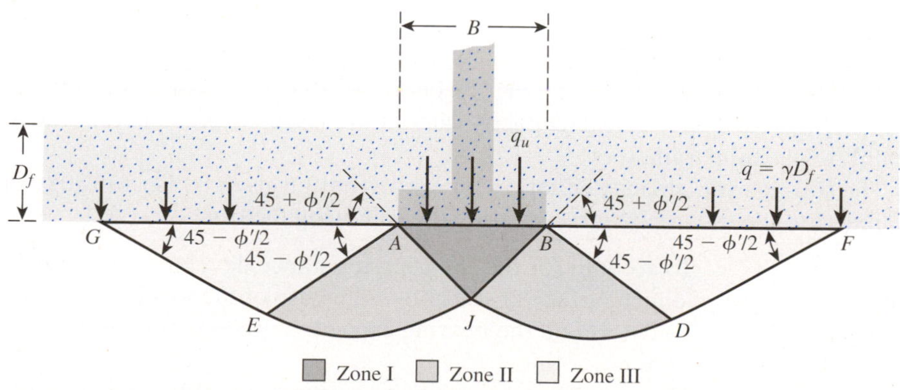

Figure 16.9 Soil-bearing capacity calculation—general shear failure

where

$$N_q = e^{\pi \tan \phi'} \tan^2\left(45 + \frac{\phi'}{2}\right) \tag{16.25}$$

Prandtl (1921) showed that

$$q_c = c'N_c \tag{16.26}$$

where

$$N_c = (N_q - 1)\cot \phi' \tag{16.27}$$

$\uparrow$

Eq. (16.25)

Vesic (1973) expressed q_γ as

$$q_\gamma = \frac{1}{2} B\gamma N_\gamma \tag{16.28}$$

where

$$N_\gamma = 2(N_q + 1) \tan \phi' \tag{16.29}$$

$\uparrow$

Eq. (16.25)

Combining Eqs. (16.23), (16.24), (16.26), and (16.28), we obtain

$$q_u = c'N_c + qN_q + \frac{1}{2}\gamma BN_\gamma \tag{16.30}$$

This equation is in the same general form as that given by Terzaghi [Eq. (16.11)]; however, the values of the bearing capacity factors are not the same. The values of N_q, N_c, and N_γ, defined by Eqs. (16.25), (16.27), and (16.29), are given in Table 16.2. But for all practical purposes, Terzaghi's bearing-capacity factors will yield good results. Differences in bearing-capacity factors are usually minor compared with the unknown soil parameters.

The soil-bearing capacity equation for a strip footing given by Eq. (16.30) can be modified for general use by incorporating the following factors:

Depth factor: To account for the shearing resistance developed along the failure surface in soil above the base of the footing

Shape factor: To determine the bearing capacity of rectangular and circular footings

Inclination factor: To determine the bearing capacity of a footing on which the direction of load application is inclined at a certain angle to the vertical

Table 16.2 Bearing-Capacity Factors N_c, N_q, and N_γ [Eqs. (16.25), (16.27) and (16.29)]

ϕ' (deg)	N_c	N_q	N_γ	ϕ' (deg)	N_c	N_q	N_γ
0	5.14	1.00	0.00	26	22.25	11.85	12.54
1	5.38	1.09	0.07	27	23.94	13.20	14.47
2	5.63	1.20	0.15	28	25.80	14.72	16.72
3	5.90	1.31	0.24	29	27.86	16.44	19.34
4	6.19	1.43	0.34	30	30.14	18.40	22.40
5	6.49	1.57	0.45	31	32.67	20.63	25.99
6	6.81	1.72	0.57	32	35.49	23.18	30.22
7	7.16	1.88	0.71	33	38.64	26.09	35.19
8	7.53	2.06	0.86	34	42.16	29.44	41.06
9	7.92	2.25	1.03	35	46.12	33.30	48.03
10	8.35	2.47	1.22	36	50.59	37.75	56.31
11	8.80	2.71	1.44	37	55.63	42.92	66.19
12	9.28	2.97	1.69	38	61.35	48.93	78.03
13	9.81	3.26	1.97	39	67.87	55.96	92.25
14	10.37	3.59	2.29	40	75.31	64.20	109.41
15	10.98	3.94	2.65	41	83.86	73.90	130.22
16	11.63	4.34	3.06	42	93.71	85.38	155.55
17	12.34	4.77	3.53	43	105.11	99.02	186.54
18	13.10	5.26	4.07	44	118.37	115.31	224.64
19	13.93	5.80	4.68	45	133.88	134.88	271.76
20	14.83	6.40	5.39	46	152.10	158.51	330.35
21	15.82	7.07	6.20	47	173.64	187.21	403.67
22	16.88	7.82	7.13	48	199.26	222.31	496.01
23	18.05	8.66	8.20	49	229.93	265.51	613.16
24	19.32	9.60	9.44	50	266.89	319.07	762.89
25	20.72	10.66	10.88				

Thus, the modified general ultimate bearing capacity equation can be written as

$$q_u = c'\lambda_{cs}\lambda_{cd}\lambda_{ci}N_c + q\lambda_{qs}\lambda_{qd}\lambda_{qi}N_q + \frac{1}{2}\lambda_{\gamma s}\lambda_{\gamma d}\lambda_{\gamma i}\gamma B N_\gamma \tag{16.31}$$

where λ_{cs}, λ_{qs}, and $\lambda_{\gamma s}$ = shape factors

λ_{cd}, λ_{qd}, and $\lambda_{\gamma d}$ = depth factors

λ_{ci}, λ_{qi}, and $\lambda_{\gamma i}$ = inclination factors

It is important to recognize the fact that, in the case of inclined loading, Eq. (16.31) provides the vertical component.

The approximate relationships for the shape, depth, and inclination factors are described below.

Shape factors The equations for the shape factors $\lambda_{cs}, \lambda_{qs}$, and $\lambda_{\gamma s}$ were recommended by De Beer (1970) and are

$$\lambda_{cs} = 1 + \left(\frac{B}{L}\right)\left(\frac{N_q}{N_c}\right) \tag{16.32}$$

$$\lambda_{qs} = 1 + \left(\frac{B}{L}\right)\tan\phi' \tag{16.33}$$

and

$$\lambda_{\gamma s} = 1 - 0.4\left(\frac{B}{L}\right) \tag{16.34}$$

where L = length of the foundation ($L > B$).
The shape factors are empirical relations based on extensive laboratory tests.

Depth factors Hansen (1970) proposed the following equations for the depth factors:

$$\lambda_{cd} = 1 + 0.4\left(\frac{D_f}{B}\right) \tag{16.35}$$

$$\lambda_{qd} = 1 + 2\tan\phi'(1 - \sin\phi')^2\frac{D_f}{B} \tag{16.36}$$

$$\lambda_{\gamma d} = 1 \tag{16.37}$$

Equations (16.35) and (16.36) are valid for $D_f/B \leq 1$. For a depth-of-embedment-to-footing width ratio greater than unity ($D_f/B > 1$), the equations have to be modified to

$$\lambda_{cd} = 1 + (0.4)\tan^{-1}\left(\frac{D_f}{B}\right) \tag{16.38}$$

$$\lambda_{qd} = 1 + 2\tan\phi'(1 - \sin\phi')^2\tan^{-1}\left(\frac{D_f}{B}\right) \tag{16.39}$$

and

$$\lambda_{\gamma d} = 1 \tag{16.40}$$

respectively. The factor $\tan^{-1}(D_f/B)$ is in radians in Eqs. (16.38) and (16.39).

Inclination factors Meyerhof (1963) suggested the following inclination factors for use in Eq. (16.31):

$$\lambda_{ci} = \lambda_{qi} = \left(1 - \frac{\alpha°}{90°}\right)^2 \tag{16.41}$$

$$\lambda_{\gamma i} = \left(1 - \frac{\alpha}{\phi'}\right)^2 \tag{16.42}$$

Here, α = inclination of the load on the foundation with respect to the vertical.

For undrained condition, if the footing is subjected to vertical loading (that is, $\alpha = 0°$), then

$$\phi = 0$$

$$c = c_u$$

$$N_\gamma = 0$$

$$N_q = 1$$

$$N_c = 5.14$$

$$\lambda_{ci} = \lambda_{qi} = \lambda_{\gamma i} = 1$$

So Eq. (16.31) transforms to

$$q_u = 5.14c_u\left[1 + 0.2\left(\frac{B}{L}\right)\right]\left[1 + 0.4\left(\frac{D_f}{B}\right)\right] + q \tag{16.43}$$

Example 16.4

A square footing is shown in Figure 16.10. Determine the safe gross inclined load (factor of safety of 3) that the footing can carry. Use Eq. (16.31).

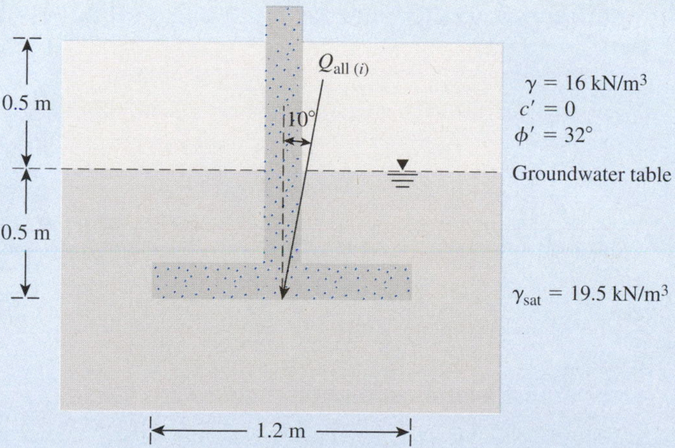

Figure 16.10

Solution

From Eq. (16.31),

$$q_u = c'\lambda_{cs}\lambda_{cd}\lambda_{ci}N_c + q\lambda_{qs}\lambda_{qd}\lambda_{qi}N_q + \frac{1}{2}\gamma'\lambda_{\gamma s}\lambda_{\gamma d}\lambda_{\gamma i}BN_\gamma$$

Because $c' = 0$,

$$q_u = q\lambda_{qs}\lambda_{qd}\lambda_{qi}N_q + \frac{1}{2}\gamma'\lambda_{\gamma s}\lambda_{\gamma d}\lambda_{\gamma i}BN_\gamma$$

From Table 16.2 for $\phi' = 32°$, $N_q = 23.18$ and $N_\gamma = 30.22$.

$$\lambda_{qs} = 1 + \frac{B}{L}\tan\phi' = 1 + \frac{1.2}{1.2}\tan 32 = 1.625$$

$$\lambda_{\gamma s} = 1 - 0.4\frac{B}{L} = 1 - 0.4\left(\frac{1.2}{1.2}\right) = 0.6$$

$$\lambda_{qd} = 1 + 2\tan\phi'(1 - \sin\phi')^2\left(\frac{D_f}{B}\right)$$

$$= 1 + 2\tan 32(1 - \sin 32)^2\left(\frac{1}{1.2}\right)$$

$$= 1.23$$

$$\lambda_{\gamma d} = 1$$

$$\lambda_{qi} = \left(1 - \frac{\alpha°}{90°}\right)^2 = \left(1 - \frac{10}{90}\right)^2 = 0.79$$

$$\lambda_{\gamma i} = \left(1 - \frac{\alpha}{\phi'}\right)^2 = \left(1 - \frac{10}{32}\right)^2 = 0.473$$

The groundwater table is located above the bottom of the footing, so, from Eq. (16.16),

$$q = (0.5)(16) + (0.5)(19.5 - 9.81) = 12.845 \text{ kN/m}^2$$

Thus,

$$q_u = (12.845)(1.625)(1.23)(0.79)(23.18) + \left(\frac{1}{2}\right)(19.5 - 9.81)(0.6)(1)(0.473)(1.2)(30.22)$$

$$= 520 \text{ kN/m}^2$$

$$q_{all} = \frac{q_u}{3} = \frac{520}{3} = 170.33 \text{ kN/m}^2$$

Hence, the gross vertical load is as follows:

$$Q = q_{all}(B^2) = 170.33(1.2)^2 = 245.88 \text{ kN}$$

Thus,

$$Q_{all(i)} = \frac{245.28}{\cos 10} = \mathbf{249\ kN}$$

16.7 Ultimate Load for Shallow Footings Under Eccentric Load (One-Way Eccentricity)

To calculate the bearing capacity of shallow footings with eccentric loading, Meyerhof (1953) introduced the concept of *effective area*. This concept can be explained with reference to Figure 16.11, in which a footing of length L and width B is subjected to an eccentric load, Q_u. If Q_u is the ultimate load on the footing, it may be approximated as follows:

1. Referring to Figures 16.11b and 16.11c, calculate the effective dimensions of the footing. If the eccentricity (e) is in the x direction (Figure 16.11b), the *effective dimensions* are

$$X = B - 2e$$

and

$$Y = L$$

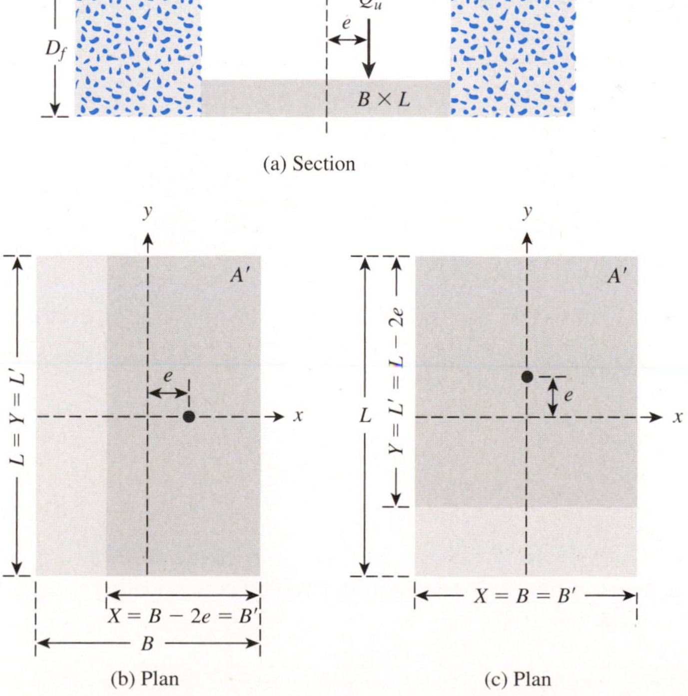

Figure 16.11 Ultimate load for shallow footing under eccentric load

However, if the eccentricity is in the y direction (Figure 16.11c), the effective dimensions are

$$Y = L - 2e$$

and

$$X = B$$

2. The lower of the two effective dimensions calculated in step 1 is the *effective width* (B') and the other is the *effective length* (L'). Thus,

$$B' = X \text{ or } Y\text{, whichever is smaller}$$
$$L' = X \text{ or } Y\text{, whichever is larger}$$

3. So the effective area is equal to B' times L'. Now, using the effective width, we can rewrite Eq. (16.31) as

$$q'_u = c'\lambda_{cs}\lambda_{cd}N_c + q\lambda_{qs}\lambda_{qd}N_q + \frac{1}{2}\lambda_{\gamma s}\lambda_{\gamma d}\gamma B'N_\gamma \tag{16.44}$$

Note that the preceding equation is obtained by substituting B' for B in Eq. (16.31). While computing the shape and depth factors, one should use B' for B and L' for L.

4. Once the value of q_u is calculated from Eq. (16.44), we can obtain the total gross ultimate load as follows:

$$Q_u = q'_u(B'L') = q_u A' \tag{16.45}$$

where A' = effective area.

Reduction factor for granular soil

Purkayastha and Char (1977) carried out stability analysis of eccentrically loaded *continuous footings on granular soil* (i.e., $c' = 0$) using the method of slices. Based on that analysis, they proposed that

$$R_k = 1 - \frac{q_{u(\text{eccentric})}}{q_{u(\text{centric})}} \tag{16.46}$$

where R_k = reduction factor

$q_{u(\text{eccentric})}$ = average ultimate load per unit area of eccentrically loaded continuous footing

$$= \frac{Q_{u(\text{eccentric})}}{B}$$

$q_{u(\text{centric})}$ = ultimate bearing capacity of centrally loaded continuous footing

$$= \frac{Q_{u(\text{centric})}}{B}$$

The magnitude of R_k can be expressed as

$$R_k = a\left(\frac{e}{B}\right)^k \tag{16.47}$$

where a and k are functions of the embedment ratio D_f/B (Table 16.3).

Hence, combining Eqs. (16.46) and (16.47) gives

$$Q_{u(\text{eccentric})} = Q_{u(\text{centric})}\left[1 - a\left(\frac{e}{B}\right)^k\right] \tag{16.48}$$

where $Q_{u(\text{eccentric})}$ and $Q_{u(\text{centric})}$ = ultimate load per unit length, respectively, for eccentrically and centrically loaded footings.

Patra, Sivakugan, Das, and Sethy (2015) conducted several model tests for *eccentrically loaded rectangular footings* on sand. Based on their test results, it was suggested that the reduction factor R_k [Eq. (16.47)] for rectangular footings can be expressed as (for all values of D_f/B),

$$R_k = a\left(\frac{e}{B}\right)^k$$

where

$$a = \left(\frac{B}{L}\right)^2 - 1.6\left(\frac{B}{L}\right) + 2.13 \tag{16.49}$$

$$k = 0.3\left(\frac{B}{L}\right)^2 - 0.56\left(\frac{B}{L}\right) + 0.9 \tag{16.50}$$

Table 16.3 Variations of a and k [Eq. (16.47)]

D_f/B	a	k
0	1.862	0.73
0.25	1.811	0.785
0.5	1.754	0.80
1.0	1.820	0.888

Example 16.5

A rectangular footing 1.5 m × 1 m is shown in Figure 16.12. Determine the magnitude of the gross ultimate load applied eccentrically for bearing-capacity failure in soil.

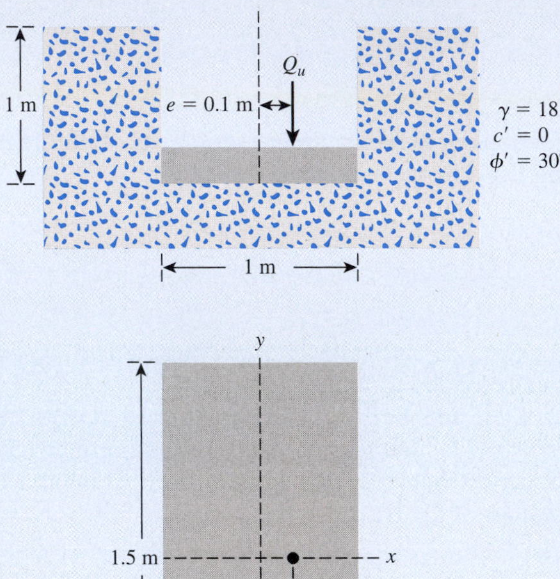

Figure 16.12

Solution

From Figures 16.11b and 16.12,

$$X = B - 2e = 1 - 2e = 1 - (2)(0.1) = 0.8 \text{ m}$$

$$Y = L = 1.5 \text{ m}$$

So, effective width $B' = 0.8$ m and effective length $L' = 1.5$ m. From Eq. (16.44),

$$q'_u = q\lambda_{qs}\lambda_{qd}N_q + \frac{1}{2}\lambda_{\gamma s}\lambda_{\gamma d}\gamma B' N_\gamma$$

From Table 16.2 for $\phi' = 30°$, $N_q = 18.4$ and $N_\gamma = 22.4$. Also,

$$\lambda_{qs} = 1 + \frac{B'}{L'}\tan\phi' = 1 + \left(\frac{0.8}{1.5}\right)\tan 30 = 1.308$$

$$\lambda_{\gamma s} = 1 - 0.4\left(\frac{0.8}{1.5}\right) = 0.787$$

$$\lambda_{qd} = 1 + 2\tan\phi'(1 - \sin\phi')^2\left(\frac{D_f}{B'}\right)$$

$$= 1 + 2\tan 30\,(1 - \sin 30)^2\left(\frac{1}{0.8}\right)$$

$$= 1.361$$

$$\lambda_{\gamma d} = 1$$

So

$$q'_u = (1 \times 18)(1.308)(1.361)(18.4) + \left(\frac{1}{2}\right)(0.787)(1.0)(18)(0.8)(22.4) = 716.53 \text{ kN/m}^2$$

Hence, from Eq. (16.45),

$$Q_u = q'_u(B'L') = (716.53)(0.8)(1.5) = \mathbf{859.8\ kN}$$

Example 16.6

Refer to Example 16.5. Determine the gross ultimate load the footing could carry by using Eqs. (16.48), (16.49), and (16.50).

Solution

$$Q_{u(\text{eccentic})} = Q_{u(\text{centic})}\left[1 - a\left(\frac{e}{B}\right)^k\right]$$

$$Q_{u(\text{centric})} = (B \times L)(q_u)$$

$$= (B \times L)\left(q\lambda_{qs}\lambda_{qd}N_q + \frac{1}{2}\lambda_{\gamma s}\lambda_{\gamma d}\gamma BN_\gamma\right)$$

$$\lambda_{qs} = 1 + \frac{B}{L}\tan\phi' = 1 + \left(\frac{1}{1.5}\right)\tan 30 = 1.385$$

$$\lambda_{\gamma s} = 1 - 0.4\left(\frac{B}{L}\right) = 1 - 0.4\left(\frac{1}{1.5}\right) = 0.733$$

$$\lambda_{qd} = 1 + 2\tan\phi'(1 - \sin\phi')^2\left(\frac{D_f}{B}\right)$$

$$= 1 + 2\tan 30(1 - \sin 30)^2\left(\frac{1}{1}\right) = 1.289$$

$$\lambda_{\gamma d} = 1.0$$

From Example 16.5, $N_q = 18.4$ and $N_\gamma = 22.4$

$$q_u = (1 \times 18)(1.385)(1.289)(18.4) + \left(\frac{1}{2}\right)(0.733)(1.0)(18)(1)(22.4)$$

$$= 739.05 \text{ kN/m}^2$$

$$Q_{u(\text{centric})} = (1 \times 1.5)(739.05) = 1108.58 \text{ kN/m}^2$$

From Eq. (16.49),

$$a = \left(\frac{B}{L}\right)^2 - 1.6\left(\frac{B}{L}\right) + 2.13 = \left(\frac{1}{1.5}\right)^2 - 1.6\left(\frac{1}{1.5}\right) + 2.13 \approx 1.51$$

From Eq. (16.50),

$$k = 0.3\left(\frac{B}{L}\right)^2 - 0.56\left(\frac{B}{L}\right) + 0.9 = 0.3\left(\frac{1}{1.5}\right)^2 - 0.56\left(\frac{1}{1.5}\right) + 0.9 \approx 0.66$$

$$R_k = a\left(\frac{e}{B}\right)^k = 1.51\left(\frac{0.1}{1.5}\right)^{0.66} = 0.253$$

Hence,

$$Q_{u(\text{eccentic})} = Q_{u(\text{centic})}(1 - R_k) = 1108.58(1 - 0.253) = \mathbf{822.57\ kN}$$

16.8 Continuous Footing Under Eccentrically Inclined Load

Figure 16.13 shows a shallow strip footing subjected to an eccentrically inclined ultimate load $Q_{u(ei)}$ per unit length. The load is inclined at an angle α to the vertical and has an eccentricity e. In order to determine $Q_{u(ei)}$, one of the following methods may be adopted.

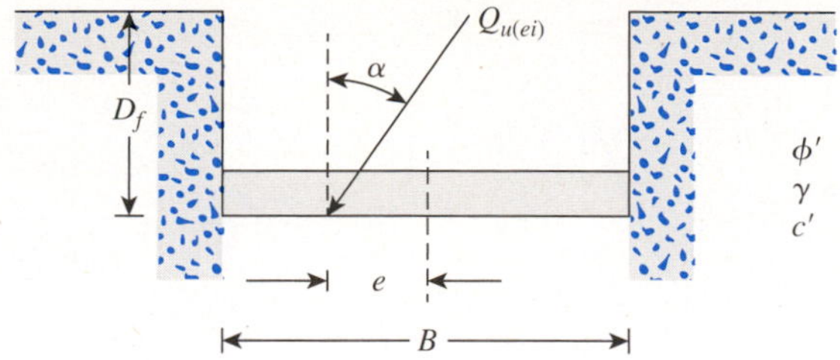

Figure 16.13 Shallow strip footing subjected to an eccentrically inclined load

Meyerhof's method

Equation (16.31) may be modified to obtain q'_u for a strip footing. For this case, the shape factors λ_{cs}, λ_{qs}, and $\lambda_{\gamma s}$ are all equal to one. Thus,

$$q'_u = c'N_c\lambda_{cd}\lambda_{ci} + qN_q\lambda_{qd}\lambda_{qi} + \frac{1}{2}\gamma B'N_\gamma\lambda_{\gamma d}\lambda_{\gamma i} \tag{16.51}$$

where B' = effective width = $B - 2e$

The *vertical component of the ultimate load* per unit length of the footing can be expressed as $Q_{u(ei)} \cos \alpha = q'_u B'$. Hence,

$$Q_{u(ei)} = \frac{B'}{\cos \alpha}\left[c'N_c\lambda_{cd}\lambda_{ci} + qN_q\lambda_{qd}\lambda_{qi} + \frac{1}{2}\gamma(B - 2e)N_\gamma\lambda_{\gamma d}\lambda_{\gamma i}\right] \tag{16.52}$$

Saran and Agarwal method

Saran and Agarwal (1991) conducted a limit equilibrium analysis to obtain the ultimate load that can be given as

$$Q_{u(ei)} = B\left[c'N_{c(ei)} + qN_{q(ei)} + \frac{1}{2}\gamma BN_{\gamma(ei)}\right] \tag{16.53}$$

where $N_{c(ei)}$, $N_{q(ei)}$, and $N_{\gamma(ei)}$ = bearing capacity factors (Figures 16.14, 16.15, and 16.16). It is important to note that Eq. (16.53) does not contain the depth factors.

Reduction factor method—granular soil

Based on about 120 model test results on dense and medium dense sand, Patra et al. (2012) have provided the empirical relationship to obtain $Q_{u(ei)}$ as

$$Q_{u(ei)} = Bq_u\left[1 - 2\left(\frac{e}{B}\right)\right]\left(1 - \frac{\alpha}{\phi'}\right)^{2-(D_f/B)} \tag{16.54}$$

where q_u = ultimate bearing capacity *with vertical centric load for a given D_f/B*.

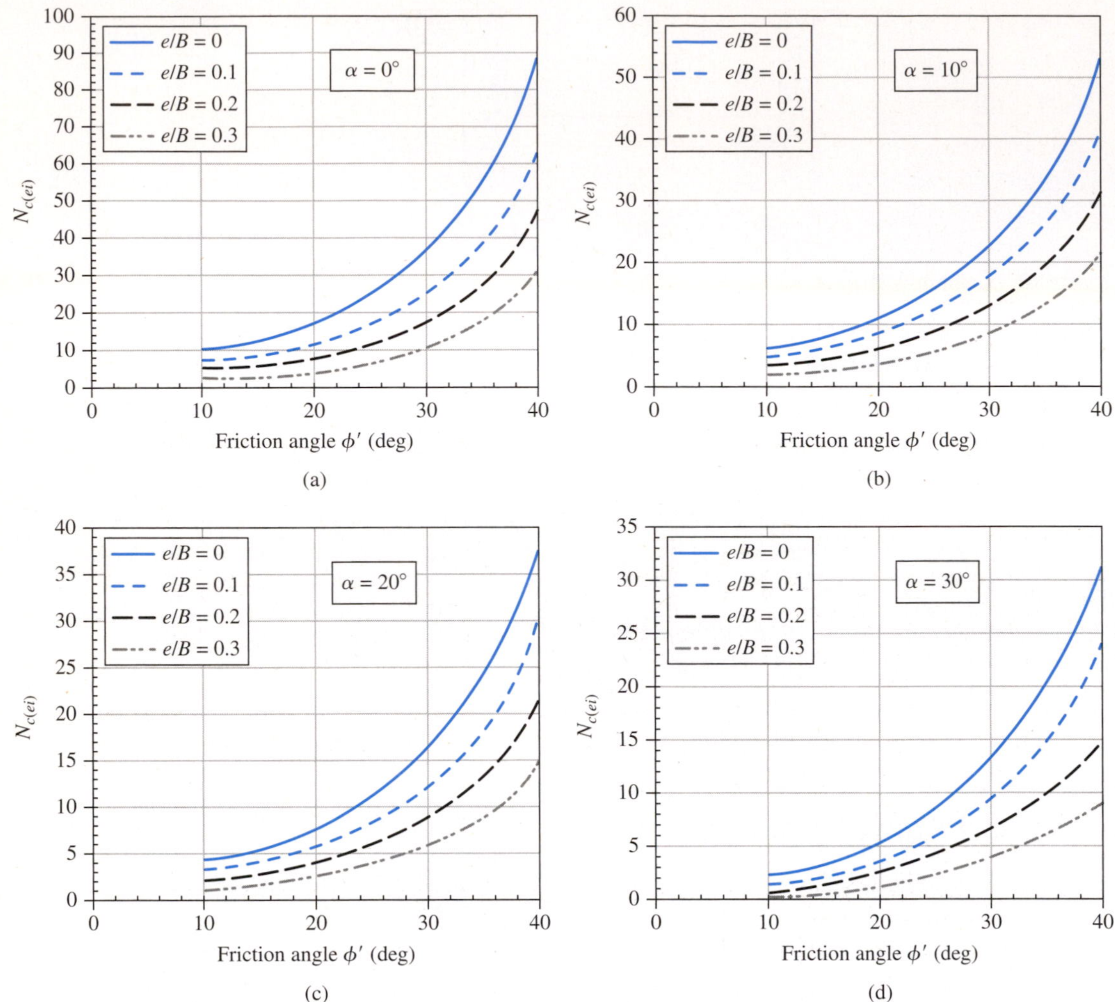

Figure 16.14 Variation of $N_{c(ei)}$: (a) $\alpha = 0°$; (b) $\alpha = 10°$; (c) $\alpha = 20°$; (d) $\alpha = 30°$

Example 16.7

Refer to Figure 16.13. A continuous footing is supported by a granular soil. The footing is subjected to an eccentrically inclined load. Given for the footing: $B = 2$ m, $e = 0.2$ m, $D_f = 1.5$ m, and $\alpha = 10°$. Given for the soil; $\phi' = 40°$, $c' = 0$, and $\gamma = 16.5$ kN/m^2.

Determine $Q_{u(ei)}$

a. Using Eq. (16.52)
b. Using Eq. (16.53)
c. Using Eq. (16.54)

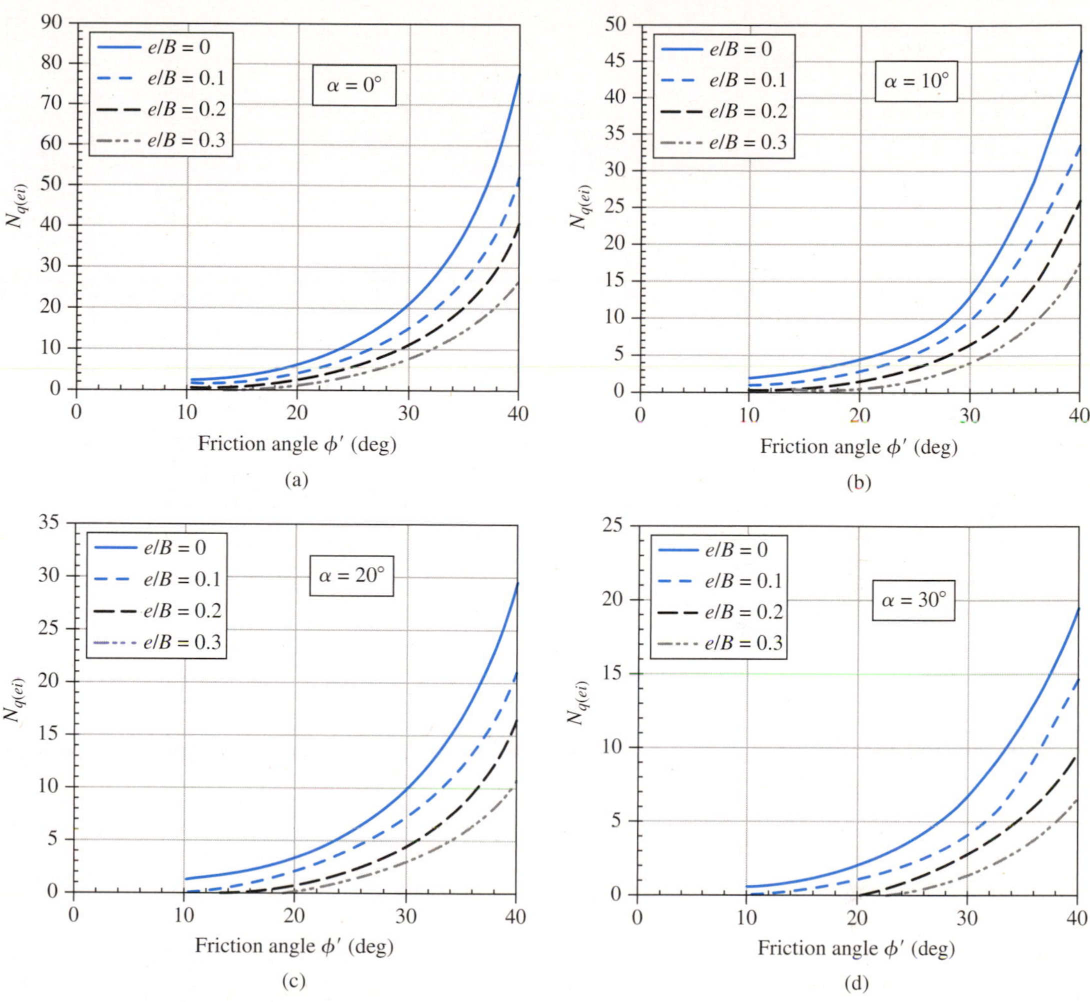

Figure 16.15 Variation of $N_{q(ei)}$: (a) $\alpha = 0°$; (b) $\alpha = 10°$; (c) $\alpha = 20°$; (d) $\alpha = 30°$

Solution

Part a

For $\phi' = 40°$ from Table 16.2, $N_q = 64.2$ and $N_\gamma = 109.41$. Hence,

$$B' = B - 2e = 2 - (2)(0.2) = 1.6 \text{ m}$$

$$\lambda_{qd} = 1 + 2\tan\phi'(1 - \sin\phi')^2 \frac{D_f}{B'} = 1 + 0.214\left(\frac{1.5}{1.6}\right) = 1.2$$

$$\lambda_{\gamma d} = 1$$

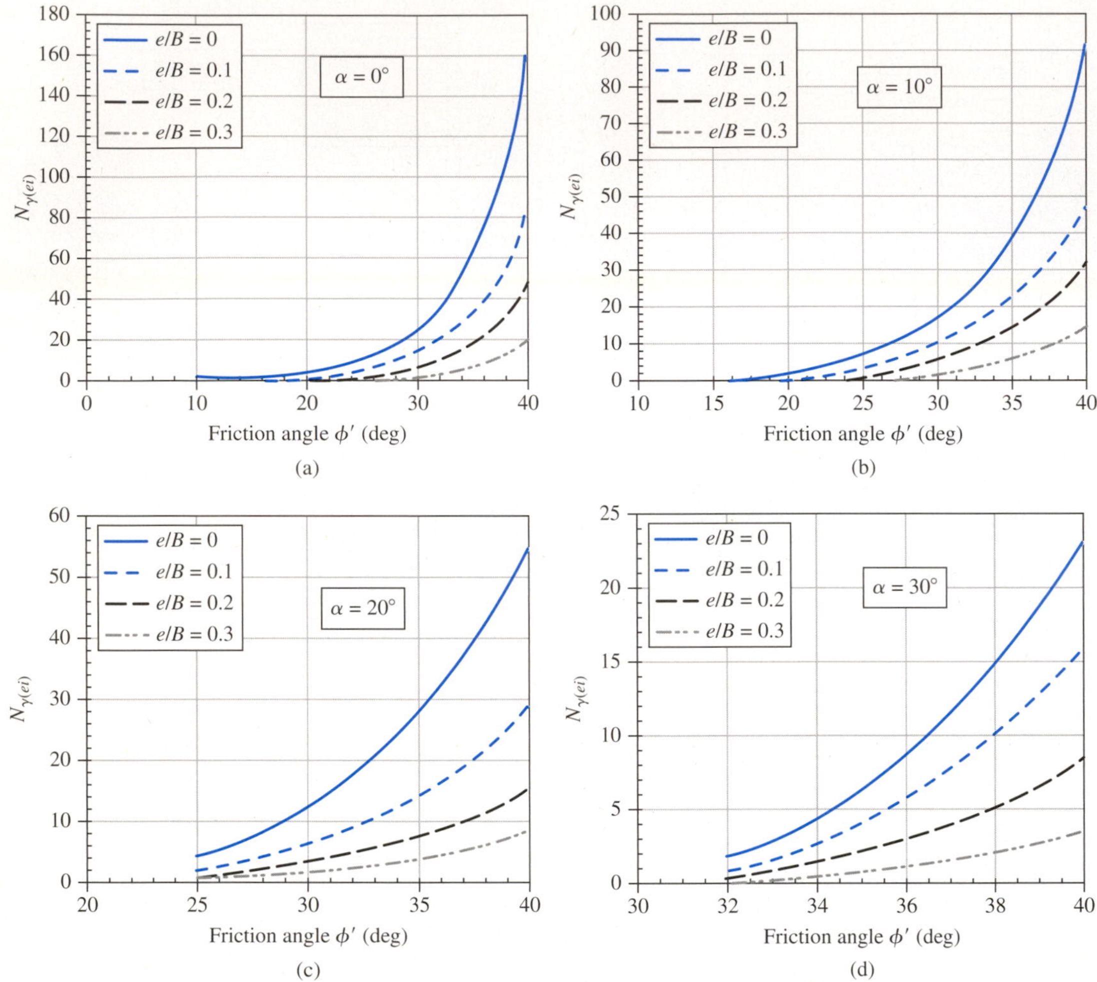

Figure 16.16 Variation of $N_{\gamma(ei)}$: (a) $\alpha = 0°$; (b) $\alpha = 10°$; (c) $\alpha = 20°$; (d) $\alpha = 30°$

$$\lambda_{qi} = \left(1 - \frac{\alpha}{90}\right)^2 = \left(1 - \frac{10}{90}\right)^2 = 0.79$$

$$\lambda_{\gamma i} = \left(1 - \frac{\alpha}{\phi'}\right)^2 = \left(1 - \frac{10}{40}\right)^2 = 0.56$$

From Eq. (16.51),

$$q'_u B' = B'\left[qN_q\lambda_{qd}\lambda_{qi} + \frac{1}{2}\gamma B' N_\gamma \lambda_{\gamma d}\lambda_{\gamma i}\right]$$

$$= 1.6\left[(1.5 \times 16.5)(64.2)(1.2)(0.79) + \left(\frac{1}{2}\right)(16.5)(1.6)(109.41)(1)(0.56)\right]$$

$$= 3704 \text{ kN/m}$$

From Eq. (16.52),

$$Q_{u(ei)} = \frac{q'_u B'}{\cos \alpha} = \frac{q'_u B'}{\cos 10} = \frac{3704}{\cos 10} \approx \mathbf{3761\ kN/m}$$

Part b

With $c' = 0$, Eq. (16.53) becomes

$$Q_{u(ei)} = B\left[qN_{q(ei)} + \frac{1}{2}\gamma B N_{\gamma(ei)}\right]$$

where $q = 1.5 \times 16.5 = 24.75$ kN/m², $B = 2$ m, $\gamma = 16.5$ kN/m³.

For this problem, $\phi' = 40°$, $e/B = 0.2/2 = 0.1$, and $\alpha = 10°$. From Figures 16.15(b) and 16.16(b), $N_{q(ei)} = 33.16$ and $N_{\gamma(ei)} = 47.48$. So,

$$Q_{u(ei)} = (2)\left[(24.75)(33.16) + \left(\frac{1}{2}\right)(16.5)(2)(47.48)\right] = \mathbf{3208.26\ kN/m}$$

Part c

From Eq. (16.54),

$$Q_{u(ei)} = Bq_u\left[1 - 2\left(\frac{e}{B}\right)\right]\left[1 - \left(\frac{\alpha}{\phi'}\right)\right]^{2-\frac{D_f}{B}}$$

$$q_u = qN_q F_{qd} + \frac{1}{2}\gamma B N_\gamma F_{\gamma d}$$

where $q = 16.5 \times 1.5 = 24.75$ kN/m², $\lambda_{qd} = 1 + 0.214(D_f/B) = 1 + 0.214(1.5/2) = 1.161$ (see Part a), $\lambda_{\gamma d} = 1$ (see Part a), $N_q = 64.2$, $N_\gamma = 109.41$, $e/B = 0.1$, and $D_f/B = 0.75$. Hence,

$$q_u = (24.75)(64.2)(1.161) + \left(\frac{1}{2}\right)(16.5)(2)(109.41)(1) = 3650\ \text{kN/m}^2$$

$$Q_{u(ei)} = (2)(3650)[1 - 2(0.1)]\left[1 - \left(\frac{10}{40}\right)\right]^{2-0.75} = \mathbf{4076\ kN/m}$$

Comments: Meyerhof's method and the method of Patra et al. yield fairly close results.

16.9 Bearing Capacity of Sand Based on Settlement

Obtaining undisturbed specimens of cohesionless sand during a soil exploration program is usually difficult. For this reason, the results of standard penetration tests (SPTs) performed during subsurface exploration are commonly used to predict the allowable soil-bearing capacity of footings on sand. (The procedure for conducting SPTs is discussed in detail in Chapter 17.)

Meyerhof (1956) proposed correlations for the *net allowable bearing capacity* (q_{net}) based on settlement (elastic). It was further revised by Meyerhof (1965) based on the field performance of footings. The correlations can be expressed as follows.

SI units

$$q_{\text{net}}(\text{kN/m}^2) = \frac{N_{60}}{0.05} F_d \left[\frac{S_e(\text{mm})}{25}\right] \quad (\text{for } B \leq 1.22 \text{ m}) \tag{16.55}$$

$$q_{\text{net}}(\text{kN/m}^2) = \frac{N_{60}}{0.08} \left(\frac{B + 0.3}{B}\right)^2 F_d \left[\frac{S_e(\text{mm})}{25}\right] \quad (\text{for } B > 1.22 \text{ m}) \tag{16.56}$$

where B = footing width (m)
S_e = settlement

In Eqs. (16.55) through (16.58),

N_{60} = field standard penetration number based on 60% average energy ratio
S_e = allowable settlement (elastic)

$$F_d = \text{depth factor} = 1 + 0.33\left(\frac{D_f}{B}\right) \leq 1.33 \tag{16.57}$$

The N_{60} values referred to in Eqs. (16.55) through (16.58) are the average values between the bottom of the footing and $2B$ below the bottom.

Comparison with field settlement observation

Meyerhof (1965) compiled the observed maximum settlement (S_e) for several mat foundations constructed on sand and gravel. These are shown in Table 16.4 (Column 5) along with the values of B, q_{net}, and N_{60}.

From Eq. (16.58), we can write

$$S_e(\text{mm}) = \frac{25\, q_{\text{net}}}{\left(\frac{N_{60}}{0.08}\right)\left(\frac{B + 0.3}{B}\right)^2 F_d} \tag{16.58}$$

As can be seen from Table 16.4, the widths B for the mats are large. Hence,

$$\left(\frac{B + 0.3}{B}\right)^2 \approx 1$$

$$F_d = 1 + 0.33\frac{D_f}{B} \approx 1$$

So

$$S_e(\text{mm}) \approx \frac{2\,q_{\text{net}}(\text{kN/m}^2)}{N_{60}} \tag{16.59}$$

Using the actual values of q_{net} and N_{60} given in Table 16.4, the magnitudes of S_e have been calculated via Eq. (16.61). These are shown in Column 6 of Table 16.4 as $S_{e(\text{predicted})}$. The ratio of $S_{e(\text{predicted})}/S_{e(\text{observed})}$ is shown in Column 7. This ratio varies from 0.84 to 3.6. Hence, it can be concluded that the allowable net bearing capacity

Table 16.4 Observed and Calculated Maximum Settlement of Mat Foundations on Sand and Gravel

Structure (1)	B (m) (2)[a]	N_{60} (3)[a]	q_{net} (kN/m²) (4)[a]	$S_{e(\text{observed})}$ (mm) (5)[a]	$S_{e(\text{predicted})}$ (mm) (6)[b]	$\frac{S_{e(\text{predicted})}}{S_{e(\text{observed})}}$ (7)
T. Edison Sao Paulo, Brazil	18.3	15	229.8	15.2	30.6	2.0
Banco de Brasil Sao Paulo, Brazil	22.9	18	239.4	27.9	26.6	0.95
Iparanga Sao Paulo, Brazil	9.1	9	306.4	35.6	68.1	1.91
C.B.I., Esplanada Sao Paulo, Brazil	14.6	22	383.0	27.9	34.8	1.25
Riscala Sao Paulo, Brazil	4.0	20	229.8	12.7	23	1.81
Thyssen Dusseldorf, Germany	22.6	25	239.4	24.1	10.2	0.8
Ministry Dusseldorf, Germany	15.9	20	220.2	21.6	22.0	1.02
Chimney Cologne, Germany	20.4	10	172.4	10.2	34.5	3.38

[a]*Compiled from Meyerhof (1965)*
[b]From Eq. (16.61)

for a given allowable settlement calculated using the empirical relation is safe and conservative.

16.10 Summary

In this chapter, theories for estimating the ultimate and allowable bearing capacities of shallow footings were presented. Estimation of the allowable bearing capacity of granular soil based on limited settlement criteria were discussed briefly.

Following is an itemized list of the important materials covered in this chapter.

- Terzaghi's ultimate bearing-capacity equations [Eqs. (16.11), (16.12), and (16.13)]
- General ultimate bearing-capacity equation [Eq. (16.31)]
- Ultimate bearing capacity of footings subjected to vertical eccentric load (Section 16.7).
- Ultimate bearing capacity of continuous footings under eccentrically inclined load (Section 16.8).
- Bearing capacity of shallow foundations based on settlement (Section 16.9)

Several building codes now used in the United States and elsewhere provide presumptive bearing capacities for various types of soil. It is extremely important to realize that they are *approximate values only*. The bearing capacity of foundations depends on several factors:

1. Subsoil stratification
2. Shear strength parameters of the subsoil
3. Location of the groundwater table
4. Environmental factors
5. Building size and weight
6. Depth of excavation
7. Type of structure

Hence, it is important that the allowable bearing capacity at a given site be determined based on the findings of soil exploration at that site, past experience of foundation construction, and fundamentals of the geotechnical engineering theories for bearing capacity.

Problems

16.1 A continuous footing is shown in Figure 16.17. Using Terzaghi's bearing capacity factors, determine the gross allowable load per unit area (q_{all}) that the footing can carry. Assume general shear failure. Given: $\gamma = 19$ kN/m^3, $c' = 31$kN/m^2, $\phi' = 28°$, $D_f = 1.5$ m, $B = 2$ m, and factor of safety = 3.5.

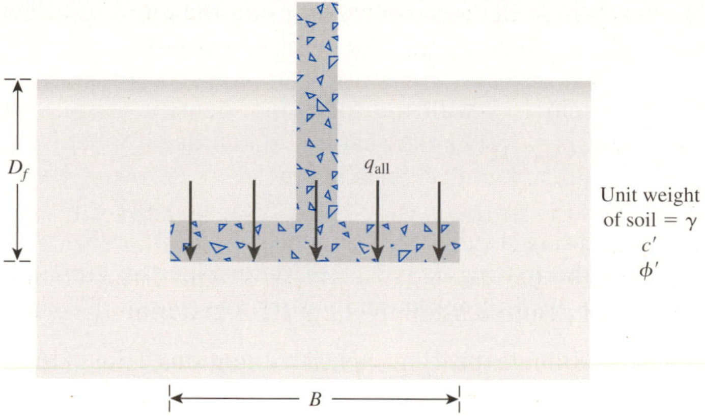

Figure 16.17

16.2. Refer to Problem 16.1. If a square footing with dimension 2 m × 2 m is used instead of the wall footing, what would be the allowable bearing capacity?

16.3 Redo Problem 16.1 with the following: $\gamma = 18.1\ \text{kN/m}^3$, $c' = 53\ \text{kN/m}^2$, $\phi' = 35°$, $D_f = 1.1$ m, $B = 1.52$ m, and factor of safety = 4.

16.4 Redo Problem 16.1 with the following: $\gamma = 16.5\ \text{kN/m}^3$, $c_u = 41\ \text{kN/m}^3$, $\phi' = 0$, $D_f = 1.5$ m, and factor of safety = 5.

16.5 Redo Problem 16.1 using the modified general ultimate bearing capacity Eq. (16.31).

16.6 Redo Problem 16.2 using the modified general ultimate bearing capacity Eq. (16.31).

16.7 Redo Problem 16.3 using the modified general ultimate bearing capacity Eq. (16.31).

16.8 Redo Problem 16.4 using the modified general ultimate bearing capacity Eq. (16.31).

16.9 A square footing is shown in Figure 16.18. Determine the gross allowable load, Q_{all}, that the footing can carry. Use Terzaghi's equation for general shear failure ($F_s = 4$). Given: $\gamma = 17\ \text{kN/m}^3$, $\gamma_{sat} = 19.2\ \text{kN/m}^3$, $c' = 32\ \text{kN/m}^3$, $\phi' = 26°$, $D_f = 1$ m, $h = 0.5$ m, and $B = 1.5$ m.

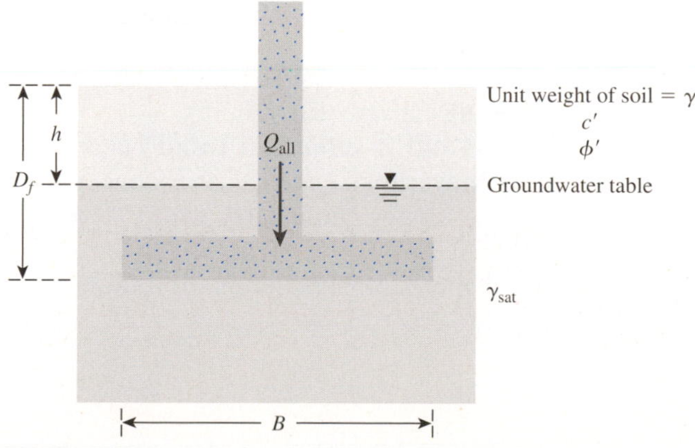

Figure 16.18

16.10 If the water table in Problem 16.9 drops down to 0.25 m below the foundation level, what would be the change in the factor of safety for the same gross allowable load?

16.11 Redo Problem 16.9 with the following: density of soil above the ground water table, ρ = 1800 kg/m^3; saturated soil density below the groundwater table, ρ_{sat} = 2050 kg/m^3; c' = 36 kN/m^3, ϕ' = 29°, D_f = 1.5 m, h = 1.5 m, and B = 2 m.

16.12 A square footing is subjected to an inclined load as shown in Figure 16.19. If the size of the footing, B = 2.25 m, determine the gross allowable inclined load, Q, that the footing can safely carry. Given: α = 12° and F_s = 3.5.

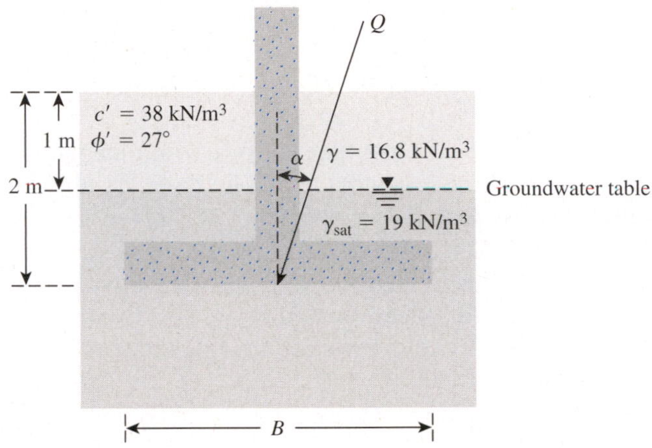

Figure 16.19

16.13 A square footing ($B \times B$) must carry a gross allowable load of 1160 kN. The base of the footing is to be located at a depth of 2 m below the ground surface. If the required factor of safety is 4.5, determine the size of the footing. Use Terzaghi's bearing capacity factors and assume general shear failure of soil. Given: γ = 17 kN/m^3, c' = 48 kN/m^2, ϕ' = 31°.

16.14 Redo Problem 16.13 with the following data: gross allowable load = 818 kN, γ = 19 kN/m^3, c' = 0, ϕ' = 26°, D_f = 2 m, and required factor of safety = 2.5.

16.15 Refer to Problem 16.13. Design the size of the footing using the modified general ultimate bearing capacity Eq. (16.31).

16.16 A square footing on sand is subjected to an eccentric load as shown in Figure 16.20. Using Meyerhof's *effective area* concept, determine the gross allowable load that the footing could carry with F_s = 4. Given: γ = 16 kN/m^3, c' = 0, ϕ' = 29°, D_f = 1.3 m, B = 1.75 m, and x = 0.25 m. Use Eqs. (16.32) through (16.42) for shape, depth, and inclination factors.

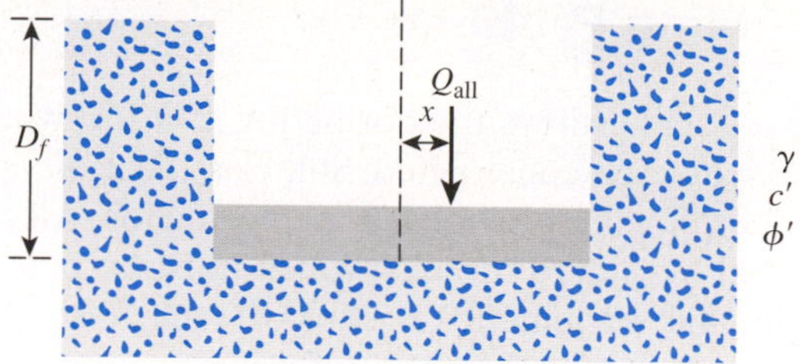

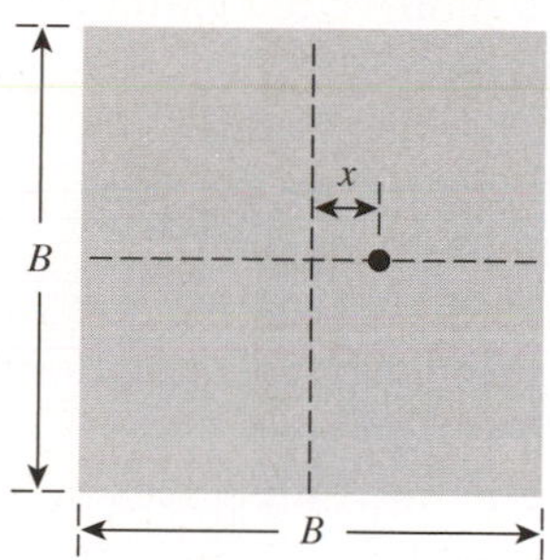

Figure 16.20

16.17 Redo Problem 16.16 with the following data: $\gamma = 18.52\ \text{kN/m}^3$, $c' = 35\ \text{kN/m}^2$, $\phi' = 33°$, $D_f = 1.37$ m, $B = 1.9$ m, and $x = 0.23$ m.

16.18 Refer to the footing in Problem 16.16. Determine the gross ultimate load the footing can carry using the Patra et al. (2015) reduction factor method for rectangular foundations given in Eqs. (16.47), (16.49), and (16.50).

16.19 Figure 16.21 shows a continuous foundation with a width of 1.8 m constructed at a depth of 1.2 m in a granular soil. The footing is subjected to an eccentrically inclined loading with $e = 0.3$ m, and $\alpha = 10°$. Determine the gross ultimate load, $Q_{u(ei)}$, that the footing can support using:

a. Meyerhof (1963) method [Eq. (16.52)]

b. Saran and Agarwal (1991) method [Eq. (16.53)]

c. Patra et al. (2012) reduction factor method [Eq. (16.54)]

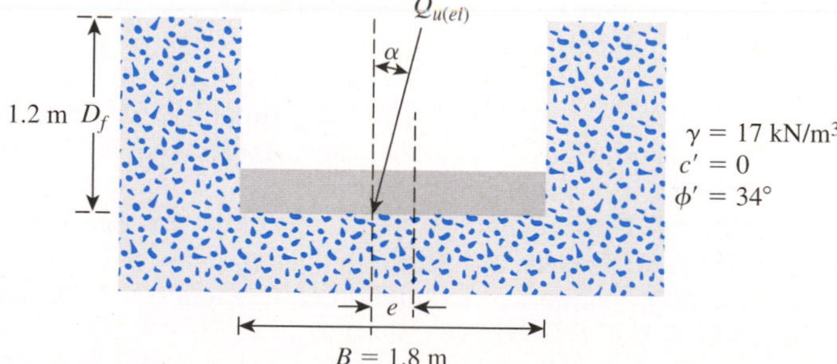

Figure 16.21

Critical Thinking Problem

C.16.1 The following table shows the boring log at a site where a multi-story shopping center would be constructed. Soil classification and the standard penetration number, N_{60}, are provided in the boring log. All columns of the building are supported by square footings which must be placed at a depth of 1.5 m. Additionally, the settlement (elastic) of each footing must be restricted to 20 mm. Since the column loads at different location can vary, a design chart is helpful for quick estimation of footing size required to support a given load.

a. Prepare a chart by plotting the variation of maximum allowable column loads with footing sizes, B = 1 m, 1.5 m, 2 m, and 3 m. Use a factor of safety of 3.

b. If the gross column load from the structure is 250 kN, how would you use this chart to select a footing size?

c. What would be the design footing size for the column in Part (b) if you use Terzaghi's bearing capacity equation? For the well graded sand, assume that $\phi' = 33°$. Use $F_s = 3$.

d. Compare and discuss the differences in footing sizes obtained in Parts b and c.

Soil type	Depth (m)	N_{60}
Well-graded sand (SW)	1	12
	2	
	3	
Sandy silts (ML)	4	7
	5	
	6	8
	7	
Gravelly sands (SP)	8	19
	9	
	10	

References

Balla, A. (1962). "Bearing Capacity of Foundations," *Journal of the Soil Mechanics and Foundations Division*, ASCE, Vol. 89, No. SM5, 12–34.

De Beer, E. E. (1970). "Experimental Determination of the Shape Factors and Bearing Capacity Factors of Sand," *Geotechnique*, Vol. 20, No. 4, 387–411.

De Beer, E. E., and Vesić, A., (1958). "Etude expérimentale de la capacité portante du sable sous des fondations directes établies en surface," *Annales des Travaux Publics de Belqique*, Vol. 59, No. 3,. 5–58.

Hansen, J. B. (1970). "A Revised and Extended Formula for Bearing Capacity," Danish Geotechnical Institute, *Bulletin No. 28*, Copenhagen.

Kumbhojkar, A. S. (1993). "Numerical Evaluation of Terzaghi's N_γ," *Journal of Geotechnical Engineering*, ASCE, Vol. 119, No. GT3, 598–607.

Lundgren, H., and Mortensen, K. (1953). "Determination by the Theory of Elasticity of the Bearing Capacity of Continuous Footing on Sand," *Proceedings*, 3rd International Conference on Soil Mechanics and Foundation Engineering, Vol. I, 409–412.

Meyerhof, G. G. (1951). "The Ultimate Bearing Capacity of Foundations," *Geotechnique*, Vol. 2, No. 4, 301–331.

Meyerhof, G. G. (1953). "The Bearing Capacity of Foundations Under Eccentric and Inclined Loads," *Proceedings*, 3rd International Conference on Soil Mechanics and Foundation Engineering, Vol. 1, 440–445.

Meyerhof, G. G. (1956). "Penetration Tests and Bearing Capacity of Cohesionless Soils," *Journal of the Soil Mechanics and Foundations Division*, American Society of Civil Engineers, Vol. 82, No. SM1, pp. 1–19.

Meyerhof, G. G. (1963). "Some Recent Research on the Bearing Capacity of Foundations," *Canadian Geotechnical Journal*, Vol. 1, 16–26.

Meyerhof, G. G. (1965). "Shallow Foundations," *Journal of the Soil Mechanics and Foundations Division*, ASCE, Vol. 91, No. SM2, pp. 21–31.

Patra, C. R., Behera, R. N., Sivakugan, N. and Das, B. M.(2012). "Ultimate Bearing Capacity of Shallow Strip Foundation under Eccentrically Inclined Load: Part I," *International Journal of Geotechnical Engineering*, Vol. 6, No. 3, 342–352.

Patra, C. R., Sivakugan, N., Das, B. M. and Sethy, B. (2015). "Ultimate Bearing Capacity of Rectangular Foundation on Sand under Eccentric Loading," *Proceedings*, XVI European Conference on Soil Mechanics and Geotechnical Engineering, Edinburgh, U.K., pp. 3153–3158

Prandtl, L. (1921). "Über die Eindringungsfestigkeit (Harte) plasticher Baustoffe und die Festigkeit von Schneiden," *Zeitschrift für Angewandte Mathematik und Mechanik*, Basel, Switzerland, Vol. 1, No. 1, 15–20.

Purkayastha, R. D., and Char, R. A. N. (1977). "Stability Analysis for Eccentrically Loaded Footings," *Journal of the Geotechnical Engineering Division*, ASCE, Vol. 103, No. 6, 647–651.

Reissner, H. (1924). "Zum Erddruckproblem," *Proceedings*, 1st International Congress of Applied Mechanics, 295–311.

Saran, S. and Agarwal, R. K. (1991). "Bearing Capacity of Eccentrically Obliquely Loaded Footing," *Journal of Geotechnical Engineering*, ASCE, Vol. 117, No. 11, 1669–1690.

Terzaghi, K. (1943). *Theoretical Soil Mechanics*, Wiley, New York.

Vesić, A. S. (1973). "Analysis of Ultimate Loads of Shallow Foundations," *Journal of the Soil Mechanics and Foundations Division*, ASCE, Vol. 99, No. SM1, 45–73.

CHAPTER 17

Subsoil Exploration

17.1 Introduction

The preceding chapters reviewed the fundamental properties of soils and their behavior under stress and strain in idealized conditions. In practice, natural soil deposits are not homogeneous, elastic, or isotropic. In some places, the stratification of soil deposits even may change greatly within a horizontal distance of 15 to 30 m. For foundation design and construction work, one must know the actual soil stratification at a given site, the laboratory test results of the soil samples obtained from various depths, and the observations made during the construction of other structures built under similar conditions. For most major structures, adequate subsoil exploration at the construction site must be conducted. The purposes of subsoil exploration include the following:

1. Determining the nature of soil at the site and its stratification
2. Obtaining disturbed and undisturbed soil samples for visual identification and appropriate laboratory tests
3. Determining the depth and nature of bedrock, if and when encountered
4. Performing some *in situ* field tests, such as permeability tests (Chapter 7), vane shear tests (Chapter 12), and standard penetration tests
5. Observing drainage conditions from and into the site
6. Assessing any special construction problems with respect to the existing structure(s) nearby
7. Determining the position of the water table

This chapter briefly summarizes subsoil exploration techniques. For additional information, refer to the *Manual of Foundation Investigations* of the American Association of State Highway and Transportation Officials (1967).

17.2 Planning for Soil Exploration

A soil exploration program for a given structure can be divided broadly into four phases:

1. *Compilation of the existing information regarding the structure:* This phase includes gathering information such as the type of structure to be constructed and its future use, the requirements of local building codes, and the column and load-bearing wall loads. If the exploration is for the construction of a bridge foundation, one must have an idea of the length of the span and the anticipated loads to which the piers and abutments will be subjected.
2. *Collection of existing information for the subsoil condition:* Considerable savings in the exploration program sometimes can be realized if the geotechnical engineer in charge of the project thoroughly reviews the existing information regarding the subsoil conditions at the site under consideration. Useful information can be obtained from the following sources:
   a. Geologic survey maps
   b. County soil survey maps prepared by the U.S. Department of Agriculture and the Soil Conservation Service
   c. Soil manuals published by the state transportation departments
   d. Existing soil exploration reports prepared for the construction of nearby structures

   Information gathered from the preceding sources provides insight into the type of soil and problems that might be encountered during actual drilling operations.
3. *Reconnaissance of the proposed construction site:* The engineer visually should inspect the site and the surrounding area. In many cases, the information gathered from such a trip is invaluable for future planning. The type of vegetation at a site, in some instances, may indicate the type of subsoil that will be encountered. The accessibility of a site and the nature of drainage into and from it also can be determined. Open cuts near the site provide an indication about the subsoil stratification. Cracks in the walls of nearby structure(s) may indicate settlement from the possible existence of soft clay layers or the presence of expansive clay soils.
4. *Detailed site investigation:* This phase consists of making several test borings at the site and collecting disturbed and undisturbed soil samples from various depths for visual observation and for laboratory tests. No hard-and-fast rule exists for determining the number of borings or the depth to which the test borings are to be advanced. For most buildings, at least one boring at each corner and one at the center should provide a start. Depending on the uniformity of the subsoil, additional test borings may be made. Table 17.1 gives guidelines for initial planning of borehole spacing.

The test borings should extend through unsuitable foundation materials to firm soil layers. Sowers and Sowers (1970) provided a rough estimate of the minimum depth of borings (unless bedrock is encountered) for multistory buildings. They can be given by the following equations, applicable to light steel or narrow concrete buildings:

$$z_b\ (\text{m}) = 3S^{0.7} \tag{17.1}$$

Table 17.1 Spacing of Borings

Project	Boring spacings m
One-story buildings	25–30
Multistory buildings	15–25
Highways	250–300
Earth dams	25–50
Residential subdivision planning	60–100

or to heavy steel or wide concrete buildings:

$$z_b\ (\text{m}) = 6S^{0.7} \tag{17.2}$$

In Eqs. (17.1) and (17.2), z_b is the approximate depth of boring and S is the number of stories.

The American Society of Civil Engineers (1972) recommended the following rules of thumb for estimating the boring depths for buildings.

1. Estimate the variation of the net effective stress increase, $\Delta\sigma'$, that will result from the construction of the proposed structure with depth. This variation can be estimated by using the principles outlined in Chapter 10. Determine the depth D_1 at which the value of $\Delta\sigma'$ is equal to 10% of the average load per unit area of the structure.
2. Plot the variation of the effective vertical stress, σ'_o, in the soil layer with depth. Compare this with the net stress increase variation, $\Delta\sigma'$, with depth as determined in step 1. Determine the depth D_2 at which $\Delta\sigma' = 0.05\sigma'_o$.
3. The smaller of the two depths, D_1 and D_2, is the approximate minimum depth of the boring.

When the soil exploration is for the construction of dams and embankments, the depth of boring may range from one-half to two times the embankment height.

The general techniques used for advancing test borings in the field and the procedure for the collection of soil samples are described in the following sections.

17.3 Boring Methods

The test boring can be advanced in the field by several methods. The simplest is the *use of augers*. Figure 17.1 shows two types of hand augers that can be used for making boreholes up to a depth of about 3 to 5 m. They can be used for soil exploration work for highways and small structures. Information regarding the types of soil present at various depths is obtained by noting the soil that holds to the auger. The soil samples collected in this manner are disturbed, but they can be used to conduct laboratory tests such as grain-size determination and Atterberg limits.

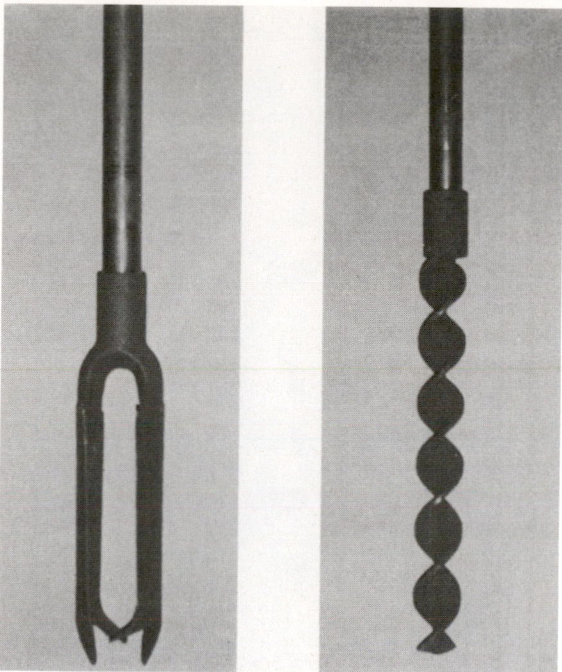

Figure 17.1 Hand augers: (a) Iwan auger; (b) slip auger *(Courtesy of Braja M. Das, Henderson, Nevada)*

When the boreholes are to be advanced to greater depths, the most common method is to use continuous-flight augers, which are power operated. The power for drilling is delivered by truck- or tractor-mounted drilling rigs. Continuous-flight augers are available commercially in 1 to 1.5 m sections. During the drilling operation, section after section of auger can be added and the hole extended downward. Continuous-flight augers can be solid stem or hollow stem. Some of the commonly used solid-stem augers have outside diameters of 67 mm, 83 mm, 102 mm, and 114 mm. The inside and outside diameters of some hollow-stem augers are given in Table 17.2.

Table 17.2 Dimensions of Commonly Used Hollow-Stem Augers

Inside diameter mm	Outside diameter mm
63.5	158.75
69.85	187.8
76.2	203.2
88.9	228.6
101.6	254.0

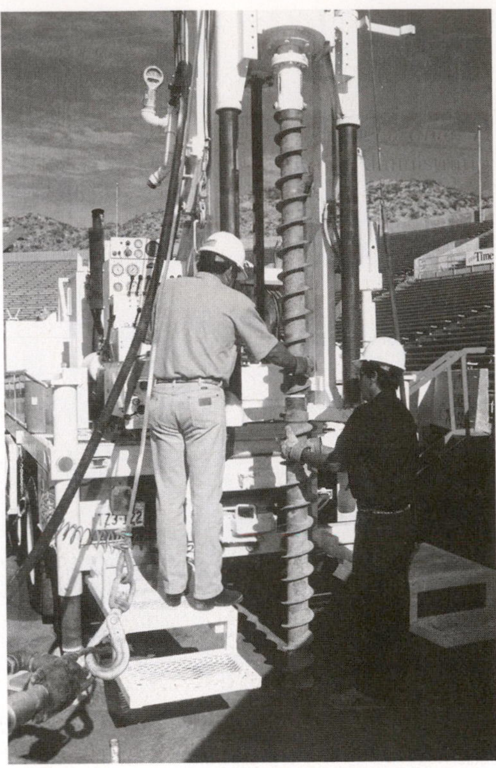

Figure 17.2 Drilling with flight augers (*Courtesy of Danny R. Anderson, PE, of Professional Service Industries, Inc., El Paso, Texas*)

Flight augers bring the loose soil from the bottom of the hole to the surface. The driller can detect the change in soil type encountered by the change of speed and the sound of drilling. Figure 17.2 shows a drilling operation with flight augers. When solid-stem augers are used, the auger must be withdrawn at regular intervals to obtain soil samples and to conduct other operations such as standard penetration tests. Hollow-stem augers have a distinct advantage in this respect—they do not have to be removed at frequent intervals for sampling or other tests. As shown in Figure 17.3, the outside of the auger acts like a casing. A removable plug is attached to the bottom of the auger by means of a center rod.

During the drilling, the plug can be pulled out with the auger in place, and soil sampling and standard penetration tests can be performed. When hollow-stem augers are used in sandy soils below the groundwater table, the sand might be pushed several feet into the stem of the auger by excess hydrostatic pressure immediately after the removal of the plug. In such conditions, the plug should not be used. Instead, water inside the hollow stem should be maintained at a higher level than the groundwater table.

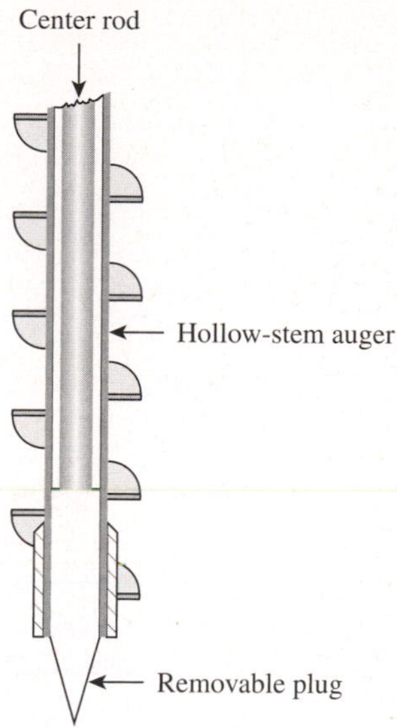

Figure 17.3 Schematic diagram of hollow-stem auger with removable plug

Rotary drilling is a procedure by which rapidly rotating drilling bits attached to the bottom of drilling rods cut and grind the soil and advance the borehole down. Several types of drilling bits are available for such work. Rotary drilling can be used in sand, clay, and rock (unless badly fissured). Water or drilling mud is forced down the drilling rods to the bits, and the return flow forces the cuttings to the surface. Drilling mud is a slurry prepared by mixing bentonite and water (bentonite is a montmorillonite clay formed by the weathering of volcanic ash). Boreholes with diameters ranging from 50 to 200 mm can be made easily by using this technique.

Wash boring is another method of advancing boreholes. In this method, a casing about 2 to 3 m long is driven into the ground. The soil inside the casing then is removed by means of a *chopping bit* that is attached to a drilling rod. Water is forced through the drilling rod, and it goes out at a very high velocity through the holes at the bottom of the chopping bit (Figure 17.4). The water and the chopped soil particles rise upward in the drill hole and overflow at the top of the casing through a T-connection. The wash water then is collected in a container. The casing can be extended with additional pieces as the borehole progresses; however, such extension is not necessary if the borehole will stay open without caving in.

Percussion drilling is an alternative method of advancing a borehole, particularly through hard soil and rock. In this technique, a heavy drilling bit is raised and lowered to chop the hard soil. Casing for this type of drilling may be required. The chopped soil particles are brought up by the circulation of water.

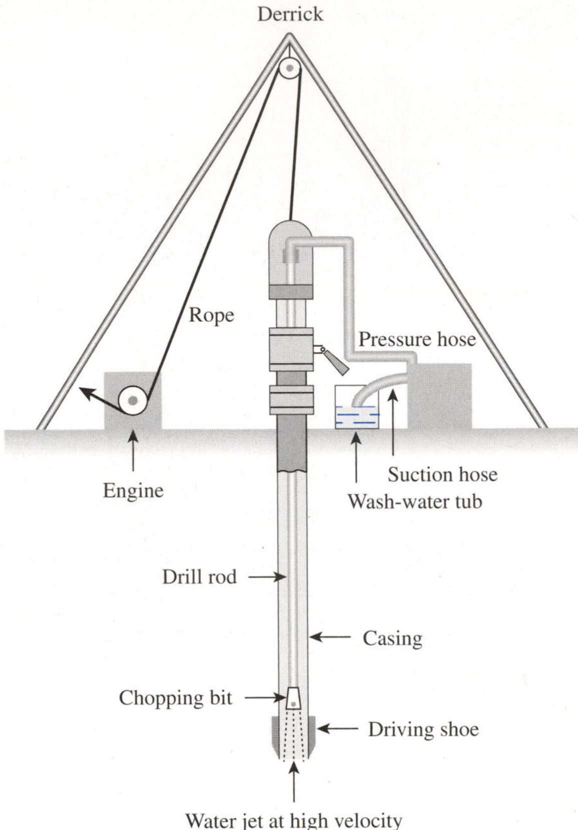

Figure 17.4 Wash boring

17.4 Common Sampling Methods

During the advancement of the boreholes, soil samples are collected at various depths for further analysis. This section briefly discusses some of the methods of sample collection.

Sampling by standard split spoon

Figure 17.5 shows a diagram of a split-spoon sampler. It consists of a tool-steel driving shoe at the bottom, a steel tube (that is split longitudinally into halves) in the middle, and a coupling at the top. The steel tube in the middle has inside and outside diameters of 34.9 mm and 50.8 mm, respectively. Figure 17.6 shows a photograph of an unassembled split-spoon sampler.

When the borehole is advanced to a desired depth, the drilling tools are removed. The split-spoon sampler is attached to the drilling rod and then lowered to the bottom of the borehole (Figure 17.7). The sampler is driven into the soil at the

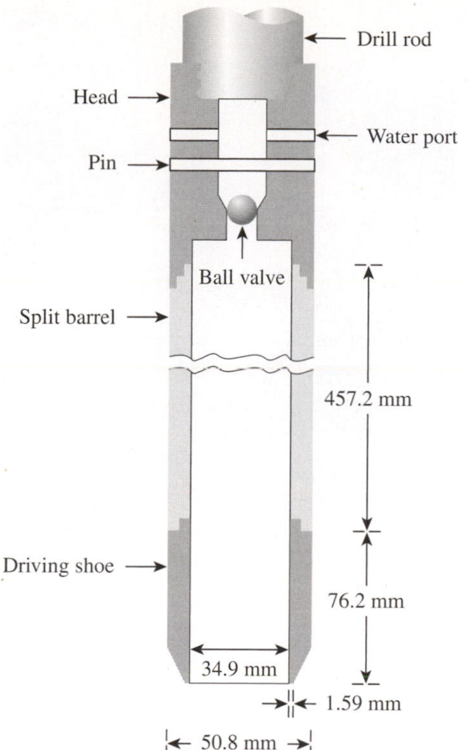

Figure 17.5 Diagram of standard split-spoon sampler

Figure 17.6 Split-spoon sampler, unassembled (*Courtesy of ELE International*)

bottom of the borehole by means of hammer blows. The hammer blows occur at the top of the drilling rod. The hammer weighs 623 N. For each blow, the hammer drops a distance of 0.762 m. The number of blows required for driving the sampler through three 152.4 mm intervals is recorded. The sum of the number of blows required for driving the last two 152.4 mm intervals is referred to as the *standard penetration number, N*. It also commonly is called the *blow count*. The interpretation of the standard penetration number is given in Section 17.6. After driving is completed, the sampler is withdrawn and the shoe and coupling are removed. The soil sample collected inside the split tube then is removed and transported to the laboratory in small glass jars. Determination of the standard penetration number and collection of split-spoon samples usually are done at 1.5 m intervals.

At this point, it is important to point out that there are several factors that will contribute to the variation of the standard penetration number, N, at a given depth for similar soil profiles. These factors include SPT hammer efficiency, borehole diameter, sampling method, and rod length factor (Seed et al., 1985; Skempton, 1986). The two most common types of SPT hammers used in the field are the *safety hammer* and *donut hammer*. They commonly are dropped by a rope with *two wraps around a pulley*.

Figure 17.7 Drilling rod with split-spoon sampler lowered to the bottom of the borehole (*Courtesy of Braja M. Das, Henderson, Nevada*)

The SPT hammer energy efficiency can be expressed as

$$E_r(\%) = \frac{\text{Actual hammer energy to the sampler}}{\text{Input energy}} \times 100 \tag{17.3}$$

$$\text{Theoretical input energy} = Wh \tag{17.4}$$

where W = Weight of the hammer ≈ 0.623 kN
h = Height of drop ≈ 0.76 m

So,

$$Wh = (0.623)(0.76) = 0.474 \text{ kN-m}$$

In the field, the magnitude of E_r can vary from 30 to 90%. The standard practice now in the U.S. is to express the N-value to an average energy ratio of 60% ($\approx N_{60}$). Thus, correcting for field procedures and on the basis of field observations, it appears reasonable to standardize the field penetration number as a function of the input driving energy and its dissipation around the sampler into the surrounding soil, or

$$N_{60} = \frac{N\eta_H\eta_B\eta_S\eta_R}{60} \tag{17.5}$$

where N_{60} = standard penetration number corrected for field conditions
N = measured penetration number

η_H = hammer efficiency (%)
η_B = correction for borehole diameter
η_S = sampler correction
η_R = correction for rod length

Based on the recommendations of Seed et al. (1985) and Skempton (1986), the variations of η_H, η_B, η_S, and η_R are summarized in Table 17.3.

Sampling by thin-wall tube

Sampling by thin-wall tube is used for obtaining fairly undisturbed soil samples. The thin-wall tubes are made of seamless, thin tubes and commonly are referred to as *Shelby tubes* (Figure 17.8). To collect samples at a given depth in a borehole, one first must remove the drilling tools. The sampler is attached to a drilling rod and lowered

Table 17.3 Variations of η_H, η_B, η_S, and η_R [Eq. (17.5)]

1. Variation of η_H

Country	Hammer type	Hammer release	η_H (%)
Japan	Donut	Free fall	78
	Donut	Rope and pulley	67
United States	Safety	Rope and pulley	60
	Donut	Rope and pulley	45
Argentina	Donut	Rope and pulley	45
China	Donut	Free fall	60
	Donut	Rope and pulley	50

2. Variation of η_B

Diameter	
mm	η_B
60–120	1
150	1.05
200	1.15

3. Variation of η_S

Variable	η_S
Standard sampler	1.0
With liner for dense sand and clay	0.8
With liner for loose sand	0.9

4. Variation of η_R

Rod length (m)	η_R
>10	1.0
6–10	0.95
4–6	0.85
0–4	0.75

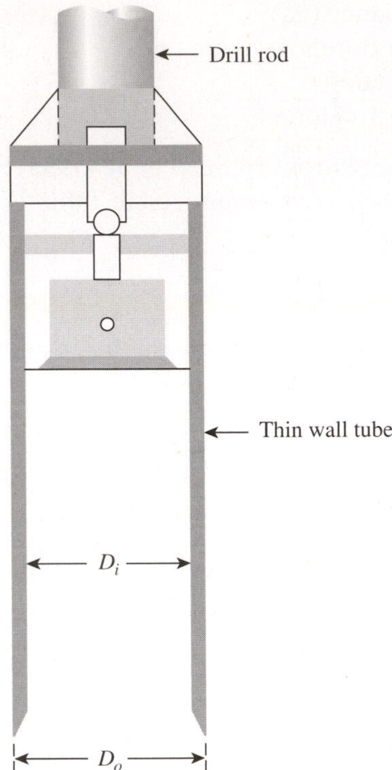

Figure 17.8 Thin-wall tube sampler

to the bottom of the borehole. After this, it is pushed hydraulically into the soil. It then is spun to shear off the base and is pulled out. The sampler with the soil inside is sealed and taken to the laboratory for testing. Most commonly used thin-wall tube samplers have outside diameters of 76.2 mm.

Sampling by Piston sampler

Piston samplers are particularly useful when highly undisturbed samples are required. The cost of recovering such samples is, of course, higher. Several types of piston samplers can be used; however, the sampler proposed by Osterberg (1952) is the most advantageous (Figure 17.9). It consists of a thin-wall tube with a piston. Initially, the piston closes the end of the thin-wall tube. The sampler first is lowered to the bottom of the borehole (Figure 17.9a), then the thin-wall tube is pushed into the soil hydraulically—past the piston. After this, the pressure is released through a hole in the piston rod (Figure 17.9b). The presence of the piston prevents distortion in the sample by neither letting the soil squeeze into the sampling tube very fast nor admitting excess soil. Samples obtained in this manner consequently are disturbed less than those obtained by Shelby tubes.

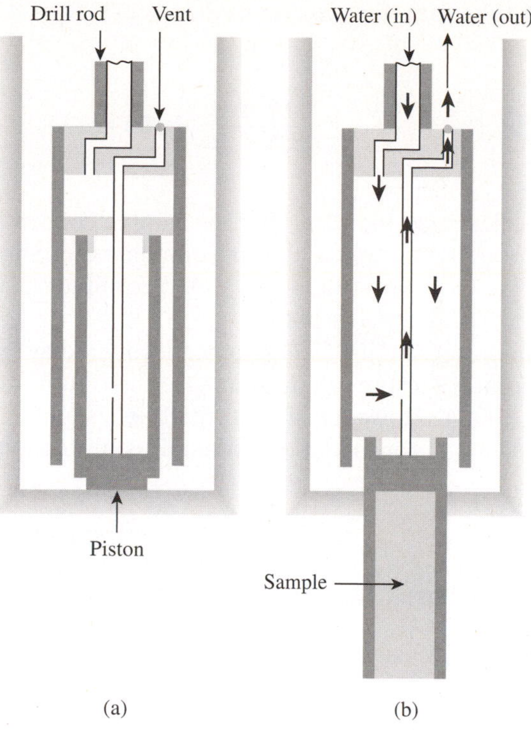

Figure 17.9 Piston sampler: (a) sampler lowered to bottom of borehole; (b) pressure released through hole in piston rod

17.5 Sample Disturbance

The degree of disturbance of the sample collected by various methods can be expressed by a term called the *area ratio*, which is given by

$$A_r(\%) = \frac{D_o^2 - D_i^2}{D_i^2} \times 100 \tag{17.6}$$

where D_o = outside diameter of the sampler
D_i = inside diameter of the sampler

A soil sample generally can be considered undisturbed if the area ratio is less than or equal to 10%. The following is a calculation of A_r for a standard split-spoon sampler and a 50.8 mm Shelby tube:

For the standard split-spoon sampler, D_i = 35.05 mm and D_o = 50.8 mm. Hence,

$$A_r(\%) = \frac{(50.8)^2 - (35.05)^2}{(35.05)^2} \times 100 = 110\%$$

For the Shelby-tube sampler (50.8 mm diameter), D_i = 47.63 mm and D_o = 50.8 mm. Hence,

$$A_r(\%) = \frac{(50.8)^2 - (47.63)^2}{(47.63)^2} \times 100 = 13.8\%$$

The preceding calculation indicates that the sample collected by split spoons is highly disturbed. The area ratio (A_r) of the 50.8 mm diameter Shelby tube samples is slightly higher than the 10% limit stated previously. For practical purposes, however, it can be treated as an undisturbed sample.

The disturbed but representative soil samples recovered by split-spoon samplers can be used for laboratory tests, such as grain-size distribution, liquid limit, plastic limit, and shrinkage limit. However, undisturbed soil samples are necessary for performing tests such as consolidation, triaxial compression, and unconfined compression.

17.6 Correlations for N_{60} in Cohesive Soil

The consistency of clay soils can be estimated from the standard penetration number N_{60}. In order to achieve that, Szechy and Vargi (1978) calculated the *consistency index* (*CI*) as

$$CI = \frac{LL - w}{LL - PL} \tag{17.7}$$

where w = natural moisture content
LL = liquid limit
PL = plastic limit

The approximate correlation among *CI*, N_{60}, and the unconfined compression strength (q_u) is given in Table 17.4.

It is important to point out that the correlation between N_{60} and q_u given in Table 17.4 is approximate. The sensitivity, S_t, of clay soil also plays an important

Table 17.4 Approximate Correlation among *CI*, N_{60}, and q_u

Standard penetration number, N_{60}	Consistency	*CI*	Unconfined compression strength, q_u kN/m²
<2	Very soft	<0.5	<25
2 to 8	Soft to medium	0.5 to 0.75	25 to 80
8 to 15	Stiff	0.75 to 1.0	80 to 150
15 to 30	Very stiff	1.0 to 1.5	150 to 400
>30	Hard	>1.5	>400

role in the actual N_{60} value obtained in the field. Based on several field test results, Kulhawy and Mayne (1990) have suggested the following correlation:

$$\frac{q_u}{p_a} = 0.58N_{60}^{0.72} \tag{17.8}$$

where p_a = atmospheric pressure ($\approx$ 100 kN/m^2)

17.7 Correlations for Standard Penetration Number in Granular Soil

In granular soils, the standard penetration number is highly dependent on the effective overburden pressure, σ'_o.

A number of empirical relationships have been proposed to convert the field-standard penetration number N_{60} to a *standard effective overburden pressure,* σ'_o, of approximately 100 kN/m^2. The general form is

$$(N_1)_{60} = C_N N_{60} \tag{17.9}$$

Several correlations have been developed over the years for the correction factor, C_N. They are given below.

In the following relationships for C_N, note that σ'_o is the effective overburden pressure and p_a = atmospheric pressure ($\approx$100 kN/m^2).

Liao and Whitman's relationship (1986)

$$C_N = \left[\frac{1}{\left(\dfrac{\sigma'_o}{p_a}\right)}\right]^{0.5} \tag{17.10}$$

Skempton's relationship (1986)

$$C_N = \frac{2}{1 + \left(\dfrac{\sigma'_o}{p_a}\right)} \text{ (for normally consolidated fine sand)} \tag{17.11}$$

$$C_N = \frac{3}{2 + \left(\dfrac{\sigma'_o}{p_a}\right)} \text{ (for normally consolidated coarse sand)} \tag{17.12}$$

$$C_N = \frac{1.7}{0.7 + \left(\dfrac{\sigma'_o}{p_a}\right)} \text{ (for overconsolidated sand)} \tag{17.13}$$

Seed et al.'s relationship (1975)

$$C_N = 1 - 1.25 \log\left(\frac{\sigma'_o}{p_a}\right) \tag{17.14}$$

Peck et al.'s relationship (1974)

$$C_N = 0.77 \log\left[\frac{20}{\left(\frac{\sigma'_o}{p_a}\right)}\right]\left(\text{for } \frac{\sigma'_o}{p_a} \geq 0.25\right) \tag{17.15}$$

Bazaraa's relationship (1967)

$$C_N = \frac{4}{1 + 4\left(\frac{\sigma'_o}{p_a}\right)}\left(\text{for } \frac{\sigma'_o}{p_a} \leq 0.75\right) \tag{17.16}$$

$$C_N = \frac{4}{3.25 + \left(\frac{\sigma'_o}{p_a}\right)}\left(\text{for } \frac{\sigma'_o}{p_a} > 0.75\right) \tag{17.17}$$

Table 17.5 shows the comparison of C_N derived using various relationships cited above. It can be seen that the magnitude of the correction factor estimated by using any one of the relationships is approximately the same, considering the uncertainties involved in conducting the standard penetration tests. Hence, it is recommended that Eq. (17.10) may be used for all calculations.

Table 17.5 Variation of C_N

	C_N						
$\frac{\sigma'_o}{p_a}$	**Eq. (17.10)**	**Eq. (17.11)**	**Eq. (17.12)**	**Eq. (17.13)**	**Eq. (17.14)**	**Eq. (17.15)**	**Eqs. (17.16) and (17.17)**
0.25	2.00	1.60	1.33	1.78	1.75	1.47	2.00
0.50	1.41	1.33	1.20	1.17	1.38	1.23	1.33
0.75	1.15	1.14	1.09	1.17	1.15	1.10	1.00
1.00	1.00	1.00	1.00	1.00	1.00	1.00	0.94
1.50	0.82	0.80	0.86	0.77	0.78	0.87	0.84
2.00	0.71	0.67	0.75	0.63	0.62	0.77	0.76
3.00	0.58	0.50	0.60	0.46	0.40	0.63	0.65
4.00	0.50	0.40	0.60	0.36	0.25	0.54	0.55

Table 17.6 Approximate Relationship between Corrected Standard Penetration Number and Relative Density of Sand

Corrected standard penetration number, $(N_1)_{60}$	Relative density, D_r(%)
0–5	0–5
5–10	5–30
10–30	30–60
30–50	60–95

Correlations for relative density

Table 17.6 shows approximate correlations for the standard penetration number, $(N_1)_{60}$, and relative density, D_r.

Cubrinovski and Ishihara (1999) proposed a correlation between N_{60} and the relative density of granular soils, D_r, in the form

$$D_r(\%) = \left[\frac{N_{60}\left(0.23 + \dfrac{0.06}{D_{50}}\right)^{1.7}}{9}\left(\frac{p_a}{\sigma'_o}\right)\right]^{0.5}(100) \tag{17.18}$$

where σ'_o = effective overburden pressure in kN/m²
D_{50} = sieve size through which 50% of soil will pass (mm)
p_a = atmospheric pressure

Meyerhof (1957) developed a correlation between D_r and N_{60} as

$$N_{60} = \left[17 + 24\left(\frac{\sigma'_o}{p_a}\right)\right]D_r^2$$

or

$$D_r(\%) = \left\{\frac{N_{60}}{\left[17 + 24\left(\dfrac{\sigma'_o}{p_a}\right)\right]}\right\}^{0.5}(100) \tag{17.19}$$

Equation (17.19) provides a reasonable estimate only for clean, medium fine sand.

Kulhawy and Mayne (1990) correlated the corrected standard penetration number and the relative density of sand in the form

$$D_r(\%) = \left[\frac{(N_1)_{60}}{C_p C_A C_{OCR}}\right]^{0.5}(100) \tag{17.20}$$

where

$$C_P = \text{grain-size correlations factor} = 60 + 25\log D_{50} \tag{17.21}$$
$$C_A = \text{correlations factor for aging} = 1.2 + 0.05\ \log\left(\tfrac{t}{100}\right) \tag{17.22}$$

C_{OCR} = correlation factor for overconsolidation = $OCR^{0.18}$ (17.23)
D_{50} = diameter through which 50% soil will pass through (mm)
t = age of soil since deposition (years)
OCR = overconsolidation ratio

Correlations for drained angle of friction

The drained angle of friction of *granular soils*, ϕ', also has been correlated to the standard penetration number. Schmertmann (1975) also provided a correlation for N_{60} versus σ'_o. After Kulhawy and Mayne (1990), this correlation can be approximated as

$$\phi' = \tan^{-1}\left[\frac{N_{60}}{12.2 + 20.3\left(\dfrac{\sigma'_o}{p_a}\right)}\right]^{0.34} \tag{17.24}$$

where p_a = atmospheric pressure (same unit as σ'_o).

The standard penetration number is a useful guideline in soil exploration and the assessment of subsoil conditions, provided that the results are interpreted correctly. Note that all equations and correlations relating to the standard penetration numbers are approximate. Because soil is not homogeneous, a wide variation in the N_{60} value may be obtained in the field. For soil deposits that contain large boulders and gravel, the standard penetration numbers may be erratic.

Example 17.1

Following are the results of a standard penetration test in sand. Determine the corrected standard penetration numbers, $(N_1)_{60}$, at various depths. Note that the water table was not observed within a depth of 10.5 m below the ground surface. Assume that the average unit weight of sand is 17.3 kN/m^3. Use Eq. (17.10).

Depth, z (m)	N_{60}
1.5	8
3.0	7
4.5	12
6.0	14
7.5	13

Solution

From Eq. (17.10),

$$C_N = \left[\frac{1}{\left(\dfrac{\sigma'_o}{p_a}\right)}\right]^{0.5}$$

$$p_a \approx 100 \text{ kN/m}^2$$

Now the following table can be prepared.

Depth, z (m)	**σ'_o (kN/m²)**	**C_N**	**N_{60}**	**$(N_1)_{60}$**
1.5	25.95	1.96	8	**≈16**
3.0	51.90	1.39	7	**≈10**
4.5	77.85	1.13	12	**≈14**
6.0	103.80	0.98	14	**≈14**
7.5	129.75	0.87	13	**≈11**

Example 17.2

Refer to Example 17.1. Using Eq. (17.24), estimate the average soil friction angle, ϕ', from $z = 0$ to $z = 7.5$ m.

Solution

From Eq. (17.24),

$$\phi' = \tan^{-1}\left[\frac{N_{60}}{12.2 + 20.3\left(\frac{\sigma'_o}{p_a}\right)}\right]^{0.34}$$

$$p_a = 100 \text{ kN/m}^2$$

Now the following table can be prepared.

Depth, z (m)	**σ'_o (kN/m²)**	**N_{60}**	**ϕ' (deg) [Eq. (17.24)]**
1.5	25.95	8	37.5
3.0	51.9	7	33.8
4.5	77.85	12	36.9
6.0	103.8	14	36.7
7.5	129.75	13	34.6

Average $\phi' \approx$ **36°**

Example 17.3

The following table gives the variation of the field standard penetration number (N_{60}) in a sand deposit:

Depth (m)	N_{60}
1.5	5
3.0	11
4.5	14
6.0	18
7.5	16
9.0	21

The groundwater table is located at a depth of 12 m. The dry unit weight of sand from 0 to a depth of 12 m is 17.6 kN/m^3. Assume the mean grain size (D_{50}) of the sand deposit to be about 0.8 mm. Estimate the variation of the relative density with depth for sand. Use Eq. (17.18).

Solution

Given: Unit weight $\gamma = 17.6$ kN/m^3; $D_{50} = 0.8$ mm. So, $\sigma'_o = (\gamma)(\text{depth})$. From Eq. (17.18),

$$D_r(\%) = \left[\frac{N_{60}\left(0.23 + \frac{0.06}{D_{50}}\right)^{1.7}}{9} \left(\frac{p_a}{\sigma'_o}\right) \right]^{0.5} (100)$$

Now the following table can be prepared.

Depth (m)	σ'_o (kN/m^2)	p_a (kN/m^2)	$\left(0.23 + \frac{0.06}{D_{50}}\right)^{1.7}$	N_{60}	D_r (%) [Eq. (17.18)]
1.5	26.4	100	0.133	5	**52.9**
3.0	52.8	100	0.133	11	**55.5**
4.5	79.2	100	0.133	14	**51.1**
6.0	105.6	100	0.133	18	**50.2**
7.5	132.0	100	0.133	16	**42.3**
9.0	158.4	100	0.133	21	**44.3**

Example 17.4

Solve Example 17.3 using Eq. (17.19).

Solution

$$D_r(\%) = \left\{ \frac{N_{60}}{\left[17 + 24\left(\frac{\sigma'_o}{p_a}\right)\right]} \right\}^{0.5}$$

Now the following table can be prepared.

Depth (m)	σ'_o (kN/m²)	p_a (kN/m²)	N_{60}	D_r (%) [Eq. (17.19)]
1.5	26.4	100	5	**46.3**
3.0	52.8	100	11	**60.9**
4.5	79.2	100	14	**62.4**
6.0	105.6	100	18	**65.2**
7.5	132.0	100	16	**57.3**
9.0	158.4	100	21	**61.8**

17.8 Other *In Situ* Tests

Depending on the type of project and the complexity of the subsoil, several types of *in situ* tests can be conducted during the exploration period. In many cases, the soil properties evaluated from the *in situ* tests yield more representative values. This better accuracy results primarily because the sample disturbance during soil exploration is eliminated. Some of the common tests that can be conducted in the field are given in the following Sections (Sections 17.9 through 17.11).

17.9 Vane Shear Test

The principles and the application of the vane shear test were discussed in Chapter 12. When soft clay is encountered during the advancement of a borehole, the undrained shear strength of clay, c_u, can be determined by conducting a vane shear test in the borehole. This test provides valuable information about the strength in undisturbed clay.

17.10 Borehole Pressuremeter Test

The pressuremeter is a device that originally was developed by Menard in 1965 for *in situ* measurement of the stress–strain modulus. This device basically consists of a pressure cell and two guard cells (Figure 17.10). The test involves expanding the pressure cell inside a borehole and measuring the expansion of its volume. The test data are interpreted on the basis of the theory of expansion of an infinitely thick cylinder of soil. Figure 17.11 shows the variation of the pressure-cell volume with changes in the cell pressure. In this figure, Zone I represents the reloading portion, during which the soil around the borehole is pushed back to its initial state—that is, the state it was in before drilling. Zone II represents a pseudoelastic zone, in which the cell volume versus cell pressure is practically linear. The zone marked III is the plastic zone. For the pseudoelastic zone,

$$E_s = 2(1 + \mu_s)V_o\frac{\Delta p}{\Delta V} \tag{17.25}$$

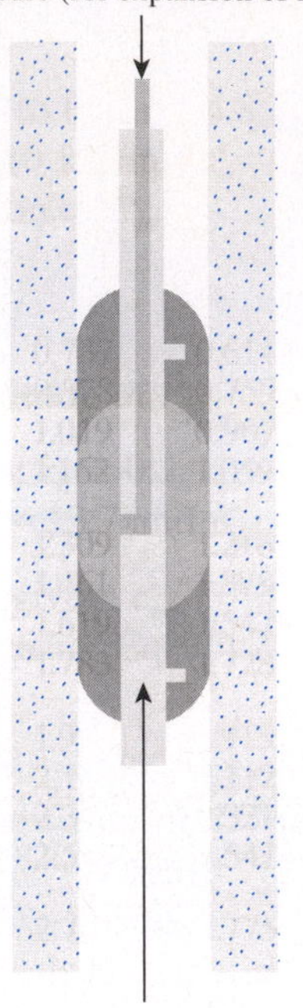

Figure 17.10 Schematic diagram for pressuremeter test

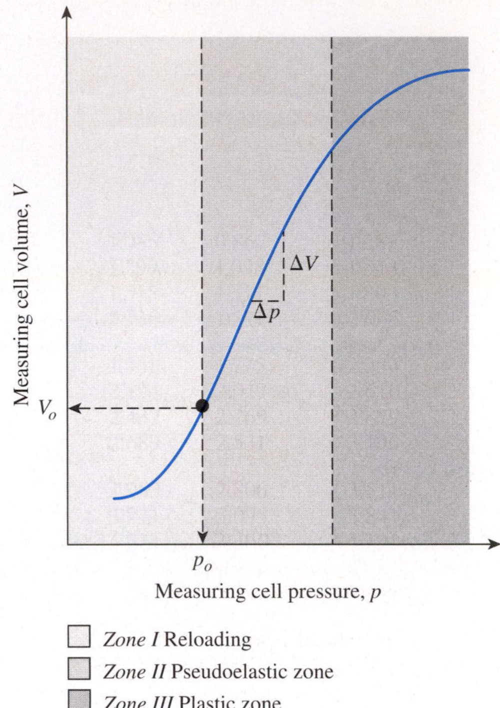

Figure 17.11 Relationship between measuring pressure and measuring volume for Menard pressuremeter

where E_s = modulus of elasticity of soil
μ_s = Poisson's ratio of soil
V_o = cell volume corresponding to pressure p_o (that is, the cell pressure corresponding to the beginning of Zone II)
$\Delta p/\Delta V$ = slope of straight-line plot of Zone II

Menard recommended a value of $\mu_s = 0.33$ for use in Eq. (17.25), but other values can be used. With $\mu_s = 0.33$,

$$E_s = 2.66\, V_o \frac{\Delta p}{\Delta V} \tag{17.26}$$

From the theory of elasticity, the relationship between the modulus of elasticity and the shear modulus can be given as

$$E_s = 2(1 + \mu_s)G_s \tag{17.27}$$

where G_s = shear modulus of soil. Hence, combining Eqs. (17.25) and (17.27) gives

$$G_s = V_o \frac{\Delta p}{\Delta V} \tag{17.28}$$

Pressuremeter test results can be used to determine the at-rest earth-pressure coefficient, K_o (Chapter 13). This coefficient can be obtained from the ratio of p_o and σ'_o (σ'_o = effective vertical stress at the depth of the test), or

$$K_o = \frac{p_o}{\sigma'_o} \tag{17.29}$$

Note that p_o (see Figure 17.11) represents the *in situ* lateral pressure.

The pressuremeter tests are very sensitive to the conditions of a borehole before the test.

17.11 Cone Penetration Test

The Dutch cone penetrometer is a device by which a 60° cone with a base area of 10 cm^2 (Figure 17.12) is pushed into the soil, and the cone end resistance, q_c, to penetration is measured. Most cone penetrometers that are used commonly have friction sleeves that follow the point. This allows independent determination of the cone resistance (q_c) and the frictional resistance (f_c) of the soil above it. The friction sleeves have an exposed surface area of about 150 cm^2.

The penetrometer shown in Figure 17.12 is a *mechanical-friction cone penetrometer*. At the present time, *electrical-friction cone penetrometers* also are used for field investigation.

One of the major advantages of the cone penetration test is that boreholes are not necessary to conduct the test. Unlike the standard penetration test, however, soil samples cannot be recovered for visual observation and laboratory tests.

Correlations for soil friction angle

Robertson and Campanella (1983) provided correlations among the vertical effective stress (σ'_o), drained soil friction angle (ϕ'), and q_c for sand. The relationship among σ'_o, ϕ', and q'_c can be approximated (Kulhawy and Mayne, 1990) as

$$\phi' = \tan^{-1}\left[0.1 + 0.38 \log\left(\frac{q_c}{\sigma'_o}\right)\right] \tag{17.30}$$

Correlations for soil modulus of elasticity

The cone penetration resistance also has been correlated with the equivalent modulus of elasticity, E_s, of soils by various investigators. Schmertmann (1978) gave the following correlations for sand to be used in elastic settlement calculations of shallow foundations:

$$E_s = 2.5q_c \text{ (for square and circular foundations)} \tag{17.31}$$

$$E_s = 3.5q_c \text{ (for foundations with } L/B \geq 10) \tag{17.32}$$

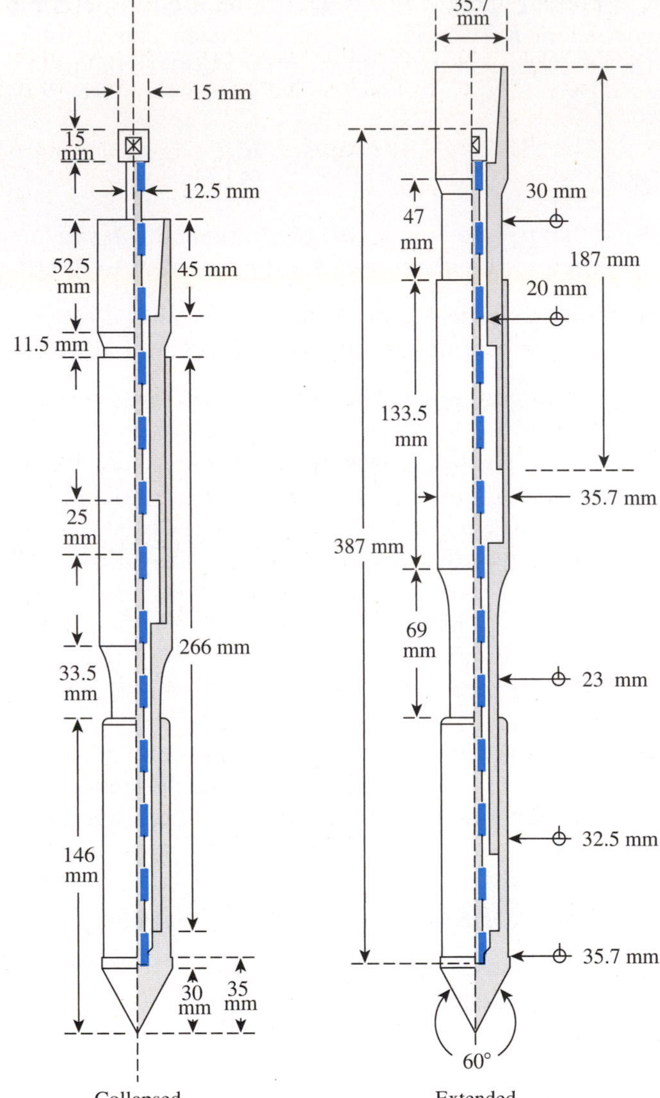

Figure 17.12 Dutch cone penetrometer with friction sleeve (From *Annual Book of ASTM Standards*, 04.08, 1991, Copyright ASTM INTERNATIONAL. Reprinted with permission.)

where L = length of the foundation
B = width of the foundation

Terzaghi, Peck and Mesri (1996) suggested the following for elastic settlement calculations in sand:

$$E_{s(L/B=1)} = 3.5q_c \tag{17.33}$$

and

$$\frac{E_{s(L/B)}}{E_{s(L/B=1)}} = 1 + 0.4\log\left(\frac{L}{B}\right) \leq 1.4 \tag{17.34}$$

Correlations for undrained cohesion of clay

Anagnostopoulos et al. (2003) have provided several correlations for the undrained cohesion (c_u) for clay soil based on a large number of field-test results conducted on a wide variety of soils.

The correlations between c_u and the cone penetration resistance (q_c) can be given as

$$c_u = \frac{q_c - \sigma_o}{N_k} \tag{17.35}$$

where σ_o = vertical total stress
N_k = bearing capacity factor ($\approx$18.3 for all cones)

Consistent units need to be used in Eq. (17.35). The values of c_u in the field tests were equal to or less than about 250 kN/m².

Similarly, the correlations between c_u and sleeve-frictional resistance (f_c) are

$$c_u = \frac{f_c}{1.26} \quad \text{(for mechanical cones)} \tag{17.36}$$

$$c_u = f_c \quad \text{(for electric cones)} \tag{17.37}$$

$$c_u = \frac{f_c}{1.21} \quad \text{(average for all cones)} \tag{17.38}$$

Correlation between q_c and N_{60}

For granular soils, several correlations have been proposed to correlate q_c and N_{60} (N_{60} = standard penetration resistance) against the mean grain size (D_{50} in mm). These correlations are of the form

$$\frac{\left(\frac{q_c}{p_a}\right)}{N_{60}} = cD_{50}^{a} \tag{17.39}$$

where p_a = atmospheric pressure ($\approx$100 kN/m²).

Table 17.7 shows the values of c and a as developed from various studies.

Table 17.7 Values of c and a [Eq. (17.39)]

Investigator		c	a
Burland and Burbidge (1985)	Upper limit	15.49	0.33
	Lower limit	4.9	0.32
Robertson and Campanella (1983)	Upper limit	10	0.26
	Lower limit	5.75	0.31
Kulhawy and Mayne (1990)		5.44	0.26
Anagnostopoulos et al. (2003)		7.64	0.26

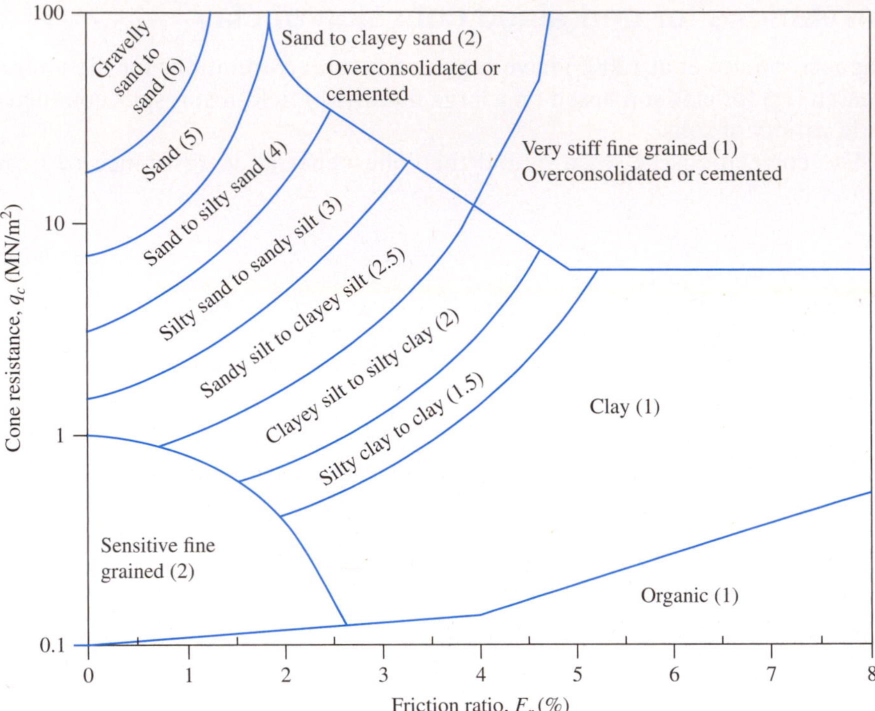

Figure 17.13 Robertson et al. correlation (1986) between q_c, F_r, and the type of soil (*Based on Robertson et al., 1986*)

Correlations of soil types

Robertson et al. (1986) provided the correlations shown in Figure 17.13 between q_c and the friction ratio [Eq. (17.40)] to identify various types of soil encountered in the field.

The friction ratio F_r is defined as

$$F_r = \frac{f_c}{q_c} \tag{17.40}$$

Example 17.5

In a deposit of normally consolidated dry sand, a cone penetration test was conducted. Following are the results:

Depth (m)	Point resistance of cone, q_c (MN/m²)
1.5	2.06
3.0	4.23
4.5	6.01
6.0	8.18
7.5	9.97
9.0	12.42

Assuming the dry unit weight of sand to be 16 kN/m³, estimate the average peak friction angle, ϕ', of the sand. Use Eq. (17.30).

Solution

From Eq. (17.30),

$$\phi' = \tan^{-1}\left[0.1 + 0.38 \log\left(\frac{q_c}{\sigma'_o}\right)\right]$$

Now the following table can be prepared.

Depth (m)	σ'_o (MN/m²)	q_c (MN/m²)	ϕ' (deg)
1.5	0.024	2.06	39.85
3.0	0.048	4.23	40.00
4.5	0.072	6.01	39.70
6.0	0.096	8.18	39.81
7.5	0.120	9.97	39.67
9.0	0.144	12.42	39.88

ϕ' (average) ≈ **39.80°**

Example 17.6

Refer to Example 17.5. Use Eq. (17.39) and Kulhawy and Mayne factors for a and c (Table 17.7) to predict the variation of N_{60} with depth. Given the mean grain size $D_{50} = 0.2$ mm.

Solution

From Eq. (17.39),

$$\frac{\left(\frac{q_c}{p_a}\right)}{N_{60}} = cD_{50}^{a}$$

From Table 17.7, $c = 5.44$ and $a = 0.26$. Also, $p_a \approx 100$ kN/m^2. Now the following table can be prepared.

Depth (m)	q_c (MN/m^2)	N_{60}
1.5	2.06	5.75 ≈ **6**
3.0	4.23	11.8 ≈ **12**
4.5	6.01	16.78 ≈ **17**
6.0	8.18	22.85 ≈ **23**
7.5	9.97	27.85 ≈ **28**
9.0	12.42	34.69 ≈ **35**

17.12 Rock Coring

It may be necessary to core rock if bedrock is encountered at a certain depth during drilling. It is always desirable that coring be done for at least 3 m. If the bedrock is weathered or irregular, the coring may need to be extended to a greater depth. For coring, a core barrel is attached to the drilling rod. A coring bit is attached to the bottom of the core barrel. The cutting element in the bit may be diamond, tungsten, or carbide. The coring is advanced by rotary drilling. Water is circulated through the drilling rod during coring, and the cuttings are washed out. Figure 17.14a shows a diagram of rock coring by the use of a single-tube core barrel. Rock cores obtained by such barrels can be fractured because of torsion. To avoid this problem, one can use double-tube core barrels (Figure 17.14b). Table 17.8 gives the details of various types of casings and core barrels, diameters of core barrel bits, and diameters of core samples obtained. The core samples smaller than the BX size tend to break away during coring.

On the basis of the length of the rock core obtained from each run, the following quantities can be obtained for evaluation of the quality of rock.

$$\text{Recovery ratio} = \frac{\text{Length of rock core recovered}}{\text{Length of coring}} \tag{17.41}$$

$$\text{Rock quality designation } (RQD) = \frac{\begin{array}{c}\Sigma \text{ Length of rock pieces}\\ \text{recovered having lengths of}\\ \text{101.6 mm or more}\end{array}}{\text{Length of coring}} \tag{17.42}$$

A recovery ratio equal to 1 indicates intact rock. However, highly fractured rocks have a recovery ratio of 0.5 or less. Deere (1963) proposed the classification system in Table 17.9 for *in situ* rocks on the basis of their RQD.

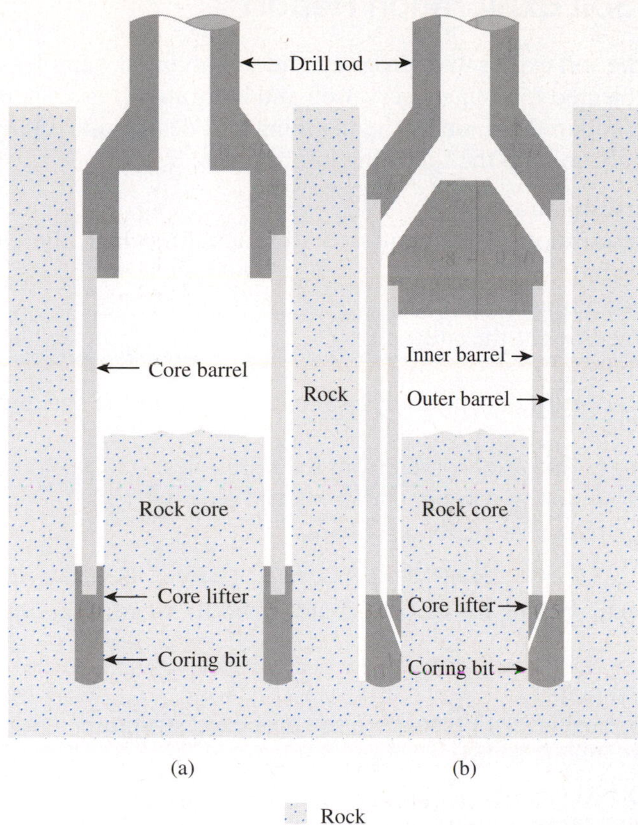

Figure 17.14 Rock coring: (a) single-tube core barrel; (b) double-tube core barrel

Table 17.8 Details of Core Barrel Designations, Bits, and Core Samples

Casing and core barrel designation	Outside diameter of core barrel bit, mm	Diameter of core sample, mm
EX	36.5	22.2
AX	47.6	28.6
BX	58.7	41.3
NX	74.6	54.0

Table 17.9 Qualitative Description of Rocks Based on *RQD*

RQD	Rock quality
1–0.9	Excellent
0.9–0.75	Good
0.75–0.5	Fair
0.5–0.25	Poor
0.25–0	Very poor

17.13 Soil Exploration Report

At the end of the soil exploration program, the soil and rock samples collected from the field are subjected to visual observation and laboratory tests. Then, a soil exploration report is prepared for use by the planning and design office. Any soil exploration report should contain the following information:

1. Scope of investigation
2. General description of the proposed structure for which the exploration has been conducted
3. Geologic conditions of the site
4. Drainage facilities at the site
5. Details of boring
6. Description of subsoil conditions as determined from the soil and rock samples collected
7. Groundwater table as observed from the boreholes
8. Details of foundation recommendations and alternatives
9. Any anticipated construction problems
10. Limitations of the investigation

The following graphic presentations also need to be attached to the soil exploration report:

1. Site location map
2. Location of borings with respect to the proposed structure
3. Boring logs (Figure 17.15)
4. Laboratory test results
5. Other special presentations

The boring log is the graphic presentation of the details gathered from each borehole.

17.14 Summary

This chapter provides a brief overview of subsoil exploration in which we have discussed the following:

- Soil exploration planning involves compilation of existing information, reconnaissance, and detailed site investigation.
- Borings are generally made with continuous-flight augers. Rotary drilling, wash boring, and percussion drilling are other methods of advancing a bore hole.
- Soil samples during boring can be obtained by standard split-spoon sampler, thin-wall tube, and piston sampler.
- Standard penetration resistance can be correlated with unconfined compression strength of cohesive soils (Section 17.6). In granular soil, it can be correlated to relative density and friction angle (Section 17.7).
- Other *in situ* tests are vane shear test, pressuremeter test, and cone penetration test (Sections 17.9, 17.10, and 17.11).
- Rock coring is done by attaching a core barrel to the drilling rod. A coring bit is attached to the bottom of the core barrel. Recovery ratio and rock quality designation are parameters to evaluate the quality of rock (Section 17.12).

BORING LOG

PROJECT TITLE Shopping center

LOCATION Intersection Hill Street and Miner Street DATE June 7, 1997

BORING NUMBER 4 TYPE OF BORING Hollow-stem auger GROUND ELEVATION 40.3 m

DESCRIPTION OF SOIL	DEPTH (m) AND SAMPLE NUMBER	STANDARD PENETRATION NUMBER, N_{60}	MOISTURE CONTENT, w (%)	COMMENTS
Tan sandy silt	0.3			
Light brown silty clay (CL)	0.6			
	0.9			
	1.2			
	SS-1 1.5	13	11	Liquid limit = 32 $PI = 9$
	1.8			
	2.1			
	2.4			
	2.7	5	24	
Groundwater table June 14, 1997	SS-2 3.0			
	3.3			
Soft clay (CL)	3.6			
	3.9			
	4.2			
	ST-1 4.5	6	28	Liquid limit = 44 $PI = 26$ q_u = unconfined compression strength = 40.7 kN/m^2
	4.8			
	5.1			
	5.4			
	5.7			
Compact sand and gravel End of boring @ 6.6 m	SS-3 6.0	32		
	6.3			
	6.6			

Figure 17.15 Typical boring log (*Note:* SS = split-spoon sample; ST = Shelby tube sample)

Problems

17.1 ASTM D 1587-08 (2014) recommends the following dimensions for thin-walled steel tube sampling of soils for geotechnical purposes:

Outside diameter, D_o (mm)	Wall thickness, t (mm)
50.8	1.24
76.2	1.65
127	3.05

Calculate the area ratio for each case and determine which sampler would be appropriate for the following soil characterization tests: grain size distribution, Atterberg limits, consolidation, and unconfined compression.

17.2 During a soil exploration program, the following choices were available for soil sampling:

- Shelby tube A: Outside diameter, $D_o = 101.6$ mm; inside diameter, $D_i = 98.4$ mm
- Shelby tube B: Outside diameter, $D_o = 89$ mm; inside diameter, $D_i = 85.7$ mm
- Split spoon sampler: Outside diameter, $D_o = 50.8$ mm; inside diameter, $D_i = 34.9$ mm

Calculate the area ratio for each case and determine the suitability of each sampler for the soil characterization tests specified in Problem 17.1.

17.3 The following are the results of a standard penetration test in sand. Determine the corrected standard penetration numbers, $(N_1)_{60}$, at the various depths given. Note that the water table was not found during the boring operation. Assume that the average unit weight of sand is 18.7 kN/m^3. Use Liao and Whitman's relationship [Eq. (17.10)]. Assume $p_a \approx 100$ kN/m^2.

Depth (m)	N_{60}
1.0	10
2.5	12
4.0	14
5.5	15
7.0	17

17.4 For the soil profile given in Problem 17.3, estimate the average soil friction angle, ϕ', using the Kulhawy and Mayne correlation [Eq. (17.24)]. Assume $p_a \approx 100$ kN/m^2.

17.5 Following are the results of a standard penetration test in fine dry sand.

Depth (m)	N_{60}
0.5	6
2.0	9
3.5	12
5.0	14
6.5	17

For the sand deposit, assume the mean grain size, D_{50}, to be 1 mm and the unit weight of sand to be 18.5 kN/m^3. Estimate the variation of relative density with depth using the correlation developed by Cubrinovski and Ishihara [Eq. (17.18)]. Assume $p_a \approx 100$ kN/m^2.

17.6 Assuming the soil in Problem 17.5 is a clean, medium fine sand, use the Meyerhof (1957) method [Eq. (17.19)] to estimate the variation of relative densities with depth.

17.7 Refer to the boring log shown in Figure 17.16. Estimate the average drained friction angle, ϕ', using the Kulhawy and Mayne correlation [Eq. (17.24)]. Assume $p_a \approx 100$ kN/m^2.

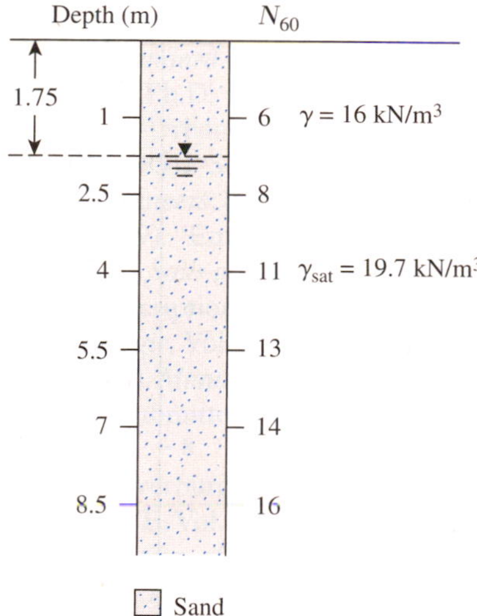

Figure 17.16

17.8 Refer to Problem 17.7 and Figure 17.16. Suppose a footing (1.5 m × 1.5 m) is constructed at a depth of 1 m.

a. Estimate the design values for N_{60} and ϕ'.

b. What is the net allowable load that the footing can carry? The maximum allowable settlement is 25 mm. Use Eqs. (16.56) and (16.61).

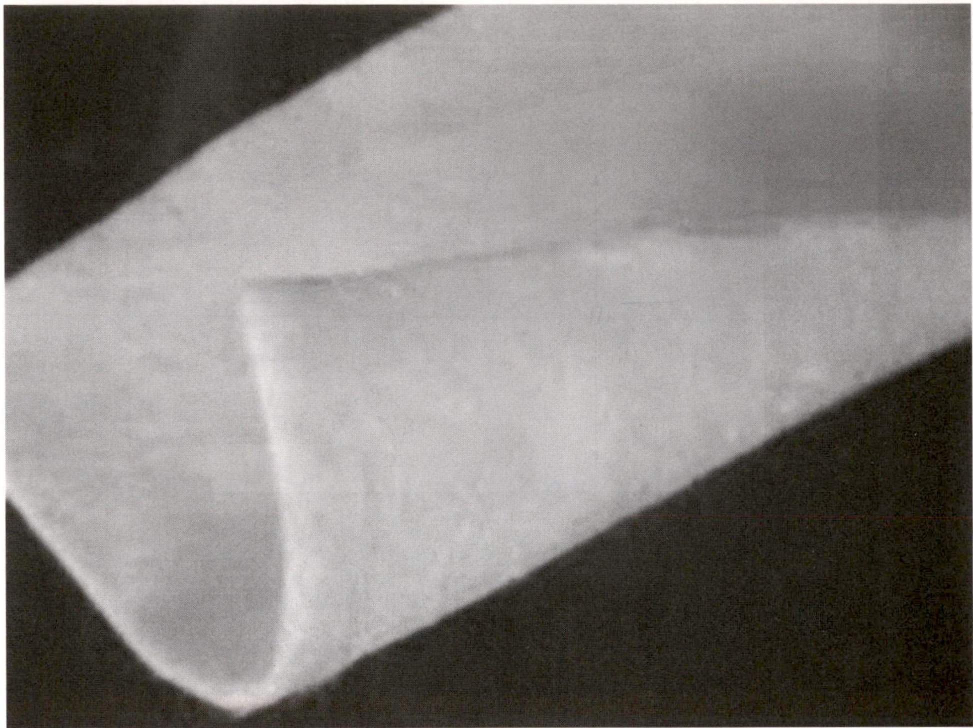

Figure 18.3 Nonwoven needle-punched geotextile (*Courtesy of Reinaldo Vega-Meyer, Beaumont, Texas*)

4. *Reinforcement*: The tensile strength of geotextiles increases the load-bearing capacity of the soil.

Geotextiles currently available commercially have thicknesses that vary from about 0.25 to 7.5 mm. The mass per unit area of these geotextiles ranges from about 150 to 700 g/cm^2.

Tensile strength

The tensile strength of geotextiles is an important parameter for design consideration, and it is generally expressed in terms of force per unit width. The tensile strength depends on several parameters such as manufacturing type, mass per unit area, thickness, and type of test. Table 18.1 gives the experimental results of some laboratory tests reported by Koerner (2005) that provide a sense of the range of parameters encountered in various geotextiles. Table 18.2 gives some typical values of the tensile properties of geotextiles.

Table 18.1 Tensile Strength and Strain at Failure for Some Geotextiles [Interpolated Values from the Results of Koerner (2005)]

Manufacturing type	Mass/unit area (g/m²)	Thickness (mm)	Tensile strength (kN/m)	Strain at failure (%)
Woven monofilament	200	0.38	37.7	21.9
Woven slit-film	170	0.25	28.2	26.3
Woven multi-filament	270	0.71	60.0	30.0
Nonwoven heat-bonded	135	0.33	24.1	68.8
Nonwoven needle-punched	200	0.63	27.3	72.5

Permeability

One of the major functions of geotextiles is filtration. For this purpose, water must be able to flow freely through the fabric of the geotextile (Figure 18.4a). Hence, the *cross-plane hydraulic conductivity* is an important parameter for design purposes. It should be realized that geotextile fabrics are compressible; however, their thicknesses may change depending on the effective normal stress to which they are being subjected. The change in thickness under normal stress also changes the cross-plane hydraulic conductivity of a geotextile. The cross-plane capability is generally expressed in terms of a quantity called *permittivity*, or

$$P = \frac{k_n}{t} \tag{18.1}$$

where P = permittivity (sec^{-1})
k_n = hydraulic conductivity for cross-plane flow (cm/sec)
t = thickness of the geotextile

The magnitude of k_n can vary from about 1×10^{-3} to about 2.5×10^{-1} cm/sec. Similarly, the magnitude of P can be in a range of 2×10^{-2} to 2.0 sec^{-1}.

Table 18.2 Typical Values of Tensile Properties of Some Geogrids

Type of geotextile	Tensile strength (kN/m)	Extension at maximum load (%)	Mass per unit area (g/cm²)
Nonwoven			
Heat-bonded	3–30	20–60	60–360
Needle-punched	7–90	30–80	100–3000
Resin-bonded	5–30	25–50	125–800
Woven			
Monofilament	20–80	20–35	150–300
Multifilament	40–1200	10–30	250–1500
Flat tape	10–90	15–25	90–250

Based on Shukla (2015)

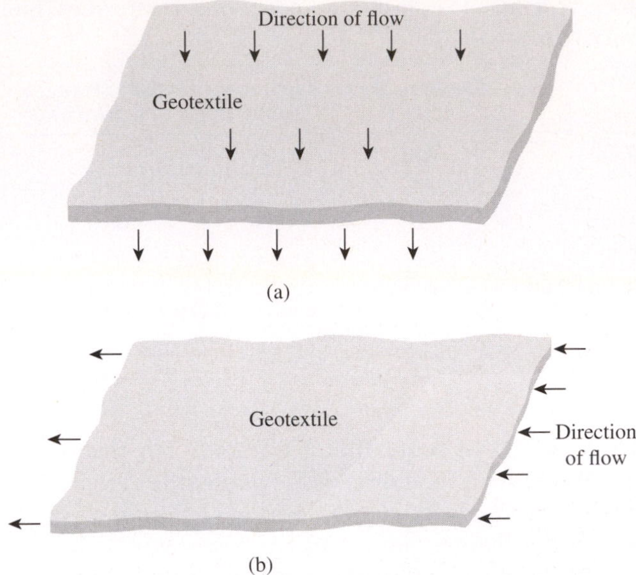

Figure 18.4 (a) Cross-plane flow through geotextile; (b) in-plane flow in geotextile

In a similar manner, to perform the function of drainage satisfactorily, geotextiles must possess excellent in-plane permeability (Figure 18.4b). For reasons stated previously, the in-plane hydraulic conductivity also depends on the compressibility and, hence, the thickness of the geotextile. The in-plane drainage capability thus can be expressed in terms of a quantity called *transmissivity*, or

$$T = k_p t \tag{18.2}$$

where T = transmissivity (m³/sec·m)
k_p = hydraulic conductivity for in-plane flow (cm/sec, m/sec)

The ranges of T and k_p are as follows (Gerry and Raymond, 1963):

- *Transmissivity T*:
  Nonwoven: 2×10^{-6} to 3×10^{-9} m³/sec·m
  Woven: 1.2×10^{-8} to 3×10^{-8} m³/sec·m
- *Hydraulic conductivity* k_p
  Nonwoven: 1×10^{-3} to 5×10^{-2} cm/sec
  Woven: 2×10^{-3} to 4×10^{-3} cm/sec

Properties to be considered for design

When a geotextile is being considered for use in design and construction, certain properties must be evaluated from tests on the geotextile to determine its applicability. A partial list of these tests follows:

1. Mass per unit area
2. Percentage of open area
3. Equivalent opening size
4. Thickness
5. Ultraviolet resistivity
6. Permittivity
7. Transmissivity
8. Puncture resistance
9. Resistance to abrasion
10. Compressibility
11. Tensile strength and elongation properties
12. Chemical resistance

Some examples of use of geotextile

There are many applications of geotextiles in the construction industry today. It is beyond the scope of this chapter to describe them all. However, Figure 18.5 shows some practical applications of geotextiles in the field. Figure 18.5a shows the use of geotextile between the subgrade and an aggregate layer in an unpaved road. The functions of the geotextile in this case are both the separation between the soft subgrade and granular layer and reinforcement. It helps reduce the aggregate thickness for a given traffic condition. Figure 18.5b shows a layer of geotextile placed for reinforcement below an embankment over a soft compressible soil layer. Geotextiles can be used to reinforce a slope to increase its factor of safety against failure (Figure 18.5c). A retaining wall with granular backfill reinforced with layers of geotextile is shown in Figure 18.5d. The geotextile layers add to the stability of the wall. Figure 18.5e shows the use of a geotextile as a filter behind a retaining wall.

18.3 Geogrid

A geogrid is defined as a polymeric (*i.e.,* geosynthetic) material consisting of connected parallel sets of tensile ribs with apertures of sufficient size to allow strike-through of the surrounding soil, stone, or other geotechnical material. Geogrids are generally made from high-density polyethylene (HDPE) and polypropylene (PP). The primary function of geogrid is *reinforcement.* Netlon Limited of the United Kingdom was the first producer of geogrid (called *extruded geogrid*). In 1982, the Tensar Corporation, presently Tensar International Corporation, introduced extruded geogrid into the United States.

Extruded geogrids are formed using a thick sheet of polyethylene or polypropylene that is punched and drawn to create apertures and to enhance engineering properties of the resulting ribs and nodes. Extruded geogrids are generally uniaxial or biaxial. A uniaxial geogrid (Figure 18.6a) is manufactured by stretching a punched sheet of extruded high-density polyethylene in one direction under carefully controlled conditions. This process aligns the polymer's long-chain molecules in the direction of the draw and results in a product with high one-directional

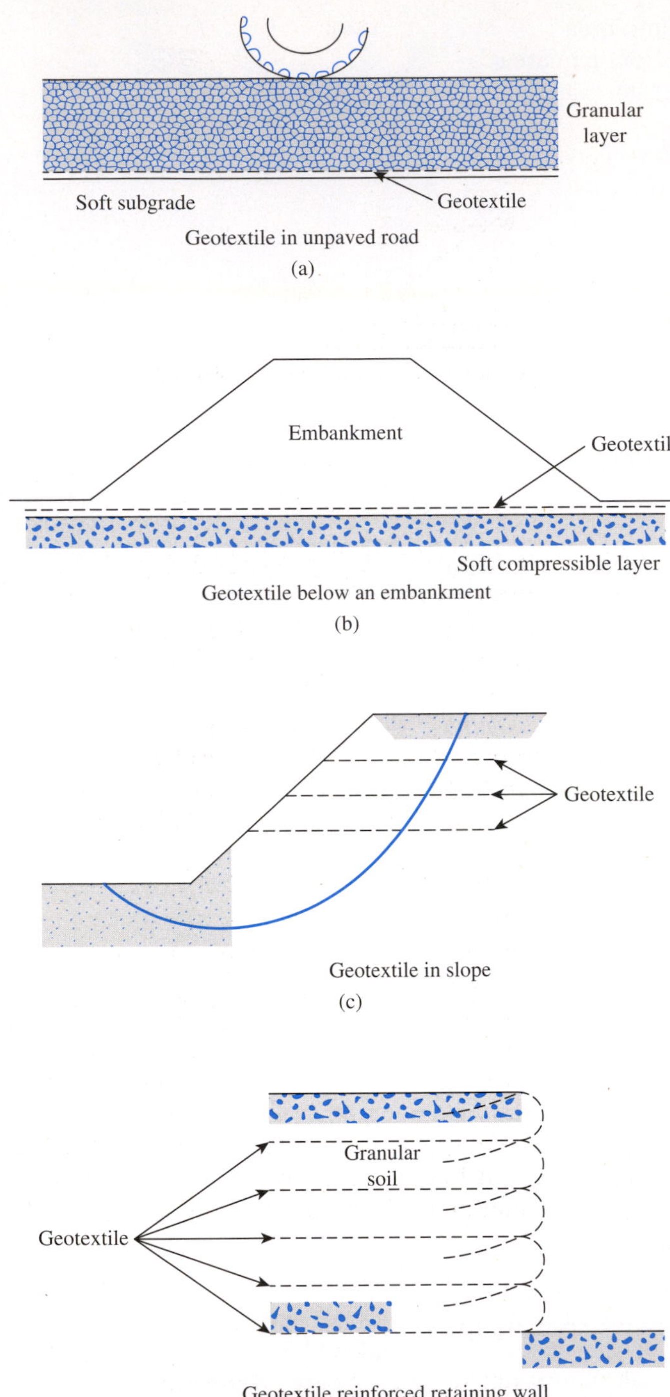

Figure 18.5 Examples of use of geotextile

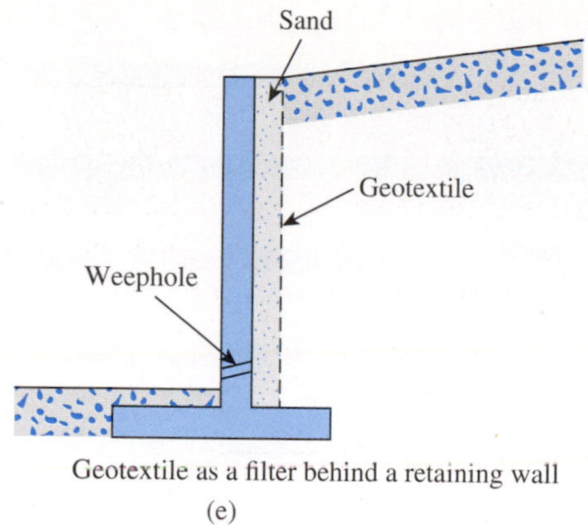

Geotextile as a filter behind a retaining wall

(e)

Figure 18.5 (*Continued*)

tensile strength and a high modulus. A biaxial geogrid (Figure 18.6b) is manufactured by stretching the punched sheet of polypropylene in two orthogonal directions. This process results in a product with high tensile strength and a high modulus in two perpendicular directions. The resulting grid apertures are either square or rectangular.

More recently, extruded geogrid with triangular apertures (triaxial) were introduced for construction purposes (Figure 18.6c). Geogrid with triangular apertures are manufactured from a punched polypropylene sheet that is then oriented in three substantially equilateral directions so that the resulting ribs have a high degree of molecular orientation.

Other types of geogrid presently available are *woven* and *welded*. A woven geogrid is manufactured by grouping polymeric strips—usually polyester or polypropylene—and weaving them into a mesh pattern that is then coated with a polymeric lacquer. A welded geogrid is manufactured by fusing junctions of polymeric strips. Figure 18.7 shows an example of a welded geogrid.

Commercial geogrids currently available for soil reinforcement have nominal rib thicknesses of about 0.5 to 1.5 mm and junctions of about 2.5 to 5 mm. The dimensions of the apertures vary from about 25 to 150 mm. Geogrids are manufactured so that the open areas of the grids are greater than 50% of the total area. They develop reinforcing strength at low strain levels (such as 2%). Extruded geogrids generally have a mass per unit area of 200 to 1100 g/m^2 with a tensile strength of 10 to 200 kN/m.

Reinforcement mechanism

As stated earlier, the major function of geogrid is *reinforcement*. They are relatively stiff. The apertures are large enough to allow interlocking with surrounding soil or rock (Figure 18.8) to perform the function of reinforcement or segregation (or both). Sarsby (1985) investigated the influence of aperture size on the size of soil particles

(a)

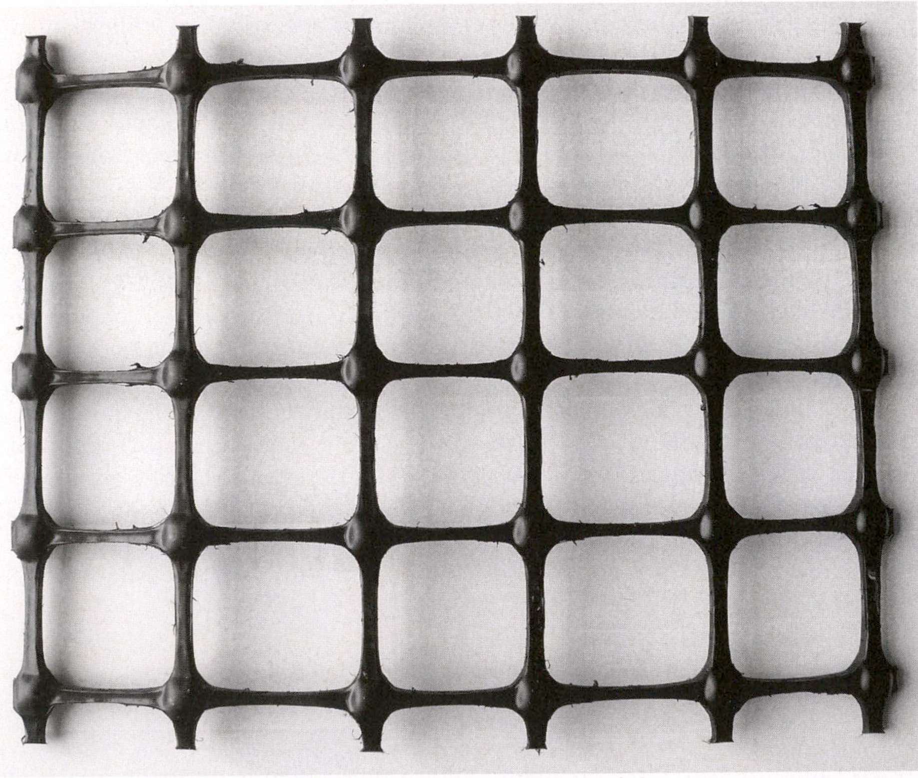

(b)

Figure 18.6 Extruded geogrid: (a) uniaxial; (b) biaxial; (c) triaxial (*Courtesy of Tensar International, Shadsworth, Blackburn, UK*)

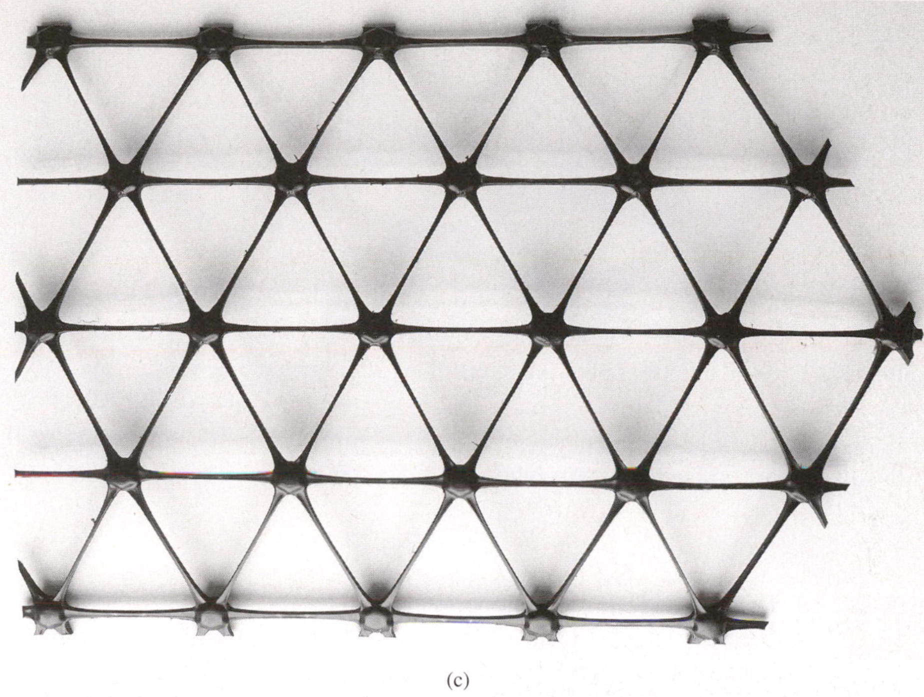

(c)

Figure 18.6 (*Continued*)

Figure 18.7 Thermo-welded geogrid (*Courtesy of Reinaldo Vega-Meyer, Beaumont, Texas*)

Figure 18.8 Geogrid apertures allowing interlocking with surrounding soil

for maximum frictional efficiency (or efficiency against pullout). According to this study, the highest efficiency occurs when

$$B_{GG} > 3.5D_{50} \tag{18.3}$$

where B_{GG} = minimum width of the geogrid aperture
D_{50} = the particle size through which 50% of the soil passes (*i.e.*, the median particle size)

Several authors have studied the reinforcement mechanisms associated with the interaction of geogrids and unbound aggregate. Perkins (1999), for example, suggested that there are four separate reinforcement mechanisms. These are shown in Figure 18.9 and are described here.

a. Confinement of the aggregate by the geogrid results in a reduction in the amount of lateral spreading.
b. Confinement results in an increase in the lateral stress within the aggregate thereby increasing its stiffness.

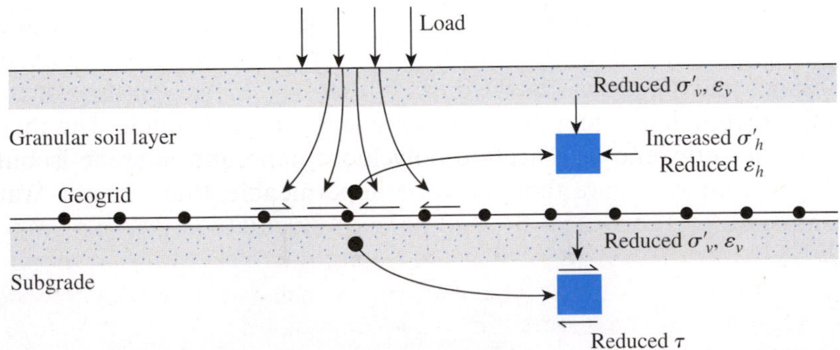

Figure 18.9 Reinforcement mechanism of geogrid in granular soil over a subgrade. (*Note*: σ'_v = vertical effective stress, σ'_h = horizontal effective stress, ε_v = normal strain in the vertical direction, ε_h = normal strain in the horizontal direction, τ = shear sress.)

c. An increased modulus of the aggregate results in an improved vertical stress distribution onto the underlying subgrade. The effect of this is that the surface deformation will be less and more uniform.
d. A reduction in the shear stress within the subgrade leading to lower vertical strain.

Mechanical properties to be considered for design

Following is a list of mechanical properties of geogrids which should be taken into consideration for design purposes:

- Strength of rib and junction
- Wide-width tensile strength
- Shear strength
- Anchorage strength from soil pullout

Examples of field application

There are several types of field application for geogrids; however, only a few of those are shown in Figure 18.10. Geogrids can be used to construct mechanically stabilized earth retaining walls as shown in Figure 18.10a. The reinforcing geogrid layers are placed in a granular backfill. Geogrids are used as reinforcement in the ballast under railroad tracks (Figure 8.10b) to reduce maintenance frequency. They can also be placed directly on weaker subgrade under railroad tracks to improve the load-bearing capacity (Figure 18.10c). There are several instances where layers of geogrid can be used to reinforce and stabilize slopes (Figure 18.10d). Ultimate and allowable bearing capacities of a shallow foundation on granular soils can be enhanced by using multi-layered geogrid reinforcement (Figure 18.10e).

18.4 Geomembrane

Geomembranes are impermeable liquid or vapor barriers made primarily from continuous polymeric sheets that are flexible. The type of polymeric material used for geomembranes may be *thermoplastic* or *thermoset*. The thermoplastic polymers include PVC, polyethylene, chlorinated polyethylene, and polyamide. The thermoset polymers include ethylene vinyl acetate, polychloroprene, and isoprene-isobutylene. Although geomembranes are thought to be impermeable, they are not. Water vapor transmission tests show that the hydraulic conductivity of geomembranes is in the range of 10^{-10} to 10^{-13} cm/sec; hence, they are only "essentially impermeable." Figure 18.11 shows several specimens of geomembrane available commercially.

Many scrim-reinforced geomembranes manufactured in single piles have thicknesses that range from 0.25 to about 0.4 mm. These single piles of geomembranes can be laminated together to make thicker geomembranes. Some geomembranes made from PVC and polyethylene may be as thick as 4.5 to 5 mm. The primary function of geomembrane is containment.

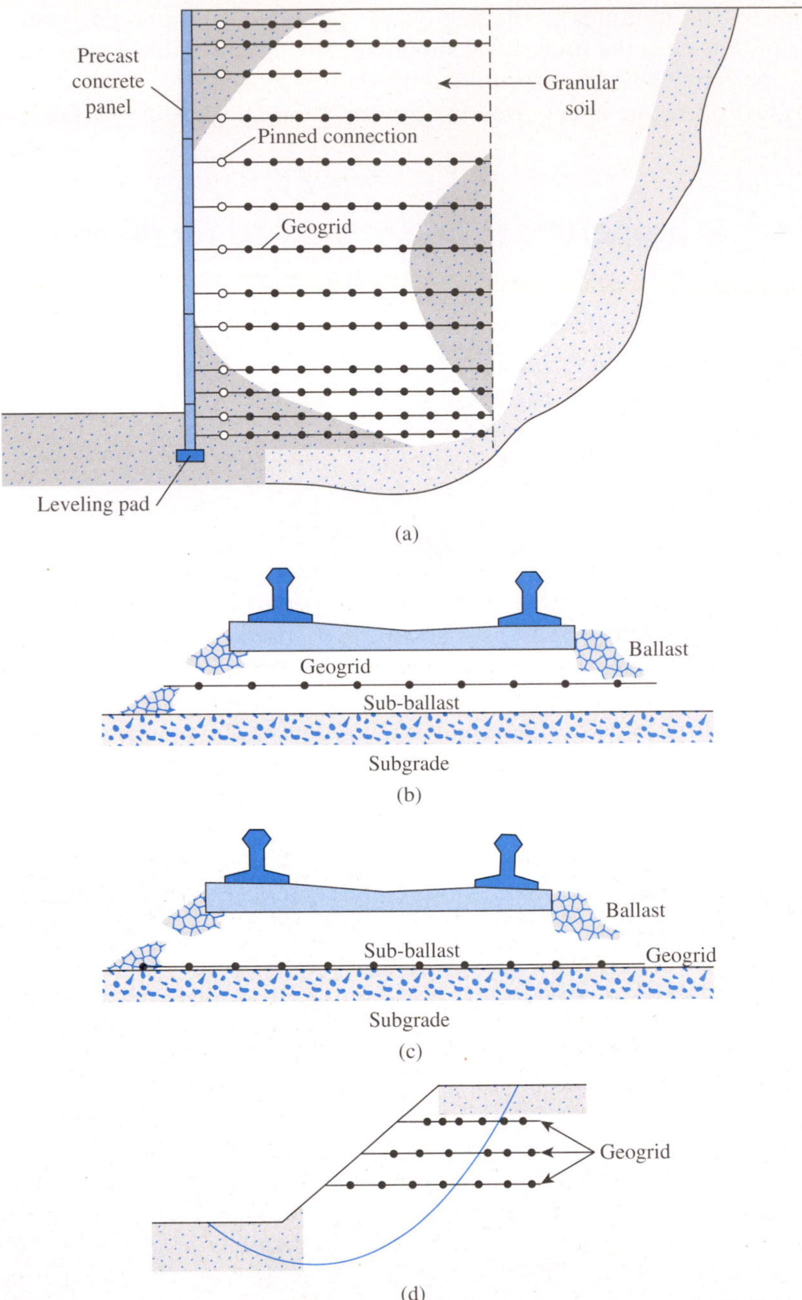

Figure 18.10 (a) Mechanically stabilized earth (MSE) retaining wall with geogrid-reinforced granular backfill; (b) geogrid reinforcement of the ballast layer under railroad track for maintenance reduction; (c) bearing capacity improvement by placement of geogrid directly on weaker subgrade under railroad track; (d) use of geogrid for slope stability; (e) use of geogrid in granular soil under shallow foundation for bearing capacity improvement

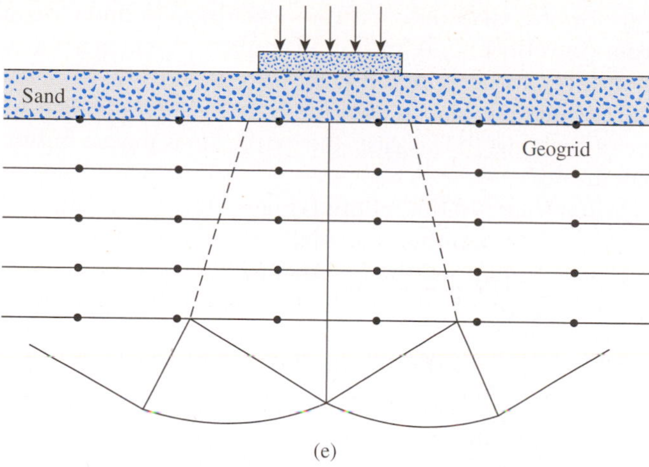

Figure 18.10 (*Continued*)

The most important aspect of construction with geomembranes is the preparation of seams; otherwise, the basic reason for using geomembrane as a liquid or vapor barrier will be defeated. Geomembrane sheets are generally seamed together in the factory to prepare larger sheets. These larger sheets are field seamed into their final position. There are several types of seams, some of which are briefly described next.

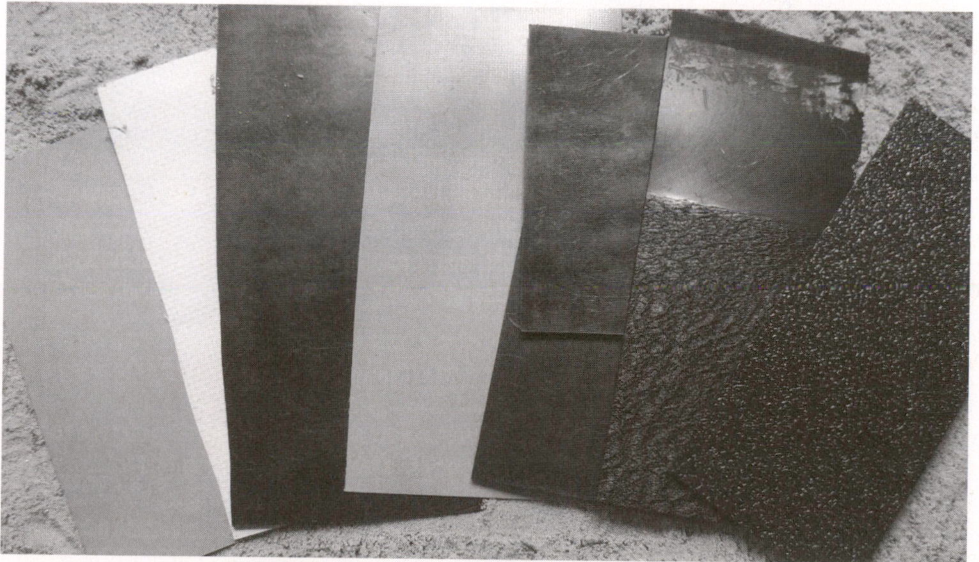

Figure 18.11 Specimens of several commercially-produced geomembranes (*Courtesy of Reinaldo Vega-Meyer, Beaumont, Texas*)

- *Extrusion seam*: Extrusion, or fusion welding, is done on geomembrane made from polyethylene. A ribbon of molten polymer is extruded between the two surfaces to be joined. Figure 18.12a shows a *fillet type* of seaming where the extrudate is placed over the edge of the seam. Figure 18.12b shows a *flat type* of seaming. In this case, the extrudate is placed between two sheets to be joined.
- *Thermal fusion seam*: In this method (Figure 18.12c), hot air is blown between the two sheets of geomembrane to melt the opposing surfaces. Following that, pressure is applied to bond the two sheets.
- *Chemical fusion seam*: In this method, a liquid solvent is applied to the edges of the two sheets of geomembrane. Pressure is then then applied for complete contact of the two edges (Figure 18.12d).
- *Adhesive seam*: For this method, bonding agents are applied to the mating surfaces of the two sheets to be joined. The two edges are then placed over each other, and pressure is applied for full contact (Figure 18.12e). This process is applicable for thermoset geomembranes.

Tensile strength

Table 18.3 provides typical values for the tensile strength of various types of geomembranes in use.

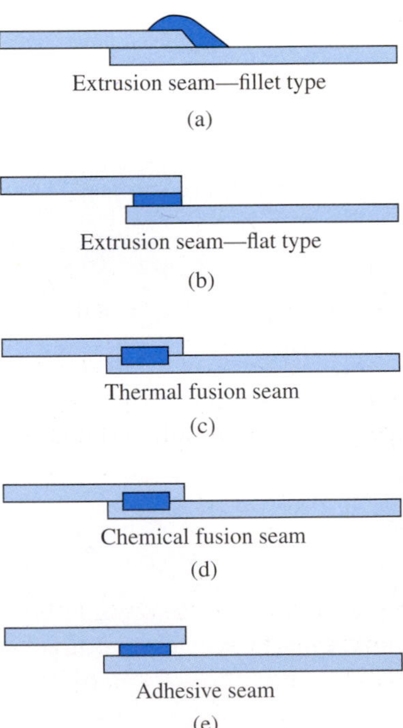

Figure 18.12 Geomembrane seams

Table 18.3 Typical Values of Tensile Properties of Geomembrane

Type of geomembrane	Tensile strength (kN/m)	Extension at maximum load (%)	Mass per unit area (g/m²)
Reinforced (made from bitumen and nonwoven geotextile)	20–60	30–60	1000–3000
Plastomeric (made from HDPE, LDPE, PP or PVC)			
Unreinforced	10–50	50–200	500–3500
Reinforced	30–60	15–30	600–1200

Based after Shukla (2015)

Properties to be considered for design

Following is a partial list of tests that should be conducted on geomembranes when they are considered for design:

1. Density
2. Mass per unit area
3. Water vapor transmission capacity
4. Tensile behavior
5. Tear resistance
6. Resistance to impact
7. Puncture resistance
8. Stress cracking
9. Chemical resistance
10. Ultraviolet light resistance
11. Thermal properties
12. Behavior of seams

Some examples of use of geomembranes

Geomembranes have been used for construction projects such as liners for seepage control in cutoff trenches, solar ponds, solid waste landfills, waterproofing within tunnel conveyance canals, seepage control in tailing dams, and others.

18.5 Geonet

Geonets are formed by the continuous extrusion of polymeric ribs at acute angles to each other. They have large openings in a net-like configuration. The primary function of geonet is drainage. Figure 18.13 is a photograph of a typical piece of geonet. Most geonets currently available are made of medium- and high-density polyethylene. They are available in rolls with widths of 1.8 to 2.1 m and lengths of 30 to 90 m. The approximate aperture sizes vary from 30 mm × 30 mm to about 6 mm × 6 mm.

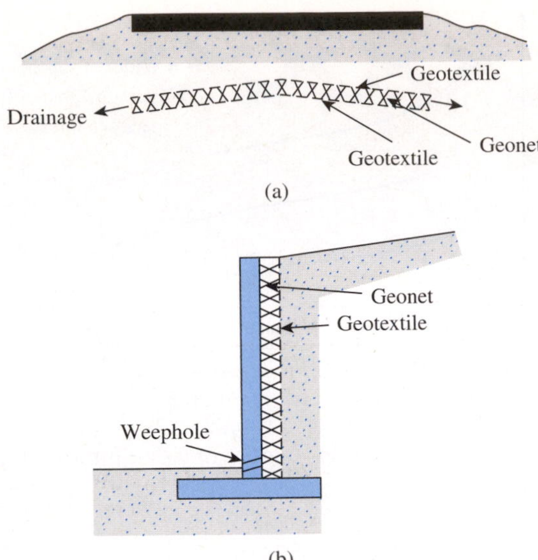

Figure 18.15 (a) Use of geonet under a highway for drainage; (b) use of geonet behind a retaining wall for drainage

geotextiles layer and the clay layer in between them. Stitch bonding can be used for reinforcement in case the woven geotextile is used to manufacture the GCL. Daniel et al. (1997) have reported the results of several permeability tests on GCL. The hydraulic conductivity ranged from about 2×10^{-9} to 2×10^{-10} cm/sec.

Figure 18.17 gives some examples for using geosynthetic clay liner. Figure 18.17a is a diagram using GCL for a canal, and Figure 18.17b shows GCL beneath a storage tank.

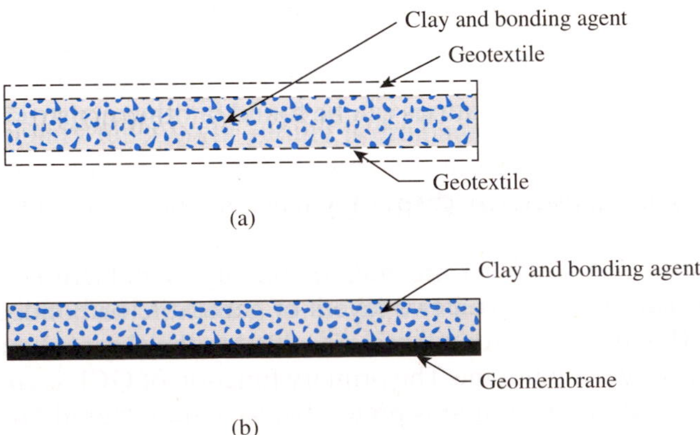

Figure 18.16 Geosynthetic clay liner: (a) clay between two geotextile layers; (b) clay on a geomembrane layer

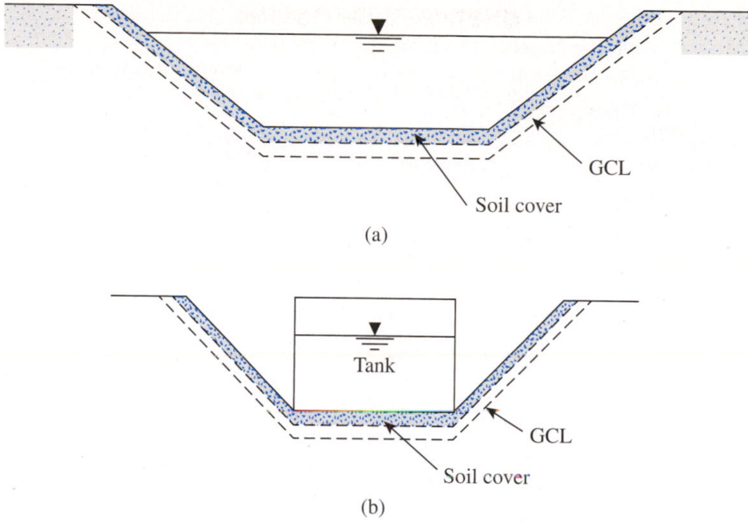

Figure 18.17 Use of geosynthetic clay liner: (a) for a canal; (b) beneath a storage tank

18.7 Summary

Geosynthetics are non-biodegradable material made from polymers such as polypropylene, polyester, polyethylene, and polyamide. Major geosynthetic products are geotextile, geogrid, geomembrane, geonet, and geosynthetic clay liner. The functions of these products are

- *Geotextile*: Separation, reinforcement, filtration, and drainage
- *Geogrid*: Reinforcement
- *Geomembrane*: Moisture barrier (containment)
- *Geonet*: Drainage
- *Geosynthetic clay liner*: Moisture barrier (containment)

The general important physical properties of these geosynthetic products have been briefly discussed in this chapter. Some examples of field applications are also provided.

Geopile, geofoam, and geocomposites are other geosynthetic products used in the construction industry; however, they have not been expanded upon in this chapter.

References

American Society for Testing and Materials (2015). *Annual Book of ASTM Standards*, Vol. 04.13, West Conshoken, PA.

Daniel, D.E., Bowders, J.J., and Gilbert, R.B. (1997). "Laboratory Hydraulic Conductivity of GCLs in Flexible-Wall Permeameters," In *Testing and Acceptance Criteria for Geosynthetic Clay Liners*, ASTM STP 1308, Ed. L.W. Well, ASTM, 208–228.

Gerry, G.S., and Raymond, G.P. (1963). "The In-Plane Permeability of Geotextiles," *Geotechnical Testing Journal*, ASTM, Vol. 6, No. 4, 181–189.

Koerner, R.M. (2005). *Designing with Geosynthetics*, 5th ed., Prentice-Hall, Englewood Cliffs, NJ.

Perkins, S.W. (1999). *Geosynthetic Reinforcement of Flexible Pavements: Laboratory-Based Pavement Test Sections*. Federal Highway Administration Report No. FHWA/MT-99-001/81838.

Sarsby, R.W. (1985). "The Influence of Aperture Size/Particle Size on the Efficiency of Grid Reinforcement," *Proceedings,* 2nd *Canadian Symposium on Geotextiles and Geomembranes*, Edmonton, 7–12.

Shukla, S.K. (2015). Personal communication.

ANSWERS TO SELECTED PROBLEMS

Chapter 2

2.1 $C_u = 4.36$; $C_c = 1.18$; well graded

2.3 Soil *A*: $C_u = 17.5$; $C_c = 5.43$; poorly graded
Soil *B*: $C_u = 8.82$; $C_c = 2.2$; well graded
Soil *C*: $C_u = 17.2$; $C_c = 2.75$; well graded

2.5 **a.**

Sieve No.	Percent finer
4	100
6	94
10	84.26
20	58.80
40	39.44
60	24.12
100	13.08
200	4.40
Pan	0

b. $D_{10} = 0.13$ mm; $D_{30} = 0.3$ mm; $D_{60} = 0.9$ mm
c. 6.92
d. 0.77

2.7 **a.**

Sieve No.	Percent finer
4	100
6	100
10	100
20	98.18
40	48.3
60	12.34
100	7.8
200	4.7
Pan	0

b. $D_{10} = 0.21$ mm; $D_{30} = 0.39$ mm; $D_{60} = 0.45$ mm
c. 2.14
d. 1.61

2.9 **b.** Gravel – 0%; Sand – 27%; Silt – 64%; Clay – 9%
c. Gravel – 0%; Sand – 32%; Silt – 59%; Clay – 9%
d. Gravel – 0%; Sand – 20%; Silt – 71%; Clay – 9%

2.11 **b.** Gravel – 0%; Sand – 35%; Silt – 30%; Clay – 35%
c. Gravel – 0%; Sand – 38%; Silt – 27%; Clay – 35%
d. Gravel – 0%; Sand – 30%; Silt – 35%; Clay – 35%

2.13 **b.** Gravel – 0%; Sand – 16%; Silt – 56%; Clay – 28%
c. Gravel – 0%; Sand – 17%; Silt – 55%; Clay – 28%
d. Gravel – 0%; Sand – 10%; Silt – 62%; Clay – 28%
2.15 0.0047

Chapter 3

3.3 $W_s = 15.6$ kN; $W_w = 2.18$ kN; $W = 17.78$ kN; $V_s = 0.591\text{ m}^3$; $V_v = 0.408\text{ m}^3$; $V_w = 0.222\text{ m}^3$
3.5 **a.** 0.657
b. 36.7%
c. 24.5%
d. 0.056 kN
3.7 **a.** 19.65 kN/m^3
b. 17.24 kN/m^3
c. 0.54
d. 0.35
e. 70.2%
f. $6.95 \times 10^{-4}\text{ m}^3$
3.9 **a.** 1.41 kN/m^3
b. 1.84 kN/m^3
3.11 **a.** 18.95 kN/m^3
b. 23.5%
c. 15.26 kN/m^3
d. 0.76
e. 84.7%
3.13 **a.** 2.72
b. 0.98
c. 18.33 kN/m^3
3.15 **a.** 0.1 kN/m^3
b. 0.44 kN/m^3
3.17 **a.** 17.6 kN/m^3
b. 0.48
c. $1.25 \times 10^{-3}\text{ m}^3$
3.19 87.6%
3.21 **a.** 2.71
b. 0.89
3.23 **a.** 14 kN/m^3
b. 27%
c. 16.3 kN/m^3
3.25 $e = 0.6$; $\gamma_d = 16.37\text{ kN/m}^3$
3.27 39.6%
3.29 17.88 kN/m^3

Chapter 4

4.1 **a.** 29.0
b. 11.5

4.3 **a.** 37.86
b. 0.304

4.5 **a.** 37.05
b. Inorganic clay of low plasticity

4.7 $LL = 33.8; I_{FC} = 43.2$

4.9 35.76

4.11 $SL = 17.63$
$SR = 2.5$

Chapter 5

5.1

Soil	Classification
A	Clay
B	Sandy clay
C	Loam
D	Sandy clay and sandy clay loam (borderline)
E	Sandy loam
F	Silty loam
G	Clay loam
H	Clay
I	Silty clay
J	Loam

5.3

Soil	Classification
A	A-2-4(0)
B	A-3(0)
C	A-2-6(0)
D	A-2-7(1)
E	A-1-b(0)

5.5

Soil	Group symbol
1	ML
2	CH
3	CL
5	ML
8	CL
10	MH

5.7

Soil	Group symbol	Group name
A	SP	Poorly graded sand
B	SW-SM	Well graded sand with silt
C	MH	Elastic silt with sand
D	CH	Fat clay
E	SC	Clayey sand

5.9

Soil	Group Symbol	Group name
1	SC	Clayey sand with gravel
2	GC	Clayey gravel with sand
3	CH	Sandy fat clay
4	CL	Lean clay with sand
5	CL	Lean clay with sand
6	SC	Clayey sand
7	CH	Fat clay with gravel
8	CH	Sandy fat clay
9	SP-SC	Poorly graded sand with clay and gravel
10	SW	Well graded sand
11	CL	Sandy lean clay
12	SP-SC	Poorly graded sand with clay

Chapter 6

6.1 78.1%

6.3

w (%)	γ_{zav} (kN/m^3)
7	22
11	20.18
15	18.65
19	17.33
23	16.18

6.5 86.8%

6.7 **a.** $\gamma_{d(\max)} \approx 19$ kN/m^3 @ $w_{opt} = 10.8\%$
b. 21.05 kN/m^3
c. 74.4%
d. 97.3%
e. 9% to 11.9%
f. 77.2%

6.9 **a.** 14.97 kN/m^3
b. 94%
c. No

6.11 **a.**

Borrow pit	Volume of soil (m^3)
I	8,165
II	8,473
III	9,749
IV	8,338

b. Borrow Pit II

6.13 $\gamma_{d(\text{field})} = 15.98$ kN/m^3; $D_r = 72.4\%$

6.15 1.44 m

6.17 $S_N = 16.35$; Rating: Good

Chapter 7

7.1 14.1×10^{-2} m^3/hr/m

7.3 **a.** 7.4×10^{-2} m^3/hr/m

b. 0.0061 cm/sec

c. 621.6 m^3

7.5 8.22×10^{-5} m^3/sec/m

7.7 26.28 cm

7.9 **a.** 8.4×10^{-13} m^2

b. 30 cm

7.11 1.0×10^{-2} cm/sec

7.13 0.05 cm/sec

7.15 0.415 cm/sec

7.17 0.035 cm/sec

7.19 6.89×10^{-7} cm/sec

7.21 278.64

7.23 3.56×10^{-5} cm/sec

Chapter 8

8.1 77.76×10^{-6} m^3/m/day

8.3 0.38 m^3/m/day

8.5 1717.5 kN/m

8.7 2.07 m^3/m/day

8.9 0.291 m^3/m/day

8.11 11.4×10^{-4} m^3/min/day

Chapter 9

9.1

	kN/m^2		
Point	σ	u	σ'
A	0	0	0
B	36.18	0	36.18
C	105.57	35.9	69.67
D	139.42	53.85	85.57

9.3

	kN/m²		
Point	σ	u	σ'
A	0	0	0
B	48	0	48
C	156	58.86	97.14
D	198.5	83.38	115.12

9.5 **a.**

	kN/m²		
Point	σ	u	σ'
A	0	0	0
B	64.8	0	64.8
C	169.2	49.05	120.15

b. 1.5 m

9.7 6.91 m

9.9 6.88 m

9.11 **a.** 781.2 cm^3/sec
b. No boiling
c. 2.77 m

9.13

	kN/m²		
Depth (m)	σ	u	σ'
0	0	0	0
3.05	54.75	$\begin{cases} 0 \\ -9.53 \end{cases}$	$\begin{matrix} 54.75 \\ 64.28 \end{matrix}$
5.48	97.34	0	97.34
10.36	188.26	47.87	140.38

9.15 2.33

Chapter 10

10.1 **a.** $\sigma_1 = 202.72\ kN/m^2$
$\sigma_3 = 117.28\ kN/m^2$
b. $\sigma_n = 177\ kN/m^2$
$\tau_n = -39.2\ kN/m^2$

10.3 **a.** $\sigma_1 = 97.2\ kN/m^2$
$\sigma_3 = 28.8\ kN/m^2$
b. $\sigma_n = 42\ kN/m^2$
$\tau_n = 27\ kN/m^2$

10.5 **a.** $\sigma_1 = 178\ kN/m^2$
$\sigma_3 = 62\ kN/m^2$
b. $\sigma_n = 176\ kN/m^2$
$\tau_n = -8\ kN/m^2$

10.7 0.885 kN/m^2
10.9 71.68 kN/m^2
10.11 9402 kN/m
10.13 572.6 kN/m^2
10.15 @ $A \rightarrow$ 395.55 kN/m^2
@ $B \rightarrow$ 348 kN/m^2
@ $C \rightarrow$ 28.68 kN/m^2
10.17 @ $A \rightarrow$ 75.2 kN/m^2
@ $B \rightarrow$ 16.5 kN/m^2
@ $C \rightarrow$ 218.72 kN/m^2
10.19

z (m)	$\Delta\sigma_z$ (kN/m^2)
0	39.24
2	29.53
4	15.54
8	5.18
10	3.41

10.21

r (m)	$\Delta\sigma_z$ (kN/m^2)
0	44.66
2	40.85
4	20.48
6	2.964
8	0.51

Chapter 11

11.1 12.1 mm
11.3 $C_c = 0.163$; $\sigma'_c = 0.7$ kg/cm^2
11.5 **b.** 125 kN/m^2
c. 5.88
11.7 0.244 m
11.9 0.065 m
11.11 7.6 cm
11.13 0.33 m
11.15 **a.** 0.43
b. 59.85 kN/m^2
11.17 1,768 days
11.19 5.37×10^{-8} m/min
11.21 **a.** 18.63 kN/m^2
b. 184 mm
c. 25%
d. 0.075 m^2/year
e. 104 mm

Chapter 12

12.1 **a.** 24.9°
b. 367.6 N
12.3 666 N
12.5 **a.** 27.5°
b. 27 kN/m^2
12.7 −96.9 kN/m^2
12.9 1376.8 kN/m^2
12.11 145.08 kN/m^2
12.13 0.548
12.15 **a.** 22°
b. 56°
c. $\sigma'_f = 331.27$ kN/m^2; $\tau_f = 133.83$ kN/m^2
12.17 **a.** $\sigma' = 179$ kN/m^2; $\tau = 70.8$ kN/m^2
b. $\sigma'_f = 214.25$ kN/m^2; $\tau_f = 119$ kN/m^2
12.19 **a.** 11.46°
b. 124.7 kN/m^2
12.21 $(\Delta\sigma_d)_f = 229.64$ kN/m^2; $(\Delta u_d)_f = -42.1$ kN/m^2
12.23 **a.** 24 kN/m^2
b. 24.21 kN/m^2
c. 57.8 kN/m^2

Chapter 13

13.1 $P_o = 393.1$ kN/m; $\bar{z} = 2.5$ m
13.3 $P_o = 649.8$ kN/m; $\bar{z} = 3.67$ m
13.5

z (m)	σ'_h (kN/m^2)
0	0
1	5.32
2	5.19
3	2.72
4	1.33
5	0.68

13.7 91.52 kN/m
13.9 $P_a = 206.5$ kN/m; $\bar{z} = 2.95$ m
13.11 $P_a = 244.16$ kN/m; $\bar{z} = 3$ m
13.13 $P_p = 535$ kN/m; $\bar{z} = 1.83$ m
13.15 $P_p = 3847.9$ kN/m; $\bar{z} = 3.2$ m
13.17 $P_a = 227.5$ kN/m; $\bar{z} = 1.97$ m
13.19 $P_a = 587.34$ kN/m; $\bar{z} = 3.72$ m
13.21 **a.** 50.72 kN/m^2
b. 177.54 kN/m^3; resultant a distance of 2.33 m from the bottom of the wall inclined at an angle $\alpha = 12°$ to the horizontal

13.23 **b.** 2.24 m
c. 334.05 kN/m
d. 434.3 kN/m; $\bar{z} = 2.086$ m
13.25 2.92 kN/m
13.27 $P_{ae} = 415.33$ kN/m; $\bar{z} = 3.45$ m

Chapter 14

14.1 2997 kN/m
14.3 3210 kN/m
14.5 3223 kN/m
14.7 1588 kN/m
14.9 692 kN/m
14.11 $P_{pe} = 9823$ kN/m; $\bar{z} = 3.12$ m
14.13 41.5 kN/m
14.15 **a.** Similar to Figure 14.15c; $\sigma_a = 64.5$ kN/m^2
b. @ $A \rightarrow$ 451.5 kN
@ $B \rightarrow$ 838.5 kN
@ $C \rightarrow$ 451.5 kN

Chapter 15

15.1 6.28 m
15.3 1.84
15.5 1.54
15.7 3.05
15.9 4.76
15.11 9.29 m
15.13 10.7 m
15.15 10.25 m; Toe circle
15.17 2.09
15.19 19.76 m; Midpoint circle
15.21 **a.** 45.9 m
b. 41.5 m
15.23 1.35
15.25 1.27
15.27 1.18
15.29 1.1
15.31 **b.** 25.74 m

Chapter 16

16.1 499.4 kN/m^2
16.3 1128 kN/m^2
16.5 534.5 kN/m^2
16.7 1150 kN/m^2
16.9 771 kN

16.11 2087 kN
16.13 $B = 1.21$ m
16.15 $B = 1$ m
16.17 3000 kN
16.19 **a.** 984 kN/m
b. 1237 kN/m
c. 1006 kN/m

Chapter 17

17.1

Tube	A_R (%)
1	5.06
2	4.47
3	4.98

All three samples are appropriate for all tests.

17.3

Depth (m)	$(N_1)_{60}$
1	23
2.5	18
4	16
5.5	15
7	15

17.5

Depth (m)	D_r (%)
0.5	94
2.0	57
3.5	50
5.0	45
6.5	44

17.7 37.6°

17.9

Depth (m)	q_c (kN/m²)
1.0	2,667
2.5	3,556
4.0	4,890
5.5	5,779
7.0	6,224
8.5	7,113

17.11 57.5 kN/m²
17.13 14
17.15 Recovery ratio = 55.5%; $RQD = 0.467$; Quality of rock—poor

INDEX

T

U

V

W

Z